Lecture Notes in Computer Science 8581

Commenced Publication in 1973
Founding and Former Series Editors:
Gerhard Goos, Juris Hartmanis, and Jan van Leeuwen

Editorial Board

Beniamino Murgante   Sanjay Misra
Ana Maria A.C. Rocha   Carmelo Torre
Jorge Gustavo Rocha   Maria Irene Falcão
David Taniar   Bernady O. Apduhan
Osvaldo Gervasi (Eds.)

Computational Science and Its Applications – ICCSA 2014

14th International Conference
Guimarães, Portugal, June 30 – July 3, 2014
Proceedings, Part III

 Springer

Volume Editors

Beniamino Murgante, University of Basilicata, Potenza, Italy
E-mail: beniamino.murgante@unibas.it

Sanjay Misra, Covenant University, Ota, Nigeria
E-mail: sanjay.misra@covenantuniversity.edu.ng

Ana Maria A.C. Rocha, University of Minho, Braga, Portugal
E-mail: arocha@dps.uminho.pt

Carmelo Torre, Politecnico di Bari, Bari, Italy
E-mail: torre@poliba.it

Jorge Gustavo Rocha, University of Minho, Braga, Portugal
E-mail: jgr@di.uminho.pt

Maria Irene Falcão, University of Minho, Braga, Portugal
E-mail: mif@math.uminho.pt

David Taniar, Monash University, Clayton, VIC, Australia
E-mail: david.taniar@infotech.monash.edu.au

Bernady O. Apduhan, Kyushu Sangyo University, Fukuoka, Japan
E-mail: bob@is.kyusan-u.ac.jp

Osvaldo Gervasi, University of Perugia, Perugia, Italy
E-mail: osvaldo.gervasi@unipg.it

ISSN 0302-9743 e-ISSN 1611-3349
ISBN 978-3-319-09149-5 e-ISBN 978-3-319-09150-1
DOI 10.1007/978-3-319-09150-1
Springer Cham Heidelberg New York Dordrecht London

Library of Congress Control Number: 2014942987

LNCS Sublibrary: SL 1 – Theoretical Computer Science and General Issues

Typesetting: Camera-ready by author, data conversion by Scientific Publishing Services, Chennai, India

Printed on acid-free paper

Springer is part of Springer Science+Business Media (www.springer.com)

Welcome Message

On behalf of the Local Organizing Committee of ICCSA 2014, it is a pleasure to welcome you to the 14th International Conference on Computational Science and Its Applications, held during June 30 – July 3, 2014. We are very proud and grateful to the ICCSA general chairs for having entrusted us with the task of organizing another event of this series of very successful conferences.

ICCSA will take place in the School of Engineering of University of Minho, which is located in close vicinity to the medieval city centre of Guimarães, a UNESCO World Heritage Site, in Northern Portugal. The historical city of Guimarães is recognized for its beauty and historical monuments. The dynamic and colorful Minho Region is famous for its landscape, gastronomy and vineyards where the unique *Vinho Verde* wine is produced.

The University of Minho is currently among the most prestigious institutions of higher education in Portugal and offers an excellent setting for the conference. Founded in 1973, the University has two major poles: the campus of Gualtar in Braga, and the campus of Azurém in Guimarães.

Plenary lectures by leading scientists and several workshops will provide a real opportunity to discuss new issues and find advanced solutions able to shape new trends in computational science.

Apart from the scientific program, a stimulant and diverse social program will be available. There will be a welcome drink at Instituto de Design, located in an old Tannery, that is an open knowledge centre and a privileged communication platform between industry and academia. Guided visits to the city of Guimarães and Porto are planned, both with beautiful and historical monuments. A guided tour and tasting in Porto wine cellars, is also planned. There will be a gala dinner at the Pousada de Santa Marinha, which is an old Augustinian convent of the 12th century refurbished, where ICCSA participants can enjoy delicious dishes and enjoy a wonderful view over the city of Guimarães.

The conference could not have happened without the dedicated work of many volunteers, recognized by the coloured shirts. We would like to thank all the collaborators, who worked hard to produce a successful ICCSA 2014, namely Irene Falcão and Maribel Santos above all, our fellow members of the local organization.

On behalf of the Local Organizing Committee of ICCSA 2014, it is our honor to cordially welcome all of you to the beautiful city of Guimarães for this unique event. Your participation and contribution to this conference will make it much more productive and successful.

We are looking forward to see you in Guimarães.

Sincerely yours,

Ana Maria A.C. Rocha
Jorge Gustavo Rocha

Preface

These 6 volumes (LNCS volumes 8579-8584) consist of the peer-reviewed papers from the 2014 International Conference on Computational Science and Its Applications (ICCSA 2014) held in Guimarães, Portugal during 30 June – 3 July 2014.

ICCSA 2014 was a successful event in the International Conferences on Computational Science and Its Applications (ICCSA) conference series, previously held in Ho Chi Minh City, Vietnam (2013), Salvador da Bahia, Brazil (2012), Santander, Spain (2011), Fukuoka, Japan (2010), Suwon, South Korea (2009), Perugia, Italy (2008), Kuala Lumpur, Malaysia (2007), Glasgow, UK (2006), Singapore (2005), Assisi, Italy (2004), Montreal, Canada (2003), and (as ICCS) Amsterdam, The Netherlands (2002) and San Francisco, USA (2001).

Computational science is a main pillar of most of the present research, industrial and commercial activities and plays a unique role in exploiting ICT innovative technologies, and the ICCSA conference series has been providing a venue for researchers and industry practitioners to discuss new ideas, to share complex problems and their solutions, and to shape new trends in computational science.

Apart from the general track, ICCSA 2014 also included 30 workshops, in various areas of computational sciences, ranging from computational science technologies, to specific areas of computational sciences, such as computational geometry and security. We accepted 58 papers for the general track, and 289 in workshops. We would like to show our appreciation to the workshops chairs and co-chairs.

The success of the ICCSA conference series, in general, and ICCSA 2014, in particular, was due to the support of many people: authors, presenters, participants, keynote speakers, workshop chairs, Organizing Committee members, student volunteers, Program Committee members, Advisory Committee members, international liaison chairs, and people in other various roles. We would like to thank them all.

We also thank our publisher, Springer–Verlag, for their acceptance to publish the proceedings and for their kind assistance and cooperation during the editing process.

We cordially invite you to visit the ICCSA website http://www.iccsa.org where you can find all relevant information about this interesting and exciting event.

June 2014

Osvaldo Gervasi
Jorge Gustavo Rocha
Bernady O. Apduhan

Organization

ICCSA 2014 was organized by University of Minho, (Portugal) University of Perugia (Italy), University of Basilicata (Italy), Monash University (Australia), Kyushu Sangyo University (Japan).

Honorary General Chairs

Antonio M. Cunha	Rector of the University of Minho, Portugal
Antonio Laganà	University of Perugia, Italy
Norio Shiratori	Tohoku University, Japan
Kenneth C. J. Tan	Qontix, UK

General Chairs

Beniamino Murgante	University of Basilicata, Italy
Ana Maria A.C. Rocha	University of Minho, Portugal
David Taniar	Monash University, Australia

Program Committee Chairs

Osvaldo Gervasi	University of Perugia, Italy
Bernady O. Apduhan	Kyushu Sangyo University, Japan
Jorge Gustavo Rocha	University of Minho, Portugal

International Advisory Committee

Jemal Abawajy	Daekin University, Australia
Dharma P. Agrawal	University of Cincinnati, USA
Claudia Bauzer Medeiros	University of Campinas, Brazil
Manfred M. Fisher	Vienna University of Economics and Business, Austria
Yee Leung	Chinese University of Hong Kong, China

International Liaison Chairs

Ana Carla P. Bitencourt	Universidade Federal do Reconcavo da Bahia, Brazil
Claudia Bauzer Medeiros	University of Campinas, Brazil
Alfredo Cuzzocrea	ICAR-CNR and University of Calabria, Italy

Marina L. Gavrilova	University of Calgary, Canada
Robert C. H. Hsu	Chung Hua University, Taiwan
Andrés Iglesias	University of Cantabria, Spain
Tai-Hoon Kim	Hannam University, Korea
Sanjay Misra	University of Minna, Nigeria
Takashi Naka	Kyushu Sangyo University, Japan
Rafael D.C. Santos	National Institute for Space Research, Brazil

Workshop and Session Organizing Chairs

Beniamino Murgante	University of Basilicata, Italy

Local Organizing Committee

Ana Maria A.C. Rocha	University of Minho, Portugal (Chair)
Jorge Gustavo Rocha	University of Minho, Portugal
Maria Irene Falcão	University of Minho, Portugal
Maribel Yasmina Santos	University of Minho, Portugal

Workshop Organizers

Advances in Complex Systems: Modeling and Parallel Implementation (ACSModPar 2014)

Georgius Sirakoulis	Democritus University of Thrace, Greece
Wiliam Spataro	University of Calabria, Italy
Giuseppe A. Trunfio	University of Sassari, Italy

Agricultural and Environment Information and Decision Support Systems (AEIDSS 2014)

Sandro Bimonte	IRSTEA France
Florence Le Ber	ENGES, France
André Miralles	IRSTEA France
François Pinet	IRSTEA France

Advances in Web Based Learning (AWBL 2014)

Mustafa Murat Inceoglu	Ege University, Turkey

Bio-inspired Computing and Applications (BIOCA 2014)

Nadia Nedjah   State University of Rio de Janeiro, Brazil
Luiza de Macedo Mourell   State University of Rio de Janeiro, Brazil

Computational and Applied Mathematics (CAM 2014)

Maria Irene Falcao   University of Minho, Portugal
Fernando Miranda   University of Minho, Portugal

Computer Aided Modeling, Simulation, and Analysis (CAMSA 2014)

Jie Shen   University of Michigan, USA

Computational and Applied Statistics (CAS 2014)

Ana Cristina Braga   University of Minho, Portugal
Ana Paula Costa Conceicao
    Amorim   University of Minho, Portugal

Computational Geometry and Security Applications (CGSA 2014)

Marina L. Gavrilova   University of Calgary, Canada
Han Ming Huang   Guangxi Normal University, China

Computational Algorithms and Sustainable Assessment (CLASS 2014)

Antonino Marvuglia   Public Research Centre Henri Tudor,
                                Luxembourg
Beniamino Murgante   University of Basilicata, Italy

Chemistry and Materials Sciences and Technologies (CMST 2014)

Antonio Laganà   University of Perugia, Italy

Computational Optimization and Applications (COA 2014)

Ana Maria A.C. Rocha   University of Minho, Portugal
Humberto Rocha   University of Coimbra, Portugal

Cities, Technologies and Planning (CTP 2014)

Giuseppe Borruso   University of Trieste, Italy
Beniamino Murgante   University of Basilicata, Italy

Computational Tools and Techniques for Citizen Science and Scientific Outreach (CTTCS 2014)

Rafael Santos   National Institute for Space Research, Brazil
Jordan Raddickand   Johns Hopkins University, USA
Ani Thakar   Johns Hopkins University, USA

Econometrics and Multidimensional Evaluation in the Urban Environment (EMEUE 2014)

Carmelo M. Torre   Polytechnic of Bari, Italy
Maria Cerreta   University of Naples Federico II, Italy
Paola Perchinunno   University of Bari, Italy
Simona Panaro   University of Naples Federico II, Italy
Raffaele Attardi   University of Naples Federico II, Italy

Future Computing Systems, Technologies, and Applications (FISTA 2014)

Bernady O. Apduhan   Kyushu Sangyo University, Japan
Rafael Santos   National Institute for Space Research, Brazil
Jianhua Ma   Hosei University, Japan
Qun Jin   Waseda University, Japan

Formal Methods, Computational Intelligence and Constraint Programming for Software Assurance (FMCICA 2014)

Valdivino Santiago Junior   National Institute for Space Research
(INPE), Brazil

Geographical Analysis, Urban Modeling, Spatial Statistics (GEOG-AN-MOD 2014)

Giuseppe Borruso   University of Trieste, Italy
Beniamino Murgante   University of Basilicata, Italy
Hartmut Asche   University of Potsdam, Germany

High Performance Computing in Engineering and Science (HPCES 2014)

Alberto Proenca   University of Minho, Portugal
Pedro Alberto   University of Coimbra, Portugal

Mobile Communications (MC 2014)

Hyunseung Choo   Sungkyunkwan University, Korea

Mobile Computing, Sensing, and Actuation for Cyber Physical Systems (MSA4CPS 2014)

Saad Qaisar   NUST School of Electrical Engineering and
                            Computer Science, Pakistan
Moonseong Kim   Korean Intellectual Property Office, Korea

New Trends on Trust Computational Models (NTTCM 2014)

Rui Costa Cardoso   Universidade da Beira Interior, Portugal
Abel Gomez   Universidade da Beira Interior, Portugal

Quantum Mechanics: Computational Strategies and Applications (QMCSA 2014)

Mirco Ragni   Universidad Federal de Bahia, Brazil
Vincenzo Aquilanti   University of Perugia, Italy
Ana Carla Peixoto Bitencourt Universidade Estadual de Feira de Santana
                            Brazil
Roger Anderson   University of California, USA
Frederico Vasconcellos Prudente Universidad Federal de Bahia, Brazil

Remote Sensing Data Analysis, Modeling, Interpretation and Applications: From a Global View to a Local Analysis (RS2014)

Rosa Lasaponara   Institute of Methodologies for Environmental
                                Analysis National Research Council, Italy
Nicola Masini   Archaeological and Monumental Heritage
                                Institute, National Research Council, Italy

Software Engineering Processes and Applications (SEPA 2014)

Sanjay Misra   Covenant University, Nigeria

Software Quality (SQ 2014)

Sanjay Misra   Covenant University, Nigeria

Advances in Spatio-Temporal Analytics (ST-Analytics 2014)

Joao Moura Pires   New University of Lisbon, Portugal
Maribel Yasmina Santos   New University of Lisbon, Portugal

Tools and Techniques in Software Development Processes (TTSDP 2014)

Sanjay Misra   Covenant University, Nigeria

Virtual Reality and its Applications (VRA 2014)

Osvaldo Gervasi   University of Perugia, Italy
Lucio Depaolis   University of Salento, Italy

Workshop of Agile Software Development Techniques (WAGILE 2014)

Eduardo Guerra   National Institute for Space Research, Brazil

Big Data:, Analytics and Management (WBDAM 2014)

Wenny Rahayu   La Trobe University, Australia

Program Committee

Jemal Abawajy	Daekin University, Australia
Kenny Adamson	University of Ulster, UK
Filipe Alvelos	University of Minho, Portugal
Paula Amaral	Universidade Nova de Lisboa, Portugal
Hartmut Asche	University of Potsdam, Germany
Md. Abul Kalam Azad	University of Minho, Portugal
Michela Bertolotto	University College Dublin, Ireland
Sandro Bimonte	CEMAGREF, TSCF, France
Rod Blais	University of Calgary, Canada
Ivan Blecic	University of Sassari, Italy
Giuseppe Borruso	University of Trieste, Italy
Yves Caniou	Lyon University, France
José A. Cardoso e Cunha	Universidade Nova de Lisboa, Portugal
Leocadio G. Casado	University of Almeria, Spain
Carlo Cattani	University of Salerno, Italy
Mete Celik	Erciyes University, Turkey
Alexander Chemeris	National Technical University of Ukraine "KPI", Ukraine
Min Young Chung	Sungkyunkwan University, Korea
Gilberto Corso Pereira	Federal University of Bahia, Brazil
M. Fernanda Costa	University of Minho, Portugal
Gaspar Cunha	University of Minho, Portugal
Alfredo Cuzzocrea	ICAR-CNR and University of Calabria, Italy
Carla Dal Sasso Freitas	Universidade Federal do Rio Grande do Sul, Brazil
Pradesh Debba	The Council for Scientific and Industrial Research (CSIR), South Africa
Hendrik Decker	Instituto Tecnológico de Informática, Spain
Frank Devai	London South Bank University, UK
Rodolphe Devillers	Memorial University of Newfoundland, Canada
Prabu Dorairaj	NetApp, India/USA
M. Irene Falcao	University of Minho, Portugal
Cherry Liu Fang	U.S. DOE Ames Laboratory, USA
Edite M.G.P. Fernandes	University of Minho, Portugal
Jose-Jesus Fernandez	National Centre for Biotechnology, CSIS, Spain
Maria Antonia Forjaz	University of Minho, Portugal
Maria Celia Furtado Rocha	PRODEB and Universidade Federal da Bahia, Brazil
Akemi Galvez	University of Cantabria, Spain
Paulino Jose Garcia Nieto	University of Oviedo, Spain
Marina Gavrilova	University of Calgary, Canada
Jerome Gensel	LSR-IMAG, France

Maria Giaoutzi	National Technical University, Athens, Greece
Andrzej M. Goscinski	Deakin University, Australia
Alex Hagen-Zanker	University of Cambridge, UK
Malgorzata Hanzl	Technical University of Lodz, Poland
Shanmugasundaram Hariharan	B.S. Abdur Rahman University, India
Eligius M.T. Hendrix	University of Malaga/Wageningen University, Spain/Netherlands
Hisamoto Hiyoshi	Gunma University, Japan
Fermin Huarte	University of Barcelona, Spain
Andres Iglesias	University of Cantabria, Spain
Mustafa Inceoglu	EGE University, Turkey
Peter Jimack	University of Leeds, UK
Qun Jin	Waseda University, Japan
Farid Karimipour	Vienna University of Technology, Austria
Baris Kazar	Oracle Corp., USA
DongSeong Kim	University of Canterbury, New Zealand
Taihoon Kim	Hannam University, Korea
Ivana Kolingerova	University of West Bohemia, Czech Republic
Dieter Kranzlmueller	LMU and LRZ Munich, Germany
Antonio Laganà	University of Perugia, Italy
Rosa Lasaponara	National Research Council, Italy
Maurizio Lazzari	National Research Council, Italy
Cheng Siong Lee	Monash University, Australia
Sangyoun Lee	Yonsei University, Korea
Jongchan Lee	Kunsan National University, Korea
Clement Leung	Hong Kong Baptist University, Hong Kong
Chendong Li	University of Connecticut, USA
Gang Li	Deakin University, Australia
Ming Li	East China Normal University, China
Fang Liu	AMES Laboratories, USA
Xin Liu	University of Calgary, Canada
Savino Longo	University of Bari, Italy
Tinghuai Ma	NanJing University of Information Science and Technology, China
Sergio Maffioletti	University of Zurich, Switzerland
Ernesto Marcheggiani	Katholieke Universiteit Leuven, Belgium
Antonino Marvuglia	Research Centre Henri Tudor, Luxembourg
Nicola Masini	National Research Council, Italy
Nirvana Meratnia	University of Twente, The Netherlands
Alfredo Milani	University of Perugia, Italy
Sanjay Misra	Federal University of Technology Minna, Nigeria
Giuseppe Modica	University of Reggio Calabria, Italy

Mario Valle	Swiss National Supercomputing Centre, Switzerland
Pablo Vanegas	University of Cuenca, Equador
Piero Giorgio Verdini	INFN Pisa and CERN, Italy
Marco Vizzari	University of Perugia, Italy
Koichi Wada	University of Tsukuba, Japan
Krzysztof Walkowiak	Wroclaw University of Technology, Poland
Robert Weibel	University of Zurich, Switzerland
Roland Wismüller	Universität Siegen, Germany
Mudasser Wyne	SOET National University, USA
Chung-Huang Yang	National Kaohsiung Normal University, Taiwan
Xin-She Yang	National Physical Laboratory, UK
Salim Zabir	France Telecom Japan Co., Japan
Haifeng Zhao	University of California at Davis, USA
Kewen Zhao	University of Qiongzhou, China
Albert Y. Zomaya	University of Sydney, Australia

Reviewers

Abdi Samane	University College Cork, Ireland
Aceto Lidia	University of Pisa, Italy
Afonso Ana Paula	University of Lisbon, Portugal
Afreixo Vera	University of Aveiro, Portugal
Aguilar Antonio	University of Barcelona, Spain
Aguilar José Alfonso	Universidad Autónoma de Sinaloa, Mexico
Ahmad Waseem	Federal University of Technology Minna, Nigeria
Aktas Mehmet	Yildiz Technical University, Turkey
Alarcon Vladimir	Universidad Diego Portales, Chile
Alberti Margarita	University of Barcelona, Spain
Ali Salman	NUST, Pakistan
Alvanides Seraphim	Northumbria University, UK
Álvarez Jacobo de Uña	University of Vigo, Spain
Alvelos Filipe	University of Minho, Portugal
Alves Cláudio	University of Minho, Portugal
Alves José Luis	University of Minho, Portugal
Amorim Ana Paula	University of Minho, Portugal
Amorim Paulo	Federal University of Rio de Janeiro, Brazil
Anderson Roger	University of California, USA
Andrade Wilkerson	Federal University of Campina Grande, Brazil
Andrienko Gennady	Fraunhofer Institute for Intelligent Analysis and Informations Systems, Germany
Apduhan Bernady	Kyushu Sangyo University, Japan
Aquilanti Vincenzo	University of Perugia, Italy
Argiolas Michele	University of Cagliari, Italy

Athayde Maria Emília
  Feijão Queiroz   University of Minho, Portugal
Attardi Raffaele   University of Napoli Federico II, Italy
Azad Md Abdul   Indian Institute of Technology Kanpur, India
Badard Thierry   Laval University, Canada
Bae Ihn-Han   Catholic University of Daegu, South Korea
Baioletti Marco   University of Perugia, Italy
Balena Pasquale   Polytechnic of Bari, Italy
Balucani Nadia   University of Perugia, Italy
Barbosa Jorge   University of Porto, Portugal
Barrientos Pablo Andres Universidad Nacional de La Plata, Australia
Bartoli Daniele   University of Perugia, Italy
Bação Fernando   New University of Lisbon, Portugal
Belanzoni Paola   University of Perugia, Italy
Bencardino Massimiliano University of Salerno, Italy
Benigni Gladys   University of Oriente, Venezuela
Bertolotto Michela   University College Dublin, Ireland
Bimonte Sandro   IRSTEA, France
Blanquer Ignacio   Universitat Politècnica de València, Spain
Bollini Letizia   University of Milano, Italy
Bonifazi Alessandro   Polytechnic of Bari, Italy
Borruso Giuseppe   University of Trieste, Italy
Bostenaru Maria   "Ion Mincu" University of Architecture and
                          Urbanism, Romania

Boucelma Omar   University Marseille, France
Braga Ana Cristina   University of Minho, Portugal
Brás Carmo   Universidade Nova de Lisboa, Portugal
Cacao Isabel   University of Aveiro, Portugal
Cadarso-Suárez Carmen University of Santiago de Compostela, Spain
Caiaffa Emanuela   ENEA, Italy
Calamita Giuseppe   National Research Council, Italy
Campagna Michele   University of Cagliari, Italy
Campobasso Francesco   University of Bari, Italy
Campos José   University of Minho, Portugal
Cannatella Daniele   University of Napoli Federico II, Italy
Canora Filomena   University of Basilicata, Italy
Cardoso Rui   Institute of Telecommunications, Portugal
Caschili Simone   University College London, UK
Ceppi Claudia   Polytechnic of Bari, Italy
Cerreta Maria   University Federico II of Naples, Italy
Chanet Jean-Pierre   IRSTEA, France
Chao Wang   University of Science and Technology of China,
                          China

Choi Joonsoo   Kookmin University, South Korea

Choo Hyunseung	Sungkyunkwan University, South Korea
Chung Min Young	Sungkyunkwan University, South Korea
Chung Myoungbeom	Sungkyunkwan University, South Korea
Clementini Eliseo	University of L'Aquila, Italy
Coelho Leandro dos Santos	PUC-PR, Brazil
Colado Anibal Zaldivar	Universidad Autónoma de Sinaloa, Mexico
Coletti Cecilia	University of Chieti, Italy
Condori Nelly	VU University Amsterdam, The Netherlands
Correia Elisete	University of Trás-Os-Montes e Alto Douro, Portugal
Correia Filipe	FEUP, Portugal
Correia Florbela Maria da Cruz Domingues	Instituto Politécnico de Viana do Castelo, Portugal
Correia Ramos Carlos	University of Evora, Portugal
Corso Pereira Gilberto	UFPA, Brazil
Cortés Ana	Universitat Autònoma de Barcelona, Spain
Costa Fernanda	University of Minho, Portugal
Costantini Alessandro	INFN, Italy
Crasso Marco	National Scientific and Technical Research Council, Argentina
Crawford Broderick	Universidad Catolica de Valparaiso, Chile
Cristia Maximiliano	CIFASIS and UNR, Argentina
Cunha Gaspar	University of Minho, Portugal
Cunha Jácome	University of Minho, Portugal
Cutini Valerio	University of Pisa, Italy
Danese Maria	IBAM, CNR, Italy
Da Silva B. Carlos	University of Lisboa, Portugal
De Almeida Regina	University of Trás-os-Montes e Alto Douro, Portugal
Debroy Vidroha	Hudson Alley Software Inc., USA
De Fino Mariella	Polytechnic of Bari, Italy
De Lotto Roberto	University of Pavia, Italy
De Paolis Lucio Tommaso	University of Salento, Italy
De Rosa Fortuna	University of Napoli Federico II, Italy
De Toro Pasquale	University of Napoli Federico II, Italy
Decker Hendrik	Instituto Tecnológico de Informática, Spain
Delamé Thomas	CNRS, France
Demyanov Vasily	Heriot-Watt University, UK
Desjardin Eric	University of Reims, France
Dwivedi Sanjay Kumar	Babasaheb Bhimrao Ambedkar University, India
Di Gangi Massimo	University of Messina, Italy
Di Leo Margherita	JRC, European Commission, Belgium

Di Trani Francesco	University of Basilicata, Italy
Dias Joana	University of Coimbra, Portugal
Dias d'Almeida Filomena	University of Porto, Portugal
Dilo Arta	University of Twente, The Netherlands
Dixit Veersain	Delhi University, India
Doan Anh Vu	Université Libre de Bruxelles, Belgium
Dorazio Laurent	ISIMA, France
Dutra Inês	University of Porto, Portugal
Eichelberger Hanno	University of Tuebingen, Germany
El-Zawawy Mohamed A.	Cairo University, Egypt
Escalona Maria-Jose	University of Seville, Spain
Falcão M. Irene	University of Minho, Portugal
Farantos Stavros	University of Crete and FORTH, Greece
Faria Susana	University of Minho, Portugal
Faruq Fatma	Carnegie Melon University,, USA
Fernandes Edite	University of Minho, Portugal
Fernandes Rosário	University of Minho, Portugal
Fernandez Joao P	Universidade da Beira Interior, Portugal
Ferreira Fátima	University of Trás-Os-Montes e Alto Douro, Portugal
Ferrão Maria	University of Beira Interior and CEMAPRE, Portugal
Figueiredo Manuel Carlos	University of Minho, Portugal
Filipe Ana	University of Minho, Portugal
Flouvat Frederic	University New Caledonia, New Caledonia
Forjaz Maria Antónia	University of Minho, Portugal
Formosa Saviour	University of Malta, Malta
Fort Marta	University of Girona, Spain
Franciosa Alfredo	University of Napoli Federico II, Italy
Freitas Adelaide de Fátima Baptista Valente	University of Aveiro, Portugal
Frydman Claudia	Laboratoire des Sciences de l'Information et des Systèmes, France
Fusco Giovanni	CNRS - UMR ESPACE, France
Fussel Donald	University of Texas at Austin, USA
Gao Shang	Zhongnan University of Economics and Law, China
Garcia Ernesto	University of the Basque Country, Spain
Garcia Tobio Javier	Centro de Supercomputación de Galicia (CESGA), Spain
Gavrilova Marina	University of Calgary, Canada
Gensel Jerome	IMAG, France
Geraldi Edoardo	National Research Council, Italy
Gervasi Osvaldo	University of Perugia, Italy

Giaoutzi Maria	National Technical University Athens, Greece
Gizzi Fabrizio	National Research Council, Italy
Gomes Maria Cecilia	Universidade Nova de Lisboa, Portugal
Gomes dos Anjos Eudisley	Federal University of ParaÃba, Brazil
Gomez Andres	Centro de Supercomputación de Galicia, CESGA (Spain)
Gonçalves Arminda Manuela	University of Minho, Portugal
Gravagnuolo Antonia	University of Napoli Federico II, Italy
Gregori M. M. H. Rodrigo	Universidade Tecnológica Federal do Paraná, Brazil
Guerlebeck Klaus	Bauhaus University Weimar, Germany
Guerra Eduardo	National Institute for Space Research, Brazil
Hagen-Zanker Alex	University of Surrey, UK
Hajou Ali	Utrecht University, The Netherlands
Hanzl Malgorzata	University of Lodz, Poland
Heijungs Reinout	VU University Amsterdam, The Netherlands
Henriques Carla	Escola Superior de Tecnologia e Gestão, Portugal
Herawan Tutut	University of Malaya, Malaysia
Iglesias Andres	University of Cantabria, Spain
Jamal Amna	National University of Singapore, Singapore
Jank Gerhard	Aachen University, Germany
Jiang Bin	University of Gävle, Sweden
Kalogirou Stamatis	Harokopio University of Athens, Greece
Kanevski Mikhail	University of Lausanne, Switzerland
Kartsaklis Christos	Oak Ridge National Laboratory, USA
Kavouras Marinos	National Technical University of Athens, Greece
Khan Murtaza	NUST, Pakistan
Khurshid Khawar	NUST, Pakistan
Kim Deok-Soo	Hanyang University, South Korea
Kim Moonseong	KIPO, South Korea
Kolingerova Ivana	University of West Bohemia, Czech Republic
Kotzinos Dimitrios	Université de Cergy-Pontoise, France
Lazzari Maurizio	CNR IBAM, Italy
Laganà Antonio	Department of Chemistry, Biology and Biotechnology, Italy
Lai Sabrina	University of Cagliari, Italy
Lanorte Antonio	CNR-IMAA, Italy
Lanza Viviana	Lombardy Regional Institute for Research, Italy
Le Duc Tai	Sungkyunkwan University, South Korea
Le Duc Thang	Sungkyunkwan University, South Korea
Lee Junghoon	Jeju National University, South Korea

Lee KangWoo	Sungkyunkwan University, South Korea
Legatiuk Dmitrii	Bauhaus University Weimar, Germany
Leonard Kathryn	California State University, USA
Lin Calvin	University of Texas at Austin, USA
Loconte Pierangela	Technical University of Bari, Italy
Lombardi Andrea	University of Perugia, Italy
Lopez Cabido Ignacio	Centro de Supercomputación de Galicia, CESGA
Lourenço Vanda Marisa	University Nova de Lisboa, Portugal
Luaces Miguel	University of A Coruña, Spain
Lucertini Giulia	IUAV, Italy
Luna Esteban Robles	Universidad Nacional de la Plata, Argentina
Machado Gaspar	University of Minho, Portugal
Magni Riccardo	Pragma Engineering SrL, Italy, Italy
Malonek Helmuth	University of Aveiro, Portugal
Manfreda Salvatore	University of Basilicata, Italy
Manso Callejo Miguel Angel	Universidad Politécnica de Madrid, Spain
Marcheggiani Ernesto	KU Lueven, Belgium
Marechal Bernard	Universidade Federal de Rio de Janeiro, Brazil
Margalef Tomas	Universitat Autònoma de Barcelona, Spain
Martellozzo Federico	University of Rome, Italy
Marvuglia Antonino	Public Research Centre Henri Tudor, Luxembourg
Matos Jose	Instituto Politecnico do Porto, Portugal
Mauro Giovanni	University of Trieste, Italy
Mauw Sjouke	University of Luxembourg, Luxembourg
Medeiros Pedro	Universidade Nova de Lisboa, Portugal
Melle Franco Manuel	University of Minho, Portugal
Melo Ana	Universidade de São Paulo, Brazil
Millo Giovanni	Generali Assicurazioni, Italy
Min-Woo Park	Sungkyunkwan University, South Korea
Miranda Fernando	University of Minho, Portugal
Misra Sanjay	Covenant University, Nigeria
Modica Giuseppe	Università Mediterranea di Reggio Calabria, Italy
Morais João	University of Aveiro, Portugal
Moreira Adriano	University of Minho, Portugal
Mota Alexandre	Universidade Federal de Pernambuco, Brazil
Moura Pires João	Universidade Nova de Lisboa - FCT, Portugal
Mourelle Luiza de Macedo	UERJ, Brazil
Mourão Maria	Polytechnic Institute of Viana do Castelo, Portugal
Murgante Beniamino	University of Basilicata, Italy
NM Tuan	Ho Chi Minh City University of Technology, Vietnam

Nagy Csaba	University of Szeged, Hungary
Nash Andrew	Vienna Transport Strategies, Austria
Natário Isabel Cristina Maciel	University Nova de Lisboa, Portugal
Nedjah Nadia	State University of Rio de Janeiro, Brazil
Nogueira Fernando	University of Coimbra, Portugal
Oliveira Irene	University of Trás-Os-Montes e Alto Douro, Portugal
Oliveira José A.	University of Minho, Portugal
Oliveira e Silva Luis	University of Lisboa, Portugal
Osaragi Toshihiro	Tokyo Institute of Technology, Japan
Ottomanelli Michele	Polytechnic of Bari, Italy
Ozturk Savas	TUBITAK, Turkey
Pacifici Leonardo	University of Perugia, Italy
Pages Carmen	Universidad de Alcala, Spain
Painho Marco	New University of Lisbon, Portugal
Pantazis Dimos	Technological Educational Institute of Athens, Greece
Paolotti Luisa	University of Perugia, Italy
Papa Enrica	University of Amsterdam, The Netherlands
Papathanasiou Jason	University of Macedonia, Greece
Pardede Eric	La Trobe University, Australia
Parissis Ioannis	Grenoble INP - LCIS, France
Park Gyung-Leen	Jeju National University, South Korea
Park Sooyeon	Korea Polytechnic University, South Korea
Pascale Stefania	University of Basilicata, Italy
Passaro Pierluigi	University of Bari Aldo Moro, Italy
Peixoto Bitencourt Ana Carla	Universidade Estadual de Feira de Santana, Brazil
Perchinunno Paola	University of Bari, Italy
Pereira Ana	Polytechnic Institute of Bragança, Portugal
Pereira Francisco	Instituto Superior de Engenharia, Portugal
Pereira Paulo	University of Minho, Portugal
Pereira Ricardo	Portugal Telecom Inovacao, Portugal
Pietrantuono Roberto	University of Napoli "Federico II", Italy
Pimentel Carina	University of Aveiro, Portugal
Pina Antonio	University of Minho, Portugal
Pinet Francois	IRSTEA, France
Piscitelli Claudia	Polytechnic University of Bari, Italy
Piñar Miguel	Universidad de Granada, Spain
Pollino Maurizio	ENEA, Italy
Potena Pasqualina	University of Bergamo, Italy
Prata Paula	University of Beira Interior, Portugal
Prosperi David	Florida Atlantic University, USA
Qaisar Saad	NURST, Pakistan

Quan Tho	Ho Chi Minh City University of Technology, Vietnam
Raffaeta Alessandra	University of Venice, Italy
Ragni Mirco	Universidade Estadual de Feira de Santana, Brazil
Rautenberg Carlos	University of Graz, Austria
Ravat Franck	IRIT, France
Raza Syed Muhammad	Sungkyunkwan University, South Korea
Ribeiro Isabel	University of Porto, Portugal
Ribeiro Ligia	University of Porto, Portugal
Rinzivillo Salvatore	University of Pisa, Italy
Rocha Ana Maria	University of Minho, Portugal
Rocha Humberto	University of Coimbra, Portugal
Rocha Jorge	University of Minho, Portugal
Rocha Maria Clara	ESTES Coimbra, Portugal
Rocha Maria	PRODEB, San Salvador, Brazil
Rodrigues Armanda	Universidade Nova de Lisboa, Portugal
Rodrigues Cristina	DPS, University of Minho, Portugal
Rodriguez Daniel	University of Alcala, Spain
Roh Yongwan	Korean IP, South Korea
Roncaratti Luiz	Instituto de Fisica, University of Brasilia, Brazil
Rosi Marzio	University of Perugia, Italy
Rossi Gianfranco	University of Parma, Italy
Rotondo Francesco	Polytechnic of Bari, Italy
Sannicandro Valentina	Polytechnic of Bari, Italy
Santos Maribel Yasmina	University of Minho, Portugal
Santos Rafael	INPE, Brazil
Santos Viviane	Universidade de São Paulo, Brazil
Santucci Valentino	University of Perugia, Italy
Saracino Gloria	University of Milano-Bicocca, Italy
Sarafian Haiduke	Pennsylvania State University, USA
Saraiva João	University of Minho, Portugal
Sarrazin Renaud	Université Libre de Bruxelles, Belgium
Schirone Dario Antonio	University of Bari, Italy
Schneider Michel	ISIMA, France
Schoier Gabriella	University of Trieste, Italy
Schutz Georges	CRP Henri Tudor, Luxembourg
Scorza Francesco	University of Basilicata, Italy
Selmaoui Nazha	University of New Caledonia, New Caledonia
Severino Ricardo Jose	University of Minho, Portugal
Shakhov Vladimir	Russian Academy of Sciences, Russia
Shen Jie	University of Michigan, USA
Shon Minhan	Sungkyunkwan University, South Korea

Shukla Ruchi	University of Johannesburg, South Africa
Silva J.C.	IPCA, Portugal
Silva de Souza Laudson	Federal University of Rio Grande do Norte, Brazil
Silva-Fortes Carina	ESTeSL-IPL, Portugal
Simão Adenilso	Universidade de São Paulo, Brazil
Singh R K	Delhi University, India
Soares Inês	INESC Porto, Portugal
Soares Maria Joana	University of Minho, Portugal
Soares Michel	Federal University of Sergipe, Brazil
Sobral Joao	University of Minho, Portugal
Son Changhwan	Sungkyunkwan University, South Korea
Sproessig Wolfgang	Technical University Bergakademie Freiberg, Germany
Su Le Hoanh	Ho Chi Minh City Technical University, Vietnam
Sá Esteves Jorge	University of Aveiro, Portugal
Tahar Sofiène	Concordia University, Canada
Tanaka Kazuaki	Kyushu Institute of Technology, Japan
Taniar David	Monash University, Australia
Tarantino Eufemia	Polytechnic of Bari, Italy
Tariq Haroon	Connekt Lab, Pakistan
Tasso Sergio	University of Perugia, Italy
Teixeira Ana Paula	University of Trás-Os-Montes e Alto Douro, Portugal
Teixeira Senhorinha	University of Minho, Portugal
Tesseire Maguelonne	IRSTEA, France
Thorat Pankaj	Sungkyunkwan University, South Korea
Tomaz Graça	Polytechnic Institute of Guarda, Portugal
Torre Carmelo Maria	Polytechnic of Bari, Italy
Trunfio Giuseppe A.	University of Sassari, Italy
Urbano Joana	LIACC University of Porto, Portugal
Vasconcelos Paulo	University of Porto, Portugal
Vella Flavio	University of Rome La Sapienza, Italy
Velloso Pedro	Universidade Federal Fluminense, Brazil
Viana Ana	INESC Porto, Portugal
Vidacs Laszlo	MTA-SZTE, Hungary
Vieira Ramadas Gisela	Polytechnic of Porto, Portugal
Vijay NLankalapalli	National Institute for Space Research, Brazil
Villalba Maite	Universidad Europea de Madrid, Spain
Viqueira José R.R.	University of Santiago de Compostela, Spain
Vona Marco	University of Basilicata, Italy

Wachowicz Monica	University of New Brunswick, Canada
Walkowiak Krzysztof	Wroclaw University of Technology, Poland
Xin Liu	Ecole Polytechnique Fédérale Lausanne, Switzerland
Yadav Nikita	Delhi Universty, India
Yatskevich Mikalai	Assioma, Italy
Yeoum Sanggil	Sungkyunkwan University, South Korea
Zalyubovskiy Vyacheslav	Russian Academy of Sciences, Russia
Zunino Alejandro	Universidad Nacional del Centro, Argentina

Sponsoring Organizations

ICCSA 2014 would not have been possible without the tremendous support of many organizations and institutions, for which all organizers and participants of ICCSA 2014 express their sincere gratitude:

Universidade do Minho
(http://www.uminho.pt)

University of Perugia, Italy
(http://www.unipg.it)

University of Basilicata, Italy (http://www.unibas.it)

MONASH University

Monash University, Australia
(http://monash.edu)

Kyushu Sangyo University, Japan
(www.kyusan-u.ac.jp)

Associação Portuguesa de Investigação Operacional
(apdio.pt)

Table of Contents

Workshop on Computational and Applied Statistics (CAS 2014)

Workshop on Computational Algorithms for Sustainability Assessment (CLASS 2014)

General Tracks

The Sustainable Limit of the Real Estate Tax:
An Urban-Scale Estimation Model

Antonio Nesticò, Gianluigi De Mare, and Marco Galante[*]

University of Salerno, Department of Civil Engineering
Via Giovanni Paolo II, 132 - 84084 Fisciano (SA), Italy
{anestico,gdemare}@unisa.it,
mgallo_1986@yahoo.it

Abstract. The equalization criterion is important in relation to the real estate tax. The amount that the taxpayer must pay to the government should be commensurate with the current market value of the asset. In several European Countries, including Italy, the real estate tax is based on cadastral incomes, which often do not express the assets real market value. The evaluation model proposed in this study, developed following the evaluation logic of typical values procedure, aims to get to tax equalization by revising tax rates according to market information properly collected from datasets provided by private or public agencies. The model is set-up following these stages: analysis of the price formation mechanisms for urban properties and choice of evaluation process; selection of real estate features needed to achieve the tax equalization aims; data collection; definition of value function; calculus of the Municipal Property Tax; model calibration. The validity of the model is demonstrated by a case study application carried out on a wide urban area in which are situated about 500 real estate units.

Keywords: municipal property tax, housing market, market value, equalization.

1 Introduction and Aim of the Study

Generally, equal distribution refers to tax redistribution based on a careful assessment of income and according to equity and distributive justice criteria. The idea is naturally extended to urban planning, thus meaning as urban equalization the assignment of equal building powers to all building lands which have the same factual and law state when the urban plan is prepared [1].

The equalization criterion retains its meaning even in relation to the real estate tax, so that the amount that the taxpayer must pay to the government should be commensurate with the current market value of the asset.

The Italian real estate tax, the Municipal Property Tax ("Imposta Municipale Propria" IMP, commonly called IMU), which has to be paid by the owners of real estate assets, including residential houses and rural buildings, is not an equalized tax. Its tax

---

[*] This paper is to be attributed in equal parts to the three authors.

B. Murgante et al. (Eds.): ICCSA 2014, Part III, LNCS 8581, pp. 1–14, 2014.
© Springer International Publishing Switzerland 2014

base, indeed, is a function of the cadastral income, which is calculated accordingly to mass appraisal criteria and not verified over time in regards to recorded prices. For this reason, today it does not provide the real market appreciation [2]. A common result of this is that, for example, urban residential units characterized by different market values, maybe because located in different areas of the city, are subjected to the same tax burden because they belong to the same cadastral category and income class as well as consistency.

The final solution of this important issue, also shared by other European Countries [3], requires the revision of the current land registry value or, more generally, a reform of the Cadastre.

While waiting for some concrete actions – some of them planned a long time ago but still not finished – we aim to introduce with this paper an economic model capable of relating the amount of the Municipal Property Tax to the most probable market value of the taxed asset, thus allowing to pursue the equalization objectives.

After studying the national and international legislation regarding taxation on real estate, the algorithm is structured by pre-identifying the variables which affect the predominant mechanisms of formation of the price of the assets. The variables are studied considering the data sources which are necessaries to work out the exact quantification of taxed properties, thus circumscribing the datasets available from the municipalities in the next phase of model implementation. Through appropriate coefficients, the model allows the operator to considerate the different weight of each feature about the real estate market value. Furthermore, in compliance with the Italian Law, the computational structure distinguishes the case of residential houses from second ones.

The model, which is tested on about 500 urban property located in the central area of the municipality of Montoro Superiore in the Province of Avellino (Italy), is developed assuming as constant the tax revenue resulting from the IMU tax burden on the whole municipal area. Nevertheless, the algorithm may be easily changed by the municipality in order to reduce or increase the citizens tax burden.

The calculations performed explicit formal relations that underlie the model, demonstrating its validity. The model described in this article is quick to be implemented and flexible to use, so it can be applied to different market conditions which are peculiar of various national territorial areas. The model is also easily adaptable with respect to different legislative frameworks, such as those in force in other countries, by simply selecting its parameters in a different way in each situation.

2   The Real Estate Taxation in International Regulations and Italian Law

The real estate tax is almost a standard practice in all Western countries [4, 5], although there are different taxation forms. In France, there are three real estate taxes: the Taxe fonciére (art. 1380 et seq. of the Code général des Impôts), a local tax paid by the real estate owner and function of cadastral income, its location (countryside or city), residential house status and taxpayer annual income; the Taxe d'habitation (art. 1407 et seq. of the Code général des Impôts), a local tax which has to be paid by the tenant and equivalent to a month of rent; the l'Impot de solidarité sur la fortune, a state tax on properties which have a value exceeding 1.300.000 € [6].

The German Constitution provides a property tax, the *Grundsteuer*. It must be paid by property owners and its revenues are given to municipalities. The tax base is calculated from the cadastral income of the property, multiplied by two coefficients, one which depends on the Land of belonging and the other one is established by municipal authorities [7].

The Spanish real estate tax system is very much similar to the Italian one. The Impuesto sobre Bienes Inmuebles (IBI), regulated by Regio Decreto Legislativo 2/2004 and its implementing regulation (417/2006), is paid by the owner of the real property rights, with a tax base equal to the cadastral value, which is about 70% of the market value. The cadastral value is a function of the size and the type (eg, luxury, elegant, civil real estate) of the property unit. The rates are set by local authorities and are generally higher in tourist areas. In addition, there is a capital levy on all properties which have a value exceeding € 700.000 [8].

In the United Kingdom the Council Tax was introduced in 1993 under the Local Government Finance Act. The taxable person is not the owner of the real estate, but the tenant. That is because it is a tax on local services rather than a pure property tax. The housing units are assessed by public administration, which indicate the market value range of the asset. In 1991, the average value, corresponding to band D, was fixed at 80.000 pounds, then revaluated between 2003 and 2006. The band A includes the urban units which value do not exceed 50% of this average value, and the band H in which properties value exceed 300% of that value (that is 320.000 pounds). The average band D is set as the basic parameter. Each local authority determines the tax rate to apply to a band D property inhabited by two adults. After that, as a result, this basic rate allows the agency to calculate the tax properties included in other bands, through constant percentage increases and decreases [9].

It is worth noticing that unlike the other mentioned property taxes, where tax base calculation is function of cadastral values, the Council Tax makes use of the real estate market value. Moreover, the Council Tax payer is the tenant, not the owner of the house, as it happens in other Countries. This occurs only in France with the Taxe d'habitation.

In Italy the first real estate tax dates back to 1992 when the Imposta Straordinaria sugli Immobili (ISI) was introduces, then replaced with the Imposta Comunale sugli Immobili (ICI). The latter has been replaced with the Imposta Municipale Propria (IMP), commonly known as IMU, established by Legislative Decree n. 23 of 14th of March 2011 "Disposizioni in materia di federalismo fiscale municipale". Specifically, the IMU is governed by Articles 8, 9 and 14 of Legislative Decree no. 23/2011; art. 13 of Legislative Decree no. 201 of 6th December 2011, ratified by the Law of 22nd December 2011, no. 214; by Articles 1 to 15 of Legislative Decree no. 504 dated 30th December 1992 and other laws applicable to municipal property taxes.

The IMU assumption of taxation is the possession of any kind of real estate, included the principal residence and rural buildings, categories which were not considered in the ICI.

The IMU amount is calculated by multiplying the tax base by the correspondent rate. There are different tax base calculating methods depending on the case in which the tax is applied to a building, a farm land or a building site.

For the buildings registered in the Cadastre, the tax base value is given by the product of the current cadastral income on 1st January of the year of taxation, revaluated by 5%, for the following multipliers:

160 for buildings belonging to cadastral group A (ordinary properties), except from categories A/10 (offices and private studies), C/2 (warehouses and storage rooms), C/6 (stables, barns and garages) and C/7 (closed or open sheds);

140 for buildings belonging to cadastral group B (ordinary properties) and categories C/3 (workshops for arts and crafts), C/4 (buildings and premises for sports services) and C/5 (bathhouses and establishment of healing waters / spas);

80 for buildings belonging to cadastral categories A/10 and D/5 (lenders, exchange and insurance institutions);

60 for buildings belonging to cadastral group D, except from category D/5; since 1st January 2013, this multiplier will be 65;

55 for buildings belonging to cadastral category C/1 (shops and stores).

The basic rate of the tax is 0,76%. The municipal authority, through a resolution of the City Council, may decide to increase or decrease the basic rate up to 0,3 percentage points.

The basic tax rate of the principal residence and its appliances is 0,4%; The municipal authority may increase or decrease the basic rate up to 0,2 percentage points. The main houses, in addition to discounts on rates, also have a 200 € additional deduction, equally divided among those who have used it as a principal residence. On top of that, there is a further 50 €discount for each child aged up to 26 years old who has the legal residence and who is habitually dwelling the housing unit (for 2012 and 2013), for a maximum of 400 € of bonus deduction. The total maximum deduction is equal to 600 €.

3   An Estimate Model for IMU Equalization

The model proposed in this paper aims to limit the uneven effects due to the exclusive use of the cadastral income for the calculation of IMU tax base, considering the average market value of the reference homogeneous zone as well as the main intrinsic features of each property [10, 11].

In the present work, the model is calibrated to the properties of cadastral group A (excluding A/10, offices and private practices), which are the core of the Italian real estate assets. Nevertheless, the calculus algorithm can be easily extended to properties of different cadastral categories. Since it is not possible to change the IMU tax base, which is strictly defined by the law, the model operates on the rate, which is the only parameter that can be varied by a local administration.

The model is set-up following these stages:

- analysis of the price formation mechanisms for urban properties and choice of evaluation process;
- selection of real estate features needed to achieve the tax equalization aims;
- data collection;

- definition of value function;
- calculus of IMU rate;
- model calibration.

1st Stage. Theoretically, for each homogeneous area and real estate type – as defined by the Italian observatory of the property market (OMI) – it should be written this function:

$$V = f[a(x_1) + b(x_2) + c(x_3) + \cdots]$$ (1)

in which the value V of the property is function of its x_i features and of the correspondents marginal prices $a, b, c, \ldots$ [12].

In practise, however, it is greatly difficult to design the function, because the marginal prices evaluation requires a high number of data and information. Therefore, it is followed the evaluation logic of *typical values procedure*, which involves the application of multiplicative differentiation coefficients of each property to average market value of a specific area. Each coefficient represents a specific real estate feature [13, 14, 15].

2nd Stage. The variables chosen for the model are:

x_1 = date of construction and maintenance status of the building;
x_2 = floor height and presence of lift;
x_3 = prevalent view.

Those features, commonly accepted in literature as fundamental for calculating assets prices, can also being easily and objectively expressed with a qualitative parameter and so translated into a numeric coefficient [12], [16], in order to facilitate a mass appraisal approach for the calculus of the real estate tax [17].

3rd Stage. It consists of finding information for each property about the location (address, cadastral references, OMI micro-zone), cadastral category, income class, consistence and land registry value, residential house *status*. Then, it is built the dataset of maximum and minimum market values registered by OMI for each micro-zone of municipal area.

Is necessary to find a link between OMI real estate typologies and cadastral categories in order to correlate the IMU rate of the specific property with market data. The study shows that OMI typologies "civilian housing" and "manors and cottages" correspond respectively with cadastral categories "A/2 civilian housing type" and "A/7 cottage housing type". Moreover, the OMI typology "economic housing" corresponds with both cadastral categories "A/3 economic housing type" and "A/4 popular housing type". The market value of the asset is also function of the three intrinsic features identified, that are date of construction and maintenance status of the building (x_1), floor height and presence of lift (x_2), prevalent view (x_3). Differentiation coefficients related to chosen features are appointed to each property. x_1 and x_2 are taken from the scale of coefficients proposed by Tamborrino [13], x_3 from the scale of coefficients proposed by Orefice [14].

4th *Stage.* The function asset value is defined. It is necessary to derive the rate to be applied to the tax base for the calculus of the tax amount.

$$Asset\ value = V_{average\ OMI} \cdot \frac{(A \cdot \overline{x_1} + B \cdot \overline{x_2} + C \cdot \overline{x_3})}{A+B+C} \cdot k = V_{average\ OMI} \cdot \alpha \cdot k \qquad (2)$$

in which:

$V_{average\ OMI}$ is the average market value of the specific OMI real estate typology, calculated as the arithmetic mean of maximum and minimum values detected in the specific micro-zone;

$\overline{x_1}$, $\overline{x_2}$ and $\overline{x_3}$ are the coefficients of differentiation normalized calculated by dividing each value by the mode of the relative sample. The relative sample is the set of properties that belong to the same OMI typology. For example, within the generic micro-zone B1, considering only economic housing, for each property $\overline{x_1}$ is calculated by dividing its specific variable x_1 by the mode of the relative sample, composed of only the economic housing located in micro-zone B1;

A, B and C are the weight of each variable, set equal to unit;

α is the ratio:

$$\alpha = \frac{(A \cdot \overline{x_1} + B \cdot \overline{x_2} + C \cdot \overline{x_3})}{A+B+C} \qquad (3)$$

Clearly, if $\alpha > 1$, it means that the property considered has better intrinsic features than the ordinary property, which has $\alpha = 1$. As a result, the property considered has a market value higher than the average property in the specific micro-zone;

k is a scale factor, equal to k_{max} or k_{min} depending on α value. If α is greater than the unit, k_{max} is used; if α is equal to the unit, then $k = 1$ and if α is smaller than the unit, k_{min} is used. Indeed there are two different scales of values, one that goes from the mean value to the maximum value and the other one that goes from the mean value to the minimum value; as a result, there are two different scale factors. k_{max} is equal to (the same applies to k_{min} replacing the maximum values with the minimum ones):

$$k_{max} = \frac{V_{max\ OMI}}{V_{max\ calculus}} \qquad (4)$$

in which $V_{max\ calculus}$ is equal to:

$$V_{max\ calculus} = V_{average\ OMI} \cdot \frac{(A \cdot \overline{x_{1max}} + B \cdot \overline{x_{2max}} + C \cdot \overline{x_{3max}})}{A+B+C} = V_{average\ OMI} \cdot \alpha_{max} \qquad (5)$$

The k factor is utilized for maintaining the asset market value inside the range registered by the OMI, by "stretching" with k_{max} the value scale which goes from the average market value to the maximum value detected and with k_{min} the scale which goes from the average market value to the minimum value registered. In fact, the calculus formula may return a maximum (or minimum) value which is different from the maximum (or minimum) value detected by OMI.

So, if the calculus value of the asset is $V_{max\,calculus}$, there is the formula:

$$Asset\ value = V_{average\,OMI} \cdot \frac{(A \cdot \overline{x_1} + B \cdot \overline{x_2} + C \cdot \overline{x_3})}{A+B+C} \cdot k = V_{average\,OMI} \cdot \alpha_{max} \cdot k_{max} \qquad (6)$$

since:

$$V_{average\,OMI} \cdot \alpha_{max} = V_{max\,calculus} \qquad (7)$$

replacing that in the formula (6) and considering the formula (4), there is:

$$Asset\ value = V_{averageOMI} \cdot \alpha_{max} \cdot k_{max} = V_{max\,calculus} \cdot \frac{V_{max\,OMI}}{V_{max\,calc}} = V_{max\,OMI} \qquad (8)$$

5th Stage. Having calculated the asset value, the new IMU rate is equal to:

$$IMU\ rate = \lambda \cdot z \cdot \frac{(A \cdot \overline{x_1} + B \cdot \overline{x_2} + C \cdot \overline{x_3})}{A+B+C} \cdot k = \lambda \cdot z \cdot \alpha \cdot k = \lambda \cdot z \cdot \Phi \qquad (9)$$

in which λ is the basic rate applied by the City, which is different between residential houses and second ones, $\Phi = \alpha \cdot k$.

The z coefficient is used to take into account the micro-zone in which the property is situated through the following relation:

$$z = \frac{V_{average\,micro-zone}}{V_{average\,TOT}} \qquad (10)$$

in which $V_{average\,TOT}$ is the average of the average market value of all the micro-zones involved in the analysis.

The z zone coefficient, which is essential when there are combined data on properties located in different micro-zones, helps us to retain the information about the difference between the total average market value and the average market value of the specific micro-zone. Indeed, the Φ coefficient expresses the percentage increase of the property price compared to the average market value of the specific micro-zone, and not in relation to the total average market value. In other words, the Φ coefficient only states the percentage difference between the market value of the property and the average market value of the specific micro-zone, while the zone coefficient z is used to reset all the values in an absolute reference scale, balanced on the total average market value of specific OMI housing typology on the entire area involved into analysis.

Obviously, tax rates are forced to remain within the ranges fixed by the Law, which are 0,2%-0,6% for residential houses and 0,46%-1,06% for the second ones.

6th Stage. Once the function that assigns the IMU rate according to the specific characteristics of the property has been produced, its cadastral category and the values actually expressed by the market, the last phase is the model calibration. Considering, according to law, a 250 € lump-sum allowance (basic deduction of € 200 plus € 50 for each dependent child under the age of 26 years old), the total tax revenue is fixed as constant. There are three subsequent phases.

First of all, is calculated the tax revenue resulting from the application of the ordinary rates. It is simple when cadastral incomes and residential house *status* are known:

$$residential\ houses\ IMU\ revenue = \sum_i[(160 \cdot 1{,}05 \cdot R_i \cdot \lambda_1) - 250\ €] \quad (11)$$

$$secondary\ houses\ IMU\ revenue = \sum_i[160 \cdot 1{,}05 \cdot R_i \cdot \lambda_2] \quad (12)$$

160 is the multiplier of R_i, the cadastral income raised by 5%; λ_1 e λ_2 are the basic tax rates of residential and secondary houses; 250 € is the standard reduction for residential houses.

In a second step, implementing the algorithms of the model, the percentage change of rate to be applied is calculated.

Since the new IMU rates cause a different total tax revenue, the last step is a goal seek in which is requested equal tax revenue by changing the basic tax rates, which are at first imposed equal to municipal ordinary rates. The calculations are performed using Excel.

In conclusion, the final IMU rate is the product between the new basic tax rate, which comes from the goal seek to achieve equality of the revenue, the z coefficient related to the micro-zone in which the property is located, and the Φ coefficient, which expresses the effect of the intrinsic asset features.

4   An Urban-Scale Application of the Model

The model is tested on the municipal area of Montoro Superiore (Italy) in which are situated 486 urban units, with the aim of assigning the IMU rate expressed by the formula (9).

Considering the cadastral data and collected information, for each property are assigned the differentiation coefficients relative to date of construction and maintenance status of the building (x_1), floor height and presence of lift (x_2), prevalent view (x_3).

Table 1 shows the maximum and minimum market values of each OMI micro-zone, then expressed in percentage of average value in order to calculate the Φ coefficient. The same table also shows the total average market values and, for each micro-zone, the percentage variations of the average value calculated in relation to total average values, necessary for the calculation of z coefficient.

For example, for civilian housings located in B1 micro-zone, Φ ranges from 0,83 to 1,17. It means that the weight of the intrinsic features is equal to 17% (in addition or decrease) of the basic tax rate λ.

Similarly, considering the different values of z coefficient in the four micro-zones, for civilian housings z varies from 88,89% to 111,70% and affects for just more than 10 percentage points the final rate, in addition and decrease.

Table 1. Market values of the four OMI micro-zones. Φ and z coefficients

OMI typologies	market value (€/m²)											
	micro-zone B1			micro-zone B2			micro-zone C1			micro-zone C2		
	min	max	average	min	max	average	Min	max	average	min	max	average
civilian housing	890	1.250	1.070	890	1.250	1.070	820	1.050	935	810	1.000	905
economic housing	810	1.050	930	810	1.100	955	680	870	775	630	890	760
cottages	1.000	1.350	1.175	1.000	1.350	1.175	890	1.150	1.020	890	1.150	1.020

OMI typologies	percentages of the maximum oscillations for the Φ coefficient (%)											
	micro-zone B1			micro-zone B2			micro-zone C1			micro-zone C2		
civilian housing	83	117	100	83	117	100	88	112	100	90	110	100
economic housing	87	113	100	85	115	100	88	112	100	83	117	100
cottages	85	115	100	85	115	100	87	113	100	87	113	100

OMI typologies	% increase of micro-zone average value compared to the total average value (z coefficient)				
	Total average market values	micro-zone B1	micro-zone B2	micro-zone C1	micro-zone C2
civilian housing	€ 995	107,54%	107,54%	93,97%	90,95%
economic housing	€ 855	108,77%	111,70%	90,64%	88,89%
cottages	€ 1.098	107,06%	107,06%	92,94%	92,94%

The formula utilized for the calculus of the basic tax rate of residential houses $\lambda_{R.H.}$ ($\lambda_{S.H.}$ is the basic tax rate of secondary houses) used to obtain the equal tax revenue, is:

$$\lambda_{R.H.} = 0,40\% + [(\lambda_{S.H.} - 0,76\%) \cdot 2/3] \tag{13}$$

in order to connect the two rates while maintaining the same percentage increment compared to average value, equal to 0,40% for residential houses and 0,76% for secondary houses; and compared to the width of the tax rate range, which is imposed by the law. The rate range for residential houses varies from 0,2% to 0,40% (so it is 0,4 percentage points wide); for secondary houses goes from 0,46% to 1,06% (so its width is equal to 0,6 percentage points).

In case study, if secondary houses tax rate passes from 0,86% (value fixed by the City) to 0,91% – with an increment of 0,15 percentage points calculated in relation to average value (0,76%), equal to a quarter of the width of the range established by the law – residential houses tax rate will raise from 0,40% to 0,50%, with the same percentage increase (equal to a quarter) of the range imposed by the law. The relations utilized for calculations are the following:

$$\lambda_{R.H..} = 0,40\% + [2 \cdot (\lambda_{S.H.} - 0,86\%)], \quad \text{if } \lambda_{S.H.} < 0,86\% \tag{14}$$

$$\lambda_{R.H..} = 0,40\% + (\lambda_{S.H.} - 0,86\%), \quad \text{if } \lambda_{S.H.} > 0,86\% \tag{15}$$

In fact, as far as secondary houses are concerned, in Montoro Superiore there is a first interval which goes from 0,46% to 0,86%, that is from the minimum to the ordinary value, and a second range which varies from the ordinary value 0,86%, to the maximum value 1,06%. The first interval is 0,4 percentage points wide, while the second one 0,2. These ranges are related with the residential houses intervals, which are both 0,2 percentage points wide: 0,2-0,4% and 0,4-0,6%. This is why the two cases must be divided: the first case in which the tax rate of secondary houses is smaller than 0,86% and its related range is the double of the other one; the second one in which the tax rate of secondary houses is bigger than 0,86%, case in which both ranges have the same width.

Table 2 shows the results of calculation expressed as total tax revenue for each of the four micro-zones. It also points out the difference between the tax revenue coming from the model application and the ordinary one coming from the application of ordinary rates already set by the local municipality. At last, the table also shows the new ordinary tax rates, which need to include the coefficients Φ and z.

Both basic rates rise of 0,022%. In particular, the rate for the residential houses goes from 0,400% to 0,422%, while that for second one goes from 0,860% to 0,882%.

Furthermore, table 3 summaries taxpayers' contributory changes.

Table 2. Tax rates and tax revenue

		tax rates	
		current	*proposed*
	residential houses	0,400%	0,422%
	secondary houses	0,860%	0,882%

		micro-zone B1			micro-zone B2		
		A/2	A/3 e A/4	A/7	A/2	A/3 e A/4	A/7
ordinary tax revenue	residential houses	€ 3.675	€ 5	€ 650	€ 522	-	-
	secondary houses	€ 26.801	€ 4.268	€ 3.615	€ 26.618	€ 14.111	-
proposed tax revenue	residential houses	€ 5.247	€ 36	€ 745	€ 748	-	-
	secondary houses	€ 29.636	€ 4.607	€ 3.970	€ 30.838	€ 14.297	-
difference in tax revenue	residential houses	€ 1.573	€ 30	€ 95	€ 226	-	-

Table 2. (*Continued*)

secondary houses	€ 2.835	€ 339	€ 354	€ 4.220	€ 186	-

		micro-zone C1			micro-zone C2		
		A/2	A/3 e A/4	A/7	A/2	A/3 e A/4	A/7
ordinary tax revenue	residential houses	€ 3.883	-	€ 522	€ 2.762	-	-
	secondary houses	€ 25.635	€ 15.085	€ 7.688	€ 21.723	€ 12.634	-
proposed tax revenue	residential houses	€ 1.928	-	€ 477	€ 2.584	-	-
	secondary houses	€ 22.382	€ 12.499	€ 7.398	€ 20.857	€ 11.950	-
difference in tax revenue	residential houses	– € 1.955	-	– € 44	– € 178	-	-
	secondary houses	– € 3.254	– € 2.586	– € 291	– € 866	– € 684	-

tax revenue	micro-zone B1		micro-zone B2	
	residential houses	secondary houses	residential houses	secondary houses
ordinary	€ 4.331	€ 34.684	€ 522	€ 40.729
proposed	€ 6.029	€ 38.213	€ 748	€ 45.135
difference	€ 1.698	€ 3.529	€ 226	€ 4.406

tax revenue	micro-zone C1		micro-zone C2			**total tax revenue**	
	residential houses	secondary houses	residential houses	secondary houses			
ordinary	€ 4.405	€ 48.409	€ 2.762	€ 34.357		ordinary	€ 170.199
proposed	€ 2.405	€ 42.278	€ 2.584	€ 32.807		proposed	€ 170.199
difference	€ 1.999	– € 6.131	– € 178	– € 1.550		difference	-

Table 3. Tax values

	micro-zone B1					
	A/2		A/3 e A/4		A/7	
	residential houses	secondary houses	residential houses	secondary houses	residential houses	secondary houses
average proposed tax	€ 131	€ 780	€ 9	€ 329	€ 373	€ 1.323
max proposed tax	€ 320	€ 1.661	€ 36	€ 560	€ 462	€ 1.671
average ordinary tax	€ 92	€ 705	€ 1	€ 305	€ 325	€ 1.205
max ordinary tax	€ 261	€ 1.504	€ 5	€ 501	€ 428	€ 1.522

Table 3. (*Continued*)

average tax difference	€ 39	€ 75	€ 8	€ 24	€ 47	€ 118
max tax difference	€ 74	€ 185	€ 30	€ 59	€ 61	€ 149

	micro-zone B2					
	A/2		**A/3 e A/4**		**A/7**	
	residential houses	secondary houses	residential houses	secondary houses	residential houses	secondary houses
average proposed tax	€ 93	€ 791	-	€ 223	-	-
max proposed tax	€ 238	€ 1.864	-	€ 564	-	-
average ordinary tax	€ 65	€ 683	-	€ 220	-	-
max ordinary tax	€ 180	€ 1.619	-	€ 589	-	-
average tax difference	€ 28	€ 108	-	€ 3	-	-
max tax difference	€ 62	€ 245	-	€ 42	-	-

	micro-zone C1					
	A/2		**A/3 e A/4**		**A/7**	
	residential houses	secondary houses	residential houses	secondary houses	residential houses	secondary houses
average proposed tax	€ 32	€ 605	-	€ 320	€ 68	€ 673
max proposed tax	€ 284	€ 1.261	-	€ 444	€ 68	€ 1.149
average ordinary tax	€ 64	€ 693	-	€ 387	€ 75	€ 699
max ordinary tax	€ 369	€ 1.446	-	€ 537	€ 75	€ 1.205
average tax difference	– € 32	– € 88	-	– € 66	– € 6	– € 26
max tax difference	-	– € 19	-	– € 15	– € 6	€ 17

	micro-zone C2					
	A/2		**A/3 e A/4**		**A/7**	
	residential houses	secondary houses	residential houses	secondary houses	residential houses	secondary houses
average proposed tax	€ 83	€ 652	-	€ 306	-	-

Table 3. (*Continued*)

max proposed tax	€ 163	€ 1.002	-	€ 459	-	-
average ordinary tax	€ 89	€ 679	-	€ 324	-	-
max ordinary tax	€ 153	€ 1.041	-	€ 466	-	-
average tax difference	– € 6	– € 27	-	– € 18	-	-
max tax difference	€ 10	– € 6	-	-	-	-

5   Conclusions

In several European Countries, including Italy, the real estate tax is based on cadastral incomes, which often do not express the assets real market value. The evaluation model proposed in this essay, developed following the evaluation logic of typical values procedure, aims to get to tax equalization by revising tax rates according to market information properly collected from datasets provided by private or public agencies.

The variables that define the asset price must be chosen carefully during the model definition phase. These features should be quick to assess in order to easily implement the calculation algorithms. These algorithms have a flexible mathematical structure, which allows the operator to set the weight of each feature which affects the tax rate. The effects of extrinsic features on market appraisal, and later on the tax burden, are considered through a zone coefficient whose value is related to actual market prices.

The validity of the model is demonstrated by a case study application carried out on a wide urban area in which are situated about 500 real estate units. Possible future ideas to further developing the model may involve the use of tools such as the Geographic Information System (GIS), which could be important to use assets information efficiently.

References

1. Morano, P.: La stima degli indici di urbanizzazione nella perequazione urbanistica. Alinea, Firenze (2007)
2. European Commission, Directorate General for Employment, Social Affairs and Inclusion: Employment and Social Developments in Europe, Belgium (2012)
3. Buoncompagni, A., Momigliano, S.: Commissione VI del Senato della Repubblica (Finanze e Tesoro). Audizione nell'ambito dell'indagine conoscitiva sulla tassazione degli immobili. Banca d'Italia, Roma (2013)
4. Hellerstein, J.H., Hellerstein, W.: State and Local Taxation, Cases and Materials, 8th edn. American Casebook Series. West Group (2001)
5. Karayan, J.E., Gupta, S., Swenson, C.W., Neff, J.: State and Local Taxation: Principles and Planning. J. Ross Publishing, U.S.A (2004)

6. Crosti, A.: Le nuove imposte patrimoniali sugli attivi posseduti in Francia. Fiscalità & Commercio internazionale, n. 6/2012, IPSOA, Milano (2012)

7. VV. AA.: La tassazione immobiliare in Italia e nei principali Paesi Europei: un confronto internazionale. Agenzia delle Entrate, Roma (2012)

8. Dallera, G.: Imposizione patrimoniale ed esperienze di altri Paesi, in The Libro Bianco. Scuola Superiore dell'Economia e delle Finanze Ezio Vanoni - Ministero dell'Economia e delle Finanze, Roma (2008)

9. Pola, G.: La Council Tax britannica. Tributi in Toscana n. 7, Firenze (2007)

10. De Mare, G., Manganelli, B., Nesticò, A.: Dynamic Analysis of the Property Market in the City of Avellino (Italy). In: Murgante, B., Misra, S., Carlini, M., Torre, C.M., Nguyen, H.-Q., Taniar, D., Apduhan, B.O., Gervasi, O. (eds.) ICCSA 2013, Part III. LNCS, vol. 7973, pp. 509–523. Springer, Heidelberg (2013)

11. De Mare, G., Nesticò, A., Tajani, F.: Building Investments for the Revitalization of the Territory: A Multisectoral Model of Economic Analysis. In: Murgante, B., Misra, S., Carlini, M., Torre, C.M., Nguyen, H.-Q., Taniar, D., Apduhan, B.O., Gervasi, O. (eds.) ICCSA 2013, Part III. LNCS, vol. 7973, pp. 493–508. Springer, Heidelberg (2013)

12. VV. AA.: Linee guida per la valutazione degli immobili in garanzia delle esposizioni creditizie. Associazione Bancaria Italiana, Roma (2011)

13. Tamborrino, M.: Come si stima il valore degli immobili. IlSole24Ore, Milano (2012)

14. Orefice, M.: Estimo civile. UTET, Torino (1995)

15. Morano, P., Tajani, F.: Estimative analysis of a segment of the bare ownership market of residential property. In: Murgante, B., Misra, S., Carlini, M., Torre, C.M., Nguyen, H.-Q., Taniar, D., Apduhan, B.O., Gervasi, O. (eds.) ICCSA 2013, Part IV. LNCS, vol. 7974, pp. 433–443. Springer, Heidelberg (2013)

16. VV. AA.: International Valuation Standards. International Valuation Standards Council, London (2011)

17. Hefferan, M., Boyd, T.: Property taxation and mass appraisal valuations in Australia. Adapting to a New Environment. Property Management 28(3) (2010)

Evaluating Environmental Risk to Technological Hazards, Using GIS Spatial Decision Making

Didier Soto, Florent Renard, and Audrey Magnon

CRGA UMR 5600 EVS, Université Jean Moulin Lyon 3.18,
rue Chevreul. Boîte 20. 69632 Lyon cedex 07, France
{didier.soto,florent.renard}@univ-lyon3.fr

Abstract. This study proposes a methodology allowing a better spatial decision making concerning the technological risks in the Greater Lyon (France). This methodology consists in three steps. Firstly, it is needed to assess the territorial vulnerability of the area, based on semi-structured interviews with experts. Secondly, it is necessary to map the exposed areas to a potential accident concerning the storage and the transport of hazardous materials. Finally, these two variables are combined to produce an environmental risk mapping. One of the main goals of this proceeding is to convert heterogeneous spatial data of assets into a mesh form thanks to a grid that homogenize the shape of the environmental stakes according to their sensibility and exposure to thermal, toxic and overpressure effects. These maps can therefore be considered as preliminary documents for the mediation of industrials, local authorities officers and territorial engineers.

Keywords: "Technological risk", "environmental assets", "GIS", "territorial vulnerability", "heterogeneous data", "homogeneous geographical information", "analytic hierarchy process", "spatial decision making".

1   Introduction

Risk management has been for a long time focused on the control of the hazards, whatever its localization or nature [19, 20]. However, over the last thirty years, an epistemological reversal has occurred. Current researches are nowadays focused on the identification of the most vulnerable areas facing natural and/or technological hazards. Social sciences have appropriated this concept and turned it into a wider spatiotemporal dimension [6].

In this study, we have been specifically interested in the territorial vulnerability [8]. We considered that there are, within any territory, human, environmental and material assets allowing it to function and to develop. However, if human and material stakes are primarily studied [13], environmental issues are hardly taken into account. In France, recent institutional studies have led to limited spatial models for the territorial management of industrial risks. Thus, the "deterministic" mapping method, used in the delimitation of the PPI (Plans Particuliers d'Intervention – particular plans of rescue), generates a dichotomous perception of risk, while the so-called "probabilistic" method, used in the

B. Murgante et al. (Eds.): ICCSA 2014, Part III, LNCS 8581, pp. 15–25, 2014.
© Springer International Publishing Switzerland 2014

statement of PPRT (Plans de Prévention des Risques Technologiques – prevention plans of technological risks) isn't sufficient for modeling the reality of risk [14]. Moreover, only the lethal effects are considered among the factors of gravity to the detriment of the others human, material and environmental assets.

Concerning the urban community of Lyon (fig. 1), the official documents [1,2,3] [9] [11] barely take into account the environmental stakes in their risk studies. Their resolution levels are modeled on a town-scale or an IRIS, which is a census block defined by the French national statistic office (INSEE). These statistical scales are insufficient and do not fit with the local constraints, mainly because of their various shapes and sizes [16].

However, environmental assets can be considered as vulnerable targets, able to strike a whole territory, especially during a technological disaster caused by the transport of hazardous materials. Such an accident may considerably impact the environment, causing irreversible damages (toxic contamination of water resources) and generating significant loss to the local community. One of these accidents already happened in the Greater Lyon, in the commune[1] of Pierre-Bénite on December 2001, when a traffic accident led to a 15 cubic meters release of hydrocarbons in a wastewater treatment station through the storm drain [4].

The aim of our study is to experiment such scenarios inside a particularly vulnerable territory regarding storage and transport of hazardous materials. Indeed, according to the Seveso II directive, the Greater Lyon concentrates 23 "high-risk" and 7 "low-risk" industrial establishments and it is noticeable that an intense circulation of hazardous materials is organized between them, by roads, railways, navigable waterways and pipelines (fig. 1). However, the agglomeration has many environmental assets. The Greater Lyon carries out 12 "Projets Nature" (Nature Projects) in the western hills, along its rivers or in the eastern plains. One of these projects ("Rhône aval-îles et lônes") is of great interest as it intersects the limits of the northern Chemical Valley PPRT perimeter.

2   Methodology

Our methodology can be subdivided into three steps. The first one has consisted in quantifying and qualifying the environmental assets vulnerability, through semi-structured interviews with local experts. The second one has led to digitize all the PPRT perimeters, in order to map the entire areas potentially exposed to an industrial risk. In a third step, we have proceeded to a combination between the spatial occurrences of hazard and vulnerability values, so as to create an urban risk mapping.

2.1   Quantifying and Qualifying the Environmental Stakes Vulnerability with a GIS

First of all, it may be useful to remember that there are three common effects to technological hazards. It can therefore be distinguished the thermal effects (a burning or a boil-over) from the toxic (accidental release of a chemical substance) and the overpressure (a blast or a Boiling Liquid Expanding Vapor Explosion-BLEVE) ones. For practical reasons, this paper is only focused on the toxic effect on environmental assets, but will be expanded to the other effects during further research.

---

[1] Smallest administrative division in France.

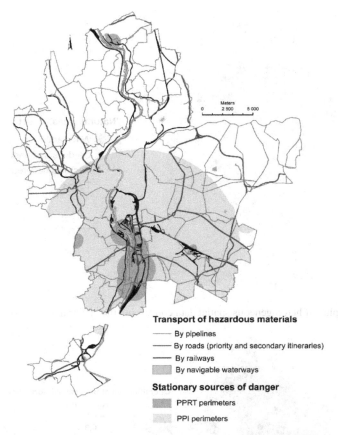

Transport of hazardous materials

‒‒‒‒‒ By pipelines
‒‒‒‒‒ By roads (priority and secondary itineraries)
‒‒‒‒‒ By railways
▮ By navigable waterways

Stationary sources of danger

▮ PPRT perimeters
▮ PPI perimeters

Fig. 1. Perimeters and pathways exposed to an accident during the transport and/or the storage of hazardous materials

Quantifying and qualifying the environmental stakes require first to work out a territorial diagnosis of the study area. Six main environmental items have been identified: "croplands", "natural areas", "protected natural areas", "water resources", "catchment areas of drinking water" and "roadside trees". In order to map these items, two main databases have been used: the ATLAS database from the Greater Lyon and the GEORHONEALPES open database, which allows downloading quantitative and qualitative data on a regional scale [15].

The main difficulty has been to convert these heterogeneous data into a standardized mapping to evaluate the environmental vulnerability. Indeed, environmental assets are displayed into punctual, linear or surface layers. Consequently, it has been necessary to convert them to create homogeneous data. For example, for the "natural areas" asset (fig.2), in a first time, it has been proceeded, thanks to a Geographical Information System (GIS)[2], to a standardization of the spatial data with a global grid

---

[2] All the geo-statistical proceedings have been realized with the *ArcGIS 10* software (Service Packs 1 & 2) developed by ESRI.

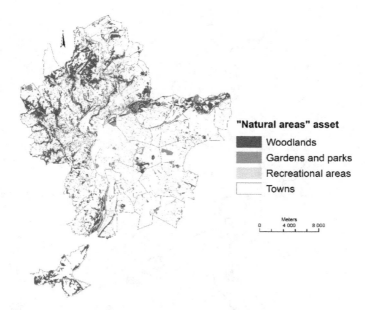

a)   Display of heterogeneous spatial data concerning the "natural areas" asset

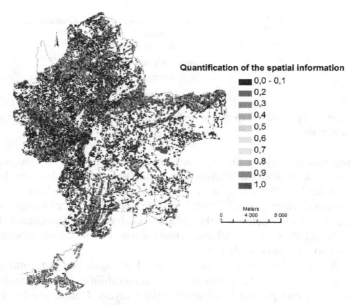

b)   Homogeneous quantitative values of "natural areas" asset after computation

Fig. 2. Result of a combination of heterogeneous zonal layers into a homogeneous quantified mesh

of 100 square meters meshes, allowing a finer view [12]. In a second time, all of the environmental targets have been weighted according to their spatial extent. A quantization factor has then been defined as a dimensionless variable, ranging from 0 to 1, "0" indicating the absence of the target in the mesh and "1" its maximal concentration. However, converting heterogeneous data to a homogeneous mesh is not subject to an automatic process. Consequently, a geo-statistical proceeding model has been created with the software application Model Builder, which consists in programming series of semi-automatic workflows. For this purpose, figure 2 displays normalized geographical information about the "natural areas" stake, resulting from a combination of different surface data.

After the standardization of the environmental information, environmental targets have been assigned weighting criteria, in order to calculate their vulnerability functions. This step relies on experts judgments. In this case [16], we focus on multi-criteria

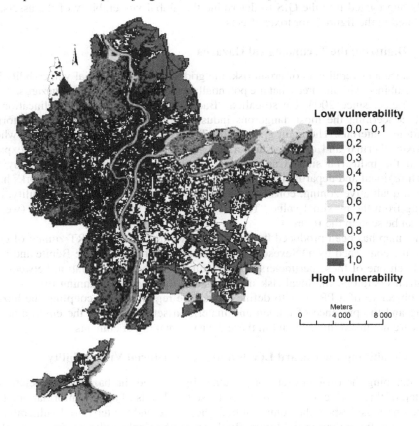

Fig. 3. Vulnerability of the environmental assets to a toxic effect

Vulnerability to a toxic effect = 0.093 "croplands" + 0.114 "natural areas" + 0.108 "protected natural areas" +0.269 "water resources" + 0.334 "catchment areas of drinking water" + 0.082 "roadside trees".

methods, in order to prioritize elements, and specifically on the Analytic Hierarchy Process, developed by T.L. Saaty, because of its easiness and its accessibility [17-18]. On this basis, the assessment of the environmental assets vulnerability has been conducted by interviewing 18 experts, familiar with this issue, chosen among local authorities risk managers, security officers, engineers, scholars and lecturers. During these semi-structured interviews, each one has carried a binary comparison of assets from a questionnaire designed with a specific software (Expert Choice), at the end of which a Consistency Ratio (CR) is calculated. All the interviews have been completed with a CR, whom value do not exceed the inconsistency threshold of 0.02 [17,18].

At the end of the interviews, we can therefore get comparison matrices, which provide a weighting value for all the environmental stakes, based on the same ranging (from "0" to "1") as their spatial concentration. These two statistical variables are finally aggregated into the GIS to determine the global vulnerability of the assets, as illustrated in the figure 3 for toxic effects.

2.2   Digitizing the Technological Hazards

To produce an overall map of urban risk, the gridding of the territorial vulnerability has to be combined with the areas that are potentially exposed to technological hazards.

In France, since 2003, the so-called "Bachelot" law provides the delineation of PPRT concerning the most dangerous industrial sites that are handling and storing hazardous materials. The main goal of PPRT is to identify a perimeter within where the toxic, thermal and overpressure effects may be visualized in case they expand beyond the industrial sites. Currently, these perimeters are being finalized by the French regional and departmental planning offices (DREAL, DDT). Their works have led to a qualitative zoning, considered as a function of intensity and probability, and ranging from the maximal value "TF+" (very strong +) to the minimal "Fai" (weak), as it can be seen in the figure 4.

This map has been produced by digitizing, into the GIS, the PPRT zoning of each commune concerned by a "Seveso" industrial site (Saint-Fons, Pierre-Bénite and Fey-zin). Each one of these perimeters is the result of an ongoing mediation between the industrial operators, the local risk managers and the regional planning officers. The main objective of a PPRT is to define a land-use regulation, by coupling the hazard zoning and the presence of human and material assets. However, the environmental stakes are not taken into account in these decision-making documents.

2.3   Combining the Hazard Levels and the Territorial Vulnerability

After mapping the different territorial vulnerabilities and the hazards perimeters, the last step of this work consists in merging these two levels of spatial information. This proceeding must respect the zoning of both the hazard and the territorial vulnerability, without modifying their initial limits. By this way, the final result (fig. 5) is not only a simple superimposition of these two layers, but rather an integrated map of the terri-torial risk. It therefore allows an operating visualization of the potential dangers. Combining hazards and vulnerability into a GIS is not a common practice – and the few existing studies could be improved, either because they try to convert the qualita-tive hazards perimeters into quantitative values [7], or because they do not take into account all of the assets [5].

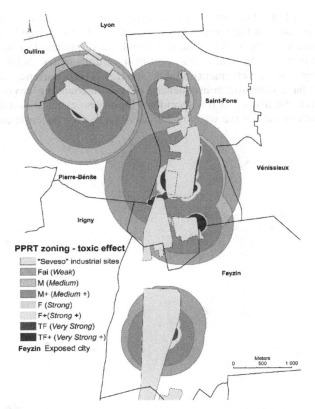

Fig. 4. Zoning of the potential areas exposed to a toxic effect in the northern part of the Chemical Valley

The combination of hazard and vulnerability is still an ongoing project concerning our study area. However, the first maps have just been produced, as it can be seen in figure 5. This one displays the coupling of the spatial extent of a toxic effect and the territorial vulnerability in the southern part of the Greater Lyon, where are located the most dangerous plants. Figure 5 displays a qualitative determination of the risk by combining the quantitative mesh of the territorial vulnerability (values ranging from to "1" to "6"[3]) and the PPRT zoning of the toxic effect (qualitative values ranging from "Fai" to "TF+"). For instance, the code "2TF" means that, for a 100 m² mesh, vulnerability value reaches 2/10, while hazard exposition may lead to a very strong toxic effect (TF).

3   Results: Assessment of the Environmental Risk to a Technological Accident

Figures 3, 4 and 5 reveal three different kinds of geographical realities. The first one concerns the territorial vulnerability. The most vulnerable places are located in the

---

[3] The ranging is the same as in figure 3, except that the values have been multiplied by 10 to facilitate encoding.

northeastern part of the Greater Lyon, corresponding to the alluvial plains of the Rhône. The main asset is the storage basin called "Le Grand Large", which can be easily seen on the map (fig.3). This one is particularly vulnerable to toxic effects and is located close to a priority road of transport of hazardous materials (N346). Moreover, this pathway crosses a channel (Canal de Jonage). The occurrence of an accident could therefore have dramatic consequences on the surrounding environmental resources. However, the history of road accidents reports no event within the "Miribel-Jonage" Park, where most of the water resources of Lyon are concentrated.

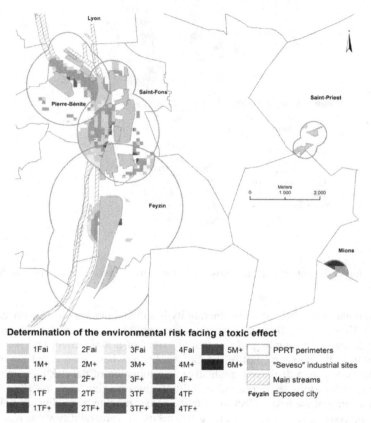

Determination of the environmental risk facing a toxic effect

1Fai	2Fai	3Fai	4Fai	5M+		PPRT perimeters
1M+	2M+	3M+	4M+	6M+		"Seveso" industrial sites
1F+	2F+	3F+	4F+			Main streams
1TF	2TF	3TF	4TF		Feyzin	Exposed city
1TF+	2TF+	3TF+	4TF+			

Fig. 5. Qualification of the environmental risk in the Southern part of the Greater Lyon

According to our results, the environmental vulnerability depends on the type of assets considered. Assets are particularly vulnerable to toxic effects, precisely the water resources (the two main streams) and the catchment areas of drinking water. However, if these environmental stakes are vulnerable, croplands and woodlands are more resistant according to the experts. Consequently, the western cultivated and woody hills [3] appear less impacted.

Figures 4 and 5 are more focused on the southern part of the Greater Lyon. In this part of the agglomeration, a huge industrial complex is located, known as the Chemical Valley, where large amount of hydrocarbons are stored and transformed. The main industrial sites involved here are Solvay, Total, Arkema and Bluestar Silicones.

For practical reasons, it has not been possible to display a whole map because sizes of the other zonings are very smaller than those of the cities of Saint-Fons, Feyzin and Pierre-Bénite. We can notice that these three towns would be almost covered by toxic fumes and/or toxic cloud if an accident occurred, according to the more "probabilistic" scenario. The "deterministic one" (fig.1) shows that more than half of the Greater Lyon would be affected.

By combining the spatial extent of the hazards with the territorial functions of vulnerability, risk is solely confined in the southern part of Lyon. Figure 5 displays a huge exposed area, corresponding with the Chemical Valley. A major environmental resource could potentially be affected: the Rhône River, running close to the plants; risk seems particularly high near "Edouard Herriot" harbor, where large quantities of oil are stored (fig.6). Risk reaches a value of "4F+", as well as the channel near the Total oil refinery.

Fig. 6. The Rhône River running close to "Edouard Herriot" harbor

Surprisingly, the commune where environmental risk reaches its highest values is Mions, in the southeastern part of the Greater Lyon. This one shelters an important catchment area of drinking water, both located under a priority road of transport of hazardous materials (A46 highway), close to a gas pipeline and in the perimeter of a PPRT (toxic effect). The combination of all these elements explains the values "5M+" and "6M+" reached by some meshes. Moreover, in this commune, an accident already happened on January 2006, when a truck, containing seven tons of zinc chloride, collapsed. It caused therefore the contamination of a cultivated field [4].

It is to be noted that this area has not been identified by the local authorities, as it is not mentioned in the most recently published official documents [1] [11]. It is undoubtedly a key advantage of our methodology to reveal new risk territories by highlighting unknown interactions between vulnerability and hazards.

4 Conclusion

This paper, aiming at evaluating the environmental vulnerability regarding technological hazards, has two main interests. The first is the use of a GIS, by which several environmental data have been compiled from many different databases. The original data were heterogeneous both in shape and size. The first task has therefore been to standardize and to quantify the environmental assets thanks to the geo-proceeding tools offered by the GIS. Thus, each stake has been converted into a mesh form of 100 meters wide. Thanks to this approach, we obtain a detailed and consistent evaluation of the territorial vulnerability of the Greater Lyon (fig.3). However, this methodology is not only restricted to the field study of this paper. It can be applied to any kind of hazards and territories. A similar study has already been done concerning the flooding hazard in the Greater Lyon [16] and is planned concerning a new area in the State of São Paulo in Brazil.

The second interest of this study is to propose a methodology combining the vulnerability and technological hazards. Each one of the PPRT covering our study area has been digitized to produce a global map (fig.4) displaying the exposure to toxic, thermal and overpressure effects, even though only toxic effects are discussed in this paper. Then, the quantitative evaluation of the vulnerability has been combined with the qualitative zoning of each effect. Thanks to our mapped results (fig.5), two areas are identified in the southern part of the Greater Lyon where environmental assets are exposed: one into the Chemical Valley industrial complex along the Rhône River; the other in the commune of Mions, exposing a large drinking water catchment area to multiple hazards. Therefore, this approach allows a better spatial decision making by highlighting still unconsidered environmental problems in previous risk studies.

However, this map, which is a first result of an ongoing project, cannot be considered as a definitive document. The first reason is that risk is only partially taken into account since the effects of an accident during the transport of hazardous materials have not yet been modeled. We only focus, in a first part, on the storage of these materials because hazard zoning has already been done within the PPRT statement. Proceeding to the same exercise with the transport of hazardous materials is much more random, according to the many scenarios and parameters which could lead to an accident. However, even if there are many constraints, taking into account this parameter would necessarily extend the areas exposed to a danger, as revealed by the history of accidents; it would therefore lead to a new map of urban risk.

The second reason is that the aggravating factors (domino effect for example) or the resilient ones (greater or lesser proximity from an emergency center for example) have not yet been considered [10] as well as the temporal rhythm, by proceeding to daily and nocturnal approaches.

Acknowledgments. We express our thanks to M. Kenji Fujiki for his advice and proofreading, and the 18 experts interviewed.

References

1. Agence d'urbanisme pour le développement de l'agglomération lyonnaise : Atlas des risques technologiques et de la vulnérabilité de l'agglomération lyonnaise (2005)
2. Agence d'urbanisme pour le développement de l'agglomération lyonnaise : Indicateurs d'exposition des populations du Grand Lyon aux risques naturels et technologiques. Observatoire du développement durable (2010)
3. Agence d'urbanisme pour le développement de l'agglomération lyonnaise : Les espaces naturels et agricoles de l'aire métropolitaine lyonnaise. Représentations métropolitaines, cartes au 1 :200 000ème (2011)
4. ARIA database, http://www.aria.developpement-durable.gouv.fr/
5. Armenakis, C., Nirupama, N.: Prioritization of disaster risk in a community using GIS. Natural Hazards 66, 15–29 (2013)
6. Becerra, S.: Vulnérabilité, risques et environnement : l'itinéraire chaotique d'un paradigme sociologique contemporain. VertigO – la Revue Électronique en Sciences de l'environnement (En ligne) 12 (2012), http://vertigo.revues.org/11988
7. Caradot, N., Granger, D., Chapgier, J., Cherqui, F., Chocat, B.: Urban flood risk assessment using sewer flooding databases. Water Science & Technology 64, 832–840 (2011)
8. D'Ercole, R., Metzger, P.: La vulnérabilité territoriale: une nouvelle approche des risques en milieu urbain. Cybergéo : European Journal of Geography, http://cybergeo.revues.org/22022
9. Grand Lyon: Cahier risques majeurs du référentiel environnemental du Grand Lyon (2004)
10. Garbolino, E., Lachtar, D.: Vulnérabilité et résilience face aux TMD dans un contexte transfrontalier. In: Bersani, C., Sacile, E. (eds.) Sécurité des Transports de Marchandises dans l'Eurorégion Alpes-Méditerranée, pp. 186–255 (2012)
11. Grand Lyon: Rapport de présentation – État initial de l'environnement – SCOT de l'agglomération lyonnaise (2010)
12. Kienberger, S., Lang, S., Zeil, P.: Spatial vulnerability units – expert based spatial modeling of socio-economic vulnerability in the Salzach catchment, Austria. Natural Hazards and Earth System Sciences 9, 767–778 (2009)
13. Leone, F.: Caractérisation des vulnérabilités aux « catastrophes naturelles » : contribution à une évaluation géographique multirisque (mouvements de terrain, séismes, tsunamis, éruptions volcaniques, cyclones). Mémoire d'Habilitation à Diriger des Recherches, Université Montpellier 3 (2007)
14. Propeck-Zimmermann, E.: Caractériser les enjeux et les vulnérabilités : de l'analyse spatiale à un mode de représentation adapté à la concertation. In: Galland, J.P., Martinais, E. (eds.) La Prévention des Risques Industriels en France 2007 – 2009, Paris (2010)
15. PRODIGE database, http://www.georhonealpes.fr
16. Renard, F., Chapon, P.M.: Une méthode d'évaluation de la vulnérabilité urbaine appliquée à l'agglomération lyonnaise. L'espace Géographique 39, 35–50 (2010)
17. Saaty, T.L.: The Analytic Hierarchy Process: Planning, Priority, Setting, Resource Allocation. McGraw-Hill, New-York (1980)
18. Saaty, T.L.: Relative measurement and its generalization in decision making. Why pairwise comparisons are central in mathematics for the measurement of intangible factors. The Analytic/Network Process. Review of the Royal Spanish Academy of Sciences, Series A, Mathematics 102, 251–318 (2008)
19. Veyret, Y., Reghezza, M.: Aléas et risques dans l'analyse géographique. Annales des Mines 40, 61–69 (2005)
20. Veyret, Y., Reghezza, M.: Vulnérabilité et risques. L'approche récente de la vulnérabilité. Annales des Mines 43, 9–13 (2006)

An Evaluation Model for the Actions in Supporting of the Environmental and Landscaping Rehabilitation of the Pasquasia's Site Mining (EN)

Fabio Naselli[1], Maria Rosa Trovato[2], and Gianpaolo Castello[1]

[1] Kore University of Enna
[2] University of Catania

Abstract. The mining activities in Sicily, over the years, have had a gradual decline until to their total abandonment, determining a degradation process for the territory and the landscape. This abandonment process, has increased the vulnerable of these sites, that often, are became incubators or sprinklers of some poisons, in them illegally allocated, becoming dangerous for the agricultural activities, the water tables, and the health of the communities nearest. Among the mining sites in Sicily's most historic and cultural interest, which in recent years has been recalled for some cases of environmental damage, there is the Pasquasia' site (EN). After the closed and the abandoned in the 1992, it was transformed into an open dump of the hazardous waste; in particular of the asbestos cement type. The study suggests, at starting to a baseline scenario as that of the remediation for the site, an evaluation model for some redevelopment integrated actions.

Keywords: Landscape mining, environmental retraining, the refunctionalization project for a mining site, Cost Benefit Analysis.

1   Introduction

The mining site Pasquasia insists on a portion of the area of approximately 40 hectares, situated in a central position of the Sicily. It is located 13 km from the town of Enna, and about 20 km from the city of Caltanissetta. The site is in front of the SS 117 bis, which connects the A19 PA-CT of the road SS 121 Caltanissetta–Agrigento. Pasquasia field was discovered in 1952, but only in 1959 the Department issued the decree No. 715, which granted the concession of the mine of mixed alkali salts of sodium, potassium, bromine, iodine mine called "Pasquasia", for a period of thirty years, to the company "Salts Trinacria" of Palermo. In the following years the mine has undergone various changes that have increased the extent and the importance from the economic point of view, and has always seen Sicily as a key player along with other private enterprises. The mining site Pasquasia was one of the largest deposits of kainite of Europe [1]: five layers covering an area of about eight kilometers. Existing industrial activities produce potassium sulphate for valuable crops, of which Italy is rich. Italy consumes about 600000 tons of potassium fertilizers, of which more

B. Murgante et al. (Eds.): ICCSA 2014, Part III, LNCS 8581, pp. 26–41, 2014.

than 380000 of 250000 of potassium chloride and potassium sulfate [1]. The Italkali SPA Company, because of the different reasons, suspended mining and processing in 1992.

2   Methodology for the Case Study

2.1   The Landscape

The study proposes an evaluation model to support the choice of redevelopment and renovation of the mining site. Elaborate scenarios arise from the aggregation of multiple functions to allocate the site and define the different ways of implementing the process of new functions that you want to implement. The aim to define the various scenarios is to mend the relationship of the component mining landscape with the territory, or to redefine a "place" in contrast to the current perception of "non-place". The term landscape assumed both in the common language and scientifically different meanings. Even at the regulatory level for a long time has not been a sufficient clear and unambiguous definition (cf. Law n.1497 of 1939 and 431 in 1985). For the first time, the "European Landscape Convention" (2000) and in Italy the recent "Cultural Heritage and Landscape Code" (Legislative Decree no. N. 42, 2004) are defined the concept of "landscape in a systematic way. Even the OECD in 2001 established the agricultural landscape: "the visible outcomes from the interaction between agriculture, natural resources and the environment, and encompass amenity, cultural, another societal values". At Europe Community level, the assessment questionnaire of the effects of actions taken in the agro environmental field by the Rural Development Program (RDP) made under Article .42 of Regulation (EC) 1795/1999 (STAR Doc. 12004/00) clearly indicates that the effectiveness of interventions in the landscape must be evaluated on the basis on their ability to improve cognitive and visual consistency of the landscape or reinforces the traditional identity elements. In summary, we can conclude that the Landscape:

1. can be considered the visible aspect of an ecological natural or anthropic system (socio-economic eco system) or a specific territory as is perceived by the people who, for various, visit it reasons;
2. the quality depends on both the objective character of the territory and from the aspirations of the population with which it comes into contact;
3. the formulation of landscape policies should be based on the public value to the landscape;
4. the subject of the political landscape and the landscape must be of high quality (to be protected) and the degraded ones (to be upgraded).

The landscape is the result of the interaction between the natural environment and the human intervention and has more historical origin, can put in some contexts the meaning of a good cultural history and as such should be subject to appropriate conservation actions.

For this reasons, the evaluation of the environment can be essentially done in two distinct criteria:

• calculating and accumulating actuality the different types of benefits that the socio-economic system is able to produce for men, in a sustainable way;

• evaluating the current state of the environment and / or its individual components according to the distance that they have compared to an ideal situation, identifiable, at least within certain limits, with the climax.

The first approach is generally used in the economic and estimate field (Costanza et al., 1997), able to provide an exhaustive classification of the benefits produced by natural ecosystems (or partially humanized) also identifying possible methods of economic evaluation [2]. The second approach, however, tends to assess the environment and the landscape primarily by the difference between its ideal state and the current situation. In this case, its value is inversely proportional to the anthropogenic interference. As above, mentioned, there is, of course poses the problem of defining what are the benefits that the landscape can produce and how they can be evaluated. The problem of definition and quantification of the benefits produced by the landscape has become particularly pressing from the moment that the Community level was earmarking for grants to improve its quality. Obviously, in order to evaluate the effectiveness and efficiency of the contributions made it became necessary to have methods of estimating the quality of the landscape both monetary and non-monetary type. In this reasons, the evaluation of the effectiveness of landscape measures was measured using quantitative assessments focused on the use of specific indicators, measurable cardinal scale, while efficiency was measured by applying feedback of monetary, with the ' cost benefit analysis. The basis of monetary valuations of the landscape are to be found mainly in the theory of the consumer and demand balance, those of non-monetary assessments are referred to both in the fields of philosophy in sociology and human ecology.

2.2   The Component of Value

To better understand which categories of benefits can produce the landscape it may be useful to refer to the classification of the value of environmental resources proposed by the Secretariat of the Convention on Biological Diversity (SCBD, 2001):

Table 1. Classification of environmental values

Values of use • current • optional • almost optional	Direct (or active) Indirect (or passive)	Extraction Non extraction
Values of non use	Existence Legacy Vicariate	

The first fundamental distinction is between use values and non-use values. The use-values depend essentially on the possibility of obtaining a personal benefit through physical interaction with the environmental good. This interaction may be voluntary or involuntary. In the case of voluntary interaction you will have direct benefits that will be kind of mining, when they are removed raw materials, flora, fauna, or whatever. This will also cover non-extraction when the users enjoy a service provided by the environment. The direct non-extractive use values are given essentially by the recreational value of the land and resources that are present there. When the use is not a volunteer type will have the benefits of indirect type that essentially assume the nature of services. The values of use, as a whole, have three components: 1) current use, 2) optional, 3) almost optional. The first category includes the benefits associated with the use of certain observable and good. The second category derives from the possibility that some subjects, although not using the asset currently intend to carry an option to be able to use in the future. The option values are a kind of payment of a deposit to make sure you have the right to use the good later in time. Finally, the values almost optional derive from the fact that some people, although they are not sure you want to use the asset in the future, they do not want to deprive themselves of the opportunity to do so. In this case, they could might mended to pay a sort of "insurance premium" to avoid the risk of failing to dispose of the environment if they need it. The non-use values refer mainly to the altruistic component of human behavior. Many people, in fact, give up a part of their income because a resource is maintained completely independent of the ability to benefit from the goods and services it can produce. This selfless attitude can affect the resource itself (existence value), the possibility that benefit future generations (bequest value) or that are to benefit others (value vicar).

As far as the landscape we will have:

• direct use benefits of non-extractive, when a person attends an area for recreation reasons with a pleasant landscape, the importance of the landscape in order to determine the recreational value of the area, however, depends mainly on the type of recreational activity carried out.

• use indirect benefits generally associated to owning a home (both for residential and tourist purposes) in an area where the landscape is more pleasant;

• benefits of non-use due to conservation of the landscape as historical – cultural good.

As mentioned above, therefore, we can gather that the value attributed to the preservation of the landscape tends to increase for people who attend more intensively a given area or that they own a house. In the case of many historical landscapes may also have some importance the conservation value, especially when they take elements of uniqueness and non-reproducibility or when there are traditional rural buildings, stone walls and other hydraulic works of the past easily recognizable.

2.3   No-monetary Values

Over the years, we have faced two different ways of relating to the landscape, which eventually affect the valuation techniques still used today. The two most used techniques are: objectivist paradigm (or physical) and subjectivist paradigm (or psychological). According to the objectivist paradigm, the value (quality) of the landscape is inherent in its components. A closer inspection of the physical features of the landscape allows you to assess their quality[3]. On the contrary, the subjectivists argue that the quality of the landscape is essentially "in the eye of the beholder." In subjectivist approach is often implied a conception of landscape understood as a visual environment in which a man works and lives. When the landscape inspires positive feelings, such as security, relaxation, pleasure, happiness, will be considered subjectively high quality. When arouse the opposite states of stress, fear, insecurity, limiting, etc. will be considered subjectively low quality (cf. Buyoff et al, 1994). Among the subjectivists can be distinguished more paradigms (cf. Friedeldey, 1995), less common, such as:

• cognitive paradigm, according to which the human mind, reworking the incentives coming from the outside, in fact determines the characteristics of the landscape.

• psychophysical paradigm, according to which, on the contrary, the objective characteristics of the landscape are crucial to the aesthetic appreciation.

• interactions paradigm, for which between man and environment there is a relationship of mutual influence. Man and environment establish mechanisms of feedback, so the quality of the landscape depends on both the objective and subjective features.

The different paradigms have led to the definition of different assessment methods and, in many ways, alternative. It can thus identify approaches:

1) Objectives (indirect)
1.1 Historical
1.2 Aesthetic
2) Subjective (visual or perceptual) (or direct)
2.1 Psychological

2.4   Monetary Values

To quantify the monetary value of the landscape can be used both approaches focusing on so-called stated preference (contingent valuation, choice experiments, etc.) and approaches based on revealed preferences (travel cost, random utility models, the hedonic pricing, etc.). Using methodologies focused on stated preferences it can be estimated, at least theoretically, all components of the value of the landscape, and using methods based on revealed preferences, we can only estimate the values of active and passive use. In particular:

• with the travel cost and random utility models is possible to calculate the value in use arising from the performance of active recreational activities carried out in a day;

• with the method of hedonic pricing is possible to estimate the value in use of active vacationers and the value in use of the person both vacationers and residents.

At the national level, the most popular method is contingent valuation, and this is thanks to the versatility of the method that deals with almost all of the issues which will be met with at the operational level in the field of landscape evaluation. The techniques of contingent valuation (CV) tend to identify directly the value of the environmental resource by simulating the existence of a market within which the user can determine their willingness to pay (WTP) [4] or accept (WTA).). It is therefore an instrument very ductile which also sees precisely in its relative simplicity of application many potential sources of error. The central moments of the application of contingent valuation are defined by the definition of the quota market and the type of approach used in order to express the user, also by means of interview, the willingness to pay (WTP) [5],[6]. Both of these steps can introduce factors affecting the goodness of the data obtained. Since the 60s, especially in the United States have been developed methodologies for the determination of the recreational value of natural resources and/or their qualitative and quantitative variations. Since these goods have no an objective market price, it has adopted the estimate of consumer surplus as a leading economic indicator of the social value of an environmental resource. To calculate the consumer surplus a number of methods have been used that, according to the scheme proposed by Mitchell and Carson (1989), can be classified as follows:

a. direct methods based on actual behavior (referendum, simulations of the market, etc.);
b. direct methods based on hypothetical behavior (Contingent Valuation - CV);
c. indirect methods based on actual behavior (Travel Cost, Hedonic Price and other approaches);
d. indirect methods based on hypothetical behavior (willingness to travel).

Among the known methods, scholarly attention has been focused mainly on the cost method of travel (Travel Cost - TC) and the contingent valuation (Contingent Valuation).

The most successful methodology (in 1994 there were more than 1,600 applications), adopted by Carson et al. [7] is the CV with tends to simulate the existence of a market, a questionnaire, in which the users can express their willingness to pay (or Willingness To Pay - WTP) in order to continue to use the resource being valued or accept a refund (or Willingness To Accept - WTA) to give it up. The application of these methods of evaluation meets in general different problems in function of the property that you want to assess and what it is able to satisfy. In any case, the procedure to follow for the assessment of environmental resource is divided into the following stages:

1. - identification of the users of the property and the definition of a sample survey;
2. - collecting the required information to the estimated value from the sample survey using a specially designed questionnaire;
3. - processing of the data collected in order to analyze the application of the good and to calculate the value.

Obtained the information referred to in paragraph 3, you can:

a. define by the Cost Benefit Analysis if any investment of enhancement and / or preservation could be attractive from a social point of view;
b. compare alternative projects of cultural and recreational development;
c. define appropriate management strategies and analyze their effect on social welfare, the local economy, the budget of the managing body and shape of the resource.

3   Results and Discussion of the Case Study

The evaluation process to support the choice of possible scenarios for the site and abandoned mining 'was conducted with the support of mixed non-monetary and monetary assessments for the evaluation of the mining landscape and for the evaluation of possible upgrading interventions.

In particular the general evaluation can be summarized as follows:

1. Finding the perceived value of the site [8];
2. Finding the wtp of users for new services in new functions [9], [10];
3. Cost benefit analysis of interventions;
4. Choice of interventions with the support of decision criteria.

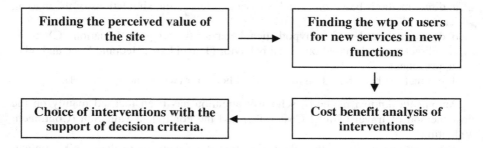

Assessment to support the identification of the perceived value of the site has been conducted in a non monetary approach by administering a questionnaire to a sample of residents in Enna. In addition, the questionnaire has been built specifically to be able to detect the perceived value of the site and wtp [11], on the part of citizens in the benefit of possible services allocated to the site. Therefore, the questionnaire has been instrumental to fulfill the first two points of the evaluation. It is composed of three different types of cards. The interviews have been divided into three age groups respectively between 18 and 35 years, between 35 and 50 years, and 50 years of age. The sample is composed of ninety individuals the first card is perceptive: there is a picture of the site and the land on which it stands, with open-ended questions with explicit reference to the landscape. The second card is cognitive and design planning and gives them total freedom to express their preferences above on the future use of the mining site of Pasquasia. The third card is aimed at identifying the willingness to pay by the given of any intended use of the site. In fact, the results of that board have been instrumental to define the demand for services related to the renovation of the site. Overall, the results were more than satisfactory, and many of the people interviewed, while noting that it was a fairly complex questionnaire, proved to be very helpful and open to the comparison, revealing a high degree of knowledge of the site in question. The data, obtained by this type of evaluation, have been reported in graphical form, divided by bands, in order to make more direct the obtained results.

The results obtained were used for the definition of the scenarios of refunctionalization of the site. It has been possible to define individual complementary and aggregates scenarios. In particular, the possible scenarios of refunctionalization identified for the site of Pasquasia are the following:

A) Reclamation
B) Protection and enhancement of the landscape-mining museum
C) University
D) Energy from renewable sources

The assumptions used are the most appropriate for a scenario like the one proposed by the site, and because they are those that were more reflected in the market and public opinion. The resumption of mining is one of the possibilities often encountered in the cards-interview assessments of non-monetary, but now, in the absence of detailed geomorphological data, it is not possible to speculate on the extractive capacity of the mine. Therefore, this alternative was not taken into consideration.

In table 2 shows the classification of costs according to the plans.

Table 2. Classification of costs

	PRIMARY	SECONDARY	INDIRECT
Reclamation	Cb1		
Museum	Cm1	Cm2	Cm3
University	Cu1	Cu2	Cu3
Alternative energy	Ce1		

Remediation of the site from polluting material will result in a direct benefit on the development of the area with an increase of the unit from the industrial point of view. The establishment of a museum on the site has resulted in a direct benefit by creating an induced touristic-cultural induced, including the specific kind of mining tourism practiced by organizations such as the ANIM.. Allocating the area a center of research and university has the effect of direct benefit to the higher unit values of the site and adjacent land (indirect benefit), and also polarizes funding derived from research and training carried out. The installation of plants for the production of alternative energy has a positive effect not only for the environment, but it involves a primary benefit directly, because it is possible to extract resources from the use and sale of produced energy.

Table 3. Classification of benefits

	principal	secondary	indirect
RECLAMATION			
Industrial upgrading	Bb1		
Reduction of environmental pollution			Bb2
MUSEUM			
Tourist and cultural activities	Bm1		
Activities of internal services		Bm2	
Activities of external services			Bm3

Table 3. (*Continued*)

UNIVERSITY			
University structure - research - facilities and services	Bu1		
University and re-search	Bu2		
Activities of internal services		Bu3	
Activities of external services			Bu4
ENERGY			
Value of energy generated	Be1		
Sale of energy gener-ated		Be2	

The characterization of the costs has been defined through market surveys (alternative energy), the results already given (remediation) and estimates point on buildings. The calculation of road works, the trimming of the areas of the bike path, furniture and many of the university buildings was performed by analytical estimates and a number of buildings, given the particular shape and type; we have made a summary index assuming a value for the redevelopment of 500.00 €/mq. Finally the complementary alternatives have been aggregated as follows:

- Reclamation; university center and museum;
- Reclamation, university, museum and alternative energy.

The Cost Benefit Analysis (CBA) is an analysis technique designed to compare the efficiency of different alternatives that can be used in a given context to achieve a well-defined objective [12]. The Cba occurs if the benefits that an alternative is able to bring to the community as a whole, the social benefits are greater than the costs (social costs). a project is considered desirable in the case where the comparison of total benefits and total costs prevailing prove the first, which is to say that the community as a whole receives a net benefit from its implementation. In the presence of alternative options for action, it is judged preferable option in which the presence of the benefits cost is greater. the logic of CBA is that the resources of a community are limited and the policy maker should be devoted to interventions that maximize the net benefit to the society The Cba is able to provide support to public policy decisions in relation to possible interventions. Traditionally Cba place ex ante and is intended to decide whether to allocate resources to a particular project or policy adjustment. In some cases, at the end of the project is a conducted Cba ex post, which provides support to evaluate the results of the same. The methodology of the Cba is characterized by the systematic characterization of the costs and benefits, for the expression of the same in terms of money and for the determination of total social net benefit of the proposal, which is the same improvement made by the current situation.

The CBA [13] is developed through a series of logical steps [14]:

1. Defining the scope of the analysis and identification of relevant costs and benefits;
2. Identification of the costs and benefits and their units of measurement;
3. Forecast in quantitative terms of costs and benefits;
4. Inter-temporal discount;
5. Aggregation of costs and benefits and the calculation of the net benefit;
6. Sensitivity analysis.

The cost benefit analysis is a technique of economic evaluation of public investments supported by the following equation:

$$\sum \frac{B}{(1+r)^n} + \sum \frac{C}{(1+r)^n} \qquad (1)$$

where B and C are the benefits and costs linked to the form, r is the discount rate or social inter-temporal discount.

For different design scenarios previously defined benefits and costs have been identified.

RECLAMATION	Activation in 2th years	Benefit			€
Industrial upgrading		Principal benefit			154.590.368,15
Environmental upgrading		Indirect benefit			5.267.068.560,00
Total benefit					5.421.658.928,15
Coefficient of actualization					0,92
Actualization B					4.987.926.213,90

MUSEUM	Activation in 6th years	Benefit	€	Investment duration	€
Cultural and turistic activities	School trips	Principal benefit	25.216,00	30 years	410.740,61
	Archeological tourism	Principal benefit	113.122,00	30 years	1.842.631,65
	Occasional tourism	Principal benefit	236.928,00	30 years	3.859.293,79
	Cruise tourism	Principal benefit	26.500,00	30 years	431.665,55
	Mining tourism	Principal benefit	60.000,00	30 years	977.333,31
Activities of internal services	Bookshop	Secondary benefit	171.715,00	30 years	2.797.046,50

		Secondary			
	Restaurant	benefit	103.029,00	30 years	**1.678.227,90**
	Bar	Secondary benefit	225.000,00	30 years	**3.664.999,92**
Activities of external services	**Transport**	Indirect benefit	171.715,00	30 years	**2.797.046,50**
Total benefit					**18.458.985,73**
Coefficient of actualization					**0,79**
Actualization B					**14.582.598,73**

UNIVERSITY	Activation in 10th years	Benefit	€	Investment duration	€
University	**Site**	Principal cost	67.166.862,28	30 years	**67 166 862,28**
University activities and research	**University fees**	Principal cost	1.206.750,00	30 years	**1 206 750,00**
Cost of internal services	**Self service**	Secondary benefit	462.000,00	30 years	**7.525.466,51**
Cost of external services	**Transport**	Indirect benefit	782.400,00	30 years	**12.744.426,40**
Total benefit					**469.860.456,90**
Coefficient of actualization					**0,68**
Actualization B					**319.505.110,69**

ENERGY	Activation in 10th years	Benefit	€	Duration of investment	€
Internal needs		Principal benefit	1.260.000,00	30 years	**20.523.999,57**
Value of energy sold		Secondary benefit	840.000,00	30 years	**13.682.666,38**
Reduction of CO_2 emissions		Indirect benefit	1.184.000,00	30 years	**19.286.044,04**
Total benefit					**53.492.709,99**
Coefficient of actualization					**0,68**
Actualization B					**36.375.042,79**

RECLAMATION	Activation in 2th years	Cost			€
Industrial and environmental upgrading		Principal cost			46.456.069,23
Total cost of upgrading					46.456.069,23
Coefficient of actualization					0,92
Actualization C					42.739.583,69

MUSEUM	Activation in 6th years	Cost	€	Investment duration	€
Cost of implementation		Principal cost	67.166.862,28	30 years	67 166 862,28
Cost of set up		Principal cost	1.206.750,00	30 years	1 206 750,00
Cost of internal services	Bookshop	Secondary cost	43.560,00	30 years	709,543,99
Cost of external services	Bar	Secondary cost	43.560,00	30 years	709,543,99
Total cost of museum					69.792.700,25
Coefficient of actualization					0,79
Actualization C					55.136.233,20

UNIVERSITY	Activation in 10th years	Cost	€	Investment duration	€
Cost of university construction		Principal cost	95.450.604,04	30 years	95.450.604
Cost of housing for students		Principal cost	1.512.000,00	30 years	24.628.799,4
Cost of internal services	Self service	Secondary cost	84.682,00	30 years	1.379.375,66
Cost of external services	Transport	Secondary cost	83.000,00	30 years	1.351.978,00
Total cost of university					122.810.757
Coefficient of actualization					0,67
Actualization C					82.283.207,3

ENERGY	Activation in 10th years	Cost	Investment duration (30)	Investment duration	€
Installation cost + service maintenance		Principal cost	17.316.576,00	30 years	17.316.576,00
Coefficient of actualization					0,67
Actualization C					11.602.105,92

The benefits and costs once identified and quantified are discounted using the discount capital. Then, applying the equation (1) we can find the net benefit generated by the intervention. In particular, this operation and has been conducted for all individual scenarios, complementary and aggregates. Below there is the table with the evaluations of all alternatives considered.

In table 4 are reported the benefits for each alternative design.

Table 4. Benefits

	Benefit
Reclamation	4 987 926 213,90 €
Museum	14 582 598,73 €
University	319 505 110,69 €
Alternative energy	36 375 042,79 €

The economically most convenient choice of interventions, can be effected in cost benefit analysis with the aid of more criteria of choice, among the most recurrent we have:

• the NPV (Net Present Value);
• the balance of benefits and costs;
• the balance of benefits and costs of costs.

The net present value (NPV), which is the discounted difference between benefits and costs, the convenience exists when the sum of the benefits is greater than or equal to the cost. A project is convenient if your van is positive and the most convenient project is with the NPV higher. Value benefit / cost discounted (RBCA), it is the relationship between benefits and costs and this ratio is related to the NPV. Value of benefits and costs on discounted costs it is the ratio of benefits minus costs of costs, this ratio is related to the VAN. For the cost benefit analysis of interventions Pasquasia mining site in order to help the decision maker called to express his opinion on different projects, and 'chose to use the selection criteria described above. The decision to adopt different kind of decision criteria was necessary because we are considering actions that trigger scenarios with characteristics and investment induced not easily comparable, so using a criterion rather than another may be detrimental to some project ideas rather than others. To this purpose, in order to avoid this, the feedback control of the different planning hypotheses through the use of multiple criteria for the decision, thus having the possibility of being able to converge the choice of the decision maker toward the alternative actually most convenient from the economic point of view limiting the distorting effect related to different ranges of application and the limitations that each type of decision criteria arise.

DECISION CRITERIA					
Aggregate actions	∑ Discounted costs	∑ Discounted benefits	VAN	B/C	(B-C)/C
Reclamation+ Museum	97.875.816	5.002.508.812	VAN 4.904.632.995	(B/C) 51,1108	(B-C)/C 50,110774
Reclamation+ Museum+ University	82.283.207	319.505.110,69	VAN 5.141.854.809	(B/C) 29,5406	(B-C)/C 28,540645
Reclamation+ Museum+ University+ Energy	11.602.105	36.375.042,79	VAN 5.166.627.835	(B/C) 27,943	(B-C)/C 26,94304

For the project proposal of the museum considered as simple action, the NPV has a negative value, since the costs will outweigh the benefits.

In the area there is not still a museum organized with territorial and extra-territorial dimension, which can support the development of this element as punctual as part of an organized structure of poly attractors. The above data obtained it was decided to proceed with the construction of the list of projects, using the ratio decision B - C / C because it takes greater account of the difference of the nature of planning hypotheses analyzed.

Whereas the hypothesis of the reclamation project is a prerequisite to the activation of any other project hypothesis, then it can be excluded from this list, which will cover all the project ideas aggregate it that incorporate. Therefore, it operates a narrowing of the field between the different designs assumptions considered.

The final choice concerns only the two design assumptions aggregated for intervention remediation, university, museum and reclamation, museum, university and energy. From the data obtained, it appears that the best condition of the project hypothesis is that when you consider the reclamation, museum and university, although the hypothesis can be considered including energy, also in relation to the value of the external wise obtained, a good investment.

RANKING OF PROJECTS	(B-C)/C		
Reclamation+Museum	(B-C/C)5	49,84%	A
Reclamation+Museum+University	(B-C/C)6	28,00%	B
Reclamation+Museum+University+Energy	(B-C/C)7	27,00%	C

RANKING OF PROJECTS	(B-C)/C		
Reclamation+Museum+University	(B-C/C)6	28,00%	B
Reclamation+Museum+University+Energy	(B-C/C)7	27,00%	C

4   Conclusions

The site Pasquasia is a complex resource, is configured with an environmental emergency. The site not only defines certain aspects of the landscape, but it represents the collective and individual history of a large portion of the community of this area. Certainly today the production capacity of this resource is open to new options far removed from the original, but it is identified as a great opportunity for this area held on the margins of the EU and national policies. The evaluation model proposed to support the choice of redevelopment and renovation of the mining site, and has been built in order to meet those goals. In particular, through the tried evaluation non-monetary to define the values perceived by the social point of view of the resource and through monetary valuations, and therefore the cost benefit analysis and' tried to identify for each scenario, the net benefit induced. The goal for the community, is 'to identify, linking non-monetary and monetary analyzes scenarios more convenient from an economic standpoint, but also those that can be better supported by the community. The non-monetary assessments have allowed, through the tabs and interviews to understand how the community perceives the presence of the mine on their territory, and in particular how it perceives and if it perceives the environmental emergency. From this analysis there are very interesting data: the sample knows very well the Pasquasia mine, and this provides an idea of the size and spatial-cultural identity of the area. In addition, interviews were instrumental in defining new scenarios for the site. Through the feedback monetary and in particular with the aid of the cost and benefits of the criteria of decisions have led to the identification of a range of alternatives that are more convenient from the economic point of view up to the identification of a planning horizon and investment which guide the choice of the decision maker. The results obtained from this last phase of the analysis points out at the non-affordability of the single alternative of the museum, because the costs outweigh the benefits accumulated and updated. In all other cases, the benefits clearly outweigh the costs, making it feasible for all possible scenarios characterized by positive economic convenience. To guide the decision maker in the choice of the most convenient scenarios from the economic point of view, tracing the information obtained from the different evaluation criteria, or from Van (net present value), B/C, B-C/C, and it has been possible to define a list of projects, and the generation of the choice of the best scenario. Among the various kind of decision criteria used, for safety reasons we have chosen to depend on the decision-making strategy by the criterion bc / c, which characterizes most marked differences between the projects and to take into account the differences of the different scenarios considered. In the end the results converge to the aggregate alternative university, museum and reclamation, providing a highest rating of just two percentage points on the alternative of the university, museum, clean and energy.

References

[1] Gabrieli, P., Trovato, J.: I misteri di Pasquasia, Ed. Lancillotto e Ginevra, Italia (2009)
[2] Signorello, G., Englin, J., Longhorn, A., De Salvo, M.: Modelling the Demand for Sicilian Regional Parks: A Compound Poisson Approach. Environmental & Resource Economics 44, 327–335 (2009)

[3] Marangon, F.: Gli interventi paesaggistico-ambientali nelle politiche regionali di sviluppo rurale, Franco Angeli, Italia (2006)

[4] Tempesta, T., Marangon, F.: Una stima del valore economico totale dei paesaggi forestali italiani tramite la valutazione contingente, Genio Rurale, n. 11. English version "The total economic value of italian forest landscapes", Italia (2004)

[5] Signorello, G.: Valutazione contingente della "disponibilità a pagare" per la fruizione di un bene ambientale: Approcci parametrici e non parametrici, Rivista di Economia Agraria (1994)

[6] Signorello, G.: La stima dei benefici di tutela di un'area naturale: un applicazione della valutazione contingente. Genio Rurale 9 (1990)

[7] Signorello, G., Cucuzza, G., De Salvo, M.: Valutazione contingente del paesaggio agrario della Costa Viola. In: Marangon F (a cura di) Gli Interventi Paesaggistico-Ambientali Nelle Politiche Regionali di Sviluppo Rurale, FrancoAngeli, Milano, Italia (2006)

[8] Sturiale, L., Trovato, M.R.: La percezione sociale a supporto della valutazione degli interventi di valorizzazione di una risorsa ambientale. Paysage Topscape 9, 365–416 (2011)

[9] Turner Kerry, R., Pearce, W., David, B.I.: Economia ambientale, Ed. Il Mulino, Italia (2003)

[10] Hoehn, J.P., Randall, A.: A Satisfactory Benefit Cost Indicator from Contingent Valuation. Journal of Environmental Economics and Management 14, 226–247 (1987)

[11] Johnston, R.J., Swallow, S.K., Weaver, T.F.: Estimating willingness to pay and resource tradeoffs with different payment mechanism: An evaluation of a funding guarantee for watershed management. Journal of Environmental Economics and Management 38, 97–120 (1999)

[12] Welsh, M.P., Poe, G.L.: Elicitation Effects in Contingent Valuation: Comparisons to a Multiple Bounded Discrete Choice Approach. Journal of Environmental Economics and Management 36, 170–185 (1998)

[13] Corners, R., Sandler, T.: The Theory of Externalities, Public Good, and Club Goods. Cambridge University Press, Cambridge (1986)

[14] Nuti, F.: L'analisi costi- benefici; il Mulino, Bologna, Italia (1987)

[15] Campbell, H., Brown, R.: Benefit-Cost Analysis. Financial and Economic Appraisal using Spreadsheets. Cambridge University Press, UK (2003)

The *Senior* as the Protagonist of the Future Economy: A Firm Case Study

Roberta Pace and Dario Schirone

Department of Political Science, Università degli Studi di Bari "Aldo Moro", Italy
darioschirone@libero.it, roberta.pace@uniba.it

Abstract. The aging of the population leads to a readjustment of society and economy and this certainly implies some costs. But the consequences on the Societies are not represented only by problems because today - more than ever before – elderly population express different needs.

The good state of health allows them to play sports, to travel and to maintain an active social life. This leads to the appearance of a new target for those who work in marketing and business. Even large firms - such as the multinational IKEA that we'll discuss in this paper - began to look with interest to the target of customers over 65 years of age, questioning their business strategy in order to capture this slice of the market, whose numerical increase seems not going to stop.

Keywords: Ageing, Marketing, Target Population, IKEA.

Introduction

The philosopher Umberto Galimerti claims:

"The West is aging badly because the values that are regulating our culture have mainly biological, economic and esthetical character in comparison to which the aging appears in all its uselessness since it is biologically decadent, economically unproductive and esthetically degrading".

Contrary to the widespread belief of a progressive population decline in the European Union, the most recent demographic projections pointed out that, from now until 2060, the low fertility rates, constant life expectancy rise and constant migration inflows will maintain almost unchanged the total population amount, although it would appear substantially more aged. As the Commission's Report discloses "[...] the EU will pass from a ratio of four persons in the working-age (from 15 to 64 years) per one person above 65 years of age to a ratio of two per one. The most considerable decline should be reported in the 2015 and 2035 period, in correspondence to the reaching of the retirement age of baby-boom children [...]". The population thus will be more and more "aged" and urbanized, with clear consequences it might have from a quantitative, qualitative and territorial point of view on a demand of goods and services [1].

In accordance with the current trends, the population under 15 years of age in Italy decreased by one third over the past 30 years. Young yield to Old. Both in the Euro

B. Murgante et al. (Eds.): ICCSA 2014, Part III, LNCS 8581, pp. 42–50, 2014.

zone and in the World, Italy is among the most eldest countries, in fact Italy is the second European Country, after Germany [2] and this condition is highlighted by the main demographic indicators of population structure that show an increasing trend year after year [3].

Indeed the National Institute of Statistics in Italy (ISTAT), has forecasted that in 2020 the population will be composed by 33.6% of old above 65 years of age and only by 12.7% of young up to 14 years of age and that the aging index (that measures the ratio between elderly and young) will rise constantly passing from 138 elderly per 100 young in 2005 to 222 in 2030 and reaching 264 in 2050 [4]. According to the ISTAT, hence, there are 3.4 elderly per one child aged less than 6 years, so the direct consequences: is the overturn of population pyramid.

The population aging entails the society and economy to readapt and it doesn't occur at zero cost.

The median age for retiring is around 65 years of age and considering the lengthening of life the providing of pensions becomes more onerous. The costs to sustain the health expenditure are rising more and more: after the age of 55, around 40% of population declares to be affected by some chronic pathology; after the age of 55 more than a half of the population uses medicines [3].

There are issues to be resolved not only because the elderly – today more than ever – express different needs. Their good health state allows them to do sports, travel and have an intense social life. This implies the creation of a new target for market operators [5].

The study will aims at pointing out how the population aging process could be an opportunity for the companies that actually were the first ones to recognize the potential that this new target might offer.

The reference case is one of IKEA Italia, a world leader in furniture retail, which in light of big and important demographic changes has face important strategic choices in order to continue being concurrent on the market.

1 Overturn of the Population Pyramid in Italy

Over the past fifty years, the Italian population lived through numerous, visible and consistent demographic transformations that are a direct consequence of the changes in habits and life styles, behaviors and choices in general. With the baby boom's years, that occurred after the Sixties, Italy recorded a demographic development entirely opposite to one from the previous years. That is: natality and fertility fall; new models of couple and family formation and dissolution; life expectancy and population aging; issues related to minority and ethnic groups and foremost those regarding migrations; different role of female population than in the past; specific issues related to infant, young and old populations [6]. So the three main factors that more than others have been causing the progressive population aging are: the fertility decline, rise of life expectancy for all age groups and the fall of mortality rates. Therefore, we can affirm that the demographic aging consists in the rise of both number and ratio of older people in respect to the other age groups; we can affirm hence that the aging is composed of absolute and relative variation of the old-age segments in the total population.

Generally looking, the current state of Italian population is not far from the scenario of other European countries [7] although it is different in respect to the timing and modes that are the result of historical and political events of our country.

The demographic phenomenon that concerns Italy as well, among others, was the overturn of the age pyramid that firstly had a large basis and a narrow top while today it has a reduced basis and a more larger and larger top [8]. Since the first population census in Italy, carried out in 1861, the total residing population practically doubled by passing from around 26 million of people to over 59 million, however in the last inter-census period (2001-2011) there was a standstill. As a matter of fact, the average number of family nucleus declined in the past 50 years from 4.0 (Census 1951) to 2.4 (Census 2011) with a proliferation of mono-component families and childless or one-child couples.

The transformation of the population structure and particularly the increase of the elderly in the total population has a clear effect on the active population: in the past fifty years the aging index (the percentage ratio between the population above 65 years of age and the total population) doubled and the forecasting has revealed that in the next fifty the increase will be exponential and the index will raise to values close to 35% [9].

Due to the demographic transition pace the percentage of the population above 65 years of age passed from 15.3% (8,700.158 persons) in 1991 to 18.7% in 2001 (10,646.874 persons) and to 20.2% (12,301.537 persons) in 2011. The increase was relevant also for the more advanced age groups: the population aged 75 years and above passed from 6.7% in 1991 (3,792.567 persons) to 8.4% in 2001 (4,762.414) to 10% in 2011.

The aging scenario can be defined also by analyzing the percentages related to the "oldest old" that is the population aged 85 years and above that passed from 1.3% in 1991 (728.817 persons) to the current 2.7% (Table 1).

Table 1. Composition of the population older than 60 years

Age	Widows/ Widowers	Male %	Female %	Total Population	% of Widows/ers on total population
60-64	298.299	48,3	51,7	**3.825.131**	6,3
65-69	399.795	47,3	52,7	**3.052.238**	5,0
70-74	648.317	45,7	54,3	**3.102.183**	5,1
75-79	826.547	42,8	57,2	**2.533.595**	4,2
80-84	901.778	38,2	61,8	**1.941.292**	3,2
85-89	715.125	32,2	67,8	**1.171.062**	1,9
90-94	261.441	26,8	73,2	**362.732**	0,6
95-99	96.640	22,1	77,9	**122.290**	0,2
100+	12.633	19,2	80,8	**16.145**	0,0

(Sources: Own elaboration. ISTAT, 2011)

The territorial analysis shows relevant geographical differences in terms of aging: the municipalities with the highest number of elderly are the small ones, followed by the mountain and big municipalities. The lowest percentage of elderly is reported in big urban centers.

2   From Aging to Business: The Slow Path

In order firstly to identify and then operate in the best way in the target markets, the marketing has been always based on the homogeneous portions subdivision. In fact, the modalities mostly used for the age groups are: less than 6 years; from 6 to 11; 12-19; 20-34; 35-49; 50-64; above 65 years [10;11].

The target age less than 49 years has always been more privileged by the marketing scholars, however in the few past years the literature (Wolfe, Snyder, 2003) theorized the ageless marketing, that is operating both for those who produce and those who consume, while not taking into account the age. The evolutions of this theory are the crossover products (cinema and TV) that are likeable to children, young and adults.

In America starting from the Sixties [12] the object of interest and studies has been the elderly consumers and senior markets [13] and the discussions began focusing on the power of the aged in the marketplace [14].

In Italy not much attention was paid on the marketing strategies oriented toward seniors, hence the road to recognizes the potential of this target in terms of profitability for companies is very slow.

Most of the literature focused on the elderly suggests to avoid offering the product whose features are not very adequate to the needs of the elderly (Umberto Colessi, 1996). The scholars invite to pay attention on comfort, convenience, simplicity, on practical and ergonomic aspects in order not to exclude the target of non young. Moreover, he urges the large organized distributions (LOD) to pay attention on the layout (the position and height of shelves) on the service availability (restrooms and benches) close to counters and on home delivery services.

A more thorough study was conducted and published between 2003 and 2005 by Francesco Casarin [15;16]. The study aimed at publishing the results of a research carried out in 2001 on some insurances, banks and distribution industries treated extensively in France envisaging the strategies that Italian firms should implement [17].

Most of the Italian marketing literature [18] is focused on distribution channels and senior's demand; it point out the advantage that the LOD would have in terms of profit if the stress was put on enhancing the simplification (as reducing the effort and the risks and eliminating the distance from the selling point to the client's home with a proper home delivery service).

The CENSIS (the Italian Center of Study and Social Investment) published a report on elderly in Italy focused on health and consumption models [19].

The mature segments are not only passive receptors that adopt the new with a delay, they can be the active protagonist and anticipators. In Italy, the marketing strategies are still largely oriented toward the young public. The elderly are still perceived as not so "attractive" – with only few sectors where the level interest for these market segments is constant and not just sporadic – both in terms of the offer composition and the distribution, so in this way the cultural marginalization is being accompanied by marketing and communication marginalization [20].

However, the 2008 global crisis seems to have speeded up the process: the phenomenon that started as purely demographic is becoming today more and more economic. The elderly, in fact, are not anymore just a minority group that sometimes is the object of interest but they are becoming an important economic figure that plays an active role both in the area of production and consumption.

3   Case Study: IKEA in Italy

Even a large distribution – such as, for example, the multinational company IKEA – started paying attention to the target over 65 years of age, questioning thus its own company strategy with an aim of conquering this new important part of the market whose numerical increase will not cease (Figure 1)

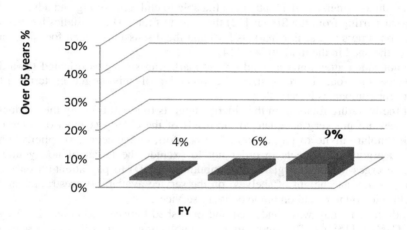

Fig. 1. Customers of the multinational IKEA over 65 years. Fiscal Years 2009, 2010 e 2011 (Sources: Own elaboration. IKEA, 2012).

A survey carried out in the multinational company IKEA on January 2012 revealed that in the past 3 years the client's profile fully reflected the scenario and trends of the Italian population aging[1].

---

[1] **Table 2.** Evolution of the Aging Index in Italy and the number of IKEA customers over 65 age (%)

	2009	2010	2011
Aging Index	143,4	144,0	144,5
Number of IKEA's customers over 65	4,0	6,0	9,0

(Sources: Own elaboration. IKEA and ISTAT, 2011)

The client's anagraphic profile has been sensibly aged: the mean age passed from 33.4 years to 39.3 years and the age group over 50 years increased from 8 to 25% while the over 65 passed from 2 to 9%; on the opposite side there is a fall of the age group 15-34 years that reduced from 43 to 34% (Figure 2). The following figure shows the percentages related to the age groups in Italy [21] and those of IKEA's clients (Internal Company survey "Brand Capital", 2011).

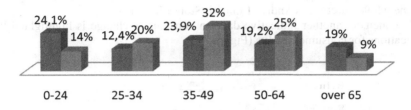

Fig. 2. Age groups of the Italian population and customer IKEA. (Sources: Own elaboration. IKEA and ISTAT, 2012).

This change of the scene has induced the company to imagine furniture styles closer to the taste of this new target. The marketing strategy, in fact, is based on the analysis of needs and tastes of local consumers and must take into account the important variations of demo-economic scenarios. The population aging has a double impact: in terms of changes in traditional type of tastes – in rural and working class geographical areas – on one side and the need to socialize and remodel their homes on another. The direct consequence is that so called "new elderly" represent the new business opportunity.

What do therefore companies do? They don't stand by and watch only. That is foremost true for those companies that usually try to anticipate the trends before the others.

A survey carried out by the Swedish multinational revealed that the adult couples, over 60 years of age, spend in average 1,000 euro per year on furniture and accessories, contrary to the average spending of a family that amounts less than 800 euro per year (Customer Satisfaction Index, IKEA, 2011). The couples over 60 years of age rarely have children at home and this contributes to further increase their purchasing power and it has been further estimated that in 2020 they will have a disposable income, with the purchasing power, higher than the national average by more than 15% [22].

It is also true that Italy is a country where the persons over 60 years of age, although being retired earlier, remain active for many years by taking care of their grandchildren, the youngest family components. In Italy, the half of the grandparents over 50 years are engaged in the care of grandchildren, younger than 13, years at least once per week; and one out of three do it on a daily basis [23]. The relative value results significantly higher in comparison to other North European Countries. Also, in Italy (as well as in Spain and Greece) even grandparents aged between 70 and 80 years have an active role in the care and education of grandchildren and, thus, can be defined as "young elderly".

"Although it is true that in terms of age the clients are aging however it is not true that these persons are spiritually "old". On the contrary, the elderly are those who today have major interests since they have more free time. It is not so rare to meet clients over 65 that use the internet easily and it is not so rare to hear that they need no for help withdrawing their purchases. However, we as retailers will not stand and watch and we are investigating the utility of adding services that would ease the purchasing process of our *ageè* clients and so it will be created, from the next fiscal year (September 2011), a new service that foresees the help to load, directly into the car, the whole purchase" (Andrea Lumiati, Store Manager).

To conclude, another indicative data on consumer's change is highlighted by the applications for consumer credits (Figure 3).

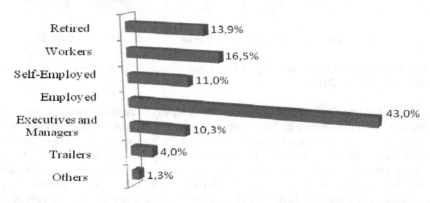

Fig. 3. Professions of those who access to credit. (Sources: Own elaboration. IKEA, 2012).

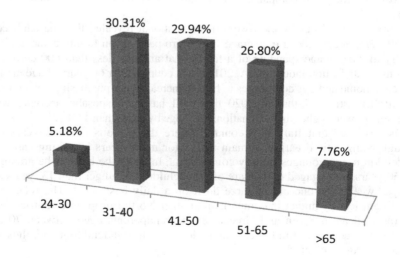

Fig. 4. Age of those who have applied for funding in 2011. (Sources: Own elaboration. IKEA, 2012).

In fact, in IKEA during the 2013 fiscal year 13,9% of those requesting and obtaining the credit for purchasing furniture and furniture accessorizes are retired; around 8% of them have more than 65 years of age (Figure 4). These data are much more higher than those of two years before when in 2009 only 8.9% were retired persons and 4.6% of them were aged above 65.

4   Conclusions

It has been stressed out how the population aging represents an unprecedented phenomenon in the history of humanity. It is well known that from today until 2050 the total number of persons aged above 60 years will exceed the number of young people; it has been rising and will continue to rise further discussions about the equanimity and solidarity among generations.

The aging is a wide phenomenon influencing all the aspects of human life: labor market, pensions, intergenerational transfers, health and assistance, family composition, dwellings, immigration and emigration, structure of political representatives; yet the study has pointed out the repercussions on savings, investments and consumption by putting in evidence how the category of over 65 years of age could be of interest for the business companies.

It is well known that the companies operate by following their goals of economic optimum, however often the sole product is not enough. The need to remain competitive and grab the target of growing clients has induced companies to modify their strategies making them not only suppliers of products but also - and even more frequently - suppliers of services. Those who follow these development opportunities in new competition arenas will be able to maintain its business position making it stable moreover in a future marked, characterized by the innovative combination of alternative products and services.

References

1. Golini, A.: Invecchiamento e Welfare. In: Racioppi, F., Rivellini, G. (eds.) Applied Demography. La Demografia per le aziende e la governance locale, pp. 11–32. Edizioni Nuova cultura, Roma (2013)
2. EUROSTAT: La situation sociale dans l'Union européenne 2005-2006. Luxembourg (2006)
3. ISTAT: Annuario Statistico (2012)
4. ISTAT: Previsioni demografiche nazionali 1 gennaio 2005 – 1 gennaio 2050 (2006)
5. Pace, R., Schirone, D.A.: Over 65enni, un nuovo target per chi fa marketing (2011), http://www.neodemos.it
6. Baldi, S., Cagiano De Azevedo, R.: La popolazione italiana verso il, Storia demografica dal dopoguerra ad oggi. Il Mulino, Bologna (2000)
7. Ferrera, M., Hemerijck, A., Rhodes, M.: The Future of Social Europe. Celta Editora,Oeiras (2000)
8. Caselli, G.: Il ruolo dei cambiamenti demografici sullo stato sociale in Italia. Rivista Italiana di Econ. Dem. e Stat. LVI, 1-2 (2002)

9. Banca d'Italia: Statistiche di Finanza Pubblica nei Paesi UE. Supplemento al Bollettino Statistico. Anno XIII, 45 (2003)
10. Kotler, P., Scott, W.G.: Marketing management. Isedi- Utet, Torino (1993)
11. Kotler, P.: Marketing management. Pearson, Milano (2007)
12. Maestow, H.L., Cosmar, S.C., Plotkin, A.: The Elderly Consumer: Past, Present and Future. Advances in Consumers Research 8, 742–747 (1981)
13. Gidlow, E.: The Senior Market, Sales Management. The Need for a New Perspective. Journal of the Academy of Marketing Science 1, 242–248(1961)
14. Klippel, R.E.: Marketing Research and the Aged Consumer. The Need for a New Perspective. Journal of the Academy of Marketing Science. 1, 242–248 (1973)
15. Casarin, F.: Marketing e domanda senior. Giapichelli, Torino (2003)
16. Casarin, F.: Canali distribuitivi e domanda senior. Micro e Macro Marketing 3, 569–579 (2004)
17. Napolitano, E.M., Vecchiato, G.: 50 plus Marketing. In: Napolitano, E.M., Scialpi, S. (eds.) Franco Angeli, Milano, pp. 11–20 (2012)
18. Collesi, U., Casarin, F.: Posizionamento e domanda senior: un'analisi empirica. In: Congresso Internazionale su Le tendenze del Marketing. Ecole Supérieure de Commerce de Paris, 21- 22 gennaio (2005)
19. CENSIS: VII Rapporto Salute. Consumi e valori degli anziani in Italia, i modelli sanitari e di consumo degli anziani (2008), http://www.censis.it
20. Salafia, P.: Chi ha paura dell'ageing attivo. Social Trends 110 (November 2010)
21. ISTAT: Annuario Statistico (2011)
22. Pizzuti, F.R.: Invecchiamento demografico, età di pensionamento e sistemi di finanziamento per la previdenza italiana. In: Cagiano De Azevedo, R. (ed.) The European Welfare in a Counterageing Society, Edizioni Kappa, Roma (2004)
23. SHARE: Survey of Health, Ageing and Retirement (2009), http://www.share-project.org

Constructing Multi-attribute Value Functions
for Sustainability Assessment of Urban Projects

Marta Bottero, Valentina Ferretti, and Giulio Mondini

Politecnico di Torino, Department of Regional and Urban Studies and Planning
{marta.bottero,valentina.ferretti,giulio.mondini}@polito.it

Abstract. This paper considers the problem of sustainability assessment in urban and territorial planning projects using the Multi Attribute Value Theory (MAVT), a particular kind of Multiple Criteria Decision Analysis method. Starting from a real case concerning the transformation of an urban area in the city of Torino (Italy), the aim of the paper is to explore the contribution of MAVT for urban planning decision making processes. In the application, several alternative projects are evaluated on the basis of different criteria and attributes, such as availability of services, urban regeneration, acoustic emissions, land consumption and so on. In the result of this approach a ranking of sustainable alternative solutions is provided.

Keywords: Multiple Criteria Decision analysis, Multi-attribute Value Theory, Strategic Environmental Assessment, Swing method.

1 Introduction

With the aim of structuring decision problems and increasing the transparency of the evaluation processes, formal methods of decision analysis are available [1]. These methods allow to better understand and communicate objectives and their expected fulfilment by different decision alternatives. This is of particular importance in decision problems in the domain of sustainability assessment of projects, plans and programmes.

It has been generally agreed that sustainable development is a multidimensional concept that includes socio-economic, ecological, technical and ethical perspectives and thus leads to issues that are simultaneously characterized by a high degree of conflict, complexity and uncertainty. When speaking about sustainability in urban and territorial planning, decision making requires consideration of trade-offs between many objectives: factors that range from the reduction of soil consumption to the optimization of the use of environmental resources, from the promotion of economic activities to the requalification of downgraded urban areas, from the endorsement of energy efficiency to the rationalization of transport systems.

To help addressing these problems, the use of Multiple Criteria Decision Analysis (MCDA) [2, 3] has gained attention in the last years.

This paper considers the problem of sustainability assessment in urban and territorial planning projects using the Multi Attribute Value Theory (MAVT) [4], a particular kind of MCDA method.

B. Murgante et al. (Eds.): ICCSA 2014, Part III, LNCS 8581, pp. 51–64, 2014.

Starting from a real case concerning the transformation of an urban area in the city of Torino (Italy), the aim of the paper is to explore the contribution of MAVT for urban planning decision making processes. In the application, several alternative projects are evaluated on the basis of different criteria and attributes, such availability of services, urban regeneration, acoustic emissions, land consumption and so on. In the result of this approach a ranking of sustainable alternative solutions is provided.

2   Methodological Background

The Multi-Attribute Value Theory (MAVT) [4] is a specific MCDA technique that can be used to address problems involving a finite and discrete set of alternative options that have to be evaluated on the basis of conflicting objectives. By being able to handle quantitative as well as qualitative data, MAVT plays a very crucial role in the field of environmental decision-making where many aspects are often intangible. Moreover, decision-making in this context is often complicated by various and conflicting stakeholder views that call for a participative decision process able to include different perspectives and facilitate the discussion.

From the methodological point of view, the multi-perspective decision process to be followed to build a MAVT model can be described as shown in Figure 1.

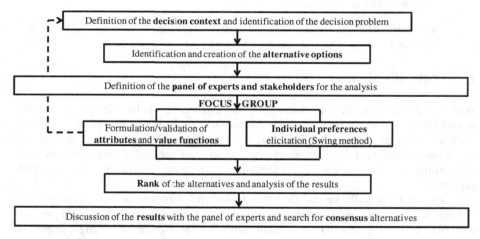

Fig. 1. Methodological steps for the development of a MAVT model making use of an experts' panel

In particular, the first steps are very crucial since they are concerned with the articulation of the decision context and the definition of the problem. This implies defining and structuring the fundamental objectives and related attributes.

Objectives are "statements of something that one desires to achieve" [5] and they depend on the problem to be analysed, on the actors involved in the decision process, and on the environment in which the decision process takes place. The degree to which objectives are achieved is measured through a set of attributes, which may be natural (they follow directly from the definition of the objective), constructed (they specify a finite number of degrees to which objectives are met), and proxy (they are only indirectly linked to the definition of the objective) [5, 6].

The second step consists in the identification and creation of alternative options.

The alternatives are the potential solutions to the decision problem. Methods and models such as visioning, problem structuring methods and scenario planning can help to promote creativity for the generation of good strategies and strategic options [7]. Once the alternative options have been identified, it is necessary to assign scores for each alternative in terms of each attribute.

The performances of each alternative specify for each attribute the outcome of the alternative. In some cases, the performances are readily available, in some other cases they have to be computed or estimated ad hoc for the problem at hand.

The next step might consist in the definition of a panel of experts for the development of the evaluation. The use of experts' panels expands the knowledge basis and may serve to avoid possible biases, which characterizes the situation with a single expert. On the other side, the use of experts' panels has a range of problems associated with it, such as the panel composition, the interaction mode between panel members and, above all, the aggregation of panel responses into a form useful for the decision [6].

In this context it is important to underline that by structuring the decision process as an iterative process the discussion with the experts during the focus group can provide useful insights for the definition of the attributes and the subsequent preference elicitation phase. In particular, bringing together experts from different disciplines allows to perform a preliminary screening of the identified attributes and to better detail the characteristics of the relevant ones [8].

The following step consists in the modelling of preferences and value trade-offs and different strategies are available for this task. The holistic scaling and the decomposed scaling strategies are the most used in practice [6]. According to the former, an overall value judgment has to be expressed of multiattribute profiles, which can be either the real alternatives or artificially designed profiles. Weights (i.e. scaling constants) and marginal value functions (which translate the performances of the alternatives into a value score representing the degree to which a decision objective is achieved) are then estimated through optimal fitting techniques (e.g. regression analysis or linear optimisation) and are the best representation of the assessor's judgments. According instead to the decomposed scaling technique, the multiattribute value function is broken down into simpler sub-tasks (the marginal value functions and the weights) which are assessed separately. The aim of decomposed scaling is to construct the multiattribute model for evaluating decision alternatives while the aim of holistic scaling is to make an inference about the underlying value functions and weights [6]. The case study illustrated in the present paper will follow the decomposed scaling approach, as explained in section 3.

The final step consists in the aggregation of the results in order to obtain the ranking of alternatives. To this end, MAVT includes different aggregation models, but the simplest and most used one is the additive model [9] as it is represented in equation (1):

$$V(a) = \sum w_i \times v_i(a_i)$$

(1)

where $V(a)$ is the overall value of alternative a, $v_i(a_i)$ is the single attribute value function reflecting alternative a's performance on attribute i, and w_i is the weight assigned to reflect the importance of attribute i.

By aggregating the options' performance across all the attributes to form an overall assessment, MAVT is thus a compensatory technique.

Finally, a sensitivity analysis is recommended in order to test the stability of the obtained results with regards to variations in the inputs. As a result, a final recommendation can be obtained and further discussed with the Decision Makers and stakeholders.

3   Application

3.1   Description of the Case Study Area and Presentation of the Alternatives

The case study that has been considered for the present application refers to the transformation of an urban area in the city of Torino (Italy). The area is located in a strategic zone of the city, close to the train station, the University campus and the most important executive headquarters buildings (Figure 2).

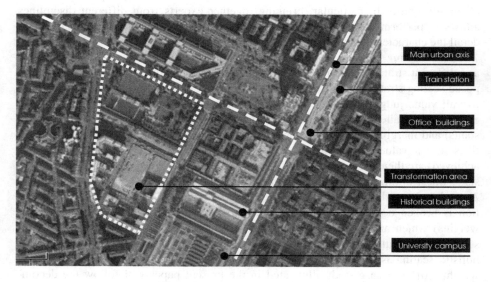

Fig. 2. The area under examination

In spite of its strategic position and intrinsic value, the area represents nowadays an urban void and different transformation projects are under consideration. The three main alternative scenarios for the area are represented in Table 1.

Table 1. Alternative scenarios for the transformation of the area

Scenario	Description
1. Conference centre	Following the recent trends of business tourism flows in the city of Torino, the main idea for the area is the construction of a large conference centre that now is missing. This scenario also includes the construction of different services, such as a luxury hotel, new commercial areas and university residences.
2. Conference centre and shopping mall	This scenario underpins the main ideas of the scenario 1 but it includes the construction of a shopping mall for providing commercial services to the area.
3. Urban park	This scenario considers to relocate the construction of the conference centre in another part of the city while the area under examination will be transformed in a urban park.

3.2   The Decision Support Process

The decision support process that we decided to follow in the present application can be structured according to subsequent steps.

We started from the definition of the decision context and the identification of the real alternatives.

The next step consisted in the definition of a panel of experts for the development of the evaluation. The use of experts panels expands the knowledge basis and may serve to avoid the possible bias which characterizes the situation with a single expert. In our case, particular attention was dedicated to the panel composition in order to have it balanced [8]. Therefore, we involved an historian, an expert in environmental engineering, a urban planner and an expert in the field of economic evaluation. In particular, we used decomposed scaling as explained by Beinat [6] and thus marginal value functions and weights were assessed separately. The overall value model was then built by combining these parts through the additive combination.

With specific reference to the assessment of the scaling constants for all the attributes, we chose the Swing Weight approach because the discussion oriented towards value ranges seemed very promising in the context of urban projects. In particular, for the lower levels of the hierarchy (i.e. the attributes level), the evaluation has been performed by individual experts in their proper field of expertise; in this case, the experts have been involved in the compilation of a specific questionnaire which is based of the Swing method approach. For the higher levels of the hierarchy (i.e., the criteria level), a focus group has been organized in order to allow a discussion among the four experts to be carried on and a common set of weights to be found.

It is important to highlight that it has been an iterative process and that the discussion with the experts during the focus group provided useful insights for the definition of the attributes and their value functions. In particular, bringing together experts from different disciplines allowed to perform a preliminary screening of the proposed attributes and to better detail the characteristics of the relevant ones.

3.3   Structuring of the Decision Problem

The aforementioned decision environment represents a complex system where the presence of interrelated elements and conflicting aspects suggests the use of a Multi-criteria Decision Analysis that is able to provide a rational base for the systematic analysis of the alternative scenarios.

A set of measurable attributes has been identified for the evaluation of the alternatives and it has been organized according to the value tree approach (Figure 3). In this case we used a top down approach for structuring the value tree meaning that we started from eliciting the ultimate objective and climbed down, asking how this objective could be achieved and thus looking for attributes.

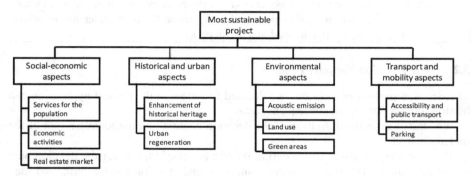

Fig. 3. The value tree for the decision problem under analysis

The main objective of our model is to determine the sustainability of the transformation scenarios for the area. To this end, 4 attributes have been considered that represent the main aspects of the decision problem. The first attribute refers to socio-economic elements of the area and takes into account the availability of services for the population, the presence of economic activities and the trends in real estate market. The second attribute considers the historical and urban factors, including the enhancement of the cultural heritage in the area and the urban requalification operations. The third attribute takes into account the environmental quality, considering the acoustic emission, the land consumption and the presence of green areas. Finally, the fourth attribute is related to transport and mobility aspects and considers the accessibility to the area by means of public transport and the creation of parking.

Following the MAVT methodology, each attribute is described by a value function which allows to scale the attributes between 0 and 1 in order to compare non-commensurable items. Table 2 describes the attributes used in the evaluation and the related value functions.

According to their performance, the three alternatives were evaluated with reference to each attribute and the evaluation is illustrated in Table 3.

Table 2. Elicitation of value functions for each attribute

Indicator	Description	Value function
1.1 Services for the population	Surface of commercial activities considered in the project (m²)	*(graph: x-axis 0 to 1000, y-axis 0 to 1)*
1.2 Economic activities	Number of jobs created by the project (n.)	*(graph: x-axis 0 to 50, y-axis 0 to 1)*
1.3 Real estate	Increase in real estate quotations of the zone (%)	*(graph: x-axis 0% to 5%, y-axis 0 to 1)*
2.1 Enhancement of cultural heritage	Levels of enhancement (low, medium and high)	*(graph: x-axis low, medium, high; y-axis 0 to 1)*
2.2 Urban regeneration	Diversity index $H' = -\sum_{j=1}^{s} p_j \, log p_j$ where s is the total number of species and p_j is proportion of s made up of the jth species	*(graph: x-axis 0 to 1, y-axis 0 to 1)*

Table 2. (*Continued*)

3.1 Acoustic emission	Noise level in the area under analysis after the project implementation	
3.2 Land use	Area subject to land reclamation measures (%)	
3.3 Green areas	Area left as green public area (%)	
4.1 Accessibility and public transport	Levels of accessibility (low, medium and high)	
4.2 Parking	Number of parking slots to be constructed (n.)	

Table 3. Evaluation of the alternatives

Indicators	Scenario 1	Scenario 2	Scenario 3
1.1	0,8	1	0
1.2	0,8	1	0
1.3	0,6	1	0
2.1	1	1	0
2.2	0,69	0,71	0
3.1	1	0	1
3.2	1	1	0,5
3.3	0	0	1
4.1	0,5	1	0
4.2	1	0	1

3.4   Weighing and Aggregation

Once the alternatives have been evaluated, it is necessary to define the importance of the different attributes of the decision problem. In this case the Swing method has been used which explicitly incorporates the attribute ranges in the elicitation question. In particular, the method asks to value each improvement from the lowest to the highest level of each attribute [7] by using a reference state in which all attributes are at their worst level and asking the interviewee to assign points (e.g. in the range 0-100) to states in which one attribute at a time moves to the best state. The weights are then proportional to these values.

In this study the evaluation has been performed by a panel of experts with expertise in the field of environmental engineering, urban planning and economic evaluation as described in section 3.2. As an example, Figure 4 shows the questionnaire they had to answer with reference to the first level attributes

The overall set of weights resulting from the elicitation procedure is shown in Table 4. The single attribute value functions have then been aggregated using the obtained set of weights and additive assumptions to calculate the total value of the specific alternatives. In particular, the global weight has been calculated for each attribute through the following equation: global weight = normalized weight of attribute x normalized weight of criterion. The calculation developed in the case under investigation provides the final priorities represented in Table 4.

From the obtained priority list it is possible to notice that scenario 2 and scenario 1 have very similar performances and that their overall scores thus present negligible differences. In any case, scenario 2 is slightly preferred to scenario 1 and scenario 3 represents the least preferred option. These results highlight the need to further investigate the alternatives by developing a sensitivity analysis on both the weights of the attributes and the original scores of the alternatives.

After obtaining a ranking of the alternatives and despite the coherence obtained in the results, it was considered useful to perform a sensitivity analysis on the final outcome of the model. The sensitivity analysis is concerned with a "what if" kind of question to see if the final answer is stable when the inputs are changed. It is of special interest to see whether these changes modify the order of the alternatives.

Elicitation tool: *Swing method*
Objective: Find the most important attributes with reference to the sustainability assessment of the project

Hypothetical alternative No.1

	Col 1	Col 2	Col 3	Col 4	Score
☺	**Socio-economic aspects** Surface of commercial activities: 1000m², number of jobs: 50, real estate: 5%				**100**
☹	↑	**Historical and urban aspects** Enhancement of cultural heritage: low, urban regeneration: 0	**Environmental aspects** Acoustic emission: high, land reclamation: 0%, green areas: 70%	**Transport and mobility** Accessibility: low, parking slots: >500	

Hypothetical alternative No.2

	Col 1	Col 2	Col 3	Col 4	Score
☺		**Historical and urban aspects** Enhancement of cultural heritage: high, urban regeneration: 1			**65**
☹	**Socio-economic aspects** Surface of commercial activities: 0m2, number of jobs: 0, real estate: 0%	↑	**Environmental aspects** Acoustic emission: high, land reclamation: 0%, green areas: 70%	**Transport and mobility** Accessibility: low, parking slots: >500	

Hypothetical alternative No.3

	Col 1	Col 2	Col 3	Col 4	Score
☺			**Environmental aspects** Acoustic emission: low, land reclamation: 100%, green areas: 100%		**90**
☹	**Socio-economic aspects** Surface of commercial activities: 0m2, number of jobs: 0, real estate: 0%	**Historical and urban aspects** Enhancement of cultural heritage: low, urban regeneration: 0	↑	**Transport and mobility** Accessibility: low, parking slots: >500	

Hypothetical alternative No.4

	Col 1	Col 2	Col 3	Col 4	Score
☺				**Transport and mobility** Accessibility: high, parking slots: 0	**80**
☹	**Socio-economic aspects** Surface of commercial activities: 0m2, number of jobs: 0, real estate: 0%	**Historical and urban aspects** Enhancement of cultural heritage: low, urban regeneration: 0	**Environmental aspects** Acoustic emission: high, land reclamation: 0%, green areas: 70%	↑	

Worst hypothetical alternative

	Col 1	Col 2	Col 3	Col 4	Score
☺					**0**
☹	**Socio-economic aspects** Surface of commercial activities: 0m2, number of jobs: 0, real estate: 0%	**Historical and urban aspects** Enhancement of cultural heritage: low, urban regeneration: 0	**Environmental aspects** Acoustic emission: high, land reclamation: 0%, green areas: 70%	**Transport and mobility** Accessibility: low, parking slots: >500	

Fig. 4. Questionnaire for the elicitation of the Swing weights for the higher level of the hierarchy

Table 4. Overall evaluation of the alternatives

Indicators	Standardized scores of the alternatives			Weights of the attributes	Weights of the criteria	Global weights of attributes	Priorities of the alternatives		
	1	2	3				1	2	3
1.1	0,8	1	0	0,42	0,3	0,13	0,10	0,13	0,00
1.2	0,8	1	0	0,38	0,3	0,11	0,09	0,11	0,00
1.3	0,6	1	0	0,2	0,3	0,06	0,04	0,06	0,00
2.1	1	1	0	0,36	0,19	0,07	0,07	0,07	0,00
2.2	0,69	0,71	0	0,64	0,19	0,12	0,08	0,09	0,00
3.1	1	0	1	0,19	0,27	0,05	0,05	0,00	0,05
3.2	1	1	0,5	0,43	0,27	0,12	0,12	0,12	0,06
3.3	0	0	1	0,38	0,27	0,10	0,00	0,00	0,10
4.1	0,5	1	0	0,71	0,24	0,17	0,09	0,17	0,00
4.2	1	0	1	0,29	0,24	0,07	0,07	0,00	0,07
Final priorities							0,70	0,74	0,28

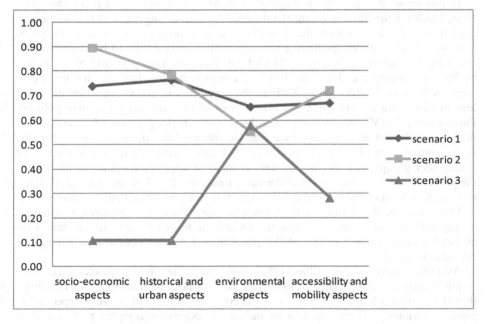

Fig. 5. Sensitivity analysis for the case under examination

In the present study the sensitivity analysis has been performed with regards to the weights of the attributes. Particularly, the weight of one attribute at time has been increased until 70% while the weights of the other three attributes have been maintained equal to 10%. The evaluation model has been run considering the new weights and the final priorities of the alternatives have been recalculated. Figure 5 represents the results of the performed sensitivity analysis. As it is possible to see, the scenario 2 has the best performances in the considered sensitivity sets of weights while the scenario 3 ranks worst in the most part of the sets. In this sense, it is possible to affirm that the model is quite stable and the rank is preserved in the most part of the considered cases.

4   Conclusions

The contribution illustrates the application of the MAVT approach in the context of the transformation of a large urban portion of the city of Torino (Italy). The results of the analysis have highlighted that the best scenario for the transformation of the area under analysis is the scenario 2, which aims at increasing the commercial function of the area under examination.

Decision-making in urban and territorial projects is usually complex because of heterogeneous stakeholder interests, multiple objectives, long planning and implementation processes, and uncertain outcomes [10]. In this context, the results of the performed analysis show that MAVT is efficient in representing the real problems of a territorial system.

In this sense, it is possible to highlight that MAVT approach has many strengths. First, MAVT helps in structuring the decision by classifying the problem in various objectives, criteria to measure the objectives and alternative options to solve the problem. MAVT also allows qualitative and quantitative information to be taken into account in the evaluation. Secondly, MAVT enhances the understanding of the policy problem by forcing the Decision Makers to compose a value function that represents their preferences. Thirdly, MAVT offers the possibility of reasoning about the problem by clarifying the strengths and weaknesses of the different alternative policies. Furthermore, MAVT strongly supports the decision-making process because it permits to clearly visualize and communicate the intermediate and final results. Finally, in the construction of the decision tree MAVT can incorporate the diverse views of stakeholder groups, considering the development of alternative options/solutions for the problem and the composition of the value functions. For these reasons, MAVT has been applied to many real-world decisions, in both the private and public sectors [11].

Moreover, the discussion oriented towards values rather than towards alternatives during both the value function construction and the level of trade-offs elicitation facilitated a better understanding of the problem and of the relationships among the considered aspects.

As a future development of the work, it would be interesting to consider uncertainty of predictions and risk attitude of DMs switching values functions to utility functions, thus developing a Multi Attribute Utility Theory model [4]. This is very important in public decisions and territorial transformation process where complex problems arise and the outcomes cannot be predicted precisely.

Secondly, mention should be made to the fact that MAVT aggregates the options' performance across all the attributes to form an overall assessment and is thus a compensatory technique. This means that the method does allow compensation of a weak performance of one attribute by a good performance of another attribute. It is interesting to notice that the compensatory approach is crucial in the field of "strong" sustainability assessment where the principle of substitution cannot be applied; in fact, according to the "strong sustainability paradigm" a good performance in one area (for example, the economic dimension) is not compensated by a poor performance in another area (for example, the environmental dimension). In this sense it would be interesting to investigate the existence of interactions among the evaluation criteria by means of the Non Additive Measures approach [12].

Thirdly, given the spatial nature of the decision problem under consideration, it would be of scientific interest to investigate the possibility of integrating the MAVT approach with Geographic Information Systems in order to develop a Multicriteria Spatial Decision Support System (MCSDSS) that will enable multi-purpose planning [13].

Acknowledgments. The paper is the result of the joint work of the three authors. Despite the overall responsibility being equally shared, Marta Bottero is responsible for paragraphs 1, 3.1 and 3.2 while Valentina Ferretti is responsible for paragraphs 2 and 3.4. Paragraph 3.3 is the result of the joint work of Marta Bottero and Valentina Ferretti while the abstract and the conclusions are the result of the joint work of the three authors.

References

1. Reicher, P., Schuwirth, N., Langhans, S.: Constructing, evaluating and visualizing value and utility functions for decision support. Environ. Modell. Softw. 46, 283–291 (2013)
2. Roy, B., Bouyssou, D.: Aide multicritére à la décision: Méthodes et case. Economica, Paris (1993)
3. Figueira, J., Greco, S., Ehrgott, M. (eds.): Multiple Criteria Decision Analysis: State of the Art Survey. Springer, New York (2005)
4. Keeney, R.L., Raiffa, H.: Decisions with Multiple Objectives: Preferences and Value Trade–offs. Wiley, New York (1976)
5. Keeney, R.L.: Value focused thinking. Harvard university press, Cambridge (1992)
6. Beinat, E.: Value functions for environmental management. Kluwer Academic Publishers, Dordrecht (1997)
7. Montibeller, G., Franco, A.: Decision and Risk Analysis for the Evaluation of Strategic Options. In: O'brien, F.A., Dyson, R.G. (eds.) Supporting Strategy: Frameworks, Methods and Models, John Wiley & Sons, Chichester (2007)
8. Ferretti, V., Bottero, M., Mondini, G.: Decision making and cultural heritage: An application of the Multi Attribute Value Theory for the reuse of historical buildings. J. Cult. Herit., http://dx.doi.org/10.1016/j.culher.2013.12.007
9. Belton, V., Stewart, T.J.: Multiple Criteria Decision Analysis: An Integrated Approach. Kluwer Academic Press, Boston (2002)

10. Bottero, M., Ferretti, V., Mondini, G.: From the environmental debt to the environmental loan: Trends and future challenges for intergenerational discounting. Environment, Development and Sustainability 15, 1623–1644 (2013)
11. Munda, G.: Multicriteria Evaluation in a Fuzzy Environment. Theory and Applications in Ecological Economics. Physical–Verlag, Heidelberg (1995)
12. Giove, S., Rosato, P., Breil, M.: An application of multicriteria decision making to built heritage: The redevelopment of Venice Arsenale. Journal of Multi-Criteria Decision Analysis 17, 85–99 (2010)
13. Malczewski, J.: GIS and Multicriteria Decision Analysis. John Wiley and Sons, New York (1999)

An Econometric Analysis of Homeownership Determinants in Belgium

Guillaume Xhignesse[1,2], Bruno Bianchet[1,3], Mario Cools[1,4], Henry-Jean Gathon[1,2], Bernard Jurion[1,2], and Jacques Teller[1,4]

[1] Research Center on City and Territory, University of Liège, Belgium
[2] Economics Department, HEC Management School, University of Liège, Belgium
{g.xhignesse,hj.gathon,b.jurion}@ulg.ac.be
[3] Geography Department, Faculty of Sciences, University of Liège, Belgium
bruno.bianchet@ulg.ac.be
[4] Architecture and Urbanism Department, Faculty of Applied Sciences,
University of Liège, Belgium
{mario.cools,jacques.teller}@ulg.ac.be

Abstract. In market economies, homeownership is associated with positive externalities. Increasing the levels of homeownership has been an objective of governments for the last decades. The analysis of the determinants of tenure status provides information to this end. This paper proposes an econometric analysis of housing tenure in Belgium. We review the main variables that have been considered in the literature as influencing housing tenure, after what we estimate a logit model. We observe a strong influence of income and age on the probability of homeownership. Couple relationship and the presence of dependent children have a positive influence, but this influence is less significant. Urban location is associated with lower probability of homeownership, compared with other areas. Our observations follow the trends described for other countries in the literature.

Keywords: Regression analysis, housing tenure, housing econometrics.

1   Introduction

The rate of homeownership varies significantly across European regions (See Fig. 1) [1]. Norway, Spain and East-European countries generally present a very high level of homeowners while Austria, Germany and Switzerland show clearly the lowest. Between these extremes, Greece, Portugal, Finland and Italy have relatively high rates of homeowners, whereas the ones of United Kingdom, the Netherlands and France are relatively low.

Belgium presents an intermediate situation, close to the average situation in the EU-28. In 2011, 72% of its total population owned and lived in its own house or flat, to 66% in 2001. These homeowners are not equally distributed among the 589 municipalities that make up the national territory (See Fig. 2) [2]; we observe a lower rate in the large and medium-size cities. Besides, homeowners are relatively rare in the five major cities of the country (Brussels, Antwerp, Ghent, Charleroi and Liège).

B. Murgante et al. (Eds.): ICCSA 2014, Part III, LNCS 8581, pp. 65–79, 2014.

Furthermore, inter-regional differences can be raised. Flanders, the most populated (6.4 million inhabitants) and richest of the three Belgian regions, counted 71% of homeowners in 2001, to 66% in Wallonia (3.6 million inhabitants) and only 40% in the Brussels-Capital region (1.1 million inhabitants).

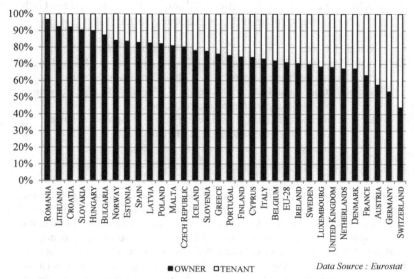

Data Source : Eurostat

Fig. 1. Population by tenure status in Europe (2011)

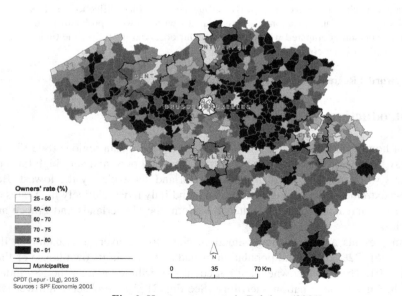

Fig. 2. Homeowners rate in Belgium (2001)

In Europe, as in the other market economies, homeownership is associated with positive externalities and socioeconomic effects [3]. For this reason, increasing the levels of homeownership has been one of the main goals of public authorities for the last decades in a lot of developed countries as well as in developing countries, and it is still the case. Such an objective manifests itself in the implementation of public policies favoring owner-occupation over the other tenure status, that is to say renting (at market or reduced price, or for free). Most of these policies use tax tools or subsidies, which are specifically studied by many authors [See, e.g., 4,5,6,7,8].

The analysis of housing tenure choice provides useful information about the elements that affect these choices for different categories of households in different situations, as a result of which such an analysis should be taken into account at the time of the definition or the renewal of housing public policies [9]. Considering the importance of such choices in the life-cycle of the individual and the social and economic consequences of aggregated homeownership level, housing tenure choice has been at the heart of numerous studies involving different fields of research. Concerning economics, this question gives rise to a prolific theoretical (for a non-technical review, see [10]) and empirical literature, the latter concerning mainly developed countries [See, e.g., 11,12,13,14,15,16], but also some developing countries [See, e.g., 17,18,19,20]. However, empirical literature exploring the continental Western Europe situation is rather poor [See, e.g., 21,22,23].

This paper proposes a classical econometric model ('classical model') of housing tenure choice for Belgium. Classical models use cross-sectional data to describe the likelihood of a household being a homeowner at a given time. Moreover, classical models use a sample of all households and, this way, model the cumulative homeownership attainment. By the way, the classical model may be justified by the fact that homeownership is a long-term decision supported with anticipated future needs as much as present needs [24]. However, it is generally admitted that classical models consider only housing tenure choice from a static perspective and then may reflect the lagged effects of earlier decisions [25,26]. Notwithstanding, the classical model seems to be the most useful model if the researchers have the ambition to describe the profile of the homeowner at a given time (what is our case) and not to explain the ownership decision per se. To the best of our knowledge, none of these models have been developed yet for Belgium. Nevertheless, Van Dam et al. [27] use a simple classical model of housing tenure choice in their study of the relationships between poverty and housing tenure in Flanders only.

This article is structured as follows. In the second section, we develop the empirical investigation framework of our research by defining briefly the main determinants of tenure status and presenting the theoretical model and data used. In the third section, the empirical estimations are presented; the accuracy and the predictive ability of the model as well as the marginal effects will be reviewed. Finally, we will discuss the main findings of our research and also provide some insights for future research.

2   Empirical Investigation Framework

2.1   Determinants of Tenure

In this part, we review briefly the main variables that have been considered in the literature as influencing housing tenure. We classify these in three well-known categories: economic, demographic and spatial variables.

Economic Variables.

Household's income is universally considered as one of the most determinant variables in the housing tenure choice literature. Its role would be mainly indirect – through borrowing constraints and relative cost of owning and renting[1] [28] – but also direct –because higher income could also be associated with higher preference for owner-occupation [12]. Most authors agree that permanent income, which describes the flows of income over a long time period according to the human capital of the household [29], is the appropriate measure of income in this context [See, e.g., 30,31]. This affirmation relies on the hypothesis that households' consumption of durable goods is determined by permanent income more than current income.

Housing prices determine, in association with income, the budget constraint for the household's consumption of housing services. They are considered in many studies as one of the most explanatory factors of homeownership. However, they remain sometimes neglected, in particular in studies focusing on household endowment factors [18]. Just like income, housing prices may intervene in models directly [See, e.g., 22] or indirectly, as a component of the relative cost of owning and renting.

Household's wealth plays an important role in the borrowing constraint and especially in the downpayment requirements. Because of the lack of available data, it remains rarely used in empirical models [22,32].

Demographic Variables.

Life-cycle variables, that is to say household size, age of the head and relationship status, reveal the household's phase in the life cycle and its mobile or stable nature. Concerning family size, larger families may be associated with later stage in life and therefore with greater ownership likelihood [33]. With regard to age, older heads are assumed to be more likely to have acquired the required savings to meet the minimum downpayment constraint [11]. Age could consequently be closely related to income and wealth and could introduce endogeneity in models [15]. As for relationship status, it could give an additional indication about the stability of the household's demand for housing [34]. However, relationship status could be problematic for the same reasons of endogeneity as age.

---

[1] That is the price-rent ratio multiplied by the household's annual user cost per euro invested in owner-occupied housing. The user cost notably includes the cost of equity, depreciation (or appreciation), mortgage interest payments, taxes and maintenance; and it depends on the household's income tax bracket [21].

The influence of *ethnicity and gender ('diversity')* of household heads on the housing tenure status is one of the most important issue in the literature. By the way, we observed that diversity considerations are normally included in housing tenure studies focussing on English-speaking countries or regions [See, e.g., 28,30,32,33,35] while it is absolutely ignored in the European literature until now.

Spatial Variables.
Residential location, which refers to the distinction between urban, suburban and rural areas, has an interesting place in some studies of housing tenure choice [14]. The introduction of the level of urbanity in the models is directly related to the urban economic theory which considers the trade-off between commununting costs and time and the household's expenditures in housing [36]. In general, higher urbanity is associated with lower ownership likelihood.

2.2 Theoretical Model and Estimation Approach

On the basis of this international review of the determinants of homeownership, we hypothesize the probability of homeownership as a function of income, life-cycle status, and location. Income and life-cycle status remain, according to the literature, the most relevant and robust [37]. In the context of an initial analysis, it seems relevant to limit the number and the complexity of the explanatory variables. Therefore, our model could be formally written as:

$$HO = f(INC, HHLFS, RLOC) \tag{1}$$

Where HO is the status of homeowner, INC is a measure of the household's income, HHLFS is a vector of life-cycle characteristics of the household, and RLOC is the residential location of the household. A definition of the variables composing the model and the expected signs of the relative parameters are presented in table 1. We precise that spatial variables are based on the classification of Belgian municipalities established by Vanneste et al. [38]. Concerning income, available information does not allow us to estimate permanent income, and current income is used as a proxy.

In order to estimate our model of housing tenure (that we consider as dichotomous), we use a logit procedure. The logistic regression model is a statistical classification model which is commonly used in discrete data analysis, and in housing tenure choice analysis in particular [14]. Specifically, the logit is defined as the natural logarithmic value of the odds in favor of a given event, that is:

$$L_i = \ln\left(\frac{P_i}{1 - P_i}\right) = \beta_1 + \beta_2 X_i + u_i \tag{2}$$

The model is based on the cumulative logistic distribution function P_i:

$$P_i = E(Y = 1 | X_i) = \frac{1}{1 + e^{-(\beta_1 + \beta_2 X_i)}} \tag{3}$$

$$P_i = \frac{1}{1+e^{-Z_i}} = \frac{e^Z}{1+e^Z} \qquad (4)$$

Where:

$$Z_i = \sum \beta_1 + \beta_2 X_i \qquad (5)$$

The ratio $P_i / (1 - P_i)$ represents the odds ratio, i.e. the odds that an event occurs $(Y = 1)$ to the odds it does not occur $(Y = 0)$; P_i is the probability that the output variable be equal to one $(Y = 1)$; X_i are the input (explicative) variables; e is the base of the natural logarithm; ß are the parameters. The estimated probabilities lie between 0 and 1 (P ranges from 0 to 1) and they are nonlinearly related to the explanatory variables. The logits are not so bounded as L goes from $-\infty$ to $+\infty$. Although L is linear in X, the probabilities themselves are not. This property is in contrast with the linear probability model [39].

Table 1. Expected influences of the independent variables

Var.	Definitions	Expected influences
Economic variables		
INC	Monthly after-tax income of the household (Categories, in euros).	Increasing positive influence with the increase of income.
Life-cycle variables (HHLFS)		
HHAGE	Age of the household's head (Categories, in years).	Increasing positive influence with the ageing of the household head.
COUPLE	Dummy variable of relationship status of the household's head: equals 1 if the head is in couple, 0 otherwise.	Positive influence.
DCHILD	Number of dependent children of the household (Continuous).	Positive influence.
Spatial variables		
URB	Dummy variable of the nature of the residential area of the household: equals 1 in case of urban area, 0 otherwise.	Negative influence.
SURB	Dummy variable of the nature of the residential area of the household: equals 1 in case of suburban area, 0 otherwise.	Positive influence

2.3   Data Description

To estimate our model, we use data collected for the Belgian Daily Mobility Survey 'BELDAM' in 2010 [40]. The data are available on request to the persons in charge of this survey. The methodology for data collection is outlined in the final report (freely downloadable on the official website). The initial database contains 8,526 individual observations and many variables relating to the daily mobility behaviour of each individual, which was the focus of this survey. Nevertheless, information about housing tenure status, life-cycle characteristics, income and location of the respondents were contained in the database and have been extracted for the need of our study. Out of these, we choose to delete observations containing at least one missing or incoherent value. At the end, 7,252 complete observations have been accepted. In addition, in order to develop a model which could also be used in a predictive way in further research, we decide to apply the technique employed by Cools et al. [41], and therefore to randomly split our sample in one primary and one secondary sample of respectively 4,835 observations (67%) and 2,417 observations (33%). This technique ensures the robustness of the model, which is estimated only on the basis of the primary sample and directly tested on the secondary sample.

We provide below some descriptive statistics of the global, the primary and the secondary samples. Table 2 shows the distribution of the global sample and the two sub-samples among the different tenure status and ensures the good representativeness of the latter over the global sample. We observe that the distribution between owners (0.697) and renters (0.303) in the global sample is close to the distribution of the population stated in the introduction of this paper. In addition, the distributions of the two sub-samples are very similar with the global sample distribution.

Table 3 shows some key statistics about the variables used in the model. These statistics concern only the primary sample which will be used to estimate the model. We can observe that homeownership is present in all categories of income and age, but is more present in the higher income and age categories. Couples are clearly more owners than tenants, while the difference is less obvious in the case of single households. In addition, owners have, on average, more dependent children than tenants. Finally, urban areas have a lower rate of homeowners than suburban or rural areas.

Table 2. Distribution of the samples

		Glob. Sample (2010)	Prim. Sample (2010)	Sec. Sample (2010)	Population (2001)[2]
Flanders	Owners	0.8055	0.8118	0.7934	0.7087
	Tenants	0.1945	0.1882	0.2066	0.2913
Wallonia	Owners	0.7538	0.7571	0.7472	0.6620
	Tenants	0.2462	0.2429	0.2528	0.3380
Brussels Capital	Owners	0.4985	0.4951	0.5054	0.3951
	Tenants	0.5015	0.5049	0.4946	0.6049
Total	Owners	0.6973	0.6991	0.6938	0.6620
	Tenants	0.3027	0.3009	0.3062	0.3380

---

[2] The detailed distribution is not available for years after 2001. The differences between the distribution of the population in 2001 and the global sample (which contains 2010 data) could be explained by the trend exposed in the introduction.

Table 3. Main descriptive statistics

			Distribution	
			Owner	**Tenant**
		Income (euros)		
	1	0 to 999	0.3540	0.6460
	2	1,000 to 1,499	0.5777	0.4223
Economic var.	3	1,500 to 1,999	0.6897	0.3103
	4	2,000 to 2,999	0.7944	0.2056
	5	3,000 to 4,999	0.8868	0.1132
	6	5,000 and more	0.8596	0.1404
		Age (years)		
	1	18-24	0.0814	0.9186
	2	25-34	0.4365	0.5635
	3	35-44	0.6650	0.3350
	4	45-54	0.7271	0.2729
	5	55-64	0.7850	0.2150
Life-cycle var.	6	65+	0.7845	0.2155
		Relationship status		
		Couple	0.8422	0.1578
		Single	0.5577	0.4423
		Dependent children		
		Mean	*0.4650*	*0.2769*
		Stand. dev.	*0.8847*	*0.7131*
		Location		
Spatial var.		Urban areas	0.6105	0.3895
		Suburban areas	0.8453	0.1547
		Rural areas	0.7885	0.2115

3   Empirical Estimation

3.1   Goodness-of-fit

The classical measurement of the goodness-of-fit (R^2) is not significative in the case of binary models. Therefore, we calculate the 'McFadden pseudo-R^2' (noted 'R_{MF}^2') which can be used as a goodness-of-fit indicator in the case of binary models, as:

$$R_{MF}^2 = 1 - \left(\frac{\log L}{\log L_0}\right)$$

(6)

Where $\log L$ is the log-likelihood of the estimated model and $\log L_0$ is the log-likelihood of the trivial model (containing only the constant term). In other words, the minimum value of the R_{MF}^2 is obtained when $\log L = \log L_0$, that is to say when the estimated model does not produce better results than the trivial model, and this value is equal to 0. On the opposite, the estimated model would be perfect if $\log L = 0$, and consequently $R_{MF}^2=1$. In our case, $R_{MF}^2=0.2195$, which is a low value compared to the perfect regression. However, pseudo-R^2 interpretations are very controversial and cannot be considered in an isolated way [42].

The goodness-of-fit of a binary model could be estimated in a simplier way by comparing the number of correct previsions of the model with the number of total observations in the sample [42]. In our case, 77.56% of the observations are correctly predicted by our model, what may be considered as a good result when we take into consideration the limited number of explanatory variables used, and their simple nature. Some relating complements are proposed in the part devoted to the predictive ability of the model.

3.2 Marginal Effects and Odds Ratios

Table 4 presents the coefficients of the logit model obtained by computing the data in Stata [43]. All the coefficients are statistically significant at 5 per cent except the SURB coefficient. However, in binary models, only the signs of the ß coefficients are interpretable as such, and the information they provide remains limited to the way (positive or negative) the probability varies compared with the given basis situation. In order to outline the amplitude of these influences, the marginal effects can be calculated. In fact, the marginal effects represent the variation of the homeownership probability in the case of a marginal change in one parameter (all other things remaining equal), still compared with the basis household.

Table 4. Estimated coefficients

	Coef.	Std. Err.	z	P>|z|
INC1	-1.203826	.1322435	-9.10	0.000
INC2	-.4719464	.1065786	-4.43	0.000
INC4	.4156449	.1165003	3.57	0.000
INC5	1.033136	.1415557	7.30	0.000
INC6	.7995854	.2166644	3.69	0.000
HHAGE1	-1.982217	.4240304	-4.67	0.000
HHAGE2	-.810933	.1308008	-6.20	0.000
HHAGE4	.3317586	.1204852	2.75	0.006
HHAGE5	.9090277	.1264432	7.19	0.000
HHAGE6	1.272747	.1220648	10.43	0.000
COUPLE	.6366912	.0911886	6.98	0.000
DCHILD	.171681	.0556677	3.08	0.002
URB	-.7424706	.1227139	-6.05	0.000
SURB	.1447139	.1398353	1.03	0.301
Constant	.5005716	.1630635	3.07	0.002

The marginal effects for our model are presented in table 5. For dummy variables, the marginal effect is calculated for a discrete change from 0 to 1. The basis household has the following characteristics: a monthly income between 1500 and 1999 euros (INC3), a head's age between 35 and 44 years (HHAGE3), single (COUPLE=0) and living in a rural area. We observe that the variables contained in the model influence the probability of homeownership in the generally expected way according to the international literature reviewed earlier. Going up the income categories has a positive influence on the probability of homeownership, while going down has a negative influence. The influence of head's age follows a similar pattern, that is to say younger households having a lower likelihood of being homeowners than older ones. A couple relationship status and dependent children have a positive influence (respectively +0.135 and +0.040). Concerning the spatial variables, urbanity has a negative influence (-0.182) on homeownership probability. On the whole, income, head's age and location have the strongest influences.

Table 5. Marginal effects

	dy/dx	Std. Error.	z	P>|z|
INC1*	-.2915026	.03003	-9.71	0.000
INC2*	-.1154378	.02583	-4.47	0.000
INC4*	.0916769	.02584	3.55	0.000
INC5*	.1999545	.02803	7.13	0.000
INC6*	.1632677	.03968	4.11	0.000
HHAGE1*	-.4374146	.06716	-6.51	0.000
HHAGE2*	-.1995671	.03138	-6.36	0.000
HHAGE4*	.0742538	.02728	2.72	0.006
HHAGE5*	.1811091	.02791	6.49	0.000
HHAGE6*	.2322762	.02814	8.25	0.000
COUPLE*	.1345831	.0195	6.90	0.000
DCHILD	.04034	.01351	2.99	0.003
URB*	-.1827752	.02902	-6.30	0.000
SURB*	.0333536	.03246	1.03	0.304
*FOR DISCRETE CHANGE FROM 0 TO 1				

However, these influences remain questionable because in such models, the marginal effects are not constant and depend on the point of the independent variable at which they are evaluated. Consequently, it could be useful to calculate the odds ratios, which constitute another (and sometimes considered as easier) way to interpret the coefficients of the model. In fact, odds ratios express the multiplicative effect between the probability, and not the probability itself. We provide these ratios in table 6. We observe that the probability of being a homeowner is almost three times superior for a household with a monthly income between 3,000 and 3,999 euros than for a household with a monthly income between 1,500 and 1,999 euros. By the way, a household with a monthly income between 1,500 and 1,999 euros has already three times more probabilities to be a homeowner than a household with a monthly income between 0 and 999 euros. Concerning head's age, the homeownership probability of young households (18-24 years) is remarkably inferior to the basis household of our model (35-44 years) – only 13% of the latter. Couples present a two times superior probability of homeownership than single households, while inhabitants of urban area have two times less probability of homeownership than inhabitants of rural areas.

Table 6. Odds ratios

	Odds ratio	Std. Error.	z	P>|z|
INC1	.3000441	.0396789	-9.10	0.000
INC2	.623787	.0664823	-4.43	0.000
INC4	1.515348	.1765385	3.57	0.000
INC5	2.809865	.3977522	7.30	0.000
INC6	2.224618	.4819956	3.69	0.000
HHAGE1	.1377635	.0584159	-4.67	0.000
HHAGE2	.4444432	.0581335	-6.20	0.000
HHAGE4	1.393416	.167886	2.75	0.006
HHAGE5	2.481908	.3138204	7.19	0.000
HHAGE6	3.570648	.4358503	10.43	0.000
COUPLE	1.890216	.1723661	6.98	0.000
DCHILD	1.187299	.0660942	3.08	0.002
URB	.4759366	.058404	-6.05	0.000
SURB	1.155709	.1616089	1.03	0.301
Constant	1.649664	.269	3.07	0.002

3.3 Predictive Ability of the Model

As mentioned earlier, we test the robustness of our model by using it on the secondary subsample. In this context, we compare the probability of homeownership given by the model for each individual with the real tenure status of the individual. The probability of homeownership for individual i is given by:

$$P(HO:X_i) = \frac{1}{1+e^{-z}}$$
(7)

$$z = \sum \beta_k X_{ik}$$
(8)

Where:

X_i = vector of given characteristics of individual i

β_k = estimated coefficients

Given the dichotomy in the tenure choice, we consider that a probability superior or equal to 0.5 involves an owner status, while a probability strictly inferior to 0.5 involves a tenant status. Based on this assumption, our model classifies correctly 76.50% of the individuals in the secondary subsample (1,849 observations correctly classified on a total of 2,417 observations). We objectively consider this result as a good result. Table 7 summarizes the predictive ability of the model and provides a comparison with the results stated in section 3.1 concerning the goodness-of-fit.

Table 7. Predictive ability of the model

	Total obs.	Correctly classified		Incorrectly classified	
		Number	%	Number	%
Prim. sample	4,835	3,750	77.56	1,085	22.44
Sec. sample	2,417	1,849	76.50	568	23.50

4   Discussion

We propose in this paper a classical econometric model of housing tenure in Belgium. To the best of our knowledge, our model is the first analysing the case of Belgium. To this end, we review in a first time the main variables that have been considered in the literature as influencing housing tenure. We find that household's income and housing prices are generally considered as the most determinant variables (notably because they determine the budget constraint of the household), while life-cycle variables (which reveal the stage of the household in the life cycle, the stability of its demand, but maybe also its wealth) and location (which refers to the trade-off between commuting cost and housing expenditures) are often also described as important. On this basis, we decide to integrate income, a vector of life-cycle characteristics and the location as the explanatory variables in our model, and we used a logit procedure to estimate it. To ensure the goodness-of-fit of the obtained model, we compare the number of correct previsions with the number of total observations in the sample. In our case, 77.56% of the observations are correctly predicted, what may be considered as a good result when we take into account the general aspect of the variables used. In addition, we test the robustness of the model with another sample. 76.50% of the individuals contained in this sample are correctly predicted by the model, what we consider also as a good result.

Regarding the expected influences of the parameters based on the literature review, our observations follow clearly the general trends. We observe a positive influence of higher income, what may be explained by a less restrictive budget constraint and/or a higher taste for privacy that involves consequently a preference for ownership from households with high income. However, we point out that the highest income group does not follow this trend and has a weaker influence on the probability of homeownership than high incomes groups. This situation could be explained by the fact that the highest income group contain very specific socio-economic and professional profiles that require sometimes high residential mobility, which is hardly compatible with homeownership. Age of the household's head influences homeownership probability in the same way as income. Actually, age may be related to wealth accumulation, what is necessary to meet the minimum downpayment constraint when acquiring a dwelling. However, the oldest groups may be expected to have a lower probability of homeowner status, because of specific needs linked to the third-age endowments. Nevertheless, we do not observe any of this in our analysis. Age could be closely related to income and would consequently introduce endogeneity in the model. Couple status and the presence of dependent children have also a positive influence on homeownership. In fact, these life-cycle variables give some indications about the stability of the household's housing demand that logically increase the likelihood of homeownership. Moreover, the presence of dependent children could involve higher requirements in terms of, amongst others, space or dwelling's characteristics, and therefore encourage the household to own its dwelling. Finally, urban location is associated, in Belgium as in other countries, with a lower probability of homeownership compared with rural or suburban areas. As mentioned above, this empirical observation may be explained with the help of the urban economic theory which develops the existing trade-off between commuting cost and housing expenditures. Furthermore, another original and recent way to explore the reasons for such a situation would be to analyse the characteristics of residential buildings that make them more likely to be owner-occupied [44].

This study presents a first insight in the field of housing tenure choice in Belgium and an additional analysis for the continental Western Europe. Moreover, the observations stated and the methodology used in this paper constitute a robust framework for in-depth research, in particular with the objective of identifying the target population which should benefit from public aid or with the aim of measuring the effects of public policies. Some extensions are proposed in the following section.

5 Perspectives

In this paper, we assess the determinants of housing tenure in Belgium with a general approach. For further research on the Belgian case, several improvements may be considered. First of all, the variables used in this approach could be precised. On one hand, economic variables such as housing prices could be introduced in the model, preferably under the form of relative prices of owning and renting. This approach seems to be the most pertinent way to analyse the influence of tax and housing policies on the choice of tenure that remains a very controversial question. On the other hand, life-cyle variables may be expanded. In particular, we observe a gap to be filled concerning the influence of diversity related variables on the tenure choice in continental European countries.

Acknowledgements. The authors wish to thank Mehdi Ould Kherroubi and Daniel Schleck for helpful discussions. This research is funded by the Government of Wallonia through the Standing Conference of Territorial Development, a multi-disciplinary research platform bringing together researchers from the three main French-speaking universities of Belgium around urban development issues.

References

1. Eurostat, http://epp.eurostat.ec.europa.eu
2. Directorate General Statistics and Economic Information of the Belgian Federal Government, http://statbel.fgov.be
3. Dietz, R.D., Haurin, D.R.: The social and private micro-level consequences of homeownership. Journal of Urban Economics 54, 401–450 (2003)

4. Bourassa, S., Yin, M.: Housing tenure choice in Australia and the United States: Impacts of alternative subsidy policies. Real Estate Economics 34, 303–328 (2006)
5. Bourassa, S., Yin, M.: Tax deductions, tax Credits and the homeownership rate of young urban adults in the United States. Urban Studies 45, 1141–1161 (2008)
6. Green, R., Vandell, K.: Giving households credit: How changes in the U.S. tax code could promote homeownership. Regional Science and Urban Economics 29, 419–444 (1999)
7. Hilber, C.A.L., Turner, T.M.: The mortgage interest deduction and its impact on homeownership decisions. Review of Economics and Statistics (forthcoming)
8. Rosen, H.S., Rosen, K.T.: Federal Taxes and Homeownership: Evidence from Time Series. Journal of Political Economy 88, 59–75 (1980)
9. Danière, A.: Determinants of tenure choice in the Third World: An empirical study of Cairo and Manilla. Journal of Housing Economics 2, 159–184 (1992)
10. Hubert, F.: The economic theory of tenure choice. In: Arnott, R.J., McMillen, D.P. (eds.) A Companion to Urban Economics, pp. 145–158. Oxford, Blackwell Publishing, United Kingdom (2008)
11. Bourassa, S.: A model of housing tenure choice in Australia. Journal of Urban Economics 37, 161–175 (1995)
12. Bourassa, S., Peng, C.-W.: Why is Taiwan's homeownership rate so high? Urban Studies 48, 2887–2904 (2011)
13. Cho, C.-J.: Joint choice of tenure and dwelling type: A multinomial logit analysis for the city of Chongju. Urban Studies 34, 1459–1473 (1997)
14. Iwarere, L.J., Williams, J.E.: A micro-market analysis of tenure choice using the logit model. The Journal of Real Estate Research 6, 327–339 (1991)
15. Li, M.M.: A Logit Model of Homeownership. Econometrica 45, 1081–1097 (1977)
16. Vernon Henderson, J., Ioannides, Y.M.: Tenure choice and the demand for housing. Economica 53, 231–246 (1986)
17. Ahmad, N.: A joint model of tenure choice and demand for housing in the city of Karachi. Urban Studies 31, 1691–1706 (1994)
18. Arimah, B.C.: The determinants of housing tenure choice in Ibadan, Nigeria. Urban Studies 34, 105–124 (1997)
19. Huang, Y., Clark, W.A.V.: Housing tenure choice in transitional urban China: A multilevel analysis. Urban Studies 39, 7–32 (2002)
20. Li, S.-M.: The housing market and tenure decisions in Chinese cities: A multivariate analysis of the case of Guangzhou. Housing Studes 15, 213–236 (2000)
21. Bourassa, S., Hoesli, M.: Why do the Swiss rent? The Journal of Real Estate Finance and Economics 40, 286–309 (2008)
22. Consuelo Colom, C., Cruz Molés, C.: Comparative analysis of the social, demographic and economic factors that influenced housing choices in Spain in 1990 and 2000. Urban Studies 45, 917–941 (2008)
23. Fischer, M.M., Aufhauser, E.: Housing choice in a regulated market: A nested multinomial logit analysis. Geographical Analysis 20, 47–69 (1988)
24. Edin, P.-A., Englund, P.: Moving costs and housing demand: Are recent movers really in equilibrium? Journal of Public Economics 44, 299–320 (1991)
25. Di Salvo, P., Ermisch, J.: Analysis of the dynamics of housing tenure choice in Britain. Journal of Urban Economics 42, 1–17 (1997)
26. Painter, G., Gabriel, S., Myers, D.: Race, immigrant status, and housing tenure choice. Journal of Urban Economics 49, 150–167 (2001)
27. Van Dam, R., Geurts, V., Pannecoucke, I.: Housing tenure, housing costs and poverty in Flanders (Belgium). Journal of Housing and the Built Environment 18, 1–23 (2003)

28. Bourassa, S.: Ethnicity, endogeneity, and housing tenure choice. Journal of Real Estate Finance and Economics 20, 323–341 (2000)
29. Goodman, A.C., Kawai, M.: Permanent income, hedonic prices, and demand for housing: New evidence. Journal of Urban Economics 12, 214–237 (1982)
30. Kain, J.F., Quigley, J.M.: Housing Market Discrimination, Home-ownership, and Savings Behavior. American Economic Review 62, 263–277 (1972)
31. Painter, G.: Tenure choice with sample selection: Differences among alternative samples. Journal of Housing Economics 9, 197–213 (2000)
32. Painter, G., Yang, L., Yu, Z.: Homeownership determinants for Chinese Americans: Assimilation, ethnic concentration and nativity. Real Estate Economics 32, 509–539 (2004)
33. Skaburskis, A.: Race and tenure in Toronto. Urban Studies 33, 223–252 (1996)
34. Carliner, G.: Determinants of Home Ownership. Land Economics, L, 109–119 (1974)
35. Gabriel, S., Painter, G.: Mobility, residential location and the American dream: The intra-metropolitan geography of minority homeownership. Real Estate Economics 36, 499–531 (2008)
36. Fujita, M.: Urban economic theory. Cambridge University Press, Cambridge (1989)
37. Raya, J., Garcia, J.: Which are the real determinants of tenure? A comparative analysis of different models of the tenure choice of a house. Urban Studies 49, 3645–3662 (2012)
38. Vanneste, I., Thomas, I., Goosens, L.: Le logement en Belgique. SPF Economie, Brussels (2007)
39. Greene, W.H.: Econometric Analysis, 7th edn. Pearson Education, Upper Saddle River (2011)
40. Belgium Daily Mobility, http://www.beldam.be
41. Cools, M., Moons, E., Wets, G.: Investigating the effect of holidays on daily traffic counts: A time series approach. Journal of the Transportation Research Board 2019, 22–31 (2007)
42. Gujarati, D.N.: Basic econometrics. McGraw-Hill, New York (2003)
43. StataCorp, http://www.stata.com
44. Coulson, N.E., Fisher, L.M.: Structure and tenure. Working Paper (2012)

A Model to Support the Decision
to Invest in Bare Dominium

Benedetto Manganelli

University of Basilicata, 10, Viale dell'Ateneo Lucano, 85100 Potenza, Italy
benedetto.manganelli@unibas.it

Abstract. This study focuses on the market of property subject to lifelong usufruct in Italy. This type of investment or disinvestment has found more widespread than in the past because of the difficulties related to the current economic
conditions that adversely affect the disposable income and access to credit of
households. This paper develops a Decision Tree Analysis model to support a
decision to invest in bare dominium. The model is perfectly suited to the analysis
of this investment, since it is able to treat the specific risk that characterizes it
that is connected to the life expectancy of the usufructuary.

Keywords: Bare dominium, Decision tree, Property investment, Real estate
valuation.

1   Introduction

In Italy, the sale of properties subject to lifelong usufruct represents more than 5% of
the transactions made in the residential real estate market. This percentage is growing
and, in 2012, it has reached the highest value of the last decade (figure 1). It will certainly increase, taking into account that the amount of homeownership is among the
highest in Europe, reaching 80% of the total. In fact, this type of contract offers advantages to the seller, who receives cash in a period of great difficulty in accessing
credit, mainly due to more and more restrictive guarantees required by the lenders,
and he has the option to use the dwelling and to reap the fruits until his death. On the
other hand, the buyer obtains both a reduction in price and a substantial saving on
transfer taxes, on the costs of management and maintenance, which continue to weigh
on the usufructuary.

The behavior of those who buy the bare dominium can more easily be compared to
that of an investor rather than to the consumer. The buyer is aware of the uncertainty
about when he will have full ownership of the property and he does not act in view of
the fulfillment of a basic need. Its behavioral model, however, does not coincide with
that of a pure investor because he is not seeking to obtain, immediately, an income, as
happens in the case of those who buy real estate, especially residential, with the aim
of selling it or lease it. He is instead pushed by reasons related to the need to preserve
his capital, or frequently by reasons of extra-economic nature; one of which is related
to the opportunity to buy a home at a price lower than the current market value, where
his children will live in the future. The purchase of the bare dominium is in effect
a bet on the life of another person, the usufructuary. This study focuses on the risk

B. Murgante et al. (Eds.): ICCSA 2014, Part III, LNCS 8581, pp. 80–89, 2014.
© Springer International Publishing Switzerland 2014

related to this bet, even though like all real estate investment, also in this case possible profit could come by the increase in property values over time.   The specific profit of the bare dominium owner may be, immediately, even very high, if the death of the usufructuary were to occur shortly after purchase, and ahead of schedule.

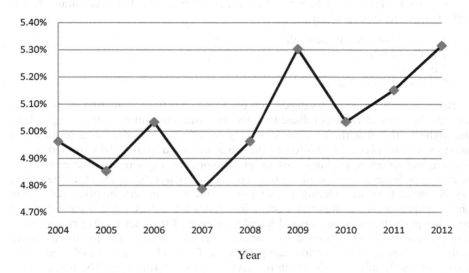

Fig. 1. Trend of bare dominium percentage of the total number of houses sold, in Italy (Source "Agenzia delle Entrate")

The bare dominium owner, on the one side, cannot enjoy the property subject to usufruct (for it he will have to wait for the death of the usufructuary); on the other hand, he has the advantage of avoiding expenditure on routine maintenance and taxes related to the availability and enjoyment of the property. Obviously, in this case the investment is free from risks arising from the management of the property (taxes, solvency of the tenants, vacancy rates, etc...).

Of no less importance is the fact that the capital gain on the difference between the prices paid and the final value of the property following the consolidation of ownership, cannot be subjected to any taxation.

The known methods for estimating the bare dominium, both those that involve the use of financial mathematics and those using coefficients defined for tax purposes, handle this risk with a deterministic approach that does not take into account the time factor and therefore the possibility of different scenarios from what is a priori considered the most probable. This paper deals with assessing the convenience to purchase a bare dominium by building a decision tree model [1] able to manage the typical risks of this investment, which is related to the life expectancy of the usufructuary.

2   The Decision Making Process

The investment value differs from market value and the comparison between these two is used to define the advisability of investing. The investment decision is the result of an evaluation of cost effectiveness. Of course, the alternative may be configured with a different investment opportunity or just with the alternative zero, if the planned activity does not happen.

The investment value provides a measure of the importance that a particular investor attaches to the flow of benefits. In the case of buying a bare dominium then it reflects the assessments about:

- The probable term of life of the usufructuary;
- The final sale price;
- Taxation;
- The financing instruments.

Because these assessments depend on economic and fiscal condition of the investors, the investment value is related to individual investor sentiment. It corresponds to the highest price that the investor is willing to offer for purchase as a result of an analysis that involves, in addition to the variables listed above, the expected profit. Modern techniques of evaluation of an investment cannot ignore the time preference. A greater or lesser extension of the time horizon has a significant effect on the result. On the other hand considering the temporal dimension in turn involves judgments about the probability of alternative scenarios. When the decision maker must evaluate complex projects, the sequence and interdependence of the choices give rise to problems cannot be solved with the use of the static net present value [2]. In this case there is the advantage that the scenarios are perfectly defined. They are, for each year of the analysis period, only two: the death or survival of the usufructuary. The probabilities of occurrence of these two possible scenarios, alternative perfectly, are exogenous data on which the assessment would not be required. In fact, the Italian National Institute of Statistics (Istat) provides them.

2.1   Decision Tree

When we know the scenarios and their probabilities, the risk of investment may be treated using probabilistic analysis techniques able to implement these data. In particular, Decision Tree Analysis perfectly reproduces the reference scheme typical of this investment. [3] [4] [5]. The logic underlying the model presupposes that the structure of the project is translated into a flowchart called "decision tree." Each branch of the tree is associated with the possible values that the project may take as a result of the occurrence of certain hypothetical scenarios, to which are assigned the precise probability. This analysis allows us to capture the interrelationships between decisions reached at different times, mutually dependent.

The analysis summarizes a chain of conditional choices, where the development hypothesis at a given stage depends on the actions taken or the possible events in the earlier stage. It is therefore necessary that the result of the interaction between the successive phases be explained at the time of evaluation. This scheme is perfectly suited to the case under study.

The principal advantage of the decision tree analysis is the opportunity to express the structure of the project through a chart that highlights the fundamental key steps (nodes) [6] [7]. Decision analysis is the process of separating a complex decision into its component parts and using a mathematical formula to reconstitute the whole decision from its parts [8] [9].

A decision tree allows assessing the impact of each decision node in relation to the effects that a decision or an event could have on the project as a whole. It is used to assess future flexibility inherent in an investment project and its convenience in different scenarios.

This methodology may therefore be seen as a first attempt to introduce into the traditional methods of capital budgeting, the concept of strategic analysis [10] [11]. The effort remains unfinished. In fact, the limit that this approach does not exceed consists in the necessity to explain the probability distribution of the scenarios that characterize the different expected results. However, what might appear a limit it is not, when as in the present case is known the exact measure of the probabilities of the possible scenarios.

The nodes in a decision tree can be seen as 'decision points', when, at that time, the strategic decision depends on the investor, or as 'chance events', which are particular phases of the development project in correspondence of which external events may occur. In general, however, the first node is configured necessarily as a decision point and is defined as the 'root point', since you can assimilate it to the roots of a tree. A very important role is played by the branches, which instead describer the possible scenarios caused by a decision or an event [6]. 'Endpoints' define the possible outcomes of the investment.

One of the tasks that require more attention by the analyst is the correct definition of the characteristics of the branches, since he must assign to each of them its probability of occurrence. The sum of the probabilities associated to the branches that result from the same node must be equal to 1 [11]. The possible scenarios in the decision tree are collectively exhaustive since you cannot add others not previously considered in the analysis.

For each chance event, it is possible to determine the flow (gain or loss) that the project takes as a function of potential scenarios and their probability of occurrence.

2.2   Model Specification

By transferring, the foregoing to investment decision on the purchase of a bare dominium was built a model for estimating the value of investment to support that decision. Obviously, the first step is the estimate of the market value of the bare dominium (V_{bd}) that defines the reference for assessing the suitability of the investment [12] [13]. This estimate is based on the complementary relationship that the bare dominium value has with the value of the real right of usufruct (V_u); it can be performed using two different approaches. Both have as a reference the market value of the property (V_m). In one case, it is calculated by capitalizing the annual yield (Y_a) on the property over the life expectancy of the usufructuary (n), from the date on which the usufruct is granted. This amount is then deducted from the market value of the property at date of sale to determine the bare dominium value. The formula for the calculation is:

$$V_{bd} = V_m - V_u = V_m - Y_a \cdot \left(\frac{(1+i)^n - 1}{i \cdot (1+i)^n} \right). \tag{1}$$

In the second case, reductive predetermined coefficients (c) for tax purposes are applied to the current market value of the property. These coefficients are according to the age of the usufructuary and the legal rate of interest (i_l). This second approach is the one most commonly used by real estate agents. In the case of lifelong usufruct, the value of this real right is inversely proportional to the age of the usufructuary. Instead, the opposite is true for the bare dominium value that reaches 90% of the market value of the property when the usufructuary age is between 93 and 99 years. The formulas used in this case are the following:

$$Y^*_a = V_m \cdot i_l \qquad V_u = Y^*_a \cdot c \qquad V_{bd} = V_m - V_u = V_m(1 - i_l \cdot c). \quad (2)$$

The subject intends to purchase the bare dominium, should be borne in mind that the potential scenarios are multiple since he does not know when there will be consolidation between the right of usufruct and bare dominium.

At each year following the year of purchase, two events are possible: one is the advantage resulting from the achievement of the full value of the property, if the usufructuary dies; the other refers the increase in value of the right to bare dominium in the event that the usufructuary survives. Below is constructed a formula able to return the value required, namely able to update automatically each year the value of the bare dominium starting from (1).

$$V_t = V_{bd}(n - t - 1) - V_{bd}(n - t) = \frac{Y_a}{(1+i)^{(n-t+1)}} \qquad for \; t = 1, 2, ..., x \quad (3)$$

with x = year-end analysis.

By placing the nodes in sequence for each specific year, you get a tree with a binomial trend.

In the construction of decision tree, the period of analysis depends on the age of the usufructuary at the time of purchase. The limit (x) is defined from the year in which the probability of survival is close to zero. Each branch, and therefore each value determined in correspondence with the possible events is associated with its probability of occurrence, the probability of survival or death of the usufructuary, which are complementary to the unit. These are provided annually by Istat.

3   Application

The model was applied to a real case. The source data are a) the market value of the property; V_m= 200,000 €, b) the annual yield; Y = 7,000 €, c) the characteristics of the usufructuary; a man of 65 years. Please note that market research conducted primarily by real estate agents showed that who offers the bare dominium usually is between 60 and 70 years old.

The investment cost, equal to the market value of property is estimated using equation (1). Life expectancy for a 65 year old man is about 18 years (source: Istat). The (1) gives the value of € 98,214.7 for $i = i_l$ = 2.5%. The latter, however, is not very different from that measured using the coefficients valid for tax purposes (€ 100,000).

Each scenario is expression of a sequence of events, connected to each other and relative to the survival or death of the usufructuary, whose probability for each year is respectively p (t) and 1-p (t). The probability of a scenario is given by the joint probability $(p(c))$ obtained by the product of all the probabilities associated with events leading to that particular node.

The development of decision tree is also related the interval between the age of the usufructuary at the time of investment; in this case we assumed that he is 65 years old. The last year covered by Istat in the construction of tables of mortality coincides with the age of 119 years. 54 + 1= 55 scenarios are therefore possible.

Table 1. Probability of death for males

Years	(1-p)/100	Years	(1-p)/100	Years	(1-p)/100
65	1.220669	84	9.404812	103	46.42319
66	1.345200	85	10.36180	104	49.76697
67	1.509628	86	11.72183	105	53.12315
68	1.642612	87	13.06181	106	56.46301
69	1.780147	88	14.23517	107	59.75802
70	1.962726	89	15.12301	108	62.98154
71	2.109858	90	16.49781	109	66.10882
72	2.342070	91	18.35475	110	69.11832
73	2.612289	92	21.43404	111	71.99192
74	2.951294	93	24.36141	112	74.71548
75	3.284192	94	26.35090	113	77.27841
76	3.647049	95	27.87837	114	79.67409
77	4.091015	96	28.45284	115	81.89926
78	4.613335	97	29.05472	116	83.95399
79	5.184270	98	29.86444	117	85.84083
80	5.930776	99	31.56170	118	87.56479
81	6.678323	100	34.84726	119 (x)	89.13248
82	7.490556	101	39.88537		
83	8.386056	102	43.12042		

The table 1 shows the probability of death related to the age of the usufructuary.

After determining the potential scenarios and their probabilities, it is possible to derive the investment value for each of them (figure 2).

The profitability indicator used for the valuation is the Internal Rate of Return (IRR) that is the rate of return that makes the net present value (NPV) equal to zero. This frees the analysis from the choice of a discount rate factor. To determine the IRR we constructed the function of total value of the investment (NPV expected) to changes in the discount rate (r).

The total value of the investment is given by the weighted sum of the values that the investment (outcomes) takes on all endpoints of the decision tree (table 2).

Table 2. Calculating the Net Present Value of cash flows at endpoints

$$NPV_1 = \left(\frac{V_m}{(1+r)}\right) - V_{bd}$$

$$NPV_2 = \left(\frac{V_1}{(1+r)^1} + \frac{V_m}{(1+r)^2}\right) - V_{bd}$$

$$NPV_3 = \left(\sum_{t=1}^{2}\frac{V_t}{(1+r)^t}\right) + \frac{V_m}{(1+r)^3} - V_{bd}$$

.

.

$$NPV_{x2} = \left(\sum_{t=1}^{x-1}\frac{V_t}{(1+r)^t}\right) + \frac{V_m}{(1+r)^x} - V_{bd}$$

$$NPV_{x1} = \left(\sum_{t=1}^{x}\frac{V_t}{(1+r)^t}\right) - V_{bd}$$

$$NPV\ expected = \sum_{t=1}^{x} NPV_t \cdot p(c_t)$$

To determine the IRR we constructed the function of the investment value (*NPV expected*) by varying the discount rate factor. The investment takes as many values (NPV) as potential scenarios (endpoints). For each outcome, the NPV is given by discounting the sequence of cash inflows preceding the endpoint (V_t) minus the purchase price equal to the bare dominium value (figure 2). The NPV calculated at a given scenario is multiplied by its probability of occurrence. All weighted NPV are finally summed to obtain the expected value of the investment. The calculation of the expected value to changes in the discount rate defines the function in the figure 3.

This function detects an internal rate of return equal to 8.54%.

This rate of return shows that it is a great opportunity taking into account 'the current state of housing market crash in which other types of investment would hardly achieve similar performance.

4   Conclusions

The result shows that the proposed model is an excellent tool to assist the decision to invest in bare dominium. The spread of this type of investment is linked to a greater awareness of rights and duties of the bare dominium owner, but above all to the better understanding of the opportunities and potential economic benefits offered.

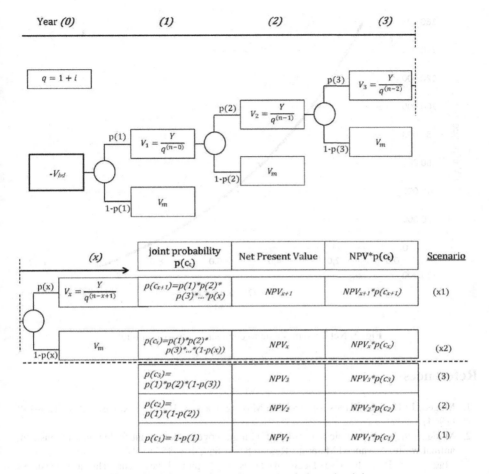

Fig. 2. Decision tree for an investment in bare dominium

The model focuses on the management of specific risk associated with this type of investment that, unlike other real estate investments, relates essentially to the life expectancy of the usufructuary. This risk is defined by a known probability, that of death or survival of an individual according to sex and age.

The economic benefit could be even greater for real estate funds or big investors capable of including in their portfolio many bare dominium. When the probability of occurrence of events is known, as in the present case, according to the law of large numbers, the average of the results obtained from a large number of opportunities affected by those events should be close to the expected value, and will tend to become closer as more investment opportunities are taken.

Further research might investigate how the convenience of investing in the bare dominium may be influenced by sex and age of the usufructuary.

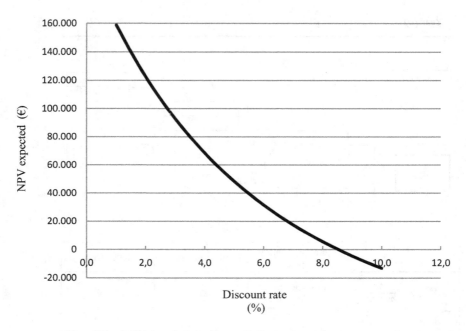

Fig. 3. Net Present Value to changes in the discount rate

References

1. Magee, J.: Decision Trees for Decision Making. Harvard Business Review 42(4), 126–138 (1964)
2. Manganelli, B.: Economic feasibility of a biogas cogeneration plant fueled with biogasfrom animal waste. Advanced Materials Research 864-867, 451–455 (2014)
3. Magee, J.: How to Use Decision Trees in Capital Investment. Harvard Business Review 2(5), 79–96 (1964)
4. De Ambrogio, W.: Programmazione reticolare. Etas Libri, Milano (1977)
5. Zaderenko, S.G.: Sistemi di programmazione reticolare. Hoepli, Milano (1970)
6. Bland, R., Beechey, C.E., Dowding, D.: Extending the model of the Decision Tree. Technical Report CSM – 162. Department of Computing Science and Mathematics – University of Stirling, Scotland (August 2002)
7. Dey, P.K.: Decision support system for risk management: A case study. Management Decision 39(8), 634–649 (2001)
8. Alemi, F., Gustafson, D.H.: Decision Analysis for healthcare managers. Health Administration Press, AUPHA Press, Chicago, Washington (2006)
9. Goodwin, P., Wright, G.: Decision Analysis for Management Judgement. John Wiley & Sons Ltd., Chichester (2004)
10. Manganelli, B.: La valutazione degli investimenti immobiliari. Franco Angeli, Milano (2013)

11. De Mare, G., Manganelli, B., Nesticò, A.: The Economic Evaluation of Investments in the Energy Sector - A Model for the Optimization of the Scenario Analyses. In: Murgante, B., Misra, S., Carlini, M., Torre, C.M., Nguyen, H.-Q., Taniar, D., Apduhan, B.O., Gervasi, O. (eds.) ICCSA 2013, Part II. LNCS, vol. 7972, pp. 359–374. Springer, Heidelberg (2013)
12. McNamee, P., Celona, J.: Decision Analysis for the Professional. SmartOrg, Inc. (2001)
13. Morano, P., Tajani, F.: Estimative Analysis of a Segment of the Bare Ownership Market of Residential Property. In: Murgante, B., Misra, S., Carlini, M., Torre, C.M., Nguyen, H.-Q., Taniar, D., Apduhan, B.O., Gervasi, O. (eds.) ICCSA 2013, Part IV. LNCS, vol. 7974, pp. 433–443. Springer, Heidelberg (2013)
14. Morano, P., Tajani, F.: Bare ownership evaluation, Hedonic price model vs artificial neural network. International Journal of Business Intelligence and Data Mining 8(4), 340–362 (2013)

A Choquet Integral Based Assessment Model
of Projects of Urban Neglected Areas: A Case of Study

Teresa Cilona and Maria Fiorella Granata[*]

University of Palermo, Department of Architecture, Palermo, Italy
{teresa.cilona,maria.granata}@unipa.it

Abstract. This paper describes a multi-criteria evaluation model supporting decisions related to redevelopment of urban residual areas, a central theme in planning practices. Renewal projects on urban or neighborhood scale are complex problems because of the social, economic and environmental implications generated on the different categories of stakeholders. In the awareness of the specific characteristics of each city, the cognitive and evaluation model is especially defined for a given urban context, although it is easily adaptable to different ones. In order to take into account the interactions among the criteria by which we compare design alternatives, the Choquet integral is implemented as aggregation function. The model applies to some alternative projects for the redevelopment of a residual urban area in Agrigento (Italy). It can be usefully employed by the local government to choose the better alternative in pursuing policy objectives or to build new projects.

Keywords: Multiple criteria decision aid, Choquet integral, neglected urban areas.

1   Introduction

Numerous studies have been conducted on multi-criteria analysis applications in various sectors [1]. In the fields of urban and regional planning applications there are a great number of studies on large complex projects, such as infrastructure works, which impacts cover wide geographical areas for a long time [2, 3, 4, 5, 6]. Although the spatial and temporal scale of impacts is usually greatly reduced, interventions of urban or neighborhood scale are also defined as complex problems, because of the implications of social, economic and environmental nature that are generated on the different categories of stakeholders [7, 8, 9, 10].

Urban planning can play a strategic role in urban sustainability [11] since it does not create merely physical spaces, but it can help creating a shared identity, a sense of social belonging [12], that is a crucial factor for local sustainability [13, 14]. Therefore, urban design needs to be addressed by making use of multi-criteria analysis

---

[*] This work is the result of an interdisciplinary collaboration of the authors, with the following contribution of T. Cilona (Chapters 4.1, 4.2, 4.3) and M.F. Granata (Chapters 1, 2, 3, 4.4, 4.5). Chapter 5 is due to both the authors.

B. Murgante et al. (Eds.): ICCSA 2014, Part III, LNCS 8581, pp. 90–105, 2014.

techniques [7], [15, 16, 17], possibly integrated with the social and financial assessments that are used in urban equalization [18, 19, 20].

The 5th Aalborg Commitment [21] ("Planning and design") recognizes the importance of "re-use and regenerate derelict or disadvantaged areas", in order to fulfill the strategic role of urban planning and design in increasing social well-being. Since every city has typical features [22], planning decisions have to be made according to these characteristics. Therefore, a unique set of indicators in evaluating the sustainability of an urban project or planning cannot be established. The specific set related to a project and to a city must meet the basic needs of local community and embrace the three pillars of sustainability (economy, environment and society) [23].

The purpose of this paper is to present a multiple criteria model for handling the refurbishment of an urban abandoned area in Agrigento (Italy), a city showing a contradiction between a high tourist potential, for the presence of the Valley of the Temples (UNESCO site), and some socio-economic difficulties. The model allows identifying some "potential actions" of urban transformation that can best meet the needs of local development. Evaluation of projects should be seen as a cognitive activity aimed at better understanding the points of view of the various stakeholders. It may also outline new design alternatives and facilitate the implementation of participative strategies [23, 24, 25].

The present study is divided into an analysis of socio-economic and environmental local context followed by a description of the project alternatives and of the multicriteria evaluation model. Then we proceed to a sensitivity analysis on the weights and, finally, we discuss the results of the evaluation.

2   Methodology

In an urban context, the feasible functional targets of neglected areas are manifold and autonomous governmental decisions are not always coherent with the needs of citizens. Given the particular characteristics of the different urban realities, we proceeded with the definition of a valuation model specifically outlined for the considered city. The model will be able to support public decisions or to provide guidance for the construction of new design alternatives.

According to the classical structure of decisional process [26], the assessment model has been constructed through the following steps: 1. Analysis of the socio-economic and environmental local system, on which base the rational formulation of the decision problem; 2. Analysis of alternative projects for the redevelopment of the residual area; 3. Identification of the criteria against which the alternatives are evaluated, on the basis of the local key sustainability concerns; 4. Identification of an appropriate procedure for assessing the overall evaluation of alternatives with respect to the adopted family of criteria; 5. Options analysis and sensitivity analysis, designed to verify the robustness of the model with respect to changes in the system of weights; 6. Formulation of final recommendations.

Dealing with sustainability, the various thematic categories and the indicators are not independent, but they influence each other [27]. The multiple forms of interaction taking place among the sub-systems that make up the complexity of towns [28] suggest the use of an aggregation operator able to seize them. The Choquet integral, which is based on fuzzy measures, is a suitable algorithm for global assessment of alternatives, since it considers the positive (synergic) or negative (redundant) interaction among criteria.

The concept of fuzzy integral [29] in the form of Choquet [30] was first applied in multiple criteria decision problems mainly in the field of industrial applications [31] and then in other various sectors, like policy capturing, analysis of root dispersal, computation of the number of citations clinical diagnosis, overall industrial performance, selection of groups of genes, evaluation of discomfort in sitting position when driving a car [32], supply chain partners and configuration selection [33], evaluation of strategies in games [34], data mining [35], choice of accommodation for travelers [36]; warehouse location in logistic problems [37]; evaluation of software quality [38]. The interest of researchers in exploring the use of the innovative aggregation operator in decision making about public concerns is quite recent and there are still few examples in literature on some specific field, like source water protection [39], evaluation of sustainability of nations [40], definition of energy regional plans [41], localization of manufacturing center [42]; evaluation of energy performance of buildings [43], redevelopment of built heritage [44], reuse of open-pit quarry [45], waterfront redevelopment designing [46]. In this work the application of Choquet integral is explored in the field of urban abandoned areas design.

3   The Aggregation Operator

In multiple criteria decision problems, the global assessment of an alternative is the result of aggregation of its partial performances with respect to each considered criteria. The most traditional approach applies the Multiple Attribute Utility Theory, according which a numerical value (utility) is assigned to each action, such that the more is the value of the global utility function the more the action is preferable. Additive aggregation operators of marginal utilities impose the constraint that any subset of criteria must be preferentially independent [47]. As remarked in the previous chapter, the independence of criteria is a drawback in modelling the preference of the decision maker about urban sustainability, where criteria are interdependent instead [27, 28] and neglecting interactions among them may give aggregate values not coherent with available information about the preferences of decision maker [31, 48].

The Choquet integral is an extension of the weighted arithmetic mean. It is able to represent dependence among criteria, as the weights related to single criterion are substituted by weights related to all combinations of criteria [49]. Two or more criteria are synergic (redundant) when their joint weight is more (less) than the sum of the weights of every single criterion of the coalition [48]. These weights can be identified by learning data or questionnaire [31].

The analytical definition of Choquet integral is now recalled. Let $G = \{g_1, g_2, ..., g_n\}$ be a finite set of cardinal criteria and $P(G)$ the power set. A fuzzy measure on G is a set function $\mu:P(G) \rightarrow [0, 1]$ that satisfies the following axioms: $\mu(\emptyset)=0$ and $\mu(G)=1$ (boundary conditions); $\forall T \subseteq R \subseteq G, \mu(T) \leq \mu(R)$ (monotonicity condition). In the contest of multicriteria decision problems, a fuzzy measure $\mu(T)$ stands for the weight of importance on a combination of criteria T. A fuzzy measure is said to be additive if $\mu(T \cup R) = \mu(T) + \mu(R)$ whenever $T \cap B = \emptyset$. It is said to be superadditive (subadditive) if

$\mu(T \cup R) \geq \mu(T) + \mu(R)$ $(\mu(T \cup R) \leq \mu(T) + \mu(R))$ with $T \cap B = \emptyset$. The superadditivity of a fuzzy measure stands for a synergy among criteria; the subadditivity implies redundancy among criteria [31]. If a fuzzy measure is additive, then it suffices to define the n coefficients $\mu(g_1)$, $\mu(g_2)$, ..., $\mu(g_n)$ to define the measure entirely. If it is super-additive or subadditive it is necessary to give the 2^n coefficients corresponding to the 2^n subsets of G.

With respect to a fuzzy measure μ, the global importance of a criterion $g_i \in G$ is not solely given by the value $\mu(g_i)$, but also by the value representing its interaction with other criteria, i.e. for every value $\mu(T \cup \{g_i\})$, $T \subseteq G \setminus \{g_i\}$. The importance index or Shapley value of criterion $g_i \in G$ with respect to $\mu(T)$ and the Murofushi-Soneda interaction index between criteria g_i, $g_j \in G$ with respect to $\mu(T)$ are suitable index to measure importance and interaction between criteria [48].

Let μ be a fuzzy measure on G. The Choquet integral of a function $f : G \rightarrow [0, 1]$ with respect to μ is defined by

$$C_\mu(f(g_1),...,f(g_n)) = \sum_{i=1}^{n} (f(g_{(i)}) - f(g_{(i-1)})) \; \mu(A_{(i)}) \tag{1}$$

where (i) stands for a permutation of the indices of the marginal utility functions and of criteria such that $0 \leq f(g_{(1)}) \leq ... \leq f(g_{(n)}) \leq 1$; $A(i) = \{g(i), ..., g(n)\}$, $i = 1, 2, ...,$ n; $f(g(0)) = 0$.

The description of interactions among all the criteria is a too hard task for the decision maker. The information on the importance of the single criteria and the interactions between couples of criteria can be given with more lucidity in many applications [48]. So, in the following application the Möbius representation of the Choquet integral with regard to 2-additive measure will be considered. Let $\mu(R)$ be a fuzzy measure; the Möbius representation is defined by the function $a : P(G) \rightarrow R$, as follows for each $R \subseteq G$:

$$a(R) = \sum_{T \subseteq R} (-1)^{|R-T|} \mu(T) \tag{2}$$

with the property: $a(\emptyset) = 0$ and $\sum_{T \subseteq G} a(T) = 1$ (boundary conditions), $\forall \, g_i \in R$ and $\forall \, R$

$\subseteq G$, $\sum_{T \subseteq G} a(T) \geq 0$ (monotonicity condition) [52]. If $R = \{i\}$ is a singleton, then

$a(\{i\}) = \mu(\{i\})$. If $R = \{i,j\}$ is a couple of criteria, then $a(\{i,j\}) = \mu(\{i,j\}) - \mu(\{i\}) - \mu(\{j\})$.

A fuzzy measure is said k-additive if $a(T) = 0$ when $|T| > k$. With respect to a 2-additive fuzzy measure, the fuzzy measure $\mu(R)$ obtained by the inverse transformation from the Möbius representation $a(R)$ is defined as

$$\mu(R) = \sum_{g_i \in R} a(\{g_i\}) + \sum_{\{g_i, g_j\} \subseteq R} a(\{g_i, g_j\}), \; \forall \, R \subseteq G \tag{3}$$

with the properties: $a(\{g_i\}) \geq 0$, $\forall \, g_i \in G$; $a(\{g_i\}) + \sum_{g_j \in T} a(\{g_i, g_j\}) \geq 0$, $\forall \, g_i \in G$

and $\forall \, T \subseteq G \setminus \{g_i\}$.

With respect to a 2-additive fuzzy measure, the Choquet integral expressed in terms of Möbius representation is given by

$$C_\mu(x) = \sum_{g_i \in G} a(\{g_i\}) u_i(g_i(x)) + \sum_{\{g_i,g_j\} \subseteq G} a(\{g_i,g_j\}) \min\{u_i(g_i(x)), u_j(g_j(x))\}, \quad (4)$$

for $x \in A$, the set of alternatives. In the same case, the importance index is given by

$$\Phi(\{g_i\}) = a(\{g_i\}) + \sum_{g_i \in G\{g_i\}} \frac{a(\{g_i,g_j\})}{2}, \quad g_i \in G \quad (5)$$

and the interaction index for a couple of criteria $g_i, g_j \in G$ is given by

$$\Phi(\{g_i,g_j\}) = a(\{g_i,g_j\}) \; [50]. \quad (6)$$

4   Evaluation Model

4.1   Analysis of the Local Socio-economic and Environmental Background

In urban planning, open space is the key element for the welfare of citizens and for ensuring a good quality of life. That is why a "culture" of urban planning should be promoted, in which public gardens and equipped green areas will offer the opportunity to enhance the local landscape heritage by providing spaces and opportunities for leisure, recreation, socialization. This will encourage the development of a civil consciousness and of a cultural heritage, and will contribute to the improvement of the psychophysical health of the inhabitants and of their life.

The study we propose refers to the city of Agrigento (Fig. 1a), on the south coast of Sicily, famous for its Valley of the Temples (World Heritage Site) and for its great historical, architectural, archaeological heritage. In the past it was also destination of great travellers, since it was at a centre of national and international trade. The city has undergone a series of transformations without following any pattern, any shape, any urban design, especially since the fifties, during the so-called speculation which caused a landslide in 1966. This event marked visibly and indelibly the area and the town [51]. Moreover, the lack of a urban planning has determined many disadvantages, as for example the lack of equipment and services, often insufficient and in some cases non-existent. We want to focus our attention and develop our study on this aspect. There are several green areas in this town which are abandoned and decaying and towards which it was not possible to create a "link" between the citizen and the place. The attempt to redevelop these spaces has often failed due to incorrect design choices and to poor maintenance. The procedure to carry out is very delicate and it is the result of a refined interpretation of the natural and environmental features that the territory owns. It needs to upgrade the green area and, where possible, increase its surface, reduce air pollution caused by traffic, make safe the "hypogeum" (underground tunnels that run through the city), and improve the aesthetic aspect of the urban landscape.

4.2   The Place as It Is

The studied area (Fig. 1*b*), of about 7.000 square meters, is situated within the historic center of Agrigento, in the district called "Rabato", between Garibaldi Street (to the north), Dante Street (south), Porta di Mare Street (east), and Finazzi Alley (west).

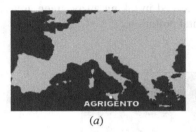
(*a*)

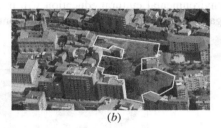
(*b*)

Fig. 1. The location of Agrigento in Italy (*a*) and a view of the "Rabado" district (*b*). The considered area is highlighted in white.

Some architectural entities are of particular value: the remains of Agugliaro palace [52], which was demolished in 2002 by the City Council for safety reasons since it had been damaged by the landslide 1966; the entrance to a "hypogeoum" [52]; and the Church of St. Catherine, now no longer in use. The area is situated on steep slopes and it is undeveloped, in fact there is a drop of about 25 meter between Garibaldi Street and Dante Street. Moreover it is in a state of neglect, with thick vegetation, except for some private lands used as storage. In the south- east a long staircase connects Dante Street and Annunciation Street, becoming the only unifying element between the historical and the contemporary city. The urban redevelopment aims at improving the quality of life and the promotion of economic and social development, both in territorial and urban areas. The intervention, therefore, must not only offer a physical and an aesthetic recovery of part of the city, but it must primarily aim at revitalizing it through the improvement of its commercial and economic potential and also its cultural and social interchange. In our specific case, the assessment of the physical and morphological characteristics, has led us to provide different solutions, in accordance with the provisions of the Town Plan, approved in 2007, which indicates the area destined partly to parks and garden and partly to parking. Note that in the following chapters the current state of the place is called "alternative 0".

4.3   Designing Proposals

The proposals below arise from the indications of the "Leipzig Charter" in 2007 on "Sustainable European Cities" [53] Intervention strategies are designed to: recover degraded areas in order to increase social cohesion; create quality and attractive public spaces and oriented-users; involve citizens in the process of recovery; strengthen the local economy and integration policies and social support; promote the environmental sustainability of existing buildings, avoiding the consumption of other green areas.

Alternative 1 – Urban Gardens and Recreational Sports Spaces. The first proposal (Fig. 2) is inspired by a national project of "Our Italy" entitled "Urban Gardens", and addresses to both private or public authorities that possess green areas and want to allocate them to the "art of cultivating" respecting the historical memory of these places and the "ethical" rules established by "OUR Italy" in accordance with ANCI (Association of Municipalities of Italy). According to this project these activities of environmental cultural education will have succeed in taking away green areas to illegal construction, speculation and environmental pollution.

Fig. 2. Some views of the redeveloped area as an urban garden and a place for recreational sport

This is why it is useful for us to realize in this part of the city an area for scientific and didactic purposes trying to identify some paths that allow learners to be able to learn in a direct and close contact with nature, expanding their knowledge in food and agriculture (with particular attention to a sustainable use of production techniques). At the same time, along these paths, fun recreational areas will be identified so as to rediscover past games, stimulate their creativity and encourage the movement. These games have been abandoned by the younger generation for a long time, such as Hopscotch game or drawings activities done with the use of colored chalks, which offer them the possibility to change the space with their signs.

The construction cost is approximately € 250,000.

Alternative 2 – Parking. The second proposal refers to the Town Plan [54]. The project includes, in the plain part of the area, the construction of a car park at ground level for 50 parking spaces (Fig. 3). This way the track, now used improperly, could be upgraded thanks to the addition of items which refer to street furniture (benches, litter bins, lighting). Moreover the flow of traffic would be simplified since it is now slowed by the presence of vehicles parked on the right and left side of the street.

The construction cost is about € 100,000.

(a)   (b)

Fig. 3. Rendering of the parking project. The redevelopment of the contiguous street (*a*) and a sight of the parking spaces (*b*).

Alternative 3 – Urban Park. Our proposal (Fig. 4) aims at the requalification of this area as a meeting and socializing place. Thanks to this intervention the entrance to the Hypogeum and the traces of the ancient Agugliaro palace will be brought to light, and this area will become a square [55].

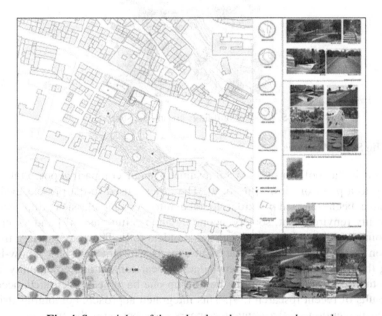

Fig. 4. Some sights of the redeveloped area as an urban park

Our main concern will be the recovery and the regulation of private spaces trying to recreate the image of a "traditional" environment choosing typical plants of the Mediterranean and the Sicilian countryside. The vegetation has an important role and plays a dual function: that of consolidation and of green lung. To create green areas for playing and recreational use, carob trees will be used. For private gardens citrus, bitter orange, mastic shrubs, oak, olive, euphorbia, broom, myrtle and rosemary will be planted. For parking use hedges or some typical aromatic essences with shielding function, will be used, such as laurel and lavender. This intervention will create public and private spaces that settle down to the morphology of the area, following the natural shape of the land, through paths, moving or fixed staircase and bioengineering systems.

The construction cost is approximately € 400,000.

Alternative 4 – Housing. The fourth proposal provides for an intervention on existing buildings which overlook on the designed area (total 3,000 square meters, including the multipurpose building the church of St. Catherine), through renovation and restoration works. Moreover, since the demand for housing is rising, we propose the construction of a new residential building with sustainable canon for 6 apartments, having the same design-shape of Agugliaro palace (Fig. 5).

The total cost is about € 3,970,000.

Fig. 5. Existing buildings around the study area

4.4   The Set of Criteria

A criterion is "a tool constructed for evaluating and comparing potential actions according to a point of view" [56, p. 9]. The criteria used in the present evaluation model have been selected in order to take into account classical dimensions of sustainability (environment, society, economy and institutions) [57] and to represent all the points of view relevant to local stakeholders [58]. They have sprung from the interaction of the authors with some local experts who not only have know-how in planning themes but also a deep knowledge of local context and community. In this way, the top-down approach and the bottom-up one have been integrated, according to the results of recent research [59]. This integrated approach, combining strictness with local perception of the context, appears coherent with the meaning of Local Agenda 21, which recognizes the local definition of sustainability, and overcomes the reductionism of top-down approaches [59]. Table 1 shows the hierarchical structure of target, macro-criteria and criteria. In it, the individual criteria are described semantically.

4.5   Option Analysis and Sensitivity Analysis

Using the Choquet integral with respect to a 2-additive fuzzy measure, the aggregation of marginal valuations of alternatives on each criterion requires the assignment of

Table 1. Hierarchical structure of target, macro-criteria and criteria

Goal	Macro-Criteria	Criteria	Description	Code
IMPROVEMENT OF URBAN QUALITY	Economic Factors	Development of economic activities	It expresses the contribution to the development of existing or new economic activities.	C1
		Tourist Attraction	It expresses the attractiveness of the city center for visitors of the Valley of the Temples.	C2
	Social Factors	Public housing	It expresses the benefit that the considered project brings to the local population in terms of availability of social housing.	C3
		Social gathering, entertainment and learning places	It concerns the creation of social gathering places and places of learning or entertainment available to school and general public.	C4
	Environmental Factors	Geological safety	It concerns the contribution to geological security of the territory.	C5
		Urban green spaces	It regards the presence of green areas in the urban context.	C6
		Urban air quality	It gives an assessment on the impact that the alternative design produces in terms of greenhouse gas emissions.	C7
		Cityscape	It regards the quality of the urban landscape generated by the implementation of the intervention.	C8
	Institutional Factors	Urban forecasts	It expresses the consistency of the intervention with urban forecast, since the need to use a variant would result in greater administrative complexity and in a longer time for carrying out the project.	C9

partial evaluation of the projects against the criteria and of an appropriate system of weights to the couples of criteria.

Table 2. Measure scale of marginal performances

Evaluation of alternatives respect to the criteria	Score
Insufficient level	0
Modest level	2,5
Medium level	5
Good level	7,5
Optimal level	10

Information retrieval at a local community is recognized as the best method to obtain valid and low cost knowledge about the same local reality [60]. Therefore, the marginal values of alternatives and the weights have been measured through the administration of some questionnaires to a panel of local experts, composed by some city planners, an agronomist and a geologist. The judgments on marginal performances have been expressed in a cardinal scale of measurement in the interval (0, 10). The value 0 expresses a totally insufficient level of the performance of the project from the considered point of view, while 10 corresponds to a fully optimal level. With the object of obtaining the partial evaluation, we asked the experts to answer the question: «How do you assess the current situation and the proposed projects for the neglected area in the district "Rabato" in Agrigento with respect to the stated criteria?». In order to assist the assignment of the points in compiling individual evaluation tables, the experts have been provided with the measure scale of marginal performance shown in Table 2.

The experts have also assigned weights to single criterion and to all the couple of criteria, answering the questions: «How do you assess the importance of the stated criteria with respect to the overall objective "quality of the city", taking into account the current living conditions in Agrigento?» and «How do you assess the importance of joint enhancement of two factors with respect to the overall objective "quality of the city", taking into account the current living conditions in Agrigento?». The weights have been expressed in a 10-point scale. In order to help the experts in the task, they have been provided with the following measure scale (Table 3).

Table 3. Measure scale of weights related to single criterion (*a*) and to couples of criteria (*b*)

Evaluation of the importance of the considered criteria	Score
insignificant	0
not much important	2,5
average importance	5
very important	7,5
extremely important	10

(*a*)

Joint evaluation of effect of pairs of criteria	Score
the joint action of the two criteria is definitely comparable	-10
the joint action of the two criteria is quite comparable	-5
the effects of the two factors simply add up	0
the joint action of the two criteria is quite synergistic	5
the joint action of the two criteria is strongly synergistic	10

(*b*)

On the grounds of acquired data, the overall evaluation table (Table 4) and system of weights (Table 5) have been constructed. In the performance matrix each row refers to a project and each column to a criterion. It describes the performance of the alternatives with respect to all criteria singly considered. The performance values and the weights correspond to the mean of the values attributed by each member of the panel of experts.

Table 4. Evaluation table

Projects	Criteria								
	C1	C2	C3	C4	C5	C6	C7	C8	C9
Alternative 0	0.8	3.3	1.0	0.0	0.8	0.0	3.2	1.5	1.3
Alternative 1	7.2	7.0	2.7	9.3	6.5	9.1	7.1	8.3	4.4
Alternative 2	3.7	2.8	3.3	3.2	4.9	3.2	1.8	2.0	9.5
Alternative 3	6.6	7.9	3.2	10.0	8.8	10.0	7.8	9.7	4.9
Alternative 4	3.5	2.5	8.3	4.5	5.3	4.2	3.2	3.3	1.3

Table 5. The system of weights

$\varphi(C1)$	9.33	$\varphi(C1,C2)$	10.00	$\varphi(C2,C4)$	8.67	$\varphi(C3,C7)$	2.83	$\varphi(C5,C7)$	3.83
$\varphi(C2)$	10.00	$\varphi(C1,C3)$	0.50	$\varphi(C2,C5)$	4.00	$\varphi(C3,C8)$	5.00	$\varphi(C5,C8)$	7.83
$\varphi(C3)$	4.08	$\varphi(C1,C4)$	9.08	$\varphi(C2,C6)$	8.33	$\varphi(C3,C9)$	0.17	$\varphi(C5,C9)$	-3.33
$\varphi(C4)$	9.00	$\varphi(C1,C5)$	1.33	$\varphi(C2,C7)$	3.83	$\varphi(C4,C5)$	3.50	$\varphi(C6,C7)$	2.17
$\varphi(C5)$	8.75	$\varphi(C1,C6)$	4.00	$\varphi(C2,C8)$	9.17	$\varphi(C4,C6)$	9.50	$\varphi(C6,C8)$	4.67
$\varphi(C6)$	8.67	$\varphi(C1,C7)$	1.00	$\varphi(C2,C9)$	6.67	$\varphi(C4,C7)$	4.33	$\varphi(C6,C9)$	0.33
$\varphi(C7)$	5.33	$\varphi(C1,C8)$	1.17	$\varphi(C3,C4)$	5.00	$\varphi(C4,C8)$	7.17	$\varphi(C7,C8)$	3.33
$\varphi(C8)$	8.50	$\varphi(C1,C9)$	-3.00	$\varphi(C3,C5)$	7.00	$\varphi(C4,C9)$	1.25	$\varphi(C7,C9)$	0.00
$\varphi(C9)$	4.58	$\varphi(C2,C3)$	0.00	$\varphi(C3,C6)$	6.67	$\varphi(C5,C6)$	6.00	$\varphi(C8,C9)$	0.00

The application of the aggregation algorithm provides the global evaluation ("comprehensive Choquet integral") of alternatives according to the comprehensive weights of the panel of experts (Table 6 - second column).

Table 6. The global performance of the projects according to the weights of the overall panel of experts (second column) and single different points of view (third, fourth and fifth column)

Alternatives	Comprehensive Choquet integral	Choquet integral		
		planners	geologist	agronomist
0 – Current state of the place	0,065	0,0659	0,0668	0,0630
1 - Urban gardens and sports spaces	0,110	0,1110	0,1098	0,1090
2 – Parking	0,086	0,0870	0,0847	0,0847
3 - Urban Park	0,111	0,1113	0,1103	0,1088
4- Housing	0,100	0,0998	0,1030	0,0982

Finally, the robustness of the preference for the scenario of transformation has been analyzed. The global evaluation of alternatives has been computed using from time to time the priorities and the interactions among the criteria expressed by the singly considered categories of experts (the planners, the geologist and the agronomist). The results show that the urban park transformation scenario is still the preferable one (Table 6 - third, fourth and fifth column) and the ranking of the projects is confirmed, since the difference resulting from the only agronomic point of view is very slight.

5   Concluding Remarks on Results

In the considered case study the local administration has an abandoned area which is susceptible of various uses and the prevalence of one alternative on another is not evident. From the study carried out we can say that it is necessary to redevelop the area as an urban park and consider it as a social place in which you can spend your free time and organize social and cultural activities outdoors. The proposed project would result in an improvement in the quality of life for both the local residents and the rest of the population of Agrigento.

The proposed methodological approach allows dealing with the choice among alternative urban projects, combining strictness with nimbleness of procedure. While classical preference modeling (outranking and additive multiple attribute utility theory methods) requires the use of independent criteria, the Choquet integral permits the aggregation of non necessarily independent information taking into consideration the perception of interactions between the criteria of local community, according to the integrated approach of sustainability. This aggregation tool gives a better representation of local community preferences, although it presents some operational difficulty, mainly pertinent to the elicitation of weights related to coalitions of criteria [48, 49]. Nevertheless, the experiment carried out seems to prove that this difficulty is not an insuperable barrier. Dealing with fuzzy integrals, another operative trouble consists in the number of coefficients involved in the model, which grows exponentially with the number of criteria [31]. Therefore, it is advisable to follow the classical rule limiting the number of considered criteria and to take into consideration only coalitions made of couples of criteria, which are more easily understandable [48].

We believe that the followed approach has permitted to face the decisional problem with greater awareness and that it may be used as a cognitive platform for explaining the decision problem in a possible successive participated decision process.

The present study is an example of generation of transparent information through which local governments can engage in constructive dialogue with citizens and build further planning decisions. Similar tools have a crucial role in increasing social capital and pursuing more sustainable development [61].

Acknowledgements. Authors are grateful to O. Campo, G. Riccobene, P. Pontei, A. Villardita e D. Vinti for completing the submitted questionnaires. They are also grateful to the anonymous referees for their constructive comments on the earlier version of this paper.

References

1. Figueira, J., Greco, S., Ehrgott, M.: Multiple criteria decision analysis. Springer, New York (2005)
2. Beinanat, E., Nijkamp, P. (eds.): Multicriteria Analysis for Land-Use Management. Kluwer, Dordrecht (1998)
3. Bana e Costa, C.A., Nunes da Silva, F., Vansnick, J.C.: Conflict dissolution in the public sector: A case study. European J. of Operational Research 130, 388–401 (2001)
4. Coutinho-Rodiguez, J., Simao, A., Antunes, C.H.: A GIS-based multicriteria spazial decision support system for planning urban infrastructures. Decision Support Systems 51, 720–726 (2011)
5. Rosso, M., Bottero, M., Pomarico, S., La Ferlita, S., Comino, E.: Integrating multicriteria evaluation and stakeholder analysis for assessing hydropower projects. Energy Policy 67, 870–881 (2014)
6. Cerreta, M., D'Auria, A., Giordano, G., De Toro, P.: Valutazione multi criterio e multi gruppo per lo studio di fattibilità del sistema di mobilità della Penisola Sorrentina. In: Giordano, G. (ed.) Pratiche di valutazione. DENAROlibri, Napoli (2004)
7. Fusco Girard, L., Nijkam, P.: Le valutazioni per lo sviluppo sostenibile della città e del territorio. Angeli, Milano (1997)
8. Giordano, G. (ed.): Pratiche di valutazione. DenaroLibri, Napoli (2004)
9. Mollica, E., Malaspina, M.: Programmare valorizzare e accompagnare lo sviluppo locale. Laruffa editore, Reggio Calabria (2012)
10. Brandon, P.S., Lombardi, P.: Evaluating sustainable development. Blackwell Publishing, Oxford (2005)
11. Forte, F., Fusco Girard, L.: Creativity and new architectural assets: The complex value of beauty. International Journal of Sustainable Development 12, 160–191 (2009)
12. Ratin, D.E.: Creative cities and/or sustainable cities: Discourses and Practices. Culture and society 4, 125–135 (2013)
13. Rizzo, F.: Il capitale sociale della città. Valutazione pianificazione e gestione. FrancoAngeli, Milano (2003)
14. Rydin, Y., Holman, N.: Re-evaluating the contribution of social capital in achieving sustainable development. Local Environment 9, 117–133 (2004)
15. Girard, L.F., Torre, C.M.: The Use of Ahp in a Multiactor Evaluation for Urban Development Programs: A Case Study. In: Murgante, B., Gervasi, O., Misra, S., Nedjah, N., Rocha, A.M.A.C., Taniar, D., Apduhan, B.O. (eds.) ICCSA 2012, Part II. LNCS, vol. 7334, pp. 157–167. Springer, Heidelberg (2012)
16. Lombardi, P. (ed.): Riuso edilizio e rigenerazione urbana. Innovazione e partecipazione. Celid, Torino (2008)
17. Cerreta, M., De Toro, P.: Assessing Urban Transformations: A SDSS for the Master Plan of Castel Capuano, Naples. In: Murgante, B., Gervasi, O., Misra, S., Nedjah, N., Rocha, A.M.A.C., Taniar, D., Apduhan, B.O. (eds.) ICCSA 2012, Part II. LNCS, vol. 7334, pp. 168–180.
    Springer, Heidelberg (2012)
18. Stanghellini, S.: La riqualificazione urbana fra leggi di mercato ed esigenze sociali. In: Fusco Girard, L., Forte, B., Cerreta, M., De Toro, P., Forte, F. (eds.) L'uomo e la città, pp. 490–501. Franco Angeli, Milano (2003)
19. Lombardi, P., Micelli, E. (eds.): Le misure del piano. Angeli, Milano (1999)

20. Stanghellini, S., Stellin, G.: Politiche di riqualificazione delle aree metropolitane: domande di valutazione e contributo delle discipline economico-estimative. Genio Rurale 7/8, 47–55 (1997)

21. European Sustainable Cities and Towns Conference, http://ec.europa.eu/environment/urban/aalborg.htm

22. Sustainable Cities International – Canadian International Development Agency: Indicators for Sustainability. How cities are monitoring and evaluating their success, Vancouver (November 2012)

23. Mondini, G.: La valutazione come processo di produzione di conoscenza per il progetto. Valori e Valutazioni 3, 5–16 (2009)

24. Fusco Girard, L., Cerreta, M., De Toro, P.: Valutazioni integrate: da "processo di apprendimento" a "gestione della conoscenza". Valori e Valutazioni 4/5, 101–115 (2010)

25. Forte, F.: I giudizi di valore nel processo di ideazione del progetto. Valori e Valutazioni 4/5, 117–125 (2010)

26. Tsoukiàs, A.: On the concept of decision aiding process. Annals of Operations Research 154, 3–27 (2007)

27. Munda, G.: Environmental economics, ecological economics, and the concept of sustainable development. Environmental Values 6, 213–233 (1997)

28. Rizzo, F.: Il territorio come organizzazione auto poietica, struttura dissipativa e sistema politico-amministrativo: una scienza del valore e delle valutazioni. In: Maciocco, G., Marchi, G. (eds.) Dimensione ecologica e sviluppo locale: Problemi di valutazione. FrancoAngeli, Milano (2000)

29. Sugeno, M.: Theory of fuzzy integrals and its applications. Doctoral thesis. Tokyo Institute of Technology (1974)

30. Choquet, G.: Theory of capacities. Annales de l'Institut Fourier 5, 131–295 (1953)

31. Grabisch, M.: The application of fuzzy integrals in multicriteria decision making. European Journal of Operational Research 89, 445–456 (1996)

32. Grabisch, M., Labreuche, C.: A decade of application of the Choquet and Sugeno integrals in multi-criteria decision aid. Annals of Operations Research 175(1), 247–286 (2010)

33. Ashayeri, J., Tuzkaya, G., Tuzkaya, U.R.: Supply chain partners and configuration selection: An intuitionistic fuzzy Choquet integral operator based approach. Expert Systems with Applications 39, 3642–3649 (2012)

34. Narukawa, Y., Torra, V.: Fuzzy measures and integrals in evaluation of strategies. Information Sciences 177, 4686–4695 (2007)

35. Wang, Z., Leung, K.S., Klir, G.J.: Applying fuzzy measures and nonlinear integrals in data mining. Fuzzy Sets and Systems 156(3), 371–380 (2005)

36. Li, G., Law, R., Vu, H.Q., Rong, J.: Discovering the hotel selection preferences of Hong Kong inbound travelers using the Choquet Integral. Tourism Management 36, 321–330 (2013)

37. Demirel, T., Demirel, N.C., Kahraman, C.: Multi-criteria warehouse location selection using Choquet integral. Expert Systems with Applications 37(5), 3943–3952 (2010)

38. Pasrija, V., Kumar, S., Srivastava, P.R.: Assessment of Software Quality: Choquet Integral Approach. Procedia Technology 6, 153–162 (2012)

39. Islam, N., Sadiq, R., Rodriguez, M.J., Francisque, A.: Evaluation of source water protection strategies: A fuzzy-based model. Journal of Environmental Management 121, 191–201 (2013)

40. Pinar, M., Cruciani, C., Giove, S., Sostero, M.: Constructing the FEEM sustainability index: A Choquet integral application. Ecological Indicators 39, 189–202 (2014)

41. Zhang, L., Zhou, D.Q., Zhou, P., Chen, Q.T.: Modelling policy decision of sustainable energy strategies for Nanjing city: A fuzzy integral approach. Renewable Energy 62, 197–203 (2014)

42. Feng, C.M., Wu, P.J., Chia, K.C.: A hybrid fuzzy integral decision-making model for locating manufacturing centers in China: A case study. European Journal of Operational Research 200(1), 63–73 (2010)
43. Lee, W.S.: Evaluating and ranking energy performance of office buildings using fuzzy measure and fuzzy integral. Energy Conversion and Management 51(1), 197–203 (2010)
44. Giove, S., Rosato, P., Breil, M.: An application of multicriteria decision making to built heritage: The redevelopment of Venice Arsenale. Journal of Multi-Criteria Decision Analysis 17, 85–99 (2011)
45. Bottero, M., Ferretti, V., Pomarico, S.: Assessing Different Possibilities for the Reuse of an Open-pit Quarry Using the Choquet Integral. Journal of Multi-Criteria Decision Analysis (2013), Published online in Wiley Online Library
46. Granata, M.F.: A multidimensional model based on the Choquet integral to evaluate performance of waterfront redevelopment projects in promoting local growth. BDC - Bollettino del Dipartimento di Conservazione dei Beni Architettonici ed Ambientali 12(1), 960–971 (2012)
47. Vincke, P.: Multicriteria Decision-aid. John Wiley & Sons, Chichester (1992)
48. Angilella, S., Greco, S., Lamantia, F., Matarazzo, B.: Assessing non-additive utility for multicriteria decision aid. European Journal of Operational Research 158, 734–744 (2004)
49. Marichal, J.L., Roubens, M.: Determination of weights of interactive criteria from a reference set. European Journal of Operational Research 124, 641–650 (2000)
50. Grabish, M.: k-Order additive discrete fuzzy measures and their representation. Fuzzy Sets and Systems 92, 167–189 (1997)
51. Martuscelli, M.: Agrigento, Relazione della Commissione d'indagine. Urbanistica 48, 31–160 (1966)
52. Miccichè, C.: Gli ipogei agrigentini tra archeologia, storia e mitologia. Industria grafica Sarcuto, Agrigento (1996)
53. European Ministers responsible for Urban Development: Leipzig Charter on Sustainable European Cities. Final Draft (May 2, 2007),
    http://ec.europa.eu/regional_policy/archive/themes/urban/
    leipzig_charter.pdf
54. Miccichè, C.: Girgenti: le pietre delle meraviglie...cadute. Osservazioni, note autentiche, documenti editi e inediti per il recupero del centro storico di Agrigento. Tipografia Arcigraf, Agrigento (2006)
55. Detailed Plan of the Historic Center of Agrigento, report and drawings of Piano Particolareggiato del Centro Storico di Agrigento (2007)
56. Roy, B.: Paradigms and Challenges. In: Figueira, J., Greco, S., Ehrgott, M. (eds.) Multiple Criteria Decision Analysis, pp. 3–24. Springer, USA (2005)
57. United Nations Commission on Sustainable Development: Indicators of sustainable development: framework and methodologies. Background paper No. 3. United Nations, New York (2001)
58. Bouyssou, D.: Building Criteria: A Prerequisite for MCDA. In: Bana e Costa, C.A. (ed.) Readings in Multiple Criteria Decision Aid. Springer, Berlin (1990)
59. Reed, M.S., Frases, E.D.G., Dougill, A.J.: An adaptive learning process for developing and applying sustainability indicators with local communities. Ecological Economics 59, 406–418 (2006)
60. Pelto, P., Pelto, G.: Anthropological Research: the Structure of Inquiry. Cambridge University Press, Cambridge (1970)
61. Kusakabe, E.: Advancing sustainable development at the local level: The case of machizukuri in Japanese cities. Progress in Planning 80, 1–65 (2013)

Clustering Analysis in a Complex Real Estate Market: The Case of Ortigia (Italy)

Salvatore Giuffrida, Giovanna Ferluga, and Alberto Valenti

Department of Civil Engineering and Architecture, University of Catania
sgiuffrida@dica.unict.it,
gio.ferluga@virgilio.it,
albvlt79@gmail.com

Abstract. Ortigia, the historic center of Syracuse, is a complex urban entity characterized by high outer homogeneity and inner heterogeneity. The evolution of its real estate market during the last decade is somehow related to the global property market one. In addition its events are connected with the evolution of the exploiting policies still ongoing. The critical observations of its features aim at providing tools able to support the decisions about subsides and local property taxes. This study continues the observations we have carried out for five years, this time involving clustering analysis, a data mining technique able to recognize different submarkets, and suitable to make the valuation pattern fit to the different market areas. For each of the latter significant characteristics have been recognized with reference to the "monetary declination" of these particular capital assets.

Keywords: Imperfect real estate markets, mass-appraisals, clustering analysis, theory of capital, income method.

1   Introduction

The real estate of Ortigia, the historic center of Syracuse, is a complex urban entity characterized by high outer homogeneity and inner heterogeneity. The first is due to its geographical, landscape, architectural and cultural identity that makes it recognizable like a brand; the second is due to its location on an islet, and therefore to its need to concentrate in itself all the urban practical and symbolic functions and activities, so that it looks like a miniaturized complex and articulated context.

The huge amount of investments made during the last fifteen years in order to boost the general renovation policy for one of the main tourist target of tourism in the Mediterranean sea, has generated a bundle of positive economic and negative social and cultural externalities during a renovation process, still ongoing, dominated by the property market for better or worst.

The real estate market observations we have carried out for the last five years, reveal the sequence of about three different and recognizable phases during which the economic, financial and monetary characteristics of property have raised playing different roles. Furthermore, the surveys and the valuations we have carried out so far,

B. Murgante et al. (Eds.): ICCSA 2014, Part III, LNCS 8581, pp. 106–121, 2014.

highlight some criticalities that don't allow to consider Ortigia as a whole market and to use linear calculation tools, mostly because of the process that has been featuring it during the last decade.

At first, during its immature stage, this heterogeneous market has been attacked by a massive trading wave affected by the gap and inconsistency between local owners who were looking backward and foreign buyers who were looking forward. The former despised some functional features, the latter appreciated the symbolic and, above all, the monetary ones. Therefore, the first stage has been characterized by the massive sale of the best architectural heritage, mostly located along the waterfront or facing the most important streets and squares.

The second phase (2005-2009) has been characterized by the success of the brand of Ortigia that confirmed it as one of the most promising real estate markets; during this phase the prices reached the top. The professional intermediaries, who ruled the assessments and the bargaining, have played an important role in this process.

The third stage (since 2010 until now), started about one and half year after the start of the economic-financial crisis, and has been characterized by the lack of liquidity, due to the crisis of credit; trades decreased significantly and many of the former investors, who purchased during the first bullish phase are now trying to sell.

This sort of bubble [14] provided some positive and negative effects: the first ones are the general improvement of the physical and functional condition of this real estate, and the fair relationship between the quality of the location and the market prices; the second ones are the filtering of the local population and the concentration of the property because of some massive investments made by professional investors or groups of real estate investors.

The evolution of this eccentric market is still ongoing and allows us to apply a general approach aimed at recognizing and connecting the multiple relationships between *value* density and tensions of *prices*.

2   Materials: Ortigia and the Real Estate Market Observations

Ortigia is the historic center of Syracuse, situated in a 50 hectares islet connected to the mainland by three bridges. Although it has been for a long time a marginal and decaying quarter, mainly due to some adverse urban-environmental conditions, an improved awareness of its extraordinary heritage has deployed, since 1990 – the date of the most important Renovation Plan by Prof. Pagnano – a huge amount of human, political and economic resources that have been involved in its infrastructural renovation and in the enhancement of its cultural-historical identity. Some other enhancement plans and laws are: the Integrated Communitary Plan, "Urban Italy" including Syracuse in 1995, the "Progetti Sponda", financed by the Province and the Region ex lege n. 433/91, providing 15 mln €; the PIT – the Mediterranean "Environmental-Museum" providing further 16 mln € the special regional laws for Ortigia n. 70/76 and n. 34/85; the Urban Recovery Plan, 5,5 mln €, the Operating program FESR 2007-2013; the Sustainable Development Plan of Syracuse; the Strategic Plan Syracuse Renewal 2020, aiming at the land enhancement and valorisation with further 115 mln € to be devolved to the harbour renovation work in Public-Private Partnership.

Syracuse (Ortigia) and the Rocky Necropolis of Pantalica have been included in the Unesco World Heritage List since 2005 (Committee Decision 29COM 8B.41). Ortigia has become one of the most important economic mover for the city of Syracuse, and the undisputed brand of the whole Province, whose economic policy is aimed at improving tourism compensating some of the worst environmental criticalities caused by the former industrial development strategy; in fact, the huge chemical pole located in the last '70s along the northern waterfront of Syracuse has destroyed and marginalized so far a significant part of the environmental heritage of one of the most beautiful Mediterranean areas.

The real estate market of Ortigia took off in 2003, after a previous depression period when prices were comparable to the marginal areas' ones, despite the incomparable architecture and landscape of this location [6].

The current market situation, that does not follow the general national trend [16], is summarized by the result of the survey carried out in 2013 by collecting a sample of 96 properties for sale and 25 for rent; two databases have been compiled and each property (record) is described by 15 attributes (field) grouped in four types of characters (location, landscape, technologic, architectural-environmental). The fields are organized in a *work breakdown structure* in which each group is divided into a certain number of characters, then in sub-characters whose importance is measured by means of a weighed score system. The heterogeneity of this real estate requires both surface area (*sqm*) and number of rooms (*r*) to be considered as quantitative characters: in fact, sometimes ancient architectural typologies have wide rooms or they are not fairly usable, so that the effective utility do not correspond to the property area [4].

The sale prices overall range is 534-4,958 €/sq.m. and 14,194-148,750 €/r; in the different areas wide ranges have been registered as well: the waterfront properties with terrace reach the maximum value, while the properties with sea view range 735-3,129 €/sq.m. and 17,647-76,191 €/room; properties facing the main streets and squares (Cavour, Rome, Matteotti, Maestranza, Vittorio Veneto, Maniace etc.) range 1,100-3,409 €/sq.m. and 22,222-93,333 €/room; along the secondary streets and alleys (Amalfitania, Alagona, Crocifisso, Dione, Mirabella, Resalibera etc.) properties range 534-2,500 €/sq.m. and 14,194-85,714 €/room (Fig. 1). The wide price ranges reveal the heterogeneity of the real estate shown for each characteristic in the diagrams of the following of Fig. 1.

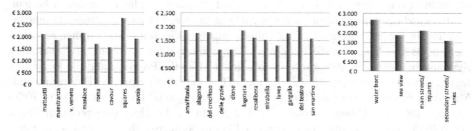

Fig. 1. Price ranges for each urban context

The overall location, landscape, technologic and architectural-environmental features of the properties, are synthesized in an appropriate analysis (Fig. 2).

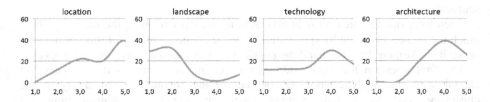

Fig. 2. Quality distribution within the sample

3   Methods and Procedures

3.1   Economic-Appraisal Features: Value and Valuation of Urban Real Estate

The study of this market has been carried out with the purpose of connecting and integrating architectural and economic features in the approach to real estate (re)production and management. In fact, the creation of value [2] in the historic centres has to be handled in order to reallocate the surplus that the property investment achieves as positive externalities due to the renovation public works.

Therefore, the main typical "three benefits" of property – 1. functional-symbolic qualities, 2. productiveness and 3. expectations – have been involved as general target of the complex behaviour of the investors, assuming that each of these benefits is considered by them as a motivation to invest or disinvest. Therefore, the three benefits can be considered located at different degrees of the goals/means scale: at the bottom we find the referential, practical and symbolic features; in the middle the economic-financial ones (costs and revenues of the investment); at the top the monetary ones, capital gains and implicit yield (Fig. 3).

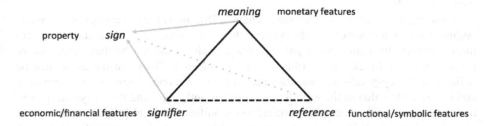

Fig. 3. The semiotic interpretation of a property and its benefits

In the proposed semiotic interpretation, properties, as signs, are the synthesis of value (as meaning) and productiveness (as signifier); the latter is the economic the economic phenomenon of the underlying physical object, the reference, that includes the bundle of functional and practical features of a property. The reference has a very weak relationship with the sign, so that the value of a property isn't strongly connected to its object features. This gap measures the eccentricity of a real estate market, that depends on how the bid-ask spread characterizes the different market segments.

Therefore, our research aims at representing on the one hand the ontology (how much the market is driven by speculation) and the phenomenology of this eccentricity (what it is like), on the other hand the possible segmentation of this property market.

The first question has been addressed by connecting some items of theory of capital with the analytic appraisal procedure, the second one by applying the clustering analysis, a data mining technique aimed at circumscribing the homogeneous submarket highlighting the three types of benefit.

The theory of capital, as addressed and defined by F. Rizzo [17], may be synthesized as follows. The income value V of an asset depends on forecasted income and capital gains: $V = k \pm |a|k$, where V is the income value, k is the supply value meant as the market current value, relevant for most operators, and $|a|$ is a factor of increasing/decreasing in value.

The income approach can be expressed as $V = I_n 1/r$, where I_n is the Net Operating Income, and r is the cap rate. If the cap rate is considered a variable, it is: $r = r' \mp |a|r'$, or $r = r'(1 \mp |a|)$, where r' is the average, or market, cap rate. Therefore, $V = I_n/r'(1 \mp |a|)$.

The differences between the capitalisation value and the market value, and between the effective cap rate and the average cap rate, are the main concern of this theory, about which each asset is considered a singularity, and its value is the result of the combination of market rules and personal determination of the investor within a speculative market.

In fact, one of the main points of a monetary approach to economy is the inversion of the relationship between value and price: although it is generally believed that price depends on value, and the former is the monetary measure of the latter, in a monetary reality, that is the reality of the globalized financial capitalism, the price tensions modify our perception of value, so that the price becomes a sort of general and abstract value [19].

Assuming the market as a "semantic chain" [18], the value (meaning) of the assets (signs of the chain) arises mostly because they communicate by their monetary features – expectations and capital gains (that are goals) – and not by their economic or functional-symbolic characters (that mostly are means). Therefore the assets can be defined as money-goods, because they play the role of money: they measure, preserve and increase the value of the wealth that they (currently) are, and that they can (potentially) become. Each speculative transaction modifies the local property market (the semantic chain) both in prices (meanings) and in characteristics (signifiers), so that the expectations arise in advance of the current prices.

When the monetary perspectives (the capital gain expectations) prevail over the practical/symbolic functions, properties become treasures to be hoarded and the market becomes asymmetric and dynamic, for better and for worst.

The monetary characterization of a speculative asset market can be appraised as the difference between the fair and the expected cap ratio. According to Hicks [11] and

Rizzo [18], differently from Fisher [3] a capital asset is in *crescendo*, *c*, or in *diminuendo*, *d*: $c \rightarrow r' > r$; $d \rightarrow r' < r$ where r' is the cap rate of a standard stream of values, and r is the cap rate of the prospective, or expected, one.

Each specific cap rate describes how the related property plays the role of a "treasure". A treasure is hoarded by itself and doesn't provide any real income, but only a psychological one. The difference between the average (standard stream) rate and the specific (perspective stream) rate measures how much a property can be considered a treasure, that depends on the prevailing of the psychological income on the real one: the former can be considered the implicit liquidity, the latter the explicit one [18].

Therefore, we can measure the degree of hoarding of a real estate market and compare different markets from this point of view.

The inverse ratio of r' or r, $1/r'$ or $1/r$, is the average period of the standard or perspective stream, defined by Hicks as "the average length of time for which the various payments are deferred from the present, when the times of deferment are weighted by the discounted values of the payments" [11].

The average period measures the "average life" of the stream. Assuming a property as a particular stream, its average life measures its certainty degree as the number of "income-years" it is able to provide. The more r is low, the longer is the average life and vice-versa. In synthesis, we affirm that the renounce to a part of the probable income (related to the standard stream rate) prolongs the average life of the property whose cap rate is lower.

The main foundations of this theory is the well known Keynesian law of the inducement to investment, in which a difference between the supply price of an asset capital and its expected value calculated by discounting the perspective yields [12] is highlighted. The marginal efficiency e is the rate that equates the expected value to the supply price of the asset capital, so the marginal efficiency can be expressed by means of the expected rate: $e = i(1 \mp |a|)$, where e is the marginal efficiency and i the interest rate. The positive difference between e and i progressively activates successive amounts of investments. Therefore, *mutatis mutandis*, cap rate can be considered a sort of marginal efficiency in the property market.

A further representation of this theory can be addressed by considering the difference between optimistic and pessimistic approach about the same asset. A transaction implies that the bid price is higher than the demand price.

As a consequence, the value (bid price) can be expressed as the result of the double projection of k (*supply price* according to Keynesian address, *demand price* in financial current uses) forward and backward: $V = k(1 + r_l)^n / (1 + r_d)^n$ in which r_l is the expected (perspective) rate of increasing in value, r_s is the observed (at cost) discount ratio, and n is the time of this projection. The result of this projection depends on the characteristics of the two (dis)investors that influence these two rates. In general, $r_l > r_s \rightarrow V > k \rightarrow a > 0$ and viceversa. The relationship between increase in value rate and discount rate rules the result of the "negotiation adventure".

Such theory of capital represents the price tensions phenomena and the consequent bid-ask spread due to different expectations as measured by the two rates. During a bull trend, bullish operators expect capital gains, $V = k(1 + |a|)$, so that purchasers easily buy, owners hardly sell; during a bear trend, bearish operators expect capital losses, $V = k(1 - |a|)$, so that purchasers hardly buy, owners easily sell. The greater the price tension within a particular market, the greater the bid-demand spread.

Appraisal science mainly concerns about the fair value; according to a semiotic approach the fair value can be considered the conventional signification that rules a normal communicative system. When the communication system gets over-communicative, the intentional signification prevails, whereas, when the system becomes under-communicative, the real signification prevails; assuming market as a communicative system, the fair value dominates in a perfect market, whereas in a speculative market, *bid prices* prevail during a bullish trend, *demand prices* prevail during a bearish trend. Within an eccentric market the individual willingness/aversion "drags and drops" the prices and jerks the whole semantic chain by modifying the expectations that influence the cap rates. Several income method valuation tools have been recently provided following the increasing/decreasing logic, both in business [8], [9] and real estate market [21], [22], [15].

The real estate of Ortigia is characterized by the heterogeneity of capitalization ratios; the survey we propose aims at defining the characteristics that mostly influence it in the different submarkets [1]. The segments can be defined from physical, economic, monetary points of view [20]. The properties are substitute units in a physical sense if they have similar characteristics, in an economic sense if they have similar productiveness abilities, in a monetary sense if they have similar rate of capitalization classes [7].

The tool we propose in order to connect the urban, architectural and landscape with monetary characteristics is the clustering analysis referred to both the referential and the semantic characteristics of the properties.

3.2   Clustering Analysis. Theoretical Frame and the Real Estate Market

The statistical study of complex social systems, such as real estate markets, identifies the data mining techniques important applications such as the cluster analysis. The real estate market of Ortigia, because of its heterogeneity and articulation, and the large dimension of the examined dataset, is a quite interesting application context that needs some additional considerations that supplement the mere application of the chosen algorithm.

In order to better understand the theoretical framework of this work, it is convenient to make a brief digression about clustering theory [5].

Recently, it has been the tendency to treat inquiries in the real estate market using statistical techniques of clustering [10], [13].

Generally, the reason that leads to apply the clustering theory to a statistical population is grouping statistical units in subsets, called clusters, the most possible internally homogeneous and externally heterogeneous. This kind of groups can be realized through clustering algorithms.

In the following, with the term partition we will mean a family of subsets of the initial sample such that two of them are disjoint and their union is the entire sample.

A clustering algorithm is a succession of steps through which is made a succession of partitions is made starting from the partition of singletons and uniting at every step only two subsets in the previous partition, until to the partition is formed by the entire sample. Subsets in various partitions are called clusters. The aggregation is realized on the basis of a certain parameter, generally linked to a method and to a metric. Happening the aggregations of clusters, the initial sample becomes more and more compact. Therefore, through a clustering algorithm, a family of partitions called hierarchy is made, having as first the one formed by singletons and as last the one formed by only the entire sample. Moreover the previous partition has the same sets as the successive one, except for the ones aggregated at the previous step in a unique set in the successive one. This means that the previous partition is less fine than the successive one.

The following definitions formalize some of the above exposed concepts.

Def. 1. *A partition P of a set U is a family of subsets U_i in U having the following properties*:

   i. $\forall i, j \ (i \neq j \rightarrow U_i \cap U_j = \emptyset)$

   ii. $\bigcup_i U_i = U$

Def. 2. *A partition P_2 is said less fine than partition P_1, symbolically $P_2 < P_1$ if:*
$$\forall U \in P_2 \ U = \bigcup_{i \in I} U_i$$
where $\{U_i : i \in I\}$ is a family of sets taken from partition P_1: in other words, if every set in partition P_2 is union of sets in partition P_1.

Def. 3. *A hierarchy G is a family of partitions $P_1, P_2, \dots, P_n$, provided with a sorting having the following properties*:

   i. P_1 is the partition of singletons

   ii. P_n is the partition formed by only U

   iii. $i < j \rightarrow P_i < P_j$

Def. 4. *A hierarchical aggregative algorithm is a proceeding such that in output provides a hierarchy, having accepted as input a population constituted by statistical units $u_1, u_2, \dots, u_n$, generally multivariate, i.e. with multidimensional vectors associated. These vectors are obtained from values assigned to variables $v_1, v_2, \dots, v_m$ accepted to describe the dataset.*

All the clustering algorithms have a common property, i.e. at every step, they aggregate among them the sets U_i, U_j minimizing a certain parameter, associated to a method. In the following table, there are the parameters associated to various methods:

Method	Parameter
Single linkage method	$min \ (D)$
Complete linkage method	$max(D)$
Average linkage method	$M(D)$
Centroids' method	$d(c_i, c_j)$

Where:

$$D = \{d(ui, uj) | \ ui \in U_i, uj \in U_j\} \tag{1}$$

c_i, c_j are the centroids in the clusters U_i, U_j and d is a metric taken from various available. d may be Euclidean metric, Manatthan metric or various other.

The algorithm used in the analysis is associated to the complete linkage method and to the Euclidean metric. So it predicts, at every step, the aggregation of two clusters of the previous partition, on the basis of the least Euclidean distance among the most distant elements.

Dendrogram. Once obtained the final hierarchy, it is possible to use a graphical object, called dendrogram, which allows us to globally visualize it, from the initial partition of singletons to the partition of the entire statistical population. In this graphic, there is an horizontal line where single elements in the sample are located. Climbing, at every level, there are the same sets of the last level with the addition of the one coming from the last aggregation. Therefore a specific partition of the hierarchy corresponds to each level.

R^2 and RMSSTD. After analyzing the dendrogram, the problem is choosing the level at which to cut the dendrogram. In other words, the problem is determining the best partition among the ones in the hierarchy, where in this context it better means the one maximizing heterogeneity between clusters and homogeneity within clusters.

To cut the dendrogram, there are two useful statistical indices that guide the choice of the best partition.

One of these indices is called R^2 and is defined as follows:

$$R^2 = 1 - \frac{W}{T} \tag{2}$$

Where:

$$W = \Sigma_{j,v} W_{jv} \qquad W_{jv} = \Sigma_{i \in I_V}(x_{ij} - \bar{x}_{jv})^2 \tag{3}$$

$$I_V = \{i : u_i \in U_v\} \qquad u_i \equiv (x_{i1}, x_{i2}, \dots, x_{im}) \tag{4}$$

Moreover:

$$T = W + B \tag{5}$$

Where:

$$B = \Sigma_{j,v} B_{jv} \qquad B_{jv} = \Sigma_V(\bar{x}_j - \overline{x_{jv}})^2 \tag{6}$$

W is known as within total variance while W_{jv} is the within variance of variable j inside cluster U_V. T is called total variance, and it is the sum of W and B, where B is the total between variance. $\overline{x_{jv}}$ is the average value of variable j relative to statistical units in the cluster U_V and $\bar{x}_j$ is the total average for variable j.

From definition of R^2, results that high values of this index indicate good partitions because these values correspond to small values of within variance W in relation to the total variance T.

The second index is known as *RMSSTD*. It is an acronym for *Root Mean Square Standard Deviation* and is defined as follows:

$$RMSSTD = \sqrt{\frac{W_V}{p(n_V - 1)}} \tag{7}$$

In this formula, p represents the number of variables investigated and n_V is the number of clusters in the current step. Moreover, in this formula, there is the within variance W_V relative to the cluster coming from the last aggregation, so it is necessary to know this value to calculate this index. Contrary to the first index, it is better having small values because RMSSTD grows up within variance of the last cluster so that great values indicate that the last cluster is very heterogeneous.

RMSSTD is not defined for trivial partition constituted by only the entire set because $n_V = 1$. Moreover it is not interesting for partition of singletons because it is the first partition so that there isn't a new entry inside it. In the cutting dendrogram, there is a simple rule predicting to stop with aggregations when this index has become much greater than the previous values.

Because it is scientifically meaningless establishing without any reference a good numeric boundary for values of R^2 and $RMSSTD$, to choose the best partition in the hierarchy arising, we will use an initial hypothesis about the maximum number of submarkets in the global market, considering only partitions with a number of clusters, corresponding to submarkets,   lesser than a maximum value. This value is linked to the dimension of the global market.

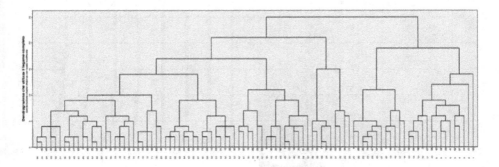

Fig. 4. The dendrogram

4   Applications and Results

The dendrogram (Fig. 4) provides a great number of possible bottom-up aggregations, so that some tests must be done in order to choose the right number of sub-markets. Therefore a spread-sheet model has been drawn up in order to represent the different segmentations. The maximum number of submarkets has been reasonably established on 8. The dendrogram shows the successive breakdowns so that the whole sample can be progressively cut (top-down) into two clusters (cut 2), three clusters (cut 3) and so on until the last cut that provides eight clusters. To do this, for each cut the first and the last identification numbers of the cluster are inserted in a triangular matrix (Fig. 5a) so that the eight segmentations are provided (Fig. 5c). The model calculates the indexes R2 and RMSSTTD (Fig. 5b) in order to choose the correct number of submarkets, in this case 5 because RMSSTD decreases more slowly the after fifth breakdown and no significant real estate differences have been recognized in successive groups. The location of the eight different segmentations (cuts) is shown in Fig. 6.

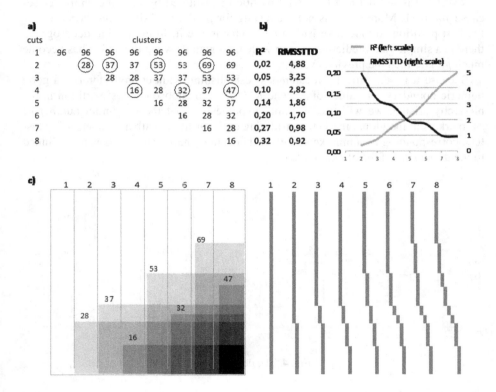

Fig. 5. Clustering application; a) each segmentation hypothesis (1 to 8) is defined by the two id. numbers shown in the column of the triangular matrix; the circle indicates the id. of each new cut; b) for each hypothesis the indexes R^2 and RMSSTTD are calculated: the graph shows their trend; c) the two histograms show the progressive clustering of each hypothesis: first column no segmentation, second column two clusters, the first from id 96 to 28, the second from 27 to 1; second column three clusters, 96-37, 36-28, 28-1, and so on; the scheme on the right differently shows the progressive clustering.

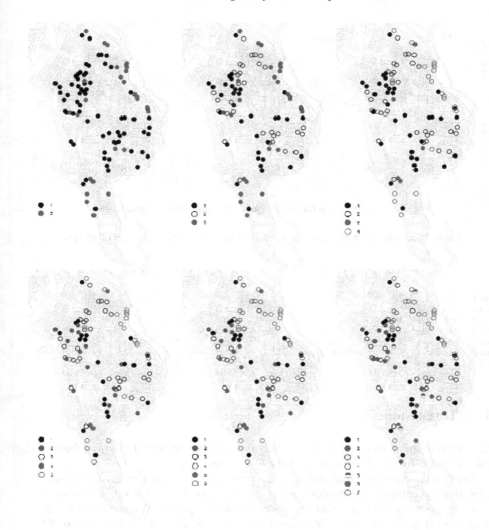

Fig. 6. Location of the different clusters for each cut (eighth cut omitted)

The next verification involves the price ranges within each cluster. This consistency analysis has been carried out by calculating the prices range (Fig. 7) and the logarithmic regression R2 index for each cluster of every cut (Tab. 1). Table one shows that after the fifth breakdown the groups are not significant (R2=1.00) in order to describe a different elasticity price/quality.

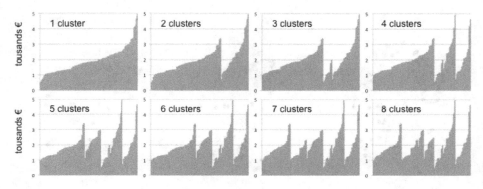

Fig. 7. Clustering price ranges for each cut (segmentation hypothesis)

Table 1. Statistic clustering consistency test: value of R^2 for each cluster of all cuts

	R^2 calculated by using room unit prices								R^2 calculated by using sq.m. unit prices							
	1	2	3	4	5	6	7	8	1	2	3	4	5	6	7	8
cut 1	0,51	-	-	-	-	-	-	-	0,49	-	-	-	-	-	-	-
cut 2	0,50	0,48	-	-	-	-	-	-	0,47	0,47	-	-	-	-	-	-
cut 3	0,41	0,91	0,48	-	-	-	-	-	0,32	0,86	0,47	-	-	-	-	-
cut 4	0,41	0,91	0,36	0,56	-	-	-	-	0,32	0,86	0,47	0,46	-	-	-	-
cut 5	0,28	0,81	0,91	0,36	0,56	-	-	-	0,15	0,45	0,86	0,47	0,46	-	-	-
cut 6	0,28	0,81	1,00	1,00	0,36	0,56	-	-	0,15	0,45	1,00	1,00	0,47	0,46	-	-
cut 7	0,18	0,31	0,81	1,00	1,00	0,36	0,56	-	0,09	0,29	0,45	1,00	1,00	0,47	0,46	-
cut 8	0,18	0,31	1,00	0,83	1,00	1,00	0,36	0,56	0,09	0,29	0,70	0,56	1,00	1,00	0,47	0,46

5   Discussions

According to the main concern of this study that involves the consistency of the monetary features, two final verifications have been carried out.

The first concerns the "bullish/bearish eccentricity" of the clusters given by multiplying the percentage of properties outside the "fair range" and the sum of the differences between the out-range and fair prices of the unit prices over and under the fair price level (Fig. 8). The more the clusters are bigger the more the indexes are significant.

The second verification concerns the cap rates. Basing of the rental survey, NOIs (I_n) have been calculated in order to find out the cap rate of each property for sale. The regression analysis carried out for each submarket allows to calculate the gross income (I_g) for each property, from which the managing expenses are deducted. According to the previous clustering, the comparison between the cap rates trend and the total quality index is displayed in Fig. 9 for the first five (the most significant) clustering hypotheses (the arrows indicate the dividing clusters that at each stage).

The clustering analysis provides some rational items for grouping the sample into five clusters. The different stages progressively distinguish ordinary and extraordinary property groups. By referring to the fifth clustering, we show the characteristics of the different groups in Fig. 10.

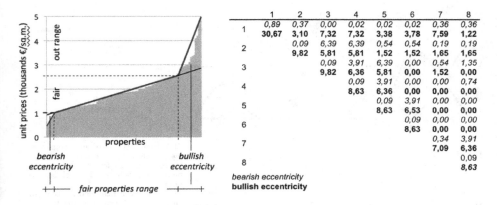

Fig. 8. Bullish/bearish eccentricity

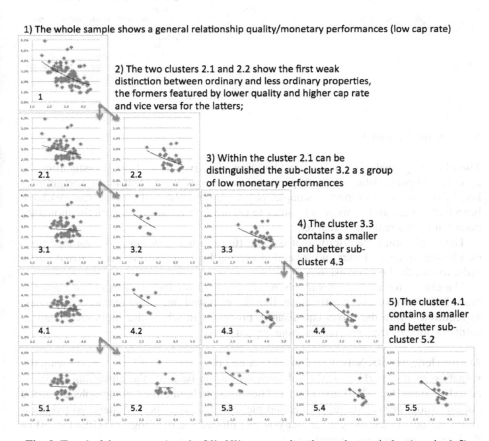

Fig. 9. Trend of the cap rate (y-axis, 0%-6%) compared to the total score index (x-axis, 1-5)

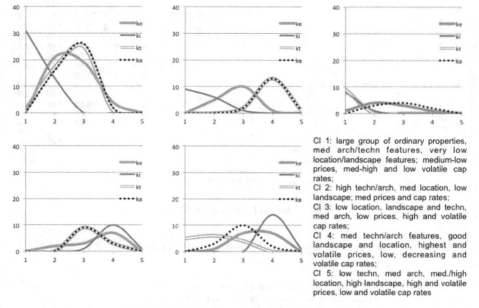

CI 1: large group of ordinary properties, med arch/techn features, very low location/landscape features; medium-low prices, med-high and low volatile cap rates;
CI 2: high techn/arch, med location, low landscape; med prices and cap rates;
CI 3: low location, landscape and techn, med arch, low prices, high and volatile cap rates;
CI 4: med techn/arch features, good landscape and location, highest and volatile prices, low, decreasing and volatile cap rates;
CI 5: low techn, med arch, med./high location, high landscape, high and volatile prices, low and volatile cap rates

Fig. 10. Description of the different clusters

6   Conclusions

The clustering analysis can be considered an effective method for investigating the monetary characteristics of an atypical, heterogeneous and eccentric real estate market. Basing on a 96 properties sample, physical, symbolic and income features have been highlighted and connected in order to define their relationship with the monetary potential of this real estate market.

The significant eccentricity, that is the difference between fair and out-range prices, changes in each cluster and several clustering hypotheses have been carried out in order to find the best consistency.

The clustering analysis is suitable to complement the income method in mass appraisals for both ordinary estimation, like the cadastral ones, and extraordinary ones, as required within the urban equalisation processes in which the supposed enhancement encourages the expectation about capital gains.

Acknowledgements. Salvatore Giuffrida edited paragraphs 1, 3.1, 4, 5, 6, drew up the valuation model, performed the calculations on the basis of the clustering analysis, edited pictures 3, 5, 7-10 and table 1; Giovanna Ferluga edited paragraphs 2 carried out the whole market survey and edited pictures 1, 2, 6; Alberto Valenti edited paragraph 3.2 performed the cluster analysis and plotted the dendrograms (like Fig. 4) as the bases to the successive analyses.

References

1. Acciani, C., Gramazio, G.: L'Albero di Decisione quale nuovo possibile percorso valutativo. Aestimum 48, 19–38 (2006)
2. Donna, G.: La creazione di valore nella gestione dell'impresa. Carocci, Roma (1999)
3. Fisher, I.: The Theory of Interest. Macmillan, New York (1930)
4. Forte, C.: Elementi di estimo urbano. Etas Kompass, Milano (1968)
5. Fraley, C., Raftery, A.E.: Model-Based Clustering, Discriminant Analysis, and Density Estimation. Journal of the American Statistical Association 97(458) (2002)
6. Giuffrida, S., Martorina, L.: Densità di valore e tensione nei prezzi in un mercato immobiliare eccentrico. Il caso di Ortigia. Valori e Valutazioni 7, 145–162 (2011)
7. Grissom, T.V., Wang, K., Webb, J.R.: The Spatial Equilibrium of Intra Regional Rates of Return and The Implications for Real Estate Portfolio Diversification. Journal of Real Estate Research 7(1), 59–71 (1991),
   http://ideas.repec.org/s/jre/issued.html
8. Guatri, L.: Trattato sulla valutazione delle aziende. EGEA, Milano (1998)
9. Guatri, L.: Il giudizio integrato di valutazione. Dalle formule al processo valutativo. Università Bocconi Editore, Milano (2000)
10. Hepşen, A., Vatansever, M.: Using Hierarchical Clustering Algorithms for Turkish Residential Market. International Journal of Economics and Finance 4, 138–150 (2012),
    http://dx.doi.org/10.5539/ijef.v4n1p138
11. Hicks, J.R.: Valore e capitale. UTET, Torino (1959)
12. Keynes, J.M.: Teoria generale della moneta, dell'interesse e dell'occupazione. UTET, Torino (2001)
13. Maitra, R.: Volodymyr, Melnykov, Soumendra, N, L.: Bootstrapping for Significance of Compact Clusters in Multi-dimensional Datasets. Jasa (2012)
14. McDonalds, J., Stokes, H.: Monetary Policy and the Housing Bubble. Journal of Real Estate Finance and Economics (2011), doi: 10.1007/s11146-011-9329-9
15. Morano, N., Manganelli, B., Tajani, F.: La valutazione del rischio nell'analisi finanziaria di Ellwood per la stima indiretta di immobili urbani. Aestimum 55, 19–41 (2009)
16. Nomisma: III Rapporto 2008. Osservatorio sul mercato immobiliare, Bologna (2012)
17. Rizzo, F.: Analisi critica della teoria delle valutazioni. Seminario Economico-dell'Università di Catania, Catania (1977)
18. Rizzo, F.: Valore e Valutazioni. La scienza dell'economia o l'economia della scienza. FrancoAngeli, Milano (1999)
19. Rizzo, F.: Dalla rivoluzione keynesiana alla nuova economia. Dis-equilibrio, tras-informazione e coefficiente di capitalizzazione. FrancoAngeli, Milano (2002)
20. Simonotti, M.: La segmentazione del mercato immobiliare urbano per la stima degli immobili urbani. Atti del XXVIII Incontro di studio Ce.S.E.T., Roma (1998)
21. Simonotti, M.: Problemi di verifica del saggio di capitalizzazione. Estimo e territorio 12, 8–14 (2009)
22. Simonotti, M.: Ricerca del saggio di capitalizzazione nel mercato immobiliare. Aestimum 59, 171–180 (2011)

The Lifestyles of Families
through Fuzzy C-Means Clustering[*]

Silvestro Montrone, Paola Perchinunno, Samuela L'Abbate, and Maria Rosaria Zitolo

DISAG, University of Bari,
Via C. Rosalba 53, 70100 Bari, Italy
{silvestro.montrone,paola.perchinunno,
samuela.labbate,maria.zitolo}@uniba.it

Abstract. The objective of this report is the analysis of the data arising from the Family Lifestyles survey conducted by the University of Bari "A. Moro" (2012-2013) through the construction of indicators of socio-economic hardship and the identification of family profiles during the current period of crisis. The approach used in this work in order to synthesize and measure the conditions of hardship of a population is based on the so-called "Totally Fuzzy and Relative" method employing a Fuzzy Sets technique in order to obtain a measure of relative incidence in a population from the statistical information provided by a plurality of indicators [1]. The subsequent step involved considering a clustering procedure (Fuzzy c-means) with the objective of outlining various profiles, not defined a priori, to be assigned to each family with different socio-economic behaviours [2]. This clustering method allows, compared to conventional methods, a set of data to belong not only to a main cluster but also to two or more clusters with "fuzzy" profiles.

Keywords: fuzzy logic, fuzzy sets, lifestyles.

1 Approaches and Methodologies for the Analysis of Socio-economic Hardship

From the 1970s to the present, a range of studies carried out on socio-economic hardship have given rise to a variety of approaches, each with corresponding methodical definitions and conceptualizations. The numerous concepts of hardship are, furthermore, attributable to a traditional distinction between absolute and relative conditions of hardship.

An approach based on the absolute concept commences from a failure to achieve the objective of a minimum level of well-being and is therefore independent of social context

---

[*] The contribution is the result of joint reflections by the authors, with the following contributions attributed to S. Montrone (chapters 1), to P. Perchinunno (chapters 3), to Maria Rosaria Zitolo (2.1 and 2.2) and to S. L'Abbate (chapters 2.3). The conclusions are the result of the common considerations of the authors.

The data used in this paper are the result of a survey funded by the Cassa di Risparmio di Puglia entitled: "Analisi statistica territoriale della povertà urbana attraverso la costruzione di indicatori di disagio socio-economico".

B. Murgante et al. (Eds.): ICCSA 2014, Part III, LNCS 8581, pp. 122–134, 2014.
© Springer International Publishing Switzerland 2014

and time. Reference is therefore made to what is known in economic theory as a basic needs approach, according to which hardship is defined as the failure to meet basic needs.

The relative approach is, however, based on the assumption that the social status of an individual cannot be defined if not by the environment in which they live or, rather, people, families and groups within the population are considered poor when living in worse conditions than the standards of the community to which they belong. They cannot therefore achieve such standard in terms of their diet, participate in activities or enjoy the living conditions and amenities which are customary or, at least, widely encouraged and approved in the societies to which they belong. The subjective approach has, for several years, arisen in a transversal framework between the relative and absolute approaches. According to the subjective approach the condition of hardship is determined by its perception by individuals or families themselves in the comparison of the perceived well-being of members of their society. The subjective approach therefore begins from the perception of the individual of a condition of social exclusion.

In order to overcome the limitations of the traditional approach it is therefore necessary to expand analysis to a wide range of indicators on living conditions and, at the same time, adopt mathematical tools which allow for accurately taking the complexity and vague nature of poverty into account. The options of scientific research were therefore oriented towards the establishment of a multi-dimensional approach, sometimes abandoning dichotomous logic in order to arrive at fuzzy classifications in which each unit belongs and, at the same time, does not belong, to a category. A multidimensional index that considers hardship as the overall condition of being disadvantaged and deprived seems the most appropriate in view of the socio-economic differential analysis of demographic phenomena.

Indeed, this has in fact created a fissure between the world of research and the official framework of the measurement of hardship arising from the need to overcome the excessively rigid and inevitably arbitrary classification in the two categories [3]. The approach used in the present work in order to synthesize and measure the condition of hardship of a population is known as Totally Fuzzy and Relative, which uses the technique of Fuzzy Sets in order to obtain a relative incidence measurement within a population, from the statistical information provided by a plurality of indicators [4].

The next step was to consider a clustering procedure (Fuzzy c-means) with the objective of outlining various profiles, not defined a priori, to be assigned to each family with different socio-economic behaviour. This clustering method, compared to conventional methods, allows a set of data to belong not only to a main cluster but also to two or more clusters with "fuzzy" profiles.

2 The Construction of Sets of Indicators of Hardship with a Fuzzy Method

2.1 Introduction

In this report the data source used in order to construct indicators of socio-economic hardship is that of the Family Lifestyles survey conducted by the University of Bari "A. Moro" (December 2012 - January 2013).

The *Family Lifestyles survey* collected significant information on income, spending behaviour, and on the use of financial loans by families with children, resident in the metropolitan city of Bari. The objective of the survey, carried out by the University of Bari was that of analyzing issues associated with the measurement of socio-economic hardship created by the difficulty of attributing a single and generally agreed definition. A methodology based on *objective variables* (those resources actually available to families) was accompanied by *subjective measurements* based on the perception of the family in terms of its social and economic condition.

In order to obtain a measurement of the level of socio-economic hardship of the families interviewed, *sets of indicators* were constructed for the detection of the possession or absence of functional goods, the ability to bear certain costs, the perception of the evolution of the economic condition of the family etc.. Such sets of indicators were used in order to obtain a fuzzy value corresponding to the level of hardship of each family.

2.2   The Fuzzy Approach in the Analysis of Socio-economic Hardship

The development of fuzzy theory initially stems from the work of Zadeh [5] and subsequently draws upon Dubois and Prade [6] and their definition of a methodological basis. Fuzzy theory develops from the assumption that every unit is associated contemporarily to all categories identified and not univocally to only one, on the basis of ties of differing intensity expressed by the concept of degrees of association. Fuzzy methodology in the field of "poverty studies" in Italy has been recently employed in the work of Cheli and Lemmi [1] who define their method "total fuzzy and relative" (TFR) on the basis of the previous contribution from Cerioli and Zani [7].

Such a method consists in the construction of a function of membership to the fuzzy totality of the poor which is continuous in nature, and able to provide a measurement of the degree of poverty present within each unit.

Given a set of **X** elements $x \in$ **X**, any fuzzy subset **A** of **X** is defined as follows:

$$A = \{X, f_A(x)\} \tag{1}$$

where $f_A(x): x \rightarrow [0,1]$ is defined as the membership function of the fuzzy subset **A** and indicates the degree of membership of x to **A**. Therefore $f_A(x) = 0$ indicates that x does not belong to **A**, while $f_A(x) = 1$ indicates that x belongs only to **A**. However, in the case of $0 < f_A(x) < 1$, x belongs partially to **A**, with a greater degree of membership the closer $f_A(x)$ is to 1.

Supposing an observation of k poverty indicators for every family, the function of membership of i*th* family to the fuzzy subset of the poor may be defined thus:

$$f(x_i) = \frac{\sum_{j=1}^{k} g(x_{ij}).w_j}{\sum_{j=1}^{k} w_j} \qquad i = 1,....,n \tag{2}$$

The w_j function in the membership function is only a *weighting system* [1], as in the generalization of Cerioli and Zani [7], whose specification is given:

$$w_j = \ln(1 / \overline{g(x_j)})$$ (3)

Theoretically, when $\overline{g(x_j)} = 1$ all families demonstrate the j-th symptom and the corresponding weight w_j results as equal to zero; when $\overline{g(x_j)} = 0$ then w_j is not defined or, rather, X_j is an inappropriate indicator for that particular set [8].

In order to avoid the issue of the distribution of frequency of the indicators of unbalanced poverty, with an elevated frequency associated to modality or to extreme values, an alternative specification of g(x_{ij}) may be used, as proposed by Cheli and Lemmi [1], where H(x_j) is the function of the division of each value x_j ordered according to an increasing risk of poverty:

$$g(x_{ij}) = \begin{cases} 0 & \text{if } x_{ij} = x_j^{(1)}; \ k = 1 \\[2em] g(x_j^{(k-1)}) + \dfrac{H(x_j^{(k)}) - H(x_j^{(k-1)})}{1 - H(x_j^{(1)})} & \text{if } x_{ij} = x_j^{(k)}; \ k > 1 \end{cases}$$ (4)

where $x_j^{(1)}, \ldots x_j^{(m)}$ represents the modality and the values of the variable X_j, ordered according to an increasing risk of poverty, so $x_j^{(1)}$ denotes the minimum risk, while $.x_j^{(m)}$ denotes the maximum risk.

2.3   The Fuzzy Approach for the Analysis of Socio-Economic Hardship

The *Total Fuzzy and Relative* (TFR) model is used in order to summarize the values emerging from analysis in a single "blurred" fuzzy value which, as described above, measures the degree of membership of an individual in the range between 0 (condition of well-being) and 1 (hardship).

The indices were chosen in order to identify levels of socio-economic hardship and were calculated so as to match the high values of the index with a high level of hardship and low values of the index with higher levels of well-being. The indices were grouped into several sets characterized by different situations:

-   **Set 1: difficulty in paying debts/instalments or buying food staples** (mortgages, other debts and taxes, utility bills, food staples);
-   **Set 2: difficulty in paying for education, health or unforeseen expenses** (costs of school meals and other subsidies for children; voucher for medical treatment in public hospitals, private medical care or other unexpected expenses);
-   **Set 3: difficulty in purchasing other goods and services** (consumption of meat or fish at least once every two days, heating or air-conditioning in the home, purchase of clothing items when needed, going to the cinema/theatre at least once a month, going on holiday for one week a year);
-   **Set 4: difficulty in participating in events** (social, religious, sporting, political, voluntary, or cultural).

When grouped into 4 classes, the fuzzy values obtained for each set of indicators indicate hardship levels between values equal to zero and values equal to one (no hardship from 0.00 to 0.25, low hardship from 0.25 to 0.50, medium hardship from 0.50 to 0.75 and uncomfortably elevated hardship from 0.75 to 1.00). The following synthetic results emerge from the application of the fuzzy model on different sets of indicators relative to families (adjusted for missing answers).

Specifically, from the data grouped into classes, 36.3% of families surveyed result as "overall fuzzy" inasmuch as they *do not demonstrate any kind of difficulty* in bearing costs (mortgages, various debts, taxes, duties, bills) or in the purchase of food staples, while 39.9% of families demonstrate conditions of low hardship. The percentages of families belonging to all of those who experience great difficulty in bearing such expenses amounted to 10.1 % (Table 1).

Table 1. Fuzzy indices relating to Set 1 (difficulty in paying debts/instalment or purchasing food staples)

Condition of...	Absolute value	%
No hardship	911	36.3%
Low hardship	1,001	39.9%
Medium hardship	343	13.7%
Elevated hardship	252	10.1%
Totale	**2,507**	**100.0%**

Source: Our elaboration on the Lifestyles archive (2012).

From the analysis of data relating to the cost of *health* or the *education of their children* it emerges that all families interviewed have greater difficulties. Indeed, only 2.7% added to 19.6% are classified within the "fuzzy whole" of those families affected by significant difficulties in bearing such expenses. The percentages of families belonging to the total of those experiencing great difficulty in bearing such expenses amounted to 27.4%. It therefore seems clear that conditions of hardship appear much more pronounced when analyzing costs relating to the purchase or payment of non-necessities (Table 2).

Table 2. Fuzzy indices relating to Set 2 (difficulty in paying for education, health or unforeseen expenses)

Condition of...	Absolute value	%
No hardship	68	2.7%
Low hardship	491	19.6%
Medium hardship	1,262	50.3%
Elevated hardship	686	27.4%
Totale	**2,507**	**100.0%**

Source: Our elaboration on the Lifestyles archive (2012).

From the analysis of data pertaining to the ability to purchase other goods or services (meat or fish at least once every two days; heating for the home; clothing; cinema/theatre at least once a month; holiday for a week away from home) it emerges that families interviewed experience difficulties similar to those emerging from the set of indicators described above. Indeed, a lower rate of families affected by great difficulty in bearing such expenses is confirmed in comparison with the first set (2.7% no hardship and 19.9% low hardship). The percentages of families belonging to the total of those who experience great difficulty in supporting such expenses amounted to 27.2% (Table 3).

Table 3. Fuzzy indices relating to Set 3 (difficulty in purchasing other goods or services)

Condition of...	Absolute value	%
No hardship	68	2.7%
Low hardship	498	19.9%
Medium hardship	1,260	50.3%
Elevated hardship	681	27.2%
Totale	**2,507**	**100.0%**

Source: Our elaboration on the Lifestyles archive (2012).

The data relating to *social hardship* also demonstrates the participation of families in religious, political and cultural events as a cost that families are likely to renounce in cases of need. Indeed, only 4.1% of families demonstrate no difficulty while more than half of families surveyed experience difficulty in maintaining participation in activities and social events (36.2 % medium hardship and 23.1% elevated hardship).

Table 4. Fuzzy indices relating to Set 4 (difficulty in participating in events)

Condition of...	Absolute value	%
No hardship	104	4.1%
Low hardship	916	36.5%
Medium hardship	908	36.2%
Elevated hardship	579	23.1%
Totale	**2,507**	**100.0%**

Source: Our elaboration on the Lifestyles archive (2012).

The following summary graphical representation of the situation resulting from the application of the fuzzy model shows the percentage composition of hardship for the different sets of indicators used.

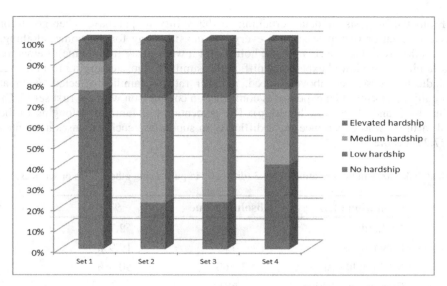

Fig. 1. Percentage of families and hardship conditions for the sets of indicators

It emerges from the data obtained through the application of the fuzzy model to the Lifestyles archive that the condition of hardship appears stronger and more greatly perceived in the case of the cost of goods and services other than the payment of debts/instalments or bills or the purchase basic necessities. Indeed, it may easily be deduced that the first expenses to be reduced in times of crisis are those relating to non-essential purchases such as going to the cinema, attending cultural and social events and performances, holidays, clothing etc..

It emerges from the data obtained through the application of the fuzzy model to the Lifestyles archive that the condition of hardship appears stronger and more greatly perceived in the case of the cost of goods and services other than the payment of debts/instalments or bills or the purchase basic necessities. Indeed, it may easily be deduced that the first expenses to be reduced in times of crisis are those relating to non-essential purchases such as going to the cinema, attending cultural and social events and performances, holidays, clothing etc..

3   Identification of Poverty Profiles

3.1   The Fuzzy C-Means Clustering

The subsequent step involved considering a clustering procedure with the objective of outlining various profiles, not defined a priori, to be assigned to each family with socio-economic behaviours.

Cluster analysis is highly advantageous as it provides "relatively distinct" (or heterogeneous) clusters, each consisting of units (families) with a high degree of "natural association". Different approaches to cluster analysis are characterized by the need to define a matrix of dissimilarity or distance between the n pairs of observations, which represent the point at which each algorithm is generated.

Fuzzy *c-means* (FCM) is a clustering method that allows a set of data to belong not only to a main cluster but also to two or more clusters. The c-means differs from the k-means objective function through the additions of the u_{ik} membership values and the fuzzifier m. The fuzzifier m determines the level of cluster fuzziness.

A fuzzy c-partition of **X**, (subset of R^n), is that which characterizes the membership of each sample point with all clusters through the identification of a membership function that assumes values of between zero and one. The sum of the memberships for each sample point must be equal to one.

This method was developed by Dunn [9] and later by Bezdek [10,11] and is based on the minimization of the following objective function:

$$J_m(U,v) = \sum_{k=1}^{n} \sum_{i=1}^{c} (u_{ik})^m (d_{ik})^2 \tag{5}$$

where:

- U is a fuzzy partition of **X**
- $v = (v_1, v_2, ..., v_c) \in \mathbf{R}^{cn}$ with $v_i \in \mathbf{R}^n$ is the centre of cluster i with $1 \le i \le c$
- u_{ik} is the degree of membership of the i-*th* fuzzy subset for the k-*th* datum
- $(d_{ik})^2 = \|x_k - v_i\|^2$ is any norm expressing the similarity between any measured data, x_k con $1 \le k \le n$ and the centre v_i .
- m is any real number greater than 1.

The FCM algorithm, via iterative optimization of J_m, produces a fuzzy c partition of the **X** data set. The steps to be followed are:

- Determine the number of clusters $2 \le c \le n$ e $m \ge 1$;
- Initialize the fuzzy c partition $U^{(0)}$, at step b=0,1,2,...
- Calculate the c cluster centres with a general equation for the i-*th* cluster centre:

$$v_i = \frac{\sum_{k=1}^{n} (u_{ik})^m x_k}{\sum_{k=1}^{n} (u_{ik})^m}$$

Subsequently updating $U^{(b)}$ the membership matrix $U^{(b+1)}$ is calculated with the equation:

$$u_{ik} = \frac{1}{\sum_{j=1}^{c} \left(\dfrac{d_{ik}}{d_{jk}} \right)^{2(m-1)}} \tag{6}$$

Finally, $U^{(b)}$ and $U^{(b+1)}$ are compared through a matrix norm, stopping if $\|U^{(b)} - U^{(b+1)}\| \le \varepsilon$, otherwise returning to calculate the c cluster centres.

3.2   An Application with Fuzzy C-Means Clustering Algorithms

The cluster analysis allowed for the identification of several profile families derived from the fuzzy applications. Applying cluster analysis on the 20 variables under observation and placing $m = 1$ and $c = 4$, only four different main clusters were obtained, each of which has its own average profile as described below and for which every family belongs exclusively to one cluster. In particular:

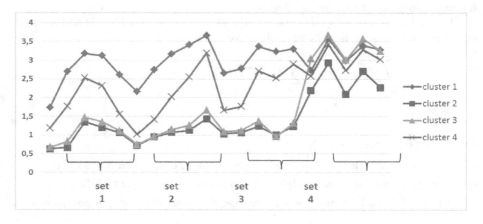

Fig. 2. Classification of families per cluster membership and level of hardship

In particular:

-    **Cluster 1:** is composed of 18.5% of families surveyed and refers to those families perceiving a situation of greater hardship in all types of expenses. It may be assumed that the poorest families fall within this cluster for which the economic crisis has only exacerbated an already existing state of hardship.
-    **Cluster 2:** is composed of 31.8% of families surveyed and refers to those families who do not perceive any hardship in any of the expenses. These families therefore belong to a social status group for which the economic crisis has not generated any significant effect as they do not demonstrate any form of economic difficulty.
-    **Cluster 3:** is composed of 24.1% of families surveyed and presents a profile of medium-high hardship, albeit diverse, with peaks corresponding only to certain expenses. Specifically, these families demonstrate greater difficulty in supporting expenses such as the payment of taxes or unforeseen expenses and the possibility of taking a holiday.
-    **Cluster 4:** is composed of the remaining 25.5% of families surveyed and demonstrates a low hardship profile in terms of bearing material costs but a high level of hardship in participating in social and cultural events or recreational activities. This cluster therefore represents those who while not demonstrating particularly evident financial difficulties experience hardship in terms of participation in social life and therefore show greater social than economic hardship.

The profiles of belonging to a single cluster or a set of clusters can, however, be identified through *Fuzzy c-means*. A simulation is carried out by placing c=4, m=1.5. Four different clusters are thus obtained, each of which with its own average profile and for which every family belongs to only one cluster and other 6 clusters of families who do not specifically belong to a well-defined cluster but belong to two or three clusters, as shown in table 5.

Table 5. Composition of clusters based on the number of families surveyed and the average value of hardship

Cluster	Absolute value	%	Value of hardship
Cluster 1	202	8.1	3.3
Cluster 2	368	14.7	1.2
Cluster 3	213	8.5	1.7
Cluster 4	201	8.0	2.2
Clusters 1,4	467	18.6	2.7
Clusters 2,3	643	25.6	1.5
Clusters 2,4	61	2.4	1.5
Clusters 3,4	156	6.2	2.2
Clusters 1,3,4	17	0.7	2.6
Clusters 2,3,4	179	7.1	1.8
Totale	**2,507**	**100.0**	

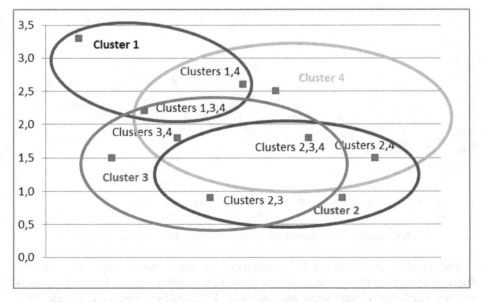

Fig. 3. Classification of families per cluster membership and level of hardship

Through a representation of the clusters on the basis of the average values of hardship, the following "fuzzy" relations between the different average profiles are obtained on the basis of membership to the different clusters.

The profile of the families belonging to a single cluster presents the only variations compared to the previous application, in cluster 3 due to the fact that this hardship profile is amplified compared to the previous simulation assuming high values for all the variables considered (from 16 to 20).

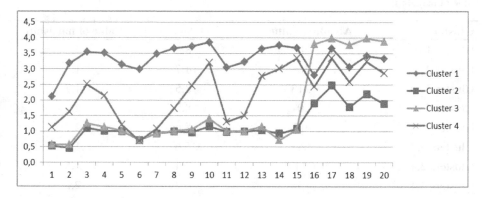

Fig. 4. Classification of families per cluster membership and level of hardship

In terms of the profiles characterized by belonging to two clusters, intermediate situations are shown between the profiles of the two clusters of membership.

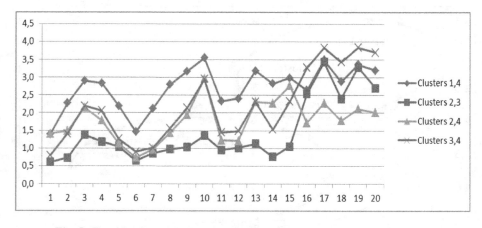

Fig. 5. Classification of families per cluster membership and level of hardship

The profiles characterized by membership to three clusters appear somewhat different from the others. The profile presented by membership to clusters 2, 3 and 4 appears particularly significant and includes 179 families who demonstrate little difficulty bearing costs, except for variable 10 (coping with exceptional unforeseen expenses) and for variables 17 and 19 (social participation). The second profile concerns clusters 1,3 and 4 and involves a negligible number of families (17) that present greater hardship than those of the previous cluster yet with an absence of hardship in terms of the possibility of going to the cinema or theatre or going on holiday once a year. Indeed, this absence of hardship appears to be linked to low-level participation in social life, as confirmed by the high values measured in the set relating to social participation. This therefore represents families demonstrating economic hardship and, above all, social hardship.

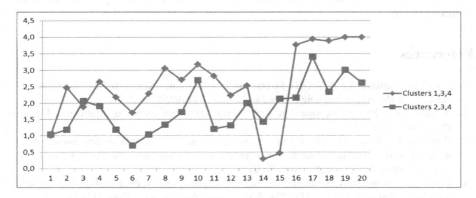

Fig. 6. Classification of families per cluster membership and level of hardship

4 Concluding Remarks

An increasing interest has, in recent years, developed in both the scientific and political fields towards the issue of poverty and, more generally, phenomena of social marginalization. Such studies are however often faced with the lack of specific statistical data. Analyses of poverty are based on surveys of a general nature and often at a "macro" level. It would be of particular interest to perform in Italy, as has already been the case for several years in the United States, surveys of "micro" areas.

The present analysis has attempted to quantify the influence of income and of family typology (number of members) in order to understand how family lifestyles may evolve. The risk of poverty estimates based on "objective" indicators, such as income or levels of debt are completely independent of the state of awareness of those directly involved. It is, however, also useful to observe the "subjective" perception of Italian people in relation to their standard of living and to the recurring causes of economic and social hardship.

The study presented seeks to overcome old classifications between the poor and non-poor by creating "blurred" profiles between those living in different circumstances.

Through the two different applications carried out in this work it is possible to:

- Analyze situations of family hardship through the synthesis of multi-dimensional sets of indicators (*Total Fuzzy and Relative method*);
- Create "fuzzy" profiles highlighting the specific peculiarities of small groups not strictly belonging to a defined profile but to a mix of different profiles (*Fuzzy c-means*).

It is hoped that the variations regarding the new family profiles emerging in general from analyses carried out with different criteria may provide a solid basis for not only a more accurate description and understanding of the phenomenon of economic hardship but also for developing indications for social policies that may contrast poverty.

References

1. Cheli, B., Lemmi, A.: A Totally Fuzzy and Relative Approach to the Multidimensional Analysis of Poverty. Economic Notes 24(1), 115–134 (1995)
2. Bezdek, J.C.: Pattern Recognition with Fuzzy Objective Function Algoritms. Plenum Press, New York (1981)
3. Betti, G., Cheli, B., Lemmi, A.: Occupazione e condizioni di vita su uno pseudo panel italia-no: Primi risultati, avanzamenti e proposte metodologiche. Working paper n. 17. Dip. di Scienze Statistiche, Padova (2002)
4. Lemmi, A., Pannuzi, N.: Continuità e discontinuità nei processi demografici. L'Italia nella transizione demografica, pp. 211–228. 4. Rubettino, Arcavacata di Rende (1995)
5. Zadeh, L.A.: Fuzzy sets. Information and Control 8(3), 338–353 (1965)
6. Dubois, D., Prade, H.: Fuzzy sets and systems. Academic Press, Boston
7. Cerioli, A., Zani, S.: A Fuzzy Approach to the Measurement of Poverty. In: Dugum, C., Zenga, M. (eds.) Income and Wealth Distribution, inequality and Poverty. Springer, Berlin (1980)
8. Lemmi, A., Pannuzi, N., Mazzolli, B., Cheli, B., Betti, G.: Misure di povertà multidimensionali e relative: il caso dell'Italia nella prima metà degli anni '90. In: Quintano, C. (ed.) Scritti di Statistica Economica, 3, Istituto di Statistica e Matematica, Istituto Universitario Navale di Napoli. Quaderni di Discussione, Curto, Napoli, vol. 13, pp. 263–319 (1997)
9. Dunn: A fuzzy relative of the ISODATA process and its use in detecting compact, well-separated clusters. Journal Cibern 3, 32–57 (1973)
10. Bezdek, J.C., Ehrlich, R., Full, W.: FCM: The fuzzy c-means clustering algorithm. Computer e Geosciences 3, 191–205 (1984)
11. Bezdek, J.C., Cannon, R.L., Dave, J.V.: Efficient Implementation of the fuzzy c-means clustering algorithms. IEEE Transactions on Patters Analysis and Machine Intelligence 8(2), 248–255 (1986)

Evaluation of the Economic Sustainability of the projects in Social Housing[*]

Francesco Tajani and Pierluigi Morano

Department of Science of Civil Engineering and Architecture
Polytechnic of Bari, Italy
francescotajani@yahoo.it,
pierluigi.morano@poliba.it

Abstract. With reference to operations of urban regeneration to be realized with the involvement of private investors, in this work a model of Operative Research is developed that allows to define, subject to the constraint of financial feasibility for the private operator, the maximum share of social housing to be borne by the private sector, the administered price of sale and/or lease of social housing and the exchange factor of the area to redevelop. The model is applied to a project of urban regeneration and social housing recently realized in an Italian Region. The research highlights the utility of the model as a tool for decision support in the planning of social housing.

Keywords: Social housing, market value, urban redevelopment, financial feasibility, operative research.

1 Introduction

The regeneration of degraded areas is among the most important issues in public policy. With the designation of degraded areas a wide range of manifestations of the degradation is indicated, that includes individual buildings or entire urban areas characterized by different location, size, use and intensity of degradation, which complex modes of action are necessary for [16].

According to European Environment Agency [7] in the period 2000-2006 about 1,000 km² of land was covered every year by artificial surfaces. Land take for urban area and infrastructure use increased between 1990 and 2000 by 5.7% across Europe, but with unequal distribution. This trend accelerated during 2000-2006: annual land take increased from 0.57% for 1990-2000 to 0.61% for 2000-2006. It is expected that by 2020 urban areas will increase their share in European land stock by approximately 1%. Urban sprawl results in an irreversible reduction of agricultural land [13] and causes damage to the ecosystem [11]. For this reason in the Lisbon Treaty and in the Europe 2020 Strategy, green growth has received more attention and will also be integrated into the next period of Cohesion Policy.

---

[*] The work must be attributed in equal parts to the two authors.

B. Murgante et al. (Eds.): ICCSA 2014, Part III, LNCS 8581, pp. 135–147, 2014.
© Springer International Publishing Switzerland 2014

On the other hand, the increase in sales prices and rents of urban real estate in recent years and the simultaneous credit crunch practiced by banks to individuals who do not have strong guarantees, are preventing access to the "home" to ever larger segments of society [21]. To contrast this phenomenon, many European countries are resorting to policies that encourage social housing [2, 4, 9, 18]. These policies provide that, in urban redevelopment or new construction to be implemented with the involvement of private investors, are fixed percentages of social housing that the private is required to realize at his expense. However, these are rates set by the Public Administration according to the housing needs or the specific regional policies, in the absence of estimates which verify the financial viability. The consequence is that the projects are not always feasible; they could be achievable only in the presence of public resources in support of the operation or with poor quality buildings.

2   Aim of the Study

In this research an evaluation model to support decisions of the Public Administrations that predispose urban regeneration and social housing to be implemented with the involvement of private operators, is developed.

Carried out with reference to an Italian regional territory, the model allows to define: the maximum amount of social housing, as a percentage of the housing planned, to be borne by the private investors; the administered price of sale and/or lease of the social housing; the exchange factor of the area to be redeveloped, with the assumption - frequently found in the Italian context - that the promoter of the initiative and the owner of the property constitute two different entities and that the owner of the area, rather than sell, agrees to commute the area with property of equivalent value that will be realized with the intervention.

Built by borrowing the principles of Operative Research, the model implements the Simplex algorithm through an appropriate software (Matematica 9.0). The Operative Research is a methodology that is frequently used in the field of assessment and/or verification of investments. Applications are found in industry, traffic, agriculture, urban planning, transport, telecommunications, corporate finance, energy and project management [1, 6, 8, 12, 14, 17, 19, 20]. The variety of topics dealt attests the remarkable efficiency of this method with which, in the present work, an easy to use and flexible model is intended to obtain, to can be applied in any territorial context.

The problem can be attributed to the determination of the best use of scarce resources that can be allocated in alternative uses. In mathematical terms, for a process that involves n possible uses for the m resources available, the generic problem of Operative Research can be summarized in Eq. 1:

$$max \text{ (o } min) \quad f(x_1, x_2, ..., x_n) \tag{1}$$

subject to the constraint system of Eq. 2:

$$\left\{ \begin{array}{l} a_{11}x_1+a_{12}x_2+....+a_{1n}x_n \le b_1 \\ a_{21}x_1+a_{22}x_2+....+a_{2n}x_n \le b_2 \\ .... \\ a_{m1}x_1+a_{m2}x_2+....+a_{mn}x_n \le b_m \end{array} \right. \qquad (2)$$

where:
- $f(x_1, x_2,..., x_n)$ is the objective function to be maximized (*max*) or minimize (*min*) in relation to the purposes of the decision maker;
- $x_1, x_2,..., x_n$ are the variables of the problem that represent the possible uses of resources;
- the generic line $\Sigma\, a_{ij}x_i \le b_i$ ($i = 1,.., m$ *e* $j = 1,.., n$) defines the i-th constraint necessary to achieve the objective, where a_{ij} is the absorption rate of the i-th resource in the j-th use, supposed unitary, and b_i is the i-th amount of the resource available to the process.

The research is organized as follows. Section 3 is divided into four parts. In the first, the case study related to the Call for "urban regeneration and social housing" banned in 2008 by the Region of Campania (Italy) is presented. Planning parameters, the percentages and types of social housing provided by the Call are specified. In the second part the economic reasons that lead to the need of social housing rates for different geographical areas are discussed. In the third part the valuation model is introduced, the assumptions, the variables, the constraints and the objective function are explained. In the fourth part the model is applied to a project presented for the Call of the Region of Campania. In Section 4 the conclusions of the work are derived.

3   The Case Study: The Urban Redevelopment and the Social Housing in the Region of Campania

The national aims of "equalized planning" indicated in the Law n. 244/2007 have found application in the Region of Campania with the Executive Decree n. 294/2008. With this disposition - whose the main principles have been incorporated into the Regional Law n. 19/2009 - the Region of Campania has issued a Call, which aims to identify public and private entities and economic operators interested in building redevelopment of urban areas and brownfield and programs of social housing.

Through balanced interventions of urban densification and recovery of the existing building, the program aims to reduce the consumption of the soil and to reutilize urban spaces for the production of residential parts integrated to services and activities.

The project proposals - eligible degraded areas or areas occupied by buildings abandoned or being disposed to redevelop and redeploy for residential purposes - should provide a mix of free residential housing and social housing, as well as services related to residence, artisanal and commercial business, parking, green areas and cultural functions. The approval of the projects is constrained to the verification of the compatibility of the proposals with the purposes and the predictions of existing planning instruments and, in any case, to the limits established by the planning and building national regulations [5].

The Executive Decree n. 294/2008 has put restrictions on the amount of social housing to be implemented in each project initiative, in proportion to the total residential floor area, divided between the types of social housing provided in the Ministerial Decree dated April 22, 2008. This Decree classifies social housing in the following three types, depending on the mode of management and the owner profile [15]:

a) social housing in public ownership. In this case, the private owner of the redeveloped area builds social housing that is then donated to the Local Authority. The management prescriptions are decreed by regional regulations;

b) social housing of private property which can be sold immediately. The criteria for determining the sale price of social housing are set by the Regional Authorities, as a result of the agreement with regional associations;

c) social housing of private property, used as permanently or temporarily rented (at least eight years). The rent is set by the Regional Authority, in consultation with regional associations, according to the different economic capacities of the beneficiaries, the composition of the families and housing characteristics. In any case, the set rent cannot be greater than that resulting from the fixed-price values determined by local territorial agreements. For temporary social housing, the selling price at the end of the rental period is defined by the Regional Authority, in accordance with regional associations.

According to the Executive Decree n. 294/2008, the minimum percentage of social housing of the project proposal must be equal to 30% of the total residential surface and divided into three types:

- type a): 5% to be donated to the municipal administration;
- type b): 10% to be sold at controlled prices (or "social");
- type c): 15% to be leased with sustainable (or "social") rent for at least eight years, with the possibility of selling with administered prices at the end of the minimum rental period.

Most of the project proposals that have participated in the Call of the Region of Campania were planned by private operators.

3.1   The Financial Feasibility of the Project Proposals: The Need of a Valuation Model

The financial benefit of the private entrepreneur, promoter of a redevelopment project, is directly proportional to the density and the marketability of the intended uses consented, as well as to the market values in the area where the intervention is placed. Conversely, the proportion of social housing to realize affect negatively the financial feasibility of the initiative. Therefore, it is evident that unique rate of 30% for all project initiatives, if it attempts to satisfy the local housing needs, does not take into account the financial aspects that may prevent the effective realization of the interventions. As location characteristics that contribute to the formation of market prices remain constant in restrained geographical areas, a same design solution has different profit margins if it is realized in a central urban area characterized by high sale prices or in a suburban area characterized by a not very dynamic real estate market [10]. Similarly, as social housing is a financial cost to the real estate developer, in terms of lower earnings for the absence of the sale of the same property in a free market, it is plausible to assume that the proportion of social housing that ensures the financial convenience of the operation in the central area, may not be financially sustainable in the peripheral zone.

For this reason, noted that the constraint of financial feasibility determines the inability to standardize on vast territories - which in the case of the cited Call correspond to the whole territory of the Region of Campania - the rate of social housing to be provided in urban redevelopment, it is appropriate that this percentage arises from a specific analysis of each project and of the corresponding items of revenue and expense. The model developed in this paper allows to define the critical threshold of social housing that preserves the constraint of financial feasibility, preliminary to the actual realization of the initiative.

3.2   Assumptions of the Model

From the estimative point of view, the market value of a buildable area - which is the case of an area subject to redevelopment - can be determined directly, recording the trades occurred recently for similar areas in the same market territory, or indirectly. The model proposed applies this second procedure, using a residual valuation.

Despite being more suitable the application of Discounted Cash Flow Analysis (DCFA) for interventions that require realization times of at least three or four years, the model is developed without considering the "time" variable. In other words, it is assumed that the implementation of the redevelopment and the sale of building products happen instantly. This simplification, while it does not involve excessive approximations in the results, allows to ignore the uncertainty related to the estimation of a specific discount rate of the cash flows for each area of intervention.

In the hypothesis that the owner of the area to be redeveloped does not coincide with the real estate developer, the market value ($V_{m,a}$) of the area is given by equation (3):

$$V_{ma} = R - K_{transf} = (R - K_{sh}) - (K_{dem/rehab} + K_{const} + K_{prof} + K_{manage} + K_{license} + K_{loan} + K_{market} + U_p) \quad (3)$$

In Table 1 the terms of equation (3) are explained.

The share of social housing leads to a reduction of the total utility of the private entrepreneur, therefore of the market value of the area to be redeveloped. In fact, given the aims of the realization of social housing, it is clear that the unit sale price (p_{sh}) and the unit rent (Ca_{sh}) of social housing should be lower than relative prices of residential free market (p_{res} e Ca_{res}). In the Region of Campania the conventional criteria for the determination of social prices expect that the sale prices and the rents are related to the construction costs of housing, through a relationship of direct proportionality. However, this relationship could not always ensure social housing prices lower than open market prices, since in areas with a depressed housing market the application of

Table 1. Items of revenue and expense in determining the market value of the intervention

$V_{m,a}$	market value of the area to be redeveloped
R'	revenues generated by the intervention, net of the cost associated to the realization of social housing surfaces
K_{transf}	total cost of the transformation of the area
R	revenues generated from the sale in the free market of all surfaces realized, assuming the absence of social housing surfaces
K_{sh}	cost/loss of revenue resulting from the allocation of part of the surfaces realized in social housing
$K_{dem/rehab}$	cost of demolition and/or rehabilitation of existing volumes on the area of intervention
K_{constr}	cost of construction of the buildings, public parking, green areas, etc.. provided by the project
K_{prof}	technical expenses
K_{manage}	operating expenses for the management of the transformation activities
$K_{license}$	concession fees
K_{loan}	financial charges, i.e. the interest on the capital borrowed for the realization of the transformation project.
K_{market}	commercialization expenses for the marketing activities of the finished product and the brokerage of real estate agencies
U_p	profit for the real estate developer

the criterion of the cost of construction could determine disproportionate social prices. In this work, therefore, it is considered more appropriate to assume a coefficient (w), less than unity, that links social prices to the residential prices in the open market that are recorded in the territory where the area to be redeveloped is located. Assuming that the differential between the prices of the social housing and the residential prices in the free market is the same both for sale and for lease, it can be written (4):

$$\begin{cases} P_{sh} = w \cdot P_{res} \\ Ca_{sh} = w \cdot Ca_{res} \end{cases} \qquad 0 \le w < 1 \qquad (4)$$

The lost revenues caused by the allocation of part of the buildings to social housing can be characterized according to the type of social housing to be realized. With reference to the categories of social housing outlined in section 4, it can be deduced that:

1) the social housing quota (x) donated by the private investor - type a) – represents a missed revenue ($K_{sh,a}$) equal to the selling price on the free market of the gross surface areas transferred to public ownership. Mathematically this cost is reflected in the expression (5):

$$K_{sh,a} = P_{res} \cdot x \qquad (5)$$

2) the social housing quota (y) that can be immediately sold by the private operator – type b) – sets a minus value ($K_{sh,b)}$) equal to the difference between the selling price on the free market and the "controlled" selling price. Equation (4) expresses the calculation of this cost item:

$$K_{sh,b)} = (p_{res} - p_{sh}) \cdot y = (1 - w) \cdot p_{res} \cdot y \tag{6}$$

3) the social housing quota (z) used for renting – type c) - generates a contraction of private benefits ($K_{sh,c)}$) that, in the case of temporary accommodation, includes two shares: i) the initial accumulation of the difference between the free market rent and the social rent, ii) the accumulation of the difference between the sale price on the free market and the social price, both being estimated at the end of the period of the lease (n). Denoting by r the discount rate for the private entrepreneur in the components i) and ii) of the social cost, the result is summarized in (7):

$$K_{sh,c)} = [\frac{(Ca_{res} - Ca_{sh}) \cdot (q^n - 1)}{r \cdot q^n} + \frac{(p_{res} - p_{sh})}{q^n}] \cdot z = [\frac{(1-w) \cdot Ca_{res} \cdot (q^n - 1)}{r \cdot q^n} + \frac{(1-w) \cdot p_{res}}{q^n}] \cdot z \tag{7}$$

The total social housing cost for the private entrepreneur is obtained from the sum of the terms (5), (6) and (7) described above, and it can be outlined by the expression (8):

$$K_{sh} = K_{sh,a)} + K_{sh,b)} + K_{sh,c)} = p_{res} \cdot x + (1-w) \cdot p_{res} \cdot y + [\frac{(1-w) \cdot Ca_{res} \cdot (q^n - 1)}{r \cdot q^n} + \frac{(1-w) \cdot p_{res}}{q^n}] \cdot z \tag{8}$$

The market value of the area to be redeveloped ($V_{m,a}$), that is determined by (3), represents the gain for the property owner. Given the ordinary market conditions, it is plausible to assume that $V_{m,a}$ must be greater than or equal to the exchange-value that the owner of the area expects from the transformation. Denoting by h the exchange coefficient, this constraint results in the expression (9):

$$V_{m,a} \geq h \cdot R' = h \cdot (R - K_{sh}) \tag{9}$$

3.3   The Model

Subject to the constraint of financial feasibility for the private developer who implements the redevelopment, the model proposed points at the same time to two objectives:

1) define the maximum rate of surfaces to be allocated to social housing, as a percentage of total residential gross floor surface planned in the project (SLP_{res}), and considering the distribution of the surfaces (x, y e z) in the three types a), b) and c) described above;

2) determine the exchange coefficient h that can be recognized to the owner of the area, considering as the upper limit (h_{sup}) the exchange coefficient that is usually applied in the territory where the intervention area is located.

Therefore, it is possible to define a model of Operative Research that translates into mathematical relationships the variables to be determined, the objective function and the constraints of the problem (Table 2). The mathematical relationships in Table 2 borrow the symbolism adopted in section 4.2. A maximum threshold for the coefficient w equal to 0.7 is assumed, i.e. a maximum deviation of 30% between selling prices on the free market and social prices.

Table 2. Explanation of the model

variables	x, y, z, w, h
objective function	$max\ (x + y + z + h)$
constraints	$V_{m,a} \geq h \cdot (R - K_{sh})$
	$x \geq 0, y \geq 0, z \geq 0, w \geq 0, h \geq 0$
	$w \leq 0.7$
	$h \leq h_{sup}$
	$x + y + z \leq SLP_{res}$

3.4   Application of the Model

The model is applied to a project proposal of the Call promulgated with Executive Decree n. 294/2008 of the Region of Campania.

The project area is in the center of a city in the Province of Salerno, in the Region of Campania, and is currently occupied by a disused industrial complex and a historical and architectural building. The area is close to the historic center or the city, primarily intended for residential and commercial functions. It is characterized by the presence of the main infrastructure and services and a particularly dynamic real estate market. The project design includes two intervention strategies. The first is aimed at the recovery and restructuring of the volume of the historic building, the demolition of the volumes in a state of decay - according to the historical and cultural constraints imposed by the relevant Institutions - and the construction of social housing, integrated services and meeting public spaces, allowing the interconnection with the surrounding urban area. The second strategy concerns the redevelopment of the territory through the demolition of the disused industrial building and the construction of a new architectural volume, adjacent to the historic building, which abuts altimetrically in line with the existing buildings on the edge of the site, redesigning harmonically the urban skyline. The new building will contain residential and commercial uses. The project idea is therefore interesting from a public point of view, as it may allow a greater seam of the city center, and from a private point of view, as it is characterized by market values that should extensively cover the transformation costs. The project involves: the construction of approximately 7,700 m^2 of residential surface, distributed in the restored historic building and the new volume to be realized, and shared in free market housing (70%) and in social housing (30%); the construction of approximately 4,000 m^2 of commercial surface; the construction of two underground floors for public parking (7,500 m^2) and the external arrangement of public spaces (approximately 8,000 m^2).

The preliminary phase of the implementation of the model has involved the determination of the revenues R generated by the realization of the project, assuming that all surfaces are sold in the open market (absence of social housing). The determination of R - so of the unit market value of the residential surfaces in the free market (p_{res}) and of commercial surfaces (p_{comm}) in the project - has been carried out through *Market Comparison Approach*, noting, in the homogeneous territory of the intervention, the prices of properties similar to those to be estimated and making the corrections imposed by the systematic process. The analysis developed have led to unit market values, for residential and commercial uses, respectively equal to 3,300 €/m² and 6,500 €/m².

Similarly, the determination of the unit rent of residential units in the open market (Ca_{res}) has been carried out. The value resulted is equal to 7.20 €/m² for month.

The terms $K_{dem/rehab}$ and K_{constr} have been estimated according to the specifics of the market of the project proposal. The analysis has provided these unit cost values: for the recovery of the historical building, equal to 950 €/m²; for the construction of residential surface, equal to 1,200 €/m²; for the construction of commercial surface, equal to 1,000 €/m²; for the construction of underground public parking surface, equal to 400 €/m²; for the arrangement of external public spaces, equal to 70 €/m². Concession fees have been determined through the coefficients tabulated by the municipality where the intervention area is located ($K_{license}$ = € 1,460,400). The other cost items of K_{transf} have been valuated as a percentage of the construction cost K_{constr} or of the revenues R' generated by the transformation. All the terms are exogenous parameters of the model, except for commercialization expenses for the marketing activities K_{market} and the entrepreneurial profit U_p, which are endogenous variables, functions of the social unknowns ($K_{market} = f(x, y, z)$ and $U_p = f(x, y, z, w)$).

The exchange coefficient that is usually applied in the territory where the intervention area is located (h_{sup}) has been estimated equal to 20%.

The number of years of temporary lease (n) of social housing of the type c) is assumed equal to eight. The private discount rate (r) which also appears in equation (7) is set equal to 7% , estimated with reference to the average annual yield of ten-year Italian BTP at December 2013 (about 6%), adjusted for expected inflation (2%) and considering a representative portion of the risk premium for real estate investment equal to 3%.

The market data and the values of the parameters for the implementation of the model are summarized in Table 3.

The results generated by the application of the model are reported in Table 4.

The output obtained allows to make the following remarks.

First of all, it is evident that the case analyzed is financially "hot": the high market values of the building products make the entrepreneurial initiative highly attractive, so that the percentage of social housing ($\%_{sh}$) calculated on the total residential surface of the project is equal to 100% and the exchange coefficient for the owner of the area is equal to the typical factor in the territory ($h = h_{sup}$). The value of the differential between prices in the free market and social prices (w), equal to 0.64, is lower than the limit assumed for the model. About the distribution of social housing in the three types considered, there is a "preference" of the algorithm for the type c) of social housing, i.e. for the realization of temporarily rented social housing, equal to 59.90%. The result is that in the case examined the estimated parameters for the prices and the

Table 3. Market data and values of the parameters for the implementation of the model

p_{res}	$3,300\ €/m^2$
p_{comm}	$6,500\ €/m^2$
Ca_{res}	$7.20\ €/m^2$ for month
$K_{dem/rehab}$	$950\ €/m^2$
K_{constr}	residential $1,200\ €/m^2$ commercial $1,000\ €/m^2$ underground parking $400\ €/m^2$ external public spaces $70\ €/m^2$
K_{prof}	$6\% \cdot K_{constr}$
K_{manage}	$3\% \cdot K_{constr}$
K_{loan}	$6\% \cdot K_{constr}$
$K_{license}$	$€\ 1,460,400$
K_{market}	$2\% \cdot [R - p_{res} \cdot (x + y + z)]$
U_p	$20\% \cdot R' = 20\% \cdot (R - K_{sh})$
SLP_{res}	$7,659\ m^2$
h_{sup}	20%
n	$8\ years$
r	7%

Table 4. - Output of the model

x	$1,098\ m^2$
y	$1,973\ m^2$
z	$4,588\ m^2$
w	0.64
h	20%

discount rate make this type of social housing the most affordable for the private entrepreneur, in a contingency of the market in which the extension of the sale period is making the formula of *rent to buy* increasingly widespread. The remaining residential surface is divided between the social housing quota of private property which can be immediately sold (25.76%) and the social housing in public ownership (14.34%).

The high degree of financial feasibility of the case analyzed and the extreme flexibility of the algorithm of the model have made it possible to implement additional analysis to determine the most advantageous solutions for the community. Figure 1 shows the output of the model by considering the coefficient w, i.e. the difference between market and social prices, as an input data with values less than 0.7 (i.e. the upper limit considered in the model).

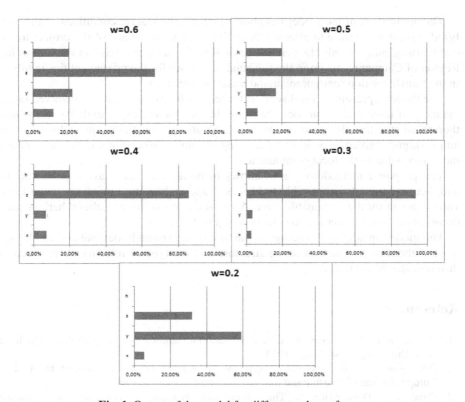

Fig. 1. Output of the model for different values of w

It could be noted that, for decreasing values of w, the percentage of social housing for rent increases, and it ranges from a value equal to 67.14 % of the total residential surface for $w = 0.6$, to a value equal to 93.75 % for $w = 0.3$, whereas the exchange coefficient for the owner of the area remains equal to the maximum value admitted by the model ($h = h_{sup} = 20\%$). These results highlight the greater convenience for the private investor to realize social housing of type c). However, for $w = 0.2$ the distribution of social housing changes radically, rewarding social housing in private property which can be sold immediately - type b) - but countering the exchange coefficient ($h = 0$). In this case, therefore, the project will be realized only if the coincidence between the economic figures of the private entrepreneur and the owner of the area occurred.

4   Conclusions

Developed with reference to urban regeneration projects which imply the involvement of private investors, the model elaborated in this work produces a compromise solution between the objectives of financial feasibility of the private operator and the collective demands of urban reorganization and realization of social housing.

The application of the model has shown that, for the market conditions of the analyzed case, it is possible to allocate the entire residential surface of the project to social housing, and not only the percentage of 30%, i.e. the share fixed in the Call of the Region of Campania, ensuring the financial feasibility for the private entities involved in the transformation (entrepreneur and owner of the area).

The model represents a valid support for the Public Administrations involved in the planning of interventions in social housing. In fact, the extreme flexibility of the mathematical structure designed allows to recalibrate the constraints, providing, for example, higher percentages of social housing in public ownership and/or social prices more accessible to the local communities.

The proposed methodology also allows to monitor the land use transformations and, because it is also applicable to historical and architectural buildings, is an innovative system for the sustainable protection and the promotion of cultural heritage as a driver of social cohesion and the creation of jobs [3].

The model, in addition to being simple to use, is extremely flexible and can be applied without major changes in the structure and the type of information required in different spatial contexts.

References

1. Beneke, R.Y., Winterboer, R.: Linear programming applications to agriculture. The Iowa State University Press, Ames (1973)
2. Boelhouwer, P.: International comparison of social housing management in Western Europe. Journal of Housing and the Built Environment 14(3) (1999)
3. Calabrò, F., Della Spina, L.: The cultural and environmental resources for sustainable development of rural areas in economically disadvantaged contexts. Economic-appraisals issues of a model of management for the valorization of public assets. In: The 3rd International Conference on Energy, Environment and Sustainable Development (ICEESD 2013). Advanced Materials Research, vol. 869-870, pp. 43–48. Trans. Tech. Publications, Switzerland (2014)
4. Cook, T., Whitehead, C.M.E.: Social housing and planning gain: is this an appropriate way of providing affordable housing? Environment and planning 34(7), 1259–1279 (2002)
5. De Mare, G., Nesticò, A., Tajani, F.: The rational quantification of social housing. In: Murgante, B., Gervasi, O., Misra, S., Nedjah, N., Rocha, A.M.A.C., Taniar, D., Apduhan, B.O. (eds.) ICCSA 2012, Part II. LNCS, vol. 7334, pp. 27–43. Springer, Heidelberg (2012)
6. Davis, I.W., Moder, J.J., Phillips, C.R.: Project Management with CPM, PERT and Precedence diagramming. Van Nostrand Reinhold (1983)
7. European Environment Agency: The European Environment. State and Outlook 2010. Copenhagen (2010)
8. Ghiani, G., Musmanno, R.: Modelli e metodi per l'organizzazione dei sistemi logistici. Pitagora Editrice, Bologna (2000)
9. Hackworth, J., Moriah, A.: Neoliberalism, contingency and urban policy: the case of social housing in Ontario. International Journal of Urban and Regional Research 30(3), 510–527 (2006)
10. Holmans, A., Kleinman, M., Royce Porter, C., Whitehead, C.M.E.: How many homes will we need? Technical Volume Shelter, 88 Old Street, London EC1V9HU (2000)

11. Houghton, R.A., Goodale, C.L.: Effects of land-change on the carbon balance of terrestrial ecosystems. Ecosystems and land use change, Geophysical, Monographs Series 153, 85–98 (2004)
12. Kerzner, H.: Project Management: a systems approach to planning scheduling and controlling. Wiley (1998)
13. Lorencovà, E., Frelichovà, J., Nelson, E., Vackà, D.: Past and future impacts of land use and climate change on agricultural ecosystem services in the Czech Republic. Elsevier (2012)
14. Manganelli, B., Tajani, F.: Optimised management for the development of extraordinary public properties. Journal of Property Investment & Finance 32(2) (2014)
15. Ministero delle Infrastrutture: Definizione di alloggio sociale ai fini dell'esenzione dall'obbligo di notifica degli aiuti di Stato, ai sensi degli articoli 87 e 88 del Trattato istitutivo della Comunità europea, Italy (2008)
16. Morano, P., Tajani, F.: The transfer of development rights for the regeneration of brownfield sites. Applied Mechanics and Materials 409-410, 971–978 (2013)
17. Olson, D.L.: Multiple Criteria Optimization: Theory, computation, and application. Wiley series in probability and mathematical statistics applied (1996)
18. Pawson, H., Kintrea, K.: Part of the problem or part of the solution? Social housing allocation policies and social exclusion in Britain. Journal of Social Policy 31(4), 643–667 (2002)
19. Roberts, F.S.: Discrete Mathematical Models with applications to social, biological and environmental problems. Prentice-Hall (1976)
20. Steuer, R.E.: Multiple Criteria Optimization: theory, computation and application. Wiley series in probability and mathematical statistics applied (1986)
21. Torre, C.M., Perchinunno, P., Rotondo, F.: Estimates of housing costs and housing difficulties: An application on Italian metropolitan areas. In: Housing, Housing Costs and Mortgages: Trends, Impact and Prediction, pp. 93–108 (2013)

Urban Renewal and Real Option Analysis: A Case Study[*]

Pierluigi Morano and Francesco Tajani

Department of Science of Civil Engineering and Architecture
Polytechnic of Bari, Italy
pierluigi.morano@poliba.it
francescotajani@yahoo.it

Abstract. The high variability of market prices and the uncertainty that, even in restrained timeframes, is characterizing the general economic situation, have led real estate operators to a prudent attitude, who tend to postpone or at least stagger the start of the initiatives on hold of more stable conditions. In this context, it is appropriate to use evaluation tools enable to enhance the investment capacity to be adapted to possible changes of the conditions initially hypothesized. In the present research Real Options Analysis (ROA) is applied to the evaluation of an investment in urban redevelopment of a former industrial complex. The result obtained shows the efficacy of the instrument.

Keywords: Real options, urban renewal, risk analysis, financial feasibility.

1   Introduction

The current historical moment is characterized by a unique combination of economic phenomena and national and international policy choices that have negatively affected the real estate sector [10, 23].

This situation is inducing a cautious attitude in the operators, who prefer to postpone or at least stagger the launch of the initiative, waiting for more favorable market conditions, due to the variability of market prices and the uncertainty about their progress in restrained timeframes [8, 20, 24]. Therefore it is appropriate to use "dynamic" assessment tools [12, 18], which allow to exploit the ability of the investment to be adapted to potential changes of the conditions initially considered [2, 3].

The Real Options Analysis (ROA), where changes in scenario assumptions are predictable, allows decision makers to adopt pricing strategies and to evaluate the effects that result from the investment choices (options), different from the initial ones [9].

The ROA is a technique for evaluating investments that can be used with success to manage the uncertainty related to possible changes of scenery. Compared with the "static" approach, that considers the present value of cash flows expected to the most likely future scenario, the ROA, when it is possible to transform the uncertainty of the project into risk, allows to carry out the risk analysis for the different options.

---

[*] The work must be attributed in equal parts to the two authors.

B. Murgante et al. (Eds.): ICCSA 2014, Part III, LNCS 8581, pp. 148–160, 2014.
© Springer International Publishing Switzerland 2014

2   Aim of the Study

With specific reference to urban redevelopment, this work aims to highlight the potential of Real Options Analysis (ROA) as a tool to support the choices of investment in real estate [4, 15, 22].

The opportunity to employ a "dynamic" process of evaluation is particularly incisive in the current economic situation, characterized by high volatility in property values and by the resulting uncertainty related to the calculation of the Net Present Value associated with the application of traditional Discounted Cash Flow Analysis (DCFA) [14]. The enhancement of the project's ability to adapt to any possible disruption of the scenario initially hypothesized is the basis for the launch of real estate operations that require a substantial financial commitment [21]. Therefore, the ROA for the verification of the financial feasibility of an investment property that provides for the urban redevelopment of a former industrial complex and the construction of a park for mixed use (residential, commercial and underground garages) is here applied.

The analytical formulations of ROA are manifold. In this work the binomial paradigm is used, which develops the changes of the initial value of the investment through probabilistic multiplicative states, that represent the evolution of the initial situation to a favorable scenario or to an unfavorable scenario [5].

The work is structured as follows. In Section 3 the case study, i.e. the urban redevelopment of a former industrial complex located in the central area of a municipality in the Province of Salerno (Italy), is illustrated. The dimensional parameters of the initiative and the prices and the costs of building products are outlined, and financial analysis is carried out using a "traditional" DCFA. In Section 4 the ROA is applied to the case study. The theoretical and practical aspects of risk analysis, of strategic analysis and quantitative analysis are developed, that identify the three phases in which the implementation of the ROA is divided. Finally, the results of the calculations are discussed and the conclusions of the work are taken.

3   The Case Study

The case study concerns a redevelopment of a former industrial complex located in the central area of Pagani, that is a town in the province of Salerno (Italy).

The area, which covers 17,000 m^2 of surface, is in fair conditions of usability and need only few interventions of reclamation. On the area insist 71,942 m^3 of abandoned buildings, mostly in poor conditions.

The specificity of the types of construction and the decay of the buildings do not recommend the conversion to new uses of the existent structures, but direct to their replacement.

The redevelopment is expected by the current planning instrument. On the basis of planning rules, in particular, 40,030.65 m^3 of new buildings can be realized, that correspond to 9,304.77 m^2 of residential gross floor area and 2,358.00 m^2 of commercial gross floor area.

The idea of transformation is schematically shown in Figure 1.

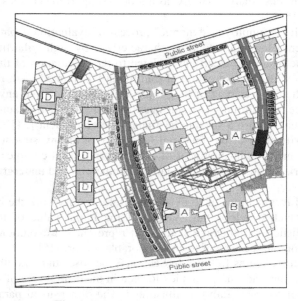

Fig. 1. Hypothesis of the transformation

Of the eleven buildings in the program, those remarked with the letters A, B and C have a height of 12.50 m (for four floors); the ground floor is intended to commercial use, the upper floors are house residences. The three buildings identified with the letter D, each of three floors with a total height of 9.45 m, are intended to residential use. The building marked with the letter E, which is also in residential use, have two floors, with a total height of 6.50 m.

Under three of the five buildings of type A and under the buildings B, D and E underground garages are planned, for a total of 75 units. The other two buildings of type A and the building C are served by parking spaces on the ground floor.

The project will be completed by the external work and the construction of a link road between two existing municipal roads located north and south of the complex.

The financial feasibility of the initiative has been verified with the traditional DCFA. The data for the construction of the business plan have been obtained by integrating the amounts reported in the official lists of the territory with the information gathered through a survey of construction companies and operators in the local real estate. The main financial data are presented in Table 1.

Table 1. Market data for the implementation of the financial analysis

COSTS	
Land purchase	150.00 €/m²
Demolition of existent buildings	17.00 €/m³
Constructions of new buildings	900.00 €/m²
Construction of underground boxes	15,000.00 €/box
External work	290.00 €/m²
REVENUES	
Residential market value	2,200 €/m²
Commercial market value	2,300 €/m²
Box market value	30,000 €/box

As basis of the evaluation the following assumptions have been considered: the time of the valuation is the second half of 2013; the analysis period, including the construction phase, is four years, divided into eight semesters; the current prices system has been assumed re-evaluated annually on expected inflation; sales have been assumed to be uniformly distributed over the eight semesters of the analysis period in the proportion of 12.50% per semester; the annual discount rate is 6.50%, which corresponds to a half-year rate equal to 3.20%.

The calculations return a Net Present Value (NPV) of the project equal to € 5,998,935.29 .

This amount, which corresponds to 25.36% of the total revenues of the entire initiative (*Revenues*), amounting to € 23,655,107.61, identifies the total profit of the operation (U_p).

However, given the current economic conditions and the significant uncertainty involved in this operation, the market survey has revealed that the profitability threshold currently set by the operators in the zone to launch similar investments ($\overline{U}$), computed in terms of the ratio between the NPV and the total revenues, is at least 27%.

Therefore the results obtained by the canonical DCFA induce to abandon the project.

4   Application of the ROA to the Case Study

The usefulness of real options emerges just in cases where the NPV is close to zero or, as in this case, it reaches values close to the threshold of acceptability of an investment [17]. In these circumstances, the identification and the translation into monetary terms of real options, i.e. the strategic opportunities that are activated with the implementation of the project and that can be captured with the actions that the entrepreneur can undertake in response to changes of scenario, can lead to recognize the convenience of initiatives that according to traditional logic would be discarded [25].

The opportunity to defer the launch of the work, due to the uncertainty related to the performance of the real estate market, as well as the ability to divide the project into three functional lots to be carried out in sequence, are the added value of the intervention that can be quantified with the use of ROA.

In the literature, different approaches are illustrated for the implementation of the Real Options Analysis. Whatever the procedure used, it is possible to distinguish three logical-operative moments that lead to the explicitation of the value of the options: Risk analysis; Strategic analysis; Quantitative analysis.

In the terminology of the ROA, the project of which is estimated the extended NPV - sum of the NPV obtained by the application of a traditional DCFA and the value of real options - constitutes the "underlying asset". The ROA developed here pertains to the so-called discrete models [7], in which the value of the underlying asset changes into specific points in time and can take only specific predetermined values. The simplicity of the analytical formulations and the schematic of the logical process make the procedure easy to understand and give an additional ease of use.

In Figure 2 the boundaries of the three functional lots hypothesized are shown. Therefore, the ROA of the 1° functional lot is developed, considering it as the "main" project (underlying asset), whose the extended Net Present Value is determined, whereas the other two lots constitute opportunities for future developments.

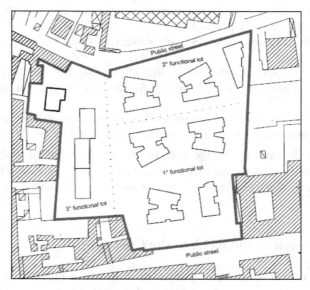

Fig. 2. Functional lots

4.1   Risk Analysis

The risk analysis allows the detection and the investigation of the uncertain variables of the initiative: the decisions related to the project depend on their evolution [13].

This phase is divided into two steps: the identification of the critical variables of the project and the valuation of the volatility.

The identification of the critical variables depends on the specific area in which the initiative pertains, the market at the time when the valuation is developed, the availability of data and information on which the analysis is set. Of these variables the connection with the underlying project is also necessary to analyze.

Even if the project involves a mix of functions consisting of residential use, commercial use and parking, in the present case the selling price of the residences has been identified as a critical variable. It is the use with the greatest weight within the framework of the realizations; the major changes in terms of price are expected for the residential destination.

The next step involves the formalization of a stochastic model that simulates the evolution of the uncertain variable.

The procedure applied in this work to estimate volatility uses the regression analysis, implemented on historical data of uncertain variables identified [16, 19]. Estimated the function that interpolates the data of the time series of selling prices, the volatility of the investment is calculated as the standard deviation of the residuals computed between the detected values and the values estimated.

The analytical evaluation of the volatility (σ) is performed using the statistical formula of the standard deviation:

$$\sigma = \sqrt{\frac{1}{n-1} \cdot \sum_{i=1}^{n}(R_i - \overline{R})^2} \tag{1}$$

where:

σ = volatility of the uncertain variable X;

n = number of years for which historical data of the variable X are available;

R_i = residual at year i expressed as a percentage, i.e. the percentage deviation of the historical data X_i with respect to the corresponding value of the interpolating function X_{ir}, deviation obtained with the equation $R_i = \dfrac{(X_i - X_{ir})}{X_{ir}}$;

$\overline{R}$ = average of residuals R_i.

The estimated statistical characteristics of the interpolating function "R-squared corrected" (0.75), the "significance of the F Fischer" (<0.05) and the "significance of the Student's t" (<0.05) have denoted a good reliability of the linear regression model adopted.

The use of the formula (1) returns a value of volatility (σ) equal to 22.03%.

4.2   Strategic Analysis

Strategic analysis concerns the identification of areas of managerial adaptability, i.e. the strategic opportunities inherent in the project and which the investor could provide.

In the literature, the options are grouped into various types [26]. In this case, the opportunities inherent in the main project are related to the possibility to defer the start of the works and to expand the size of the initial operation.

A compound option is considered [11], that consists of the following opportunities in chronological order: a) the option to defer the launch of the works of the project in three years, b) the option to expand the implementation of the 2° functional lot, viable from the second year and by the end of the fifth semester of the analysis period considered, c) the option to expand the realization of the 3° functional lot, feasible from the second year. The volumes and gross floor surfaces (SLP) pertaining to each functional lot considered are shown in Table 2.

Table 2. Sizing and economic parameters for the functional lots

		Volume [m³]	SLP [m²]	SLP [n]	Market value [€/m²]	Market value [€/n]
Main project 1° lot	Residential	19,003.37	4,075.41		2,200	
	Commercial		1,358.00		2,300	
	Underground parking			50		30,000
2° lot	Residential	13,975.75	3,000.00		2,200	
	Commercial		1,000.00		2,300	
	Underground parking		-			-
3° lot	Residential	7,051.53	2,229.36		2,200	
	Commercial				-	
	Underground parking			25		30,000

It must be noted that the listed opportunities to expand and to defer are sequential options. In fact, every viable option changes the economic characteristics of the main project, thus defining a new underlying asset which is related to the possible activation of the subsequent options.

4.3   Quantitative Analysis

In the quantitative analysis the value of real options is determined. The analytical formulation adopted borrows the logic of dynamic programming for solving optimization problems of decisions. This technique involves the construction of a binomial model, according to which the value of the underlying asset is a stochastic variable that changes over time evolving through only two probabilistic multiplicative states, defined by the coefficients $u > 1$ and $d < 1$. These coefficients are, respectively, the evolution of the initial state to a favorable scenario (u) or to an unfavorable scenario (d). The method of solution develops the possible values of the underlying asset over the life of the option and then discounts at the present time the value of optimal decisions in the future.

In the quantitative analysis of the binomial model, there are three steps:
1) valuation of the NPV of the underlying project;
2) modeling of the stochastic process through the scenarios tree;
3) construction of the decision tree and determination of the extended NPV.

4.3.1   Valuation of the NPV of the Underlying Project

The use of a traditional DCFA returns the Net Present Value of the project. In this step the evaluator has to discern the revenues and the costs of the intervention, the analysis period, the discount rate and other parameters that influence the outcome of the evaluation.

In order to consider only the voices characterized by components of uncertainty for further analytical processing, the NPV with the exception of the investment costs has been determined, as these items are routinely characterized by deterministic nature. The value obtained (*VA*) is a stochastic variable that does not yet take into account the flexibility inherent in the investment.

The Net Present Value of the underlying project is determined by considering the same parameters used in case the DCFA is applied to the entire project solution and quantifying revenues and costs related to the different size of the intervention. The analysis period is still four years, divided into six-month intervals ($dt = 0.5$) and it is assumed that the sales are spread evenly over the eight semesters. The result is a NPV of the main project (first functional lot) equal to € 2,682,295.

The value (*VA*) of the underlying project is formed by the accumulation at the time of the assessment of revenues generated from the sales of building products.

The subsequent processing requires the development of a DCFA for the two options to expand identified (2° and 3° functional lots). Table 3 summarizes the results in terms of present value of the revenues and present value of the costs of the three functional lots evaluated individually.

Table 3. Revenues and costs of the functional lots

	Present value of revenues [€]	Present value of costs [€]
Main project	11,424,130	8,741,835
2° functional lot	7,481,970	5,556,610
3° functional lot	4,753,650	3,362,370

4.3.2   Modeling the Stochastic Process through the Scenarios Tree

In the binomial model, the mathematical expressions for the calculation of the two possible states of evolution (*u* and *d*) of the value of the underlying project are:

$$\begin{cases} u = e^{\sigma \cdot \sqrt{dt}} \\ d = e^{-\sigma \cdot \sqrt{dt}} \end{cases} \tag{2}$$

where:

e = Euler number;

σ = riskiness of the underlying project revenues (volatility);

dt = time interval between two successive periods.

Depending on the values of u and d, VA of the underlying project changes in each time interval with a "jump" evolution, represented by a typical tree structure (*scenarios tree*).

The formulas of (2) are used to define the evolution of VA of the 1° functional lot (*main project*). The parameters required for the construction of the scenarios tree are summarized in Table 4.

Table 4. Parameters for the construction of the scenarios tree

σ	22.03%
u	1.169
d	0.856
dt	0.5

4.3.3   Construction of the Decision Tree and Determination of Extended NPV

Starting from the scenarios tree, the binomial model enhances the options inherent in the underlying project by constructing a tree diagram (decision tree) for each option identified.

A specific maximization function is associated with each class of options. The maximization function works by comparing, in the exercise period (m) of the option and for every possible scenario of that period, the value of the activation of the option with the alternative "zero" (non-exercise of the option), and selecting the major of the two values. The maximization algorithm, applied in all scenarios n of the period in which the option is feasible, allows to determine a vector of the majors, with size $n \cdot 1$.

Unlike the scenarios tree, which is built from left to right, i.e. from the initial moment until the final instant of the time period of evaluation, the evolution of the decision tree is defined backward, moving from the period on which the option is exercised (m) up to the year zero. The operation is carried out by weighting the elements of the vector of the majors determined, through coefficients called risk neutral probability, determined by (3):

$$p = \frac{e^{(r_f - \delta)dt} - d}{u - d}$$

(3)

where:
$e =$ Euler number;
$r_f =$ risk free discount rate;
$\delta =$ loss in value of the underlying project[1].

---

[1]   "The loss of value resulting from cash flows or returns of convenience occurred between two points of possible decision are a characteristic of real assets, which must be taken into account in an evaluation model of real options. [...] For option pricing models it is very convenient to treat the cash flows or returns of convenience as a constant percentage of the value of the underlying asset" [1]. The parameter δ is representative of a constant proportionality between the cash flows of each reporting period and the present value of the underlying asset.

The result is then brought to the current moment through a risk-free discount rate and because of the time interval dt between two successive periods.

In the case of the activation of sequential options during the analysis period, the operative logic of the algorithm requires the verification of the global effect of the options from the furthest one chronologically, then gradually implementing the procedure for the other options closer to the time of the assessment. The mathematical expressions that lead to the value of the compound option are built in such a way that the value of the i-th option "has memory" of the value of the chronologically later option [6].

In the present case, it is necessary to determine in advance the expansion coefficients for the second and the third functional lots (respectively, a_2 and a_3) and the strike prices of the three options (I_{a2} and I_{a3} for the options to expand and I_d for the option to defer), in order to implement the related optional algorithms.

Developed the DCFA for the two options to expand, the corresponding coefficients are considered to be equal to the percentage increase of the revenues of the main project determined by the exercise of each option to expand considered. The strike prices are given, for the two options to expand, by the investment costs of the second and the third functional lots; for the option to defer, by the investment cost of the main project.

In Table 5 the main parameters that contribute to the estimation of risk neutral probability using the formula (3) and to the definition of the algorithms of the binomial model are explained.

Table 5. Parameters of the algorithms of the binomial model

σ	22.03%
u	1.169
d	0.856
r_f	3.00%
δ	12.50%
p	0.313
Dt	0.50
a_2	0.65
a_3	0.42
I_{a2}	€ 5,556,610
I_{a3}	€ 3,362,370
I_d	€ 8,741,835

The mathematical formalization of the maximization functions of the sequential options identified (option to defer and two options to expand) is shown in Table 6.

In Table 7 the meaning of the terms that appear in the equations is explained, using the apex $i = (a3, a2, d)$ to indicate the i-th option among those in analysis (to expand 3° lot, to expand 2° lot, to defer).

The analysis returns a value of the compound option considered (V_{option}) amounted to € 3,998,052.

The extended NPV of the main project is equal to the sum of the NPV of the underlying asset and the total value of the sequential options identified:

$$extended\ NPV = NPV + V_{cption} = 2,682,295 + 3,998,052 = €\ 6,680,347 \qquad (4)$$

The return for the entrepreneur (U_p), equal to the ratio between the value of transformation and total revenues of the initiative is the following:

$$U_p = \frac{extended\ NPV}{Re\ venues} = \frac{6,680,347}{23,655,107.61} = 28.24\% > \overline{U} = 27\% \qquad (5)$$

Therefore, the entrepreneur's profit is greater than the threshold of financial acceptability fixed. The difference in value between the application of ROA and the development of a traditional DCFA - difference of about € 700,000 - allows to appreciate the financial feasibility of the project analyzed.

Table 6. Algorithms of the sequential options identified

Option to expand 3° lot	
$t = 8$	$VA_t^{a3} = \max(a_3 \cdot VA_t - I_{a3}),0]$
$3 \le t < 8$	$VA_t^{a3} = \max[(a_3 \cdot VA_t - I_{a3}), \dfrac{p \cdot VA^{a3+}_{t+dt} + (1-p) \cdot VA^{a3-}_{t+dt}}{e^{r_f \cdot dt}}]$
$0 \le t < 3$	$VA_t^{a3} = \dfrac{p \cdot VA^{a3+}_{t+dt} + (1-p) \cdot VA^{a3-}_{t+dt}}{e^{r_f \cdot dt}}$

Option to expand 2° lot	
$t = 5$	$VA_t^{a2} = \max(a_2 \cdot VA_t - I_{a2} + VA^{a3}_t),0]$
$3 \le t < 5$	$VA_t^{a2} = \max[(a_2 \cdot VA_t - I_{a2} + VA^{a3}_t), \dfrac{p \cdot VA^{a2+}_{t+dt} + (1-p) \cdot VA^{a2-}_{t+dt}}{e^{r_f \cdot dt}}]$
$0 \le t < 3$	$VA_t^{a2} = \dfrac{p \cdot VA^{a2+}_{t+dt} + (1-p) \cdot VA^{a2-}_{t+dt}}{e^{r_f \cdot dt}}$

Option to defer	
$t = 5$	$VA_t^d = \max(VA_t - I_d + VA^{a2}_t),0]$
$2 \le t < 5$	$VA_t^d = \max[(VA_t - I_d + VA^{a2}_t), \dfrac{p \cdot VA^{d+}_{t+dt} + (1-p) \cdot VA^{d-}_{t+dt}}{e^{r_f \cdot dt}}]$
$0 \le t < 2$	$VA_t^d = \dfrac{p \cdot VA^{d+}_{t+dt} + (1-p) \cdot VA^{d-}_{t+dt}}{e^{r_f \cdot dt}}$

Table 7. Meaning of the terms in the algorithms of the binomial model implemented

VA_t^i	value of the i-th option at time t
VA_t	value of the underlying asset (1° functional lot) at time t
VA^{i+}_{t+dt}	value of the i-th option at time $t+dt$ and assuming favorable evolution of the value at time t
VA^{i-}_{t+dt}	value of the i-th optionat time $t+dt$ and assuming unfavorable evolution of the value at time t

5   Conclusions

In the field of real estate investments, the potentialities of the ROA have been enucleated applying the technique to the financial evaluation of a urban redevelopment project of an abandoned industrial complex. Enhancing the factor of "wait" to start the initiative and identifying the relationship of complementarity between the different functional lots that compose the project, the options identified return an added value that enhances the financial feasibility of the investment. Moreover, the binomial approach employed is easy to use and to read for the operator.

The "finance static", even if it is complemented by simulations on the sensitivity of some input variables or by probabilistic scenarios, is not able to enhance the flexibility in the management decisions. The real options belong to the "dynamic finance" because their value and their exercise depend on the evolution of the uncertain variables: the task of management is to control and to exercise these options at the best time.

References

1. Amram, M., Kulatilaka, N.: Real Options: Strategie d'investimento in un mondo dominato dall'incertezza. Etas, Milano (2000)
2. Bauer, M.: An analysis of the use of discounting cash flow methods and real options to value flexibility in real estate development projects. Lambert Academic Publishing, Berlin (2009)
3. Brennan, M.J., Trigeorgis, L.: Project Flexibility, Agency and Competition. Oxford University Press, New York (2000)
4. Bulan, L., Mayer, C., Tsuriel Somerville, C.: Irreversible investment, real options, and competition: Evidence from real estate development. Journal of Urban Economics 65, 237–251 (2009)
5. Copeland, T., Antikarov, V.: Opzioni reali: Tecniche di analisi e valutazione. Il Sole 24 Ore, Milano (2003)
6. Copeland, T., Keenan, P.T.: Making Real Options Real. The McKinsey Quarterly (3) (1998)
7. Cox, J., Ross, S., Rubinstein, M.: Options Pricing: a simplified approach. Journal of Financial Economics 7 (1979)
8. Dixit, A.K.: Investment and hysteresis. Journal of Economic Perspectives 6, 107–132 (1992)

9. Dixit, A.K., Pindyck, R.S.: Investment under uncertainty. Princeton University Press, Princeton (1994)
10. Eurostat: Euro-area house prices down by 2.5% - third quarter 2012 compared with third quarter 2011 (2013)
11. Geske, R.: The valuation of compound options. Journal of Financial Economics (1) (1979)
12. Guarini, M.R., Battisti, F., Buccarini, C.: Rome: Re-Qualification Program for the Street Markets in Public-Private Partnership. A Further Proposal for the Flaminio II Street Market. In: Civil, Structural and Environmental Engineering (GCCSEE 2013). Advanced Materials Research, vol. 838-841, pp. 2928–2933 (2014)
13. Hertz, D.: Risk Analysis in Capital Investment. Harvard Business Review 42 (1964)
14. Kulatilaka, N., Markus, A.J.: Project valuation under uncertainly: where does DCF fail? Journal of Applied Corporate Finance 5(3) (1992)
15. Manganelli, B., Morano, P., Tajani, F.: Risk assessment in estimating the capitalization rate. WSEAS Transactions on Business and Economics 11 (2014)
16. Micalizzi, A.: Opzioni reali. Logiche e casi di valutazione degli investimenti in contesti di incertezza. Egea, Milano (1997)
17. Milne, A., Whalley, A.E.: Time to build, option value and investment decisions: A comment. Journal of Financial Economics 56, 325–332 (2000)
18. Morano, P., Tajani, F.: The transfer of development rights for the regeneration of brown-field sites. Applied Mechanics and Materials 409-410, 971–978 (2013)
19. Mun, J.: Real Options Analysis. Tools and techniques for valuing strategic investments and decisions. John Wiley & Sons, New Jersey (2006)
20. Pacheco-de-Almeida, G., Zemsky, P.: The effect of time-to-build on strategic investment under uncertainty. Rand Journal of Economics 34, 166–182 (2003)
21. Rocha, K., Salles, L., Alcaraz Garcia, F.A., Sardinha, J.A., Teixeira, J.P.: Real estate and real options - a case study. Emerging Markets Review 8, 67–79 (2007)
22. Schatzki, T.: Options, uncertainty and sunk costs: an empirical analysis of land use change. Journal of Environmental Economics and Management 46, 86–105 (2003)
23. Standard & Poor's: Europe's recession is still dragging down house prices in most markets (2013)
24. Titman, S.: Urban Land Prices Under Uncertainty. The American Economic Review 75(3), 505–514 (1985)
25. Triantis, A.J., Hodder, J.E.: Valuing flexibility as a complex option. Journal of Finance 45, 549–565 (1990)
26. Trigeorgis, L.: Real Options. Managerial Flexibility and Strategy in Resource Allocation. The MIT Press, Cambridge (1996)

Urban Redevelopment: A Multi-criteria Valuation Model Optimized through the Fuzzy Logic[*]

Pierluigi Morano[1], Marco Locurcio[2], Francesco Tajani[1], and Maria Rosaria Guarini[2]

[1] Department of Science of Civil Engineering and Architecture,Polytechnic of Bari, Italy
`pierluigi.morano@poliba.it, francescotajani@yahoo.it`
[2] Department of Architecture and Design, University "La Sapienza", Rome, Italy
`locurciomarco@yahoo.it, mariarosaria.guarini@uniroma1.it`

Abstract. In decision-making processes related to urban redevelopment, the clarity and the transparency play a primary role. In these contexts, the multi-criteria techniques, despite having a wide application, are not always adequate to represent and to quantify the quality effects of the urban initiatives, as well as to compare the alternatives for choosing the best solution, phases in which the logical rules followed by the decision-maker are not usually explicited. In the present work, with reference to a multi-criteria model recently developed for the municipality of Rome (Italy) to streamline and make more transparent the definition of urban regeneration projects, a solution to these issues is proposed, through the use of a fuzzy logic system. Using linguistic variables and expressions of ordinary language, logical rules followed by the decision-maker in performing the evaluations have been formalized. The result is a decision-making process clear and easy to understand, with positive effects on the legitimacy of the decisions of the Public Administration.

Keywords: Urban redevelopment, decision support models, multi-criteria analysis, fuzzy logic systems.

1   Introduction

The regeneration of degraded or disused areas is among the most important issues in the field of urban policies of the major European countries. The possibilities to revive the image of entire parts of the city, to correct - with measures of quality - the damage caused by the urban sprawl, to meet the needs of residential buildings and public spaces avoiding the consumption of new areas, to generate significant economic impacts on the territory through the multiplier effects triggered by investments in the construction industry, are connected to the physical and functional recovery and the enhancement of degraded or disused areas [9, 16, 17].

These observations, generally applicable, are reflected particularly in the city of Rome (Italy), which is characterized by a long history and a strong potential, but also by significant situations of degradation.

---

[*] The work must be attributed in equal parts to the authors.

B. Murgante et al. (Eds.): ICCSA 2014, Part III, LNCS 8581, pp. 161–175, 2014.
© Springer International Publishing Switzerland 2014

Starting from these premises, the municipality of Rome has entrusted in 2011 to the Urban Disused Areas Association (AUDIS) the task of developing an instrument to support the preparation of urban regeneration plans and projects.

The AUDIS has participated for more than a decade in the European debate on urban quality. In Italy, the Association has carried out the monitoring of various operations of urban transformation [3] and has developed the "AUDIS Urban regeneration charter" [1].

Through the comparison with public and private operators, experts and social actors, the AUDIS has come to the drafting of the "Urban quality Protocol of the capital city of Rome" [2]. The Protocol is an operational tool that aims to streamline and make more transparent the process of design and construction of urban regeneration projects to be implemented with the involvement of private investors. The objectives include the increase of the quality of the areas to be redeveloped, the decrease in programming and approval times of plans and projects, the reduction of the discretionary in the choices of the Public Administration.

The Protocol consists of the "Identity Card Project", which describes the regeneration project and identifies it briefly, and the "Matrix of Urban Quality", which is the instrument of implementation of the Protocol. The Matrix identifies a checklist of targets, translated into evaluation criteria explained with suitable parameters, which the private investor must take into account to prepare the redevelopment project. Through these elements, the ability of the project to improve the urban, architectural, social, economic and environmental qualities and the energy and landscape aspects of the urban territory on which the initiative is planned, are examined.

The process that occurs is the following. The Public Administration fills in the "targets" column of the Matrix, setting the level through the specification of the relevant criteria and parameters; the private operator fills in the "data" column, showing how the proposed project pursues the objectives; the Public Administration, with the support of experts (planners, valuators, sociologists, etc.), in the "evaluation" column expresses for each parameter a synthetic judgment (good, sufficient or insufficient), which indicates how the proposed project achieves the objectives set. A polar graph provides a schematic overview of the complex quality of the project.

Although this procedure is certainly more effective than the evaluation methods currently used by the Public Administration, as it leads to a transparent and timely examination of the initiatives, it has some limitations that should be highlighted. First of all, it is designed to evaluate a single project proposal: therefore it does not permit to explore alternative solutions of the same project or to compare proposals from different promoters. Secondly, in the analysis the same weight is given to each objective, whereas, for technical and political reasons related to the programmatic purposes of the Public Administration or to contingent motivations, it may be appropriate to assign a different importance. Finally, the stages in which the judgment on the project's ability to pursue the objectives set is expressed and the best solution is chosen are quite cryptic, since nothing is said about the logical rules followed in performing the evaluations, limiting the transparency of the process.

2   Aims

The present work aims to improve the transparency and the effectiveness of the evaluation procedure of the Urban quality Protocol through the use of a fuzzy logic model. With the use of linguistic variables, this model allows to make more transparent the phase of the evaluation of project priorities by the Public Administration, as well as to introduce the possibility to attribute different weights to the parameters which explain the objectives of the regeneration, and to carry out the comparison between the projects submitted by different operators. The use of fuzzy logic permits to formalize the logical rules of the decision-maker, employing expressions of ordinary language, so that the decisional criteria adopted are clear and easy to understand for all stakeholders, with positive effects on the legitimacy of the choices.

The research is organized as follows. In Section 3 brief remarks on the theory of fuzzy logic systems are given. In Section 4 the case study, concerning the urban and environmental renewal of a neighborhood in the periphery of Rome, are illustrated. In Section 5, the evaluation procedure developed in Urban quality Protocol is improved through a fuzzy logic system. In Section 6 the conclusions of the work are derived.

3   Overview on Fuzzy Logic Systems

The improvement of the evaluation procedure of the Urban quality Protocol has been carried out through the implementation of a fuzzy logic system [10, 13]. This is a system in which the inputs and the outputs are fuzzy sets, and the output is determined as a function of the input, applying the fuzzy rules [7].

The fuzzy rules are the core of the system [25]. A fuzzy rule represents the association of an input linguistic relationship, obtained by the connection of several linguistic variables to an output linguistic expression, that also typically consists of multiple linguistic variables [20]. The fuzzy rules make it possible to translate into formal models, using natural language, the mechanism which the decision-maker applies to assume the choice [15, 21].

The fuzzy rules are built through the directions of experts of the problem and reflect their behavior in those circumstances [4]. Basically, a fuzzy rule describes, in words, the rational but intuitive process that a person follows to produce the action to be taken, that is to reach a final decision starting from qualitative and quantitative information on the phenomenon and on the basis of similar experiences that he has already addressed [8, 18, 24].

In this paper, the stages of the construction of the logic fuzzy system are directly developed with reference to a case study of redevelopment of the Tor Fiscale district in Rome.

4   The Case Study

Located in the south-east quadrant of Rome, the Tor Fiscale district has a surface of about 29 hectares and a population of 2,174 inhabitants. The social structure is popular, and in recent years there has been a sharp increase in foreign nationals, mostly without permission. The area in analysis consists of a vast green zone of public property and an inhabited core. It is bounded by three parks, which are part of the broader Appia Antica Regional Park, characterized by numerous archaeological finds, among which tombs, the Tower of Fiscale, the remains of Roman villas of the imperial era, the remains of Claudio and Felice Roman aqueducts and the ancient route of the Latina Street, are to be reported.

The area is crossed by the Appia Nuova Street that, despite improving its accessibility, represents at the same time an element of discontinuity which is added to the railway Rome-Naples. In the north there is the Metro Station A of "Arco di Travertino". The urban structure looks disorganized and fragmented: small industrial and craft activities, shacks and illegal buildings, popular and luxury housing coexist. The commercial activities are almost entirely absent.

The district is characterized by widespread situations of degradation. In fact, as well as the lack of public services and functions, there are several illegal dumps that disfigure the green areas, whereas the two public houses, recently renovated by the Public Administration to be allocated to services for the Park, have been occupied by a community of immigrants.

Among the strengths of the district, as areas of action for the enhancement, there are the location near the city center, the good accessibility, the large presence of the green and the existence of archaeological remains.

4.1   Aims of the Public Administration

The objectives of the Pubic Administration are derived from the analysis of the existing planning instruments, in particular the Urban Plan of Rome (Art. 14, 52, 53, 55, 63 and 72 of the Technical Implementing Rules). In summary, the following objectives can be identified:

- planning aims: the increase of the services supply and the integration of the internal mobility, with particular attention to the connection with the subway;
- public space aims: the improvement of the security and the elimination of illegal building;
- social aims: the certain usability of public spaces; the encouragement of the integration among the inhabitants of the district; the increase of the supply of commercial and social district services; the improvement of   the attractiveness of the park;
- economic aims: the minimization of the public contribution to the initiative and the encouragement of the private involvement to finance public works;
- environmental aims: the limitation of the soil sealing;
- cultural aims: the promotion of the recovery and the restoration of the existing heritage with cultural activities;
- landscape aims: the enhancement of the existing naturalness and the overcoming of the ruptures determined by the Appia Nuova Street and the railway Rome - Naples.

Known the objectives, it is possible to identify the criteria (C_i) and the parameters (P_i) which can represent them in an effective manner. With reference to the case study of the present work, the most consistent criteria and parameters have been enucleated from the Matrix of Urban Quality[1] (Table 1).

---

[1] www.audis.it.

Table 1. Parameters of the Protocol considered

1. Planning quality

C01	Typology of the urban plant	P01. Morphological/urban design
		P02. Conformation of public space
C02	Mix of functions	P06. Distribution
C05	Internal mobility	P12. Distribution of parking in function of the incentive to "slow" mobility

3. Quality of Public Space

C14	Usability, accessibility and safety of the public space	P31. Usability, accessibility and safety of the public space

4. Social quality

C16	Composition and diversity of residential supply	P37. Accessibility to the residential supply
		P39. Presence of structures dedicated to specific classes of citizens
C18	Availability of services to individuals and families	P41. Public/private educational and cultural services
		P43. Centers of social aggregation
C19	Availability of district services	P44. Commercial services
		P46. Sport services
C20	Availability of urban services	P49. Urban services

5. Economic quality

C25	Sustainability of the initiative for the Municipal Administration	P61. Sustainability in the operating phase

6. Environmental quality

C27	Consideration and management of heat islands	P65. Relationship between permeable and impermeable materials

8. Cultural quality

C38	Number and quality of transformed/preserved items and their justification	P88. Type of the initiative
C39	Modalities of using the preserved heritage	P89. Conservation of the functions

9. Landscape quality

C40	Total perception of the landscape	P91. Consideration of existing landmarks
		P93. Construction/development of new landmarks
C41	Accessibility and visual enjoyment of the landscape	P95. Coverage/darkening of improper elements
		P96. Integration of the landscape (visual cones, etc.)

4.2   Project A: Maximization of the Development of Public Services

The project proposal A aims to increase the services and meeting places of the area. It involves the creation of a road that, tracing the route of the Latina Street, connects the park of the Latin Tombs with the Tower of Fiscale. The area will be characterized by low-rise buildings to not obstruct the view of the Tower of Fiscale. In the buildings closest to the park, cultural and leisure activities will be included, that integrate the characteristics of the neighboring environment. In areas facing the Appia Nuova Street the private executive buildings will be located.

4.3   Project B: Maximization of Building Development

To minimize the public contribution in the initiative, in the project proposal B the building capacity of the area is fully exploited by implementing the maximum building index provided by the Urban Plan of Rome. The reduction of land consumption is achieved with 8-10 storey residential buildings overlooking the park, with shops on the ground floor. A part of these buildings will be devoted to social housing. In peripheral areas multi-storey car parks and executive buildings will be located, which will also provide a barrier of protection against the noise originated by the intense traffic on the Appia Nuova Street.

4.4   Project C: Maximization of Naturalistic Development

The project proposal C aims to enhance the archaeological and natural character of the area. It provides these interventions: the construction of a museum, in order to host the most significant finds; the incentive and the relocation of polluting activities through the payment of compensation in development rights to be materialized in the productive areas; the dismantling of slums and the construction of small volumes; the construction of a green area with adjacent parking, to mitigate the noise of roads and railways.

5   The Fuzzy Model

The improvement of the procedure defined by the Urban quality Protocol through elements of fuzzy logic leads to the multi-criteria fuzzy evaluation model whose the phases are summarized below:

1. *division of the decision-making process in the parts that compose it.* The decision-making process can be summarized as follows: Public Administration states the objectives to be pursued with the redevelopment by defining the corresponding parameters; are fixed the weights of these parameters; the private operators illustrate how the proposed projects meet the targets set; Public Administration, through a panel of experts, values how each proposed project reaches the parameters of the objectives; Public Administration defines the priority of each project; the best project is identified.

2. *identification of input and output variables of the decision-making process.* The input variables of the process are represented by the weights that the Public Administration attributes to the parameters of each objective and by the ability of the solution in analysis to pursue those aims. It should be noted that the attribution of priority to the parameters of the objectives, provided in the proposed model, identifies an additional step compared to what is planned in the Protocol. It is hypothesized that the importance of each parameter is given by a score ranging from 1 (low importance) to 10 (maximum importance). In this way, the values of the first input variable (v_{in1}), which express the "importance of the parameters", are identified. These values are shown in the second column of Table 2.

On the basis of information provided by private developers, the experts (planners, estimators, sociologists, etc.) that operate in support of the Public Administration characterize the ability of each solution in analysis to pursue the parameters that make explicit the goals set by the Public Administration. This ability is expressed as a score between 0 (lack of achievement) and 10 (maximum achievement). In this way the values of the second input variable ($v_{in2,i}$) are defined, which indicates the "capacity of the solution in analysis to pursue the parameters". These values are shown in columns from three to five ofTable2.

On the basis of the importance attributed to the various parameters and the degree of achievement of the parameters, the next steps will define through the fuzzy logic system the output variable ($v_{out,i}$) that identifies the "priority of the solutions in analysis for each parameter ".

3. *definition of the membership functions.* The *set of definition*, the *shape* and the *labels* of the membership functions must be defined. In the present work, it has been assumed that, for the input variables, the set of definition coincides with the scale of points awarded to each of them; for the output variable, a score from 0 to 10 can be considered, that detects an output value, as the input value is defined in linguistic terms in the fuzzy rules defined in the next step. The shape is assumed triangular type, simple to treat analytically but capable to offer a good approximation for the representation of the phenomenon.

Table 2. Inputs and outputs of the model

Parameter	v_{in1}	$v_{in2,A}$	$v_{in2,B}$	$v_{in2,C}$	$v_{out,A}$	$v_{out,B}$	$v_{out,C}$
P01	8	9	7	4	8.8	7.5	6.5
P02	9	6	5	4	7.1	7	6.5
P06	9	9	7	4	9.4	7.5	6.5
P12	4	4	7	10	4.1	5.2	6.5
P31	3	4	5	9	3.5	3.8	5.8
P37	7	6	8	4	6.4	7.5	5.4
P39	8	6	8	3	7.1	8.8	5.9
P41	9	8	7	5	8.8	7.5	7
P43	9	8	4	6	8.8	6.5	7.1

Table 2. (*Continued*)

Parameter	v_{in1}	$v_{in2,A}$	$v_{in2,B}$	$v_{in2,C}$	$v_{out,A}$	$v_{out,B}$	$v_{out,C}$
P44	5	5	9	1	5	7	3
P46	4	4	9	2	4.1	6.4	2.9
P49	8	2	4	9	5	6.5	8.8
P61	9	4	7	2	6.5	7.5	5
P65	5	5	3	10	5	3.9	7
P88	9	8	4	7	8.8	6.5	7.5
P89	6	6	5	4	5.9	5.5	5
P91	7	7	4	9	6.6	5.4	7.5
P93	3	8	7	10	5.8	5	5.8
P95	4	3	7	2	3.5	5.2	2.9
P96	7	10	3	7	7.5	5	6.6
				$v_{out,med}$	6.4	6.3	6

The labels (e) are three in number (low = B, medium = M, high = A) for the input variables and five (low = B, moderately low = MB, medium = M, moderately high = MA, high = A) for the output variable, in order to ensure a better level of detail for the expression of the judgment of the decision maker. In Figures 1, 2 and 3 the membership functions of the input and output variables of the model are plotted and the respective formal expressions are summarized.

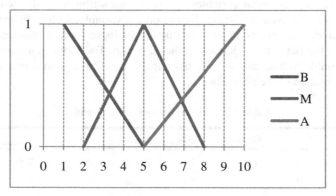

Fig. 1. Membership function of the first input variable

$$f_{in1,B} = \frac{5-v_{in1}}{4} \text{ for } 1 \leq v_{in1} \leq 5$$

$$f_{in1,M} = \frac{v_{in1}-2}{3} \text{ for } 2 \leq v_{in1} \leq 5$$

$$f_{in1,M} = \frac{8-v_{in1}}{3} \text{ for } 5 \leq v_{in1} \leq 8$$

$$f_{in1,A} = \frac{v_{in1}-5}{5} \text{ for } 5 \leq v_{in1} \leq 10$$

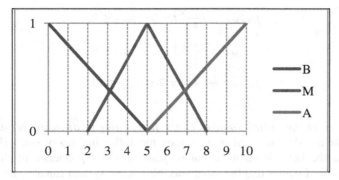

Fig. 2. Membership function of the second input variable

$$f_{in2,B} = \frac{5-v_{in2}}{5} \text{ for } 0 \leq v_{in2} \leq 5$$

$$f_{in2,M} = \frac{v_{in2}-2}{3} \text{ for } 2 \leq v_{in2} \leq 5$$

$$f_{in2,M} = \frac{8-v_{in2}}{3} \text{ for } 5 \leq v_{in2} \leq 8$$

$$f_{in2,A} = \frac{v_{in2}-5}{5} \text{ for } 5 \leq v_{in2} \leq 10$$

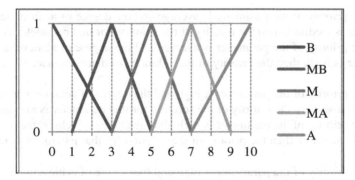

Fig.3. Membership function of the output variable

$$f_{out,B} = \frac{3-v_{out}}{3} \text{ for } 0 \leq v_{out} \leq 3$$

$$f_{out,MB} = \frac{v_{out}-1}{3} \text{ for } 1 \leq v_{out} \leq 3$$

$$f_{out,MB} = \frac{5-v_{out}}{2} \text{ for } 3 \leq v_{out} \leq 5$$

$$f_{out,M} = \frac{v_{out}-3}{2} \text{ for } 3 \leq v_{out} \leq 5$$

$$f_{out,M} = \frac{7-v_{out}}{2} \text{ for } 5 \leq v_{out} \leq 7$$

$$f_{out,MA} = \frac{v_{out}-5}{2} \text{ for } 5 \leq v_{out} \leq 7$$

$$f_{out,MA} = \frac{9-v_{out}}{2} \text{ for } 7 \leq v_{out} \leq 9$$

$$f_{out,A} = \frac{v_{out}-7}{3} \text{ for } 7 \leq v_{out} \leq 10$$

4. *clarification by the decision-maker of the fuzzy rules in linguistic terms.* It is assumed that the decision-maker expresses the following fuzzy rules (R_i), which summarize the behavior that he put in place in correspondence of the possible combinations of values that the input variables of the system can assume [18]:

R_1: if the priority of the parameter is low and the degree of achievement of the parameter is low, then the priority of the solution for that parameter is low;

R_2: if the priority of the parameter is low and the degree of achievement of the parameter is medium, then the priority of the solution for that parameter is moderately low ;

R_3: if the priority of the parameter is low and the degree of achievement of the parameter is high, then the priority of the solution for that parameter is average;

R_4: if the priority of the parameter is average and the degree of achievement of the parameter is low, then the priority of the solution for that parameter is moderately low;

R_5: if the priority of the parameter is average and the degree of achievement of the parameter is medium, then the priority of the solution for that parameter is average;

R_6: if the priority of the parameter is average and the degree of achievement of the parameter is high, then the priority of the solution for that parameter is moderately high;

R_7: if the priority of the parameter is high and the degree of achievement of the parameter is low, then the priority of the solution for that parameter is average;

R_8: if the priority of the parameter is high and the degree of achievement of the parameter is medium, then the priority of the solution for that parameter is moderately high;

R_9: if the priority of the parameter is high and the degree of achievement of the parameter is high, then the priority of the solution for that parameter is high.

The encoding of these rules is shown in Table 3 .

5. *fuzzification of the data of the input variables.* The fuzzification takes place through the membership functions of the input variables of the model [14, 6]. As the membership functions are overlapped, most membership functions, in general, correspond to each value of the input variable. In the case study, the overlap occurs at most between two membership functions and therefore two degrees of membership for each input value can be "activated" at most. The results of the fuzzification of the values of the input variables are shown in Tables 4 and 5.

Table 3. Fuzzy rules

Code	e_{in1}	e_{in2}	e_{out}
R_1	B	B	B
R_2	B	M	MB
R_3	B	A	M
R_4	M	B	MB
R_5	M	M	M
R_6	M	A	MA
R_7	A	B	M
R_8	A	M	MA
R_9	A	A	A

Table 4. Fuzzification of the scores of the first input variable

v_{in1}	f_{in1}		
	B	M	A
1	1	0	0
2	0.75	0	0
3	0.50	0.33	0
4	0.25	0.67	0
5	0	1	0
6	0	0.67	0.20
7	0	0.33	0.40
8	0	0	0.60
9	0	0	0.80
10	0	0	1

6 *inference and composition.* In the inference phase the "approximate" reasoning coded in the fuzzy rules is implemented. The values of the output variable are associated with the corresponding pair of fuzzy values obtained with the fuzzifica-tion, using the mechanisms summarized in the fuzzy rules activated by the pair of fuzzy values. For example: the priority to be attributed to the project B for the parameter "P12. Distribution of parking in function of the incentive to the slow mobility" is to be determined; the input values of the two input variables are, respec-tively, 4 and 7. The fuzzification of the score 4 of the degree of preference actives (see Table 4) the fuzzy set corresponding to the label "low", with degree of mem-bership 0.25, and the label "average" with degree of membership 0.67, whereas

Table 5. Fuzzification of the scores of the second input variable

v_{in2}	f_{in2}		
	B	M	A
0	1	0	0
1	0.80	0	0
2	0.60	0	0
3	0.40	0.33	0
4	0.20	0.67	0
5	0	1	0
6	0	0.67	0.20
7	0	0.33	0.40
8	0	0	0.60
9	0	0	0.80
10	0	0	1

the fuzzification of the score 7 of the degree of achievement actives at the same time (see Table 5) the fuzzy sets corresponding to the label "average", with degree of membership 0.33, and the label "high", with degree of activation 0.4. Thus the linguistic rules are: $R_2: BM \rightarrow MB$, $R_3: BA \rightarrow M$, $R_5: MM \rightarrow M$ and $R_6: MA \rightarrow MA$.

Identified the active rules, the logical operator to be applied should be chosen, in order to obtain the degree of membership of the output fuzzy set starting from the degrees of membership of the input sets. For the inference, the logical operator MIN is here applied, which identifies, as degree of membership of the output set, the smallest degree of membership of the input fuzzy sets. It thus ensures that, in the determination of the degree of activation of the output set, the input variables of the rule that is being implemented act at the same time, without neglecting the variable associated with the smallest value of the degree of membership [23].

The composition has the task of combining fuzzy sets that result from the fuzzification into a single fuzzy set. In this work, the composition is carried out through the logical operator MAX. The set that results associates the maximum value of the fuzzy sets obtained by the inference with each point of the output values of the output variable, so that, even if a rule has a marginal influence on the action to be taken , may also exerts its effect [22].

6. *defuzzification*. The defuzzification has the purpose of obtaining the representative value of the set obtained with the composition [11, 12]. In fact, the fuzzy shape obtained from the composition is not useful for practical purposes and must be converted into a deterministic value. The operator here chosen for defuzzification is the Center of Gravity (COG), which identifies as output the abscissa of the center of gravity of the compound set. Compared to other operators, the use of

COG ensures greater "sensitivity" of the model, because it considers the contribution of all the output variables [5]. Given the possible j values of the variable $v_{out,j}$, the output value relative to the i-th parameter will be:

$$v_{out,i} = \frac{\Sigma_j v_{out,j} \cdot f_{out,j}}{\Sigma_j f_{out,j}} \tag{1}$$

The repeated application of the fuzzy logic system for all parameters of the project leads to the priorities of the solutions ($v_{out,i}$), contained in columns from six to eight of Table 2.

7. *identification of the best project*. The priorities of the projects (six to eight columns of Table 2) show that there isn't a solution with an absolute predominance on the others, that is a project having at the same time the priorities in all the components. In these situations, the best project is the solution that allows the best compromise between the priorities of all the components in analysis [19].
Considering that each of the six to eight columns of Table 2 identifies the priorities of a project, the compromise solution can be found by determining, as summary index of the goodness of the solution itself, the arithmetic mean ($v_{out,med}$) of the scores generated by the defuzzification (last line of Table 2):

$$v_{out,med} = media(v_{out,i}) \tag{2}$$

The result obtained allows to order the solutions and to identify the project capable to pursue the highest level of the objectives, "weighted" with the importance assigned by the Public Administration to the corresponding parameters. In this case study the dominant solution is the project A, with $v_{out,med} = 6.4$.

6   Conclusions

The rationalization of the evaluation model developed by AUDIS in the "Protocol of urban quality of the capital city of Rome" through the fuzzy logic highlights many benefits that arise from the prerogative and the structure of fuzzy logic systems.

The graphical representation of the membership functions reduces the vagueness and the ambiguities related to the qualitative judgments and provides a baseline for the phases in which the comparison between the experts that support the decisions to be taken by the Public Administration is performed.

The clarification of the valuation criteria of the decision-maker through logical rules, which are real codes of behavior expressed in terms of natural language, allows the immediate understanding of the criteria themselves by all the partners involved in the decision-making process. The retraceable process of the model legitimates the decisions of the Public Administration and serves as mechanism for verifying the consistency of the results with respect to the objectives of the initiative.

The structure of the model is flexible, able to adapt to the specificities of the case study and to the changes that - over time - could arise in the evaluation mechanism of the decision-maker. The model can be adjusted to the changed conditions which may occur, modifying the logical rules through the addition or deletion of linguistic variables, the update of the form and the values of the membership functions and the increase/decrease in the number of "labels" of the states of the system. In this way, the model can also be adapted to the degree of approximation related to the phases of planning and to the specificities of the territorial area to be redeveloped.

References

1. Audis: Carta Audis della rigenerazione urbana (2008)
2. Audis: Il Protocollo della qualità di Roma Capitale. Definire e valutare la qualità dei progetti urbani complessi. Roma Capitale, Risorse per Roma (2012)
3. Audis: Monitoraggio della rigenerazione urbana attraverso indicatori condivisi. Ricerca AUDIS per la Regione Emilia Romagna (2010)
4. Bagnoli, C., Smith, H.C.: The Theory of Fuzz Logic and its Application to Real Estate Valuation. Journal of Real Estate Research 16(2) (1998)
5. Bai, Y., Wang, D.: Fundamentals of Fuzzy Logic Control – Fuzzy Sets, Fuzzy Rules and Defuzzifications. In: Advanced Fuzzy Logic Technologies in Industrial Applications. Springer (2006)
6. Bandemer, H., Siegried, G.: Fuzzy sets, fuzzy logic, fuzzy methods: With applications. Chichester. Whiley (1992)
7. Cammarata, S.: Sistemi a logica fuzzy. Come rendere intelligenti le macchine. ETAS Libri, Milano (1997)
8. Chen, S., Hwuang, C.L.: Fuzzy multiple attribute decision making: Methods and applications. Lecture notes in Economics and mathematical systems. Springen, Berlin (1992)
9. De Mare, G., Nesticò, A., Tajani, F.: Building investments for the revitalization of the territory: A multisectoral model of economic analysis. In: Murgante, B., Misra, S., Carlini, M., Torre, C.M., Nguyen, H.-Q., Taniar, D., Apduhan, B.O., Gervasi, O. (eds.) ICCSA 2013, Part III. LNCS, vol. 7973, pp. 493–508. Springer, Heidelberg (2013)
10. Dubois, D., Prade, H.: Fuzzy sets and systems: Theory and applications. Academic Press, New York (1980)
11. Filev, D., Yager, R.: A generalized defuzzification method under BAD distributions. Int. J. Intell. Syst. (1991)
12. Jiang, H., Eastman, J.R.: Application of fuzzy measures in multi-criteria evaluation in GIS. International Journal of Geographical Information Science (2010)
13. Kaufmann, A.: Theory of fuzzy subset, vol. I. Academic Press, New York (1975)
14. Klir, G., Folger, T.: Fuzzy sets, uncertainty, and information. Prentice Hall, Englewood, Cliffs (1988)
15. Kosko, B.: Fuzzy Thinking: The New Science of Fuzzy Logic. Hyperion (1993)
16. Morano, P., Tajani, F.: Break even analysis for the financial verification of urban regeneration projects. Applied Mechanics and Materials 438-439, 1830–1835 (2013)
17. Morano, P., Tajani, F.: The transfer of development rights for the regeneration of brownfield sites. Applied Mechanics and Materials 409-410, 971–978 (2013)
18. Munda, G.: Multicriteria evaluation in a fuzzy environment. Physica (1997)
19. Orlowski, S.D., Kacprzyk, J.: Optimization model using fuzzy sets and possibility theory (1987)
20. Ross, T.J.: Fuzzy Logic with Engineering Applications, 2nd edn. John Wiley & Sons, Ltd. (2004)

21. Seo, F., Sakawa, M.: Multiple criteria decision analysis in regional planning – Concepts, methods and application. D. Reidel publishing company (1996)
22. Terano, T., Asai, K., Sugeno, M.: Fuzzy Systems Theory and its Applications. Academic Press Inc., San Diego (1992)
23. Tong, R.M., Bonnisone, P.P.: A linguistic approach to decision-making with Fuzzy Sets. IEEE Transaction on System, Man and Cybernetics smc-10(11) (1980)
24. Veronesi, M., Visioli, A.: Logica Fuzzy. Fondamenti teorici e applicazioni pratiche. Franco Angeli Editore, Milano (2003)
25. Zadeh, L.A.: Calculus of fuzzy Restriction-Fuzzy sets and their application to cognitive and decision process. Academic Press, New York (1975)

Architectural Heritage Restoration Management:
The Case of the Borgo of Cassibile (Italy)

Salvatore Giuffrida[1], Angelita Bellissimo[2], and Pietro Copani[3]

[1] Department of Civil and Environmental Engineering and Architecture - University of Catania
sgiuffrida@dica.unict.it
[2] bellissimoangelita@libero.it
[3] Ministero Beni Culturali Ambientali Regione Puglia
pietro.copani@beniculturali.it

Abstract. Recovery and renovation are two opposite management approaches to minor cultural heritage, when the lack of public funding exposes it to risk of abandonment and ruin. Referring to an architectural complex located in a minor urban centre of the Syracuse Province, this paper provides a project management valuation pattern, based on parceling the buildings in standard elementary units suitable to be handled by the means of a digital appraisal cost model. Successively, a business plan model, based on the same database, is applied in order to assess financial sustainability and economic convenience for the investments involving the different buildings and their different functions. This model contains the analysis, referring to each single component, work, and project choice and facilitates the maximization of the ratio economic and architectural qualities.

Keywords: architectural heritage economy restoration/recovery numeric modelling financial sustainability economic convenience scenario analysis.

1 Introduction

1.1 Economy and Architecture in Cultural Property Heritage

Economy and architecture have been for a long time separate and opposed research areas although, on the one hand architectural activity designs and handles capital assets and, on the other, social-economic surplus privileges the long run stocks – architecture, indeed – both in the private (real estate) and in the public (social and infrastructural capital) forms.

Cultural heritage can be considered the best investment in terms of social capital, but some distinctions have to be made in order to address restoration and its different approaches. Restoration involves both monumental and minor constructions, public and private goods, urban and land locations.

The case of minor cultural heritage, in peripheral locations and owned privately, like the present one, involves public values and choices as well as major, urban and public monuments. However, in this case social values must be distinguished in order to instruct and integrate both private and public points of view. Private investments

B. Murgante et al. (Eds.): ICCSA 2014, Part III, LNCS 8581, pp. 176–191, 2014.

and public boosting, contribute to enhance both kinds of capital in terms of income and stock, so that a general method carries out measuring costs, prices and values in order to highlight efficiency, effectiveness and fairness of the renovation of the architectural complex.

The general restoration approach does not consider economic concerns as well as recovery and renovation ones that, on the contrary, assume functional and economic features as conditions of the intervention feasibility. Therefore, restoration considers the economic features as a financial constraint, whereas renovation and recovery assume it as condition and opportunity. Monumental goods must be restored if funding is available; minor architectural goods could be recovered if doing it is convenient. In the first case material, as a vehicle of the architectural value, is assumed as a target, so that money is a constraint; in the second, money is the goal, whereas material is a constraint [17].

1.2   Cultural Heritage as Money Goods

A new and different approach has been addressed by F. Rizzo [18], who defines cultural-environmental heritage as money goods: in this way, it works like money being required by itself. In fact, as well as money, it measures, saves and increases the value of wealth of the cultural developed civilizations; oppositely, its abandon measures, wastes and decreases the richness of a civilization announcing its next decaying. The social-politic willingness to pay for – and, above all, to invest in – cultural heritage is the civilization index of a social (national or local) community: different approach have been carried out in order to address and apply these concepts.

A general approach to the government of the conservation of cultural heritage is due to M. Finocchiaro et al. [7] about the efficiency of the conservation activity by the analysis of the performance of the Regional (Sicilian) Heritage Authority; a more specific approach has been carried out by E. D. Massimo et al. [14], proposing a parametric cost standardization approach basing on GIS tools in urban public and private property heritage; a detailed study has been carried out by B. Manganelli [12] who proposes a valuation and decision model of single investment basing on the comparison of different utilities and on the restoration cost.

2   Materials

2.1   The Ancient Borgo of Cassibile

Cassibile is an ancient village about 14 km in the south of Syracuse, along the SS 114 near Manghisi River. It marks the boundary between Siracusa and Avola. It is surrounded by several substantial touristic centers. Human settlements have existed since the Bronze Age. There is significant evidence that an extended cemetery near Cassibile existed.

For several centuries, the Cassibile property was owned by many important Sicilian families (Arezzo, Speciale). Currently, the Loffredo family owns this property. Between 1840 and 1870 the Marquise of Cassibile built a workers' village known as "Borgo di Cassibile" [11].

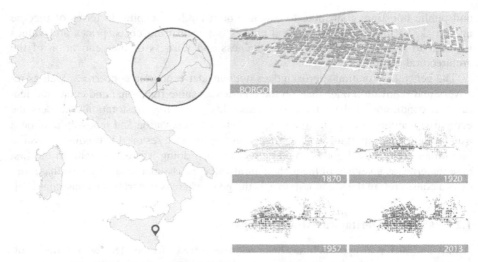

Fig. 1. The Borgo of Cassibile: Location and historical development

Nowadays the owner is Silvestro Ferdinando Gutkowski Pulejo Loffredo, whose properties extend from the Cassibile River to the mountainous area of Noto.

During the second world war the 'short armistice' was signed in Cassibile in 1943; the consequence was that the hostility ended between Italy and its allies [13].

The main structure of the village is composed of low-rise houses in two parallel lines, whose facades are characterized by the regular succession of doors and windows framed into doorposts and lintel. These are made of typical white local stone, and light pink plaster, made of the traditional mixture of lime and pounded bricks. The seriality of the architectural complex has been helpful to the standardization of the costs calculation model.

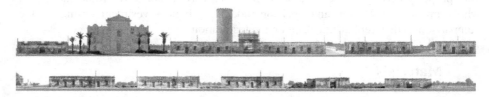

Fig. 2. Facades of the Borgo of Cassibile

The first segment of the architectural complex is a wide square. In the background, stands the vertical facade of the church dedicated to the Sacred Family. The village was built by Silvestro Loffredo for his workers and there was a succession of houses near their work site. Nowadays, the big farm is well-known for the production of almonds livestock and dairy products [1].

Fig. 3. Perception of the village in the current condition

The strict house type consists of single cell, which are well-lighted by a door and a window. Sometimes, the house includes a compartment in the back, for the kitchen which has an oven. In some cases, there is a second entrance in the back. In the original scheme the village also had lodgings and barracks. According to this analysis, the plan supposes to preserve the typological layout of the original masonries, and to locate within each building flexible and interchangeable uses, as more as possible. As a consequence, the economic and financial structure has changed as well.

In the main street there is a great neoclassical gate where "Via dei Campi" starts from, connecting the village and the marquise's mansion. The Borgo shows the owner's entrepreneurial attitudes. The village was a community for farm workers and a temporary accommodation as well.

Currently, the village is collapsing since it lost its former role and identity because of the rapid development of a new town, nearby. Some of its blocks are used as warehouses, some others as non-EU citizens' houses. An organic-food shop for the marquises' farm products, is established within one of the blocks facing the square.

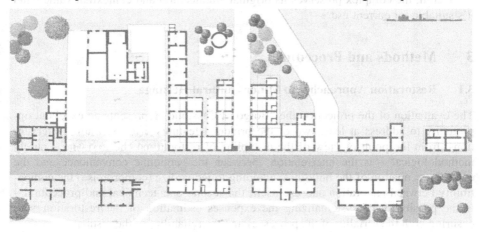

Fig. 4. Plan of the ancient village in its current condition

2.2   Buildings Characteristics and Conditions

Quite similar features characterize the blocks. Their geometric shape and profiles display their original simple function. The dwellings are at ground floor. The units are made of a standard cell, whose area is always about 18 sq. m. Each unit has a door and a window.   The village consists of 10 blocks and a church. Only three buildings are ruined. Two of them, were the original lodgings and the barracks, which were two floor buildings, possibly belonging to a different period. The study of the technical features of the buildings, has allowed us to identify six types of stonework.

The roof is formed by inclined surfaces arranged in simple horizontal warping made of fir-tree beams, sealed by typical Sicilian shingles, laid over the traditional diagonal sticks insulated surface. Instead of the gutter, the facades are protected from rain water by the characteristic "cappuccina". It's a protrusion made of shingles on the top of the outer wall. The shell is composed of an inclined roof from a median ridge beam.

The general conditions demonstrate the state of negligence of most part of the architectural complex. However, in the ruined ones, the ground floor planking levels are still present in all the buildings, but there isn't any flooring.

The wooden floors of the two-story buildings are inexistent, nevertheless some traces of them can be observed on the top of the walls. They were made of a simple warping of beams and a continuous surface of planks, on which the flooring was laid. There aren't any traces of staircases. The original shingle completely collapsed in the ruined buildings, whereas in the blocks they are still intact. The plasters and frames are inexistent.

Today, the static conditions are quite critical, only in reference to the two-story buildings. The shingle and floors have collapsed and the walls are full of cracks and some local collapses are noticeable. The one-story buildings are almost intact from a structural point of view, no cracks are visible and only general renovation is necessary.

Overall, the complex preserves its original architectural and contextual value   and it's suitable for current use.

3   Methods and Procedure

3.1   Restoration Approaches to Minor Cultural Heritage

The evaluation of the project on the "Borgo of Cassibile" represents an excellent opportunity to address at least two of the principal features of conservation interventions, from the point of view of the discipline of restoration. The first one – purely methodological – is the interrelation   between the economic convenience and the end-use/occupation of the buildings, avoiding to jeopardize (compromise) the compatibility between restoration and use itself; the second one–technical and procedural – is the possibility of systematizing the expenses estimation of the restoration yard ensuring both the seriality of the processes and the reliability of the results.

These two themes are intriguing and they have always marked the main problems of the restoration discipline, but they have different path of historiography. The first one regards one of the basic requirements to ensure the conservation of cultural heritage and it is recognized as such since the early experiences of theorization of restoration, starting with the Athens Charter for the Restoration of Historic Monuments (1931). In fact, the Charter "recommends that the occupation of buildings, which ensures the continuity of their life, should be maintained but that they should be used for a purpose which respects their historic or artistic character", [16] and the same requirement has been confirmed in all the principal international documents about Restoration .

The second one has been always debated by scholars and theorists of the discipline, due to the extreme complexity of all processes within the restoration yards, including the economic ones.

In methodological terms, the relationship between restoration and economic disciplines involves several issues related to the aims of the restoration itself, which often causes "uneconomic" choices by definition. Nevertheless, it is easy to show that in many cases a choice that seems uneconomic during one of the early stages of the project or the yard can become convenient in the medium or long term. Indeed, as we saw in the introduction, the advantages vary from savings in maintenance costs, increase of the value, and indirect spin-off on the society. Moreover, if the analysis includes the cultural aspects of the intervention (which often overlap or include the socio-economic impact), it is much easier to get a positive balance by restoring a property that belongs to the national architectural heritage.

This assumption corresponds to the possibility that a conservative restoration can be a convenient choice from a strictly economic point of view, and this is one of the basic elements used in the analysis of the value of the Borgo of Cassibile. However, in our instance the conservative approach can't be replaced with other ones, because the group of buildings is protected by Italian law since it is part of the national cultural heritage. Therefore, works such as demolition of significant parts or the alteration of the original architectural characteristics are not permitted.

Considering the above, the inclusion of the restoration principles within the economic evaluations can be seen as a real challenge. Transformation costs must be calculated on the basis of project choices that are conformed to the principles of conservation and that regard buildings that already have an intrinsic value as cultural goods.

To respect the conservative approach is a main aim of any restoration intervention, but in the case of analysis of the Borgo it takes further meanings, since this approach should be included in a comprehensive evaluation analysis and, above all, its principles have to be respected during the stage of costs estimation, by using modeling that include the standardization of processes, at least partially.

Therefore, it is useful to take into account, among the potential occupations to be included in the evaluation, only those ones that are considered compatible with the requirements of the good's conservation. From this point of view, the Borgo has significant potential, firstly because it was built to accommodate different but complementary functions: although the plan was essentially productive, linked to the exploitation of agricultural lands of the marquisate, in the Borgo there are homes, warehouses and special buildings for commercial activities and even a church. Moreover, this is not the only feature that makes the Borgo highly flexible and adaptable to accommodate different functions: the nature of the individual buildings makes them ductile, as they have regular shapes and almost all of them have only one floor above the ground.

It is not a coincidence that the chosen occupation for the village – the most convenient one – is multi-functional and includes various activities (housing, offices, commercial and service activities), none of which contemplates destructive works or incompatible with the preservation of the property.

In technical and procedural terms, the restoration of the Borgo suggests the usual issue related to the process of standardization of the rules within a whole program of recovery. This topic is often debated by scholars of the restoration of historic centers, because it regards the contrast between the conservation requirements for a specific single object and the need of ensuring the minimum quality requirements in large-scale programs. On the one hand, it is quite obvious that the operations of restoration must always adapt to the unique characteristics of each individual good, in order to achieve the best outcome in any conditions of conservation of the same good; on the other hand, to provide common rules for large-scale interventions means - inevitably – that the same rules have to be necessarily approximated, as minimum as possible. In our case, the common rules are works that correspond to the calculation processes, single or grouped, which transform the good from the initial state to the standards required to host the chosen occupation, in order to calculate the costs of transformation.

This methodology becomes more complex during the stage of checking of each building: first of all, because the initial state of the preservation of the buildings has to be considered, also because it determines considerably different restoration costs as well as increases of up to 1500%. However, it is possible to extend the systematization of the individual cases (each type of intervention consists of 5 classes, from best to worst conditions) to the whole panorama of the restoration that has to be performed: also in this sense, the seriality of the buildings of the Borgo has helped, both because of the building types said above, both because these latter - built with homogeneous materials and in the same period - have similar conservation issues, so they can be easily considered as standardized. For instance, the fact that none of the buildings has an appropriate insulation of the earth-floor, has forced to provide for a ventilated crawl space, with the same characteristics in all the buildings, removing the existing floor (not particularly valuable) while installing a new one, that has to be suitable to the characteristics of the nineteenth century buildings.

The comprehensive catalogue of the potential interventions was elaborated by the survey of some building samples; it has provided a schedule that can be adapted to almost all of the buildings of the Borgo. As noted above, thanks to the seriality of the buildings very good result have been reached, despite premises were problematic. But the same method has enormous potential in other areas of the restoration, with the necessary modifications and limitations that would be considered as acceptable. For instance, it can be used in the analysis of historical city centers as a whole, which usually have serial aspects in constructive features and also architectural ones; again, most of the historical centers have been built and transformed according to simple rules of construction shared by an entire community; so these centers, since their formation, are ideal areas to apply serial processes, both in analysis and project.

The supposed possibility of extending these studies to a certain number of cases, such as in the Borgo of Cassibile or in many historical centers of European or Mediterranean area, makes the evaluation method a very versatile tool for economic analysis, especially in large scale projects. Nevertheless, in the case of analysis of a cultural good, the extension of the range of the method corresponds to the increase of the variables related to the preservation of the same good. For this reason, the analysis of the value of a cultural good must always start from the assumption that culture itself is value, and it can be increased through the preservation of its features, with the support of evaluations of both material and non-material goods.

3.2   The analysis, Valuation and Planning Pattern

The purpose to harmonize the planning restoration process and entrepreneurial approach stimulates a general methodology suitable to adapt ends to means, and to verify the correspondence between results and aims.

According to the previous theoretical premises, we assume material as a value in itself and the assurance of the transmission of the heritage information message, so that interventions aim to preserve the traditional material values and to integrate contemporary functions into the original "genius loci" [2].

The general methodological scheme is synthesized in fig. 5.

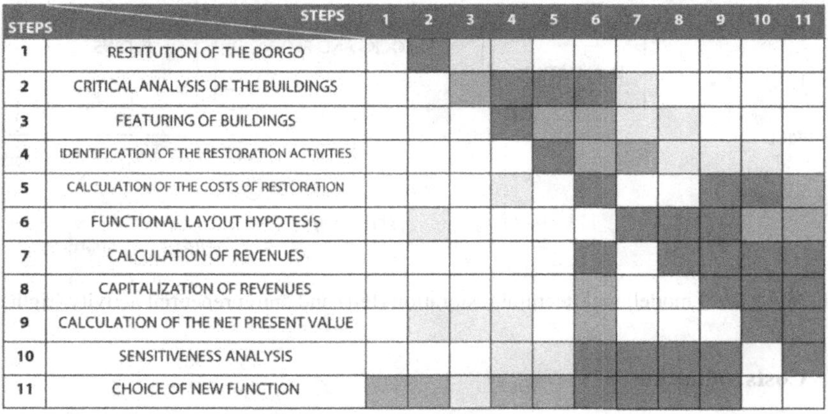

STEPS	STEPS	1	2	3	4	5	6	7	8	9	10	11
1	RESTITUTION OF THE BORGO		■									
2	CRITICAL ANALYSIS OF THE BUILDINGS		■	■	■							
3	FEATURING OF BUILDINGS			■								
4	IDENTIFICATION OF THE RESTORATION ACTIVITIES				■		■					
5	CALCULATION OF THE COSTS OF RESTORATION					■		■				
6	FUNCTIONAL LAYOUT HYPOTESIS						■			■		
7	CALCULATION OF REVENUES						■	■				
8	CAPITALIZATION OF REVENUES								■			
9	CALCULATION OF THE NET PRESENT VALUE							■		■		
10	SENSITIVENESS ANALYSIS						■				■	
11	CHOICE OF NEW FUNCTION	■				■						

Fig. 5. General methodological scheme

The main point of this study is the current value of the Borgo according to the restoration constraint we assumed. The proposed model aims at appraising the capital value of the heritage real estate according to the different plan options. A more transformative option could increase the economic convenience of the investments reducing the historical cultural value and vice-versa. Therefore, the options must be contained within the typological and material conservation (the constraint) so that the most convenient of them generates the highest current value of the architectural complex as a whole. The model uses some business accounting tools based on the cash flow approach and sensitiveness analysis. Therefore, it verifies the economic performances of the supposed destinations, and the consistency with the restoration instances as claimed by the specific characteristics of the buildings. According to the well-known appraisal extraction method, the current value V_i of each ith building of the Borgo is:

$$V_i = \frac{\frac{I_n}{r} - C(1+wacc)^m(1+r\prime)^n}{(1+wacc)^m(1+r\prime)^n} \tag{1}$$

where: I_n is the Net Operating Income (NOI) of each building, the difference between the gross income I_g and the managing expenses, r is the cap rate, *wacc* is the *weighed average cost of capital*, $r\prime$ is the business risk premium rate, m the short loan life and n the transformation period. Many other variables, concerning the long term loans are included into the model as later shown. Therefore, the proposed pattern integrates two sections, the first aimed at costs, the second at revenues (Fig. 6).

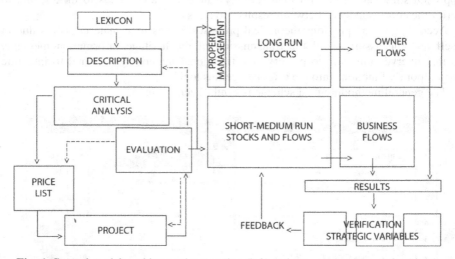

Fig. 6. General model: architectural restoration (left) and entrepreneurial activity (right)

3.3   Costs valuation

The cost valuation's main tool is a matrix in which the elementary building components are listed as rows and the elementary renovation and integration activities are listed as columns. The description of the buildings involves some different elements:

1. the list of the components, so that they can be inserted as standard types; the elementary components are: compartments, walls, openings; roofs, planking levels;
2. the topologic id. numbers that indicate the collocation of each element within the major one (e.g. an opening within a wall, and the wall within a compartment):
3. dimensions: to each element is associated the k dimensions required by the unit by which the price of the possible works are expressed; unit prices are assumed from specialized price lists [4]; each type of dimension is defined as a vector $\overrightarrow{d_k}$;

4. qualities β_i: provide standardized information about materials and techniques of each element;
5. status γ_i: standard types of maintenance status are listed and associated to each element.
6. The prescription of the restoration activities includes:
7. the list of the works w_j, selected by the type of element, its quality and condition $(w_j = f(\beta_i, \gamma_i))$;
8. their unit prices p_j, each of which selects its specific $\overrightarrow{d_{kj}}$ vector;
9. the quantities q_j of every components affected by each work, given as $q_j = \overrightarrow{d_{kj}} \cdot \overrightarrow{a_j}$, as next explained;
10. the resulting partial costs of each work, $c_j = p_j \cdot q_j$;
11. as for the result, the total cost of the project $C = \sum_j c_j$.

A matrix handles elements (rows) and works (columns). The generic element a_{ij} contains the percentage of the ith element interested by the jth work, so that a_{ij} "turns on" a certain quota of the jth intervention supposed for the specific ith building component (e.g, the partial reconstruction of a component). For each work the vector $\overrightarrow{a_j}$ (column) is defined as the list of the percentage of the $\overrightarrow{d_{kj}}$ elements involved in the jth work (Fig. 7).

The database allows totaling the different works by elements in order to assess the quota of the cost of the different elements or groups of elements referred to each building and work or groups of them [6]. Therefore, some parametric costs have been calculated for each maintenance status and for each element, in order to drawing the relationship between an aggregate maintenance status index and an aggregate cost index, in order to extend the specific valuation to every similar building [15].

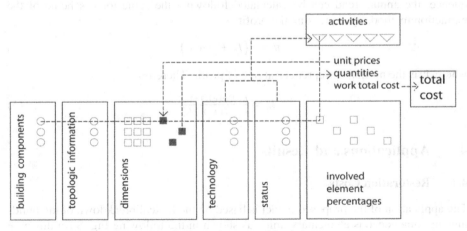

Fig. 7. General cost model scheme

3.4   Revenues Valuation

The revenues of a real estate investment depend on the market value of the renovated buildings. The Borgo can't be considered ordinary real estate whose value can be elicited from the local market price, because its value is strictly connected with the specific purposes and function of the buildings and their relationships. Therefore a market comparison approach can't be assumed because of the inactivity of the market in a segment not actually defined. Consequently, the market value should be calculated by applying the income method. The gross income is the rent that the tenants and/or the firms pay for the productive use of the buildings, so that an analytic business plan approach has been carried out in order to calculate the cash flow of the different supposed economic activities [19]. The cash flow analysis' main point is to turn each stock value into annual amounts so that they can be compared with the typical balance sheet amounts. By doing this it is possible to integrate appraisal and business approach as a solid foundation for the restoration aims. Each stock amount C is turned in annual amount y by the formula:

$$y = C \frac{r''q^n}{q^n - 1} \tag{2}$$

where r" is the interest rate, depending on n, that is the duration of the considered stock, $q = (1 + r)$. The annual gross income Ig is:

$$I_g = R - y - e - \pi \tag{3}$$

where R are the annual revenues, y the depreciation rates, e the annual expenses, π the entrepreneurial profit. The last expression shows the conflict between rent and profit and more generally between real estate owners and entrepreneurs. As consequence, the annual rend can be calculated following the same logic scheme of the extraction method, by expressing the profit as:

$$\pi = \bar{r}(I_g + y + e) \tag{4}$$

where $\bar{r}$ is the normal entrepreneurial rate of profit. Therefore:

$$I_g = \frac{R - (y + e)(1 + \bar{r})}{(1 + \bar{r})} \tag{5}$$

4   Applications and Results

4.1   Restoration Costs

The application of the proposed model is based on the logical breakdown of the buildings in homogeneous elementary units, as shown in the following Fig. 8 (in this case the total number is 1014).

From the analytic calculation of the cost of the three most representative buildings some parametric costs for each status degree have been deducted for each element restoration as shown in the following Tab. 1.

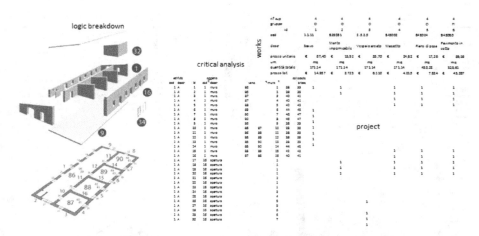

Fig. 8. Application of the general cost model scheme (Fig. 6)

Table 1. Parametric costs for each component and for each status degree

degree	mesonries		shells		openings		plankings	
	unit cost	score	unit cost	score	unit cost	score	unit cost	score
1	€ 28,86	5,0	€ 130,06	5,0	€ 20,91	5,0	€ 17,26	5,0
2	€ 40,67	4,8	€ 190,07	1,0	€ 185,50	3,5	€ 95,12	3,4
3	€ 80,24	4,3			€ 250,00	3,0	€ 213,04	1,0
4	€ 307,90	1,4			€ 306,91	2,5		
5	€ 342,16	1,0			€ 326,96	2,3		
6					€ 350,06	2,1		
7					€ 469,91	1,0		

Successfully the partial and total costs for each building have been calculated (Fig. 9, left histogram). Furthermore, the analysis of the relationships between status degree and dimensions of the different components and between these and the unit and total costs (Fig. 9, top table showing the first five buildings results) allowed us to calculate some different indexes of maintenance status (Fig. 9 right diagram x-axis) [10], and indexes of costs (y-axis) that have been compared for each building referring to volume unit (red points) and area unit (blue points). The relationship confirms the consistency of status degree and cost so that the parametric valuation can be considered reliable.

4.2   Revenues and Business Issues

The proposed business plan model is articulated as shown in two main sections, land-architectural and business, and in two further subsections, the owner's and the entrepreneur's balance sheets [20].

buildings	A				B				C				D				E			
	masonries	openings	plankings	shells	masonries	openings	plankings	shells	masonries	openings	plankings	shells	masonries	openings	plankings	shells	masonries	openings	plankings	shells
total element score	80	43	85	107	215	33	42	75	209	26	40	72	861	138	179	350	313	23	51	95
medium element score	5,0	2,8	5,0	5,0	5,0	2,8	5,0	5,0	2,8	5,0	5,0	2,3	3,4	2,0	1,0	3,4	3,0	1,0	1,0	1,0
element volume	341	43	85	107	215	33	42	75	209	26	40	72	861	138	179	350	313	23	51	95
total building volume	576	576	576	576	364	364	364	364	346	346	346	346	1527	1527	1527	1527	483	483	483	483
element cost	9,8	16,9	4,9	36,7	6,2	13,0	2,4	25,6	6,0	10,2	2,3	24,5	210,1	46,7	115,2	136,2	47,3	8,7	36,5	47,7
total element cost	68	68	68	68	47	47	47	47	43	43	43	43	508	508	508	508	140	140	140	140
weight1 elem vol/totvol 1	0,59	0,07	0,15	0,19	0,59	0,09	0,11	0,21	0,60	0,07	0,12	0,21	0,56	0,09	0,12	0,23	0,65	0,05	0,11	0,20
weight2 elem cost/totcost 2	0,14	0,25	0,07	0,54	0,13	0,28	0,05	0,54	0,14	0,24	0,05	0,57	0,41	0,09	0,23	0,27	0,34	0,06	0,26	0,34
weight 3 elem unit cost 3	0,29	0,15	0,17	0,39	0,29	0,15	0,17	0,39	0,29	0,15	0,17	0,39	0,29	0,15	0,17	0,39	0,29	0,15	0,17	0,39
weight 4 aver prev weights 4	0,34	0,16	0,13	0,37	0,34	0,17	0,11	0,38	0,34	0,15	0,11	0,39	0,42	0,11	0,17	0,30	0,42	0,09	0,18	0,31

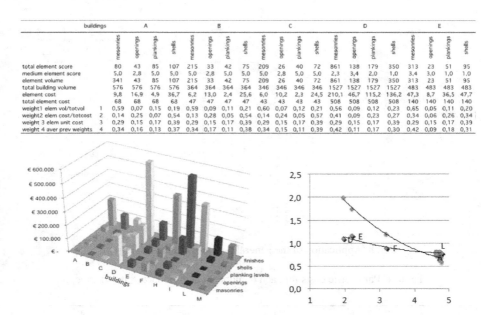

Fig. 9. Comparison of the total costs of the different buildings

4.3   Revenues and Business Issues

The business plan model is split in two main sections, architectural and firm, and in two subsections, the owner's and the entrepreneur's balance sheets [17] (fig. 10).

The architectural one (upper-left) concerns the capital asset value and the rent as depending on the different transformation options, in this case shops, accommodation, catering, offices, places for events and exhibitions. It involves the different choices about the fair combinations of: architectural transformation/preservation works, public/private areas; intensive/extensive functions. These combinations shouldn't involve only intense function: in fact the larger the experience range, the higher is the willingness to pay. Therefore, the supposed ratio intensive/extensive function is around 60%.

The firm features concern efficiency and convenience of the entrepreneurial activities settled within the complex, and assumes profit as the main criterion. It involves: the dimensioning of the most productive and appraised compartments; the definition of the financial horizon that influences the kind of lease contract and the kind of furniture and equipment to implement; the expected revenues from each function; the financial condition as implemented into the model (Fig. 10). The valuation results (Tab. 2) refer to four different quality/revenues hypotheses [8,9].

For the main distributive variables and financial features, some sensitiveness analyses have been carried out in order to compare their different elasticity (Fig. 11).

The sensitiveness analysis shows that the most influent distributive variables are cap rate -3,03% and long term interest rate, -2,79%, whereas a longer loan life has a significant positive effect whereas, on the contrary, a longer work period reduces the convenience and, as a consequence, the remainder presents value of the site.

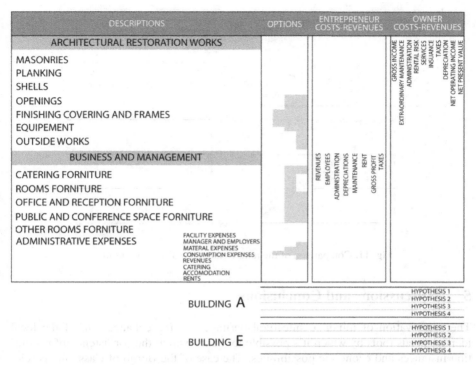

Fig. 10. General business plan scheme with the integration of the option module

Table 2. Distributive variables and evaluative results of different options

		distributive variables and factors	Entrepreneur		1		2		3		4
2,0%	rcap	cap rate	business revenues	€	29.046	€	29.585	€	30.123	€	30.662
2,0%	ra	depreciation interest rate	employees	€	6.367	€	6.367	€	6.367	€	6.367
1,0%	rn	entrepreneurial profit rate	administration	€	7.800	€	8.190	€	8.580	€	8.970
1,0%	ro	opportunity cost of equity	depreciations	€	4.011	€	3.869	€	3.615	€	3.615
6%	rot	owner profit rate	maintenance	€	704	€	685	€	633	€	633
3	dr	works period	rent	€	9.876	€	10.181	€	10.630	€	10.773
1,02	qa	depreciation factor	gross profit	€	288	€	293	€	298	€	304
1,01	q	opportunity factor	taxes	€	95	€	97	€	98	€	100
5,1%	ral	long run interest rate	**Owner**		**1**		**2**		**3**		**4**
1,051	qal	long run factor	gross income (rent)	€	9.876	€	10.181	€	10.630	€	10.773
1,5%	wacc	weighed average cost of capital	extraordinary maintenance	€	296	€	305	€	319	€	323
10,1	il	loan life	building depreciation	€	-	€	-	€	-	€	-
3%	qm	ordinary maintenance rate	administration	€	99	€	102	€	106	€	108
33%	tr	taxes rate	rental risks	€	198	€	204	€	213	€	215
3%	ms	extraordinary maintenance rate	services	€	-	€	-	€	-	€	-
1%	am	administration rate	insurance	€	99	€	102	€	106	€	108
2%	al	rental risk rates	taxes	€	3.259	€	3.360	€	3.508	€	3.555
0%	se	services expenses rate	other depreciation	€	-	€	-	€	-	€	-
1%	as	insurance rate	Net Operating Income	€	5.926	€	6.108	€	6.378	€	6.464
50%	D/C	leverage	capitalization value	€	296.292	€	305.423	€	318.888	€	323.188
			building costs	€	224.736	€	216.564	€	210.927	€	207.885
			normal profit	€	17.089	€	17.616	€	18.393	€	18.641
			Extraction value	€	**54.466**	€	**71.243**	€	**89.569**	€	**96.662**
			Net Present Value	€	**98.144**	€	**116.670**	€	**137.403**	€	**145.298**

	variations		elasticity
cap rate	2,00%	€ 119.936	-3,03%
	2,02%	€ 116.307	
depreciation rate	2,00%	€ 119.936	-0,15%
	2,02%	€ 119.756	
long run interest rate	5,00%	-€ 181.385	-2,79%
	5,05%	-€ 186.449	
leverage	50,00%	€ 174.688	-0,46%
	50,50%	€ 173.888	

	variations		elasticity
works period	3,00	€ 119.936	-0,17%
	3,01	€ 119.731	
loan life	10,0	€ 174.688	1,19%
	10,1	€ 176.768	

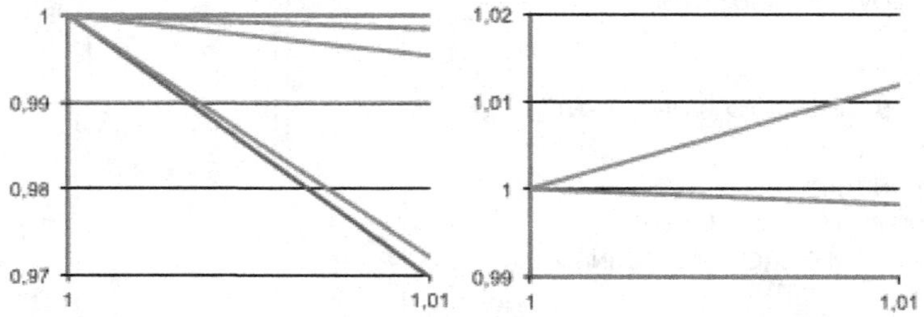

Fig. 11. Comparison of the total costs of the different buildings

5   Discussions and Conclusions

The preservation of minor architectural heritage and the enhancement of the local identity needs tools by which it's possible to make explicit the consistency of restoration instances and economic possibilities. The case of the Borgo of Cassibile, is relevant because its high potential due to its general morphological consistency and the proposed valuations confirm that, despite its status, the most part of the total cost of restoration of the architectural complex is due to the surface works as finishing, frames and covering, and to the equipment, so that no significant transformation needs. As the consequence the preservation of the original materials and techniques can be considered consistent with the productive function of the buildings that can be considered as flexible containers for different alternative productive uses [3].

This approach can be extended over similar complexes in which is not clearly defined the relationship between intrinsic values, future perspectives and economic opportunities. The model could be successfully enhanced by a 3D GIS extension helpful to connect the building topologic structure and the general database into a single tool.

This study confirms the possibility of a fair integration of architectural and economic sciences and applications and of historical values, social-economic decision making and individual creative initiative [5]. Some advices can also be assumed:

— the connection between stock long run features (capital private and social values) and short run cash flows as results of dynamic activities that often implement a "colonizer approach" that adversely affects the local traditional features;
— the role played by the financial and distributive variables that (as perceived with the sensitiveness analysis) often modify the convenience of the investment; cultural heritage economy is a long run activity which takes great advantages by a favorable financial context where uncertainty and risk are reduced by low interest rates;

— the role played by the capitalization rate: if heritage investments get surer a greater number of investors will be attracted by the potentiality of long run investment assuming as underlying singular or rare assets whose value is expected to increase.

Acknowledgement. Salvatore Giuffrida edited paragraphs 1, 3.2, 3.3, 3.4, 4, 5, drew up the cost model, business plan and economic calculations. Angelita Bellissimo edited paragraph 2, did all the restitution of the buildings, the critical analysis and the characterization. Pietro Copani edited paragraph 3.1 and supervised the restitution.

References

1. Accascina, M., et al.: Le arti in Sicilia nel '700: studi in memoria di Maria Accascina, pp. 87–98. Assessorato dei Beni Culturali e Ambientali e della Pubblica Istruzione, Palermo (1992)
2. Boscarino, S., Prescia, R.: Il restauro di necessità. FrancoAngeli, Milano (1992)
3. Caponnetto, R.: La valutazione del decadimento prestazionale di monumenti tipici del contesto mediterraneo. Editecnica, Palermo (2006)
4. DEI : Recupero, ristrutturazione e manutenzione. DEI, Roma (2013)
5. Donna, G.: La creazione di valore nella gestione dell'impresa. Carocci, Roma (1999)
6. Fattinanzi, E.: La valutazione della qualità e dei costi nei progetti residenziali. Il brevetto SISCo. In: Valori e Valutazioni. Anno IV n. 7, pp. 49–66. DEI, Roma (2011)
7. Finocchiaro Castro, M., et al.: Public Intervention on Heritage Conservation And Determinant of Heritage Authorities' Performance: A Semi-Parametric Analysis. Int. Tax Public. Finance 18, 1–16 (2011)
8. Guatri, L.: Trattato sulla valutazione delle aziende. EGEA, Milano (1998)
9. Imperatori, G.: Il project financing. Il sole 24 ore, Milano (1995)
10. International Society of Parametric Analysist: Parametric Estimating Handbook. ISPA, Vienna (2008)
11. Lippi Guidi, A.: Masserie e vecchi manieri nel siracusano. Lombardi, Milano (1990)
12. Manganelli, B.: Valutazioni economico-estimative nella valorizzazione di edifici storico-architettonici. Aestimum 51, 21–42 (2007)
13. Mangiameli, R.: Società locale e guerra totale. Lombardi, Siracusa (1995)
14. Massimo, D.M., et al.: Derivazione di costi parametrici di recupero edilizio da stime dettagliate con l'uso dei GIS. Asita II, pp.1505–1510 (2007)
15. Mattia, S.: Momenti della valutazione preventiva dei costi nel processo edilizio. In: Mattia, S. (ed.) Progetto qualità e decisione. Over, Bergamo (1993)
16. Participants of the First International Congress of Architects and Technicians of Historic Monuments: The Athens Charter for the Restoration of Historic Monuments. Athens (1931)
17. Rizzo, F.: Economia del patrimonio architettonico ambientale. Franco Angeli, Milano (1989)
18. Rizzo, F.: Il mercato e la valutazione dei quadri. In: La scienza estimativa e il suo contributo per la valutazione dei beni artistici e culturali, Atti al VII Incontro del Centro Studi di Estimo, pp. 157–176. Le Monnier, Firenze (1978)
19. Sahlman, W.: How to write a great Business Plan. In: Harvard Business Review, pp. 98–108. Cambridge (July/August 1997)
20. Singler, A.: Businessplan. 3. Auflage, 128 S. Haufe-Lexware, München (2010)

Family Lifestyles through the Construction of a Structural Equation Modeling

Silvestro Montrone, Paola Perchinunno, and Alessandro Rizzi

DISAG, University of Bari,
Via C. Rosalba 53, 70100 Bari, Italy
{silvestro.montrone,paola.perchinunno,alessandro.rizzi}@uniba.it

Abstract. The construction of a structural equation modeling can be used to identify the latent variables underlying a determined phenomenon. Developed thanks to the impulses of the statistician psychometrician Karl Joreskog, such models are increasingly applied in the social field. The current study aims at identifying the latent variables underlying the change in family lifestyles caused by the economic recession which has spread in all countries of the European Union and continues to make its effects felt. More specifically, a research carried out by the University of Bari has been taken into account in order to monitor how the lifestyles of the families living in Bari have evolved in time of crisis.

Keywords: Lifestyles, SEM, latent variables, fitting.

1   Structural Equation Modeling

Structural Equation Modeling (SEM) is a valid tool to evaluate the interrelations among not directly measured, that is, latent variables. This justifies its increasingly wide use above all in the social field, since it is an interesting point of synthesis between factorial analysis and path analysis.

Developed as a method to decompose the correlation into different pieces to interpret its effects, path analysis is very closely linked to the concept of multiple regression, as it implies the analysis of simultaneous multiple regression equations. Unlike multiple linear regression models that take into account only the direct relationships between the independent variables and the dependent variables, path analysis allows to highlight the direct, indirect and total effect of one variable on the other, in which as direct effects are understood those effects that are not mediated by any other variable, unlike the indirect effects that require the intervention of a variable precisely defined as mediator variable.

Since within a structural equation modeling a variable can appear as an independent variable in some equations and as a dependent variable in others, it is appropriate to distinguish between:

- *Exogenous variables*, i.e., the variables external to the model, which appear in the same model always and only as independent variables;

- *Endogenous variables*, that are those variables appearing in the model alternately as dependent variables and as independent variables.

B. Murgante et al. (Eds.): ICCSA 2014, Part III, LNCS 8581, pp. 192–207, 2014.

The above mentioned definition shows that the exogenous variables can be either deterministic or stochastic, unlike the endogenous variables that can be only stochastic. In this context, Wooldrige's contribution should be mentioned, as it distinguishes between an endogenous variable or an exogenous one, depending on whether it is correlated or uncorrelated with the term of error of the model [1].

Using the symbolism of path analysis it is possible to graphically represent a structural equation modeling according to the following criteria:

- The latent variables are enclosed in a circle or in an ellipse if they are exogenous, in a rectangle or in a square, if they are endogenous. The stochastic errors are indicated by the corresponding letters, but without being circled.
- The causal link between two variables is indicated by a one-direction arrow which starts from the cause (independent) variable and ends at the effect (dependent) variable; the association between two variables (covariation, correlation) is indicated by a bidirectional arrow. The absence of arrows indicates the lack of relations between the variables.
- The strength of these relationships is indicated by reporting in corresponding of the arrow the value of the regression coefficient in the case of causal relationships, or the value of the coefficient of correlation in the case of association relationships. In symbolic terms, the structural parameter presents two indices which refer respectively to the variable of arrival and to the variable of departure.

Each structural equation modeling can be decomposed into a structural (or inside) model and a measurement (or external) model.

The internal model allows to study the causal relationships between the latent variables, unlike the external model that does not deal with the problem of causality, but with the problem of the relationships between the latent variables and their indicators (or manifest variables).

Each of these parts is summarized in a "basic equation" that, written in matrix form, can be translated into the following system

$$\begin{cases} \eta = B\eta + \Gamma\xi + \varsigma \\ Y = \Lambda_y\eta + \varepsilon \\ X = \Lambda_x\xi + \delta \end{cases} \tag{1}$$

in which the first equation refers to the structural model, whereas the second and the third equation refer to the measurement models related respectively to the endogenous and exogenous variables.

Let us dwell on the first basic equation given by:

$$\eta = B\eta + \Gamma\xi + \varsigma \tag{2}$$

It contains:

1) The vector η of the endogenous latent variables of (m x 1) dimension, where m is the number of the endogenous latent variables; the vector ς of (m x 1) dimension related to the errors of the endogenous latent variables and, finally, the vector ξ of the exogenous latent variables of (n x1) dimension, where n is the number of such variables;

2) The matrix **B** of the structural coefficients between the endogenous latent variables of (m x m) order, whose main diagonal is always composed by elements equal to zero, because the regression coefficients of each variable on itself are null; the matrix Γ of the structural coefficients between the exogenous and endogenous latent variables of (m x n) order.

3) To the purpose of a correct specification this part of the model needs two more matrices: the symmetric matrix $\boldsymbol{\Phi}$ (phi) of (n x n) order which contains the covariances between the exogenous variables and the symmetric matrix $\boldsymbol{\psi}$ (psi) of (m x m) order which contains the covariances between the errors ς. This matrix is very important, because it allows to insert into the model the effects of those variables which have been excluded from it, but operating really on the observed data.

The second basic equation

$$Y = \Lambda_y \eta + \varepsilon \qquad (3)$$

is composed by:

1) The vector **Y** of the observed endogenous variables of (p x 1) dimension, where p is the number of the observed variables Y the vector ε of the errors of the observed variables Y of (p x 1) dimension; the vector η of the endogenous latent variables of (m x 1) dimension;

2) The matrix $\boldsymbol{\Lambda_y}$ of (p x m) order, related to the structural coefficients between the observed endogenous variables Y and the endogenous latent variables η;

3) The covariance matrix of the errors ε, $\boldsymbol{\theta^\varepsilon}$, which is squared, symmetric and of (p x p) order.

Finally, the third basic equation

$$X = \Lambda_x \xi + \delta \qquad (4)$$

is composed by:

1) The vector X of the observed exogenous variables of (q x 1) dimension, where q is the number of the observed variables X; the vector δ of the errors of the observed variables X of (q x 1) dimension; the vector   of the exogenous latent variables of   (n x 1) dimension;

2) The matrix Λ_x of the structural coefficients between the observed variables X and the latent variables ξ of (q x n) order.

3) The covariance matrix of the errors δ, θ^δ, which is squared, symmetric and of (q x q) order.

2   The Application of the Model

The evolution of the lifestyles of the families residing in the metropolitan city of Bari as the result of the current economic recession was assessed through the project "Lifestyles in time of crisis". The survey was addressed to young couples with children in preschool and school age. The questionnaires were distributed at the educational Institutes that joined the project. The innovative importance of the survey can be attributed to the joint use of both objective variables and subjective measurements. Just these last ones allows to explore the perception of poverty: As a matter of fact, it is a poor man not only who has not got the minimum resources necessary to enjoy a decent standard of living, but also, and above all, who feels poor in the perception of his own life condition.

The questionnaire submitted to the families is divided into 5 sections:

1. General news of the respondent (gender, age, educational level, marital status, place of residence, housing conditions);
2. Evaluation of the employment status of the respondent and his relatives;
3. Assessment of the degree of social integration (participation in religious, political, sporting events, etc ...);
4. Evaluation of the difficulty in sustaining certain types of expenses;
5. Judgment of the interviewee on his own condition.

The construction of a structural equation model (SEM) achieved the purpose of identifying those latent factors of the lifestyles of 4,400 interviewed families. As the questionnaire consisted of 61 variables, before the construction of the model, an exploratory factor analysis was performed to simplify the phenomenon by transforming the number of the starting variables in a more simple structure [2].

To this aim, the previous quantification of each variable was carried out through an appropriate procedure of "optimal scaling" to assess, with respect to each variable, the categories of the variables themselves, as well as the score for the answers provided by each interviewed family.

Specifically, the exploratory factor analysis was conducted by selecting among the various procedures ALSOS [1] (Alternative Least Square Optimal Scaling), the CATPCA (Categorical Principal Component Analysis), a non-parametric algorithm which uses as a method of quantifying the main components of p variables transformed and optimized within the factor p-dimensional space, with p <m [3,4].

It was considered appropriate to exclude from further analyses the variables with a communality less than 0.55, by reducing the analysis to 41 variables on which it was carried out an exploratory factor analysis to identify their latent factors.

Eight factors were thus obtained. An oblique rotation was carried out upon them: the methods of oblique rotation, starting from an orthogonal rotation of VARIMAX type, operate a transformation of the factor loadings to increase the large ones in absolute value, and diminish the importance of the smaller weights. In particular, in order to improve the separation among factors and to obtain a matrix of rotated weights with a single dominant factor in correspondence with each original variable, a rotation PROMAX was carried out. In this way, all the elements of the matrix obtained by the rotation VARIMAX are raised to a power k> 1 (not necessarily whole). Consequently, the eight factors to which the following labels:

---

[1] The methods ALSOS are exploratory techniques that are preferable to traditional methods of scaling, as they are free from parametric assumptions, such as the normal distribution of the latent variables.

-   **F1: Purchases difficulties**
-   **F2: Loans for spending**
-   **F3: Supports for the study**
-   **F4: Social participation**
-   **F5: Payments difficulties**
-   **F6: Availability of appliances**
-   **F7: Family incomes**
-   **F8: Family status**

are associated, account for 67.70% of the total variance.

More specifically, it was observed that **Factor 1** underlies all options of question 27 "*Has your family had difficulties in recent months to afford these expenses?*", excluded the option relating to the payment of the school canteen.

Factor 2 summarizes the reply options to question 18 "*In the last year, did your family need to resort to loans for consumption?*" excluded the response 18.8 (loans for other reasons).

Factor 3 summarizes the availability of various supports for the study through question 21 "*Have your sons got the availability at home of any of the following supports?*" excluded the options 21.a (availability of a room just for them), 21.b (a quiet place to study), and 21.c (a desk for homework).

The social participation, investigated by question 16 "*In the last six months, have you noticed a participation of your family in events / organizations?*", is condensed in the **Factor 4**, which ignores however the participation in religious events.

Factor 5 concerns the difficulties that families have experienced in the payment of taxes and duties, rates of other debts, bills and food staples (questions 23-26). We would like to emphasize that this factor does not take into account the difficulties of the families in the payment of loan installments.

Question 19 "*Has your family currently the use of the following goods?*" has allowed us to probe the possession of the families of a number of appliances. **Factor 6** reflects the availability of three appliances which are refrigerator, washing machine, and TV, respectively.

Factor 7 summarizes the information which can be derived from questions 13-14-15 of the questionnaire regarding the number of earners in household income, of relatives in working age and those who, although they are not of working age, contribute to the family income.

Finally, **Factor 8** "family status" takes into account the respondent's residence in Italy and the eventual spouse.

Confirmatory factor analysis was performed with the module AMOS (Analysis of Moment Structures) of the SPSS statistical package through which it was possible to construct a structural equation modeling.

Therefore, we can refer to a SEM model (shown in the following page) consisting of 8 latent variables, *six* of which are endogenous and *two* exogenous. The exogenous latent variables are:

- ξ_1: Family Status
- ξ_2: Family incomes

while the endogenous latent variables are:
- η_1 : purchases difficulties
- η_2: loans for consumption
- η_3 : supports for the study
- η_4 : social participation
- η_5 : payment difficulties
- η_6 : availability of appliances

The structural model can be expressed by the following equations system:

$$\eta_1 = \eta_3\beta_{13} + \eta_5\beta_{15} + \varsigma_1$$
$$\eta_2 = \eta_3\beta_{23} + \xi_1\gamma_{21} + \xi_2\gamma_{22} + \varsigma_2$$
$$\eta_3 = \eta_4\beta_{34} + \xi_2\gamma_{32} + \varsigma_3 \qquad (5)$$
$$\eta_4 = \xi_2\gamma_{42} + \varsigma_4$$
$$\eta_5 = \eta_2\beta_{52} + \eta_3\beta_{53} + \eta_4\beta_{54} + \eta_6\beta_{56} + \varsigma_5$$
$$\eta_6 = \eta_2\beta_{62} + \eta_3\beta_{63} + \xi_1\gamma_{61} + \varsigma_6$$

It should be very appropriate, even before analyzing the structural model, to focus on the extent of the regression weights associated to each latent variable. The latent variable "family income" shows high enough regression weights for each indicator associated with it: if it is not surprising the high regression weight associated with the variable "relatives in working age" (0.88), it must be underlined how the regression weight associated with the "relatives not in working age income earners" (0.73), could be considered as indicative of the necessity to continue to carry out regularly remunerated work activities, when the working age has been exceed, as well as the need of the family members not yet in working age to support family budget.

The second exogenous latent variable *"family status"* is marked by very low regression weights: they start from a minimum of 0.20 for the variable "spouse's residence in Italy" up to a maximum weight of regression equal to 0.57 for the "spouse's working type".

Observing the regression weights associated to the endogenous latent variable *"purchases difficulties"*, it can be observed that the highest values concern the ticket payment (0.83), the maintenance of private medical care (0.80), and the purchase of clothing articles (0.80). Although the right of health is insuppressible, it is ever growing the number of families forced to deal with a systematic difficulty in paying the ticket, as our research has largely documented. On the other, the inefficiencies of the national health system as well as the exceeding bureaucracy, such as long waiting lists, lead to a growing number of users who turn to private health that is most certainly faster, but also more expensive.

Not surprisingly, the lowest regression weights are read in correspondence on the purchase of movies tickets (0.62), of magazines and newspapers (0.59), and of holidays (0.59): obviously, given that such expenses are not related to the satisfaction of the insuppressible needs, they are the first to be eliminated in case of economic difficulties.

To this regard, the latent variable *"supports for the study"* includes a series of tools, whose eventual availability will improve without any doubt the quality of the study of youngsters: the highest regression weights are to be read in relation to books (0.78), classic literature (0.73), encyclopedia (0.72). The low values, however, in relation to the supports such as computer (0.56) and internet connection (0.66) should not be misled.

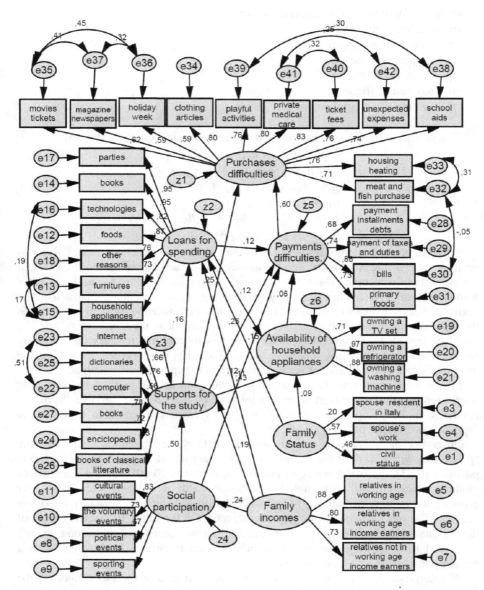

Given that the majority of the interviewed families reported to own a computer and a connection to the Internet, we notice that whenever these supports are not present, such absence is not automatically attributed to merely "*economic*" motives, but to the fact that, since such supports are being used within the school context, the families think not to buy them.

The latent variable "*social participation*" is important, as it allows to monitor the participation of families in social and political life: as already mentioned, a poor man is the one who, feeling relegated to the margins of society, is excluded from itself.

Within this context, it is important to note the high weight of regression associated with the participation in cultural events provides a further confirmation of the importance that families recognize and attribute to the factor "culture." In addition, he lowest regression weight (0.66) is registered in relation to the participation in political demonstrations: a direct consequence of the widespread disaffection with politics, which is witnessed by the ever higher abstentionism during the political or administrative elections. The same value of the regression weight concerning the participation in sporting events can be read as a witness of the difficulties of the families to take part to these events, often due to high costs.

As for the "*payment difficulties*", families have had to deal essentially with the difficulty in the payment of bills related to a series of consumptions (0.86), taxes and charges (0.74), and primary food (0.73), outlining thus a framework consistent to what is being daily shown by media.

Finally, the last η_6 endogenous latent variable concerning the *availability of appliances* presents very high regression weights especially with respect to the owning of the refrigerator (0.97) as well as washing machine (0.88).

By analysing the structural model, it can be observed how the only fully endogenous variable (that is, influenced by other variables, without affecting any of them) is "*purchases difficulties*" which is directly influenced by the "payment difficulties" (0.60) and "supports for the study" (0.25).

As a matter of fact, it is quite obvious that the difficulty in certain payments (rates of debts and taxes, bills) has repercussions on the criticality when making a series of purchases (private medical care, ticket fees, clothing, school aids), it is also evident that when the families decide to make available to their children some supports for the study to improve its quality, they have to come across greater difficulties in performing a series of purchases. The indirect relationships that link the latent variable η_1 with the other latent variables are less evident, but not less important. For example, let us take into account the path:

Family incomes → Social participation → Supports for study → Loans for spending → Payment difficulties →Purchases difficulties

It can be reasonably interpreted as follow: the participation in the social life of the country involves the bearing of some costs; therefore, the presence of average family incomes, or non-low, influences directly (0.24) the participation in a series of events. Particularly, the participation in events and/or cultural performances makes aware the families to the importance of a good education for their sons, as shown by the significant direct effect of the "*social participation*" on the availability of "*supports for the study*" (0.50).

To this regard, it should be noted how the family incomes influence the availability of teaching aids both directly (0.19) and indirectly through the variable "*social participation*" that is a mediator variable (0.12).

The families of the current study are not particularly well-off; thus, it is evident that when family incomes are used both to participate to a greater extent in the social life of the country, and to provide to a greater extent the supports for the study (in a direct and an indirect way), they are forced to seek to loans for spending to meet a variety of needs, and this could explain the direct relationship "*supports for the study*"

- "*loans for spending*" (0.16). Also in this case the variable "*family incomes*" influences the demand for "*loans for spending*" directly (0:11) and indirectly through the variables "social participation" and "supports for the study," although very weakly (0.05).

When families get a loan for spending, they engage to give back the amounts received in accordance with a predetermined amortization schedule; the need to comply with this commitment can cause the onset of difficulties in making payments concerning a number of utilities, rates of other debts, taxes and charges (0.12) leading thus to a greater difficulty to make a series of purchases (0.60). It should be noted how the demand for "*loans for spending*", does not influence "*difficulty in purchasing*" directly, but indirectly through the mediator variable of "*payment difficulties*" (0.08).

There is a key aspect to consider, which cannot be ignored: education is an investment, though, that produces results in a long period, certainly not in a short one. The evidence of this fact lies in the immediate investments in education that would probably lead the families to a worsening of economic difficulties and because of those investments families are forced to seek loans for spending and to face greater difficulties in complying with a series of payments.

The chain of the above described causal relationships can be considerably simplified by taking into account as the starting point of the path not the variable "family incomes", but the variable "social participation" or "supports for the study."

By considering the second exogenous latent variable of our model ξ_1 "family status", we notice that it influences, although indirectly, the difficulty in purchases through the two following paths:

1)   *Family status* → *Loans for spending* → *Payments difficulties* → *Difficulty purchases*

2)   *Family status* → *Availability of household appliances* → *Payments Difficulties* → *Difficulty purchases*

As a matter of fact, It is quite obvious that the family status, whose indicators are marital status, spouse's work and the residence in Italy of the respondent's spouse, directly influence both the demand for loans for spending (0.15) and, although weakly, the availability of certain appliances such as TV, washing machine and refrigerator (0.09).

3   Adaptation of the Model to the Data

After specifying the model, it is necessary to verify its fit to the data on the basis of the comparison between the theoretical covariance matrix (that produced by the model) and the covariance matrix of the observed data; then the functions of the residual or the difference between S and Σ needs to be referred to.

This comparison allows us to reach the **not falsification of the model**. This concept needs to be studied in depth. In fact, as the observed data can be compatible with different models, at the end of this phase, the conclusion according to which the model is true will be never reached, but the statement, the model is not false, or it is not contradicted by the data, will be simply obtained.

The problem of formulating this difference in terms of a known statistical distribution, then, rises, so that, in the evaluation of the difference $(S - \Sigma)$, the stochastic fluctuations of sampling can be excluded.

When the model is right and the sample is large enough, the standard deviation $(S - \Sigma)$ tends to be distributed according to a chi-square (χ^2) with a number of degrees of freedom (df) equal to:

$$df = \frac{1}{2}(p+q)(p+q+1) - t \tag{6}$$

where t is the number of free parameters to be estimated, p is the number of the observed endogenous variables Y, and q is the number of the observed exogenous variables X.

The degrees of freedom measure the *parsimony* of the model, i.e. the greater is the number of the degrees of freedom, the lower is the number of the parameters to be estimated and, consequently, the better is the model.

These aspects are so important that they have to be considered in the evaluation process of the model, in the sense that the researcher, faced with the choice between two models, should always opt for the more parsimonious one, i.e., with the highest number of degrees of freedom.

On the basis of what has been previously said, we have taken into account the value of the χ^2 divided by its degrees of freedom, which should assume a value between one and three as a measure of the model fit to the data, so that the model cannot be falsified.

From a practical point of view, we have to test the hypothesis of the identity between the observed covariance matrix S and the theoretical one Σ.

$$\begin{cases} H_0: S = \Sigma \\ H_1: S \neq \Sigma \end{cases} \tag{7}$$

To test the null hypothesis, it is necessary to consider the minimum value of the function of empirical discrepancy $\hat{C}$. . If this value is found to be less than the tabulated one $\chi^2_{g,\alpha}$, it means that the model is not falsified by the data and that the deviation $(S - \Sigma)$ is due to stochastic fluctuations. On the contrary, should the calculated value of the statistic test will be greater than the tabulated one, we can easily conclude that the model cannot be considered compatible with the data, since the difference $(S - \Sigma)$ cannot be attributed to stochastic fluctuations.

However, the adaptation measures that refer to the χ^2 have the disadvantage of being highly sensitive to the sample size, so that the same residue $(S - \Sigma)$ could give either a not significant value of χ^2 , or a significant one, depending on whether the sample consists of a limited number of cases, with the following risk of committing an error of the type 1, i.e., the error to reject a true model.

Indeed, if the sample is very large, even a good model, almost close to reality, gives a significant value of the χ^2 that will lead the researcher to reject the null hypothesis, thus making an error.

This problem is overcome by evaluating the fit of the model to the data through a series of statistics which do not seem to be sensitive to the sample size, although they refer to the function of the above mentioned discrepancy. The most widely used of them are:

1) **Goodness of Fit Index (GFI)**

$$GFI = 1 - \left[\frac{\hat{C}}{\hat{C}_k}\right] \tag{8}$$

The Goodness-of-Fit statistic (GFI) was created by Jöreskog and Sorbom as an alternative to the Chi-Square test and calculates the proportion of variance that is accounted for by the estimated population covariance and it shows how closely the model comes to replicating the observed covariance matrix.

A value of this index close to 0.9 is indicative of the good adaptation of the model to the data, being the maximal value of this index equal to 1 in the case of a perfect adaptation of the model to the data (saturated model). However, simulation studies have shown that when factor loadings and sample sizes are low a higher cut-off of 0.95 is more appropriate.

2) **Adjusted Goodness of Fit Index (AGFI)**

$$AGFI = 1 - \left(\frac{k}{df}\right)(1 - GFI) \tag{9}$$

This index takes into account the distortion due to the size of the sample, where k is the number of variances covariances produced by the model equal to $\frac{1}{2}(p + q)(p + q + 1)$ which varies between zero and one with the same meaning of the previous index. Values of the AGFI greater than 0.9 indicate a good fit of the model to the data.

3) **Root Mean-Square Residuals (RMR) and Standardized RMR (SRMR)**

This index is obtained by performing the square root of the average of the squared deviations between the sample variance and its obtained estimate assuming that the model is correct. Intuitively it could happen that the smaller is the value of this index, the better is the fit of the model to the data.

The range of the RMR is calculated based upon the scales of each indicator, therefore the RMR becomes difficult to interpret. The standardized RMR (SRMR) resolves this problem and is therefore much more meaningful to interpret.

4) **Normed Fit Index (NFI)**

This index was proposed by Bentler and Bonnet [5] and its expression is given by:

$$NFI = 1 - \left[\frac{\hat{C}}{\hat{C}_b}\right] \tag{10}$$

where $\hat{C}_b$ is the value of the discrepancy function of the null model (of independence). The null/independence model is the worst case scenario as it specifies that all measured variables are uncorrelated .Also this index, whose range of variation is given by the interval [0.1], must be greater than 0.9 so that the model can be considered adequate.

5) Relative Fit Index (RFI)

This index was proposed by Bollen [6] in 1986 and is obtained by correcting the Normed Fit Index with the degrees of freedom of the two values of discrepancy:

$$RFI = 1 - \left[\frac{\hat{C}}{\hat{C}_b}\right] * \left[\frac{g_b}{g}\right] \tag{11}$$

It is an index that, considering the degrees of freedom of the two values of discrepancy, pays more attention to the parsimony of the model and, as the previous models, varies between zero and one.

6) Incremental Fit Index (IFI)

Bollen [7] proposed in 1989 an index relatively insensitive to sample size. Values that exceed .90 are regarded as acceptable, although this index can exceed 1. To compute the IFI, first the difference between the chi square of the independence model, in which variables are uncorrelated, and the chi-square of the target model is calculated. Next, the difference between the chi-square of the target model and the df for the target model is calculated. The ratio of these values represents the IFI

$$IFI = \frac{(\hat{C}_b - \hat{C})}{(\hat{C}_b - g)} \tag{12}$$

7) Comparative Fit Index (CFI)

Developed by Bentler in 1990 [8], it is an index that can be considered as a revised and correct form of the NFI that also takes into account the amplitude of the sample. This index also assumes that the latent variables are uncorrelated and compares the covariance matrix of the sample with that of the null model. A value close to one indicates a good fit of the model. A cut-off criterion of CFI $\geq$ 0.90 was initially advanced however, recent studies have shown that a value greater than 0.90 is needed in order to ensure that misspecified models are not accepted.

$$CFI = 1 - \frac{\max(\hat{C} - g; 0)}{\max(\hat{C}_b - g_b)} \tag{13}$$

8) Tucker –Lewis Index (TLI)

Also known as "Non-Normed Fit Index," this index calculated by Tucker and Lewis [9] in the context of the analysis of the moments, has a typical range of variation between zero and one, although it has no upper limit. To compute TLI:
   a) First, divide the chi-square for the target model and the null model by their corresponding df-value which generates relatives chi-square for each model.
   b) Next, calculate the differences between these relative chi-squares
   c) Finally, divide this difference by the relative chi-square for the null model minus one.

Its expression is given by the following formula:

$$TLI = \left[\frac{\left(\hat{C}_b / g_b \right) - \left(\hat{C} / g \right)}{\left(\hat{C}_b / g_b \right) - 1} \right] \tag{14}$$

9) Expected Cross Validation Index (ECVI)

Given by the relationship between the Akaike information criterion and the number of the degrees of freedom, this index is a measure of the discrepancy the covariance matrix implied by the model, and the expected covariance matrix in another sample of equivalent size extracted from the same population. It must be as close as possible to the index related to the saturated model.

10) Root Mean Square Error of Approximation (RMSEA)

Developed by Browne and Cudek [10] in 1993, it is an index that takes into account the complexity of the model and is obtained from the square root of the relationship between $(\hat{C} - g)$ and g, or zero and g if $(\hat{C} - g) < 0$

$$RMSEA = \sqrt{\max(\hat{C} - g, 0)/g} \tag{15}$$

The RMSEA tells us how well the model, with unknown but optimally chosen parameter estimates would fit the populations covariance matrix . In other words, the RMSEA favours parsimony in that it will choose the model with the lesser number of parameters.

The ideal matching is achieved when this index should be equal to zero, but in reality a value of this index equal to 0.08 indicates a reasonable matching of the data to the model.

After this overview of the most commonly used indices to evaluate the fit of the model to the data, we can then evaluate the goodness of the model

Table 1. Indices of model fit to the data

Index	Value
GFI	0.940
AGFI	0.931
RMR	0.038
NFI	0.953
RFI	0.949
IFI	0.960
CFI	0.960
TLI	0.956
ECVI	1.211
RMSEA	0.036

In our case the values of the indices show an excellent fit of the model to the data.

4   Decomposition of the Model into Sub-groups

As previously mentioned, the SEM model was constructed by referring to the 4.400 families, whose children were enrolled at one of the seven Schools involved in the project.

In a following phase it was decided to split our model in four sub-models according the four following groups:

1) XVII° Circolo Didattico Poggiofranco – XIV° Circolo Didattico Re David – Istituto Comprensivo Mazzini – Modugno
2) IX° Circolo Didattico Japigia 1 – Istituto Comprensivo Balilla Imbriani
3) Istituto Comprensivo Don Milani
4) Istituto Comprensivo Gabelli

The first aspect, which our attention has been focused on, was to verify whether the path of causal relations was confirmed within each single group.

Family incomes → Social participation → Supports for the study → Loans for spending → Payments difficulties → Purchases difficulties

As a matter of fact, it has been verified that the above mentioned path would show as a mirror-like way in three of the four considered groups, because only in the Comprehensive Institute Gabelli the acquisition of supports for the study has no influence on the demand for loans for spending, but exclusively on the payments difficulty. Therefore, in this Institute the following sequence of causal relations was:

Family incomes → Social participation → Supports for the study → Payments difficulties → Purchases difficulties

Unlike the baseline model, in which the supports for the study influenced the payments difficulty both directly and indirectly through the mediating variable "Loans for spending", in the area of Santo Spirito the indirect effect disappears and the direct one remains in the model itself. Thus, in this context the demand for loans for spending is influenced only by the level of family incomes.

However, the most interesting aspects have emerged when we have focused on "triangle" of causal relations

1)   Family incomes →   Social participation
2)   Social participation → Supports for the study
3)   Family incomes → Supports for the study

Table 2. Causal relations and direct effects

Causal relations	Groups			
	Poggiofranco Re David Mazzini	Japigia 1 Balilla	Don Milani	Gabelli
Family incomes Social participation	**0.17**	0.26	0.35	0.24
Social participation Supports for the study	**0.24**	**0.54**	**0.39**	**0.51**
Family incomes Supports for the study	0.16	0.15	0.23	0.19

As previously seen in the baseline model, the medium-high or not low family incomes influences the family participation in the social life of the Country and it is just through this participation that families become aware of the importance of a good education for their sons.

Within this framework, Table 2 allows us to capture interesting aspects since the families considered in the 1st group (Poggiofranco – Re David – Mazzini) are the wealthiest; for them the social participation in the Country life is something unavoidable, certainly influenced by the level of the family incomes, although in a lesser extent compared to what happens somewhere else. On the other hand, this proposition emerges in the fact that the direct effect of the family incomes on social participation seems to be the lowest in the four groups considered (0.17). Whatever has been said so far, it can be realized even better by observing how this consideration is valid only for the group in which social participation influences, although weakly, both the demand for loans for spending (0.10) and the purchases difficulty (0.13). That is, these families, even if they did not renounce to participate actively in social life, would be willing to renounce to the satisfaction of needs for more essential requirements. Moreover, the families of this group, characterized by high academic qualifications and stable occupations, are aware of the importance of a good education for their children as well as of the participation in social life. As a matter of fact, the direct effect of social participation on the purchase of supports for the study is the lowest (0.24), compared with a value equal to 0.51 at Santo Spirito and to 0.54 in Balilla-Japigia group. This datum supports what earlier hypothesized, i.e. this group of families has no need to participate in social life in order to understand the importance of a good education for their children, because it is aware of it independently from this participation.

These are only some of the aspects that emerge from such comparisons: in the current study we have preferred to focus particularly on the importance of the participation in the social life, as the right approach to examine the change of the family life styles consists in monitoring how families perceive their own condition: the poor is not only the one who has not a lot of economic resources to conduct a respectable life, but also and above all is the individual who is marginalised in the society, since he/she considers himself more a burden rather than as a resource.

5   Conclusions

The analysis has allowed us to highlight in detail the critical times that the families of Bari are going through. In most cases, they had to revise their habits of spending, not only eliminating expenses that, although important, are unrelated to the irrepressible needs (for ex., week-long holidays, movie tickets and/or theatre tickets), but also significantly reducing the spending for basic necessities.

In some cases, the access to such goods has been made possible only through the use of loans, which, however, have also a downside: to force the family to take on other commitments and to face other deadlines laid down by the plans of debt repayment.

Although the outlined framework is far from promising, there are positive aspects need to be highlighted: the families recognize without any doubt the great importance of a good education for their children (as a matter of fact, it has been observed how people recourse to consumer loans for the purchase of books and dictionaries). This reflects the firm belief that a good education can act as the "engine" of the social mobility, being, therefore, a very important lever through which people can free themselves from a situation of poverty or indigence.

References

1. Wooldridge, J.M.: Econometric Analysis of Cross Section and Panel Data. The M.I.T. Press, Cambridge (2002)
2. Fabbris, L.: Statistica multivariata. Analisi esplorativa dei dati. McGraw- Hill, Milano (1997)
3. De Leuwe, J., Meulman, J.J.: A special jackknife for multidimensional scaling. Journal of Classification 3, 97–112 (1986)
4. Meulman, J.J., Heiser, W.J.: Categories 10.0. SPSS Inc., Chicago (1999)
5. Bentler, P.M., Bonnet, D.C.: Significance Test and Goodness of Fit in the Analysis of Covariance Structures. Psychological Bulletin 88(3), 588–606 (1980)
6. Bollen, K.A.: Sample size and Bentler and Bonnet's non normed fit index. Psycometria 51, 375–377 (1986)
7. Bollen, K.A.: A new incremental fit index for general structural equations models. Sociological Method and Research 17, 303–316 (1989)
8. Bentler, P.M.: Comparative Fit Index Structurals Models. Psychological Bulletin 107(2), 238–246 (1990)
9. Tucker, L.R., Lewis, C.: A reliability coefficient for maximum likelihood factor analysis. Psycometrika 38, 1–10 (1973)
10. Browne, M.W., Cudeck, R.: Alternative ways of assessing model fit. In: Bollen, K.A., Long, J.S. (eds.) Testing Structural Equations Models, pp. 136–162. Sage, Newbury Park (1993)

Benchmarking Multi-criteria Evaluation: A Proposed Method for the Definition of Benchmarks in Negotiation Public-Private Partnerships

Maria Rosaria Guarini and Fabrizio Battisti

Department of Architecture and Design (DIAP), Faculty of Architecture,
University of Rome, Sapienza
Via A. Gramsci 53, 00197 Rome (RM)
{mariarosaria.guarini,fabrizio.battisti}@uniroma1.it

Abstract. In Italy, new processes of settlement transformation based on negotiation-type public-private partnerships (PPPN) have been standardised to cope
with the degradation of many urban areas. However, these standards have not
provided for *benchmarks* referring to the contents of partnerships or assessment
procedures aimed at assessing the initiatives undertaken with respect to public
utility objectives. This has often led to redevelopment initiatives geared more
towards the satisfaction of private rather than public interests. The proposed
methodology, structured on the integration of a *Benchmarking* process with
multi-criteria evaluation techniques known as *Benchmarking Multi-criteria
Evaluation* (BME) enables the definition of *benchmarks* through a participatory
process of the different *Stakeholders* involved in a PPPN to which the BME is
applied. The *benchmarks* can be used both for renewing the planning of the
PPPN concerned and for verifying the quality of the initiatives within the same
PPPN process.

Keywords: Appraisal, Multi-Criteria Analysis, Benchmarking, Public-Private
Partnership, Stakeholders, Governance.

1 Introduction

In Italy, activities related to territorial administration (for both redevelopment and
development), as currently (2014) implemented by the majority of local governments,
are based on the search for a balance between public and collective interests, of which
the Public Administration (PA) is the carrier, and the interests of which private entities are the carriers. Exclusively public intervention in redevelopment and recovery
initiatives for settlements has become an "extreme" *modus operandi* creating major
problems related to the identification of investment and management resources in
particular [1].

In the 1990s, and then more recently during the current economic downturn (2008-
2014), scientific debate (both European and Italian) was aimed at identifying action
strategies to address the problems regarding the "urban dimension"; in particular, this
debate has focused on the definition of new procedures able to generate growth,

B. Murgante et al. (Eds.): ICCSA 2014, Part III, LNCS 8581, pp. 208–223, 2014.

competitiveness and physical renewal of the territory through urban redevelopment and limiting the use of public resources. In this context, the subject of the Public-Private Partnership (PPP) has assumed particular importance. In this respect, the European Union has introduced instruments, which have then been implemented by the Member States, which provide for recourse to PPPs, both to activate wider negotiation-type processes for territorial redevelopment (PPPN)[1] and to undertake traditional works of public interest (PPPT)[2].

With reference to the territorial redevelopment process, starting in the 1990s, Italy followed the European experience and issued "innovative" standards, introducing new planning instruments: the so-called Complex Programmes[3] (Integrated Intervention Programmes, Urban Redevelopment Programmes and Urban Rehabilitation Programmes) pertaining to the PPPN. These instruments are more flexible than the traditional authoritative territorial government models used by the PA[4]; in initiatives under the Complex Programmes, the PA can in fact "soften" the exercise of its urban planning authority by negotiating proposals submitted also by private entities, as an exception to municipal planning instruments [2].

At national and regional level, these instruments have been standardised, with legislative devices containing general principles; implementing regulations[5] and/or memoranda regarding the method of preparation, evaluation (*ex ante*, *in itinere* and *ex post*), implementation and management of initiatives to be activated with these instruments have almost never been issued. Consequently, specific indicators by which to measure and compare the expected effects of the initiatives to be activated or that have been activated have not been identified.

Due to technical difficulties and/or the discretion that the policy-maker sometimes reserves in implementing territorial redevelopment programmes based on PPPN, local governments have rarely developed protocols of direction containing specific benchmarks to transparently verify both the equity of treatment among private parties and,

---

[1] Urban Pilot Projects Urban I and Urban II are among the PPPN experiences carried out at European level.
[2] The use of the following instruments is indicated for implementation of PPPT at European level: Design Build Finance Transfer, Service Contracts, Management Contracts, Build Lease Transfer, Design Build -Finance Operate, Concession, Build Operate Own, etc.; in the specific Italian situation, the following have been implemented: Project Financing, Leasing in Building, Building and Management Concessions.
[3] The Complex Programmes were established by Law no. 179/1992, and were subsequently implemented by the Regions with specific standards.
[4] The objective of PAs is to increase collective benefits, while private parties are oriented towards maximising revenues, profits and extra-profits through the opportunity to implement initiatives through the same variant of the planning instrument.
[5] At the regional level, there is only one significant experience conducted by the Marche Region (2009), which issued the Implementing Regulation of Regional Law no. 16 of 2005 concerning "Regulation of urban redevelopment projects and strategies for ecologically equipped productive areas". The Regulation provides both criteria for determining the level of public interest in urban redevelopment projects and criteria and parameters to comprehensively assess initiatives related to redevelopment programmes. The Regulation also provides *benchmarks*, or performance thresholds, that the initiatives must follow.

especially, the suitability of the proposed initiative in terms of collective and social objectives[6] [3].

In this context, the first PPPN experiences launched in Italy, promoted almost exclusively by private operators and built on "generic" references to the law, were therefore used more to act as driving forces for low-risk financial profit, without commitments and special guarantees for the public, rather than achieving public interest objectives [4].

However, within a complex and overall vision (multiplicity of objectives, types of interventions, stakeholders involved) of the modalities of planning and implementing PPPN processes, particular importance should instead be given to the distribution of commitments and conveniences between public and private operators [5].

In fact, on the Italian scene this is one of the critical situations found in initiatives planned and implemented with such programmes and is also attributable to the lack and/or inappropriate use of assessment instruments supporting the choice of decisions to be taken.

Useful indications for responding to this problem can be found in several EU directives which have recognised the use of assessment techniques and instruments[7] to support decisions related to the planning of complex territorial redevelopment processes. Among these, assessment techniques such as Benchmarking[8] (BCM) [6], [7], and Multi Criteria Analysis[9] (MCA) are or particular interest [8], [9], [10].

Already during the late 1990s, the European Commission recognised the BCM as an "integrator of quality processes and stimulator of learning processes both in industries and in different situations, such as Public Administrations" [11]. In this sense, the BCM was intended as an instrument to cyclically and continuously improve the performance of a company or a PA, insofar as it allows a comparison of social behaviour, business practices, market structures, and public, national, regional, sectoral and corporate institutions, also in the light of the continuous evolution of the social and economic framework within which the same PA and/or businesses operate.

---

[6] A particular case is that of the City Council of Rome which, in 2011, entrusted the task of preparing a study to define a "Protocol for the Urban Quality of Rome Capital" to the Urban Brownfield Areas Association (AUDIS) (City Council Memorandum 6830, 5 May 2011). The work of defining the "Protocol for the Urban Quality of Rome Capital", conducted through the comparison of public parties, private operators, scholars and social stakeholders involved in the urban redevelopment plans and projects was based on the "Charter of the Quality of Urban Renewal" published by AUDIS in 2008. The " Protocol for the Urban Quality of Rome Capital. Define and evaluate the quality of complex urban projects" was presented in Rome on 22 May 2012 in the Pietro da Cortona room of the Campidoglio.

[7] The main assessment techniques and instruments indicated in European directives are: Multicriteria analysis, Cost-benefit analysis, SWOT Analysis, Benchmarking, Techniques for participation, interaction and conflict resolution, and Fuzzy Analysis.

[8] It is possible to consult some guidelines for the use of Benchmarking, promoted by the European Commission, based on the Benchmarking methodology at
http://ec.europa.eu/internal_market/securities/benchmarks/index_en.htm.

[9] It is possible to consult some cases for use of the MCA, promoted by the European Commission, at
http://ec.europa.eu/europeaid/evaluation/methodology/tools/
too_cri_som_en.htm

In 2005, the European Commission recognised the MCA as a useful assessment instrument to be used in complex situations involving *Stakeholders* with different objectives, roles, positions, interests and opinions[10] [12].

These assessment instruments assume greater effectiveness when all "decision-maker" Stakeholders are considered and when methods and techniques are used to encourage their participation through consultation and interaction and for the resolution of conflicts which may arise among the expectations of different stakeholders [13].

These assessment methods and techniques, widespread at European level, have only found marginal recognition in Italian legislation[11] and practice. Nevertheless, in scientific circles there are many different proposals for the application and experimentation of these techniques (BCM and MCA).

However, an integrated and joint application of these two inclusive[12] and complex[13] assessment procedures has never been proposed [8], [9], [10], [14], [15], [16], [17].

It is believed that the contextual and integrated application of procedures relating to the BCM and the MCA can be used for the participatory definition of objective and shared performance references (benchmarks) of levels of quality aimed at improving efficiency, effectiveness and transparency in negotiation-type partnership territorial redevelopment processes.

With reference to the redevelopment instruments based on the PPPN, the benchmarks resulting from a process of this type may provide useful elements for combining private and public viewpoints.

2   Objectives

This document is part of the debate mentioned with the explicit aim of proposing a *Benchmarking Multi-criteria Evaluation* (hereinafter BME) method that enables identification of *benchmarks* (in relation to *ex post* assessments of experiences, preferably virtuous, already carried out) to guide the planning of PPPN-based processes and/or training for and/or assessment (*ex ante*) of new initiatives.

The BME, structured on the BCM and the MCA, will act as support to improve the transparency and effectiveness of the administrative activities of the PA in the territorial government (planning processes, training, validation and implementation of settlement transformation initiatives), by focusing on the interaction between public and private partners regarding their conveniences (balance of resources used and produced, risks and guarantees).

---

[10] The European Commission specifies the cases in which the different MCA techniques can be used in assessments (*ex ante*, *in itinere* and *ex post*) and the methods and conditions for the best expression of their potential and for their correct use.

[11] In the Italian legislation system, the use of some MCA techniques is only contemplated in Legislative Decree No. 163 of 2006 in order to select the most economically advantageous offer in the tenders of public works.

[12] Inclusive, because they contemplate the active participation of various stakeholders.

[13] Complex, because they take into account different and heterogeneous aspects.

The BME is calibrated in this text, which illustrates the methodological proposal, and in its application verification (already developed but not shown here) [15] to be applied, in the Italian context, to PPPN-based urban redevelopment processes, but it could also be used in settlement transformation processed based on the PPPT. Moreover, it is believed that the BME can also be used with reference to other European and international contexts.

The method proposed is shown below, and described in sections 3.1. Purpose and Use; 3.2. Structure; 3.3. Method of Implementation of the different phases; 4. Conclusions.

3   The Proposed Methodology: Benchmarking Multi-criteria Evaluation (BME)

3.1   Purpose and Use

In the BME, the MCA is implemented within a BCM model, the *Deming Cycle*, to activate a multi-dimensional learning and change process as part of PPPN-based territorial redevelopment processes and initiatives.

The BME is aimed at defining benchmarks (performance references) designed to improve quality standards (in an objective way that is shared among the Stakeholders involved) of any new planning of such processes and of the initiatives through which they are implemented.

The benchmarks are performance for which Stakeholders express sufficient satisfaction at the very least. Therefore, they can be parameters useful to the PA[14] (provincial, regional and local) that must both define guidelines for the planning of these processes and be expressed when the contents of the individual initiatives proposed are approved.

In particular, the benchmarks resulting from the application of BME can be used:

- by PAs with planning and authorisation responsibility for initiatives related to these instruments. For this purpose, explanatory and policy-making instruments must be structured and organised by the PAs (e.g. guidelines, implementing regulations, strategy memoranda);

- by promoters of initiatives relating to individual processes in order to align their proposals to the references assumed as a direction from the PAs.

A municipal PA could make use of the BME to evaluate its actions with respect to any specific PPPN processes and to define the trends that must be followed in order to promote initiatives in its territory aligned with quality standards (restricted territorial scope). However, at the local level, a significant and satisfying number of initiatives of the same process for which the BME was implemented may not have been started; this would result in the low significance of the results of this method.

---

[14] Without prejudice to the reform of Local Authorities currently (2014) before the Italian Parliament, in the framework of the activities for which the PAs are responsible: i) Presidential Decree No. 8 of 15 January 1972 and Presidential Decree No. 616 of 14 July 1977 gave Regions legislative powers in relation to urban planning; ii) with reform of Chapter V of the Constitution (from 2001), Provinces were given territorial and provincial planning responsibilities, iii) pursuant to chapter V of the Constitution (from 1948), Municipalities were given local territorial government powers.

Therefore, use of the method at provincial level, but above all regional level (vast territorial scope), assumes greater interest; in vast but nevertheless homogenous territorial areas, it may be the starting point for stimulating local administrations towards "learning" and "improvement" of their assets, by borrowing the best practices promoted at their territorial level (homogenous), insofar as it permits a comparison of the work of different local PAs, being able to consider a significant sample of initiatives.

Cyclic application of the BME may enable the achievement of increasingly better and shared performance appropriate to the PPPN reference framework for which the same BME was implemented.

3.2   Structure

The BME is structured as a *Deming Cycle* in which the assessment nodes are resolved through MCA techniques (in order to permit the assessment of different, and sometimes heterogeneous, aspects that characterise the PPPN processes) as well as through logical and mathematical functions.

The BME is an assessment instrument that can be used by a PA that considers it necessary to perfect the activity related to a specific territorial redevelopment process based on the PPPN. Better results are achieved if the BME is applied to processes in which a large number of significant initiatives are implemented.

Once the specific PPPN to be assessed is identified, it is possible to launch the BME, which is divided into four macro phases:

1) The *Plan* phase, designed to identify the key elements for subsequent implementation of purely assessment phases (*Do* and *Check*): survey and analysis of the initiatives (adopted and approved) in the specific PPPN process taken into consideration, and recognition of the sample of initiatives on which to implement the BME (alternatives); identification of Stakeholder categories to engage;

2) The *Do* phase, designed to define and explain all variables related to the alternatives to be considered in the assessment: construction of the impact matrix (formulation of criteria, sub-criteria, indicators) and the matrix of viewpoints (weighting sub-criteria and objective functions); insertion of data (*input*) in the matrices;

3) The *Check* phase, designed to define the benchmarks: collection of opinions through appropriate MCA techniques to obtain the quality and hierarchical orders of the alternatives; identification of the most significant (*best in class*[15]) alternatives in relation to the level and quality of interaction and satisfaction of the main public and private partners and other parties involved; identification of benchmarks derived from best performance found among the alternatives;

4) The *Act* phase, designed to organise benchmarks into guidelines so that they become useful references for the forecasts and decisions to be made in new initiatives similar to those analysed in the BME process by bringing efficiency and effectiveness

---

[15] It is considered appropriate to briefly summarise the difference between best in class and best practices, terms that are sometimes used interchangeably. As part of a limited sample of initiatives, best in class means those initiatives that are positively the most significant; generally, best practices includes the most significant positive experiences with reference to a non-limited and broader set of initiatives. The difference lies in the size of the framework in which the significance of a certain experience/initiative is recognised.

to the PPPN process to be improved: re-assembly of the set of benchmarks in a document, in line with the applicable legislation and regulation of the type of PPPN process to which the BME is applied.

3.3   Implementation

The following briefly describes the operating procedures for implementing the various BME phases (Table 1).

Table 1. BME phases

Phases	Action	Output
Plan	Analysis of the initiatives relating to the type of PPPN	Highlight the most significant aspect
	Identification of alternatives	Select significant and representative initiatives
	Identification of Stakeholder (Stk)	Choise representative Stk for number and categories
Do	Impact Matrix	Criteria (envronmental, procedural, socio-economic, etc)
		Sub-criteria relating to criteria
		Indicators relating to sub-criteria
		Imput data
	Viewpoint Matrix	Weights
		Objective Function
	Other viewpoint	N. of alternatives to be considered best in class
		Satisfaction of the performance for each sub-criterion
Check	Aggregate input data	Appraisal score
	Classifications of alternatives	Preference order of alternatives for each Stk category
	Identification of performance of the best in class	Average, maximum, modal performance
	Definition of a benchmark for each sub-criterion considered	Related to level of satisfaction of best in class's performance
Act	Pre Act: organisation of the benchmarks identified	Guidelines or regulatory framework
	In Act: approval and institutionalisation of the BME results	BME and benchmarks: protocol, standard, procedural

Plan. The following is carried out in the Plan phase:

1) Analysis of the initiatives relating to the type of PPPN process to which the PA has decided to implement the BME; according to criteria (urban planning, social and procedural) through which to enable a thorough reading of the initiatives in order to highlight the most significant aspects;

2) Identification of alternatives *(An)* to be considered in application of the BME: verify comparability of the initiatives activated in the PPPN process to which the BME is applied and select a significant and representative sample of initiatives (alternative). It is appropriate to select this sample if there is a large number of initiatives which, if all of them have been taken into consideration, can make the BME process particularly long or complex, or if the initiatives were activated during broader time horizons. A necessary condition for the selection of the *set* of alternatives is the choice of initiatives that have not only been initiated (and under investigation), but that have already been approved by the competent bodies and are covered by an urban planning agreement. In fact, they become legally valid only after approval and, in particular at the time of "agreement" between the PA and private operator.

3) Identification of Stakeholder *(STn)* categories to engage in the *benchmark* definition process. Depending on the type of PPPN process and level of detail to be given to the BME, the composition of the category and number of representatives to be interviewed will be different. Normally, because they are affected by the PPPN-based redevelopment processes, the Stakeholder categories that may be engaged in the BME are: institutions (state, regional and local), business owners, property owners, economic

operators, residents, workers and tourists. The plurality of Stakeholder viewpoints is essential to give the process a suitable degree of participation and horizontal government (governance), in line with current European trends. Once the Stakeholder categories are defined, it is also necessary to specify the number of subjects to be interviewed for each different Stakeholder category, and to identify the contact method to be used to carry out the interviews for obtaining the opinions of the same Stakeholders.

Do. The following is carried out in the Do phase:

1) Construction of the impact matrix (Table 2), into which the input data representative of the performances *(i)* related to each of the initiatives considered (significant sample) and constituting the assessment alternatives *(An),* will be inserted ordinately. Construction of this matrix may take place following collection, analysis and processing of data regarding: i) objectives, constraints, requirements, guidelines and specific guidelines set out in the legal systems that govern the PPPN process covered by the BME; ii) the alternatives considered (dimensional, financial, procedural, economic, social, administrative, etc.).

The information derived from previous analysis and processing makes it possible to:

a) Select/identify assessment criteria *(Cn)*, sub-criteria *(SCn)* and related indicators *(In)* specific to the type of PPPN process in question. The criteria and sub-criteria must be formulated and calibrated as the assessment proceeds in relation to the type of process to which the BME is being applied. With reference to the indications of the European Commission [12]:

- there are usually 5 categories of criteria against which to define the sub-criteria (in sufficient numbers to significantly express the transformations generated by the initiative): environmental, financial, socio-economic, procedural and technical;

- The *set* of sub-criteria should be a *corpus* that is in line with the assessment purposes and balanced (the indications must be unequivocal and, therefore, a special interest should not be measured by more than one sub-criteria) through which concrete and credible results can be achieved;

- The indicators are to be defined and explained on the basis of the possibility of: i) using existing data that are easily and statistically comparable; ii) making complex issues understandable; iii) integrating with other assessment indicators. In assessment practice, it is known that the indicators can usually refer to different scales and measures, and that they should still be related to what it is intended to describe. They may be divided into four macro-categories: i) quantitative with legal standards, referring to all quantitative data that can be quantitatively measurable and comparable with a threshold defined by law; ii) the quantitative without *legal* standards, referring to data that can be assessed through thresholds defined *ad hoc*; iii) qualitative (with possible quantitative elements), iv) cartographic;

b) Entry of the input data of each sub-criterion for each of the alternatives in the impact matrix *[i(SCn;An)]*. These are deduced from all collected documents (project and administrative) regarding the alternatives considered;

Table 2. Impact matrix (with the inclusion of criteria, sub-criteria and indicator examples)

Criteria (Cn)		Sub-criteria (SCn)	Indicators (In)	Alternatives (An)		
				a	b	n
Financial	F.1	Advantages for the proposer: non-repayable grants	% on the total amount of the intervention coming from public funding	i(F.1;Aa)	i(F.1;Ab)	i(F.1;An)
	F.2	Financial benefits for the public involved in the initiative	Ratio (%) between plus-public value (in €) and plus-private value (in €)	i(F.2;Aa)	i(F.2;Ab)	i(F.2;An)
	F.3	Guarantees for the public	% of the amount corresponding to the plus-public value paid or guaranteed by a surety	i(F.3;Aa)	i(F.3;Ab)	i(F.3;An)
	F.n	…	…	i(F.n;Aa)	i(F.n;Ab)	i(F.n;An)
Procedural	P.1	Reliability of promoters	Years of business of the company proposing the initiative	i(P.1;Aa)	i(P.1;Ab)	i(P.1;An)
	P.n	…	…	i(P.n;Aa)	i(P.n;Ab)	i(P.n;An)
Socio-economic	S.1	Residential capacity	% established residents/total resident in the City	i(S.1;Aa)	i(S.1;Ab)	i(S.1;An)
	S.2	Population density	Inhabitants / ha	i(S.2;Aa)	i(S.2;Ab)	i(S.2;An)
	S.3	Workforce during operation	% number of works to be used in the initiatives (public and private) provided for in the PII/number of unemployed people in the municipal area (ISTAT data)	i(S.3;Aa)	i(S.3;Ab)	i(S.3;An)
	S.n	…	…	i(S.n;Aa)	i(S.n;Ab)	i(S.n;An)
Environmental	A.1	Pollution level	Quality - Pollution risk (air, water, soil, subsoil, acoustic)	i(A.1;Aa)	i(A.1;Ab)	i(A.1;An)
	A.n	…	…	i(A.n;Aa)	i(A.n;Ab)	i(A.n;An)
Urban	U.1	Total private use building potential	Index of private territorial building suitability	i(U.1;Aa)	i(U.1;Ab)	i(U.1;An)
	U.2	Level of functional mix (residential)	% residential cubic meters on total private cubic meters	i(U.1;Aa)	i(U.1;Ab)	i(U.1;An)
	U.3	Level of public-private volume	% public cubic meters on private cubic meters	i(U.1;Aa)	i(U.1;Ab)	i(U.1;An)
	U.n	…	…	i(U.n;Aa)	i(U.n;Ab)	i(U.n;An)

c) (Possible) standardisation of input data (heterogeneous) inserted into the impact matrix through: i) linear normalisation functions; ii) logical-mathematical functions (zero-max, min-max, max-max, vector, line total, zero mean); iii) value and utility functions. The choice of the normalisation function should be made in relation to heterogeneity/homogeneity, scrap value, impact matrix data input. This operation permits making the impact matrix homogeneous for the subsequent collection of opinions in the *Check* phase;

d) (Possible) implementation of the dominance analysis, needed if there are some "dominate" alternatives, in the Paretian sense, and, therefore, immediately recognisable as "not satisfactory" at this stage, even before the collection of reviews;

2) Viewpoint matrix (Table 3) to explain the preferences of a significant number of subjects ordinarily representative of the categories of Stakeholder to be considered in the assessment. To construct this matrix, the viewpoints of a sample of Stakeholders should be pointed out in order to proceed with the: i) weighting of the criteria and sub-criteria *(p)*; ii) definition of the objective functions *(fo)*; iii) collection of data useful for

the implementation of the next *Check* phase (definition of the best in class selection criteria and demarcation of the satisfaction level perceived by the Stakeholder categories in relation to certain performances).

Table 3. Viewpoints matrix

| Criteria (Cn) | Sub-criteria (SCn) | Indicators (In) | Category of Stakeholders *(STn)* | | | |
| | | | a | | n | |
			Weights	Objective function	Weights	Objective function
Financial	F.1	Advantages for the proposer: non-repayable grants	% on the total amount of the intervention coming from public funding	$p(F.1;STa)$	$fo(F.1;STa)$	$p(F.1;STn)$	$fo(F.1;STn)$
	F.2	Financial benefits for the public involved in the initiative	Ratio (%) between plus-public value (in €) and plus-private value (in €)	$p(F.2;STa)$	$fo(F.2;STa)$	$p(F.2;STn)$	$fo(F.2;STn)$
	F.3	Guarantees for the public	% of the amount corresponding to the plus-public value paid or guaranteed by a surety	$p(F.3;STa)$	$fo(F.3;STa)$	$p(F.3;STn)$	$fo(F.3;STn)$
	F.n	…	…	$p(F.n;STa)$	$fo(F.n;STa)$	$p(F.n;STn)$	$fo(F.n;STn)$
Procedural	P.1	Reliability of promoters	Years of business of the company proposing the initiative	$p(P.1;STa)$	$fo(P.1;STa)$	$p(P.1;STn)$	$fo(P.1;STn)$
	P.n	…	…	$p(P.n;STa)$	$fo(P.n;STa)$	$p(P.n;STn)$	$fo(P.n;STn)$
Socio-economic	S.1	Residential capacity	% established residents/total resident in the City	$p(S.1;STa)$	$fo(S.1;STa)$	$p(S.1;STn)$	$fo(S.1;STn)$
	S.2	Population density	Inhabitants / ha	$p(S.2;STa)$	$fo(S.2;STa)$	$p(S.2;STn)$	$fo(S.2;STn)$
	S.3	Workforce during operation	% number of works to be used in the initiatives (public and private) provided for in the PII/number of unemployed people in the municipal area (ISTAT data)	$p(S.3;STa)$	$fo(S.3;STa)$	$p(S.3;STn)$	$fo(S.3;STn)$
	S.n	…	…	$p(S.n;STa)$	$fo(S.n;STa)$	$p(S.n;STn)$	$fo(S.n;STn)$
Environmental	A.1	Pollution level	Quality - Pollution risk (air, water, soil, subsoil, acoustic)	$p(A.1;STa)$	$fo(A.1;STa)$	$p(A.1;STn)$	$fo(A.1;STn)$
	A.n	…	…	$p(A.n;STa)$	$fo(A.n;STa)$	$p(A.n;STn)$	$fo(A.n;STn)$
Urban	U.1	Total private use building potential	Index of private territorial building suitability	$p(U.1;STa)$	$fo(U.1;STa)$	$p(U.1;STn)$	$fo(U.1;STn)$
	U.2	Level of functional mix (residential)	% residential cubic meters on total private cubic meters	$p(U.1;STa)$	$fo(U.1;STa)$	$p(U.1;STn)$	$fo(U.1;STn)$
	U.3	Level of public-private volume	% public cubic meters on private cubic meters	$p(U.1;STa)$	$fo(U.1;STa)$	$p(U.1;STn)$	$fo(U.1;STn)$
	U.n	…	…	$p(U.n;STa)$	$fo(U.n;STa)$	$p(U.n;STn)$	$fo(U.n;STn)$

The collection of viewpoints (synthetic) of the Stakeholder categories may be conducted through use of the Stakeholder Analysis by interviewing a significant sample of representatives from the various Stakeholder categories involved in the BME.

Therefore, through the Stakeholders Analysis, and processing data from interviews with arithmetic averages, it is possible to:

a) Assign weights (an indication of the importance assigned by Stakeholders and/or decision-makers) to each criterion and sub-criterion *[p(SCn;An)]*. For this purpose, the following techniques are used, which are to be chosen in relation to the number of respondents and the level of "robustness" to be achieved in the weighing of the criteria and sub-criteria: direct assignment; pairwise comparison; *Paired Comparison Technique; Delphi Method*; methods based on a single order. This construction of a hierarchy makes it possible to define an order of importance among various criteria and/or sub-criteria;

b) Define the objective function with the aim of representing the optimal expectation or preferable performance level for each sub-criterion, especially for homogeneous stakeholder categories *[fo(SCn;An)]*. Satisfaction in relation to the performance of the indicator is detected by interviewing subjects in relation to each sub-criterion. Three orientations of the objective function are usually identified: i) maximisation of performance (increase in the value expressed by the indicator); ii) minimisation of the performance (decrease in the value expressed by the indicator); iii) inclusion of performance within predetermined values (value expressed by the indicator within a pre-established range).

c) Define the number of alternatives to be considered *best in class* as well as the levels of satisfaction (very high, high medium, low and no satisfaction) of the *performance* for each sub-criterion. This is necessary to implement the *Check* phase.

Check. The following is carried out during the Check phase:

1) Aggregate input data (possibly made homogeneous) of the impact matrices and viewpoints to obtain the appraisal score *(as)* (output data) from which to define the hierarchy from among the alternatives considered for each category of Stakeholders *[as(An;STn)]*. Opinions may be aggregated using various systems such as AHP, REGIME, ANP, MAUT, TOPSIS, NAIADE, etc. [16], [17]; the technique should be chosen on the basis of a number of alternatives, criteria, sub-criteria and subjects to be considered in the assessment as well as the possible legal recognition of opinion collection systems in regulations relative to the PPPN to which the BME is applied;

2) Formation of classifications indicating the preference order *(pos)* of alternatives for each Stakeholder category *[pos(An;STn)]* (Table 4).

Table 4. Single group rankings and definition of the *best in class* (with sample positioning)

Alternatives	Stakeholder a			Stakeholder b			Stakeholder n		
	Appraisal score	Classification		Appraisal score	Classification		Appraisal score	Classification	
Aa	as(Aa;Sta)	pos(Aa;Sta)	6	as(Aa;Stb)	pos(Aa;Stb)	4	as(Aa;Stn)	pos(Aa;Stn)	6
Ab	as(Ab;Sta)	pos(Ab;Sta)	5	as(Ab;Stb)	pos(Ab;Stb)	5	as(Ab;Stn)	pos(Ab;Stn)	2
Ac	as(Ac;Sta)	pos(Ac;Sta)	4	as(Ac;Stb)	pos(Ac;Stb)	6	as(Ac;Stn)	pos(Ac;Stn)	5
Ad	as(Ad;Sta)	pos(Ad;Sta)	2	as(Ad;Stb)	pos(Ad;Stb)	3	as(Ad;Stn)	pos(Ad;Stn)	4
Ae	as(Ae;Sta)	pos(Ae;Sta)	1	as(Ae;Stb)	pos(Ae;Stb)	1	as(Ae;Stn)	pos(Ae;Stn)	3
Af	as(Af;Sta)	pos(Af;Sta)	3	as(Af;Stb)	pos(Af;Stb)	2	as(Af;Stn)	pos(Af;Stn)	1
An	as(An;Sta)	pos(An;Sta)	...	as(An;Stb)	pos(An;Stb)	...	as(An;Stn)	pos(An;Stn)	...

N.B.: Featured best in class (criteria of selection: top alternatives - n - (3) of each classification "single-group" at least for several - n - (2) category of Stakeholder of every (3) category Stakeholders

The collection of opinions permits sorting the alternatives according to a qualitative-hierarchical order (different "single group" classifications for each Stakeholder category); once the classifications are defined, and the results of the Stakeholders Analysis are retrieved (see Do phase, section 2, point c), it is possible to define the best in class or the n alternatives preferred by the n Stakeholders;

3) Identification of performance average, maximum and possibly modal of the best in class (pre-benchmark range). With the best in class performance, summary schemes are created (one for each sub-criterion) in which these values are given; subsequently, it is possible to revise such values in order to identify the average *[Vmed(SCN)]*, maximum *[Vmax(SCN)]* and possibly the modal *[Vmod(SCN)]* of all the best in class of each sub-criterion (Table 5);

Table 5. Definition of the *pre-benchmark range* (with examples of *pre-benchmarks*)

Criteria (Cn)		Sub-criteria (SCn)	Inticators (In)	Average value		Maximum value (predominant objective function)		Modal value	
Financial	F.1	Advantages for the proposer: non-repayable grants	% on the total amount of the intervention coming from public funding	Vmed(F.1)	0,00%	Vmax(F.1)	0,00%	Vmod(F.1)	n.p.
	F.2	Financial benefits for the public involved in the initiative	Ratio (%) between plus-public value (in €) and plus-private value (in €)	Vmed(F.2)	58,10%	Vmax(F.2)	66,76%	Vmod(F.2)	n.p.
	F.3	Guarantees for the public	% of the amount corresponding to the plus-public value paid or guaranteed by a surety	Vmed(F.3)	0,00%	Vmax(F.3)	0,00%	Vmod(F.3)	n.p.
	F.n	...	...	Vmed(F.n)		Vmax(F.n)		Vmod(F.n)	
Procedural	P.1	Reliability of promoters	Years of business of the company proposing the initiative	Vmed(P.1)	21,6	Vmax(P.1)	44	Vmod(P.1)	n.p.
	P.n	...	...	Vmed(P.n)		Vmax(P.n)		Vmod(P.n)	
Socio-economic	S.1	Residential capacity	% established residents/total resident in the City	Vmed(S.1)	1,73%	Vmax(S.1)	0,00%	Vmod(S.1)	n.p.
	S.2	Population density	Inhabitants / ha	Vmed(S.1)	56	Vmax(S.1)	93	Vmod(S.1)	n.p.
	S.3	Workforce during operation	% number of works to be used in the initiatives (public and private) provided for in the PII/number of unemployed people in the municipal area (ISTAT data)	Vmed(S.1)	0,68%	Vmax(S.1)	1,79%	Vmod(S.1)	n.p.
	S.n	...	...	Vmed(S.n)		Vmax(S.n)		Vmod(S.n)	
Environmental	A.1	Pollution level	Quality - Pollution risk (air, water, soil, subsoil, acoustic)	Vmed(A.1)	Low	Vmax(A.1)	Void	Vmod(A.1)	n.p.
	A.n	...	...	Vmed(A.n)		Vmax(A.n)		Vmod(A.n)	
Urban	U.1	Total private use building potential	Index of private territorial building suitability	Vmed(U.1)	1,29	Vmax(U.1)	0,67	Vmod(U.1)	n.p.
	U.2	Level of functional mix (residential)	% residential cubic meters on total private cubic meters	Vmed(U.2)	58,89%	Vmax(U.2)	0,00%	Vmod(U.2)	n.p.
	U.3	Level of public-private volume	% public cubic meters on private cubic meters	Vmed(U.3)	26,17%	Vmax(U.3)	74,52%	Vmod(U.3)	n.p.
	U.n	...	...	Vmed(U.n)		Vmax(U.n)		Vmod(U.n)	

n.p. = not present

4) Definition of a benchmark *(B)* for each sub-criterion considered *[B(SCn)]*. Prior to the definition of the benchmarks, the degree of satisfaction *(g)* of the *Stakeholder* category must be verified (defined within the Stakeholders Analysis, see the *Do* phase paragraph 2 point c) with the average *(gVmed)*, maximum *(gVmAX)* and modal *(gVmod) values* of the best in class. This permits verification of the satisfaction level for each Stakeholder category *[g(SCn;STn)]* and consequently the

'"acceptability" of the references identified. Once this verification is completed, the definition of the Benchmark may be direct (the average, maximum, modal, or intermediate value of these becomes the benchmark) when: i) the satisfaction expressed by the Stakeholder category (average, maximum and modal values of the best in class) is at minimum "average" (best if "high" or "very high"): the benchmark may be the average value, the maximum value, the modal value or, alternatively, an interpolated value between any of the above depending on the level of satisfaction expressed by the Stakeholders. It is best to assume the value for which there is greatest satisfaction expressed by the Stakeholders as the benchmarks); ii) there is consistency with the legislation that governs the process type for which the BME has been implemented (Table 6).

Table 6. Verify the *Stakeholders'* acceptability of the average and maximum values (with examples of satisfaction) and *benchmarks* proposal (examples of *benchmarks*)

Sub-criteria (SCn)	Average value	Satisfaction (g)		Maximum value (predominant objective function)	Satisfaction (g)		Benchmark proposal
		g(STa)	g(STn)		g(STa)	g(STn)	
F.1	0.00%	gVmed(F.1;STa;) Very low	gVmed(F.1;STn;) Very low	0.00%	gVmax(F.1;STa;) Very low	gVmax(F.1;STn;) Very low	0.00%
F.2	58.10%	gVmed(F.2;STa;) High	gVmed(F.2;STn;) Alta	66.76%	gVmax(F.2;STa;) Very high	gVmax(F.2;STn;) High	60.00%
F.3	0.00%	gVmed(F.3;STa;) Very low	gVmed(F.3;STn;) Very low	0.00%	gVmax(F.3;STa;) Very low	gVmax(F.3;STn;) Low	10.00%
F.n	...	gVmed(F.n;STa;)	gVmed(F.n;STn)	...	gVmax(F.n;STa;)	gVmax(F.n;STn;)	...
P.1	21.6	gVmed(P.1;STa;) Medium	gVmed(P.1;STn) Medium	44	gVmax(P.1;STa;) Very high	gVmax(P.1;STn;) Very high	25
P.n	...	gVmed(P.n;STa;)	gVmed(P.n;STn;)	...	gVmax(P.n;STa;)	gVmax(P.n;STn;)	...
S.1	1.73%	gVmed(S.1;STa;) Medium	gVmed(S.1;STn;) High	0.00%	gVmax(S.1;STa;) Very high	gVmax(S.1;STn;) Very high	2.00%
S.2	56	gVmed(S.1;STa;) Low	gVmed(S.1;STn;) Medium	93	gVmax(S.1;STa;) Low	gVmax(S.1;STn;) Very low	100
S.3	0.68%	gVmed(S.3;STa;) High	gVmed(S.3;STn;) High	1.79%	gVmax(S.3;STa;) Very high	gVmax(S.3;STn;) Very high	0.70%
S.n	...	gVmed(S.n;STa;)	gVmed(S.n;STn;)	...	gVmax(S.n;STa;)	gVmax(S.n;STn;)	...
A.1	Basso	gVmed(A.1;STa;) High	gVmed(A.1;STn;) High	Void	gVmax(A.1;STa;) Very high	gVmax(A.1;STn;) Very high	Void
A.n	...	gVmed(A.n;STa;)	gVmed(A.n;STn;)	...	gVmax(A.n;STa;)	gVmax(A.n;STn;)	...
U.1	1.29	gVmed(U.1;STa;) Medium	gVmed(U.1;STn;) Medium	0.67	gVmax(U.1;STa;) Very high	gVmax(U.1;STn;) Medium	1.00
U.2	58.89%	gVmed(U.2;STa;) High	gVmed(U.2;STn;) High	0.00%	gVmax(U.2;STa;) Very high	gVmax(U.2;STn;) Very low	60.00%
U.3	26.17%	gVmed(U.3;STa;) High	gVmed(U.3;STn;) High	74.52%	gVmax(U.3;STa;) Very high	gVmax(U.3;STn;) Very high	30.00%
U.n	...	gVmed(U.n;STa;)	gVmed(U.n;STn;)	...	gVmax(U.n;STa;)	gVmax(U.n;STn;)	...

If the benchmarks obtained are inconsistent according to the latter, the causes of the discrepancy should be investigated by checking if the BME has been properly applied, and, in spite of the correct application of the BME, if the value identified show the actual non-fulfilment of the Stakeholders' expectations.

If there are BME application inconsistencies, the application of the method should be repeated with the necessary corrections; if the values defined with the BME are still not satisfactory for the Stakeholders, the party responsible for directing the BME process may propose an alternative Benchmark by changing the minimum expectation expressed by the same during data collection; however, it is important that the defined Benchmark be effectively achievable and does not represent an unreachable expectation that is detached from reality.

Act (Pre Act and In Act). This phase can be divided into two sub-phases:
- *Pre Act*, organisation of the benchmarks identified as guidelines or regulatory frameworks;
- *In Act*, approval and institutionalisation of the BME results.

In the *Pre Act* phase, the benchmarks are contextualised differently by defining the guidelines, which are considered as a set of recommendations developed systematically, and integrated and coordinated with each other.

In the *In Act* phase, the guidelines are approved and organised in a regulatory device thus making them "protocol" and/or "standard" and/or "procedure"[16] for the authority promoting the BME.

Thus, the guidelines formulated are an expression of behaviour and *modus operandi* shared by different PPPN stakeholders, and represent a starting point for subsequent planning regarding the type of process subjected to the BME, and may represent references to be followed during the validation phases (adoption and approval) of new initiatives.

4   Conclusions

The proposed methodology, called *Benchmarking Multi-criteria Evaluation* (BME), created by placing two analysis and evaluation approaches into a new and unique system: the MCA and BCM, can be applied to any type of PPPN-based redevelopment process.

The BME is closely linked to the best experiences that have already taken place and whose contents can actually be replicated, if satisfactory to the Stakeholders.

In a specific PPPN process, if certain framework conditions (of a defined spatial area at a specific point in time) have enabled some initiatives (best in class) to achieve certain performance levels for which the Stakeholders have expressed satisfaction, and if those framework conditions persist, new initiatives with performance levels comparable (if not superior) to past initiatives, which were considered the best and most satisfactory, can and should be provided. Consequently, the new PPPN programming and the relative initiatives should provide for the attainment of the same

---

[16] The term "protocol" refers to a pre-defined behaviour pattern within an activity that describes a rigid sequence of behaviours. It is a document that provides a sequence of actions that must be carried out to achieve the given objective. In this case, "standard" refers to values expressed by an indicator with which the frequency of activities or services rendered or the *performances* of the intervention identified are defined by using a scale as a reference measure. This term has a normative meaning associated with an explicit quality opinion. Procedure means a set of professional actions aimed at the set objective, or a sequence of actions that are more or less rigid that describe individual process phases to harmonise activities and behaviours by reducing individual discretion.

satisfactory performance levels. Therefore, the benchmarks defined with the BME are interconnected to the framework in which the method is applied, and represent a synthesis between practice, legislation, expectations and priorities of Stakeholders (specific to a given context, at a specific time). Thus, as already mentioned, the BME enables the synthesis of different interests expressed through a "rational" observation of reality and not referred to abstract expectations. Hence, the outlined methodological approach permits translating these expectations into reference elements to build benchmarks that represent a balanced composition and that may allow the PA (local, provincial and regional) a more virtuous management of settlement redevelopment and recovery processes based on the PPPN. So, benchmarks are clear, objective and democratic elements through which a phenomenon can be represented and shaped and which, in this case, pertains to settlement redevelopment and recovery. Consequently, when forms of PPPN are used in settlement redevelopment processes, the proposed method may be an opportunity for PAs to have a greater guarantee of producing development, being competitive, boosting the economy and improving the quality of life of a community [15].

References

1. Curti, F.: Lo scambio leale. Negoziazione urbanistica e offerta privata di spazi e servizi pubblici, Officina, Rome (2007)
2. Morano, P.: La compensazione urbanistica nell'acquisizione consensuale non onerosa di risorse ad uso pubblico. In: Bentivegna, V., Miccoli, S. (eds.) Valutazione e Progettazione Urbanistica. Metodologia e applicazioni, pp. 175–206 (2010) ISBN: 978-88-496-0401-6
3. Stanghellini, S.: Il negoziato pubblico privato nei progetti urbani. Principi, metodi e tecniche di valutazione, Dei, Rome (2012)
4. Urbani, P.: Territorio e poteri emergenti. La politica di sviluppo tra urbanistica e mercato, Giappichelli, Turin (2007)
5. Morano, P., Tajani, F.: Break Even Analysis for the financial verification of urban regeneration projects. In: Applied Mechanics and Materials, vol. 438-439, pp. 1830–1835. Trans Teach Publications Ltd. (2013), doi:10.4028/www.scientific.net/AMM.438-439.1830 ISSN: 1660-9336
6. Karloff, B., Ostblom, S.: Benchmarking: A signpost to excellence in quality and productivity. Wiley, Chichester (1993)
7. Camp, R.: Business process Benchmarking: Trovare e migliorare le prassi vincenti, Editoriale Itaca, Milano (1996)
8. Nijikamp, P., Rietveld, P., Voogd, H.: Multicriteria evaluation in physical planning. North Holland Publ., Amsterdam (1990)
9. Keeney, R.L., Raiffa. H.: Decisions with multiple objectives: Preferences and value tradeoffs. John Wiley, New York (1993); Republished by Cambridge University Press, New York
10. Lichfield, N., Barbanente, A., Borri, D., Khakee, A., Pratt, A.: Evaluation in spatial planning: Facing the challenge of complexity. Kluwer Academic Publishers, Dordrecht (1998)
11. European Commission (1997), Benchmarking: Implementation of an Instrument available to economic actors and public authorities, Communication from the Commission to the Council, the European Parliament, the Economic and Social Committee and the Committee of the Regions. COM (97) 153 final, April 16, 1997, Bruxelles (1997)

12. European Commission, Europe Aid Cooperation Office, Linee guida per l'analisi multicriteria (UE), Quando e perché l'analisi multicriteria dovrebbe essere utilizzata, Bruxelles (2005)
13. Guarini, M.R., Battisti, F.: Evaluation and Management of Land-Development Processes Based on the Public-Private Partnership. Advanced Materials Research 869-870, 154–161 (2014); online available at, http://www.scientific.net (since December I3, 2013), doi: 10.4028/www.scientific.net/AMR.869-870.154 ISSN: 10226680, ISBN 978-303785975-9
14. Fusco Girard, L., Nijkamp, P.: Le valutazioni per lo sviluppo sostenibile della città e del territorio, Franco Angeli, Rome (2012)
15. Roscelli, R.: Misurare nell'incertezza. Valutazioni e trasformazioni territoriali. Celid, Turin (2005) ISBN: 88-7661-664-0
16. Battisti, F.: Valutazioni comparative per lo sviluppo dei processi di riqualificazione del territorio a partenariato pubblico-privato: una proposta di metodo. Sperimentazione sui Programmi Integrati di Intervento della Regione Lazio. PhD Thesis in Doctoral School of "Riqualificazione e Recupero Insediativo" (XXIV ciclo), Tutor Prof.ssa M.R. Guarini, Facoltà di Architettura, Sapienza University of Rome (2012), http://padis.uniroma1.it/handle/10805/2088
17. Mattia, S.: Costruzione e valutazione della sostenibilità dei progetti, vol. 1 & 2, Franco Angeli, Rome (2007)

A DSS to Assess and Manage the Urban Performances in the Regeneration Plan: The Case Study of Pachino

Maria Rosa Trovato[1] and Salvatore Giuffrida[2]

Department of Civil Engineering and Architecture - University of Catania
{mrtrovato,sgiuffrida}@dica.unict.it

Abstract. The historic centers, by using a large sustainability concept, should be the places where the quality and the symbolic values may prevail on the quantitative and functional ones, but often, the poor resources and the low quality of the management do not allow to promote a sustainable retraining process for them. The plan decision makers should verify the impact of the technical choices and then the sustainability of the actions, but also the impact of the planning policies on the economic performance and then on the economic sustainability of these actions. Therefore they must be helped to identify the planning actions, their funds and the value system they want to promote. In this regard, this study proposes a model to support the management of the retraining plan for Pachino's historic center. The proposed model is a DSS that is developed using the MAUT and the IMO-DRSA tool.

Keywords: sustainable retraining plan MAUT·IMO-DRSA economic performance equalization.

1 Introduction

Nowadays town planning strongly focuses on the different types of sustainability, especially in the old towns, where qualitative and symbolic values prevail over the quantitative and functional ones. In contrast, pressing budget constraints don't allow the historic centers to be unproductive places to which devolve unlimited resources. In the ordinary urban planning the supposed interventions, as ruled by the Detailed Renewal Plan, give rise to a set of positive and negative externalities that, as far as possible, an equalized plan should avoid and/or adequately compensate. As a result, the plan should be designed and applied as a supportive process in which wealth – property market value, architectural quality and functional performances – increases the public benefit. The economic-evaluative structure can be considered nowadays *the big issue* both in urban planning and in estimation theory and practice. The main concern are the internalization of the real estate value surplus due to the social urban capital enhancement, and the connected negotiation procedures that have to be made explicit in order to encourage the participation [4], [14, 17]. An effective organization of the public consensus depends on the combination of a strong technical approach and a flexible preferences modelling pattern able to help the local administration to: increase its skills, that means the capability to produce a wide range of planning options; reduce its responsibility about the results of the planning process, by delegating

B. Murgante et al. (Eds.): ICCSA 2014, Part III, LNCS 8581, pp. 224–239, 2014.

a relevant part of the decisions. One of the most important points of equalization, can be considered the equilibrium of restrictions and concessions, to be obtained by means of different types and practices of development transfers and fees. Their dimension depends on the appraisal of the value of the private and public wealth that is created and/or destroyed at the same time. In fact, the concept of value includes many different meanings, and/or depends on different issues: among them the abstract ones, expressed by money, and the concrete ones, scarcely expressible by money. They several conflicts and, above all, make participation and communication hard. A technological approach defines top-down the meaning and the measure of value, basing on skill, whereas the social and political one establishes a wide range of ways of consultation to rise bottom-up the frame of the preferences.

2   Materials: The Historic Center of Pachino

Pachino (Fig 1, left), is a town located in the south-eastern Sicily, near the southern boundary of the province of Syracuse, two kilometres from the sea, standing upon a hill from which it dominates a wide territory comprising a relevant landscape and environmental heritage feature. The town was founded by concession of Ferdinand IV, in 1760, in order to populate the Feud of Scibbini.

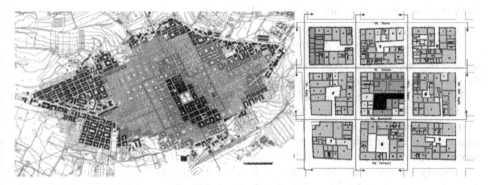

Fig. 1. Pachino old town historical development

Its urban frame is a grid of districts about 40 x 40 m in size, laying on a hill on the top of which is the wide square Vittorio Emanuele. The urban fabric have evolved through this framework to the last '70s when the grid matched the diagonal directions of the peripheral roads. As a result, the urban fabric is scarcely characterized by urban spaces and voids; the urban quality depends on some valuable architecture and fifteen civil and religious monuments, dating back since XVIII to early XIX century. The most important one is the main Church dedicated to the Holy Crucifix, built in in 1790.

The town performs low location tension and the real estate market prices are very close to the cost of construction. This circumstance legitimates the implementation of regeneration strategies, rather transformative, that must be legitimated by the most appropriate decision making tools. Our concern involves the relationship between the production of real estate value and the formation of social urban capital. For the case study of Pachino a strong application of MCDM involved nine "districts" including Tasca Palace, one of the main monumental building of Pachino. Tasca Palace, as well as other important   buildings or public areas characterized by a high potential social value can be taken as the targets of renovation processes within overall strategies involving the minor buildings. The renovation actions will be fitted referring to the urban-architectural values that should be explicit. The current Master Plan (1988) includes a Detailed plan that rules the historic center conservation. We assume the plan rules as general purposes of the value based meta-strategic tool we propose. The studied context is the area between Vittorio Emanuele square, Anita and Rome streets, Ferrucci street and includes 98 Architectural Units (AU); Fig. 1 right). All the characteristics of the UA are expressed by using a standard lexicon so that and the information can be handled for statistical purposes (Tab. 1). Only some significant results of the wide and detailed analysis are shown in the following applications.

3   Methods and Procedures

The model to support the evaluation, the selection and the economic performance management of retraining plan has been developed as a DDS, on the basis of the composition between two approaches of MCDA type:

-   *MAUT (Multi-attribute utility Theory)*
-   *IMO-DRSA (Interactive Multiobjective Optimization)*

The two approaches allow to develop a robust model that meets the needs of the various process stakeholders.

In this regard, we report a brief summary of the two approaches.

MAUT. The Multi-attribute utility Theory is based on the main Hypothesis that every decision maker tries to optimize, consciously or implicitly, a function which aggregates all their points of view. therefore the decision maker's preferences can be represented by a function, called the utility function U [12]. This function is not necessarily known at the beginning of the decision process, so the decision maker needs to construct it first. The utility function is a way of measuring the desirability or the preference of objects, called alternatives. The utility score is the degree of well-being those alternatives provide to the decision maker. The utility function is composed of various criteria which enable the assessment of the global utility of an alternative. Each alternative of set A is evaluated on the basis of function U and receives a "utility score" $U(a)$. The issue of the incomparability between two alternatives, as in the outranking methods, does not arise since two utility scores are always comparable. So, the preference relation on set A is based on the utility scores I transitive. This means that if alternative a is better than alternative b, which in turn is better than alternative c, we can conclude that a is also better than c basing on the utility score.

Table 1. List of the characteristics and the valuation of four AU

				1	2	3	4
identification	id AU (part)			1	2	3	4
	id BI			1	1	1	1
	Id. AU (prog)			1	2	3	4
	type/denomination			Palazzo Santucci	palazzetto	palazzetto	palazzetto
dimensions	n rooms	n		28	6	7	6
	ground area	sq.m		420	36	76	65
	real estate area	sq.m	curr	840	108	227	130
		sq.m	exp	840	144	302,6666667	260
	open area	sq.m		0	0	0	0
	n floors	n	curr	2	3	3	2
		n	exp	2	4	4	4
	main front width	m		36	6	7	7
	other fronts width	m		12	0	7	7
outer positional features or location	location			Piazza Vittorio E	Via Roma	Via Roma	Via Roma
	accessibility	cod	curr	accesibility, park	accesibility, park	accesibility, park	accesibility, park
	functional value	cod	curr	rare function an	mono destinatio	mono destinatio	mono destinatio
	simbolic value	cod	curr	greatest	invalid	none	none
	first floor function	cod	curr	commercial	commercial	residential	resdential
	upper floors functions	cod	curr	commercial	residential	residential	residential
	uses quality	cod	curr	elevated	normal	none	none
	use	cod	curr	used	used	used	used
	panoramic	cod	curr	good	good	good	good
inner positional features	view	cod	curr	square	secondary street	secondary street	secondary street
	well-lighted	cod	curr	good	decent	decent	decent
	finishing	cod	curr	colored plaster a	ordinary plaster	ordinary plaster	ordinary plaster
technological features	frames	cod	curr	wood/aluminiun	anodized alumin	anodized alumin	anodized alumin
	conservation of building	cod	curr	no damage	decoration degr	light crack	light crack
	renovations	cod	curr	total recent	none	none	none
architectural features	age	cod	curr	previous 1900	1981-2000	1981-2000	1981-2000
	language element	cod	curr	decorations	no definition	no definition	no definition
	historical value	cod	curr	elevated	none	none	none
	decorum	cod	curr	without degrada	system, boiler	addition	no degradation
assessment	outer features (ke)	score	curr	5,00	2,83	2,53	2,53
		score	exp	5,00	3,04	2,77	2,77
	inner features (ki)	score	curr	5,00	3,85	3,85	3,85
		score	exp	5,00	5,00	5,00	5,00
	thecnological features (kt)	score	curr	4,70	2,80	2,00	2,00
		score	exp	4,79	5,00	5,00	5,00
	architectural features ka	score	curr	5,00	1,30	1,00	2,20
		score	exp	5,00	5,00	5,00	5,00
	totale average weighed	score	curr	4,88	2,72	2,28	2,52
	score (K)	score	exp	4,92	4,61	4,55	4,55
	unit market value (area)	€/sq.m	curr	€ 1.070	€ 529	€ 419	€ 479
		€/sq.m	exp	€ 1.079	€ 1.002	€ 989	€ 989
	unit market value (rooms)	€/room	curr	€ 43.800	€ 22.150	€ 17.750	€ 20.150
		€	exp	€ 44.160	€ 41.085	€ 40.545	€ 40.545
	total market value (area)	€	curr	€ 898.800	€ 57.105	€ 95.056	€ 62.238
		€	exp	€ 906.360	€ 144.306	€ 299.224	€ 257.043
	total market value (rooms)	€	curr	€ 1.226.400	€ 132.900	€ 124.250	€ 120.900
		€	exp	€ 1.236.480	€ 305.672	€ 406.531	€ 454.104
	average market value (area; rooms)	€	curr	€ 1.062.600	€ 95.003	€ 109.653	€ 91.569
		€	exp	€ 1.071.420	€ 224.989	€ 352.878	€ 355.573
	surplus %	%	curr	-13%	-29%	-12%	-24%
		%	exp	-13%	-26%	-13%	-22%
project hypotesis	intervention category			restoration	new building + r	new building + e	new building + e
	conservation cost	€		€ 8.568	€ 0	€ 0	€ 0
	rebuilding cost	€			€ 28.800	€ 60.533	€ 104.000
	gross surplus	€		€ 8.820	€ 129.987	€ 243.224	€ 264.005
	net surplus	€		€ 252	€ 101.187	€ 182.691	€ 160.005

The utility function U can be defined in several different ways. The most common approach is the additive model, where, F is the set of q criteria $f_j (j = 1, \dots, q)$.

The evaluations of the alternatives $f_j(a_i)$ $(i = 1, \dots, s)$ are first transformed into marginal utility contributions, denoted by U_j, in order to avoid scale problems. The general additive utility function can be written as follows:

$$\forall a_i \in A : U(a_i) = U\left(f_1(a_1), \dots, f_q(a_s)\right) = \sum_{j=1}^{q} U_j\left(f_j(a_i)\right) \cdot w_j \qquad (1)$$

Where $U_j(f_j) \geq 0$ *is usually a non-decreasing function, and* w_j *represents the weight of the criterion* f_j. *They generally satisfy the normalization constraint:*

$$\sum_{j=1}^{q} w_j = 1 \qquad (2)$$

The weights represent trade-off, that means the amount a decision maker is ready to trade on one criterion in order to gain one unit on another criterion [12]. When using additive functions, some conditions such as the preferential independence one between the criteria need to be respected. The marginal utility functions are such that the best alternative on specific criterion has a marginal utility score of 1 and the worst alternative, on the same criterion, a score of 0. If the weights are normalized, the utility score of an alternative is always between 0 and 1 [13].

DRSA. The DRSA tool belongs to the "rough set" family [15, 16]. The original rough set approach is not able, however, to deal with preference-ordered attribute domains and decision classes. Solving this problem was crucial for the application of the rough set approach to the multi criteria decision analysis (MCDA). For this reason, Greco [6] has proposed an extension of the rough sets theory that is able to deal with inconsistencies typical to exemplary decisions in MCDA problems. This innovation is mainly based on replacement of the indiscernibility relation with a dominance relation in the rough approximation of decision classes [8]. An important consequence of this fact is a possibility of inferring from exemplary decisions the preference model in terms of decision rules being logical statements of the type "*if ..., then ...*" [7]. The information about objects of the decision (also called actions) is represented in the form of an information table. The rows of the table are labeled by objects, whereas the columns are labeled by attributes, and the entries of the table are attribute-values. Formally, a data table is the 4-tuple $S = \{U, Q, V, f\}$, where U is a finite set of objects (universe), $Q = \{q_1, \dots, q_m\}$ is a finite set of attributes, V_q is the domain of the attribute q, $V = \bigcup_{q \in Q} V_q$ and $f : U \times Q \to V$ is a total function such that $f(x, q) \in V$ for each $q \in Q, x \in U$, called information function. The set Q is, in general, divided into a set C of condition attributes and a set D of decision attributes. The information table for which condition and decision attributes are distinguished is called *decision table* [8]. The condition attributes with domains (scales) ordered according to decreasing or increasing preference are called criteria. For criterion $q \in Q, \succcurlyeq_q$ is a weak preference relation (also called outranking relation) on U such that $x \succcurlyeq_q y$ means "*x is at least as good as y with respect to criterion q*". We suppose that $\succcurlyeq_q$ is a total preorder, i.e. a strongly complete and transitive binary relation, defined on U on the basis of evaluations $f(., q)$. We assume, without loss of generality, that the preference is increasing with the value of $f(., q)$ for every criterion $q \in C$. Furthermore, we assume that the set of decision attributes D is a singleton $\{d\}$. Decision attribute d makes a partition of U into a finite number of decision classes, $Cl = Cl_t, t \in T$, $T = \{1, \dots, n\}$, such that each $x \in U$ belongs to one and only one class $Cl_t \in Cl$. We suppose that the classes are preference-ordered, i.e. for all $r, s \in T$, such that $r > s$, the objects from Cl_r are preferred to the objects from Cl_r. The sets to be approximated are called upward union and downward union of classes, respectively:

$$Cl_t^{\geq} = U_{s>t} Cl_s, \ Cl_t^{\leq} = U_{s<t} Cl_s; \ t = 1, ..., n \tag{3}$$

The statement $x \in Cl_t^{\geq}$ means "x *belongs to at least class* Cl_t", while $x \in Cl_t^{\leq}$ means "x *belongs to at most class* Cl_t". The key idea of the rough set approach is representation (approximation) of knowledge generated by decision attributes, by "granules of knowledge" generated by condition attributes. In DRSA, where condition attributes are criteria and decision classes are preference ordered, the represented knowledge is a collection of upward and downward unions of classes and the "granules of knowledge" are sets of objects defined using a dominance relation [8]. We say that x dominates y with respect to $P \subset C$ (shortly, x P-dominates y), denoted by xD_py, if for every criterion $q \in P$, $f(x,q) \geq (f,y,q)$. P-dominance is reflexive. Given a set of criteria $P \subset C$ and $x \subset U$ the "granules of knowledge" used for approximation in DRSA are:

- a set of objects dominating x, called P-dominating set:

$$D_P^+(x) = \{y \in U : yD_px\} \tag{4}$$

- a set of objects dominated by x, called P-dominated set:

$$D_P^-(x) = \{y \in U : xD_py\} \tag{5}$$

Let us recall that the dominance principle (or *Pareto principle*) requires that an object x dominating object y on all considered criteria (i.e. x *having evaluations at least as good as* y *on all considered criteria*) should also dominate y on the decision (i.e. x *should be assigned to at least as good a decision class as* y). The P-lower approximation of $Cl_t^{\geq}$, denoted by $\underline{P}(Cl_t^{\geq})$ and the P-upper approximation of $Cl_t^{\geq}$ denoted by $\overline{P}(Cl_t^{\geq})$ are defined as follows $t = 1, ..., n$:

$$\underline{P}(Cl_t^{\geq}) = \{x \in U : D_P^+(x) \subseteq C_t^{\geq}\} \tag{6}$$

$$\overline{P}(Cl_t^{\geq}) = \{x \in U : D_P^-(x) \cap C_t^{\geq} \neq 0\} = U_{x \in Cl_t^{\geq}} D^+(x) \tag{7}$$

Analogously, one can define the P-lower approximation $Cl_t^{\leq}$ and the P-upper approximation $Cl_t^{\leq}$.

The P-lower and P-upper approximations so defined satisfy the following inclusion properties for each $t = 1, ..., n$ and for all $P \subset C$:

$$\underline{P}(Cl_t^{\geq}) \subseteq Cl_t^{\geq} \subseteq \overline{P}(Cl_t^{\geq}) \tag{8}$$

$$\underline{P}(Cl_t^{\leq}) \subseteq Cl_t^{\leq} \subseteq \overline{P}(Cl_t^{\leq}) \tag{9}$$

For every $P \subset I$, the quality of approximation of sorting Cl by a set of criteria P is defined as the ratio of the number of objects P-consistent with the dominance principle and the number of all the objects in U.

$$\gamma_P(Cl) = \frac{|U - (\bigcup_{t \in \{1,...,m\}} Bn_P(Cl_t^{\geq})) \cup (\bigcup_{t \in \{1,...,m\}} Bn_P(Cl_t^{\geq}))|}{|U|} \tag{10}$$

$\gamma_P(Cl)$ can be seen as a degree of consistency of the sorting examples, where P is the set of criteria and Cl is the considered sorting. Each minimal (in the sense of inclusion) subset $P \subset I$ such that $\gamma_P(Cl) = \gamma_I(Cl)$ is called a reduct of sorting (Cl), and is denoted by $REDCl$. Let us remark that for a given set of sorting examples one can have more than one reduct. The intersection of all reducts is called the core, and is denoted by $CORECl$.

IMO-DRSA. Interactive Multiobjective Optimization is based on the application of a logical preference model built using the Dominance-based Rough Set Approach (DRSA) [10]. The method is composed of two main stages that alternate in an interactive procedure. In the first stage, a sample of solutions from the Pareto optimal set (or from its approximation) is generated. In the second stage, the Decision Maker (DM) indicates relatively good solutions in the generated sample. From this information, a preference model expressed in terms of "if ..., then ..." decision rules is induced using DRSA [9]. These rules define some new constraints which can be added to original constraints of the problem, cutting off non-interesting solutions from the currently considered Pareto optimal set. A new sample of solutions is generated in the next iteration from the reduced Pareto optimal set. The interaction continues until the DM finds a satisfactory solution in the generated sample. This procedure permits a progressive exploration of the Pareto optimal set in zones which are interesting from the point of view of DM's preferences. (Fig. 2).

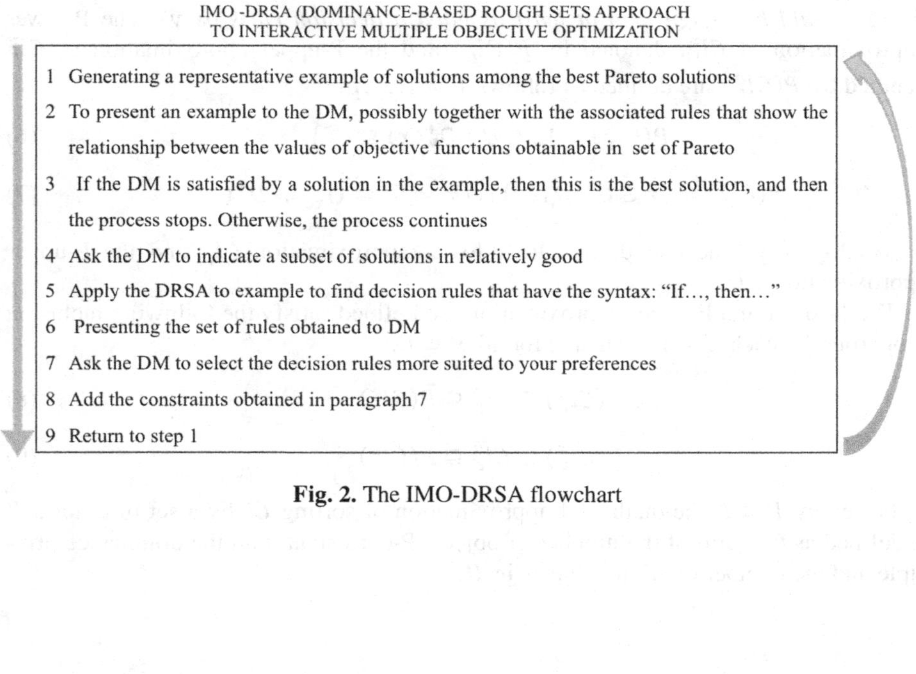

IMO -DRSA (DOMINANCE-BASED ROUGH SETS APPROACH
TO INTERACTIVE MULTIPLE OBJECTIVE OPTIMIZATION

1 Generating a representative example of solutions among the best Pareto solutions

2 To present an example to the DM, possibly together with the associated rules that show the relationship between the values of objective functions obtainable in set of Pareto

3   If the DM is satisfied by a solution in the example, then this is the best solution, and then the process stops. Otherwise, the process continues

4 Ask the DM to indicate a subset of solutions in relatively good

5 Apply the DRSA to example to find decision rules that have the syntax: "If..., then..."

6   Presenting the set of rules obtained to DM

7 Ask the DM to select the decision rules more suited to your preferences

8 Add the constraints obtained in paragraph 7

9 Return to step 1

Fig. 2. The IMO-DRSA flowchart

4   Application

4.1   General Approach

In the case of the retraining plan for the Pachino's old town, the stakeholders are helped to identify the system of the plan actions that best reflect the technical, economic, social and political sustainability.

In particular, the general objective is to define a DDS to support the plan's stakeholders into the identifying of the simple or compound alternatives that best meet the economic performance which they want to reach. The supporting phase to the identification of the actions plan and of their economic performance is instrumental to the management of the retraining process which it want to implement

In particular, the correlation between the economic performance and the actions plan is perceived differently by the plan's stakeholders

Among plan stakeholders, it is certainly is possible to identify two groups: the technicians on the one hand and the politicians and the economic actors on the other hand.

The former are interested in evaluating and choosing among the actions those that reflect the technical, regulatory and economic sustainability, while the latter are interested in evaluating and choosing among the actions those that reflect the social, political and economic sustainability. The language of the first group is different from that of the second one, the latter have more difficulty to understand the effects of the plan actions and to use the mathematical and technical language.

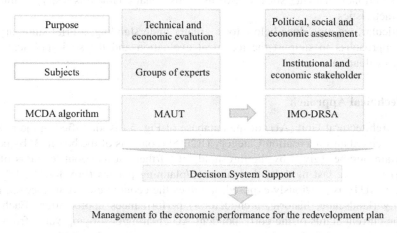

Fig. 3. Decision System Support. The economic sustainability

In this regard, this study proposes two stages of evaluation and selection of the retraining plan actions: one techno-economic type, and a political-social-economic type, which aim at the identification of the economic performance for the plan that best meets their points of view, so promoting better overall management of the process. In this regard, this study proposes a model of DDS [1, 2], that can be defined by using a hybrid approach (MCDA) and that can be used to manage the level of the economic performance which the stakeholders want to implement with the retraining plan of the Pachino's old town (Fig. 3). By the synthesis of the solutions from a technical viewpoint and from the political and economic viewpoint it is possible to define the structure of the redevelopment plan that all the stakeholders want to implement. The analysis was conducted at the technical level with a model that allows to evaluate on the basis of the real estate characteristics of the blocks of the Pachino's old town, the interventions types and urban equalization scheme that offer within the constraints of sustainability (technical, normative and equalization type) a certain level of economic performance.

The considered actions for the plan are those identified by the group of technicians. They are the following: a) integration volume; b) routine maintenance; c) emergency maintenance; d) restructuring; e) restoration; f) demolition; g) routine maintenance and new construction; h) emergency maintenance and new construction; i) restructuring and new construction; j) demolition and reconstruction with an in cubage increase.

Interventions from point a) to f) are evaluated in a gradual way from the single action on the architectural unit to the global action for the plan. The action types were obtained by combining in a gradual manner some simple shares. From the single action on the architectural unit have been defined 352 alternatives, which range from the most conservative type, namely the restoration and routine maintenance, to the most transformative type, namely the demolition and/or cubage increase.

The criteria by using the multi-criteria analysis for the two MCDA approaches are the following: 1. age; 2. architectural language; 3. decorum; 4. testimonial value; 5. intrinsic positional characteristics; 6. extrinsic positional characteristics; 7. technological characteristics .

In particular, we reproduce below for a greater understanding, a brief summary of the two approaches to support the technical evaluation and the socio-political and economic evaluation.

4.2   Technical Approach

To each Architectural Unit (AU) of the database of Fig. 5 (as identified in the rows), can be associated an Intervention Category (IC) [5] (columns of the box n. 3) by passing a certain number (given by the threshold) of "if/then" tests about le status of its characteristics. The last module, the valuation/planning pattern (box four of Fig. 5) based on MAUT, as previously exposed, involves the economic (revenues-costs) and the quality (landscape, historic, architectural) performances appreciation. Each IC affects the current status of the correspondent AU creating/destroying value from the four different points of view. A good economic performance due to the *Cubage increase* associated to AU n. 4 produces bad quality performances (Fig. 4).

The progressive reduction of the conservative thresholds activates transformative interventions, first to the less qualified AU and progressively to the more qualified ones. In order to define and display a trade off function, or Pareto frontier (Fig. 7), between quantitative and qualitative performances a successive threshold reduction pattern has been arranged by supposing 352 possible combinations of threshold level.

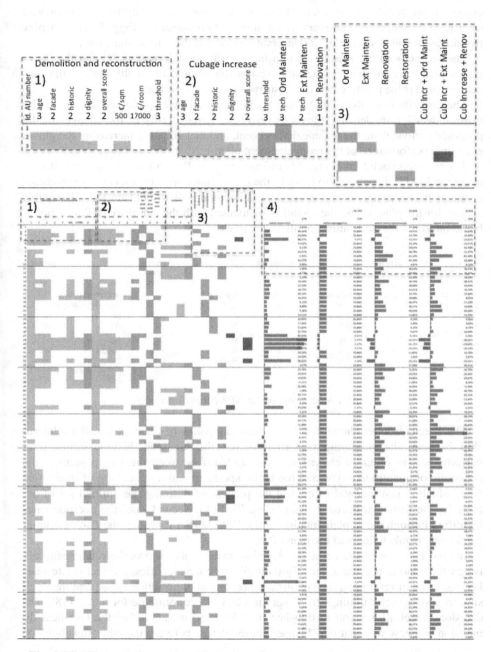

Fig. 4. IC-AU associations and multicriteria performances evaluation: conservative pattern

Each combination displays the structure of the conditions that determine the association of the suitable IC to each AU, so that everyone can verify the consistency of each IC-AU association for every given threshold combination. Figg. 5 and 6 display two different solutions, the most conservative one and the intermediate one.

The red cells in the central column indicate the transformative IC and the histograms of the second half of the table shows the amount of wealth (value) that has been created (blue positive bars) or destroyed (red negative bars).

The analysis from the political and social-economic viewpoint was conducted by using a decision model that allows to verify the sustainability of the changes of the proposed actions from the technical viewpoint, and that aimed at increasing the global economic performance of the redevelopment plan.

The two approaches allow the identification of the Pareto frontier. In the first case using the MAUT is possible to identify the utility function with respect to the technical group from the action on the single architectural unit to until on the all possible actions for the plan. Using the model implemented with the aid of MAUT, it is possible to build the Pareto frontier for all alternatives.

Fig. 5. Assessment of the different performances

The Pareto frontier shows for the various categories of intervention, the relationship between the economic performance and the landscape-historic-architectural performance, in order to characterize the trade-off between the two objectives the economic of quantitative type and the landscape-historical-architectural of qualitative type, which is instrumental to control the process of defining the optimal set of actions for the technical decision maker (Fig. 7). In the second case, the decision rules identified through the implementation of the IMO-DRSA represent the Pareto frontier for the other stakeholders of the plan (Tab. 2).

In this regard, we have defined a decision problem that shows the relationship between the condition criteria, namely the building characteristics, the real estate market, the costs of retraining, the equalization regime and the action types, and the decision to assign such buildings to a specific class of economic performance [18]. The decision model allows you to translate the identified model within the technical dimension in a new way that may be used for the debate between the other stakeholders.

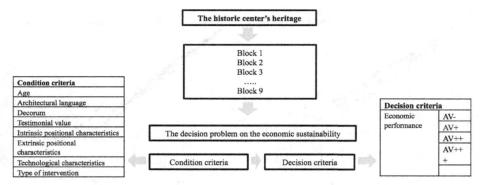

Fig. 6. The decision problem about the economic sustainability

4.3   Social Political Economic Approach

The analysis from the political and social-economic viewpoint was conducted by using a decision model that allows to verify the sustainability of the changes at the proposed actions from the technical viewpoint, and that has aimed at increasing the global economic performance of the redevelopment plan (Fig. 6).

In this regard, we have defined a decision problem that shows the relationship between the condition criteria, namely the building characteristics, the real estate market, the costs of retraining, the equalization regime and the actions types, and the decision to grant such buildings to a specific class of economic performance [18]. The decision model allows you to translate the identified model within the technical dimension in a new way that may be used for the debate between the other stakeholders.

5   Results and Discussion

Technical approach. As a result, the simulation about the different strategies shows a discontinue Pareto frontier as related to the typical heterogeneity of an urban context assumed as a discrete system. The technical assessment model to verify the sustainability of the interventions (technical, legal, and economic equalization), has been implemented with the algorithm of DRSA. It is the example sample on whose basis, the various stakeholders are called upon to express their views.

The information table associated to the sample is presented on the Fig. 6 (information able is omitted due its big size). The decision rules obtained from the sample of example are presented in the Tab. 2.

Next, we have identified the core and the reduced for the approximation, which allow you to identify the criteria that most influence the economic performance of the plan actions. In particular, the expert group attributes most importance to the shown criteria in the table that follows. In this case the core coincides with the only reduced of the approximation. Once identified the decision rules of the sample it is possible to classify all the buildings of the blocks for the Pachino's old town, assigning them to a specific class of economic performance within the placed constraints at the base of

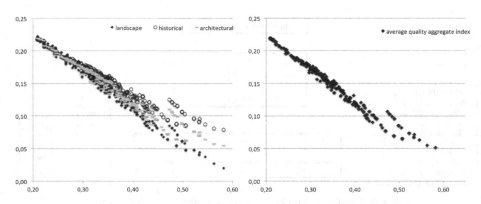

Fig. 7. Economic (x-axis) and qualitative performances (y-axis): the Pareto frontier for disaggregate (left) and aggregate (right) quality performances

Table 2. The condition criteria and the decision criteria

n.	Decision rules
1	*If* the judgment for the technological characteristics are discreet and the intervention is the new construction-increase in cubature- maintenance, *then* you will arrive at economic performance 6% < DV< 25%
2	*If* the judgment for the decorum is no degradation and that for the intrinsic positional characteristics is good, *then* you will arrive at economic performance 0 <DV <15%
3	*If* the age of the building is between 1901 and 1945 and the Extrinsic positional characteristics is discrete, the technological characteristics is discrete, *then* you will arrive at economic performance of 0 <DV <15%
4	*If* the judgment for the extrinsic positional characteristics is sufficient, that for the intrinsic positional characteristics is discreet and the intervention is the maintenance, *then* you will arrive at economic performance DV <0
5	*If* the judgment for the testimonial value is not expressed, that for the extrinsic positional characteristics is sufficient, that for the technological characteristics is sufficient and the intervention is new construction+building+increase in cubature+ maintenance, *then* you will arrive at economic performance DV <0
6	*If* the judgment for the decor is heavily modified, that for the intrinsic characteristics is discrete, *then* you will arrive at economic performance 0 <DV <15%
7	*If* the judgment for the testimonial value is not expressed , that for the extrinsic positional characteristics is not expressed, that for the technological characteristics is sufficient, *then* you will arrive at economic performance 0< DV< 15%
8	*If* the judgment for the testimonial value is not expressed , that for the extrinsic positional characteristics is sufficient and that for the intrinsic positional characteristics is discrete, *then* you will arrive at economic performance 0< DV< 15%
9	*If* the judgment for the extrinsic positional characteristics is sufficient, *then* you will arrive at economic performance 0< DV< 15%

the technical operators. In fact, this structure represents the law in terms of the economic performance which the technical group wants to implement in order to make the plan effective. Once identified the decision rules for the plan from the technical point of view, they may be proposed to stakeholders (Tab. 3).

If the solution in terms of the economic performance is acceptable, the process can be considered concluded, but the DMs want to increase the level of the economic performance for the plan, often, without being aware of the effects of their choice. The defined model by means of the algorithm of the IMO-DRSA enables to examine the effects on the real estate assets generated by a change of some actions, that are aimed at increasing the level of the economic performance. In particular, in this case, the DMs have requested to raise the level of the overall economic performance,

Table 3. Core and *Reduct* of the approximation of the experts group

	Core		Reduct
1	Testimonial value	**1**	Testimonial value
2	Intrinsic positional characteristics	**2**	Intrinsic positional characteristics
3	Extrinsic positional characteristics	**3**	Extrinsic positional caracteristics
4	Technological characteristics	**4**	Technological characteristics
5	Interventions	**5**	Interventions

proposing three changes, i.e. the increase of airspace, the lowering of the threshold provided by the technicians for the demolition and their combined effect [19].

Table 4. The changes of the economic performance according to the stakeholders' preferences

n. Block	Building	Economic performance class on the basis of the stakeholder judgement (decrease the threshold for the demolition+ increase cubature)	Economic performance class on the basis of the stakeholder judgement (only increase cubature)	Economic performance class on the basis of the judgement of the experts gruop
	54	AV+	AV+	AV+
	55	AV+	AV++	AV++
	56	AV+	AV++	AV-
	57	AV+	AV++	AV-
6	58	AV+	AV+	AV-
	59	AV+	AV++	AV+
	60	AV++	AV++	AV++
	61	AV++	AV++	AV-
	62	AV++	AV++	AV-

In all cases the decision rules have been modified and the economic performance classes for all buildings have been reclassified. We show in the following Tab. 4, only the results for the block 6 (table for all blocks is omitted because of its size). The identified decision rules correlate the increase of the economic performance with the change of the intervention, but also, with all the other condition criteria [19]. Thus, the DMs are helped to choose the best solution – i.e., what among the proposals they perceive as more sustainable way – and to control the trade-off between the economic performance and the actions, the real estate heritage and the equalization scheme rules.

6   Conclusions

The manner in which the actions for the redevelopment plan are implemented, is the result of a mediation between the stakeholders. The stakeholder's process has often different goals, some communication difficulties, and the lack of the awareness of the effects of their choices. They do not have a technical education, but a political one, so they need to be coordinated in the knowledge and choice phases of the actions supporting the retraining plan in a sustainable way. The study proposes two stages of the evaluation and the selection of actions in support of the redevelopment process: one Technical-economic one and a political-social-economic one. Both stages are aimed at identifying the economic performance of actions that best meet the views of the stakeholders, in order to encourage better management of the process. In this regard, we proposed a DSS of hybrid type, which uses two different approaches to MCDA, which best meet the needs of those involved in the process:

— MAUT approach, which is based on the main hypothesis that each decision-maker seeks to optimize, consciously or implicitly, a function which aggregates all their points of view, i.e. the utility function. The utility function is a way to measure the desirability or the preference on some objects, that are called alternatives. The weights represent trade-off, that is, the amount a decision maker is ready to trade on one criterion in order to gain one unit on another criterion.
— IMO-DRSA approach that uses a language that best interfaces with the political and economic decision-maker; it is capable to produce the law for the management of the plan, offering the ability to make some changes taking into account the effects of their choices on the plan.

Finally, the group of experts who have built a data base, during the first phase of the evaluation, can use it as a Data Mining, recalling the verification information on the concessions, the increased airspace, the permits demolition, through some "if ..., then ..." rules.

Acknowledgements. Salvatore Giuffrida edited paragraphs 1, 2, 4.2, drew up the valuation model by using the MAUT approach and the data mining in the final phase, edited pictures 1, 4 and table 1; Maria Rosa Trovato edited paragraphs 3, 4.1,4.3, 5, 6, drew up the valuation model by using the IMO-DRSA approach, edited pictures 2, 3, 6 and tables 2, 3, 4.

References

1. Belton, V., Branke, J., Eskelinen, P., Greco, S., Molina, J., Ruiz, F., Słowiński, R.: Interactive Multiobjective Optimization from a Learning Perspective. In: Branke, J., Deb, K., Miettinen, K., Słowiński, R. (eds.) Multiobjective Optimization. LNCS, vol. 5252, pp. 405–433. Springer, Heidelberg (2008)
2. Comes, T., Wijngaards, N., Schultmann, F.: Intelligent technologies for decision support and preference modelling. In: Doumpos, M., et al. (eds.) Multiple Criteria Decision Aid and Artificial Intelligence, pp. 45–72. Wiley, United Kingdom (2013)

3. Dyer, J.S.: MAUT. In: Figueira, J.R., et al. (eds.) Multiple Criteria Decision Analysis: State of the Art Surveys, pp. 265–295. Springer, Berlin (2005)
4. Giuffrida, S.: Evaluation for the new ethics of planning process. In: Fusco, G.L., et al. (eds.) L'uomo e la città: Verso uno sviluppo umano e sostenibile, Milano, FrancoAngeli (2003)
5. Giuffrida, S., Ferluga, G., Gagliano, F.: Social Housing nei Quartieri portuali storici di Siracusa. Un modello WebGIS per la valutazione e la programmazione 11, 121–154 (2013)
6. Greco, S., Matarazzo, B., Słowiński, R.: Rough approximation of a preference relation by dominance relations. European J. Operational Research 117, 63–83 (1999)
7. Greco, S., Matarazzo, B., Słowiński, R.: Rough sets theory for multicriteria decision analysis. European J. of Operational Research 129, 1–47 (2001)
8. Greco, S., Matarazzo, B., Słowiński, R.: Dominance-based rough set approach to knowledge discovery – (I) general perspective (II) extensions and applications. In: Zhong, N., Liu, J. (eds.) Intelligent Technologies for Information Analysis, pp. 513–612. Springer, Berlin (2004)
9. Greco, S., Matarazzo, B., Słowiński, R.: Decision rule approach. In: Figueira, J., Greco, S., Ehrgott, M. (eds.) Multiple Criteria Decision Analysis: State of the Art Surveys, pp. 507–563. Springer, Berlin (2005)
10. Greco, S., Matarazzo, B., Słowiński, R.: Dominance-Based Rough Set Approach to Interactive Multiobjective Optimization. In: Branke, J., Deb, K., Miettinen, K., Słowiński, R. (eds.) Multiobjective Optimization. LNCS, vol. 5252, pp. 121–155. Springer, Heidelberg (2008)
11. Ishizaka, A., Nemery, P.: Multi-Criteria decision analysis, pp. 81–104. Wiley, U K (2013)
12. Figueira, J.R., Greco, S., Mousseau, V., Słowiński, R.: Interactive Multiobjective Optimization Using a Set of Additive Value Functions. In: Branke, J., Deb, K., Miettinen, K., Słowiński, R. (eds.) Multiobjective Optimization. LNCS, vol. 5252, pp. 97–119. Springer, Heidelberg (2008)
13. Micelli, E.: La perequazione urbanistica. Marsilio, Venezia (2004)
14. Keeney, R.L., Raiffa, H.: Decision with Multiple Objectives: Preference and Value Tradeoffs. John Wiley and Sons, New York (1976)
15. Pawlak, Z.: Rough Sets. Kluwer, Dordrecht (1991)
16. Pawlak, Z.: Rough sets. International Journal of Computer and Information Sciences 11, 341–356 (1982)
17. Stanghellini, S.: Il negoziato pubblico privato nei progetti urbani. DEI, Roma (2012)
18. Trovato, M.R.: DRSA-IMO approach to support at a decision model for the social, architectural, urban and energetic retraining planning for the old town of Mazara del Vallo. 71 Meeting of the EWG MCDA. Politecnico di Torino (2010)
19. Trovato, M.R.: Information and Communication Technologies (ICTs) and Participatory Values to Support of the Territorial Governance Processes. In: Society, Integration, Education: Utopias and Dystopias in Landscape and Cultural Mosaic, Udine, pp. 273–284 (2013)

The Impact of Intelligent Building Technologies on the Urban Environment

Sergio Selicato and Daniela Violante

Polytechnic University of Bari, Bari, Italy
{s.selicato88,violante.daniela}@gmail.com

Abstract. The aim of this paper is to suggest guidelines for sustainable planning, understood as the union of actions relating to the building and the city. The strong connection between these two subjects requires a planning that should not travel on parallel tracks, as often happens, but it must be formed by a rational consideration of both. They have been analyzed for two types of urban fabric and, in relation to their characteristics, have been highlighted different problems. So we suggested some project solutions and use of materials that simultaneously improve the interior comfort of buildings and the urban comfort.

Keywords: technologies, building, façade, urban, places, comfort, guidelines.

1 Introduction

More and more the subject of environmental sustainability impregnates urban planning and the methods of building design. However, in both areas large steps forward have been made during the last decade, the two areas have mostly travelled in parallel. This paper aims instead to demonstrate how the design of sustainable buildings and the design of urban spaces are closely interrelated and influence each other.

This gap between planning and architectural design has been protracting for years, leading in most cases, on the one hand, to inconsistent plan forecasts compared with the context and mostly characterized by mere choice of location, and on the other hand, to the design buildings, sometimes architecturally significant, detached from the surrounding. The rapprochement among architectural design, evolved in the last decade through bioclimatic approach, and urban design is the first step from which start to build comfortable urban places, able to generate greater sense of belonging in the local population.

In construction, the current challenge is to be able to enforce all the essential innovations to ensure that every building reduce its environmental impact, energy consumption and emissions during the various stages of construction, maintenance and management. This will be possible through a more efficient management and diffusion of techniques and solutions that are part of clean energy. In this sense, the bioclimatic approach is pushing more and more to exploit as much as renewable energy sources (such as solar energy, wind energy, energy provided from biomass and waste production), in the implementation of cutting edge technological systems and in the integration of innovative smart materials within the building shell. Just the building

B. Murgante et al. (Eds.): ICCSA 2014, Part III, LNCS 8581, pp. 240–252, 2014.

shell – the architectural element on which we will focus attention later – marks the boundary between the built environment and the urban space. It contributes significantly to defining the perception that people have of each place. It seems clear, in fact, as a façade covered mainly by plant species may influence the quality of urban space , both in terms of climate (reduction of the heat island effect, reduction of the humidity rate etc.) and in a perceptive way (greater sense of urban quality), and it may influence the comfort inside the building ensuring a low thermal transmittance and a great mismatch of the thermal wave.

So, although it appears so obvious that a design that follows the principles of bio-architecture influences also the urban environment, until now urban design has been limited, in the most virtuous of cases, in the determination of guidelines within planning tools. It has oriented in a more detailed way the design of green areas, an observance of the permeability of surfaces and other parameters, able to ensure a minimum level of quality in public spaces.

The specific objective of this paper is to aim a criteria design, able to consider simultaneously buildings and urban space. In detail, we will consider two type of boundary context – the compact urban fabric and the marginal suburban fabric –in order to provide design guidelines to obtain that urban comfort and the indoor quality.

2   Literary Review

2.1   Planning, Quality and Urban Comfort

The debate on urban and local planning has produced a great deal of thinking on methods and contents, which have found feedback in the normative-institutional apparatus and in practical experimentation. The theoretical reflection on the nature and role of planning have expanded, in particular, the areas of interest including, as well as the spatial organization, also the economic and social aspects. Have assumed great importance, over the past few decades, the approach to follow, without ignoring the logic of rationality, essentially aims to define choices as widely shared on the social sphere [1].As for the contents, increasingly the planning aims to ensure an awareness in the use of available resources and to mitigate conflicts between different instances, enabling a stable growth rate [2]. Therefore, the planning becomes to coordinate actions through an "aware" effort of the community, thus assuming the collective aspect as " activities regulation of the individual from the community" [2].

The central theme has become urban quality [3], although its forms are multiple, as well as the knowledge that is involved. What clearly emerges in the ongoing long time debate on these topics is nothing more than the need for a joint vision in urban planning by the side of all those disciplines which, for different reasons and in different ways, dealing with settlement, social, environmental and territorial issues, addressing specifically the issue of the quality of urban space [1].

In fact, it is clear that the objective of the quality of urban life is a multi-disciplinary and multi-dimensional concept, and as such must be represented by a ratio of various layers, whereas the quality of urban life is the result of the relationship between these dimensions, different in relation to places, social and economic contexts [4].

As suggested in the literature, you can rely seven main dimensions that contribute to the quality of urban life: environmental aspects related to the area's natural; physical, referring to the structures of the urban fabric; furniture, in relation to accessibility, traffic and transport issues; social, linked to integration of individuals and their participation in social life; psychological, on the sense of membership to the city life; economic and finally political. These values are all interrelated and dependent on each other, as summarized in the "Shape Heptagon " in Fig. 1 [4] .

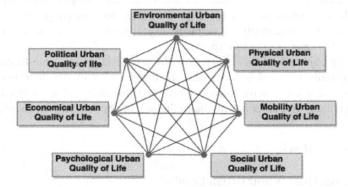

Fig. 1. Urban quality of life dimensions – Heptagon Shape (the researcher, 2012)

These seven parameters explained previously, have set a focus on the physical and environmental characteristics of urban quality, features closely related with the definition of urban comfort [5] and bioclimatic planning applied to the urban scale, through 'introduction of bioclimatic criteria for the design of the system, open spaces, materials, green, infrastructure [6].With the expression of urban comfort we refer to thermal comfort, comfort refers to the quantity of wind, solar radiation and air quality, these parameters can be taken as indicators of the urban microclimate [7]. However it should be pointed out that the conditions of urban comfort are not totally controllable since you have to evaluate a mental and physical aspect of 'user [8] linked to, idea of the quality of the living spaces of the city.

From a physical point of view, the evaluation of the comfort depends on several parameters and the combination thereof [9]: air temperature, relative humidity and wind speed.

In turn, air temperature is determined by contribute to several local factors that can be identified as: building density, the structure and urban morphology, the peculiarity of buildings components of buildings and open spaces and the incident solar radiation.

Building density varies as a function of the thermal capacity of each sub-area or urban sector to retain and store heat, and, consequently, varies the amount of thermal energy emitted into the environment, which is proportional to the energy absorbed.

The urban morphology and the building structure of each urban sector, with particular reference to the relationship of height / distance between the buildings and then the geometry of the so-called "urban canyons" [10], are key factors in the dynamics of the phenomena of reflection and absorption of solar radiation . In fact among the facades of buildings are generated phenomena of multiple reflection-absorption, depending on the characteristics of albedo (reflection coefficient), "sky view factor" between the facades and between the building and the sky, as well as the gaps between them [11].

The "urban canyons", characterized by a difference of the buildings height compared to the width of "canyon" (distance between the two built-scenes that delimit a road), determine a greater absorption of thermal energy and a subsequent increased heat radiation in environment [10].

The factor of the albedo of the ground has an important role in relation in determining the amount of shortwave solar energy reflected and therefore of high quantity of energy stored or irradiated as long-wave thermal radiation.

The sky view factor (SVF), expressed in a scale from 0 to 1, consists in measuring of the solid angle of sky view from an urban space and determines the radiant heat exchange between the city and the sky, resulting in the strong correlation between the data, the incident solar radiation on surfaces and temperatures [12].

The wind is one of the most important factors that affect the comfort conditions in outdoor open spaces but, at the same time, it is difficult to predict and control an environment exposed to the wind because it is influenced by several factors of global, regional and local levels. On a global scale the wind originates from the movement of air from areas of high pressure to areas of low pressure, the speed and direction of the wind depend on global factors but are subsequently influenced by the type of the local landscape. 'therefore is important understand that there are considerable differences in the environment exposed to the wind from side to side of the city, or even at the micro scale, from side to side of an open space [13].

The direct effects of the wind can be divided into two broad categories: mechanical effects and thermal effects [14]. The mechanical effects of wind can be perceived at a wind speed of more than 4-5 ms-1: above 10 m/s is generally unpleasant to walk and above 15 m/s there is a real risk of accidents [15]. As for the thermal effects you can use the criteria of comfort 5ms-1 present in standard tables (used for air temperatures above 10 °C) if it is assumed that people adapt their behaviour and clothing to the season [16].

2.2   The Sustainable Buildings

What has been said about the necessity of careful planning in order to ensure high standards of urban comfort must necessarily be expressed in the definition of a building that, in an aspect that descends from the general to the particular, knows how to explain all parameters mentioned, and, rather, it adds new and unavoidable parameters.

The building has long since projected into the so-called "sustainable building", that in addition to have as its objective the creation of a more comfortable and healthy environment for their occupants, trying to have a less damaging impact (and also should have ameliorative effects) for external surrounding environment. The use of materials, technologies and design solutions leads to improving the performance of building and mitigate the phenomena related to urban space as, for example, heat island (Fig. 2) [17].

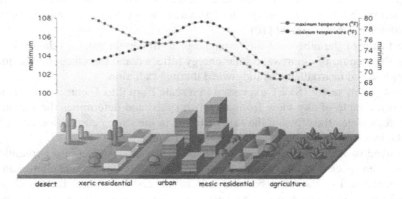

Fig. 2. Representation of urban heat island

A first aspect that affects a building in the fields mentioned before, is certainly intended the building envelope, understood as an architectural element that defines and encloses the perimeter of the building structure [18].

New materials for the "skin" [18] of buildings are proposed to mitigate environmental temperature and protect the interior environment. In this regard, the cool materials represent a solution that obtain increasing interest: they are characterized by high values of solar reflectance and high emissivity in the infrared. The first size reduces the absorption of solar radiation by building materials, so that limit the temperature rise in the presence of a high solar radiation during daylight hours. These materials also have high emissivity in the infrared to emit into the sky during the night and thus dissipate the collected heat, without transferring it inside the buildings. [19]. The usual construction materials, except metals, have a high emissivity but generally have a high solar absorption which causes a considerable rise of temperature of the material, beyond the values of air temperature. A parameter used to express the material's ability to stay cool under the solar radiation is the index SRI (Solar reflectance Index) calculated in steady state conditions. This is in relation to the solar reflectance and to the emissivity [20].

Excellent examples of vernacular architecture attest to the use of white colour finishes, which reflect solar radiation during the day and also dispose of, at night, as infrared radiation, the heat stored in buildings in the Mediterranean area. So the concept is not new, but it is interesting to analyse the combination of old concepts and new technologies; in particular, by analysing the reflectance spectrum of the cool material between 300 and 2500 nanometres, is known as the value of this is much higher than the typical materials of construction in the infrared range, while maintaining the typical value of the colour on the visible range [21]. Some studies have investigated the potential of cool material to reduce thermal profiles in the city, and found an improvement in the external thermal comfort conditions and a reduction of requirements for building air-conditioning [22]. In general, it can be said that the absorption of the sun's radiation by opaque surfaces of buildings can also be minimized by using reflective materials, so simplifying, characterized by very light colours.

Another large context that can be explored to improve the performance of the building that produces ameliorative effects at the level of the urban environment is the integration of green: vertical gardens and green roofs.

The "Natural Skin", in fact, interacts with the environment by changing climate by reducing the impact of pollutants and, where it becomes excessive, the discomfort caused by wind and noisy sources. The location of garden in coverage or in correspondence of the vertical surfaces, the type of plant cover, the level of permeability, the rapidity of growth, in relation to climatic context stressing, define the energy performance of the casing towards the internal environment and the outer space. [23]. The energy that affects on vertical surface partly is reflected , partly is transmitted , partly is absorbed (consumption for photosynthesis process) and partly is dissipated in the form of heat. While facade components and surfaces that complete the traditional roofing systems, when they are hit by direct radiation, tend to increase its surface temperature, the plant cover, however, through the process of evapo-transpiration, maintains a surface temperature near to temperature of external air [23]. The deciduous shrubs placed in a vertical garden stop the 'Excessive incoming sunlight during the summer months and lose their leaves in winter to allow solar radiation to heat the building, so, in a natural way, by changing the transmittance of building [24]. In this regard it is useful to discern two types of green wall: one that involves the integration of plants in the wall stratigraphy, the other in which the wall is vine-covered. In the first the selected essences are placed on a layer of felt, resting on a PVC layer anchored to a metal support structure detached from the wall to allow the transpiration. A system of irrigation pipes nourishes and maintains wet the felt layer, using gravity from the top of the structure. So, the second type of green-wall is cheaper and easier assembling than the first.

During the summer months, the lowest temperature in the layer closest to the green cover, increases the efficiency of HVAC (Heating, Ventilation and Air Conditioning) through the adoption of cooling air intakes, reducing the average daily demand of energy for the cooling of interior spaces [25].

At least, a solution able to operate on the urban environment, reducing the amount of pollution in the air, is the use of materials that use the Honda-Fujishima effect. This phenomenon is related to catalysis exhibited by plants: the principle has been studied and reproduced though photocatalytic coatings, based on titanium dioxide (a colourless crystalline powder), which, exposed to light, converts the oxygen in very oxidizing chemical compounds, able to react and decompose organic and inorganic pollutants. The ideal applications in the construction sector are the vertical external surfaces: the tests give to the material an average duration of ten years and so these are great solutions because they reduce the need of maintenance.

3   Research Methodology

As always, the growth of urban areas, the high incidence of economic dynamics on building process, the increase of levels of comfort have meant that the current built heritage came to be defined by whole inadequate, inefficient and energy-consuming portions, often significantly larger than the remaining parts.

The research that we want to carry out, otherwise the definition of design guidelines that consider the bioclimatic aspects applied to the building and the complementary design of urban spaces, is based on a preliminary methodological definition of urban settlement fabric, through a distinction of it in different morphologic and typological characterizations.

The analysis of the existing building fabric, which constitute a large part of Italian cities, directly affect on the energy behaviour, paying attention on the interrelations established among the energy behaviour of the built environment and urban fabric's morphology: the relationship between energy and form involved relevant fields of scientific research, leading to demonstrate the nature and entity of the relationship, which exist not only in the context of living systems (plants or animals), but also in built systems [26, 27].

Not today the energy and environmental requirements have been undertaken in the context of urban planning, as well as the energy performance have been often analyzed in conjunction with the various density parameters; the present attempt is to establish a defined scientific knowledge about the relationship between shape and energy in the urban environment, reducing the field to the assumptions and quantitative analyzes and allowing for a process of rational and scientific knowledge as possible, by many considered essential to tackle the tasks currently assigned to the discipline of architecture [28].

If, from the theoretical point of view it appears that the research of interactions between construction and quality of urban areas are constantly evolving, from a purely legal point of view it is not. In Italy, in fact, the urban legislation has a complex structure, as well as the building legislation, but both are moving on parallel tracks. The result is an approach that doesn't allow to get the opportunities arising from the interaction between the respective laws. More recently it has gained an increasing recognition of this interaction, although the attention from who is responsible for ensuring compliance with the rules seem to favour the procedural aspects rather than the content.

Urban form is defined by three fundamental physical elements: buildings and their related open spaces, plots or lots, and streets (Loconte et al. 2012) [29]. Specifically, the issue has been faced through the experimentation conducted on two case studies, able to demonstrate adequately the realities present in almost all cities: the compact urban fabric and the urban peripheral.

When we speak about compact fabric we refer to a high-density settlement, with not too high and mixed-use buildings, concentrated on housing units, large enough to offer a range of social and economic benefits, and with walking distances from homes. In addition, compact fabrics settlements have a greater accessibility of services, which are readily available to residents, an increased vitality of spaces and places of urban centres; a functionality of the residential streets that enforces the sense of places through the game of children, the encounter between people (rather than being only paths for the transit to another place), and a greater sense of community that facilitates social interaction[30].

However, the peripheral urban areas are characterized by urban spaces sufficiently disarticulated and disconnected, often without a clear relationship among morphological and typological-environmental attitude. In this context we refer to high buildings, in which the comparison between the side and horizontal surface is high.

The dispersion of the city in rural land appears to a lot of people as one of the worst effect of urban sprawl: a phenomenon that can generate a perverse settlement system very complex and expensive to manage, both from the point of view of transport and mobility, and from that of the maintenance of the buildings and urbanized spaces. In fact, the system of sprawl-town is too energy-intensive and its consumption are unsustainable. In this sense, the phenomenon of sprawl is considered the principal reason of the progressive destruction of the identity landscape as well as of the natural resources and local economies.

The main issues , related to urban comfort on the one hand, and on the other side of the design quality of the existent building, have been identified for the two areas of study. The analysis was conducted looking at the same time both factors, taking especially the interactions occurring between them.

In the case of the compact urban fabric, with regard to the open spaces, was immediately gained a deficit of lighting and ventilation, because they are characterized by alleys with a very low section, that limits the permeability of inner zones to the prevailing wind. Furthermore, the lack of green areas emerges above all in the very compact fabric, due the compactness of the urban settlement. However, if on one hand the presence of very close buildings obstructs a correct sunshine of urban spaces and of the interior of the buildings, on the other hand it is a factor of cooling (due to shading) especially in summer Mediterranean climates. In order to get a detailed analysis of the phenomenon, would be more appropriate have a mapping of shading during the year, since solar radiation and shading are two of the most important factors to be evaluated during the project. To map the shade that protects from the sun means to collect data about shadows for each our of a day once a season and add this images to obtain an annual profile of shading regarding the site in question[31].

Analysing the buildings comes to light that most of them are energy inefficient. In a such context it is necessary to define improvement actions that could be applied on existing buildings, going to carry out an energy-restoration intervention. One of the possible interventions, able to work in both areas of analysis, may be represented by the construction of roof gardens, able to increase the thermal insulation of roofs, and at the same time to reduce the heat island effect. Moreover, an intervention of this kind would also lead to benefits from the point of view of the absorption of humidity presents in the air. Analyzing the buildings of the compact fabric, the burning off surfaces that most influence in the building energy balance are the horizontal (because they are low-rise buildings with one or two vertical burning off fronts). That's why an intervention of this type appears to be more convenient than the same applied to a suburban building. In the city of Bari, for example, thanks to the project Shagree (Green Shadows Program) was provided the construction of green roofs over 12 terraces, in order to reduce external noise and to contribute to the filtration of pollution. The new green roof will be born in neighbourhoods of the compact city and will also help to improve the energy performance of the buildings on which will arise. Nevertheless it is possible to operate on the facade through the use of photocatalytic or light coatings. These actions, in addition to reducing domestic consumption, solve the problem of the natural lighting exterior lack, reflecting the solar radiation. Furthermore, the photocatalytic materials, thanks to their chemical composition, reduce the pollution in the air.

However, in an urban settlement like this we don't consider appropriate the use of green facades. In particular, the reasons that lead us to this choice are the lack of proper lighting and ventilation for the growth of the plants, and the worsening of sunlight reflection by buildings (remembering that, in the compact fabric, the low natural light of urban spaces is one of the most relevant problem). In addition to this, there is the difficulty of relating the vegetation with the ground when the green is integrated in the stratigraphy of the vertical wall.

In peripheral urban fabric, however, there are problems related to a high incident solar radiation and to the presence of highways.

The first phenomenon is measurable through the SVF (Sky View Factor), or rather the portion of the sky visible from a point on the earth's surface considering an hemispherical cap of radius defined by the user (this parameter varies in a range between 0, in case of complete darkening of the sky, and 1 in case of full view of the same). A high value of SVF means an increase of the surface temperatures of urban open spaces and of building skill. It is amplified if the surfaces are made of coarse and dark material.

The high-traffic flow characterizing the roads is measured by the hourly flow rate, i.e. the number of vehicles passing in an hour in a given section of road (vehicles/hour). A high value of this parameter leads to many disadvantages, including an increase of pollution in the atmosphere, a boost of outside temperature and of the heat island phenomenon, and other phenomena related to acoustic discomfort.

In this context, focusing the attention on the building, there are often high and/or isolated buildings in which the dispersants prevalent surfaces are the vertical walls. For this reason, the interventions on the roof are less effective on the heat balance of building. So we recommend the interventions on the walls.

A first possible intervention is the realization of clear and with a reduced roughness external surfaces. These material characteristics, in fact, contribute to obtain a reduced value of albedo, or rather a low value of the incident solar radiation that is reflected in all directions. A similar intervention, but still not popular, is the use of photocatalytic plaster as the Hotel de Police in Bordeaux that reduces the action of pollutants effects, typical of urban areas.

Another possible intervention, which involves higher costs but also further benefits from the point of view of the interior comfort and of the climatic perception of the urban environment, is the integration of green surfaces in the façade. There are many advantages that this solution presents for the designer:

- Thermal regulation due to thermal transpiration of plants that cools the air;
- Purification of the air through the absorption of carbon dioxide and the production of oxygen;
- Filtration and purification of air pollutants;
- Noise and glare reduction through the absorbing of the sound and light waves.

Fig. 3. Bosco Verticale, Boeri Studio

A successful example of this is the "Bosco Verticale©" in Milan, designed by Boeri Studio: an urban forestation project. The presence of trees, shrubs and green walls, continuously distributed on the building facades, helps to regenerate the environment and the urban biodiversity limiting the expansion of the city. It consists of two towers of 110 and 76 meters; it was realized in the centre of Milan within the "Porta Nuova" project, on the edge of the neighbourhood "Isola", and will host about one 1000 trees (up to 9 meter high) as well as numerous shrubs and flowering plants, as well as about 2300 m2 of green wall (Fig. 3).

4   Conclusion

The main theme and purpose of the paper was to highlight how within the contemporary city is possible to identify homogeneous scenarios, which define guidelines that take in account both the building and urban quality. These two objectives don't travel on parallel tracks, as can be seen instead by most of the examples of past design, but are closely related to a condition of reciprocity.

The analysis was conducted for the compact and peripheral urban context, defining as first the characteristics, and then the issues related to environmental interventions. As a result of these considerations we obtained the data by which drive project interventions. Consider the two extremes context that characterize the cities, highlights the relevance of a preliminary study on urban morphology. Only after this characterization it is possible identify the problems and the subsequent solutions, often different from each other.

So, the goal of the research is to direct the designer to a joint planning between the urban and the built environment. It is necessary to define, starting from the stage of planning and scheduling, an intervention strategy and not numerical parameters that may be inefficient and extremely binding. These project activities will involve not only merely urban space, but all the buildings that with their heights, materials, technological solutions influence on it, not less of the classical used parameters (density, land use, infrastructure and services).

The paper also refers to the laws that govern the project, hoping to overcome the gap between the present Italian legislative apparatus and technological development that involves the construction field. Too often, in fact, the excessive bureaucratic process compromises the laws effectiveness, even when these show character of innovation. The attention is only for the compliance of the rules: in this way long periods of time have spent and the quality of planning decreases. Downstream of this, legislation should not be a constraint, but rather should direct the designer to choose the best solutions for smart development and quality of urban and built environment. The legislator's goals must be to provide a framework within which the designer can navigate and choose the solutions considered most appropriate in order to the rule of the art. At the same time, the design team (engineers, architects, landscape architects, planners etc.) must be free to investigate, from time to time, the best solution provided by technologies, depending from the context of reference and from the needs of the case. The purpose of an efficient legislative system should be provide the right tools in order that people involved could find one the best alternative, downstream of the design process, rather than simply identify the only possible way.

References

1. Pavan, V.M.: I requisiti di qualità ambientale nel progetto urbano. Università degli Studi di Cagliari (2008)
2. Landaver, C.: Theory of National Economic Planning. University of California Press, Berkeley (1947)
3. Abis, E.: Piani e politiche per la città. Metodi e Pratiche. Franco Angeli Editore. Milano (2003)
4. Serag El Din, H., Shalaby, A., Elsayed, F.H., Elariane, S.A.: Principles of urban quality of life for a neighborhood. HBRC Journal (2013)
5. Krüger, E.L., Minella, F.O., Rasia, F.: Impact of urban geometry on outdoor thermal comfort and air quality from field measurements in Curtiba, Brasilia. Universitade Tecnologica do Paraná, Curtiba, Brazil (2010)
6. Calace, F.: La sostenibilità nelle pratiche della progettazione urbana. INU (Istituto Nazionale Urbanistica) (2011)
7. Carmeliet, J., Concherio, J., Steinemann, R.: Sustainableurban design, interview with Jan Carmeliet (2012)
8. Dessì, V.: Progettare il comfort urbano. Soluzioni per un'integrazione fra società e territorio. Gruppo editoriale Esselibri-Simone (2007)
9. Benedetti, C.: Cmfort urbano. Le guide pratiche del Master CasaClima, vol. 8. Bolzen-Bolzano University Press (2013)
10. Adolphe, L.: A simplified model of urban morphology: Application to an analysis of the environmental performance of cities. School of Architecture de Toulose, France (2000)
11. D'Olimpio, D.: Morfologia Urbana e microclima locale: strumenti e metodi per lo studio e la valutazione. EcoEdility (2009)
12. Dettori, S.: Il Comfort ambientale negli spazi aperti. Università degli Studi di Sassari (2011)
13. Kofoed, N., Gaardsted, M.: Considerazione sul vento negli spazi urbani. Estratto da: Progettare gli spazi aperti nell'ambiente urbano: un approccio bioclimatico. Esbensen Consulting Engineers Ltd., Danimarca (2004)
14. Penworden, A.D., Wise, A.F.E.: Wind environment around buildings. Department of the Environment BRE. Her Majesty's stationary office, London (1975)
15. Bjerregaard, A.D., Nielsen, F.: Wind environment around building. Danish Buildings Research Institute, Hørsholm (1981)
16. Davenport, A.G.: An approach to human comfort criteria for environmental wind condition. Swedish National Building Research Institute, Stockholm (1972)
17. Rosenfeld, A.H.: A mitigation of urban heat island: Materials, utility programs, updates. Elsevier (1995)
18. Herzog, T., Kripper, R., Lang, W.: Atlantedellefacciate. UTET, Torino (2006)
19. Fanchiotti, A., Carnielo, E.: Impatto di cool material sulla mitigazione dell'isola di calore urbana sui livelli di comfort termico negli edifici. UniversitàdegliStudi Roma Tre (2011)
20. Berdahl, P., Bretz, S.E.: Preliminary survey of the solar reflectance of cool roofing materials. Energy and Buildings 25 (1997)
21. Zinzi, M.: Cool material and cool roofs: Potentialities in mediterranean buildings, vol. 4. Earthscan (2010)

22. Synnefa, A., Karlessi, T., Giatoni, N., Santa Marius, M., Assimakopulos, N.D., Papakatsi-kas, C.: Experimental testing of cool colored thin layer asphalt and estimation of its potential to improve the urban microclimate. Building and Environment 46 (2011)
23. Greco, A., Quagliarini, E.: L'involucro edilizio. Una progettazione complessa. Editrice Alinea (2007)
24. Akbori, H., Konopacki, S.: Calculating energy-saving potentials of heat island reuction strategies. Elsevier (2005)
25. Killingsworth, B., Lemay, L., Peng, T.: The urban heat island effect and concrete's role in mitigation, part I. Concrete in Focus, National Ready Mixed Concrete Association (2011)
26. Knowles Ralph, L.: Energia e forma un approccio ecologico allo sviluppo urbano, Le scienze dell'artificiale. Muzzio, Padova (1981)
27. Olgyay, V., Mancuso, G.: Progettare con il clima. Un approccio bioclimatico al regiona-lismo architettonico. Franco Muzzio Editore (1990)
28. Cecere, C., Roura, H.C., Morganti, M., Clementella, G.: Dalla riqualificazione energetica al recupero sostenibile. Un metodo di analisi energetica dei tessuti della città compatta. Ricerche e progetti per il territorio, la città e l'architettura (2012)
29. Loconte, P., Ceppi, C., Lubisco, G., Mancini, F., Piscitelli, C., Selicato, F.: Climate altera-tion in the metropolitan area of Bari: temperatures and relationship with characters of urban context. In: Murgante, B., Gervasi, O., Misra, S., Nedjah, N., Rocha, A.M.A.C., Taniar, D., Apduhan, B.O. (eds.) ICCSA 2012, Part II. LNCS, vol. 7334, pp. 517–531. Springer, Heidelberg (2012)
30. CPRE, Campaign to protect rural England: High density doesn't necessarily mean high-rise, or low-quality council estates. Excerpt from the report "Compact Sustainble Communities" (2006), http://www.cpre.org.uk
31. Nikolopoulou, M.: Designing the open spaces in the urban environment: a bioclimatic ap-proach (2004)

Time Scheduling Post Earthquake Reconstruction Ethics and Aesthetics into a WebGIS Interface

Salvatore Giuffrida and Filippo Gagliano

Department of Civil Engineering and Architecture - University of Catania
sgiuffrida@dica.unict.it,
fmgagliamo@gmail.com

Abstract. One of the most relevant questions concerning the reconstruction planning after the 2009 earthquake in Abruzzo is the funding allocation pattern and its consequences on the directions for the rebirth of the damaged towns. This paper presents the experience of Villa Sant'Angelo, whose Reconstruction Plan has been conceived as an integrated platform including analysis, valuation and plan. The valuation stage overcomes the monetary one, the only one required in reality, and extends to fairness and urban quality. The Spreadsheets and the connected WebGIS tool we have performed to control all phases of this process, include these valuation functions and arrange, in particular, the funding time schedule. For this purpose, they allow to implement a lot of different strategies displaying at real time the spatial funds allocation in the different periods. The tool can be used by everyone admitted to the participation practice as managed by the local authority.

Keywords: Post earthquake reconstruction, Web-GIS tools, quali-quantitative evaluation, funding schedule, participation.

Introduction

The earthquake of Abruzzo, in 2009, involved an area, the seismic crater, of 2387 sq.km, that includes 56 municipalities and the administrative centre, L'Aquila, and about 67,000 people, accommodated first in hotels and houses for rent and then in new specifically built districts.

This huge economic effort reduced the currently available resources for the reconstruction; furthermore the complex administrative procedures, that must be followed, lengthen the time required.

One of the main requirements is the objectification of the criteria for the allocation of funding. These criteria must be shared in order to build a significant and remarkable cost/merit function for each intervention to be funded.

The complexity of questions and needs of a wide and deeply damaged area, pushed the regional authorities to involve the planning practice and to resort to have the reconstruction plans edited by the engineering departments of each municipality. This process is subject to the procedural steps of the ordinary planning system.

The general objectives of the management of a large-scale reconstruction, mostly due to landscape, urban and architectural values [2], have configured the Reconstruction Plans as programming tools for future funding.

B. Murgante et al. (Eds.): ICCSA 2014, Part III, LNCS 8581, pp. 253–268, 2014.

Materials

Villa Sant'Angelo, with the nearby hamlet of Tussillo, is one of the cities most damaged by the earthquake, within the Valley of the River Aterno. It is approximately 16 km from L'Aquila and it has 423 inhabitants.

The Reconstruction Plan (RP) [1], [5] covers the area of the two towns bounded under DCDR 3/2010, comprising an area of 88 183 sq., a built a volume of 196.300 cu.m, 417 Architectonic Units (AU) (Fig. 1a-b) joined in 100 Blocks.

The architectural heritage have been subject to considerable damage, whose extent can be measured on the basis of the distribution of the *Usability Classes* (UC) (Fig., 1b) (*A: Usable* for buildings with minor damage, *B: Temporary unusable* the edifice temporarily uninhabitable, *C: partially unusable, D: Religious buildings* with a specification of the damage degree; *E: unusable buildings* due to severe damage or collapse) recognized by the Emergency Management Office and reported into the AeDES sheets.

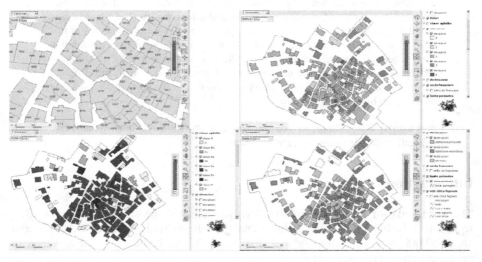

Fig. 1. WebGIS queries: a. AU identification, b. dimensions; c. Usability Classes, d. functions

One of the main functions of the PDR was the calculation of the costs of reconstruction [3, 4]. For this purpose it was first necessary to verify the title to the funding by ascertaining: *private property* - the property status (primary or secondary residence) and some functional characteristics; *public buildings* - direct (actual damage caused by the earthquake) and indirect (substantial damage in the reconstruction process) causal link with the earthquake. Therefore, private and public works to be funded by the State were distinct from those differently funded.

The general purpose to start a consistent and transparent funding program produced a wide legislative framework [3] that rules the plan procedure and the methodology of calculation of quantities and cost as well.

Municipalities are the recipients of the funding of public works, mainly involving: the disposal of the debris; repair or reconstruction of networks of underground services; repair of the surfaces of open spaces and public parks, the reactivation of public lighting. The owners are the recipients of private financing; they must form consortia when, as happens in most cases, their dwellings are part of a *Block*.

The technical-economic tools requested for the RP admission are: the *Technical Economic Balance Sheet* (TEBS), the *Monitoring Synthesis* (MS), the *Time Schedule* (TS). All of them result from the whole information system as formalized in the general database. The former is composed of records that are the single homogeneous urban objects, which can be considered the "minimum *information bearer* units" [13], and fields that include all their significant features. The relevance of these features is defined with reference to the whole planning process so that the database we drew up is extended to economic evaluation and to reconstruction planning and financing programming as well.

Two types of minimum units have been considered: the Architectonic Unit (AU) and the Urban Unit (UU). An AU is the minimum building unit which can be assumed as homogeneous by several viewpoints such as constructive, functional, architectural, morphological and typological [6], [9], [12]. The UU is the Block.

The database is composed of different sections:
- the first one concerns the identification (Block id., AU id.) and the dimensions of each AU; the dimensions have been accurately and carefully surveyed referring to the different aims they are useful for. In fact the cost has to be computed by using different surfaces or cubage for each different UC: number of properties for A; sq.m of Total Area (TA) for B and C; sq.m for E (Tab. 1);
- the second one includes some qualities that are relevant for calculating the contribution (function, public/private ownership, strategic function, primary or secondary dwelling, architectural value, any heritage bond, UC, physical condition) (Tab. 1);

- the third section contains the calculation of the costs for each AU (Tab. 2);
- the fourth section calculates the additional (technical) expenses (Tab. 2);
- the fifth section includes the urban object, as previously defined (tab. omitted).

1. TEBS. The Technical Economic Balance Sheet (TEBS) of the two urban centres, Villa Sant'Angelo and Tussillo, is the basic sheet for calculating the reconstruction cost of the works prescribed by the Reconstruction Plan. TEBS includes both ordinary (reconstruction) and extraordinary (social and economical development) works. The expenses are aggregated into some reports: a general sheet including the aggregate amounts (Tab. 3); private buildings amount; public residential building amount; public roads, open spaces and equipment amounts specified for each subcategory (all tables are omitted). Special sections include the demolitions of ruins and the transport of wreckage to landfill (table. omitted).

Table 1.   General database: synthesis of sections 1 and 2

id Block	id Architectonic Unit	GA ground area (sqm)	floors (n)	underground floors (n)	TA total area (actic excl) (sqm)	AA actic area (20%GA-16%)	NUA net usable area including AA (sqm)	GUA gross usable area (including not res area and AA	NRA not res area (sqm)	OA overall area (GUA-0.4xNRA+0.6xAA)	V volume	function	public/private ownership	strategic (yes/not)	primary or secondary dwell	architectonic value	% increase	bond	SGE usability	1 = UA in BI; 0= single UA	properties (n)	in BI with B/C	in BI with E
1	1	133	3		398	22,3	425	334	0	348	1242	dwell	private		second res				C -	1	4	1	1
1	2	40	0		0	0,0	0	0	0	0	0	-	private		second res					1	0	1	1
1	3	162	2		323	27,1	355	271	0	288	1027	dwell	private		second res				A -	1	3	1	1
1	4	281	2	1	652	47,3	709	548	75	546	1791	dwell	private		primary res				E -	1	6	1	1
1	5	233	3		698	39,1	745	586	0	610	2178	dwell	private		primary res			y	E -	1	7	1	1
1	6	51	1		51	8,6	61	43	0	48	172	dwell	private		second res			y	E -	1	1	1	1
2	7	27	2		57	4,8	62	48	24	41	180	mix dwell	private		second res				E -	1	1	0	1
2	8	56	2		113	9,5	124	95	0	100	358	dwell	private		second res				E -	1	2	0	1
2	9	90	2		180	15,1	198	151	0	160	572	dwell	private		second res				E -	1	2	0	1
2	10	78	2		156	13,1	172	131	0	139	497	dwell	private		second res				E -	1	2	0	1
2	11	105	2		212	17,8	233	178	0	188	673	dwell	private		second res				A -	1	2	0	1
59	12	56	2		112	0,0	112	94	47	75	336	-	private		second res				B -	1	2	1	0
3	13	551	2	1	1440	86,5	1543	1209	344	1123	3274	dwell	private		primary res	av	39%		E -	1	14	1	1
3	14	28	1		28	4,7	33	23	0	26	94	dwell	private		second res				E -	1	1	1	1
3	15	52	1		52	8,7	62	44	0	49	175	dwell	private		second res				B -	1	1	1	1
3	16	51	3	1	201	8,6	211	169	41	158	476	dwell	private		primary res				A -	1	2	1	1
3	17	47	2		95	8,0	104	80	0	84	301	dwell	private		second res				A -	1	1	1	1
3	18	10	2		19	1,6	21	16	0	17	61	dwell	private		second res				E -	1	1	1	1
3	19	78	2		155	13,0	171	130	0	138	493	dwell	private		second res				E -	1	1	1	1
3	20	112	2		223	18,8	246	188	0	199	710	dwell	private		primary res				E -	1	3	1	1
60	21	20	1		20	3,3	24	17	0	20	67	dwell							A -	0	1	0	0
28	22	106	2		211	17,8	233	178	0	188	672	dwell	private		primary res				B -	1	2	1	0
28	23	49	1		49	8,2	58	41	0	46	163	dwell	private		second res				B -	1	1	1	0
28	24	41	2		82	6,9	90	69	0	73	260	dwell	private		primary res				B -	1	1	1	0
28	25	40	2		79	6,7	87	67	0	71	252	dwell	private		second res				B -	1	1	1	0
28	26	14	1		14	2,4	17	12	0	13	48	dwell	private		second res				B -	1	1	1	0
61	27	173	1		173	0,0	173	146	146	87	520	-	private		second res				A -	1	2	0	0
61	28	108	2		216	0,0	216	181	91	145	647	-	private		second res				A -	1	2	0	0
31	29	103	2		206	17,3	227	173	0	183	655	dwell	private		second res				E -	1	2	0	1
30	30	215	3		646	36,2	689	543	0	564	2015	dwell	private		primary res				E -	1	6	0	1
51	31	60	1		60	0,0	60	51	51	30	181	-	private		second res				E -	0	1	0	1
52	32	72	1		72	0,0	72	60	60	36	216	-	private		second res				A -	0	1	0	0
32	33	162	2		325	27,3	357	273	0	289	1033	dwell	private		second res				E -	1	3	0	1
55	34	93	2		185	15,6	204	156	78	134	589	mix dwell	private		prima casa				A -	1	2	1	1
24	35	166	2		333	28,0	366	280	140	240	1058	mix dwell	private		primary res				A -	1	4	0	1
24	36	98	2		195	16,4	215	164	82	141	621	mix dwell	private		second res				E -	1	2	0	1
24	37	123	3		368	20,6	392	309	103	280	1147	mix dwell	private		second res				E -	1	4	0	1
17	38	40	2		79	6,6	87	66	0	70	251	dwell	private		second res				E -	1	1	0	1
17	39	31	3		94	5,2	100	79	0	82	292	dwell	private		second res				E -	1	1	0	1
17	40	90	2		180	15,1	198	151	0	160	573	dwell	private		primary res				E -	1	2	0	1
17	41	70	2		140	11,8	154	118	0	125	446	dwell	private		primary res				E -	1	2	0	1
17	42	36	2		71	6,0	79	60	0	64	227	dwell	private		primary res				E -	1	1	0	1
17	43	31	2		62	0,0	62	52	26	52	187	public bui	public r		second res				E -	1	1	0	1
17	44	43	2		85	7,2	94	72	0	76	271	dwell	private		second res				E -	1	1	0	1
17	45	56	2		112	9,4	124	94	0	100	357	dwell	private		second res				E -	1	2	0	1
17	46	32	2		64	5,4	71	54	0	57	204	dwell	private		primary res				E -	1	1	0	1
15	47	45	2		90	7,5	99	75	0	80	286	dwell	private		second res				E -	1	1	0	1
15	48	41	2		82	6,9	91	69	0	73	262	dwell	private		second res				E -	1	1	0	1
15	49	25	2		49	0,0	49	41	21	33	148	other use	private		other uses				E -	1	1	0	1
46	50	47	1		47	0,0	47	40	40	24	142	-	private		second res				E -	0	1	0	1

2. MS. The *Monitoring Synthesis* records general information about the consistency between the RP and the former planning tools, and aggregates all the costs in order to compare them with the main dimensions of the Municipality: total area, population, and so on. This way some indicators can be assumed in order to compare the unit cost of each municipality with the others' ones (table omitted);

Table 2. General database: synthesis of section 3 and 4

Id Block	Id Architectonic Unit	Usability class	A-UC: base (€/property)	B-UC: base (€/sqm)	D-UC base (€/cu.m)	E-UC base (€/TSsqm)	A-UC structural enhancement if with E (€/sqm)	B/C-UC structural enhancement (€/sqm)	% increase for bond	General total cost A, B/C, D, E	additional expenses: consortium president fee	additional expenses: geotecnical surveys	additional expenses: project management	additional expenses: taxes	% additional expenses
1	1	C - partially unusable	€	600				€ 195		€ 316.507	€ 4.155	€ -	€ 45.090	€ 23.887	23%
1	2									€ -	€ -	€ -	€ -	€ -	0%
1	3	A - usable					€ 2.500			€ 60.795	€ -	€ -	€ -	€ -	0%
1	4	E - unusable				€ 1.277				€ 697.466	€ 12.131	€ 12.912	€ 131.654	€ 69.747	34%
1	5	E - unusable				€ 1.277			100%	€ 1.557.383	€ 27.088	€ 28.832	€ 293.972	€ 155.738	33%
1	6	E - unusable				€ 1.277			100%	€ 122.989	€ 2.139	€ 2.277	€ 23.215	€ 12.299	33%
2	7	E - unusable				€ 1.277				€ 52.347	€ 1.317	969	€ 9.881	€ 5.235	35%
2	8	E - unusable				€ 1.277				€ 127.995	€ 3.221	€ 2.370	€ 24.160	€ 12.799	34%
2	9	E - unusable				€ 1.277				€ 204.610	€ 5.150	€ 3.788	€ 38.622	€ 20.461	34%
2	10	E - unusable				€ 1.277				€ 177.783	€ 4.474	€ 3.291	€ 33.558	€ 17.778	34%
2	11	A - usable					€ 2.500			€ 39.914	€ -	€ -	€ -	€ -	0%
59	12	B - temp unusable		250				€ 150		€ 44.800	€ 705	€ -	€ 5.285	€ 2.800	20%
3	13	E - unusable				€ 1.277				€ 1.993.468	€ 35.153	€ 36.905	€ 376.287	€ 199.347	33%
3	14	E - unusable				€ 1.277				€ 33.474	€ 590	€ 620	€ 6.318	€ 3.347	33%
3	15	B - temp unusable		250				€ 195		€ 23.140	€ 229	€ -	€ 2.454	€ 1.300	17%
3	16	A - usable	€ 10.000				€ 2.500	€ 2.500		€ 44.221	€ -	€ -	€ -	€ -	0%
3	17	A - usable					€ 2.500			€ 18.142	€ -	€ -	€ -	€ -	0%
3	18	E - unusable				€ 1.277				€ 21.825	€ 385	€ 404	€ 4.120	€ 2.183	33%
3	19	E - unusable				€ 1.277				€ 176.192	€ 3.107	€ 3.262	€ 33.258	€ 17.619	33%
3	20	E - unusable				€ 1.277				€ 253.943	€ 4.478	€ 4.701	€ 47.934	€ 25.394	33%
60	21	A - usable					€ 2.500			€ -	€ -	€ -	€ -	€ -	0%
28	22	B - temp unusable		250				€ 150		€ 84.560	€ 1.330	€ -	€ 9.976	€ 5.285	20%
28	23	B - temp unusable		250				€ 150		€ 19.440	€ 306	€ -	€ 2.293	€ 1.215	20%
28	24	B - temp unusable		250				€ 150		€ 32.720	€ 515	€ -	€ 3.860	€ 2.045	20%
28	25	B - temp unusable		250				€ 150		€ 31.680	€ 498	€ -	€ 3.737	€ 1.980	20%
28	26	B - temp unusable		250				€ 150		€ 5.720	€ 90	€ -	€ 675	€ 358	20%
61	27	A - usable					€ 2.500			€ 5.000	€ -	€ -	€ -	€ -	0%
61	28	A - usable					€ 2.500			€ 5.000	€ -	€ -	€ -	€ -	0%
31	29	E - unusable				€ 1.277				€ 234.164	€ 5.893	€ 4.335	€ 44.201	€ 23.416	34%
30	30	E - unusable				€ 1.277				€ 720.355	€ 18.130	€ 13.336	€ 135.974	€ 72.035	34%
51	31	E - unusable				€ 1.277				€ 38.863	€ 978	€ 719	€ 7.336	€ 3.886	35%
52	32	A - usable								€ -	€ -	€ -	€ -	€ -	0%
32	33	E - unusable				€ 1.277				€ 369.207	€ 9.292	€ 6.835	€ 69.691	€ 36.921	34%
55	34	A - usable	€ 10.000				€ 2.500			€ 43.058	€ -	€ -	€ -	€ -	0%
24	35	A - usable	€ 10.000				€ 2.500			€ 67.412	€ -	€ -	€ -	€ -	0%
24	36	E - unusable				€ 1.277				€ 180.207	€ 4.535	€ 3.336	€ 34.016	€ 18.021	35%
24	37	E - unusable				€ 1.277				€ 357.316	€ 8.993	€ 6.615	€ 67.447	€ 35.732	35%
17	38	E - unusable				€ 1.277				€ 89.801	€ 2.260	€ 1.662	€ 16.951	€ 8.980	34%
17	39	E - unusable				€ 1.277				€ 104.390	€ 2.627	€ 1.933	€ 19.705	€ 10.439	34%
17	40	E - unusable				€ 1.277				€ 204.837	€ 5.155	€ 3.792	€ 38.665	€ 20.484	34%
17	41	E - unusable				€ 1.277				€ 159.368	€ 4.011	€ 2.950	€ 30.082	€ 15.937	34%
17	42	E - unusable				€ 1.277				€ 81.162	€ 2.043	€ 1.503	€ 15.320	€ 8.116	34%
17	43	E - unusable				€ 1.277				€ 66.702	€ -	€ -	€ -	€ -	0%
17	44	E - unusable				€ 1.277				€ 97.048	€ 2.443	€ 1.797	€ 18.319	€ 9.705	34%
17	45	E - unusable				€ 1.277				€ 127.767	€ 3.216	€ 2.365	€ 24.117	€ 12.777	34%
17	46	E - unusable				€ 1.277				€ 72.977	€ 1.837	€ 1.351	€ 13.775	€ 7.298	34%
15	47	E - unusable				€ 1.277				€ 102.077	€ 2.569	€ 1.890	€ 19.268	€ 10.208	34%
15	48	E - unusable				€ 1.277				€ 93.757	€ 2.360	€ 1.736	€ 17.698	€ 9.376	34%
15	49	E - unusable				€ 1.277				€ 42.380	€ 1.067	€ 785	€ 8.000	€ 4.238	35%
46	50	E - unusable				€ 1.277				€ 30.370	€ 764	€ 562	€ 5.733	€ 3.037	35%

3. TS. The *Time Schedule* is the main point of this study because it expresses the fair combination between the features of the urban-architectural context, as represented by the technical restitutions, and the strategy of the local government that expresses, by means of it, its value scale.

258   S. Giuffrida and F. Gagliano

Table 3. Overall Technical Economic Balance Sheet

Commissario Delegato per la Ricostruzione
Presidente della Regione Abruzzo

Piani di Ricostruzione ex art. 14, co. 5bis, Legge 77/2009
Comune di VILLA SANT'ANGELO
Comune di Villa Sant'Angelo - Aree interne alla perimetrazione

QUADRO TECNICO ECONOMICO DI RIEPILOGO

TIPOLOGIA	IMPORTI TOTALI	
EDILIZIA PRIVATA	€	76.549.038
EDILIZIA RESIDENZIALE PUBBLICA	€	-
EDILIZIA PUBBLICA E PER IL CULTO	€	6.922.104
di cui 1. Edifici di interesse strategico € 1.053.096 2. Edifici non di interesse strategico € 380.580 3. Edifici per il culto € 5.488.428		
RETI E SPAZI PUBBLICI	€	9.770.647
di cui 1. Rete servizi € 5.347.474 2. Rete viaria € 877.847 3. Spazi pubblici € 3.545.326		
ESPROPRI	€	877.449
DEMOLIZIONI, MACERIE, MESSA IN SICUREZZA	€	3.414.078
IMPORTO TOTALE FINALE	€	**97.533.316,72**

Methods and Procedures

The *Time Schedule* is a planning document that describes the reconstruction funding allocation during the period of time conventionally supposed by the decision makers – six years, in this case. The funding is assigned to each Block managed by the overlying Consortium that edits the project and submits it to the Technical Office of the Municipality.

Each Blok (the corresponding Consortium) "competes" to achieve the funding as soon as possible, so that a fair and objective ranking system has to be arranged in order to avoid any discretional and not supported decision.

The purposes of the Municipality are various and, as a consequence, conflicting and they are more or less consciously connected with *fairness, effectiveness, convenience* and *efficiency*, as required by the Regional Mission Structure for Reconstruction. As the consequence the criteria for ranking the blocks to be reconstructed do their best to interpret the four points addressed as follows.

Fairness. One of the main concerns of the general reconstruction plan is to give back the primary residences to the evacuated people, now hosted in temporary dwellings; as a consequence the blocks to be rebuilt could not be suitable from many other viewpoints.

Effectiveness. A second general concern, somehow conflicting with the previous one, is to reactivate the main urban functions by rebuilding soon the central areas – in this case around the main squares and along the main streets – in order to achieve a sort of Minimum Urban Structure, able to encourage the population to move back. In fact, the long period of desertion of the city has gradually changed the perception of places by citizens and has weakened their emotional bond. Therefore, it could be hard to say how much the reconstruction is a local or general concern, and whom the reconstruction is useful to.

Convenience. The architectural heritage has different attributes and it's differently worthy. As a consequence the first and/or the second concerns could be not able to take into account the creation of property value and its relationship with the reconstruction cost as a relevant point; in fact, neither cost nor value are assumed as ranking criteria in funding.

Efficiency. The last point is the organization and coordination of the works. Building sites should be organized according to the main access roads and should be avoided heavy vehicles should not be allowed again to transit through the reconstructed areas and damage the new pavements and underground services. Another point of efficiency is the ratio result/cost so that, in order to maximize quantitative efficacy (rebuilt-houses/budget) less expensive work should come first.

The question of the best funding annual allocation can be addressed as the constrained optimization of a multiattribute function [15], for each available annual budget. The maximization is based on the ranking of the buildings that have been recognized eligible for funding, the Blocks B_i. Each block is composed of k $(1, 2, \dots, m)$ AUs; the detached AUs $(m = 1)$ have been considered Blocks as well. Each Blok has been characterized from the viewpoint of four attributes a_g $(g = 1, 2, \dots, 4)$ that specify the previously described four main points (fairness, effectiveness, convenience and efficiency). The attributes have been assigned to each Architectonic Unit, A_{ki} included in B_i, so that the score S_i (ranging on a scale from 0 to 2), depends on the scores of all the AUs, s_{ki}. Furthermore, S_i depends on the way of aggregating s_{ki}, so that the score of the block should be specified as $S_{ih} = f_h(s_{ki})$; f_h $(h = 1, 2, \dots, 4)$ is one of the four ways of aggregating s_{ki}. A weight λ_g $(\sum_g \lambda_g = 1)$ is associated to each attribute, so that $s_k = \sum_g a_g \lambda_g$.

The four ways of aggregating are:

$S_{i1} = \sum_k s_k \bar{v}_{ki}$, in which $\bar{v}_{ki} = v_{ki}/V_i$, where v_{ki} is the volume of the kth AU of the ith Block and V_i is the total volume of the ith block;

$S_{i2} = \sum_k s_k \bar{V}_k$, in which $\bar{V}_k = v_{ki}/V$, where V is the total volume;

$S_{i3} = S_k$, in which $S_k = s_k/k$;

$S_{i4} = \sum_k s_k G_k$, in which $G_k = g_{ki}/G$, where g_{ki} is the ground area of the kth AU of the ith Block and G_i is the total ground area of the ith block.

The consistency of each different aggregation of the AU scores is displayed in the graphs of Fig. 2, in which the Blocks have been sorted in descending order by A_g score.

The AU attributes are: *Social value* a_1, depending on the percentage of primary houses in each AU; *Urban value* a_2, depending on the average weighed "Manhattan distance" of each AU from the main squares and roads, measured with the GIS spatial analysis functions; *Property value* a_3, including the criteria generally used for real estate appraising (destination, dimension, typology) except the previously considered location: *Logistic* a_4, connected with the technical reconstruction feasibility, given by the distance from the main access roads.

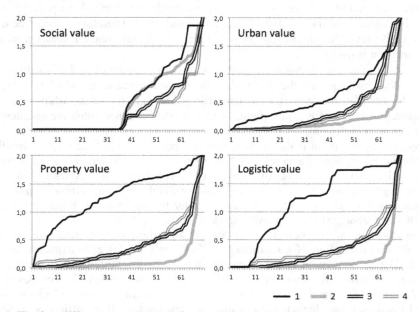

Fig. 2. The four different ways (1 black, 2 grey, 3 blue, 4 red) of aggregating the AU scores A_g: A_1 Social (upper left), A_2 Urban (upper right), A_3 Property (lower left), A_4 Logistic (lower right)

The AU attributes are: *Social value* a_1, depending on the percentage of primary houses in each AU; *Urban value* a_2, depending on the average weighed "Manhattan distance" of each AU from the main squares and roads, measured with the GIS spatial analysis functions; *Property value* a_3, including the criteria generally used for real estate appraising (destination, dimension, typology) except the previously considered location: *Logistic* a_4, connected with the technical reconstruction feasibility, given by the distance from the main access roads.

An overall score S_i is associated to each of the 70 B_i (Tussillo is omitted in this synthesis):

$$\forall B_i \in B, \exists f_1(B_i) = S_i; 1 \le B_i \le 70; 0 \le S_i \le 2$$

From the TEBS the total reconstruction cost C_i is associated to each B_i:

$$\forall B_i \in B, \exists f_2(B_i) = C_i$$

For each weigh system hypothesis a S_i decresing ranking of B is given. For each annual budget $W_y, y = (1, 2, ..., 6)$ a subset $D \subset B$ is defined; D contains the Block in the top of the ranking and the total cost of the Blocks of D can not exceed W_i.

$$\forall W_i \exists D \subset B: \sum_{i=1}^{m} C_i \leq W_i$$

The WebGIS Time Scheduling Platform

One of the main concerns the local authority has expressed is the transparency of the decision making about time scheduling in order to not privilege anyone on where the reconstruction works should start from.

The WebGis pattern we have provided allows to display at real time each type of information of the Reconstruction Plan and to connect their different thematic layers. All restitutions and critical analysis are included in this relational spatial database [14] so that each kind of valuation can be explained by going back toward the upper information.

The datasets and the calculation functions are written into a geodatabase pattern within a WebGIS frame so that everyone can gain access to the system. The system allows the implementation of pieces of information from different workstations at the same time, so that the whole information system is part of a unique data source.

The WebGIS planning platform [16], realized with a *pmapper* interface, supports APACHE Web Server and the Geodatabase POSTGRES, that has a POSTGIS spatial extension.

The WebGIS interface consists of a dialog box divided in: a wide area displaying the map, the information contents layers on the right, grouped by consistent categories and visible by checkboxes; some tools allow to zoom, refresh and so on, and by clicking on the specific item an information sheet appears. The management of information and the definition of the graphic attributes for each layer are processed by the *CGI* module of *Map Server* by means of the parameters of the *mapfile* and the variables of the *template* file, performed and edited by the operator.

The left frame displays the time schedule input interface: the weigh system, the annual budget for each year, the aggregating system choice, and some other technical parameters whose description can be omitted. Once inserted the variables the WebGIS module performs the calculation by using the functions of the geodatabase, extracts and displays the requested output as processed within the browser.

The geodatabase's logic architecture (Fig.3) is based on work breakdown structures that is the most simple and fast way to achieve information by the GiST indexing system, that allows to group data by means of their specific topological relationships between the items that can be: proximity, overlapping, inclusion, managed by the command line:

```
CREATE INDEX id ON agg_vsa USING GIST (the_geom);
```
whereas the command line:
```
CLUSTER the_geom_index ON agg_vsa
```
groups the Blocks by means of a mobile index.

The ranking, for each weigh system, is done by a *trigger* that arranges certain space-time events like the temporary storage of the last ranking from which the most worthy Blocks are extracted:

```
CREATE OR REPLACE FUNCTION aggregate_update() RETURNS trigger AS
$$
  BEGIN
  UPDATE agg_vsa_history
        (gid, id, name, v_soc, v_log, v_urb, v_fun, bud_1,
bud_2, bud_3, bud_4, bud_4, bud_6, geom, created, created_by)
  VALUES
        (NEW.gid, NEW.id, NEW.name, NEW.v_soc, NEW.v_log,
NEW.v_urb, NEW.v_fun, NEW.bud_1, NEW.bud_2, NEW.bud_3,
NEW.bud_4, NEW.bud_5, NEW.bud_6, NEW.geom, current_timestamp,
current_user);
    RETURN NEW;
  END;
$$
LANGUAGE plpgsql;

CREATE TRIGGER agg_vsa_trigger
AFTER INSERT ON agg_vsa
FOR EACH ROW EXECUTE PROCEDURE aggregate_update();
```

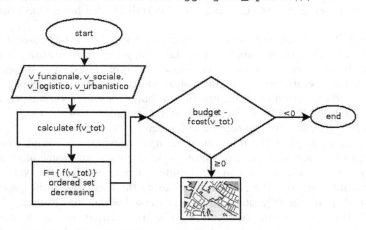

Fig. 3. The logic structure of the WebGIS platform

The ranking function code is synthesized as follows:

```
$str_conexao='host=95.110.192.251 dbname=postgis port=5432
user=xxx password=xxx';
$conexao=pg_connect($str_conexao) or die('A conexão ao banco de
dados falhou!');
$consulta=pg_exec($conexao,"update pesi_tu_vsa SET p_costo =
($v_costo)"); //urbanistico
...
$consulta=pg_exec($conexao,"update agg_vsa SET fu_agg_ =
fu_agg_d*p_ricavo from pesi_tu_vsa ;");
```

```
...
if(!$query = @pg_query("create or replace view max as SELECT
MAX(tot_agg) as maxx FROM agg_vsa ;")); // max
...
if(!$query = @pg_query("create or replace view budget_vsa as
select SUM(budget_euro) AS temp3 FROM agg_vsa ;"));
...
$consulta=pg_exec($conexao,"update agg_vsa SET budget_in =
$budget_a where tot_agg_=  (SELECT max(tot_agg_) FROM
agg_vsa);");
...
$consulta=pg_exec($conexao,"update vsa_ord set budget_in =
$budget_a ;");
for ($u= 1; $u<94; $u++) {
$consulta=pg_exec($conexao,"update vsa_ord set budget_costo =
budget_in - $z[$u] where idd = $u ;");
$consulta=pg_exec($conexao,"select budget_costo from vsa_ord
where idd = $u ;");
...
$consulta=pg_exec($conexao,"CREATE  table o_vsa AS SELECT idd,
tot_agg_, budget_in, id_agg, the_geom, budget_costo,
budget_costo_b, budget_costo_c, budget_costo_d, budget_costo_e,
budget_costo_f, budget_costo_g, costo, intervallo, int_b, int_c,
int_d, int_e, int_f, int_g, b_a, b_b, b_c, b_d, b_e, b_f, b_g
FROM vsa_ord ;");
```

Sinthesizing, the WebGIS pattern allocates the fundings as follows: once earmarked the annual budget, the best Blocks (the highest ones in the ranking the sum of which cost is covered by the budget) are selected for the first year and their id. numbers are plotted into a *temporary view*; the costs of reconstruction are transferred to the Geodatabase that plots them on the WebGIS; the difference between the annual budget and the cost of the Blocks selected in that year is added to the budget of the next year. This routine is repeated up to the last year or until the entire budget is exhausted if the funding period can be reduced or must be extended.

The temporary view is updated from time to time with the results of the following years and the process restart every time times the weights change or a different budget is earmarked or a different way of aggregating the AU scores is chosen.

Applications and Results

The long dialoguing with the Municipality of Villa Sant'Angelo has allowed us to realize some general concerns and expectations of both administration and citizens. A rough layout of the possible scheduling by the Municipality has been found out by connecting the different pieces of information, although sometimes a little confused, we obtained. The following two-year periods general scheme includes the expectations and the preferences of the Municipality.

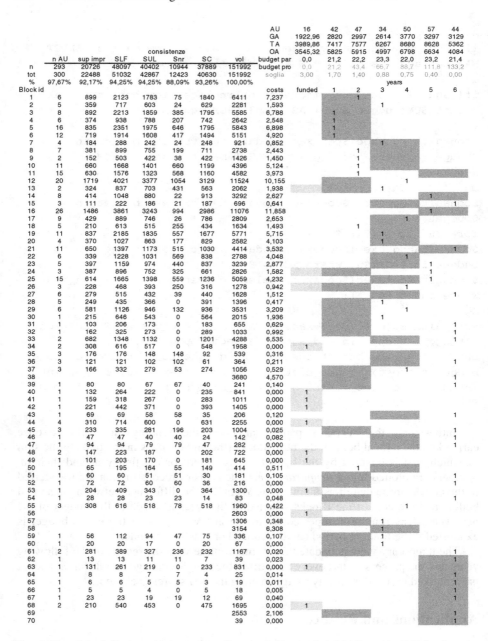

Fig. 4. Time scheduling comparison: the cells containing number **1** indicate the Blocks that are supposed to be funded in each year of the reconstruction period; the dark grey cells indicate the two-year period funding program planned by the Municipality

Fig. 4 shows the relevant differences between technical assessment and the wishes of the Municipality and the tool we proposed has been helpful in order to fit the two visions. Different weight systems have been successfully inserted in order to explore the economic (cost) and qualitative (social, urban, property, logistic) structure of the two town as filtered by the general and specific points of the reconstruction (Fig. 5-8.).

Fig. 5. Time schedule for weigh system hypothesis 1

Fig. 6. Time schedule for weigh system hypothesis 2

Fig. 7. Time schedule for weigh system hypothesis 3

Fig. 8. Time schedule for weigh system hypothesis 4

Discussions and Conclusions

The post earthquake reconstruction in Abruzzo is being performed by means of the Reconstruction Plans. The criticalities of the damaged land and town and the articulated complex of values involved in the decision making process claimed a decision system support aimed at making the funding of the reconstruction fair and reliable.

The huge dataset and the restitutions available from the Reconstruction Plan allowed us to realize a strict calculation pattern for cost accounting and qualitative valuations.

The funding time scheduling ranks the Blocks by their complex (social, urban, property, logistic) value, as represented by an aggregate score, and then allocates the designed annual budget.

The criticalities that in general affect each qualitative valuation and the consequent decision making process, have been here magnified. The conflict between the different concerns hasn't been totally reduced. In fact, the main objective of the local administration is an organic reconstruction plan starting from the main urban centres; the model validates this program only partially, because the social point cannot be ignored or underrated. Furthermore, the program, concentrates most of the funds into the first half of the funding period so that it could result unfeasible.

Of course, such ductile politic approach cannot totally fit the rigid structure of an objective tool. The tool has the main purpose to provide the guidelines from which some exception can be discussed and accepted.

The attempt to match the two approaches, in fact, allowed to measure how the different criteria were weighed by the authority to perform the hoped strategy and this confirms the utility of similar tools within a decision making process.

Another constraint of a reconstruction participative plan is the conflict between the individual refunding practice (connecting the State and the owner) and the collective social political approach affirmed by the use of planning and programming. The former does not allow any flexibility and is more coherent with a technical approach, the latter is based on a concept of value depending on the spatial, temporal and social context.

The management of the plan basically will face, for better and for worst, two constructive and partly conflicting needs: on the one hand, the recovery of local identity undermined by the earthquake; on the other, the added value of the urban and property capital asset (as an aggregate entity) that the reconstruction will generate, and that will foster the repopulation of the damaged towns [7, 8, 9].

The proposed communication and participation platform, arranging technical and social communication tools aims at making ethics and aesthetics two undistinguishable features [17] of the future management of reconstruction.

Acknowledgements. The Reconstruction Plan has been edited within an Agreement between the Municipality of Villa Sant'Angelo and the Department of Architecture of the University of Catania. All the contents and the materials have been edited by the working group led by the Scientific Director, prof. Caterina Carocci. The authors used these datasets and surveys for the technical and economic valuations and for the elaboration of the WebGIS tools provided for the time schedule sheet. Salvatore Giuffrida edited paragraphs 1, 2, 3.1, 4 and 5 and drew up the TEBS, the qualitative valuation model and the time schedule pattern. Filippo Gagliano edited paragraphs 3.2, figg. 1, 3, 5-8 and drew up the WebGIS tool.

References

1. Andreani, F., Carocci, C.F.: Urban fabric, construction types and the art of citybuilding. Approaches and methods for post earthquakere construction plans. IJPP Italian Journal of Planning Practice III(1), 69–89 (2013)
2. Boscarino, S., Prescia, R.: Il restauro di necessità. FrancoAngeli, Milano (1992)
3. Carbonara, S.: Il sisma abruzzese del 2009: la previsione di spesa per la ricostruzione. Valori e Valutazioni 11, 67–85 (2013)
4. Carbonara, S., Cerasa, D., Spacone, E.: Una proposta per la stima sommaria dei costi nella ricostruzione post-sismica. Territori 17, 67–85 (2014)
5. Carocci, C.F.: Piani di Ricostruzione post sisma tra conservazione e rigenerazione urbana. I casi di Villa Sant'Angelo e Fossa (AQ). In: Castagneto, F., Fiore, V. (eds.) Recupero, Valorizzazione Manutenzione nei Centri Storici, LetteraVentidue, Siracusa, pp. 114–117 (2013)
6. Carocci, C.F., Vitale, M.R.: Criteri, norme e line guida per gli interventi nei Piani di Ricostruzione post sisma di Villa Sant'Angelo e Fossa (AQ). In: Castagneto, F., Fiore, V. (eds.) Recupero, Valorizzazione Manutenzione nei Centri Storici, pp. 118–121. Letteraventidue, Siracusa (2013)
7. Carocci, C.F., Circo, C.: Le debolezze della città storica. Effetti sismici sul tessuto edilizio murario. In: Blasi, C. (ed.) Architettura Storica e Terremoti. Protocolli Operativi per la Conoscenza e la Tutela, pp. 153–175. Wolters Kluwer, Italia (2013)
8. Carocci, C.F.: Small centres damaged by 2009 L'Aquila earthquake: On site analyses of historical masonry aggregates. Bulletin of Earthquake Engineering (2011)
9. Carocci, C.F., Cattari, S., Circo, C., Indelicato, D., Tocci, C.: A methodology for approaching the reconstruction of historical centres heavily damaged by 2009 L'Aquila earthquake. Advanced Materials Research 133-134, 1113–1118 (2010)
10. Carocci, C.F., Lagomarsino, S.: Gli edifici in muratura nei centri storici dell'Aquilano. Progettazione Sismica 3, 117–134 (2009)
11. Fistole, R.: Il Piano digitale. Verso un nuovo governo delle trasformazioni urbane e territoriali, Urbanistica digitale, Edizioni Scientifiche italiane, Napoli (2008)
12. Giuffrida, S., Carocci, C.F., Gagliano, F.: Qualità urbana ed equità sociale nel finanziamento della ricostruzione in Abruzzo. In: Castagneto, F., Fiore, V. (eds.) Recupero, Valorizzazione Manutenzione nei Centri Storici, pp. 122–125. Letteraventidue, Siracusa (2013)
13. Giuffrida, S.: Proposta del modello multicriteriale per l'individuazione della stategia di conservazione. In: Boscarino, et al. (eds.) Petralia Soprana. Ipotesi di Restauro Urbano e Studi di Analisi Multicriteriale, Medina, Palermo, pp. 76–103 (1994)
14. Maguire, D.J., Batty, M., Goodchild, M.F. (eds.): GIS, Spatial Analysis, and Modelling, Redlands. ESRI Press, CA (2005)
15. Malczewski, J.: GIS and Multicriteria Decision Analysis. John Wiley, Hardcover (1999)
16. Murgante, B.: L'informazione geografica a supporto della pianificazione territoriale. Franco Angeli (2008)
17. Rizzo, F.: Etica dei valori economici ed economia dei valori etici. FrancoAngeli, Milano (2004)

A Spatial Econometrics Analysis for Road Accidents in Lisbon

Paula Simões[1], Sílvia Shrubsall[2], and Isabel Natário[3]

[1] Instituto Superior de Engenharia de Lisboa and CMA, Rua Conselheiro Emídio Navarro, 1, 1959-007 Lisboa Portugal
paulasimoes@adm.isel.pt
[2] CESUR, Instituto Superior Técnico - Technical University of Lisbon, Portugal
silviashrubsall@gmail.com
[3] Faculdade de Ciências e Tecnologia (UNL) and CEAUL, Quinta da Torre, 2825-114 Caparica, Portugal
icn@fct.unl.pt

Abstract. This paper presents a spatial econometrics analysis for the number of road accidents with victims in the smallest administrative divisions of Lisbon, considering as a baseline a log-Poisson model for environmental factors. Spatial correlation on data is investigated for data alone and for the residuals of the baseline model without and with spatial-autocorrelated and spatial-lagged terms. In all the cases no spatial autocorrelation was detected.

Keywords: Spatial Econometrics, Moran's I, Spatial Autoregressive Model (SAR), Spatial Error Model (SEM), Lagrange Multipliers tests, Road Accidents.

1 Introduction

Lisbon, the capital city of Portugal, has one of the Europe's highest number of road accidents per inhabitant [9]. In the year of 2007, for example, 2149 road accidents with victims were registered in an universe of about 565,000 inhabitants and an area of $85Km^2$ (Project SACRA, Spatial Analysis of Child Road Accidents, PTDC/TRA/66161/2006, and Instituto Nacional de Estatística).

The occurrence of road accidents is often conditioned by many factors, from those intrinsic to the accident, as the driver's specificities, type of vehicles or weather conditions, to road specific variables including road physical state and traffic signs, as well as traffic conditions. However, frequently, external factors that also contribute for the happening are not taken into consideration, such as all the environmental surroundings of the incident. For example, if we are talking about a city neighbourhood with many children, as expected around school grounds, or with many elderly as expected around hospitals or medical centers, it may constitute a place of higher accident risk. Additionally, if not accounted for, these environmental factors may induce some degree of spatial dependence in the risk of accident.

B. Murgante et al. (Eds.): ICCSA 2014, Part III, LNCS 8581, pp. 269–283, 2014.
© Springer International Publishing Switzerland 2014

Under this perspective, with spatially autocorrelated risk, a relatively recent way for modelling the number of the occurrences of road accidents is considered here to circumvent the limitations of classical methods when dealing with spatial data samples. When sample data is of this type, two scenarios have to be addressed, spatial autocorrelation between observations, and spatial heterogeneity in relations. Under these, fundamental assumptions in traditional statistic methods, that data values are derived from independent observations or a single relationship with constant variance exists across the sample data, are no longer guaranteed[11]. Spatial econometrics is an adequate alternative, that can be used when dealing with observations that describe geographic phenomenons or events.

A spatial econometrics approach allows to assess the magnitude of the space influences in accident risk, taking into account the environmental factors, by introducing a specific weighting scheme, in which relationships among spatial areas are specified. The topology or spatial pattern of the data are carried out by the choice of a spatial weights or contiguity matrix, commonly denoted by the letter W, and represents our comprehension of spatial association among spatial units [10]. Two different spatial econometrics models are considered, the spatial autoregressive model (SAR), and the spatial error model (SEM), both estimated by maximum likelihood.

Considering the number of road accidents in Lisbon in each of its smallest administrative division of Lisbon city, called *freguesia*, [13] studied the most significant environmental factors affecting road accident risk in Lisbon in 2007, but has not made an attempt to incorporate or evaluate the possible spatial dependence in that risk, for these aggregated data setting.

Through an empirical study, standard spatial econometrics techniques are used to look for spatial dependence in the number of accidents, determining common association measures, previously to the inclusion of the environmental factors, and afterwards. Several neighborhood structures are considered in this quest.

The paper is organised as follows: section 2 reviews, describes and explains some of the spatial econometrics methods that are used in the following section for spotting and modelling the road accidents risk in Lisbon in 2007. Section 3 presents the results and section 4 discusses them.

2   Materials and Methods

2.1   Spatial Dependence

Data for which location attributes are an important source of information, when taken into account, a spatial perspective yields. The recognition of spatial dimension can give more meaningful results than an analysis that ignores it, however observations that constitute geographic phenomenons may be spatially associated.

Spatial association, also referred to as spatial autocorrelation, corresponds to situations where observations or spatial units are non-independent over space, that is nearby spatial units are associated in some way. Such association can be identified in a number of ways, using a scatter-plot and plotting each value against the mean of neighbouring areas, the **Moran's scatter plot**, or using an spatial autocorrelation statistic, such as **Moran's I** and **Geary's C**.

Both of these statistics require the choice of a spatial weights or contiguity matrix, usually denoted by the letter W, that represents the topology or spatial arrangement of the data and represents our understanding of spatial association among all areas units [10]. Usually, $w_{ii} = 0$, $i = 1, ..., n,$, where n is the number of spatial units, but for $i \neq j$, w_{ij}, the association measure between area i and area j, can be defined in many different ways, usually as a minimum distance between areas, [8].

The referred association measures are described as follows:

Moran's scatter plot

Moran's scatter plot is a graph that allows to visually explore the spatial autocorrelation. Considering a spatial weights matrix, W, this graph has in the x axis the standardized values of a variable Y_i and in the $y axis$ the weighted mean (by $w_{i,j}$) of all the others elements. In case that matrix W is a contiguity matrix, the y axis correspond to the average of the standardized values of the neighbours of each spatial unit [8].

Moran's scatter plot with points essentially in the odd quadrants, indicates the presence of positive correlation, that is, high or low values tend to cluster in space; points in the even quadrants indicates the presence of negative correlation, because locations tend to be surrounded by neighbours with very dissimilar values of the same variables; for no spatial correlation, the points will be around the origin.

Moran's I Statistics

The Moran's I statistics is one of the most known statistics to measure spatial association. Considering a regression model, $Y = \beta X + \mu$, this statistics can be applied directly to the dependent variable, or to the regression residuals obtained by ordinary least squares, and it is formally given by

$$I = \frac{n \sum_i \sum_j w_{i,j}(Y_i - \overline{Y})(Y_j - \overline{Y})}{(\sum_{i \neq j} w_{i,j}) \sum_i (Y_i - \overline{Y})^2}$$

(1)

For statistics I a test for the null hypotheses of spatial independence can be built under two different situations: a randomized statistics distribution or a normal approximation. A significantly positive value of I indicates the presence of direct spatial correlation, a significantly negative value an inverse spatial correlation and when I is close to zero the absence of spatial correlation. Note that for relatively small values of n, the I distribution can be far away from the normal distribution and the randomized test is preferred.

Using Monte Carlo simulation, the position of the observed Moran's I statistics is situated among the I values simulated from the distribution under the null, they are compared and the empirical p value is calculated, [8].

Geary's C Statistics

Another statistics used to measure the spatial association is Geary's C statistics, given by:

$$c = \frac{(n-1)\sum_i \sum_j w_{i,j}(Y_i - Y_j)^2}{2(\sum_{i \neq j} w_{i,j})\sum_i (Y_i - \overline{Y})^2}$$

This statistics is always positive, with one for the expected value, values of c less than the expected value indicates the presence of direct spatial association, and otherwise, the presence of inverse spatial association [8].

When spatial autocorrelation is identified, due to its distinct nature, a specialized set of methods is needed. In order to capture dependencies across spatial units, spatially correlated variables are introduced in the model specification, which are weighted averages of the neighbours, where the definition of neighbours is carried out through the specification of the spatial weights matrix W. These variables, depending on the problem, can be included in the dependent response as well as in the explanatory variables data set or in the error terms [5].

2.2 The W Matrix

The definition of the spatial weights matrix W, where the spatial relationships among spatial units are specified, is very important since estimation results depend on the choice of this matrix. There are several different approaches to define spatial relations between two locations or spatial units, but they can essentially be classified into two main groups, spatial contiguity approach, and the distance based approach. Typical types of neighbouring matrices, for spatial contiguity approach, are the linear, the rook, the bishop, the queen contiguity matrices W, and for the distance approach, we have the k-nearest neighbours, inverse distance and distance decay function matrices. Given the ongoing study, some of these approaches are presented:[1]

- **Contiguity Matrix [11]:** Represents a $n \times n$ symmetric matrix, where $w_{i,j} = 1$, when i and j are neighbours and 0 when they are not. By convention, the diagonal elements are set to zero. W is usually standardized so that all columns sum to one, $\widetilde{w_{i,j}} = \frac{w_{i,j}}{\sum_j w_{i,j}}$, and operations with the W matrix are an average over neighbouring values.
  - **Rook contiguity:** Two regions are considered neighbours if they share a common border, and for these $w_{i,j} = 1$.

---

[1] To see the development of others approaches see [11].

- **Queen contiguity:** Regions that share a common border or a vertex are considered neighbours, and for these $w_{i,j} = 1$.

– **Distance Approach [6]:** makes direct use of the latitude-longitude coordinates associated with spatial data observations.
   - **Critical Cut-off Neighborhood:** Two regions i and j are considered neighbours if $0 \leq d_{i,j} < d^*$, with $d_{i,j}$ the appropriate distance adopted between regions, and d^* representing the critical cut-off or threshold distance, beyond which no direct spatial influence between spatial units is considered.
   - **k-Nearest Neighbor:** Given the centroid distances from each spatial unit i to all units $j \neq i$ ranked as, $d_{i,j}(1) < d_{i,j}(2) < ... < d_{i,j}(n-1)$, for each $k = 1, 2, ..., n-1$, the set $N_k(i) = \{j(1), j(2), ..., j(k)\}$ contains the k closest units to i, and for each given k, the k-nearest neighbor matrix, has the form: $w_{ij} = 1$, $j \in N_k(i)$, and is zero otherwise.

2.3 Spatial Econometrics Models

Spatial autoregressive econometrics models are used to model spatial data and provide a relatively complete treatment from a classic perspective [11]. Two of the several models available are described, the Spatial Autoregressive Model (SAR) and the Spatial Error Model (SEM).

Spatial Autoregressive Model. A first-order spatial autoregressive model, in its simplest form is given by,

$$y = \rho W y + \varepsilon$$
$$\varepsilon \sim N(0, \sigma^2 I_n)$$

where y and ε are $n \times 1$ random vectors, the ε vector is i.i.d., y correspond to the dependent variable, spatially autocorrelated, and W is an $n \times n$ spatial contiguity matrix. This model tries to explain variation in y as a linear combination of neighbouring units with no other explanatory variables and it is frequently used for checking residuals for spatial autocorrelation, where ρ represents the autoregressive parameter.

The ordinary least squares estimation is inappropriate for a model that includes spatial effects. Applying least squares to this model produces a biased estimate for the spatial autoregressive parameter ρ since with

$$\widehat{\rho} = (y'W'Wy)^- 1 y'W'y,$$

$$E(\widehat{\rho}) = E[(y'W'Wy)^- 1 y'W'(\rho W y + \varepsilon)] =$$
$$= \rho + E[(y'W'Wy)^- 1 y'W'\varepsilon]$$

and that gives inconsistency estimates, once $E(\widehat{\rho}) \neq \rho$.

In this model to estimate ρ we should use the maximum likelihood estimator, using a "simplex univariate optimization routine" to find a value of ρ that maximizes the likelihood function,[11].

Using this estimate of ρ, the estimate for the parameter σ^2 is provided by,

$$\hat{\sigma}^2 = \tfrac{1}{n}[(y - \hat{\rho}Wy)'(y - \hat{\rho}Wy)]$$

An extension of this spatial model is,

$$y = \rho Wy + X\beta + \varepsilon$$
$$\varepsilon \sim N(0, \sigma^2 I_n)$$

where X is a $n \times k$ matrix of explanatory variables and parameters β reflects the influence of this covariates on the y variation over the spatial sample. This model is also named "mixed regressive-autoregressive model" [11] because it combines the standard regression model with a spatially dependent variable model. As in the previous model, a maximum likelihood iterative estimation is carried out in order to obtain the autoregressive parameter, ρ, that maximizes the likelihood function, and consequently allows the compution of $\hat{\beta}$ and $\hat{\sigma}_{\varepsilon}^2$.

Spatial Error Model. The spatial error model, a regression model with spatial autocorrelation in the residuals,

$$y = X\beta + u$$
$$u = \lambda Wu + \varepsilon$$
$$\varepsilon \sim N(0, \sigma^2 I_n)$$

where, y is an $n \times 1$ vector of the dependent variable, X is a $n \times k$ matrix of explanatory variables with parameter β reflecting the influence of this variables on variation of variable y. W is a known $n \times n$ spatial contiguity matrix and the parameter λ is a coefficient on the autocorrelated residuals.

Spatial autocorrelation in the least-squares residuals can be detected by an appropriate statistical test, like, for example, that based on the Moran's I statistics.

2.4   Lagrange Multipliers Tests

The Moran's I test for spatial autocorrelation is not suited for choosing which spatial autocorrelation model is better for the specific form of spatial dependence in data. For this the Lagrange Multiplier (LM) test, based on the residuals e of a gaussian linear regression model, has proven to be more adequate. This is because the used test statistics takes a different form, whether the alternative hypothesis to the non-existence of spatial autocorrelation is a spatial error or a spatial lag model [1], [2], [3].

LM test for spatial error: the test statistics is

$$LM_{\text{error test}} = \frac{e^T W e}{tr(W^T W + W^2)},$$

where W is the weight matrix and $tr(\cdot)$ is the matrix trace. Under the null hypothesis, this statistics is approximately qui-squared distributed with 1 degree of freedom.

LM test for spatial lag dependence: the test statistics is

$$\text{LM}_{\text{lag test}} = \frac{\left(e^T W y e / \hat{\sigma}^2\right)^2}{(WX\beta)^T M(WX\beta)/\hat{\sigma}^2 + tr(W^T W + W^2)},$$

where $M = I - X(X^T X)^{-1} X^T$. Under the null hypothesis, this statistics is approximately qui-squared distributed with 1 degree of freedom.

Robust LM test for spatial error: This is a robust version of the LM test for spatial error robust to the presence of spatial lag dependence. The test statistics is

$$\text{LMR}_{\text{error test}} = \text{LM}_{\text{error test}} - \text{LM}_{\text{lag test}}.$$

Under the null hypothesis, this statistics is approximately qui-squared distributed with 1 degree of freedom.

Robust LM test for spatial lag dependence: This is a robust version of the LM test for spatial lag dependence robust to the presence of spatial error. The test statistics is

$$\text{LMR}_{\text{lag test}} = \text{LM}_{\text{lag test}} - \text{LM}_{\text{error test}}.$$

Under the null hypothesis, this statistics is approximately qui-squared distributed with 1 degree of freedom.

LM test for spatial error and spatial lag dependence: the test statistics is

$$\text{LM}_{\text{SARMA}} = \text{LMR}_{\text{error test}} + \text{LMR}_{\text{lag test}}.$$

Under the null hypothesis, this statistics is approximately qui-squared distributed with 2 degree of freedom.

3   Results

This section presents and describes the data and investigates if it displays spatial autocorrelation. They have been previously analysed in [13] and a Poisson log-linear model has been estimated, including the most significant variables. In this study we have addressed the spatial part, which has not been considered before for these data.

3.1   Data

The available data is a comprehensive data set of all road accidents with victims in the city of Lisbon for the years of 2004 to 2007. These data have been geo-referenced and, under project SACRA (Spatial Analysis of Child Road Accidents, PTDC/TRA/66161/2006), they have been further organized into the smallest city administrative divisions, called *freguesias*. Data for several environmental factors have also been gathered and selected in order to better estimate risk. For more details on data and covariate detailed description see [13].

For this work we have focused on the 2007 data. The panel on the left of Figure 1 depicts the map of all the registered accidents by freguesia.

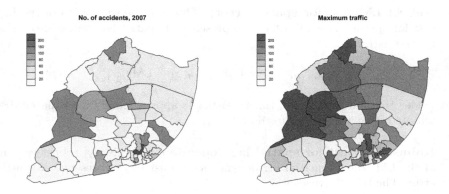

Fig. 1. Number of all road accidents with victims in Lisbon's freguesias in 2007 (left) and maximum traffic in each freguesia (right)

For this year approximately 41 accidents in average per freguesia have occurred, 50% of the freguesias have between 12 and 55 accidents, with a median value of 23, and a positive asymmetric distribution of the number of accidents per freguesia is evident - see the data histogram and boxplot, respectively, in the left and in the right panel of Figure 2. There are a couple of freguesias with a number of accidents oddly higher then the remaining, higher than 120, but that most probably is associated with the freguesia dimension and corresponding traffic, which has to be considered in the modelling.

The number of accidents in each freguesia was then previously modeled through a log-Poisson model [13]. Because the number of accidents is naturally proportional to the amount of traffic, this information is included in the model as an offset so that, in fact, what is modelled is the rate of accidents per traffic unit. The information used on this referes to the maximum number of cars circulating in a certain road section, per hour, in the morning rush hour and is depicted in the right panel of Figure 1.

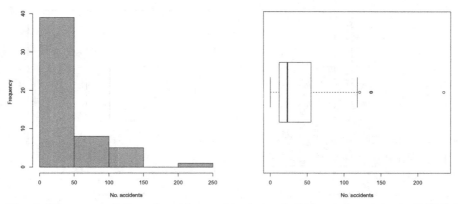

Fig. 2. Histogram of all road accidents with victims in Lisbon's freguesias in 2007 (left) and corresponding boxplot (right)

The most significant factors were found to be the number of hospitals and health centers per freguesia, the number of schools per freguesia, the proportion of resident elderly, the proportion of resident population using car, the proportion of the resident population working in the same freguesia, the proportion of resident population with no school education and the monthly housing charges.

From these factors, the number of schools and the number of the resident population working in the same freguesia are the ones most strongly positive correlated with the number of accidents (0.8), followed by the proportion of resident population with no school education (0.56), the number of resident population using car (0.46) and the number of hospitals and health centers (0.49), these with a moderated correlation.

To proceed with the evaluation of possible spatial dependence for this data, by making use of the tests described in Section 2.4 which rely on a linear model, it was considered the log transformation of the number of accidents per traffic unit, whose histogram and respective boxplot are presented in Figure 3 - clearly symmetric. A linear model was fitted to this transformed data, but in order to allow comparisons with the Poisson log-linear model fitted before the covariates had also to be transformed by a factor of $\left(1 + \frac{1}{2 \times \text{traffic}}\right)$ - for details on this approximation see the Appendix.

3.2   Spatial Correlation and Spatial Lag

In this subsection spatial dependencies on data were investigated. Given that all spatial analysis are conditional on the choice of the spatial weights matrix, several weight matrices were considered according to the neighbour structure - first and second order queen neighbourhood (lisboaqueen and lisboaqueen2), 5 nearest neighbours (lisboa5neigh) - given that 5 is the average number of contiguity

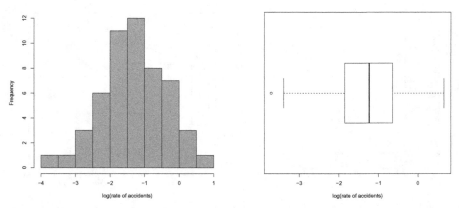

Fig. 3. Histogram of the transformed variable number of accidents per amount of traffic (left) and corresponding boxplot (right)

neighbours - and distance criteria (lisboa.min.Dist, lisboa.2min.Dist) - corresponding to the minimum distance necessary to meet some neighbour and twice that distance. The rook neighbourhood structure was not considered because is basically coincided with the queen one. Table 1 summarizes the percentage of non-zero weights and the average number of neighbours corresponding to each matrix.

Table 1. Summary measures of the several weight matrices considered

Weight matrices	% non-zero weights	No. Neighbours
lisboaqueen	9.8	5.1
lisboaqueen2	17.2	9.1
lisboa5neigh	9.43	5
lisboa.min.Dist	33.3	17.7
lisboa.2min.Dist	65.7	34.8

All the results presented bellow were computed using package spdep [7] of R-project software, following [4].

Considering first the lisboaqueen W matrix, using Moran's I statistic (1), for two sided test, both under normality ($I = -0.050$, $p = 0.71$) or considering a randomization distribution of the statistic ($I = -0.050$, $p = 0.70$), resulted in a clear non-rejection of the spatial independence hypothesis. Left panel of Figure 2 depicts the Moran's I Permutation Test Plot, based on 999 Monte Carlo simulations of the statistics permutation distribution. Here it can be seen that the observed value of the statistic (in blue) is right in the middle of the estimated statistic distribution, leading to the non-rejection of the null. This is further confirmed in the right side of Figure 2, where the Moran's scatter plot is presented.

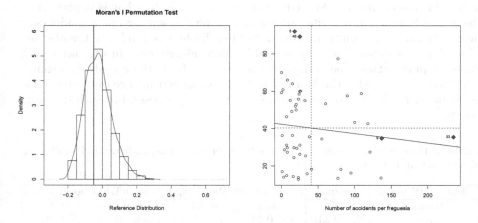

Fig. 4. Moran's I Permutation Test Plot (left) and Moran's scatter plot (right), for lisboaqueen W matrix

For all the other W matrices considered the results were similar and are presented in Table 2.

Table 2. Observed Moran's I and corresponding p-values for two sided test of spatial independence, both considering a randomization distribution for I and under normality

Weight matrices	Randomization	Normality
lisboaqueen	$I = -0.019$ $(p = 0.9989)$	$I = -0.019$ $(p = 0.9989)$
lisboaqueen2	$I = 0.042$ $(p = 0.3091)$	$I = 0.042$ $(p = 0.3096)$
lisboa5neigh	$I = -0.03$ $(p = 0.8849)$	$I = -0.03$ $(p = 0.885)$
lisboa.min.Dist	$I = 0.042$ $(p = 0.3504)$	$I = 0.042$ $(p = 0.3509)$
lisboa.2min.Dist	$I = -0.017$ $(p = 0.9434)$	$I = -0.017$ $(p = 0.9435)$

Although spatial autocorrelation was not found in data, we further investigated the spatial autocorrelation on the residuals of the log-Poisson model fitted in [13] and, of course, none was found: using a randomization distribution of the statistic and a two sided test, $I = -0.01$ $(p = 0.91)$, for lisboaqueen, $I = -0.10$ $(p = 0.19)$ for lisboaqueen2, $I = -0.03$ $(p = 0.93)$ for lisboa5neigh, $I = 0.04$ $(p = 0.35)$ for lisboa.min.Dist and $I = -0.002$ $(p = 0.49)$ for lisboa.2min.Dist.

3.3 Testing for Spatial Error and Spatial Lag Dependence

In this subsection the Lagrange Multipliers tests are used for spatial autocorrelation, based on the linear model fitted to the log number of accidents per unit traffic, including the transformed covariates according to what is explained in the end of Subsection 3.1, that approximates the Poisson log-linear model considered before. The results are summarized in Table 3 and indicate the absence of spatial error dependence and spatial lag dependence, even in the case of the existence of the other type of dependence, and also for both dependencies simultaneously. Additionally, the Moran's test for spatial dependence of the residuals of the linear model is also presented for comparison purposes, but with the same results.

Table 3. Results of Lagrange Multipliers tests for spatial dependence and spatial lag dependence and of the Moran's test

Test \ W matrix	lisboaqueen	lisboaqueen2	lisboa5neig	lisboaminTrh	lisboa2minTrh
Moran's test	0.014 (p=0.33)	0.028 (p=0.23)	0.004 (p=0.38)	0.059 (p=0.11)	-0.005 (p=0.30)
LM error test	0.025 (p=0.87)	0.012 (p=0.91)	0.067 (p=0.80)	0.053 (p=0.82)	0.078 (p=0.96)
LMR error test	0.177 (p=0.67)	0.003 (p=0.96)	0.222 (p=0.64)	0.048 (p=0.83)	0.225 (p=0.89)
LM lag test	0.002 (p=0.96)	0.566 (p=0.45)	0.363 (p=0.55)	0.927 (p=0.34)	0.929 (p=0.63)
LMR lag test	0.684 (p=0.41)	0.013 (p=0.91)	1.020 (p=0.31)	0.347 (p=0.56)	1.030 (p=0.60)
LM SARMA test	0.019 (p=0.89)	0.334 (p=0.56)	0.002 (p=0.96)	0.318 (p=0.57)	0.337 (p=0.85)

4   Discussion

This paper constitutes the first steps of a PhD project. Within the scope of the spatial econometrics methods for road accident with victims data in Lisbon, spatial-correlation was not found and the addition of spatial structure to the model did not improve estimation. This was a bit unexpected, but maybe we ought to look for spatial heterogeneity instead, which constitutes further work to be done. Another reason for this might be the degree of data aggregation that we have used, which was the smallest administrative division of Lisboa, that may have masked the expected data spatial correlation. This will be tested in future work, as we have these data geo-referenced.

When running the analysis it was realized that the available tests for spatial autocorrelation heavily depend on a linear model assumption, which is frequently not appropriated. It is probably needed a generalization of the tests for a vaster class of models as, for example, the generalized linear models [12].

This study further comprehends the analysis of only of those accidents that were considered severe - because someone has died as a consequence of the accident within 30 days of its happening. No spatial structure was also find in that data set, so the corresponding analysis was omitted here.

Acknowledgments. This work is financed by National Funds through FCT - Fundação para a Ciência e a Tecnologia (Portuguese Foundation for Science and Technology) - in the scope of project PEst-OE/MAT/UI0006/2014 and project PEst-OE/MAT/UI0297/2014 (Centro de Matemática e Aplicações). The authors acknowledge the anonymous referees for very useful and pertinent comments on a previous paper version.

References

1. Anselin, L., Rey, S.: Properties of tests for spatial dependency in linear regression models. Geographical Analysis 23, 112–131 (1991)
2. Anselin, L., Florax, R.: Small sample properties of tests for spatial dependence in regression models: Some further results. In: Anselin, L., Florax, R. (eds.) New Directions in Spatial Econometrics Geographical Analysis. Springer, Berlin (1995)
3. Anselin, L., Bera, A., Florax, R., Yoon, M.: Simple diagnostic tests for spatial dependence. Regional Science and Urban Economics 26, 77–104 (1996)
4. Anselin, L.: Spatial Regression Analysis in R - A Workbook. Center for Spatially Integrated Social Sciences (2007)
5. Anselin, L.: Thirty Years of Spatial Econometrics. Papers in Regional Science 89, 3–25 (2010)
6. Arbia, G.: Spatial Econometrics Statistical Foundations and Applications to Regional Convergence. Springer (2006)
7. Bivand, R., et al.: spdep: Spatial dependence: weighting schemes, statistics and models (2014)
8. Carvalho, M.L., Natário, I.: Análise de Dados Espaciais. Sociedade Portuguesa de Estatística (2008)
9. European Transport Safety Council. 5th Road Safety PIN Report. Road Safety Target Outcome: 100,000 fewer deaths since 2001. Brussels (2011), http://www.etsc.eu/documents/pin/report.pdf (downloaded in May 2012)
10. Fischer, M.M.: Spatial Analysis. Springer (2006)
11. LeSage, J.: The teory and Practice of Spatial Econometrics. University of Toledo (1999)
12. McCullagh, P., Nelder, J.: Generalized Linear Models, 2nd edn. Chapman and Hall/CRC, Boca Raton (1989)
13. Nunes, A.R.: Modelação Espacial de Acidentes Rodoviários na Cidade de Lisboa. Master Thesis. FCT-UNL (2011)

Appendix: From a Poisson Log-linear Model to a Linear Log Model

Consider the Poisson log-linear model

$$Y \sim \text{Poisson(offset.var} \times \exp(\beta_0 + \beta_1 x_1 + \ldots + \beta_k x_k)),$$

for which

$$E[Y] = \mu_Y = \text{offset.var} \times \exp(\beta_0 + \beta_1 x_1 + \ldots + \beta_k x_k) \Leftrightarrow$$

$$\Leftrightarrow \log\left(E\left[\frac{Y}{\text{offset.var}}\right]\right) = \beta_0 + \beta_1 x_1 + \ldots + \beta_k x_k$$

Consider now the usual log transformation to be performed to the Poisson data in order to achieve symmetry:

$$W = \begin{cases} \log(Y), Y > 0 \\ 0, \quad Y = 0 \end{cases}$$

Focus on the first branch of W definition. Here:

$$W = \log(Y) = \log(\mu_Y) + \log\left(1 + \frac{(Y - \mu_Y)}{\mu_Y}\right) \approx (\text{2nd order Taylor expansion of the log})$$

$$\approx \log(\mu_Y) + \frac{(Y - \mu_Y)}{\mu_Y} - \frac{(Y - \mu_Y)^2}{2\mu_Y^2}$$

So,

$$E[W] = E[\log(Y)] \approx \log(\mu_Y) + E\left[\frac{(Y - \mu_Y)}{\mu_Y}\right] - E\left[\frac{(Y - \mu_Y)^2}{2\mu_Y^2}\right] = \log(\mu_Y) - \frac{1}{2\mu_Y} =$$

$$= \log(\text{offset.var} \times \exp(\beta_0 + \beta_1 x_1 + \ldots + \beta_k x_k)) -$$

$$- \frac{1}{2\text{offset.var} \times \exp(\beta_0 + \beta_1 x_1 + \ldots + \beta_k x_k)} =$$

$$= \log(\text{offset.var}) + (\beta_0 + \beta_1 x_1 + \ldots + \beta_k x_k) -$$

$$- \frac{\exp(-\beta_0 - \beta_1 x_1 - \ldots - \beta_k x_k)}{2\text{offset.var}} \approx (\text{1st order Taylor expansion of the exponencial})$$

$$\approx \log(\text{offset.var}) + (\beta_0 + \beta_1 x_1 + \ldots + \beta_k x_k) - \frac{1 - \beta_0 - \beta_1 x_1 - \ldots - \beta_k x_k}{2\text{offset.var}}$$

$$= \log(\text{offset.var}) + \underbrace{\left\{\beta_0\left(1 + \frac{1}{2\text{offset.var}}\right) - 1\right\}}_{\text{Intercept}} + \beta_1 x_1\left(1 + \frac{1}{2\text{offset.var}}\right) +$$

$$+ \ldots + \beta_k x_k\left(1 + \frac{1}{2\text{offset.var}}\right)$$

Consequently, and because in this particular application $P(Y = 0)$ is very small (there are always accidents!) so that $E[W]$ can be well approximated for the value above, the coefficients of the linear model are approximately the same as the coefficients of the Poisson log-linear model provided that the values of the covariates $x_1, x_2, \ldots, x_n$ are multiplied by $\left(1 + \frac{1}{2\text{offset.var}}\right)$.

Sketching Smart and Fair Cities
WebGIS and Spread Sheets in a Code

Salvatore Giuffrida and Filippo Gagliano

Department of Civil Engineering and Architecture - University of Catania
sgiuffrida@dica.unict.it,
fmgagliamo@gmail.com

Abstract. The operative context of the urban regeneration is nowadays strictly connected to the information systems by means of which the urban wealth surplus is reallocated. One of the most relevant points of the regeneration policies is the lack of a general tool that allows to roughly "sketch the city", that means, that produces significant options about physical, qualitative and monetary contents of the plan. The pattern we propose allows to implement detailed planning options that are valued real time so that the final solution can be progressively fitted until the best one is found. The pattern works by constantly comparing with and without project options, so that the qualitative and quantitative surplus production can be monitored. The unification of the input system makes the system more reliable.

Keywords: Urban regeneration, transfer development rights, Gis-Spreadsheets model, plan appraisal, Web planning.

1 Introduction

Smart cities are nowadays the big issue of urban redevelopment and sustainability [18], and the condition of a possible *information oriented* democracy. Information Technologies (IT) widely contribute to increase and enhance communication. Cable cities, E-Government, and Intelligent cities are progressive stages of a more general trend especially in the property market and in energy efficiency and environmental economy [4] that confirms the new economics approach based on the new value theory focused on *shape surplus*: "value is a creative combination of matter, energy and information" [13].

The concept of Smart City is generally associated to a hi-tech and comfortable place, capable to manage the energy and entropy fluxes and, above all, allowing total access to information [2]. This concept can be extended involving information technologies and communication theory according to the concept of urban social capital [12].

A smart city is a unity (strongly featured) articulated, self-referential and self-conflicting in which contradictions can be reduced to the general spirit that makes it recognizable. A smart city is a valuable city if that information (as *shape attribute*) is

B. Murgante et al. (Eds.): ICCSA 2014, Part III, LNCS 8581, pp. 284–299, 2014.

assumed as value substance. Therefore, information is relevant as semantic (not mathematic) information. Semantic information can be easily transferred and assimilated if the interpretation codes of recipient and issuer match. As a consequence, the semantic one is the information best suited to the communicative processes. Therefore, social systems are formed as a consequence of the evolution of communication processes. The social macro-system communicates within itself and arises if it distinguishes itself from the environment [10]. Its communication code selects what is relevant to the system and what, instead is relevant to the environment. Similarly some sub-systems arise within the macro-system by communicating by means of a different (more specific) (sub-)code. A sub-system is formed when it shares a code, that is a value system and a selective criterion. The social urban communication assumes a hierarchy of selection codes: at the top stay the "generalized symbolic media" we can consider as meta-values.

Sketching cities has two meanings.

The first, more general, refers to the concept of connotation. A sketch, with a few strokes, captures the characteristics of a person whose essence would not otherwise be understandable. The sketch (and consequently, the work of art) is the form of reality that would otherwise be unordered, chaotic and incomprehensible. According to the general information theory a sketch selects the semantic information within an *equiprobable source*. According to the sociological luhmannian approach these saliences (outstanding features) can be considered a specification of the generalized symbolic media. An axiological approach [8] assumes different specification of these *media* as valuation criteria that selects or validates planning actions. Therefore, a Smart City a high-tech social framework able to manage huge information amounts among which to select the most communicative quota, with the purpose of generating the fair communication code system: in fact, if the system has less codes, less sub-systems, it can get self-referential, if it has too many codes and sub-systems it can lose its unity.

The second meaning of *sketching cities* is applied by the proposed methodology that integrates evaluation in planning practice [17] by the means of a GIS code that connects economic-qualitative calculation to the practice of digital design. The information system is a code that selects only the relevant information so that it is, by definition, a planning tool. On the one hand the amount of information and its resolution is definitely important, but on the other, the capability of selecting and arranging this information makes the difference [15].

As a conclusion, a smart city is a fair city as well. The relationship between information and democracy has been addressed within the UNECE *Convention on Access to Information, Public Participation in Decision-making and Access to Justice in Environmental Matters*, usually known as the *Aarhus Convention*, signed in 1998 and entered into force on 30 October 2001 and nowadays collecting 47 parties, 46 states and the European Union. The Aarhus Convention grants the public rights regarding access to information, public participation and access to justice, in governmental decision-making processes on matters concerning the local, national and trans-boundary environment. It focuses on interactions between the public and public authorities [14].

The proposed contribution focuses on the formation of an operational tool that contains and unifies information, assessment and decisions within a computer code that allows to draw directly topological objects with CAD and generate the corresponding logical entities on the spreadsheet. The same entities can be changed using both interfaces. The instrument contains all of the terms of assessment that are used to verify the results on the basis of well-established principles of urban equalization tools.

2   Methods and Procedures

2.1   General Items

The methodology for the formation of an urban management tool referring to the technologies applied to the smart city, concerns three integrated but conceptually and instrumentally distinguishable functions: 1. the descriptive function, 2. the evaluative function, 3. the project function.

1. The descriptive function includes the techniques for the digital representation of a spatial context, and in particular of its physical-geometrical and objective realm. It requires as a first phase the decomposition of the latter in minimal homogeneous unities constituting the logical-descriptive model, and subsequently their re-composition in the systemic form of the relations between these parts, that form the different thematic fields: topologic, functional, environmental, social, economic, administrative, recreational, communicative etc.

2. The *evaluative function* concerns the individuation of the relations between the objects and the transformation of these relational links in terms of values necessary for the construction of specific functions of replacement between themselves or between the functions they perform. Among the primary functions required by the evaluative model there are those necessary for the delimitation of sub-sets or sub-systems of the represented territorial-urban complex, and therefore for the formation of the hierarchies between the entities in relation to specific concerns and aims. The relation between the characteristics of status and value of each element is described under the several aspects considered relevant. The formation of a set of terms of judgment grouped in a unitary model with which the system as a whole is evaluated in each configuration assigned to it, is an essential part of the evaluative function. This evaluative model systematizes the partial evaluations relating to the status of the different single elements, and general evaluations concerning the system as a whole in relation to indicators and criteria at the adequate scale.

3. The *project function* is carried out by the decision makers through the availability of tools with which they can modify the status of the entities (elements and sub-systems) on the point of view of the characteristics by means of which they are described. The obtained results contribute to indicate the other data which may be useful to deepen the evaluations, and the other functions which may be added to the project module. The result is a circular model in which the demand of data is selected and specified by the valuations; the valuations (methods and contents of value) are selected by the choices which must be made; the choices are made less discretional by the extension and depth of the informative support. The instrumental base of the model includes a Geo-database containing the whole set of the spatial objects; specific functions are implemented to connect the elementary objects reconstructing the

topologic unity of the represented urban context. These functions may also concern the relations between objects and/or functions, and may be of provisional type inside the limits of the available knowledge and the experimented causal models. The geo-database is currently one of the most versatile platforms for the implementation of complex, mixed and numerical calculation functions, and therefore is useful to combine representation, valuation and decision.

The representation of the real system is provided through the reorganization of the logic-attribute data into homogeneous groups placed on independent layers, each constituting an informational level. If connected according to specific search keys, the different levels of information may select additional information, or, conversely, combine and structure the information as a whole.

The method used in the proposed application included the following steps:

1. the formation of a spatial database, based on the standardization of data and spatial information to which they relate;

2. the formation of a general map of spatial homogeneous objects (record), bearers of information (fields) and recipients of transformation;

3. the implementation of planning regulations: the system recognizes admissible transformations associated with objects, and suggests the limits and conditions;

4. the implementation of performance features relating to the spatial objects;

5. the implementation of functions for the evaluation of economic and functional performances;

6. the formation of a system of weighting factors of the evaluation criteria, by means of which it is possible to make the strategy of the decision maker explicit;

7. the implementation of the models of decision-making.

The WebGIS interface allows you to engage with stakeholders and to verify the connection between the social value system, emerging in the political negotiation (ethical and aesthetic values), and the system of individual preference (hedonic value).

The model allows to perform sensitivity and scenario analysis with which the most influential variables can be identified and the conformation of the plan can be adapted to the aggregate preferences of the stakeholders.

The pattern includes two sets of input for each row of the database and for each characteristics that can be friendly handled by the means of the WebGIS extension. As a consequence, the wide range of scenarios coming from the combination of the different variable, mostly the quantitative and the functional ones, stimulate the participation and the debate about the main points of the urban policy

2.2   The Stages of the Planning Process and the Code

The relational database integrates GIS and spreadsheet functions within a software architecture that allows them to be used as input, output and results verification.

We used the POSTGRES database with spatial extension POSTGIS platform MAPSERVER. QGIS is a GIS application used for each new spatial unit. The code creates a new record in the database and also modifies the fields of the existing entities when an attribute is modified by GIS. Conversely, each new record inserted is converted to a graphic entity and positioned by means of the coordinates of the vertices; characteristics entered in the fields appear in real time on the map [9].

The functions ST_Intersects(), ST_Buffer(), ST_Union(), ST_Distance(), ST_Overlaps(), ST_Touches() allow the different steps of planning:

1) Perimeter of the area: the boundary action includes all objects in the database and analyses their spatial relationships. The objects are: buildings, open spaces, road segments, street furniture, etc.; the building units (BU) are described according to their geometric characteristics (perimeter, area, height, top and bottom) material, functional, distinguishing private buildings (ZTO, cadastral category) and equipment;

2) diagnosis (benchmarking): the system checks the consistency between the existing entities and the general Plan prescription (density, urban standards, distances etc.);

3) modifications: according to the results of the analysis, some qualitative and quantitative modifications can be applied to the entities with the purpose of raising the standard of the equipment or the quality of the buildings; the size or the area occupied by the existing buildings, the roads, and urban design may be changed as well.

4) further investigation: the economic and financial verification is based on an archive of pre-established parametric costs and market prices associated with the supposed IC and functions.

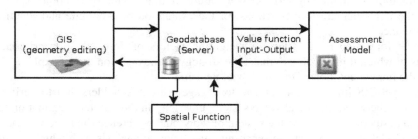

Fig. 1. Rules for assigning each AU to the appropriate Intervention Category

The Geodatabase uses a GiST spatial index (Generalized Search Trees) that groups the data in: things on one side, things which overlap, things which are inside.

Some trigger functions update the tables of Geodatatabse as well as of the spreadsheet model by storing the previous planning hypothesis. Fig. 1 synthesizes the logic scheme of the described pattern, and the main part of the code is displayed as follows.

The following strings contain the code that calculates the surface of ZTO as they fall within the building units by calling the function space ST Intersection.

```
Trigger della funzione spaziale ST_Intersection()
...
CREATE OR REPLACE FUNCTION spatial_update() RETURNS trig-
ger AS
$$
  BEGIN
    UPDATE spatial_history
```

```
ST_Intersection(prg.the_geom, edifici.the_geom) As
the_geom, id, geniozto, cod, zto, p_area_mq, area_mq,
perc FROM prg INNER JOIN prg ON
ST_Intersects(prg.the_geom, edifici.the_geom) where
id_edificio
    VALUES
...
ST_Intersection(prg.the_geom, edifici.the_geom) As
the_geom, NEW.id, NEW.zto, NEW.cod, NEW.zto,
NEW.p_area_mq, NEW.area_mq, NEW.perc FROM prg INNER JOIN
prg ON ST_Intersects(prg.the_geom, edifici.the_geom)
where NEW.id_edificio;
    RETURN NEW;
...
END;
$$
...
LANGUAGE plpgsql;
CREATE TRIGGER spatial_trigger
AFTER UPDATE ON spatial
FOR EACH ROW EXECUTE PROCEDURE
spatial_update();
```

3   Materials and Application

3.1   The Case Study

The port area of Catania [7] is a wide and complex site – nowadays characterized by a high physical and social degradation – the Masterplan individuates as a strategic area for the economic development of the historical center of the city [1] (Fig. 2).

The Masterplan supposes a bundle of interventions ranging from conservative to transformative. The historic urban fabric is supposed to be preserved by intervention categories of restoration and maintenance; in rare cases building replacements are allowed. On the contrary, the area closer to the harbor is supposed to be substantially transformed by the plan that prefigures the radical change of the general function and the modification of shapes and cubage, especially in the parts occupied by old factories and damaged buildings. (Fig. 2).

The analysis of the context included a survey of the status of the Building Units (BU) included in the area. For each BU a sheet containing some pictures and the most important quantitative and qualitative data has been completed; those data have been turned into non dimensional scores using specifically drawn utility functions.

The proposed model integrates two different planning approaches: the general one and the detailed one. Both converge to define a best (fair) plan configuration, as verified by the evaluation tools.

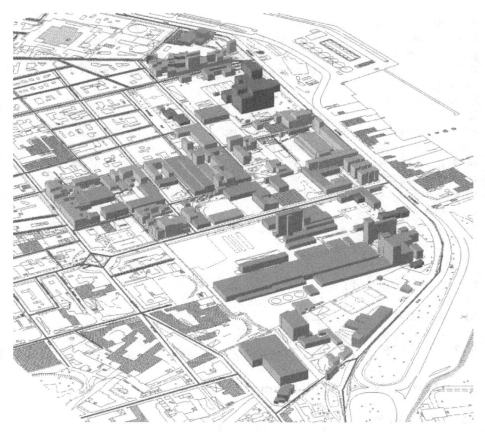

Fig. 2. Current state of the studied area

3.2   The Overall Approach: Decision Making Tool for the Intervention Categories

The Decision System Support [3] that can be applied within an equalized Urban Regeneration Plan, associates the appropriate Intervention Category IC as chosen within a certain set associated to each BU. The database is made of records, the BUs, and fields, the features. The model transforms the feature into scores ranging from 1 (low quality) to 5 (high quality) by the means of a utility function $b_h = f(u_h)$, $(h = 1, 2, ..., n)$ performed for the hth feature, so that each BU is defined by a score vector $\vec{b} = \{b_1, ..., b_n\}$; therefore, to each BU is associated the appropriate IC (Restoration Re, Ordinary maintenance Mo, Extraordinary maintenance Ms, Renovation Ri, Demolition D and reconstruction, Cubage increase Cu) sorted in this case starting from the most conservative (a different order can be given).

The IC is associated to the BU if the following generic condition is true:

$$IC \leftrightarrow \exists i_1, ..., i_k (k \leq t, t \leq 1) \mid b_{i_j} \geq s_{i_j} \forall j = 1, ..., k \tag{1}$$

where s_{i_j} represents the threshold of the jth criterion, k is the number of the criteria by means of which $s_{i_j} \geq \theta_{i_j}$ is verified and t is the minimum threshold;

furthermore: if more than one IC are selected the higher in the sorted list G_m prevails. Sorting of G_m can be chosen according to the kind of strategy (conservative or transformative). The IC *Cubage increase* is generally associated with the conservative IC of the building to be maintained. As a consequence, the strategy depends on the thresholds system and on the IC sorting (Fig. 3).

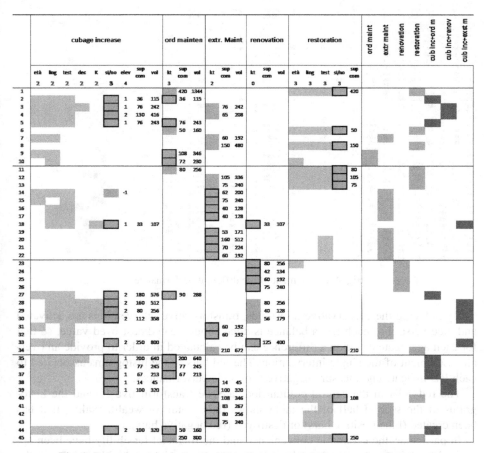

Fig. 3. Rules for assigning each AU to the appropriate Intervention Category

The last module, the valuation/planning pattern, involves the economic (revenues-costs) performance appraisal as meant in an equalization pattern [5] (Fig. 4). Each IC affects the current status of the correspondent AU creating/destroying value from the four different points of view. A good economic performance due to the *Cubage increase* associated to AU n. 4 produces bad quality performances.

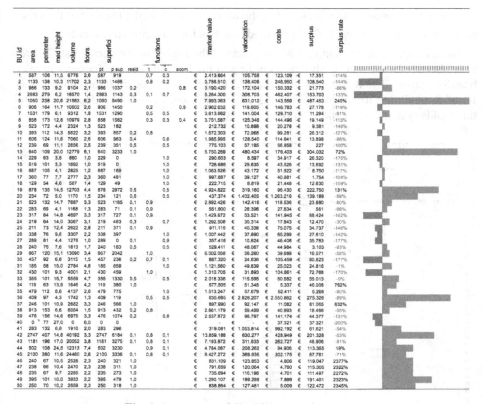

Fig. 4. Assessment of the different performances

By reducing the conservative thresholds, transformative interventions are activated and vice versa. For each BU a balance is done about created/destroyed value, so that for each aggregate criterion a difference can be calculated in order to provide an overall assessment of the single intervention. The red bars of the histograms associated to each BU indicate and measure negative economic performances.

The red cells in the central column indicate the transformative IC and the histograms of the second half of the table shows the amount of wealth (value) that has been created (blue positive bars) or destroyed (red negative bars).

In particular, the economic convenience and the financial feasibility have been appraised applying the extraction method, an evaluation criterion highly helpful in the valuation of the real estate development rights [16].

The ICs, as generally assigned, create and destroy different kinds of value: real estate value is generally produced by transformation and cubage increases that destroy landscape value.

An IC creates value when the intervention gives rise to a surplus S, as calculated according to the well-known appraisal extraction method, the surplus of the investment supposed by each IC of each BU is:

$$S = \frac{V^* - v - k(1+\bar{r})^m (1+r\prime)^n}{(1+\bar{r})^m (1+r\prime)^n} \qquad (2)$$

where: V^* is the final real estate value (after transformation), v is the current real estate value, k is the transformation/preservation cost (building cost), $\bar{r}$ is the well known *wacc*, $r\prime$ is the business risk premium rate, m the short loan life and n the transformation period [6].

3.3   Spatial and Topological Analysis Functions for WebGIS Design

The conceptual structure of the analysis, valuation and planning is composed of "entities" or elements of reality, "objects" or elements represented in the geo-database, and "symbols" reproducing the real object/entities. The previously exposed valuations are based on queries relating to localization (where), thematic (what) and temporal (when) characteristics of the BUs. The alphanumeric data or attributes, that are filed as tables in which each line corresponds to a distinctive entity, and each column or field corresponds to an attribute, are represented by qualities of the geographical entities and/or by quantitative measurements, expressible on nominal, ordinal, cardinal, relational scales. Once the conceptual structure, and subsequently the entities and/or the classes of entities chosen to represent the real geographical realm, the number of independent layers useful to represent the various classes of entities, and the number of tables for the alphanumeric filing are set up, the setting speed of the data bank should be checked. This operation is carried out through the study of an entity/relation model (E/R). The E/S model bases on the concepts of Entity, Class of entity, Attribute, Relation, Event, key and hierarchy, for the logical construction of the data relational banks [8].

The process of assigning the appropriate IC to the each BU is based on these queries and it can be considered the basic, automatic and generalized, planning step. The next step, the more detailed one, supposes the modification of the different IC in order to achieve best specific economic and qualitative results the spatial function are not able to do. The following Fig. 5-8 show some functions of the proposed WebGIS-Spreadsheet pattern by the means of which it is possible to correct the previous more general pattern. The modifications of the quantitative and qualitative characteristics of each BU are assumed by the economic and qualitative pattern and displayed as evaluative results.

The possibilities of the pattern are enhanced by the WebGIS interface. The WebGIS interface consists of a dialog box including: a wide area displaying the map, the information contents layers on the right, grouped by consistent categories and visible by checkboxes; some tools allow to zoom, refresh and so on. By clicking on the specific item an information sheet appears. The management of information and the definition of the graphic attributes for each layer are processed by the CGI module of Map Server by means of the parameters of the *mapfile* and the variables of the template file, performed and edited by the operator.

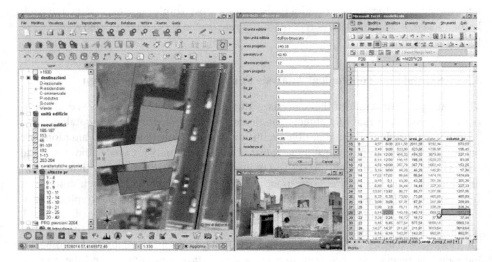

Fig. 5. Modification of the height and update of the volume

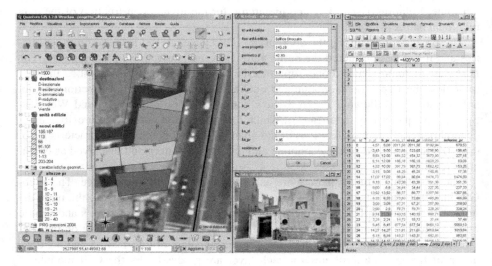

Fig. 6. Geometric modification (area and perimeter)

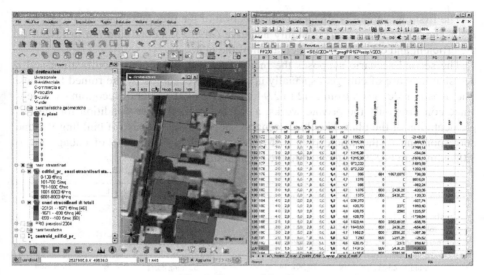

Fig. 7. Uses modification

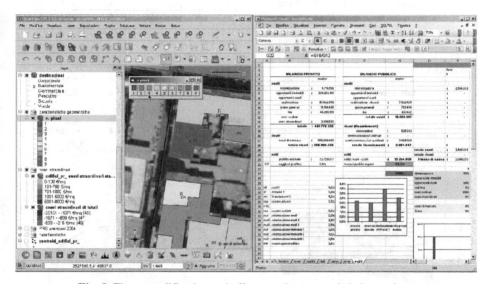

Fig. 8. Floors modification and effects on the economic balance sheet

4   Results and Discussions

As a result of the iterative mechanism of implementation of the prescriptions, in order to reach the fair shape of convenience/sustainability of the plan, three different scenarios have been drawn.

The first one supposes a general conservation of the present cubage and uses. As a result the convenience ratio is highly negative (-493%), a condition not in favour of the efficient and effective implementation of the plan.

The second scenario supposes the increasing of the functions modification and a decisive increase of cubage, especially upon the site occupied by the ruined factories. The density index goes from 2.42 to 2.94cu.m/sq.m (+21%) but the economic convenience concerns only the private works, a fair but non effective condition for the implementation of the plan. The third scenario supposes the increase of building volume corresponding to the whole plan self-financing with a further 3% of cubage up to 3.08cu.m/sq.m (+27% compared to the original index) (Fig. 9).

PRIVATE BALANCE		
costs		
soil preparation	€	4.995.375
property opportunity costs	€	202.564.922
soil opportunity costs		
building costs	€	175.502.145
general expenses	€	17.550.214
taxes	€	20.503.680
ord. concession fees		
extraord. concession fees	€	1.056.667
total cost		422.173.004
revenues		
property market value	€	443.281.654
total revenues		443.281.654
results		
normal profit	€	21.108.650
profit rate		5,0%

PUBLIC BALANCE		
costs		
soil preparation		
property opportunity costs		
soil opportunity costs		
building costs	€	25.635.884
general expenses	€	2.563.588
taxes	€	5.164.660
total costs	€	33.364.133
revenues - funding		
concessions		
ord. concession fees	€	-
extraord. concession fees	€	1.056.667
total funding	€	1.056.667
esiti		
net value	-€	32.307.466
% public cost funding		3,2%

Fig. 9. Private and public balance sheets and total amounts composition

The sensitiveness analysis we have carried out by running some routines implemented in the model of economic calculation allowed us to notice (Fig 8 upper-right graph):

the inverse relationship between the self-financing index and the normal profit rate; the high elasticity factor, -25,4% that indicates the importance of the normal profit rate in the bargain between public and private subject; the importance of the financial and monetary variables (Fig 8, lower right graph): the rate of increase in value of properties, that is supposed to vary from -2% to 2% giving rise to an increase of 33% (1 year), 57% (2 years) and 82% (3 years) of the self-financing index; the investment duration and the consequent loan life span from 1 to 3 years, that can be considered the more decisive parameter for the bargains, as shown in the last graph.

The evaluation of the plan financial sustainability – also represented through the self-financing index, that is the ratio between the extraordinary concession fees and the cost of public works – has been carried out comparing private and public balances for each scenario. In Fig. 10 are shown: the composition of private cost, in order to underline the two main quotas, the opportunity cost of properties and the building costs; the grouped components of cost and revenues for the two bargainers and in particular the comparison between public costs and extraordinary concession fees calculated. A spatial simulation of the cubage increase is provided in Fig. 11.

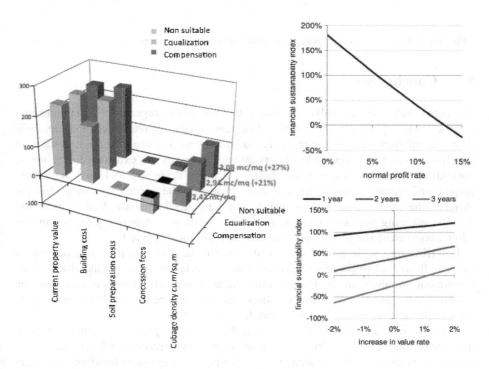

Fig. 10. Sensitiveness and scenarios analysis

Fig. 11. Hypothesis about the cubage increase for economic convenience of the regeneration

5   Conclusions

The paper addresses an analytic approach to smart urban regeneration planning in a historic context [11]. An integrated tool combines logic (spreadsheet) and topologic (GIS) calculations and allows to apply: as first, general decision support system, in which the rules for assigning the Intervention Categories are extended to the whole context; as second, on the previous basis, a detailed designing tool uses the double graphic and numeric input system, in order to specify the general IC and modify the shape of the single building units included in the general database. Both general and detailed planning systems converge to define the most sustainable plan option and allow easy scenario analyses whose results are real time displayed.

Referring to a wide central district of Catania, the application of an equalization pattern allowed us to perceive the impact of the variation of the financial and time-investment variables on the final results. These results have been connected to the cubage index, and the need for a volume increase has been calculated in order to achieve the balance between created and destroyed value.

The most relevant possible conclusion is that the actual practice of urban regeneration needs some rigid conditions to adjust to the characteristics and exigencies of a complex context like a fringe area in which the historical quota of buildings is as relevant as the potential of strategic land resources.

On the basis of the scenario analyses it has been possible to affirm that the creation of urban economic wealth through the instrument of urban regeneration and densification must be handled with care, and the model we carried out, as shown, seems particularly suitable to this concern.

Acknowledgements. Salvatore Giuffrida edited paragraph 1, 2.1, 3.2, 4, 5, figures 3, 4, 9, 10 and drew up the spreadsheet model and the economic evaluation. Filippo Gagliano edited paragraphs 2.2, 3.1, 3.3, figures 1, 2, 5-8, 11 and drew up the Web-GIS model and the spatial calculations.

References

1. Amata, G. (ed.): La Sicilia e il trasporto integrato. Un approccio per la modernizzazione del sistema di trasporti, Troina (EN), Città aperta (2005)
2. Dirks, S., Keeling, J.: How smart is your city? Helping cities measure progress, Somers (2009)
3. Dyer, J.S.: MAUT. In: Figueira, J.R., et al. (eds.) Multiple Criteria Decision Analysis: State of the Art Surveys, pp. 265–295. Springer, Berlin (2005)
4. Fregonara, E.: Architettura sostenibile, risparmio energetico e mercato immobiliare. Territori 10, 16–25 (2012)
5. Giuffrida, S.: Programmi di recupero a Napoli. La valutazione del progetto integrato e la prassi del processo perequativo. Genio Rurale – Estimo e territorio 4 (2003)
6. Giuffrida, S.: Dalla teoria del capitale alla prassi della perequazione urbanistica: il caso del PRG di Cercola (NA). In: Giordano, G. (ed.) Pratiche di Valutazione, pp. 145–187. Denaro Libri, Napoli (2004)
7. Giuffrida, S., Gagliano, F.: Information and Creativity of the Port City Plan. A Gis-Based Assessment Model for Catania's Harbor area. BDC Università di Napoli 721-735 (2012)
8. Kammeier, H.D.: New tools for spatial analysis and planning as components of an incremental planning-support system. Environ. Plan. B: Plan. Design 26, 365–380 (1999)
9. Li, X., Yeh, A.G.O.: Modelling Sustainable Urban Development Bythe Integration of Constrained Cellular Automata and GIS. International Journal of Geographical Information Science 14(2), 131–152 (2000)
10. Luhmann, N.: Sistemi sociali. Fondamenti di una teoria generale. Il Mulino, Bologna (1990)
11. Mocerino, C.: Tecno efficienza nella smart rigenerazione urbana. In: Castagneto, F., Fiore, V. (eds.) Recupero, Valorizzazione Manutenzione nei Centri Storici, pp. 262–265. Letteraventidue, Siracusa (2013)
12. Rizzo, F.: Il capitale sociale della città. Franco Angeli, Milano (2003)
13. Rizzo, F.: Valore e valutazioni. La scienza dell'economia o l'economia della scienza, Franco Angeli, Milano (1999)
14. Rodenhoff, V.: The Aarhus convention and its implications for the 'Institutions' of the European Community. Review of European Community and International Environmental Law 11, 343–357 (2003)
15. Sanchez, L., et al.: Integration of Utilities Infrastructures in a Future Internet Enabled Smart City Framwork. Sensors 13 (2013)
16. Stanghellini, S.: Il plafond perequativo del Prg di Catania, Urbanistica Informazioni 197 settembre-ottobre, Roma, Inu Edizioni (2004)
17. Torre, C.M., Bonifazi, A.: l'integrazione metodologica tra piano e procedure valutative. La VAS del Piano Urbanistico Generale di Monopoli. In: Bentivegna, V., Miccoli, S. (eds.) Valutazione, Progettazione, Urbanistica, DEI, Roma (2010)
18. Vattano, S.: Una rigenerazione smart per i centri storici. In: Castagneto, F., Fiore, V. (eds.) Recupero, Valorizzazione Manutenzione nei Centri Storici, pp. 266–269. Letteraventidue, Siracusa (2013)

An Application of Analytic Network Process in the Planning Process: The Case of an Urban Transformation in Palermo (Italy)

Grazia Napoli and Filippo Schilleci

University of Palermo, Department of Architecture.
Viale delle Scienze, Edificio 14, 90128 Palermo, Italy
(grazia.napoli,filippo.schilleci)@unipa.it

Abstract. The primary objective of this study is to test the multicriteria analysis application in favor of a selection process among alternative transformations of an urban area in the city of Palermo. The choice is referred to as a strategy-oriented one aiming to create "new urban centralities" able to redraw all urban structures that start to activate renewal processes within the existing city. The application of multicriteria analysis technique, such as the Analytic Network Process (ANP) – BOCR model, is due to the need to represent the complexity of the decision problem characterized by interrelations among several elements described by many indicators from different levels. The case study is also an opportunity to verify how the multicriteria analysis can contribute to the decision in the presence of mixed information, such as quantitative and qualitative ones. The application of the sensitivity analysis allows knowing and considering the conditions that can modify the earlier ranking of alternatives.

Keywords: Analytic Network Process, Multicriteria Analysis, Urban Planning, Decision Making; Palermo (Italy).

1   Introduction

Historically, the city often has a growth process that has gradually changed its image over time. Until the nineteenth century, the cities are still contained within their walls. The main changes are related to the need for new housing projects for the growing number of population. However, no particular attention is given to the issue of services and public spaces.

From the mid-nineteenth century, the above-mentioned transformations are clearly intended to provide the cities with main infrastructures, public services, and public housing. However, over the second half of the twentieth century, the projects on the city have focused more on redevelopment than on enlargement issues. However, the common theme has always dealt with the urban transformation. The processes have been managed by many different principles. Renewal, renovation, restoration, and rehabilitation are indeed related to the idea of urban space improvement and citizen's life.

Quite apart from the process name identification, the whole interventions are still managed by a planning instrument where the decisions are taken. In Italy, there are

B. Murgante et al. (Eds.): ICCSA 2014, Part III, LNCS 8581, pp. 300–314, 2014.

many instruments related to the urban transformation so that, at times, it has led to confusion and overlap of responsibilities. However, the main reference point remains to be the Master Plan as a framework plan, which would provide developing guidelines with regard to the cities. Indeed, renewal arguments have often been managed by another planning instrument family, known as the "Complex Program", generated by EU funding. «They are conceived to support the development and implementation of specific projects through cooperation between public and private sectors. The objective is to implement development that will promote and guide economic and social development» [1, p.80].

In this respect, it is about cross-sectoral instruments whose origin refers to the public housing discipline and deals with the residential asset maintenance-related issues attempting to improve settlement livability by reducing soil consumption.

In fact, the implementation of an urban renewal is to produce not only temporary benefits arising from the buildings but also long-term results to institute a domino effect that would lead to the transformation of brownfields and marginal areas in urban spaces able to attract economic activities such as commercial, tourist, cultural, recreational, and innovative services.

These instruments are called "complex" either because of their economic aspect, due to the contraction of public resources or the public and private stakeholders involved in the practices, or because of a wide range of financial resources affecting the interventions, which can be taken into account as a result of a multiple and different destination of uses.

It is highly necessary to examine these issues to remedy the widespread imbalances in urban planning, housing, and social services.

In this scenario, the decision (public) maker has to choose among the different planning instruments, whose infrastructures and public services better succeeded in pursuing the objective of increased urban quality. Otherwise, this decision must take into account the involvement of the private investments and the constraint of the financial feasibility.

The multicriteria analisys can support the decision maker because they are able to represent the complexity of the elements involved that belong to different levels: environmental, economic, financial, cultural, and social. The multicriteria models, that can be applied to make a decision, belong to various "families" [2,3,4,5] and are numerous, such as AHP and ANP [6,7], Macbeth [8], Electre [3] [9], Promethee [10], Rough sets [11] and many others. The choice of the Analytic Network Process (ANP) depends on the demand to apply to this case study a model able to manage both the multi-dimensionality of the decision, and the complexity of the interactions and quantitative and qualitative data.

2   Scenarios of Urban Transformation in Palermo

As previously mentioned, in Italy, there are many instruments aimed at the administration of urban transformation. It is not often easy to understand their specific role and relation among each other; to clarify their roles and relations, we have to divide them into two main sets: the plans and the projects. In its various forms, the plan is the urban development control instrument. The program is a recent Italian form of implementation planning acting on parts of the city, linked to specific calls as a consequence of European funds for renewal and limited to some parts in the south of Italy. If the plans provide a general system of rules, it is possible that the programs could not be in accordance with them, because, in many cases, they have decision-making autonomy by law.

About urban transformations, in the first set, we can find the main planning instrument for the management of the territory: the Master Plan. Particularly, through detailed plans, it has the task of thinking about the new interventions to organically act on the existing city. Generally, in the Master Plan's process, the social participation activities are limited in some phases. The second set includes special and complex programs that, starting from the last decade of the twentieth century, have been gradually established in accordance with specific laws arising from EU directives. We can observe that in this second set, regardless of the different name, all the programs are linked to the purpose of renewal of parts of the city characterized by physical and social decay with an ecological and sustainable approach [12]. More recently, we can find a new form of this typology of program, the "Piano Strategico", which is an integration of the different instruments of national and community planning activated in some specific areas, and it is considered as a general programming framework [13,14]. This program is linked to a succeeding idea to plan a territory, imagined as a strategic vision finalized to coordinate all the actions even if resulting from different funding and is characterized by real social participation activities.

It has been almost ten years since Palermo adopted a Master Plan. It is also a city where many of the programs previously mentioned have been experimented in several urban areas. Obviously, the case study area, more than being planned by the Master Plan, is intended as a strategic node for urban regeneration by the municipal administration. In fact, a part of the area was included in one of the projects under the "Programmi di Riqualificazione e Sviluppo Sostenibile del Territorio" (PRUSST), but no actions followed. However, the whole area has been identified as one of the urban centralities, named "International Polo City of Culture", on which the "Piano Strategico" of Palermo, approved in 2011, is based. In particular, the design of this area has been chosen as a pilot project between the 12 so-called "Area di Trasformazione Integrata" (ATI) to assess the effects of the processes of regeneration, development, and advancement of the city [15]. In the ATI process, one of the structural elements has been the participatory approach.

The following is a brief description of three different scenarios of use/transformation pointed out previously and related to the urban area in matter:

Status quo. The area is located in the central part of the city. It is characterized by the axis that goes from "Fossa di Danisinni" to Notarbartolo Station, crossing the former Lolli Station and including some nodes of urban centrality. It is a quadrangle extending from Via Dante—an important street that connects the center of the nineteenth century with the ring road—to Via Notarbartolo—considered one of the main access roads to the city. The whole area is characterized by the presence of a railway axis, still largely used to enter the city from Trapani, and of some historical and modern buildings—some of which are related to the railway area and without any use, some are abandoned areas, and some are tree-lined squares. Actually, there are only a few actions of renewal such as a project focused on the system of local mobility (connected to the more general forecasts of the Master Plan) and the restoration of the historical building of the Lolli rail station (purchased by private investors with the provision of cultural activities). These actions, if not complemented, cannot start a real redevelopment of the entire area (Fig. 1).

Master Plan. The Master Plan actually in force has been definitively approved in 2004 and provides that the area must be submitted to a detailed planning because of the unity of the functions and the strategic location. The indications to develop in the detailed planning are only related to mobility and public services related to education, culture, green area, and offices. The plan also provides some services inherent to the railway station, which is expected to be strengthened and rationalized. No indication is expected to increase housing or social housing (Fig. 2).

ATI2. The ATI2, called "Danisinni-Lolli-Notarbartolo", «is set up as a node strategic thinning system of urban mobility and to locate new urban centrality that act as attractors, as activators of economies and producers of quality» [15, p.31]. There is an urban project[1] in an area that goes from the Notarbartolo train station to the Lolli train station. The proximity of some important monuments, such as the "Cantieri Culturali della Zisa" and the "Castello arabo-normanno della Zisa", suggests providing a new and strategic system of accessibility, accommodation, and production spaces. «The project provides the renewal of the railway station area with the cover of the trench and the project of a green public area and services of about 160,000 square meters that will contain management and commercial functions and the new "gateway" of the city: a large public building with different functions as railway station, multimedia of contemporary, special accommodation activities» [15, p.33]. As to the first project of the Strategic Plan, particular attention has been placed on the financial feasibility in accordance with the economic sustainability and public-private partnership. «The basic idea was a strong desire not to distort the area by inserting high profitability services with a low coherence with the cultural and creative nature of the area. Another basic idea was the awareness that cultural services in a city like Palermo fail to produce income to compensate the costs of management. For this purpose it is discarded the idea of the exclusive use of the "concessione di gestione" to suggest to grant "diritti edificatori" to private with a rigorous and continuous controls by the municipal administration» [15, p.34]. The project has stated to pay greater attention to reduce the lower class' difficulties in accessing housing through social housing buildings (Fig. 3). The main characteristics of the three alternatives are summarized in Table 1.

---

[1] The urban project and the financial feasibility have been realized by the Municipality of Palermo; Town Planning department; Arch. coordinator Vincenzo Polizzi, because of a "Ministero delle Infrastrutture e dei Trasporti" financial operation obtained in 2004.

Fig. 1. Status quo

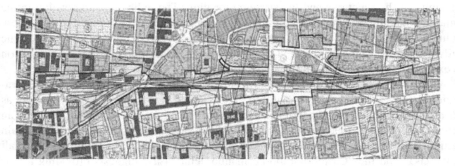

Fig. 2. Master Plan

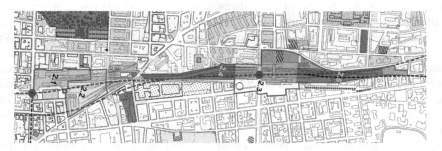

Fig. 3. ATI 2

Table 1. Characteristics of the alternatives

Status quo	New mobility is designed in relation to the railway link and to the future system of mobility to connect the center with the outskirts: some new streets are being realized, and a few of them have a different path from the original project because of a variant of Master Plan. There are abandoned areas and buildings, but no urban renewal or projects of improving housing are in act.
Master Plan	The Master Plan provides for a detailed planning in the area and for museums and cultural facilities, offices and business centers, school equipment, public and green spaces and parking, and a new and more functional mobility design, but it does not provide for any housing.

Table 1. (Continued)

ATI 2	The ATI2 project provides for the realization of the following: public services, underground parking garages, green public areas, new buildings for housing, offices and business centers, and a reorganization of the mobility system with a new design of the streets. The project decides to facilitate the social housing (40% of buildings for housing).

3   The ANP and the Structure of the Model

The ANP has been selected among the different multicriteria models because its holistic approach allows the expression of the complexity of a decision problem and makes explicit all the relations among the elements of a system.

The ANP method developed by Saaty [7] is a multicriteria discrete model, or otherwise, it is a model in which a finite number of alternatives are analyzed according to many criteria. Its structure is a network formed by clusters of elements and founded on pairwise comparison measurements in accordance with a ratio scale expressed by the Decision Maker (DM) and/or specialists; moreover, its outcome is the ranking of the alternatives with their corresponding scores. The peculiarity of the ANP is the possibility to express feedback and dependences within and between the clusters of the network elements. The model can be a simple or a structured network, where a hierarchy of control can be created, according to Saaty's suggestion [16], by means of the application of the Benefits, Opportunities, Costs, and Risks (BOCR) model; in this case, four subnetworks are built and their results are consequently synthesized.

The ANP has been already applied in many cases to solve different decision problems as regards urban and land transformations, such as: the location of a new waste incinerator plant [17], the highway corridor planning [18], the urban and land transformations [19,20], and the use of an artic coastal area [21].

In this case study, the decision problem concerns the choice among three alternative scenarios about an urban renewal and land use transformation of a central area in Palermo, which is strategic for the following development of the city, to reach the goal of maximizing the urban quality.

The analysis of these scenarios ("Status quo", "Master Plan" and "ATI2") is carried out according to four principal levels on which the differences/divergences of the alternative can be compared: social-cultural, economic-financial, infrastructural and administrative. The most important matters particularly concern the trade-off between public and private land use, the distribution of the ground rent obtained by the urban renewal, the balancing between private profit and social welfare, and the finding of financial resources needed to realize the public services.

At first point, the ANP methodology requires the development of the network that better delineates the decision problem, establishing the following: the DM's objective, the clusters (groups) of nodes (elements) that define the aspects of the decision problem, and the alternatives. Furthermore, the relations (feedback and dependences) within and between the clusters of the network elements are incorporated.

The comparison among the alternatives is based on a model belonging to the integration between the BOCR model of the ANP and the analysis of the urban renewal previously mentioned, articulated in the social-cultural, economic-financial, infra-structural, and administrative levels that are used as control criteria. The model is shown in Fig. 4.

Inevitably, in these levels, there are topics of political nature concerning the social system involved in the urban renewal; and topics of technical nature and mixed (polit-ical and technical) nature, and for this reason, the development of the model structure has required various experts' opinions interacting in a focus group[2]. Each expert has expressed technical opinion about his specific field of interest, and all experts have collectively stated the priorities among the control criteria.

Fig. 4. The BOCR model

The BOCR method application has allowed the identification of each alternative regarding certain positive and negative concerns in the short-medium term (Benefits and Costs) and the negative and uncertain concerns in the long term (Opportunities and Risks).

The network structure shows that every subnetwork is composed of 2 or 3 control criteria that represent the main levels of analysis of the decision problem. The nodes, intended as the relevant decisional aspects, are defined for each control criterion, as well as the elements (the number of elements under every control criterion is between 1 and 6), and finally, the relations are outlined. According to the ANP methodology, the type of relation between all the elements and the alternatives, introduced in the model, is the feedback type, as Saaty wrote «Not only does the importance of the criteria determine the importance of the alternatives as in a hierarchy, but also the importance of the alternatives themselves determines the importance of the criteria» [21, p.5]. There are further relations among the elements of the clusters: under the "Social-cultural" control criterion (subnetwork Benefits), the element "Cultural ser-vices" influences the "Quality of the public space", and the elements "Social partici-pation in renewal" and "Restoration of historical buildings" influence "Cultural

---

[2]   The group is composed of Giulia Bonafede as expert in the administrative field, Grazia Napoli in the economic-financial, Marco Picone in infrastructural, and Filippo Schilleci in social-cultural.

services", and there is a loop between the two elements of the "Social cohesion" cluster (Fig. 5).

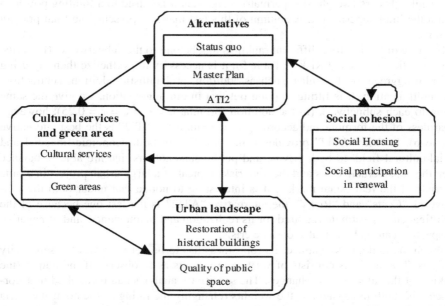

Fig. 5. Clusters and elements under the "Social-cultural" control criterion (subnetwork Benefits)

With reference to some works existing in the literature [17,18], there are 24 estimated elements using quantitative or qualitative indicators to facilitate the judgments of preference among the alternatives, and obviously, benefits and opportunities have to be maximized, whereas costs and risks have to be minimized. The clusters and the elements in the decision subnetworks are shown in Table 2 and some of them are described in Table 3.

Getting to this stage, pairwise comparisons are made by the DM/experts to express their judgments on the elements in accordance with another element of the system to establish their relative importance. These pairwise comparisons are made at the clusters level with respect to the control criteria and among the elements of the same cluster or between those elements linked to each other. All these judgments have a technical nature and are expressed by the corresponding experts, whereas the judgments among the control criteria (in each subnetwork) are collectively expressed by the focus group. This multicriteria model could take into account social groups' point of view, especially with reference to the relative importance of the control criteria and the clusters, in case the municipality is willing to adopt a wider participatory planning approach because a social participation is already a part of the planning instrument process (see §2). The judgments are expressed in the Saaty's fundamental scale (ratio scale of 1-9) and are filled in the paired comparison matrix that gives the priorities.

Resulting local and global priorities are illustrated in Table 4. All these priorities constitute a supermatrix (unweighted supermatrix) that is multiplied by the vector of the weight clusters (weighted supermatrix) and then it is raised to a limiting power to obtain the limit supermatrix. A column of this last matrix represents the final priority vector[3].

Because of having four different rankings that belong to the subnetworks (Benefits, Opportunities, Costs and Risks) (Table 5), it is necessary to synthetize them by using different aggregation formulas, such as addition, probabilistic addition, subtraction, and multiplication. It is fitting to point out that in our aggregation, we give the same rating to each subnetwork (in the addition formula, $b = o = c = r = 0,25$); however, regardless of the formula and according to the ranking, "ATI2" is the best alternative (Table 6). In fact, "ATI2" provides many benefits, in both economic-financial and social-cultural field, to both private and public stakeholders that are able to balance even the considerable costs and the high risks typical of a large urban project in a time of crisis of the real estate market. It is interesting to notice that under both the subnetworks "Costs" and "Risks", the alternative "ATI2" is the last one because of the high financial investment required for its realization and maintenance, and of negative perspective caused by actual economic crisis.

When the ranking is obtained, its robustness has to be verified with the sensitivity analysis. This analysis consists of varying the inputs as to observe if the output (the ranking of the alternatives) changes. The sensitivity analysis can be applied in accordance with both the system of the weights (changing the rating of the control criteria or the rating of the clusters) and the intensity of the influence of an element in each pairwise comparison.

In this paper, the consequences depending on the rating variation of the 4 subnetworks (Benefits, Oppurtunities, Costs, and Risks) have been verified to discover if one of the alternatives dominates the others or, otherwise, to make clear which conditions generate the variation of the ranking. The result is that the alternative ATI2 dominates the others, and it is also a robust choice because it can be verified in Fig. 6 where the x-axis represents the weight of each subnetwork (independent variable) and the y-axis represents the priorities of the alternatives (dependent variable): the ATI2 remains the best alternative for every value of the ratings.

The final ranking of the alternatives can also be obtained by using strategic criteria, such as "Creation of urban centralities", "Social consensus", and "Economic development", to weigh the four subnetworks and to emphasize the prevailing intrinsic political nature in a decision making concerning urban renewal. The same ratings of Benefits, Costs, Opportunities, and Risks with respect to the goal applied above should be initially replaced with the weighing of the strategic criteria with respect to the goal and, second, with defining the BOCR weights with respect to the strategic criteria. These weights could be applied in the aggregation formulas previously mentioned to confirm or deny that the previous priority is robust and that the ATI2 is the preferable alternative.

---

[3]   The data were processed using the software available in the website
www.superdecision.com.

Table 2. Control criteria, clusters, and elements in the decision subnetworks

BOCR	Control Criteria	Clusters	Elements (Indicators)		Unit	Status quo	Master Plan	ATI2
Benefits	Economic-financial	Real estate	RB	Residential buildings	m³	0	0	286.105
			CB	Commercial buildings	m²	1.000	1.000	40.440
			OB	Office buildings	m²	1.688	1.688	83.821
		Financing	FA	Feasibility analysis	Expert judgement	low	low	medium
	Infrastructural	Accessibility	RS	Local road system	Classes	low	medium	high
			PA	Parking areas	m²	4.866	19.011	114.471
	Social-cultural	Cultural services and green areas	CS	Cultural services	Classes	low	medium	medium
			GA	Green areas	m²	5.024	15.786	42.975
		Urban landscape	RH	Restoration of historical buildings	m²	800	800	1200
			PS	Quality of public space	Classes	low	medium	high
		Social cohesion	SH	Social housing	m³	0	0	114.442
			SP	Social participation in renewal	Classes	low	low	medium
Opportunities	Economic-financial	Economic interdependence	PI	Increase of private investments	Expert judgement	none	low	high
		Real estate	IV	Increase of real estate values	Expert judgement	none	low	medium
	Infrastructural	Accessibility	IT	Intermodal transport node	Classes	none	low	medium
	Social-cultural	Cultural services and green areas	LP	Links of parks system	Classes	none	low	medium
Costs	Economic-financial	Financing	PI	Public investment	Expert judgement	none	low	medium
			PR	Private investment	Expert judgement	none	high	medium
		Efficiency	TR	Time of realization of project	Classes	none	high	medium
	Administrative	Efficiency	TA	Time of adopting plans	Classes	none	medium	high
Risks	Economic-financial	Financing	FI	Difficulties in finding investments	Expert judgement	low	medium	high
		Losses	CM	Maintaining costs	Classes	low	medium	high
			RC	Unsold real estate	Expert judgement	none	none	high
	Administrative	Efficiency	GT	Planning management	Classes	none	high	medium

Table 3. Opportunity subnetwork

Indicators	Description
Increase of private investments	This indicator is referred to the interdependence that exists among the several sectors of an economic system so that an investment in a sector activates the production in other sectors linked, generating economic value added (EVA).
Increase of real estate values	The real estate values of the neighborhood can increase because of the renewal project that appreciably improves the urban quality. Even if this potential increase can be badly affected by the crisis of the real estate market.
Intermodal transport node	The realization of an intermodal transport node makes a contribution to the design of a new mobility system that better connects the city center with the outskirts.
Links of parks system	The realization of green areas allows the linking of urban parks and constituting a system.

Table 4. Local and global priorities

BOCR	Control Criteria	Clusters	Elements	Local priorities	Global priorities
Benefits	Economic-financial 0,30	Alternatives	ATI2	0.78359	0.35821
			Master Plan	0.14971	0.06844
			Status quo	0.06670	0.03049
		Real estate	Residential buildings	0.65744	0.07514
			Commercial buildings	0.19229	0.02198
			Office buildings	0.15027	0.01717
		Financing	Feasibility analysis	0.10000	0.42857
	Infrastructural 0,06	Alternatives	ATI2	0.78539	0.33660
			Master Plan	0.14882	0.06378
			Status quo	0.06579	0.02820
		Accessibility	Local road system	0.67940	0.38823
			Parking areas	0.32060	0.18320
	Social-cultural 0,64	Alternatives	ATI2	0.72915	0.27977
			Master Plan	0.20075	0.07703
			Status quo	0.07010	0.02690
		Cultural services and green areas	Cultural services	0.80343	0.22410
			Green areas	0.19657	0.05483
		Urban landscape	Restoration of historical buildings	0.44470	0.08625
			Quality of public space	0.55530	0.10771
		Social cohesion	Social housing	0.54621	0.07834
			Social participation in renewal	0.45379	0.06508

Table 4. (*Continued*)

Opportunities	Economic-financial 0,56	Alternatives	ATI2	0.74829	0.37415
			Master Plan	0.17618	0.08809
			Status quo	0.07553	0.03776
		Economic interdependence	Increase of private investments	0.75000	0.37500
		Real estate	Increase of real estate values	0.25000	0.12500
	Infrastructural 0,35	Alternatives	ATI2	0.31849	0.31850
			Master Plan	0.12914	0.12914
			Status quo	0.05237	0.05237
		Accessibility	Intermodal transport node	0.50000	0.50000
	Social-cultural 0,09	Alternatives	ATI2	0.31849	0.31849
			Master Plan	0.12914	0.12914
			Status quo	0.05237	0.05237
		Cultural services and green areas	Links of parks system	0.50000	0.50000
Costs	Economic-financial 0,83	Alternatives	ATI2	0.09585	0.04382
			Master Plan	0.26694	0.12203
			Status quo	0.63721	0.29130
		Financing	Public investment	0.40864	0.17513
			Private investment	0.59136	0.25344
		Efficiency	Time of realization of project	0.10000	0.11429
	Administrative 0,17	Alternatives	ATI2	0.09790	0.09790
			Master Plan	0.15541	0.15541
			Status quo	0.24669	0.24669
		Efficiency	Time of adopting plans	0.50000	0.50000
Risks	Economic-financial 0,67	Alternatives	ATI2	0.10711	0.04897
			Master Plan	0.26893	0.12294
			Status quo	0.62396	0.28524
		Financing	Difficulties in finding investments	0.52698	0.28608
			Maintaining costs	0.26249	0.14250
		Losses	Unsold real estate	0.21053	0.11429
	Administrative 0,33	Alternatives	ATI2	0.09420	0.09420
			Master Plan	0.36532	0.36532
			Status quo	0.04048	0.04048
		Efficiency	Planning management	0.50000	0.50000

Table 5. Priorities for alternatives under BOCR

Alternatives	Benefits			Opportunities		
	Ideals	Normals	Raw	Ideals	Normals	Raw
ATI2	1.000000	0.747680	1.000000	1.000000	0.694761	1.000000
Master Plan	0.245317	0.183418	0.245317	0.310420	0.215668	0.310420
Status quo	0.092153	0.068901	0.092153	0.128924	0.089572	0.128924
	Costs			Risks		
	Ideals	Normals	Raw	Ideals	Normals	Raw
ATI2	0.355779	0.182398	0.355779	0.282746	0.158378	0.229119
Master Plan	0.594788	0.304931	0.594788	1.000000	0.560143	0.810337
Status quo	1.000000	0.512671	1.000000	0.502513	0.281479	0.407205

Table 6. Final ranking of alternatives (normalized form)

Alternatives	Additive synthesis bB+oO-cC-rR	Multiplicative synthesis (B*O)/(C*R)
ATI2	0,6958	0,9646
Master Plan	-0,2943	0,0124
Status quo	-0,0099	0,0230

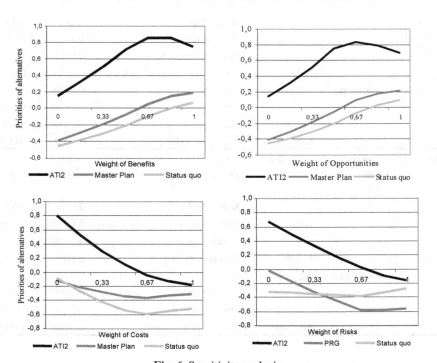

Fig. 6. Sensitivity analysis

4   Conclusions

This work has analyzed the application of ANP-BOCR method relative to the choice of the best alternative transformation of a central area in the city of Palermo.

The application has allowed emphasizing several issues:

-   in the phase of model construction, the focus group has been very usefull and profitable for the development—delineated in a detailed way—of the political, social, and technical complexity of the decisional problem that is distinctive of the projects of urban renewal;
-   in the phase of expression of the judgments concerning the priorities (of political and social nature), in the focus group, some diverging opinions have emerged but they have reached a compromise. However, it is recognized that specific Group Decision Making methods should be applied for the management of the collegial judgments (Saaty [21]);
-   in the phase of the technical judgment expression, using a system of measurable indicators has emphasized the different performances of the alternatives and has made the experts' job easier;
-   in the phase in which the outcome has been obtained, the sensitivity analysis has allowed the verification of the robustness of the ranking and the role that the elements of the model recover in the choice of the alternative.

In every case, even if the MCDM methods, whose output is a ranking, do not allow knowing how much the solution is near to the best condition but only which is the better alternative among the ones analyzed, the fact that every step of the model application is explicit, endorsed, and declared, constitutes itself an incontrovertible quality of the ANP especially when it is applied in composite and conflictual context as the urban transformation.

References

1. Lo Piccolo, F., Schilleci, F.: Local Development Partnership Programmes in Sicily: Planning Cities without Plans. Planning Practice & Research 20(1), 79–87 (2005)
2. Bana e Costa, C.A.: Readings in Multiple Criteria Decision Aid. Springer, Berlin (1990)
3. Figueira, J., Greco, S., Ehrgott, M. (eds.): Multiple Criteria Decision Analysis. State of the Art Survey. Springer, New York (2005)
4. Bouyssou, D., Marchant, T., Pirlot, M., Tsoukiàs, A., Vincke, P.: Evaluation and Decision Models: Stepping Stones for the Analyst. Springer, Berlin (2006)
5. Ishizaka, A., Nemery, P.: Multi-Criteria Decision Analysis. Wiley, Chichester (2013)
6. Saaty, T.L.: The Analytic Hierarchy Process. McGraw Hill, New York (1980)
7. Saaty, T.L.: Fundamentals of Decision-making and Priority Theory With the Analytic Hierarchy Process. RWS Publications, Pittsburgh (2000)
8. Bana e Costa, C.A., Vansnick, J.-C.: Applications of the MACBETH approach in the framework of an additive aggregation model. Journal of Multi-Criteria Decision Analysis 6(2), 107–114 (1997)
9. Roy, B., Bouyssou, D.: Aide multicritére à la décision: Méthodes et case. Economica, Paris (1995)

10. Brans, J., Vincke, P.: A preference ranking organization method: The Promethee method. Management Science 31, 647–656 (1985)
11. Greco, S., Matarazzo, B., Slowinski, R.: Rough sets theory for multicriteria decision analysis. European Journal of Operational Research 129, 1–47 (2001)
12. Avarello, P., Ricci, M. (eds.): Politiche urbane: dai programmi complessi alle politiche integrate di sviluppo urbano. Inuedizioni, Roma (2000)
13. Curti, F., Gibelli, M.C. (eds.): Pianificazione strategica e gestione dello sviluppo urbano. Alinea, Firenze (1996)
14. Archibugi, F.: Introduzione alla pianificazione strategica in ambito pubblico. Alinea, Firenze (2005)
15. ANCE: Riqualificare le città per arginare il declino. L'area Lolli Notarbartolo. Quaderni Opus Concretum, Palermo (2011)
16. Saaty, T.L.: Theory and Applications of the Analytic Network Process. RWS Publications, Pittsburgh (2005)
17. Bottero, M., Ferretti, V.: An Analytic Network Process-based Approach for Location Problems: The Case of a New Waste Incinerator Plant in the Province of Torino (Italy). Journal of Multi-Criteria Decision Analysis 17, 63–84 (2011)
18. Piantanakulchai, M.: Analytic network process model for highway corridor planning. In: Proceedings of the 8th International Symposium on the Analytic Hierarchy Process, Honolulu (2005)
19. Bottero, M., Lami, I.M., Lombardi, P.: Analytic Network Process. La valutazione di scenari di trasformazione urbana e territoriale. Alinea, Firenze (2008)
20. Bottero, M., Ferretti, V.: Integrating the analytic network process (ANP) and the driving forcepressure-state-impact-responses (DPSIR) model for the sustainability assessment of territorial transformations. International Journal of Management of Environmental Quality 21(5), 618–644 (2010)
21. Saaty, T.L.: Decision Making With the Analytic Network Process. Economic, Political, Social and Technological Applications with Benefits, Opportunities, Costs and Risks. Springer Science, New York (2006)

VGI to Enhance Minor Historic Centres and Their Territorial Cultural Heritage[*]

Pierangela Loconte and Francesco Rotondo

Polytechnic of Bari, via Orabona, 4, 70124, Bari, Italy
{pierangela.loconte,francesco.rotondo}@poliba.it

Abstract. The advent of Web 2.0 and the development of information and communication technologies represent an opportunity to reduce the spatial distances and enhance fast access to any type of information.

In particular, in recent years, the spread of Smart Mobile Applications and an increasing accessibility to social networks has provoked a growth of exchanged information and shared knowledge. Among the multiple applications, for spatial planning disciplines, it is interesting those related to the city and to urban and regional development's processes.

The processes of e-planning and e-participation for enhancing minor historic centers and their territorial cultural heritage have a particular interest.

This paper will try to understand what are the new strategies for development and enhancement of cultural territorial systems through the use of these tools and what are the perspectives for development.

In particular, in the first part of the work we try to understand what are the new tools available for the construction of knowledge and the enhancement of cultural heritage, while in the second part we define possible uses for these tools and why these technologies can be an opportunity for minor historic centers and their surrounding landscapes.

Finally, in the last part we bring the reader's attention to a series of international case studies of particular interest in order to understand whether these new tools can be an effective support for decision-making processes and for the preparation of plans and projects of regeneration and enhancement of minor historic centers and their territorial cultural heritage.

Keywords: ICT, minor historic centers, participation, volunteered geographic information, urban and regional planning.

1   Introduction

As highlighted by the European Landscape Convention and is now widely discussed and settled in the national and international urban debate, the "Landscape designates a certain part of the territory, as perceived by people, whose character derives from the natural and / or human interrelationships " [1].

---

[*] The paper is the result of a joint work of the authors, also if chapter 1 and 5 has been written by Francesco Rotondo, the others by Pierangela Loconte.

B. Murgante et al. (Eds.): ICCSA 2014, Part III, LNCS 8581, pp. 315–329, 2014.

The complexity of the territorial system is due to the variety of relational systems that constitute it and their constantly change over time.

For this reason it seems preferable testing new approaches based on their characteristics more then to define from the outset correct methods for the construction of landscape's protection and development policies.

As widely claimed in the literature, landscape management is any measure introduced to steer changes brought about by economic, social or environmental needs and should be dynamic and adaptive and to seek to improve landscape quality on the basis of the local population's expectations [1] [2] [3].

In particular, the sense of belonging to a community in their living environment is reflected in the landscape.

It is not only the outcome of processes of social identification, but also and especially a "value continuously built by the will of those who live and use the land"[4].

The landscape management and enhancement policies, then, are intended to promote the development of local systems and, at the same time, contribute to the construction of a common identity structured on a shared heritage of signs and images handed down from the past.

The structure of the landscape and territory requires new approaches able to translate the physical wealth and relational complexity in resources for development [40] [41].

In particular, contrary to what happened in the past, there is now the need to start from the expectations of the resident population and build on these policies able to increase the levels of well-being and capable of activating pathways of development.

The territory is understood as essential support to increase the levels of well-being and at the same time meeting the basic needs through the "pursuit of virtuous relationships between environmental , social, territorial, economic, political sustainability that makes coherent basic needs, self -reliance, eco-development "[5].

The introduction of Web 2.0 and the emergence of new smart technologies represent a new opportunity and can support the creation of these virtuous relationships and improving levels of territorial knowledge to build a participated project of development.

According with Wang et al. [6] e-Planning can enable easy access to information, guidance and services that support and assist planning applicants, and streamlined means of sharing and exchanging information among key players.

In particular, ICT methods can provide fundamental support for communication needs in multi-agent decisional processes, increasing data access and levels of information, thereby enhancing knowledge levels of the specific issues and process dynamics [7] [8] .

We concentrate, therefore, on the role that these new tools can have in the processes of territorial development and enhancement, with particular reference to historical landscapes characterized by the presence of small urban centers that need to be valued.

We try, therefore, to understand what might be the role of the population within these processes and, consequently, which input can be given by the Information and Communication Technologies (ICT) and Volunteered Geographic Information (VGI).

2   ICT and VGI to Enhance the Value of Cultural Heritage in Minor Historic Centers

In recent years, the mobile smart technologies have been widely circulated leading to an increase in the ability of people to communicate, share and improve knowledge levels.

The evolution of this type of technology has also led to social change, facilitating virtual socializing on social networks and simplifying processes of involvement of the population.

This was certainly facilitated by the increase number of people able to access the network and the resulting change in the behavior of publics linked to the use of the internet and interactive tools, with particular reference to social networks.

The cultural change introduced by the arrival of Web 2.0 has enabled it to overcome the barriers of space, shortening distances in space and time and the change in the type of user and their needs and expectations linked to the network.

In particular, the web users have been transformed from simple consumers of information and online services to producers of contents, helping to increase levels of knowledge in any field of life.

The role of these technologies can be considered particularly significant in all those areas characterized by high levels of complexity, where it becomes difficult to recognize a unique, shared knowledge and where it becomes difficult to make a synthesis of existing relational systems, often changing with the territorial context.

Based on the above, the use of ICT can support the processes of development and enhancement of the landscape, characterized by high degrees of complexity and relatedness.

People play an important role in shaping landscapes because they have always adapted their environment to better fit their changing needs [3][9] and their behavior have been changed by the place character, so that the use of ICT technologies can help to make people more involved in the transformation of their territory, contributing to the construction of shared development processes, in a fast changing society, as the contemporary one.

In this perspective, today, between different forms of ICT it is particularly significant and interesting the use of Volunteered Geographic Information (VGI) and Public Participation Geographic Information Systems (PPGIS).

Both tools involve the investigation and identification of locations that are important to individuals [10].

The term VGI refers to geospatial data that are voluntarily created by citizens who are untrained in the disciplines of geography, cartography or related fields [11].

PPGIS also aim to strengthen public participation through the use of interactive tools but PPGIS projects are often implemented to inform planning and policy issues while VGI systems may have no explicit purpose other than participant enjoyment [12]

Both are configured as a simplified method (PPGIS are not always simple) for the involvement of the population in decision-making processes for the construction of plans and projects for urban development and regional planning.

In this way, the governance of relations between different actors with the aim of promoting cooperation and social learning [13] can be greatly simplified.

In fact, thanks to increased levels of virtual discussion and sharing is now easier to be able to involve the population in e-planning and e-partecipation improvement.

In particular, the use of VGI and PPGIS allows people to move from interactivity to interaction and, contributes to the building of networks, which are no longer based on information exchange alone but also on knowledge sharing.

This type of content interaction (location-based) changes the processes used to produce, update and distribute geographic information [14].

In this way, therefore, it is not only possible to collect or disseminate a large amount of information but also to define the users information available spatially, facilitating the implementation of knowledge at territorial level.

These collaborative geospatial tools have approached the web-users in the manner of Geographic Information Systems, helping to create new types of links with the expert knowledge.

The VGI has completely changed the perception and use of geographic information, and provides a growing number of users with the possibility to interact with maps and data [11].

At the same time, it has found new resources and new ways of implementation of the instruments being able to improve bottom-up participatory processes.

As claimed by Garnero et al. [15] VGI is the primary information source standing at the basis of all the modern web applications connected with communities, social networks and contributive databases.

So the contributions that can be provided by the use of this type of technology are remarkable.

In particular:

- Allow you to build new frameworks of knowledge or implementing and updating existing ones;
- Activate voluntary participatory processes, shortening the distance between the decision maker and the end user, enabling new forms of leadership and interaction at different levels;
- Facilitate the involvement of the population in the planning process;
- Allow to collect and display maps and photographs through a series of information that the planner would be hardly able to collect;
- Enable us to understand what is the perception that people have of their own territory, such as the resources, opportunities and threats, the ideas for its development.

According to the claims, these innovative tools can represent an interesting resource in the construction of a participatory government of the territory.

At the same time, however, we find new problems, including:

- The need for a wide spread of the Internet and technologies for network access available for the entire population because it is a virtual participatory processes;
- The difficulty of reaching all segments of the population: these instruments, in fact, are able to engage more readily to young people, available to share personal information online while they may find some difficult with older people and who not use technological tools (the well known digital divide);
- The consistency and reliability of information put by users involved in the process;
- The need to have an expert figure able to structuring and managing participatory processes, who owns, therefore, a specific education on the use of ICT, geographic information systems and urban and regional planning and participatory processes. Moreover, this "expert" should be able to integrate these participatory tools with more traditional ones so as to collect the greatest number of contributions for the construction of plans and projects seeking to overcome critical issues described in the use of ICT and VGI.

The outlook for discussion and innovation open by the use of ICT and VGI in planning are particularly significant, especially in the case of the construction of strategies for enhancement of homogenous territorial areas based on the recognition of existing resources and the appropriation of place and local identity.

In general, tools like Open Street Map, Wikimapia e Google Map Maker enable experts and amateur enthusiasts alike to create and share limited, theme oriented geospatial information [16].

In addition to these tools, social networks take on a new and interesting role, especially if applied in synergy with the new Smart Mobile technologies through the use and implementation of Application.

3   Building Landscape Valorisation Strategies

The complexity described in the European Landscape Convention [1] and the variety of existing relational systems within can be briefly described in the Fig. 1.

As shown in the figure, the landscape is the set of natural and cultural landscapes, have territorial, environmental, cultural, economic and social structures and conformations, and value for society than within them performs its daily life.

In particular, in 1972, UNESCO defines like Cultural Landscapes the duty of ensuring the identification, protection, conservation, presentation and transmission to future generations of the cultural and natural heritage (art. 4) [18], in which there in a great relationship between history, culture and population.

Within the cultural landscapes, historic towns have a particularly significant role.

At one time they were considered only as a place characterized by a high amount of historical and architectural heritage while today have taken on a new meaning, both in urban than territorial level.

Today, the historic centers, with particular reference to those smaller, may be considered a privileged place in which you can see traces of the past, the sedimentation of the functions, the concentration of the peculiarities that characterize a given territorial area.

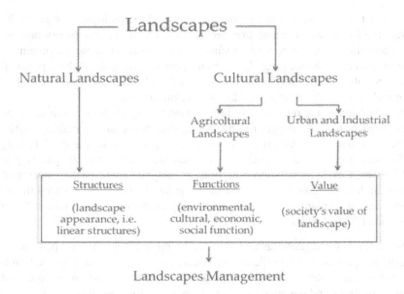

Fig. 1. Synthesis scheme of landscape structure and its characters ([17], modified)

In this view, the historical center is the place where you can read the cultural, social and economic characters.

The historical centers are not separable from their landscape and can be considered as territorial points of reference in which focuses on tangible heritage and intangible asset [40] [41].

They represent a common heritage and, albeit with considerable difficulties due to changes in lifestyles, yet maintain a close relationship with their environment, in contrast to what happens for the larger centers.

Due to the processes of modernization and globalization, sometimes the smaller historic centers have lost their role in the local context.

They are now often characterized by poor quality of urban life, high levels of abandonment of existing buildings and lack of services that force the population to move to larger centers, which can better meet the needs of contemporary life [42].

There is, therefore, the need to give small historical centers new opportunities of development through the activation of the processes of regeneration and enhancement that are no longer concentrated on making single physical interventions but who are able to propose an integrated system of actions both at urban and territorial level.

For this reason the historic centre in an opportunity for the formulation of a recomposition project [19].

What is missing are the models of development that are able to look at the system in a unified way, to understand the role that each part has in them and to respond to emerging critical issues.

To do this, there is a need to begin by the resident population, by the social structure and local economy, by the ability to respond to real needs, for building pathways to enhancement and successful development.

The historical and cultural heritage is a starting resource, but it needs to be gone beyond the concept of a "museum of historical and architectural heritage" to make them functional to the new needs of the resident and non-resident population, to became again a contemporary place of life.

Implement intersectoral and shared actions may represent an opportunity to urban, social, cultural and economic revival able to make again the historic centers a territorial reference point, as in the past.

The cultural heritage represented by city / ancient sites, cultural landscapes, etc. are able to stimulate the production of new relational values (social capital) that are given, and, therefore, indirectly promote economic activities to the extent that it is managed as a common good [20].

There is a need to define strategies to protect and value all existing landscapes within a given territory, whether natural or manmade.

The management of the landscape and its enhancement goes through the definition of a precise strategy of development able to define as the determination of the basic long-term goals and objectives and the adoption of courses of action and the allocation of resources necessary for carrying out these goals [21].

In particular, the intent of the strategy must be to understand how the structures, the functions and values that make up the landscape should be put in relation with each other and with the external environment in order to achieve the development goals, both in the short term and in the medium and long term.

Obviously the achievement of the general and specific objectives is closely connected with the knowledge of the area and its material and immaterial characteristics.

What has been said converges in the definition of the UNESCO concept of Historic Urban Landscape based on the need of improvement and integration of landscape and historical centers and on the centrality of local values in the dynamics of sustainable development.

In particular, "the HUL approach aims at preserving the quality of human environment and enhancing the productivity of urban spaces. It integrates the goals of urban heritage conservation with the goals of social and economic development" [39].

According to this logic, then, the exploitation of minor historical centers is inextricably linked to the development of the landscape in which they are and there is the need to find new tools for documenting and mapping the heritage of the communities as well as to protect natural and environmental resources.

Plans and development programs capable of improving the quality of life of the population and configure new opportunities in relation to the construction of sustainable forms of tourism could be built on a shared knowledge.

The concept of Historic Urban Landscape has become a necessity of concrete through the definition of specific operational methods.

One of these may be constituted by the strategic approach that is the basis of construction management plans of UNESCO sites [22] for their definition of enhancement actions both for minor historical centers and their local contexts with a unified approach.

This approach is particularly interesting, and applicable to any territorial area through the definition of:

- strategies for knowledge, related to the continuous monitoring of resources;
- conservation strategies, which are reflected in the procedures to achieve a coordinated and systematic organization;
- participation strategies for the involvement and growth of identity values;
- developing strategies for the enhancement of the local level;
- marketing strategies and territorial communication.

In this logic processes of e-planning, e-partecipation is able to support each of these five strategies providing an opportunity for the valorization and regeneration of minor historic centers.

This becomes possible through the construction of future scenarios with the support of the people.

The role of participation in design and planning is a mechanism to support the identification of the issues, interests, priorities and wishes of those who will make use of the site, as well as those who may be affected by the future development [23].

In this perspective, then, the Information and Communication Technologies and, in particular, Volunteered Geographic Information (VGI) or social networks might constitute interesting tools for supporting the processes of development and enhancement of minor historic centers and their territories.

4   ITC and VGI in Practice for the Enhancement of Minor Historical Centers and Their Territories

As claimed by Modica et al. [24] understanding the diversity of perception and values related to the landscape is fundamental [3] and descends from a thorough consideration of the dualistic relationship between people and the places they live [25].

The complexity of the territory, of the relations between its parts and the definition of the role of minor historical centers raises the need to implement different kinds of strategies that can achieve goals and overcome existing weaknesses.

As argued by Wang et al. urban planning is a complex task requiring multidimensional urban information (spatial, social, economic, etc.).

The need for assistance in performing urban planning tasks has led to the rapid development of urban information systems, especially "e-Planning" systems, with the support of government policy and emerging information and communication technologies (ICT) [6].

Although the use of tools such as VGI, the PPGIS and in general all the tools of ICT can be an exciting opportunity to launch and support the processes of enhancement, it is important to say that these are just one of the means that can be used to achieve objectives and that their use should be assessed in relation to each individual situation.

If place-based conservation management is to take root, methods must be developed to identify places that integrate the geography of place with the human perception of place [26].

It then attempts, in this context, to look at in a series of cases in which these tools have proven and continue to be an excellent support for the construction of shared development strategies. In particular, we analyzed Italian and international best practices in which these tools have been used in support of the five development strategies identified in the precedent chapter.

With regard to the widening of the knowledge system through the web-users tools such as OpenStreetMap, Wikimapia, Google Map Maker, we can say that they are already well known and sufficiently used. They allow any registered user to enter information of any kind through the deployment possibilities with smart mobile applications.

The selected examples of international good practice show how it is possible to implement multiple strategies using these tools.

The Italian Sardinia and Apulia regions, in the drafting of their new Regional Landscape Plan have created web platforms to support the process of participation and, at the same time collect impressions, comments and reports on the state of the landscape.

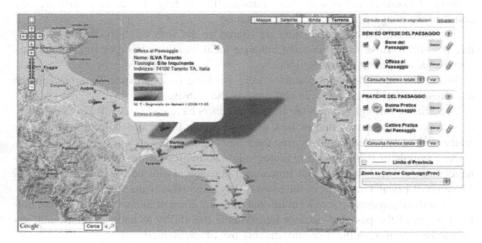

Fig. 2. Landscape Observatory of Apulia Region

SardegnaGeoBlog [33] and the Landscape Observatory of Apulia Region [34] are two examples of Volunteered Geographic Information in which the user can locate places on the interactive maps, make reports and post comments, contributing to the sharing of knowledge and the intangible heritage that is the basis of the processes of regeneration and enhancement.

At the same time, they can make judgments about the quality of the landscape, of good or bad practice and have the opportunity to make suggestions for the territories in which they live.

These tools can both respond to the strategy for the implementation of knowledge and at the same time are able to give ideas to discuss policies of development and protection.

Tools like this are an implemental tool of the indications of the European Landscape Convention in which acceding states undertake to "integrate landscape into policies for land use planning, in urban, cultural, environmental, agricultural, social and economic planning, as well as in others that may have direct or indirect impact on the landscape "

Another interesting experience in the construction of knowledge, but also for the dissemination of culture and local knowledge, with particular reference to minor historical centers and their landscapes, is the project LAB.net plus [35] linked to the creation of Cross-border network for the enhancement of landscapes and local identities[2] that involved face the regions of Sardinia, Tuscany, Liguria and Collectivité Regional de Corse.

The goal of the project is to develop cooperation and promoting the creation of networks for the enhancement of local culture oriented to deepening the relative skills to the valorization of landscapes.

This was made possible through the creation of a cross-border network of laboratories that have worked in the territories trying to involve the population in the development of Local Development Plans, through the drafting of guidelines and projects and the following pilot projects implementation.

The work carried out (the results are shared on the web site http://labnet-plus.eu/) had the aim to raise awareness and enhance local identities in order to promote sustainable development of the territory.

While the network of laboratories were working in the respective areas, all the documents were shared online in real time on the website.

In addition, through the geo-tag instrument users have had (and still have) the ability to report an item of value or lack of value of the landscape, helping to build the framework of knowledge and sharing their perception of the area.

An example of ICT supporting conservation strategies for the development and enhancement of local material and immaterial heritage is offered, in Italy, by MUVIG Museum of Local Traditions of the town of Viggiano (Basilicata Region) [36].

What distinguishes it from a traditional museum located in a small village that collects traces of the past and peasant traditions is the ability to have an additional service to be able to visit the museum, making use of augmented reality.

The use of QR codes give the opportunity to benefit from having information available to the mobile and internet technologies.

In addition, the web-site and on the Facebook page of the museum users have the ability to interact in real time and to contribute to the evolution of the museum and the dissemination of information related to local traditions, the rural past and the objects contained within the museum.

In this way, the museum has become the heritage of all, encouraging anyone among the inhabitants of Viggiano to share with the museum community the historical relics of a long-gone country reality of which they possess.

---

[2]   Programm funded by the European program 2007-2013 Operational Programme Italy-France "Maritime", Axis IV, Objective 1.

Fig. 3. Muvig Facebook page

In this way, the museum seeks to preserve and enhance a heritage that would otherwise be forgotten, encouraging the community to look for traces of his own story and tell it.

The use of new technology helps to engage in this process of "building of community and identity" the younger generation or tourists also.

Other interesting examples structured through the involvement of local people in order to define new strategies of valorization are developed by the Landscape Institute Values & PPGIS [37].

The PPGIS research has led to interesting examples of the involvement of the population in the processes of knowledge, enhancement and development of the region through the creation of interactive maps with Google Maps.

Public participation geographic information systems (PPGIS) are methods that seek to democratise spatial information and technology, often through mapping at local levels of social organisation to produce knowledge of place [26].

Particularly interesting are the cases of the construction of Public Land Values in Victoria (Australia) e del PPGIS project in Southland Region of New Zealand prepared in collaboration with New Zealand Department of Conservation (DOC).

In both cases, through the involvement of thousands of users (i.e. 1900 Participants in the case of Victoria) has allowed the construction of thematic maps able to show the distribution of selected public land values and management preferences.

The aim was to identify strategies for development through public participation in order to integrate natural and historical resources with the demand for tourism and facilities.

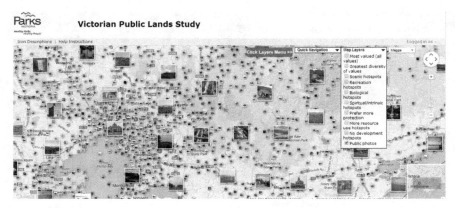

Fig. 4. http://www.landscapevalues.org/

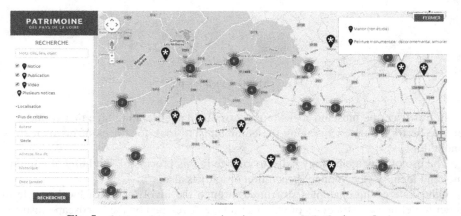

Fig. 5. http://www.patrimoine.paysdelaloire.fr/

PPGIS data collection consisted of two parts: (a) spatial attribute mapping using a custom Google Maps application; and (b) general survey questions assessing participants' familiarity with conservation areas in the regions and selected socio-demographic information. [26].

In both cases, the participants had the opportunity to report scenic spots and natural areas of particular value, assets and areas of historical, recreational points or undeveloped areas or degraded by helping to identify locations of significant value to preserve and enhance.

In this way, PPGIS systems have increasingly exploited Internet technology to identify spatial attributes from local and regional populations [27] [28] [29] [30] [31] [32].

Finally, with regard to the marketing strategies and communication planning, one of the Pays de la Loire is an interesting example of the application of integrated social networks, smart mobile applications for tourist use of the territory and for the dissemination of current activities and services existing in the Région Pays de la Loire [38] who were able to create a web-based system that allows any user to get and share information about the region, its towns, its history and culture through all the most popular social networks and through special app easily downloadable.

An example is the app that will improve the use of public transport and optimize travel throughout the geographical area; one for the identification of existing assets and cultural resources; one linked to the tourist trail by bike that can be implemented by all users.

5   Conclusions

As explained above, the Voluntereed Geographic Informations can be very useful in building a shared understanding of the local cultural heritage present in minor historical centers and their surrounding landscapes of which are an integral part.

ICT can contribute significantly in contemporary society, to inform and create communication between services, users, residents, stakeholders.

In fact, the first permit to deepen the expert knowledge through residents and users of cultural heritage contained in the historical centers and landscapes in which they are integrated for centuries.

Through the use of VGI is possible to integrate and deepen the expert knowledge to the detail characteristics of the perception of residents and users of cultural heritage thus making it more suitable to their needs.

In particular, the use of these tools can give substance to the concept of Historic Urban Landscape and can be an interesting resource for its construction.

The decrease of the distances between landscape, planning and population, implemented through the use of ICT and VGI, may represent a new opportunity for social development and economic development.

It seems really relevant for urban planning the possibility to transform final users of geographic information in its producers

It remains difficult to assess the consistency and reliability of the information entered. What tools we have to certify the information inserted or the evaluation of services offered on the web.

Self-control of the accuracy of the information entered by the users of these systems to build territorial identity self-managed by users seems the only remedy is seriously viable, coherently with examples in other fields such as, just as an example, the trip advisor success in sharing travel experiences.

References

1. Council of Europe: European Landscape Convention. Strasbourg (2000),
   http://conventions.coe.int/Treaty/en/Treaties/Html/176.htm
2. Rescia, A.J., Willaarts, B.A., Schmitz, M.F., Aguilera, P.A.: Changes in land uses and management in two Nature Reserves in Spain: Evaluating the social–ecological resilience of cultural landscapes. Landscape and Urban Planning 98, 26–35 (2010)

3. Aretano, R., Petrosillo, I., Zaccarelli, N., Semeraro, T., Zurlini, G.: People perception of landscape change effects on ecosystem services in small Mediterranean islands: A combination of subjective and objective assessments. Landscape and Urban Planning (112), 63–73 (2013)
4. Clementi, A.: Revisioni di paesaggio, Meltemi, Roma (2002)
5. Magnaghi, A.: The Urban Village: A Charter for Democracy and Local Self-sustainable Development. Zed Books, London (2005)
6. Wang, H., Song, Y., Hamilton, A., Curwell, S.: Urban information integration for advanced e-Planning in Europe. Government Information Quarterly (24), 736–754 (2007)
7. Shiffer, M.J.: Towards a collaborative planning system. Environ. Plan. B Plan. Des. 19, 709–722 (1992)
8. Rotondo, F.: The U-City Paradigm: Opportunities and Risks for E-Democracy in Collaborative Planning. Future Internet 4, 563–574 (2012), doi:10.3390/fi4020563
9. Antrop, M.: Why landscapes of the past are important for the future. Landscape and Urban Planning 70, 21–34 (2005)
10. Tulloch, D.: Is VGI participation? From vernal pools to video games. GeoJournal 72(3), 161–171 (2008)
11. Goodchild, M.F.: Citizens as sensors: The world of volunteered geography. GeoJournal 69, 211–221 (2007)
12. Brown, G., Kelly, M., Whitall, D.: Which 'public'? Sampling effects in public participation GIS (PPGIS) and volunteered geographic information (VGI) systems for public lands management. Journal of Environmental Planning and Management (2013), http://dx.doi.org/10.1080/09640568.2012.741045
13. Healey, P.: Collaborative planning: Shaping Places in Fragmented Societies. University of British Columbia Press, Vancouver (1997)
14. Roche, S., Mericskay, B., Batita, W., Bach, M., Rondeau, M.: WikiGIS Basic Concepts: Web 2.0 for Geospatial Collaboration. Future Internet 4, 265–284 (2012)
15. Garnero, G., Corrias, A., Manigas, L., Zedda, S.V.: VGI, Augmented Reality and Smart Web Application: Projects of Development in the Territory of the Sardinia Region. In: Murgante, B., Misra, S., Carlini, M., Torre, C.M., Nguyen, H.-Q., Taniar, D., Apduhan, B.O., Gervasi, O. (eds.) ICCSA 2013, Part IV. LNCS, vol. 7974, pp. 77–92. Springer, Heidelberg (2013)
16. Rotondo, F., Selicato, F.: E-Democracy in Collaborative Planning: A Critical Review. In: Murgante, B., Gervasi, O., Iglesias, A., Taniar, D., Apduhan, B.O. (eds.) ICCSA 2011, Part II. LNCS, vol. 6783, pp. 199–209. Springer, Heidelberg (2011)
17. Organization of Economic Co-Operation and Development: Environmental Indicators for Agriculture Methods and Results, OECD, Parigi (2001)
18. UNESCO: Convention Concerning the Protection of the World Cultural and Natural Heritage (1972)
19. Bonfantini, G.B.: Planning the historic centres in Italy: For a critical outline. Planum. The Journal of Urbanism (25), 1–19 (2012)
20. Fusco Girard, L.: Processi valutativi per lo sviluppo sostenibile. In: Brunetta, G., Pistoia, R. (eds.) (a cura di), Trasformazioni, coesioni, sviluppo territoriale. Temi emergenti nella scienze regionali; Associazione Italiana di Scienze Regionali, Franco Angeli, Milano, vol. (38) (2006)
21. Chandler, A.D.: Strategy and structure. Chapters un the history of the American industrial enterprise. The MIT Press, Cambridge (1962)
22. Ministero per i Beni e le Attività Culturali: Progetto di definizione di un modello per la realizzazione dei Piani di Gestione dei siti UNESCO (2005)
23. Kelsey, C., Gray, H.: The citizen survey process in parks and recreation. American Alliance for Health, Physical Education, Recreation and Dance, Reston, VA (1986)

24. Modica, G., Zoccali, P., Di Fazio, S.: The e-Participation in Tranquillity Areas Identifica- tion as a Key Factor for Sustainable Landscape Planning. In: Murgante, B., Misra, S., Carlini, M., Torre, C.M., Nguyen, H.-Q., Taniar, D., Apduhan, B.O., Gervasi, O. (eds.) ICCSA 2013, Part III. LNCS, vol. 7973, pp. 550–565. Springer, Heidelberg (2013)

25. Tress, B., Tress, G.: Capitalising on multiplicity: A transdisciplinary systems approach to landscape research. Landscape and Urban Planning 57, 143–157 (2001)

26. Brown g., Weber D.: A place-based approach to conservation management using public participation GIS (PPGIS). Journal of Environmental Planning and Management, 1–19 (2012)

27. Carver, S., et al.: Public participation, GIS, and cyberdemocracy: Evaluating on-line spatial decision support systems. Environment and Planning B: Planning and Design 28, 907–921 (2001)

28. Kingston, R.: Public participation in local policy decision-making: The role of web-based mapping. Cartographic Journal 44(2), 138–144 (2007)

29. Beverly, J., et al.: Assessing spatial attributes of forest landscape values: An internet based participatory mapping approach. Canadian Journal of Forest Research 38(2), 289–303 (2008)

30. Brown, G., Reed, P.: Public participation GIS: A new method for national forest planning. Forest Science 55(2), 166–182 (2009)

31. Brown, G., Weber, D.: Public participation GIS: A new method for use in national park planning. Landscape and Urban Planning 102(1), 1–15 (2011)

32. Pocewicz, A., et al.: An evaluation of internet versus paper-based methods for public partici- pation geographic information systems (PPGIS). Transactions in GIS 16(1), 39–53 (2012)

33. Apulia Region: Landscape Observatory (2009),
    http://paesaggio.regione.puglia.it/index.php/osservatorio/
    introduzione.html (accessed February 6, 2014)

34. Sardinia Region: Sardinia geoblog (2011),
    http://webgis.regione.sardegna.it/sardegnageoblog/?page_id=2
    (accessed February 6, 2014)

35. Sardinia Region: LAB.net plus. Cross-border network for the enhancement of landscapes and local identities (2011), http://labnet-plus.eu/ (accessed February 6, 2014)

36. Viggiano Municipality: MUVIG. Museum of Local Traditions,
    http://www.darteq.eu/muvig/ (accessed February 6, 2014)

37. Landscape Values & PPGIS Institute: http://www.landscapevalues.org/ (ac- cessed February 6, 2014)

38. Région Pays de la Loire: http://www.patrimoine.paysdelaloire.fr/ (accessed February 6, 2014)

39. UNESCO: Recommendation on the Conservation of Historic Urban Landscape, 1st Draft for discussion (2010)

40. Attardi, R., Pisani, G., Selicato, S.: Scenario Workshop as tools for Planning the Redeve- lopment of Historic Territorial Assets. In: Scutelnicu, E.D., Rotondo, F., Varum, H. (eds.) Recent Advances in Engineering Mechanics, Structures and Urban Planning. WSEAS Press, Cambridge (2013)

41. Rotondo, F., Loconte, P.: Innovations in practice. Regional regeneration strategies for muni- cipalities networks. In: Scutelnicu, E.D., Rotondo, F., Varum, H. (eds.) Recent Advances in Engineering Mechanics, Structures and Urban Planning. WSEAS Press, Cambridge (2013)

42. Torre, C.M., Perchinunno, P., Rotondo, F.: Estimates of housing costs and housing difficulties: An application on Italian metropolitan areas. In: Housing, Housing Costs and Mortgages: Trends, Impact and Prediction, pp. 93–108. Nova Science Publications, New York (2013)

Marginality and/or Centrality of the Open Spaces in the Strategies for Upgrading the Energy Efficiency of the City

Claudia Piscitelli, Francesco Selicato, and Daniela Violante

Department of Civil Engineering and Architecture, Polytechnic of Bari
Via Orabona 4, 70125 Bari, Italy
{claudia.piscitelli,francesco.selicato}@poliba.it,
violante.daniela@gmail.com

Abstract. The control of the institutions and planners on the expansion processes has been recently lacked, above all in terms of quality and functionality. Institutions and planners have not be able to build social spaces that had the same strong identity of the historic and consolidated city, though recognizing that the search for socializing has changed and the meaning of urban spaces are changing.

Nowadays the need of producing clean energy should take a central role in urban planning as well as the need of social and efficient public space. Due to the scarcity of financial resources it is necessary to identify the priority implementation on public spaces and the best strategies to start a redevelopment process. The paper aims to reflect on these problems through the consideration obtained from the experiments conducted on the city of Bari, in the Apulia region, taken as a case study.

Keywords: Public space, GIS, Regeneration.

1 Introduction

Since the middle of the last century, the traditional planning has focused its attention, firstly, on the housing demand and on the need for a greater healthiness space settlement, then, on the services and public spaces outfit, and even after, on mobility and infrastructure. Finally, in the end of the nineties, the traditional planning reach to focus its attention on the consumption of resources and on the new environmental themes, in line with the new sustainability's development goals.

However, in contrast with the expectation, this has produced a boundless land consumption and a strong settlement diffusion around the consolidated city, generating places where the existing urban structure has been disappeared or transformed totally. In fact, the city has increased in a disorganized way, with anonymous spaces where the new standardized and repetitive building is too often temporary, and its duration is determined by the remuneration of the investments.

Therefore, the growth of the city has led to a loss of identity and values, all along recognized as significant attributes of space settlement. The marked idea of social

B. Murgante et al. (Eds.): ICCSA 2014, Part III, LNCS 8581, pp. 330–344, 2014.

nature of the historic city, which associated the architectural value of squares, palaces, parks, fountains and churches to their institutional and symbolic identities, was the key to understand how the community could feel and perceive urban spaces as their own living environment [1][2]. In fact, these spaces, physically defined by their architectural value, were carriers of an identity and were known and designed in measure or in opposition to the man; these spaces were also related to each other, forming an urban fabric.

The reasons that led to these forms of dissolution of the contemporary city, which summarizes all its drama in "Cosmopolis" by Sandercock [3], are many. The devolution of the city and the settlement diffusion are also an expression of the lifestyle and social issues changes. These are different from the past [4], but institutions and planners have not be able to properly manage the change.

Urban growth is certainly happened in a too frenetic way and in a very short period of time to be sufficiently metabolized. The control of the community institutions and planners of expansion processes is lacked, above all in terms of quality and functionality. Institutions and planners have not be able to build social spaces that had the same strong identity of the historic and consolidated city, though recognizing that the search for socializing has changed and the meaning of urban spaces are changing. It should also be noted that in traditional planning, the approach to the problems solution, related to the specific topic of interest – housing needs, mobility, space for the community, the environmental issue – was often partially practiced. Often, a global view of the discussed issues is lacked.

For some years and today, additional instances has been arisen, largely determined by the energy-intensive nature of the contemporary city (due to high energy consumption, due to new construction methods, due to the forms of settlement and different lifestyles), and related to the need for a more climatic and energy comfort, both in indoors and in the open spaces. It is clear that the two areas cannot be separately considered; the quality performances on both themes can undoubtedly improve if we take in account their mutual influence, and the urban design, even before the building design, must create the conditions [5].

Consequently, the space settlement's climatic comfort should take a central role in urban planning, like it happens in the legislation of the Apulia Region[1], located in the south of the Country. The need that these new instances are promoted and tested, in an integrated manner with those strictly referred to building, is not least, in order to avoid the mistakes of the past. It is a question to reflect on a variety of design criteria, recognizing that the complexity of the issue must inevitably lead to make choices in terms of priority interests.

The paper aims to reflect on these problems through the consideration obtained from the experiments conducted on the city of Bari, in the Apulia region, taken as a case study.

---

[1] Regional Document of the General Arrangement for making Executive Urban Plans (Delibera di Giunta Regionale n. 2751 del 14.12.2010).

2   The Redevelopment of the Open Spaces as a Tool for the Urban Regeneration of a Quarter

The perception of degradation in a given context is configured on the urban fabric, for its buildings, but not least for its voids. The sense of neglect that many public spaces - considered here as places of sociability, "the site of a long and lively debate" [6] - shown in the modern city suburbs, it helps to convey the sense of decline that is evident from an external point of view , but not only. The disaffection of citizens for the urban context in which they live - as soon as they cross the threshold of their home - is also the result of the low reception that many public spaces offer [7]. And this same disaffection becomes easily cause of further degradation, feeding so a dangerous vicious circle that ends up to keep away any possible investor [8]. In fact, "the neglect the ground project in the second half of the twentieth century is not a secondary cause of poverty semantics of the contemporary city, which results in modes of use of urban space and land more and more reductive and a less attractiveness of consolidated city compared to other forms of development such as those represented by the urban sprawl. "[9].

Public space plays a particularly important role in the urban context, because it inevitably impacts on the daily lives of those who for any reason has to cross it. But its strength mainly lies in its being multiform: the public space can separate or unite [10], stimulate or relax, be loud or silent, provided that it has its own characterization; otherwise it will carry out its worst function: to convey a sense of neglect and indolence, which will be reflected on everything and everyone around it.

On the other hand, the public administration has to have the task of giving back or providing the citizens with the spaces that can meet their unconscious urban needs. To achieve this goal it is necessary to give each public space its own connotation, which reports directly to the context and users, creating processes in the medium to long term, so as to reach a total regeneration of spaces.

2.1   Experiences of Regeneration Starting from the Public Space

In the regeneration policies of the last decades, the public space has become a key element for revitalizing an area more or less wide and to give a new incentive and a new look - not just formal - to a whole part of the city. In fact there are many interventions that focuses attention on urban design of small-scale included in an urban vision more generally, with particular reference to the redevelopment "within a predefined plot of public spaces, open spaces, interstices of constructed, connecting pedestrian and vehicular paths between the various places of living" [11]. The New Deal for Community Programme, launched in 1998 by the British government, is one of the first successful experiences more oriented in this direction.

A neighbourhood that has benefited from the program was the London district of Islington. An in-depth study based on technical analysis and community participation, which lasted about three years, has allowed us to highlight the problems that gave a poor urban quality to the district: low quality of public spaces, poor allocation of spaces for pedestrians and activities for socialization, large paved surfaces and lack of green areas, many residual spaces between those public and private, areas with a physical configuration that lent itself to the phenomena of crime, many rough surfaces that allow a mobility difficult for the weaker sections of the population (elderly, children, disabled). The strategy that ensued saw a long period of cooperation between the Commission of the New Deal and the Council of the District of Islington, from the identification of those areas that needed more substantial implementations and that, being in a strategic position in the district, represented the connective tissue between its single parts, as well as between them and the surrounding districts [12]. The map of the analysis shows the increased emphasis given to those spaces that connected residences, commercial activities, parks and facilities. The main structure of the strategy was also represented by a continuous green mesh, designed with the triple function to create a greater sense of presence of green and open spaces in a medium - high density area, which helps to give an identity to the whole district and finally defines a basic framework to guide possible future project activities.

3   Climatic and Environmental Issues in Redeveloping Public Spaces

Europe has undertaken an ambitious road that have to aim the sustainable development and the fight against climate change and puts it on world leader in difficult international discussion. It is using a strategy that searches all possible alliances for a development model that ensures global sustainability and preserve current levels of well-being and environmental quality. The cities are associated in programs that move forward the fight against climate change in favor of a new kind of quality of life. Prototypes of this concept are the Covenant of Majors[2], independent initiative of the municipalities of Europe regarding the most challenging objectives of mitigation of climate-change emissions[3] and the Smart City Network Program sponsored by the Strategic Energy Technology Plan called "SET-Plan"[4]. This SET-Plan defines the concept of Smart Cities, as "cities and metropolitan areas that are taking appropriate measures to the 40% reduction of greenhouse gas emissions by 2020, through the use and production sustainable energy. The main components of the measures to be taken regard implementations on buildings, local energy networks and transport system "[13].

This definition is very focused on the energy component to the detriment, perhaps, other peculiar aspects of the concept of Smart Cities, such as ones related to the ecological improvement of the urban environment (Eco-Towns), or ones related to social

---

[2] 3721 cities had signed in March 2012, the Covenant of Mayors to increase energy efficiency and use of renewable energy sources in order to exceed the EU target of 20% reduction in CO2 emissions by 2020 .

[3] The veto of Poland, which has 95% of electricity generation from coal, has prevented in March 9, 2012 at the Council to adopt the most stringent emission reduction aims beyond 2020 provided by the Roadmap 2050.

[4] SET-Plan, adopted in 2008, the technology cornerstone of the EU's climate and energy policies.

issues (Social Sustainability). In the opinion of several researchers [14], it may include at least two essential characters of the Smart Cities concept:

The widespread use of ICT, such as infrastructure that conveys intangible flows of information and knowledge;

The regard of social capital, in other words the increase of competences, creativity and social inclusion of citizens, through their involvement.

The idea of a smart city should not be a utopian aim to reach at all costs and in any way; it should not be the only driving force of a complex planning process of the city. Thus, an abuse of this concept could cause even a failure of urban planning. Pursuing the only objective of energy-efficient city could move away the designer from the idea of the "city of man". This idea includes some considerations that may also be in contrast with the idea beforehand exposed.

A clear example is precisely expressed in this paper: the public space. The assessments for an energy upgrading will be very different if we consider as a priority the social centrality that a space has to assume. It has to be considered at the same time aspects that refer to the production of energy from renewable sources and the physical and morphological factors that will make the space a place for socializing. There are two categories of factors that affect the relationship between the energy and the public space:

    parameters related to the user:
-   biological-physiological conditions (sex, age ..);
- Sociological conditions;
- Psychological conditions;
environmental parameters related to the physical space:
- morphology of space;
- materials used;
- introduction of urban green spaces. [15]

In order to make a project oriented to the production of clean energy, in a preliminary assessment of the project it is evident that the second category of parameters have been taken in account and studied through a climatic analysis, in order to obtain the best production efficiency of the implementation. The climatic analysis aims to identify the level of sustainable energy efficiency that the site could express. The main climatic parameters on urban scale are:

    shape parameters;
    Sky View Factor (SVF) and Morphologic Protection Index (MPI);
    H/D/L (respectively, height, width and distance of the buildings that are on an urban space).

It is therefore necessary to characterize the site under consideration in relation to environmental parameters that affect the conditions of the urban space and energy efficiency, in relation to the project elements that interfere with these ones.

4   Defining Priority Areas of Implementation: Urban Planning and Energy Efficiency

With regards to the importance of the role of public spaces in the regeneration of a neighbourhood, it is necessary to select the priority areas to be tackled, which can actually play a strategic role. This evaluation may be assigned primarily to a qualitative research on all urban aspects which interact with the public space system in order to establish a new urban quality of an area, associated with a new physical, social and pedestrian district. The qualitative-urban considerations can be linked with considerations and quantitative analysis related to the potential energy of the same open spaces, with the aim of projecting the intervention towards sustainability and clean energy production. The results about the priority areas of intervention for each of the two analyzes could be overlapped or otherwise totally different.

4.1 Urban Issues

One of the aspects of urban planning from which the analysis of the criticalities and potentialities of an area referring to the public space is certainly mobility. The public space to be implemented, in fact, has to be a link between existing routes and / or potential ones, especially in bicycle and pedestrian path. Furthermore, even if many contemporary suburbs were designed according to the vehicular traffic, the public space can represent today the primary tool which can allow the citizen - and the pedestrian - to reclaim urban spaces that are truly usable and liveable, rather than only suitable to a quick and careless crossing by car. In function of this, it is also important to analyze the size of those urban voids currently present and that represent potential public spaces, in order to understand which already today, for its configuration and localization , lend themselves to a redesign of the open space . The others, however, due to their large size, could require priority of an implementation of a volumetric infill in order to compose with the new volumes a new configuration of the empty spaces, which will then be designed only at a later time. The paths of soft mobility must intercept the open spaces that are so valued as real potential object of implementation, and must create a connection between them and the residential areas, often nowadays introverted and enclosed. The aim should be to increase the permeability of the entire pedestrian area, linking them with the currently mono-functional areas, pursuing the goal of the integration between contexts and functions. In this context , it becomes essential to analyze also the current existing service areas, commercial facilities , public green areas, and , in general, any aggregation point that represents today a centre of attraction in the neighbourhood or in the urban scale, and that is another potential point of the system of the public spaces of the project.

The single implementation that includes the transformation of an urban void in a public space in this way becomes truly strategic, as contextualized and immersed in a logic of larger scale, which has been identified as a starting point to be tackled as a priority, and from which it is possible to derive in a long-term vision the whole system of district assumed.

4.2 Energy Issues

The identification of areas to be implemented concerning the preferential energetic character, in which the choice of a redevelopment of the urban fabric is exclusively based on factors related to energy production, is entrust to several criteria and instruments.

GIS (Geographic Information System) is an efficient instrument to acquire the data necessary for the selection of the best areas for the energy improvement. If we talk about a energy requalification for the production of electricity through solar radiation, the criteria, or rather the different layers that are extracted from the planning software, to choose the areas, can be summarized in:

Surface temperature (°K) with respect to time dates that identify the year seasons;

Sky View Factor (SFV), which represents the portion of sky visible from a point on the earth's surface considering a semi-spherical cap with a radius defined by the user;

Morphometric Protection Index (MPI), related to protection that the building elements observe towards an open space.

Using GIS software, the work is based on a grid that can have different sizes depending on the accuracy of the data. A value of a single feature correspond to each cell of the grids. Usually the open areas to be studied have an area greater than that of the reference cell, and therefore an open space is defined by multiple outputs of the same parameter. For this it is necessary to calculate the weighted average according to the area intersected by spaces; this in order to obtain a single coefficient related to the open space for the individual parameters.

Another criteria of analysis is the study of the solar incident radiation, qualitative and quantitative, on open spaces. Thanks to simulation software it is possible to check the movement of the sun with its shadow mapping in relation to the urban fabric and quantify the power compared to the time unit of the solar incident radiation on the areas in question. The simulation is performed with respect to the temporal dates and the most important times of the day defined by the designer. It highlights areas which benefit from direct radiation and therefore which areas are the best to accommodate solar collection elements.

The solar quantitative analysis is defined, instead, by measuring the intensity of the radiation incident on the surfaces of the analysis model. Every open space, ideally configured by a plane, is questioned about the power of the radiation, usually referring to the monthly total, divided for each hour of the day.

These criteria generate a new idea of planning, and urban spaces may represent appropriate sites for the implementation of pilot projects aimed at upgrading the energy efficiency and integration of technologies for the use of renewable sources.

5   The Case Study: Bari-Poggiofranco

5.1   The Context

The case study chosen is the neighbourhood Poggiofranco in Bari, the Regional capital the Apulia Region (Italy). The shape of the current city is primarily related to three significant historical periods: the period of the nineteenth century, the modern period that starts from the early twentieth century until after World War II and the contemporary period. The neighbourhood Poggiofranco is located in the last one. It is characterized by the high speed of realization, poorly reflective and more related to intentionality speculative [16].

The neighbourhood is divided in two parts by an important arterial road and a large urban void. The high area of the district, closer to the city center, is more orderly; the lower zone has no rules instead: uneven shapes with high buildings with irregular forms stand without any relationship with the surrounding context.

5.2   The Case Study: Analysis of Urban Parameters

The district Poggiofranco, especially its south-part, looks like a typical product of the Modernist school. Indeed, it seems like a transposition into the reality of the "absence of a meaningful experience and systematic open space [...] enormously dilated", so they seem "into dust in a series of episodic fragments connected to each other by spaces devoid of a clear Statute" [17].

Analyzing the urban fabric we see a strong heterogeneity of morphologies and directions: individual actions are in fact often unrelated to each other, disconnection further emphasized by the fences that help to isolate each batch from its context, accentuating the introverted and independent blocks, totally detached from the context adjacent and with no intent to give rise to paths continuity. The compactness of the morphology, as well as the density, is decreasing from north to south, where the residential fabric seems to disintegrate gradually. If the north area has a recognizable street pattern that attempts to interface with the built environment and public spaces create some linear pedestrian paths, the vast distances that separate the sections of road built in the south deny any attempt to connection on a human scale.

The entire path of roads seems designed on the basis of vehicular traffic at the expense of pedestrians and cyclists . The system is based on a considerable section of the main roads on which there are the secondary access to the residences. The spaces and paths appear dilated, considerable distant each other, partly because of the significant size of the lots and single-purpose residential, which cannot be crossed by pedestrians if they are not residents, force pedestrians to use only the roads that surround them and make the whole area impervious to fruition. The pedestrians are also disadvantaged because of the high speed of the vehicles which is also allowed by the large stretches of the roads. Although the Municipal Urban Plan included for this neighbourhood a provision of minimum services for residents - a few years before the drafting of the Plan it has been approved the law DI 1444/68 which compelled the plans to verify the minimum standard of 18 square meters per inhabitant for facilities -, its implementation has revealed an almost total ineffectiveness. The facilities are in fact undersized for the number of residents because most of those ones included in the plan were never realized. Those ones realized are characterized by a substantial mono-functionality - they are mostly schools. The facilities are concentrated in a few points (with reference to the offices) while commercial activities, which are necessary intermediaries between the public space and the built environment, are deficient and fail to create real continuous paths with high usability, because of their sporadic presence and the enormous distances interposed. Furthermore, the broad streets and numerous parking lots allow people to go near the shops with their own private car, make their purchase and get away with the same private vehicle. As a consequence, there are not activities and interactions between citizens and public space [18] - where the public space is also intended as the street and the sidewalk - that contribute to the lack of pedestrian vitality and usability of the district in its entirety.

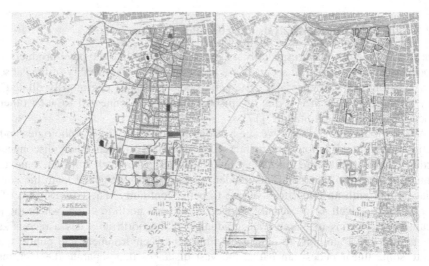

Fig. 1. Open spaces and commercial activities in Poggiofranco neighbourhood in Bari

The qualitative analyses have focused priority of implementation on some residual spaces within the core of the district, between the macro-residential blocks, characterized by a restricted size and proximity to business activities, offices, existing green areas and public spaces, which may be all involved in a system of pedestrian paths that can also effectively cover the areas currently open and wide as well as the private areas in the buildings surroundings.

5.3   The Case Study: Analysis of Energy Parameters

The analysis about the best spaces to produce solar energy starts with the identification of all open spaces in the area under consideration. In order to outline all the areas it has been used the overlap of the digital map of the Apulia Region and the orthophoto, assigning a code more readable to every localized area. In addition to the residual spaces squares, the gardens of the neighbourhood have been also mapped; that is for a completeness of data collected (fig.2).

After the survey operation, we proceeded with the characterization of the spaces from the energy point of view. We have considered the following parameters: surface temperature compared to four temporal dates (12.02.2006, 09.05.2003, 06.07.2001, 08.09.2001); Sky View Factor (SFV) and Morphometric Protection Index (MPI). For each mapped open space, we obtained a numerical data for each single parameter, extracting it from the GIS software. The use of these quantitative parameters is

Fig. 2. Map of the open spaces analyzed in Poggiofranco neighbourhood in Bari

dictated by the current availability of data. There is a clear shortage of these factors and therefore it is necessary an action for the periodic monitoring of these.

In the next step all the values have been sorted in descending order, with the exception of the MPI, which follows the opposite order[5]. The aim is to obtain some areas that fall regularly in all the top positions of the six different parameters. We looked for an objective assessment to delimit the range of "top positions", leading to the following method: for each parameter, we calculated the mean value and the range to consider includes all values above average (for MPI, we evaluated the data below the mean value). The Table 1 summarizes all the values.

The areas identified are close each other. These areas are decentralized from the dense built fabric and are in proximity to an important highway of the city. It emphasizes that the intense traffic flow could play a significant role on the high temperature (fig.3).

This assessment is combined with the solar qualitative and quantitative analysis. In relation to the areas identified by the first analysis, we have studied the maps of shading and calculated the incident solar radiation expressed in kW/h. These considerations have been generated by the modular software Autodesk Ecotect Analysis. The first step in using the tools of software Ecotect, consists in realizing the three-dimensional model of the structure that we want to investigate, inputting the materials and geometries characteristics of the buildings. In the case under analysis, we considered a volumetric schematic model, without windows, with average heights and materials properties dictated by the user. For this input what is useful for solar studying is the reflection coefficient of the external surfaces of buildings. These coefficients were retrieved from the document "Protocollo Itaca Puglia"[6].

---

[5] A high temperature and SVF correspond to a low protection that buildings make on an open space.

[6] Edited with the assistance of technical and scientific ITC – CNR, iiSBE Italia and Environment Park.

Table 1. Sorted data. The range above the mean value is identified by the frame in black.

Cod	T 12.02.06	Cod	T 09.05.03	Cod	T 06.07.01	Cod	T 08.09.01	Cod	MPI	Cod	SVF
P_6	302,52	P_6	311,50	P_15	313,23	P_28	311,24	P_28	0,1580	P_15	0,9954
P_28	288,00	P_15	309,92	P_23	312,62	P_23	310,10	P_12	0,1954	P_23	0,9953
P_23	287,86	P_19	309,33	P_22	312,07	P_15	309,29	P_15	0,2069	P_13	0,9942
P_15	287,75	P_21	309,12	P_28	311,96	P_22	308,56	P_23	0,2282	P_28	0,9936
P_27	287,72	P_11	309,03	P_17	311,74	P_1	308,42	P_11	0,2435	P_21	0,9929
P_24	287,04	P_1	308,94	P_24	311,67	P_27	308,27	P_21	0,2572	P_12	0,9925
P_19	287,00	P_27	308,63	P_18	311,58	P_19	308,09	P_13	0,2584	P_11	0,9917
P_21	286,96	P_23	308,59	P_1	311,45	P_24	308,01	P_9	0,2611	P_17	0,9911
P_22	286,91	P_10	308,57	P_10	311,31	P_7	307,68	P_1	0,2705	P_6	0,9898
P_17	286,80	P_14	308,55	P_19	311,23	P_10	307,54	P_20	0,2773	P_20	0,9874
P_10	286,80	P_20	308,35	P_11	311,22	P_11	307,47	P_17	0,2792	P_9	0,9858
P_7	286,72	P_7	308,29	P_27	311,14	P_20	307,33	P_6	0,2833	P_7	0,9854
P_16	286,66	P_22	308,24	P_21	310,96	P_12	307,08	P_24	0,2837	P_5	0,9835
P_1	286,64	P_28	308,18	P_7	310,84	P_9	307,07	P_27	0,3006	P_22	0,9820
P_14	286,60	P_12	308,12	P_16	310,80	P_17	306,96	P_7	0,3229	P_24	0,9818
P_20	286,58	P_18	307,84	P_20	310,68	P_16	306,87	P_5	0,3277	P_3	0,9803
P_11	286,50	P_2	307,81	P_9	310,48	P_13	306,78	P_10	0,3335	P_2	0,9793
P_13	286,48	P_16	307,67	P_14	310,39	P_18	306,77	P_4	0,3498	P_10	0,9711
P_18	286,43	P_9	307,66	P_13	310,38	P_21	306,75	P_14	0,3502	P_1	0,9696
P_9	286,42	P_17	307,65	P_25	310,28	P_6	306,69	P_2	0,3508	P_25	0,9694
P_26	286,26	P_5	307,46	P_6	310,08	P_14	306,05	P_19	0,3570	P_16	0,9649
P_5	286,19	P_4	307,45	P_4	309,37	P_5	305,81	P_22	0,3629	P_26	0,9626
P_2	286,10	P_13	307,41	P_5	309,35	P_25	305,25	P_18	0,3671	P_4	0,9595
P_25	285,87	P_3	307,35	P_26	309,34	P_26	305,01	P_3	0,3965	P_19	0,9593
P_3	285,87	P_25	306,95	P_12	309,06	P_2	304,81	P_25	0,4037	P_27	0,9579
P_12	285,64	P_24	306,52	P_3	309,04	P_4	304,72	P_16	0,4112	P_14	0,9355
P_4	285,57	P_26	306,45	P_2	308,72	P_3	304,39	P_26	0,4438	P_18	0,9342
P_8	285,45	P_8	305,76	P_8	308,29	P_8	303,90	P_8	0,4945	P_8	0,9230

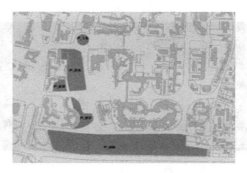

Fig. 3. Extract of open spaces with the best energy characteristics of the neighborhood

The simulation of solar radiation with its shadows has been performed with respect to the four dates obtained from thermal maps (12:02, 09:05, 06:07 and 08:09), in the most significant times of the day, 9:00, 12:00, 16:00, 18:00 (Fig.4).

The shadows study points out as the open spaces mapped result to be sufficiently distant from the built and then they benefit throughout the year of the direct radiation.

For the incident radiation relative to the total monthly, the software generates a colour diagram and a table of data for each open space: for the first one there are eleven intervals, identified by a different colour which coincides to a value in Wh (Fig.5); in the data table there are the numerical values for the different months of the year.

Fig. 4. Representation of the shadows at certain times of the days considered: 12 february hours 9.00, 09 may hours 12.00, 06 july hours 16.00, 08 september hours 18.00

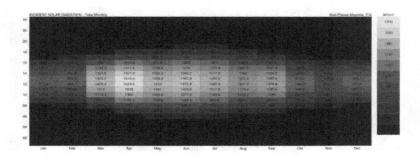

Fig. 5. Sun exposure related to space P_16

We can summarize the overall data for each space object of simulation in Table 2.

Table 2. Summary exposure data of the areas affected by the calculation of the incident radiation

AREA	TOT. INCIDENT RADIATION to mq [KWh/mq/year]	SURFACE [mq]	TOT. INCIDENT RADIATION [KWh/anno]
P_16	1247,882	11.342,55	14154157,77
P_22	1276,897	5.349,71	6831028,65
P_23	1308,572	24.489,99	32046915,19
P_24	1297,272	11.786,42	15290188,75
P_27	1262,771	11.036,79	13936933,29
P_28	1302,951	103.855,82	135319044,5

The values of the radiation incident on the areas are also the basis for the design of the energy production elements. With kWh / m² per year we have the opportunity to choose photovoltaic technologies and design the urban elements that will house these technologies and assess the productivity of regenerated areas.

Thanks to the analysis we can conclude that the best areas for energy improvement are the spaces of large dimensions, which appear to be marginal to the built environment. This aspect is correlated with the urban morphology: a fabric more open, where the buildings are less dense, it is usually closer to the countryside, away from the core of the city. This feature is essential for upgrading the energy efficiency because it is necessary to have a high SVF and therefore low MPI (and consequently high temperatures) to achieve high values of incident solar radiation. It must be emphasized, however, that this characterization of the environmental elements of open spaces (guideline of methodology) is the antithesis of the criteria of urban comfort and well-being of users.

6   Conclusions

The growing interest related to the themes of suburbs regeneration has led researchers and professionals to focus on this subject. While suburban area can mean high social and physical exclusion - so not necessarily placed at the edge of the city from a geographical point of view (LR21/2008 Apulia Region) - on the other hand it is often found that the areas which are more affected by degradation, are located just behind physical margins between the town and the country. The open and residual spaces in these areas are presented with highly heterogeneous features: on the one hand spaces between built volumes, without a precise configuration and identity, without a defined user and an intentionally designed shape; on the other hand wide open spaces through which the city overlooks the countryside. The regeneration of urban areas today cannot ignore the revitalization of open spaces and the Administrations are increasingly faced with the opportunity to leverage public funds - primarily deriving from the European Union - to redevelop parts of city. The scarcity of current resources also forces themselves authorities to carefully select the areas on which to operate, in an attempt to achieve maximum effectiveness in the narrow context but also positive implications on the entire scale of the neighbourhood, as well as on the urban context. The evaluations that can be made are of different nature: on one hand, the benefit must be closely linked to the quality of urban life meant as the liveability of the context, occupant comfort, usability of space; on the other hand we cannot ignore considerations that invest the energy-environment issue. The redevelopment of open spaces is, in fact, an opportunity for energy and smart reform of the city, an opportunity to undertake initiatives to see the city as a resources producer. The two dimensions of evaluation of open spaces to the intervention priorities (the qualitative-urban planning and the quantitative-energy one) sometimes lead to conflicting results, as demonstrated by the case study analyzed. From a urban point of view it is shown, in fact, how it is appropriate that the regeneration process originates from the redevelopment of areas in the immediate proximity to buildings on a human scale, which can be interconnected to pedestrian activities. On the contrary, the greater energy efficiency is ensured from the most marginal and large size spaces, far from the shading of the buildings and in which the temperature and solar radiation are more effective. In conclusion, the case study demonstrates how the energy requalification can be better interfaced with strategies for regeneration of physical urban margins, strategically considering their effectiveness and their role in an urban scale rather than on a neighbourhood scale. The project of these spaces allows to create dialogue between city and country to implement cost-effective implementations in terms of resources allocated to the energy production. So rather than conflict we can speak simply of different strategic dimensions of the intervention, in which before defining the priority areas of implementation, we must set the objectives: on one hand the energy redevelopment of huge marginal public spaces concerning the urban scale, on the other hand the improvement of the social system on the neighbourhood scale. Certainly this does not exclude an interaction between the two dimensions, but we intend to focus the attention on how interventions should be the result of analyses finalized to a specific objective that, depending on the context, can change in order to not waste resources.

Notes. This paper is the result of authors' joint work. Contributions come as follows: By Claudia Piscitelli: Sections 2-4-4.1-5.2-6; by Francesco Selicato: Sections 1; by Daniela Violante: Sections 3-4.2-5.1-5.3.

References

1. Magnaghi, A.: Il territorio dell'abitare, Franco Angeli, Milano (1994)
2. Magnaghi, A.: Il progetto locale. Verso la coscienza di luogo, Bollati Boringhieri, Torino (2000)
3. Sandercock, L.: Towards Cosmopolis. Wiley, New York (1998)
4. Hall, P.: Megacittà, città-mondo e città globali. Urbanistica (116), 11–28 (2001)
5. Loconte, P., Ceppi, C., Lubisco, G., Mancini, F., Piscitelli, C., Selicato, F.: Climate alteration in the metropolitan area of bari: Temperatures and relationship with characters of urban context. In: Murgante, B., Gervasi, O., Misra, S., Nedjah, N., Rocha, A.M.A.C., Taniar, D., Apduhan, B.O. (eds.) ICCSA 2012, Part II. LNCS, vol. 7334, pp. 517–531. Springer, Heidelberg (2012)
6. Purini, F.: Spazio pubblico e conflitto, Lectio Magistralis presso Master di secondo Livello in Progetto dello spazio pubblico, Università di Pisa (2011)
7. Gehl, J.: Vita in città, Maggioli Editore, Santarcangelo di Romagna (2012)
8. Carmona, M.: The Value of Public Space, Cabe Space, London (2004)
9. Secchi, B.: Progetto di suolo 2 in Aldo Aymonino, Valerio Paolo Mosco Spazi pubblici contemporanei. Architettura a volume zero, Skira, Milano (2006-2008)
10. Gabellini, P.: Tecniche Urbanistiche, Carocci, Roma (2001)
11. Rotondo, F., Selicato, F.: Progettazione Urbanistica. McGraw Hill, Milano (2010)
12. Kessler, L.: Public Space Strategy. A plan for parks, streets and estates in the EC1 New Deal for Communities area, Report, Islington, London (2004)
13. Sadler, B., Verheem, R.: Strategic Environmental Assessment Status, Challenges and Future Directions. Ministry of Housing, Spatial Planning and the Environment WP 54, The Hague, Netherlands (1996)
14. Partidário, M.R.: Elements of an SEA framework. Improving the added-value of SEA. Environmental Impact Assessment Review 20(6), 647–663 (2000)
15. Sheate, W.R.: Tools, techniques & approaches for sustainability. World Scientific Publishing, Singapore (2010)
16. Selicato, F.: Bari. Morfogenesi dello spazio urbano, Adda Editore, Bari (2003)
17. Secchi, B.: La città del ventesimo secolo, Editori Laterza, Bari (2005)
18. Gehl, J.: Kaefer, Reigstad (2006)

Sensitivity Analysis of Spatial Autocorrelation Using Distinct Geometrical Settings: Guidelines for the Urban Econometrician

Antonio Manuel Rodrigues and Jose Antonio Tenedorio

e-GEO - Faculdade de Ciencias Sociais e Humanas
Universidade Nova de Lisboa
amrodrigues@fcsh.unl.pt,
http://www.fcsh.unl.pt/e-geo/?q=en

Abstract. Inferences based on spatial analysis of areal data depend greatly on the method used to quantify the degree of proximity between spatial units - regions. These proximity measures are normally organized in the form of weights matrices, which are used to obtain statistics that take into account neighbourhood relations between agents. In any scientific field where the focus is on human behaviour, areal datasets are immensely relevant since this is the most common form of data collection (normally as count data). The method or schema used to divide a continuous spatial surface into sets of discrete units influence inferences about geographical and social phenomena, mainly because these units are neither homogeneous nor regular. This article tests the effect of different geometrical data aggregation schemas on global spatial autocorrelation statistics. Two geographical variables are taken into account: scale (resolution) and form (regularity). This is achieved through the use of different aggregation levels and geometrical schemas. Five different datasets are used, all representing the distribution of resident population aggregated for two study areas, with the objective of consistently test the effect of different spatial aggregation schemas.

Keywords: Spatial Autocorrelation, spatial weights matrix, spillover effects.

1 Introduction

The common aspect between any geographical phenomena is the possibility to identify its location relative to any model which represents the surface of the Earth. Moreover, any individual or collective action is conditioned (and conditions) actions taken by agents located nearby [15]. Hence, we may speak of a contagious or spillover effect which result in clear spatial patterns. Distance is a chief component in any geographical model, because it quantifies the relation in space between intervening agents.

In quantitative analysis of geographical data, it is common for distance between intervening agents to be compiled in the form of a weights matrix wich

B. Murgante et al. (Eds.): ICCSA 2014, Part III, LNCS 8581, pp. 345–356, 2014.

quantifies neighbouhood relations [5,2,12,9,10,13,3]; yet, there is still muc work to be done in the field in order fully undersand the theoretical and practical impications of choosing particular specifications of the weigths matrices in detriment of others [11]. The present article aims at providing a clear understanding of the impact of distinct specifications when two variables - scale and form - vary.

Most human activities are continuous in nature. Individual actions and in general spatial individual attributes (eg. where someone works, lives or goes shopping) take place in a specific point-location. Yet, in the Geographical Information Sciences (GIS), it is common for spatial information to be compiled and made available for a set in spatial units/regions [2]. Available datasets aggregate such attributes according to pre-defined geometrical regional settings. Geography is technically absent from these datasets as each record is independent [9]. Because of this, a set of Exploratory Spatial Data Analysis (ESDA) tools has been developed which measure the relations between these units [2]. The backbone of ESDA statistics is the use of spatial weights matrices where each cell represents the geographical relation between a pair of spatial units [9,4,3]. In recent years, the field of spatial autocorrelation and spatial weights have gained attention in ecological studies with the greater availability of spatial datasets [7,14].

The same variable, the same spatial phenomenon may be studied using different levels of aggregation. This is analogous to the concept of image resolution. Hence, the term *geographical* or *spatial resolution* may be used as referring to how detailed aggregates are. In practice, for a fixed atudy-area, resolution is higher the greater the number of spatial units for which counts are compiled.

The consequence of variations in spatial resolution and level of regularity is that an exploratory analysis of the same phenomenon may yield different results. Although the aggregation level may lead to distinct conclusions, these are not necessarily wrong; nonetheless, awareness for such differences is important as spatial diffusion occurs at different speeds according to the level of detail. In effect, diffusion speed may be the same, only distances are different.

There are several problems related to these aggregation exercises and the resulting geometrical schemas: first, using summary statistics which refer to the average behaviour of individuals or the sum of some demographic attribute may lead to mis-interpretations of social phenomena since the internal area of each spatial unit is not homogeneous. This is knnown in the literature as *ecological fallacy* [16]. In practice, aggregation means that part of the information is lost. Moreover, different regional schema may result in different analysis.

The second problem highlights the consequences of using different geometrical schema and is known in the literature as the Modifiable Areal Unit Problem MAUP [2,8]. But why is there a multitude of geometrical boundaries which aggregate data for the same geographical plane? Different criteria or motivations normally result in different sets of boundaries and spatial forms. Regions may be defined as natural units (ex. a river basin), functional units (ex. a metropolitan area), anthropic units (ex. agricultural landscapes) or administrative units (ex. Nomenclature of Territorial Units for Statistics NUTS).

Geographical scale (resolution) and in particular form (regularity) should be taken into account in any strategy which aim at choosing an adequate specification of neighbourhood relations. This is so because the degree in which (human) phenomena is clustered or spread-out in space - the degree of *compacity* - affects contagious behaviour. The compacity properties of a region can be measured as a proportion to its equal-area circle [1]. As Angel, Parent and Civco write in their 2010 essay, although there are many ways of measuring compactness, many of these properties are highly correlated [1]. Within the present work, it is suffice yet important to understand that geometric regularity as a form of compacteness affects geographical similarity and hence spatial autocorrelation measures. If we understand the circle as the most compact form in nature then, as recognized long ago, the hexagonal grid is the most regular collection of contiguous geometrical entities (in a two-dimensional world) [6]. Hence, the hexagon is used in the present study and a reference shape representing greatest regularity.

One principle which is of paramount importance to this whole theme field is the assumption that individual attributes and actions are in general not independent in space; there are spatial patterns associated with each phenomenon which may be quantified using spatial autocorrelation metrics. These quantify the relation between observed quantities (aggregated for a set of spatial units) and the weighted mean of the same variable for a set of neighbours - spatial lags. These exploratory measures are highly dependent on the concept of neighbourhood used or the way distance between units is quantified.

Methods used to measure distances range from the existence of contiguity to the quantification of time-distances between centroids or urban nodes. The fact that there are no clear and formal methodologies which can assist in the choice of the most adequate proximity measurement hampers robustness of statistical analysis. For this reason, as mentioned above, this is a field where further research is needed. This article examines the effect of distinct spatial data aggregation schemes on measures of spatial autocorrelation. Hence, the main objective is to test the stability of spatial autocorrelation measures to different methods of spatial data aggregation and to the use of different distance metrics. Using different specifications for the spatial weights matrix, sensitivity to geographical scale and to geometric (ir)regularity will be tested using sets of administrative regions and regular hexagonal grids overlaid on the study-areas. Observed variations in global spatial autocorrelation statistics contribute as guidelines for further research.

The next section starts by describing the study-area and datasets used, as well as the study area. This is be followed by a brief discussion of the types of spatial weights matrices used as well as the global spatial autocorrelation statistics. Section 3 describes the results. Section 4 concludes

2   Methodology

The methodology developed was focused on the objective of quantifying the sensitivity of inferences made in respect to any anthropic or socio-economic variable, given the geographic concentration/dispersion of data. The study of spatial phenomena is particularly sensible to the way proximity between regions is measured.

The focus of the present research are two distinct geographical settings: Continental Portugal (mixed urban-rural) and the Lisbon municipality (urban). For each, the distribution of resident population (Census 2001 data) is the variable of interest. Resident population, rather than population density was chosen since one objective of the study is to examine how regional irregularities and geographical scale affect spatial analysis.

2.1 Study-Area and Data

The spatial independence of observations will be tested in order to identify existing spatial patterns. For each study area, the distribution of the variable of interest will be examined for different geographical levels of aggregation (resolutions): in respect to Continental Portugal, the datasets used correspond to the NUTS 3 regions (28 spatial units) and the municipalities - or local councils - (278). An extra dataset will be used, which results from the aggregation of the data from the highest disaggregation level (over 170 thousand census tracts) according to a set of 278 regular hexagons (figure 1) . In respect to the Lisbon area, two geographical datasets will be compared: one related to the local authorities - freguesias, and a second which results also from the aggregation of resident population from census tracts according to a hexagonal grid (figure 2).

Sensitivity of global autocorrelation measures result in potentially distinct spatial patterns. While using different administrative levels allows testing the effect of geographical scale (resolution), the use of an hexagonal grid allows testing the effect of geometrical (ir)regularities (MAUP).

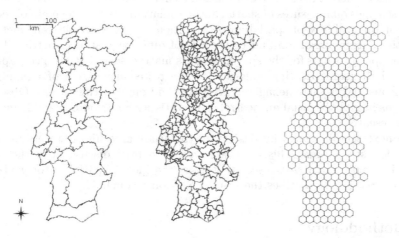

Fig. 1. Continental Portugal: NUTS3 (n=28), municipalities / local councils (n=278), hexagonal grid (n=278)

The objective of using hexagonal grids is to compare how two different geometrical schemas - with the same number of spatial units - yield different spatial autocorrelation patterns. Data from the 2001 Census was allocated to each hexagon based on the centroid of each census tract. In the case of Continental Portugal, the spatial surface is divided into over 170 thousand, whilst in the case of the Lisbon municipality this number is 4389. The large number of census tracts ensure spatial allocation based on the centroids rather than on each polygon does not carry significant distortions. Other methods could have been used for the allocation of data from census tracts according; however, the high spatial resolution of raw data guarantees robustness of the results.

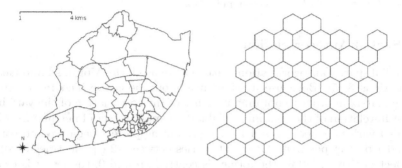

Fig. 2. Lisbon municipality: local authorities (n=53), hexagonal grid (n=53)

2.2   Spatial Weights

As mentioned, The study of the spatial structure associated with the variable of interest is strongly dependent on the form chosen of the spatial weights matrix (W), where the w_{ij} element represents the proximity between spatial units i and j. Two common ways of defining a proximity relationship between regions are through a contiguity matrix (W^f) and through a nearest neighbours matrix (W^K). In terms of contiguity matrices, it is possible to choose a binary specification, where $w_{ij} = 1$ when i and j are contiguous or zero otherwise (values in the main diagonal are also set to zero). This binary specification is a special case of a more general kind of contiguity matrix, where each connection may be represented in a number of forms, the most informative of which is that where each element different from zero is equal to the length of the common border between neighbours. Formally, the weights matrix takes the form:

$$W^f = \begin{cases} w_{ij} = 0 & \text{if } i = j \\ w_{ij} = f_{ij} & \text{if } f_{ij} \neq 0 \\ w_{ij} = 0 & \text{if } f_{ij} = 0 \end{cases} \tag{1}$$

where f_{ij} represents the common border between regions i and j. In the present study, a binary specification was chosen, since one of the spatial structures analysed is a regular hexagonal grid, in which case contiguity measures are equal across the spatial plane. For the same reason, the chosen nearest neighbours matrix is a binary matrix, where elements w_{ij} is equal to one when j belongs to the set of k neighbours. Formally:

$$W^k = \begin{cases} w_{ij} = 0 & \text{if } i = j \\ w_{ij} = 1 & \text{if } j \in k \\ w_{ij} = 0 & \text{if } j \notin k \end{cases} \tag{2}$$

where k is the set of i's nearest neighbours.

2.3 Spatial Autocorrelation

Two global spatial autocorrelation statistics are used: the Moran's I and Geary's C -equations 3 [5]. Both use a set of neighbourhoods specified in the spatial weights matrix in order to obtain a measure of the co-variation of the variable of interest in relation of the spatially weighted mean of the set of neighbours. While Moran's I varies between -1 and 1 (-1 representing perfect negative autocorrelation and 1 perfect positive), Geary's C varies between -2 (perfect negative) and 0 (perfect positive). While the former is centred around 0, the latter is centred around 1.

$$I = \frac{n}{\sum_{i=1}^{n}\sum_{j=1}^{n} w_{ij}} \frac{\sum_{i=1}^{n}\sum_{j=1}^{n} w_{ij}(x_i - \bar{x})(x_j - \bar{x})}{\sum_{i=1}^{n} w_{ij}(x_i - \bar{x})^2} \tag{3}$$

$$C = \frac{(n-1)}{2\sum_{i=1}^{n}\sum_{j=1}^{n} w_{ij}} \frac{\sum_{i=1}^{n}\sum_{j=1}^{n} w_{ij}(x_i - x_j)}{\sum_{i=1}^{n}(x_i - \bar{x})^2} \tag{4}$$

3 Results

The visualization of spatial patterns in respect to the variable of interest is dependent on scale and form. Figures 3 and 4 represent the geographical and statistical distribution of resident population respectively for Continental Portugal and Lisbon municipality. In relation to the former, data is aggregated for 278 municipalities, whilst in relation to the latter, the geometrical schema used - local parish level - is made of 53 spatial units.

One common aspect related to the size of administrative units is the fact that generally the area of spatial units increases as distance from the main urban nodes rises. In relation to the distribution of population in the Northern area of the country, since municipalities are larger as we move to the eastern/inland side of the country, the sum of resident population is higher, in spite of the smaller population density.

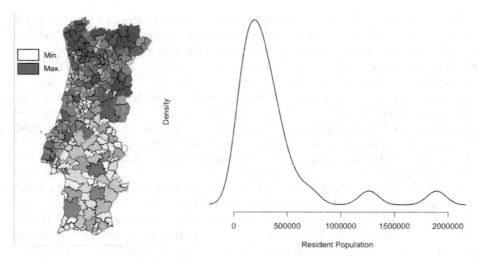

Fig. 3. Spatial distribution of resident population (Continental Portugal)

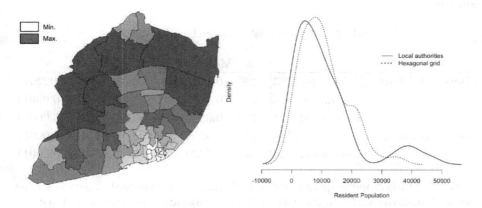

Fig. 4. Spatial distribution of resident population (Lisbon municipality)

In relation to the Lisbon municipality, it is clear that there is a core of smaller local authorities which correspond to the older part of the city. This fact potentially influences the value of spatial autocorrelation measures, as the size of regions greatly varies along the spatial surface.

Figure 3 and 4 also include the graphical representation of the empirical density functions which show in both cases the existence of a small group of regions with large resident population numbers. In relation to Continental Portugal, it is interesting to note the existence of an intermediate group of medium to large size municipalities. Figure 4 also shows the empirical density associated with resident population values for the hexagonal grid; eliminating differences in terms of form does not eliminated the skewness in the data.

Next, the two global spatial autocorrelation statistics, Moran's I and Geary's C were calculated for the two study areas using a contiguity spatial weights matrix. In relation to Continental Portugal, the statistics were computed for the distribution of resident population for the 28 NUTS 3 regions, the 278 municipalities and the 278 regular hexagons. For the NUTS 3 regions, there is significant weak positive autocorrelation in the data, which can be explained by the non-functional nature of this regional aggregation level and the large size of each region. NUTS regions were created for statistical purposes and in most cases lack physical, social and/or urban coherence. When data is disaggregated at the municipal level, spatial autocorrelation rises (higher Moran's I and lower Geary C). Hence, for the same variable, it is shown that spatial patterns are considerably different as the geographical level of analysis changes. Still related to Continental Portugal, as mentioned above, the same data was aggregated according to an hexagonal grid with the number of spatial units equal to the number of municipalities. In this case, positive spatial autocorrelation decreases according to both statistics. Again, regular hexagonal grids lack functional coherence, although the values are higher than at the NUTS 3 level. As for both indicators, the number of spatial units (n) is in the numerator, this fact is not surprising.

Table 1. Global spatial autocorrelation statistics

		n	Moran's I	Geary C
Continental Portugal	NUTS3	28	0.1995 (0.009)	0.7119 (0.0226)
	Municipalities	278	0.5688 (0.0036)	0.4482 (< 0.0001)
	Hexagonal grid	278	0.3582 (< 0.0001)	0.5692 (< 0.0001)
Lisbon municipality	Parishes	53	0.5359 (< 0.0001)	0.5374 (< 0.0001)
	Hexagonal grid	53	0.3429 (< 0.0001)	0.6393 (< 0.0001)

For the Lisbon municipality, a similar behaviour is observed. There is strong positive autocorrelation in the data at the administrative level (local authorities), whilst the spatial pattern is weaker when an hexagonal grid is used. Still for this study-area, it was thought as important to analyse, using a Moran scatterplot, the influence of outliers. This was so given the observed differences in terms of area of spatial units as distance from the older urban core increases.

The Moran scatterplot represents on the horizontal axis the variable of interest and on the vertical axis the spatially weighted mean value of the same variable. Figure 5 shows that there is a group of administrative units whose value of resident population and of the lagged variable is well above the rest. Moreover, all these regions are located away from the old urban core. In order to test the influence of heterogeneity in terms of regional size, global spatial autocorrelation was computed for a set of units whose area is below the median (figure 6a). This represented a total of 27 spatial units, only one of which was located away from the core; this one was excluded from the dataset as contiguity would be lost otherwise. Finally, in order to test the effect of spatial units' size, two datasets

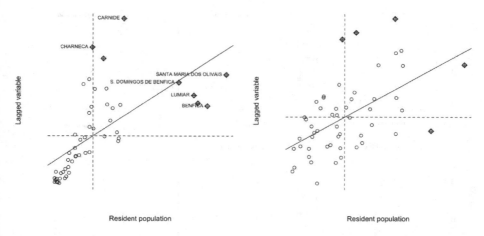

Fig. 5. Moran scatterplot (administrative units and hexagonal grid) Lisbon municipality

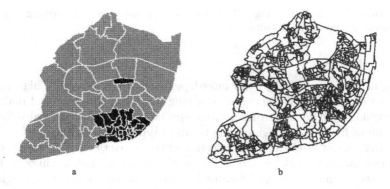

Fig. 6. (a) Set of the 27 smaller administrative units; (b) Set of 728 spatial units Lisbon municipality

were merged: one representing the 26 urban core local authorities and the other the aggregated census tracts 1. This resulted in a dataset with 728 spatial units (figure 6b).

The spatial autocorrelation statistics for the old urban core, in spite of the small number of spatial units (26), is very high (Moran's $I = 0.62$ and Geary's $C = 0.26$), both significant at the 99% level. This demonstrates that for a small set of small functional urban regions, homogeneous in terms of area, spatial patterns are very strong. For the merged dataset, consisting of 728 spatial units covering the whole of Lisbon municipality, autocorrelation is weaker, in spite of the much higher number of regions (Moran's $I = 0.42$ and Geary's $C = 0.26$). This highlights a behaviour which is consistent in all datasets: that functional long established administrative limits condition self-organizing urban growth around central administrative nodes; it stresses the importance of the administrative function in

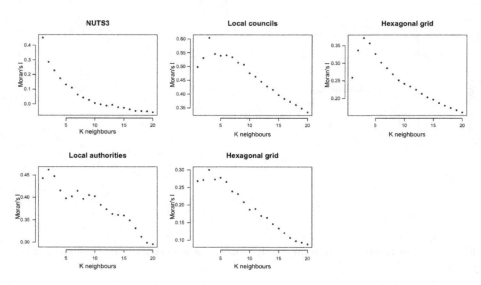

Fig. 7. Moran's I statistic using W^k spatial weights matrices with varying number of neighbours - Monte Carlo results

shaping urban form. This greater interdependence represented by high positive spatial autocorrelation is the evidence of long-established regularity. Finally, this regularity highlighted by strong positive spatial autocorrelation is also likely to be a result of internal homogeneity in terms of type of construction.

The final objective of the paper was to test the impact of an extra form for the spatial weights matrix; W^k matrices with a varying number of neighbours. Figure 7 shows the variation of the Moran's I statistic as the number of neighbours increases. The first point to be made from this Monte Carlo simulation results is that the statistic reaches its maximum when k is equal to or less than three, which is important to note considering that the mean number of contiguous regions in all datasets is between five and six. Also, for the NUTS 3 datasets, Moran's I is highest with K equal to one, which results from the large size and non-functional nature of this geographical level. Also, and confirming prior results, autocorrelation is stronger when using administrative regions (municipalities and local authorities). This confirms that regularity comes second to function.

4   Conclusion

This article intended to demonstrate how sensible are measures of global spatial autocorrelation to the geographical level of aggregation as well as to the regularity of spatial units in terms of size and shape (spatial resolution and form). Using several datasets for two study-areas, it was shown that: first, consistency in terms of self-organizing human activity around urban nodes at the national level result in stronger spatial patterns. This factor is more important than regularity as the values for the autocorrelation statistics are smaller when using hexagonal grids. Also, at the urban level, regularity of the urban core, centred around administrative units also result in higher values for the statistics. This result is amplified because of the weaker patterns observed when considering a large dataset of administrative regions and census tracts combined. This shows that census tracts lack functional homogeneity.

The results of this study are important in terms of providing some guidelines when global spatial autocorrelation measures are used. Particularly, when including spatial dependence in statistical models, one should take into consideration the nature of geographical units. In other words, it is important to study the regularity of the spatial units and their functional homogeneity. Ultimately, results help acknowledging the effect of differences in terms spatial aggregation methodologies and contribute to construct a more general set of rules for choosing the correct set of regions.

References

1. Angel, S., Parent, J., Civco, D.L.: Ten compactness properties of circles: measuring shape in geography. Canadian Geographer / Le Géographe Canadien 54(4), 441–461 (2010), http://doi.wiley.com/10.1111/j.1541-0064.2009.00304.x
2. Anselin, L.: Spatial Data Analysis With GIS: A Introduction to Application in the Social Sciences (1992)
3. Bivand, R.: The Problem of Spatial Autocorrelation: Forty years on (2011)
4. Bivand, R.S., Pebesma, E.J., Gómez-Rubio, V.: Applied spatial data analysis with R, vol. 747248717. Springer (2008)
5. Cliff, A.D., Ord, K.: Spatial Autocorrelation: A Review of Existing and New Measures with Applications. Economic Geography 46(Suppl.), 269–292 (1970)
6. Dacey, M.F.: The Geometry of Central Place Theory. Geografiska Annaler 47(2), 111–124 (1965)
7. Dormann, C., McPherson, J., Araújo, M., Bivand, R., Bolliger, J., Carl, G., Davies, R., Hirzel, A., Jetz, W., Daniel Kissling, W., Kühn, I., Ohlemüller, R., Peres-Neto, P., Reineking, B., Schröder, B., Schurr, F., Wilson, R.: Methods to account for spatial autocorrelation in the analysis of species distributional data: A review. Ecography 30(5), 609–628 (2007), http://dx.doi.org/10.1111/j.2007.0906-7590.05171.x
8. Fotheringham, A.S., Brunsdon, C., Charlton, M.: Geographically weighted regression: The analysis of spatially varying relationships. John Wiley & Sons (2003)
9. Getis, A.: Reflections on spatial autocorrelation. Regional Science and Urban Economics 37, 491–496 (2007)
10. Getis, A.: A history of the concept of spatial autocorrelation: A geographer's perspective. Geographical Analysis 40(3), 297–309 (2008), http://dx.doi.org/10.1111/j.1538-4632.2008.00727.x
11. Getis, A.: Spatial Weights Matrices. Geographical Analysis 41, 404–410 (2009)
12. Getis, A., Ord, J.K.: The Analysis of Spatial Association by Use of Distance Statistics. Geographical Analysis 24(3), 189–206 (1992)
13. Goodchild, M.F.: What Problem? Spatial Autocorrelation and Geographic Information Science. Geographical Analysis 41(4), 411–417 (2009), http://doi.wiley.com/10.1111/j.1538-4632.2009.00769.x

14. Koenig, W.D.: Spatial autocorrelation of ecological phenomena. Trends in Ecology & Evolution 14(7, 1) (2014), http://www.cell.com/trends/ecology-evolution/abstract/ S0169-5347(98)01533-X
15. Miller, H.J.: Tobler's first law and spatial analysis. Annals of the Association of American Geographers 94(2), 284–289 (2004)
16. Rogerson, P.: Statistical Methods for Geography. SAGE Publications (2001)

Fighting in the Shadows:
Health Policies and Data Availability about HIV/AIDS
A Case Study in Apulia (Italy)

Dario Antonio Schirone, Thaís García Pereiro, and Roberta Pace

Department of Political Science, Università degli Studi di Bari "Aldo Moro", Italy
{t.garcia.pereiro,roberta.pace,d.schirone}@uniba.it

Abstract. Despite the interventions enacted by the EU in 2009/2013 to fight HIV/AIDS, the latest Italian budget measures contain important cuts to health expenditure. These cuts are not allowing local authorities to develop health policies to fight against the epidemic in an efficient way by limiting both prevention and care. The main purpose is to analyze the incidence and policies regarding HIV/AIDS in Italy through a case study in the Apulia Region. First, are examined the national policy strategies and the modifications introduced by local authorities to face HIV/AIDS. Subsequently, are described the incidence and the main features of HIV/AIDS in Italy and Apulia; identifying the socio-demographic profile of an HIV-positive-individuals sample. The Italian situation regarding HIV/AIDS is still surrounded of shadows. The study of the infected population gives significant clues about the interventions that are needed, and constitutes the basis for public health policy and planning at a local level.

Keywords: AIDS, HIV population, Health Policies, Italy, Apulia.

1   Introduction

In Europe, HIV/AIDS prevention, treatment and care demand the attention of all public sectors. EU Member States are facing an increasing number of cases and the resulting medical, social and economic consequences. Over 50.000 newly diagnosed HIV cases in the EU in 2007 and an estimated two million infected people illustrate the serious dimension of the situation [1].

In the European Union, considerable treatment advances have prolonged and improved lives of people who contract the infection but, unfortunately, such advances contrasts with a diminished focus on prevention and the re-emergence of high-risk sexual behaviors [2]. The European Commission has institutionally established that HIV infection is still a threat to public health.

The surveillance of HIV infection in Italy indicated that in 2009 there were 4.5 newly diagnosed positive HIV cases per 100,000 Italian residents and 22.2 newly positive HIV cases per 100,000 foreign residents, with a median age of 39 years for males and 35 years for females. The incidence is higher in the Center and Northern

B. Murgante et al. (Eds.): ICCSA 2014, Part III, LNCS 8581, pp. 357–365, 2014.
© Springer International Publishing Switzerland 2014

regions of the country. In Italy there are between 143,000 and 165,000 HIV-positive persons, of which more than 22,000 with AIDS. One in four HIV-positive individuals is unaware about it.

Despite the suggested interventions of the European Union to fight against the spread of HIV/AIDS and its absolute priority in health policies, the budget package recently introduced by the Italian Government further cuts welfare and health expenditure which, in the country, are already below the EU and OECD countries' average. Such measures consider Populations' Health and Welfare only as a "cost" rather than an "investment" for the future and the sustainable, intelligent and inclusive growth of the country. "Investing in health today to save tomorrow" should be the main strategy of a forward looking government, however, in the era of federalism, cuts do not allow the regions and local authorities to develop accurate and effective social and health policies for an extensive and incisive combat of the epidemic, undermining thus the right of all citizens to the prevention and treatment of HIV/AIDS.

Within this Italian ambiguous context characterized by complex interrelations among health policies and HIV/AIDS [3]; the central aim of this paper is to analyze each one of these spheres through a case study of the Apulia region. First, we study the general policy strategies adopted in the country and, more specifically, the adjustments made by the local authorities in order to cope with HIV-positive patients. Secondly, we describe the incidence and main features of HIV in Italy and in Apulia in order to give a contextualization for the analysis of our sample in which we identify the socio-demographic profile of HIV-positive individuals attending medical consultation.

2   Data and Methodology

In the first part of the paper, devoted to health policies, the information is drawn: at the European level, from the Guide-lines given by the European Union in the document entitled The fight against HIV/AIDS in the European Union and neighboring countries, 2009-2013; at the country level, from both most recent National and Regional Plans for Health. The main purpose of this first part is to highlight the discordance between what has been recommended by the EU and what has been done by the local Italian authorities responsible for health services [4].

The second part responds to the formal-institutional measurement of HIV/AIDS incidence in Italy and Apulia, through the tracking and surveillance Registry of Infective Diseases collected by the correspondent Ministry and the Regional Epidemiological Observatory. Currently we just count on publications made by the observatory.

As it will be demonstrated in the following sections, one of the greatest obstacles that the socio-demographic research on HIV/AIDS faces in Italy is the scarcity of data or, at least, the impossibility of gaining access to data. Thus, with the purpose of coping with this gap, the idea was to gather information about the HIV-positive population in Italy, which remained a very ambitious objective. Nonetheless, the final compilation results are more than satisfying despite being drawn from just one region (Apulia).

The last part of the paper is dedicated to the analysis of a Survey conducted in collaboration with the Hospital of Bari and the Infectious Diseases Clinic. It was designed an anonymous questionnaire supplied during medical consultation to a number of 150 HIV-positive individuals during a six months period (April /September 2011).

3   European Health Policies Regarding HIV/AIDS

The overall guides of the EU strategy are to reduce the number of new HIV cases in all European countries by 2013; to improve access to prevention, treatment and care; and, finally, to improve the quality of life of people living with HIV / AIDS [4].

There is no correspondence between the guides that, according the EU, need to be absolutely followed concerning HIV/AIDS and the policy lines that the Italian Government, in general, and the Apulia Region, specifically, have developed and are implementing at the local level. Concerning the Italian case, Regions play a central role in the planning, organization and management of health services; and are called to ensure its effective delivery based on specific territorial needs.

Within the content of the Apulia's Regional Health Plan (2008-2010) there is an entire section dedicated to the prevention and vigilance of infective diseases. The focus is on vaccination as the most important public health measure to reduce the incidence of infective diseases incidence, but there is no specific reference to HIV or related. Nonetheless, one of the goals of the educational program is to inform about the prevention of sexually transmitted diseases and low-risk behaviors regarding health, which constitutes indirectly a policy measure that could cope the situation by slightly improving HIV numbers in the future. The lack of a direct declaration of HIV/AIDS in Apulia's health policies illustrates the "invisibility" of the disease not only at a private, but also at a public level.

According to the Euro HIV Index (2009), both care and conditions for people living with HIV/AIDS in Italy need to improve radically. This first survey ranked Italy 27th out of 29 countries. Italy's performance is very irregular in every sub-discipline of the Index, as one of the main problems being the lack of national data availability. The report also highlights the urgency to implement an effective monitoring and evaluation system. As a proven prevention method, Italy could introduce mandatory sexual education in schools and education campaigns for the general public in order to relieve the frequent discrimination faced by patients, which forces them live completely anonymous lives.

4   The Incidence of HIV/AIDS: Data Challenges

4.1   Italy: Collection Systems and Data Reports

In Italy, the systematic collection of data on cases of AIDS began over sixty years ago, back in 1982. Two years after it was institutionalized in a National System of Surveillance in which all clinically diagnosed cases in the country have being registered. After November, 28th 1986 Decree (DM n. 288), AIDS in Italy became an infectious disease of mandatory notification. Since 1987, the Surveillance System of AIDS is managed by the AIDS Operative Center (COA) of the Superior Institute of Health (ISS). In collaboration with regional institutes, the COA is in charge of the collection, analysis and periodic data publication and dissemination of annual reports.

Following the publication of a Decree of the Ministry of Health on March 2008, many Italian regions have established a monitoring system of new HIV infections, joining other regions and provinces that had already been organized independently for data collection for several years. In addition, to obtain a more accurate picture of the HIV epidemic, some regions have decided to retrieve information even before the beginning of the monitoring system. Therefore, time series regarding HIV infections vary greatly among regions (Table 1).

Table 1. Data availability of AIDS cases by Italian regions. Time series 1985-2009. (Sources: Own elaboration. Superior Institute of Health)

Regions	1985															2009									
	85	86	87	88	89	90	91	92	93	94	95	96	97	98	99	0	1	2	3	4	5	6	7	8	9
1 Lazio	x	x	x	x	x	x	x	x	x	x	x	x	x	x	x	x	x	x	x	x	x	x	x	x	x
2 Veneto	-	-	-	x	x	x	x	x	x	x	x	x	x	x	x	x	x	x	x	x	x	x	x	x	x
3 Friuli-Venezia Giulia	x	x	x	x	x	x	x	x	x	x	x	x	x	x	x	x	x	x	x	x	x	x	x	x	x
4 Piedmont	-	-	-	-	-	-	-	-	-	-	-	-	-	-	x	x	x	x	x	x	x	x	x	x	x
5 Liguria	-	-	-	-	-	-	-	-	-	-	-	-	-	-	-	-	x	x	x	x	x	x	x	x	x
6 Apulia	-	-	-	-	-	-	-	-	-	-	-	-	-	-	-	-	-	-	-	-	-	-	x	x	x
7 Marche	-	-	-	-	-	-	-	-	-	-	-	-	-	-	-	-	-	-	-	-	-	-	x	x	x
8 Emilia-Romagna	-	-	-	-	-	-	-	-	-	-	-	-	-	-	-	-	-	-	-	-	-	x	x	x	x
9 Aosta Valley	-	-	-	-	-	-	-	-	-	-	-	-	-	-	-	-	-	-	-	-	-	-	-	x	x
10 Lombardy	-	-	-	-	-	-	-	-	-	-	-	-	-	-	-	-	-	-	-	-	-	-	-	-	x
11 Calabria	-	-	-	-	-	-	-	-	-	-	-	-	-	-	-	-	-	-	-	-	-	-	-	-	x
12 Umbria	-	-	-	-	-	-	-	-	-	-	-	-	-	-	-	-	-	-	-	-	-	-	-	-	x
Provinces																									
13 Trento	x	x	x	x	x	x	x	x	x	x	x	x	x	x	x	x	x	x	x	x	x	x	x	x	x
14 Bolzano	x	x	x	x	x	x	x	x	x	x	x	x	x	x	x	x	x	x	x	x	x	x	x	x	x
15 Sassari	-	-	-	-	-	-	-	-	-	-	-	-	x	x	x	x	x	x	x	x	x	x	x	x	x
16 Catania	-	-	-	-	-	-	-	-	-	-	-	-	-	-	-	-	-	-	-	-	-	-	x	x	x
17 Pescara	-	-	-	-	-	-	-	-	-	-	-	-	-	-	-	-	-	-	-	-	-	x	x	x	x

Note: x data available; - without data.

The Superior Institute of Health collects, manages and analyzes the resulting reports and ensures the return of the collected information to the Ministry of Health [5]. Essentially, to the Surveillance System are reported only first HIV infection diagnoses, regardless of the presence of symptoms. The cases reported by the provinces and regions participating in the system certainly do not represent all cases of newly diagnosed HIV infections occurring in Italy, but can provide a useful indication of the spread and progress of the infection in the country. In 2009, regions and provinces where the surveillance system was activated represented almost three quarters (72.1%) of the total Italian population. Since the beginning of the epidemic in 1982 and until 2009, in Italy have been reported about 63,000 AIDS cases, of which nearly 40,000 died.

The data from this surveillance system indicate that 6 new HIV-positive cases per 100,000 residents were diagnosed in 2009, placing Italy among the countries of Western Europe with an upper-middle HIV incidence [6]. However, the underreporting of AIDS cases influences the real dimensions of the phenomenon, studies allowed to estimate for Italy a rate of not reported cases close to 10%.

In the period 1985-2009, 45,707 new diagnoses of HIV infections (71% males) were reported in 17 Italian regions collaborating with the system. Regarding the calculated incidence[1] of new diagnoses, the medium value for the whole country in 2009 is 6.0 for 100,000 residents. However, there are some regional differences that have to be necessarily highlighted: the values oscillate from 1.6 in Calabria to 9.3 in Emilia-Romagna [5], showing a large regional heterogeneity that could be due to a real difference on the incidence or a greater rate of underreporting in some of these regions.

4.2   A Regional Look: Apulia

Since 1996 the Regional Epidemiological Observatory maintains the Register of the Regional AIDS cases, collecting reports of all AIDS cases residing in Apulia. Since the mid nineties there has been a downward trend strongly influenced by the spread of highly effective antiretroviral therapies. In recent years the number of reported cases is around 60 and the regional distribution is quite homogeneous.

In the Apulia region the surveillance system of HIV infection was activated in 2000. This system is based on reports of both public and private laboratories that perform HIV testing and the regional coverage stood only at 50% (half of the information is missing). Laboratories have been collecting data on the subject in full compliance with privacy legislation and sent quarterly a summary of the total number of screening tests performed, the number of HIV-positive cases found and confirmed and a card with an encrypted code, essential to minimize the possibility of double reporting. However, this flow of information was done on voluntary basis and only slightly more than 50% of the regional laboratories surveyed (which declared to test for HIV) have regularly sent the requested data.

With the Decree of 2008 (Gazzetta Ufficiale No. 175 of 28/07/08) [7], the Italian Health Ministry established the National Surveillance System for newly diagnosed HIV infections, which became subject to mandatory notification. Up to 2008, only AIDS notification was mandatory. Since 2009, in Apulia the new surveillance system based on notifications from all the clinical centers of the region (U.U.O.O. Surgeries and Infectious Diseases) has been implemented. By filling out a special form of case detection, Clinics report all new cases of HIV infection. The information flow involves transmitting these notifications to the Epidemiological Observatory of Apulia (OER-Puglia) which acts as a Regional Center for the coordination, monitoring and dissemination of the data collected to the Operative Center AIDS (COA-ISS) and the Councillorship of Regional Health Policies (Resolution Service Territorial Prevention - Apulia Region, Prot No. 24/16943/1 of 18/11/2008). In the region, retrospectively, it was possible to collect data on new diagnoses also referring to the years 2007 and 2008 [8].

---

[1] The incidence rate is calculated by the COA relating the new HIV diagnoses with the population residing that year on the regions and provinces considered.

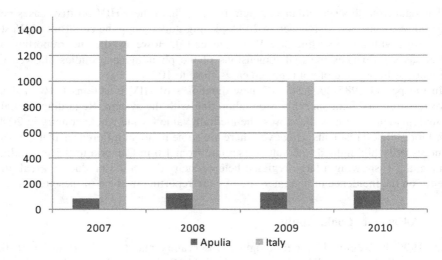

Fig. 1. New diagnoses of HIV in Italy and Apulia. 2007-2010.
(Sources: Own elaboration. Superior Institute of Health & Epidemiological Observatory Apulia Region.)

In the period 2007-2010, 484 new HIV cases were reported in Apulia, 10.6% of all new diagnoses registered in the country for the last four years. Most part of diagnoses regards men and women around 30 and 41 years old, with a sex ratio of 3 to 1 favoring males and a considerable proportion of foreigners that reaches 12%.

Regarding the evolution of the infected population, the trends in both Apulia and Italy diverge: while the new diagnoses in the first are highly increasing, in the second are decreasing. In Apulia a steady rise in the new HIV diagnoses is observed, which increased from 83 in 2007 to almost 142 in 2010, with an incidence rate also on the rise and superior to 3.3 for 100,000 persons residing in Apulia in the last year under observation (Figure 1). A constant phenomenon in Apulia, as well as in the rest of the country, regards the predominant route of HIV infection which consists on risky sexual behaviors, mostly heterosexual [8].

4.3   About Our Sample

As it has been proved here, the epidemic information system in Italy is not in compliance with the criteria that the EU expect from its members. There is a significant lack of data regarding the HIV/AIDS infected population that begins at a local level and, as a consequence, cannot certainly covers the national level. In Apulia, the reached covering is limited, but this is not the only deficit, further information regarding the socio-demographic characteristics of this population is not available for scientific research. Undoubtedly, that is the reason that gives scientific meaning to quantitative data collected from HIV-positive patients residing in Apulia.

On the first try, a large list of public practitioners at the national level were contacted in order to gather some voluntary information about their patients, the answers were almost all inexistent or negative. But a door was opened in the Apulia Region. Initially, the expectations were broader than those compressed in the final question-naire because the condition for its administration was: brief with only a few items. The designed questionnaire included some questions regarding the socio-demographic profile (age, gender, educational attainment, nationality, place of residence, religiosi-ty, among others) of 150 HIV-positive individuals who attended medical consultation during a six months period (April-September 2011). It is not possible to know exactly the representativeness of our sample considering the real number of HIV-positive population in the Region (unknown), but surely allows developing a first approach to the specific knowledge of such population and its implications for public policies.

Practitioners of Bari's Hospital and Infectious Diseases Clinic administered the questionnaire to a random sample of 150 persons with HIV who were currently under usual care. Even if random, the sample could be not necessarily epidemiologically representative because did not include all possible infected individuals but only those under treatment.

There are two alternatives for establishing the representativeness of the data col-lected: 1) if compared to newly diagnosed HIV cases in the region for the period 2007-2010, the sample covers 31% of this population; 2) if compared to the preva-lence of cases in Apulia in 2010, the percentage is around 17%.

The Socio-Demographic Profile
The sample shows a more balanced sex structure of the HIV-positive population in Apulia in 2010, of which 58.1% of the respondents are men, with a sex ratio equal to 1.4. With respect to age, the minimum and maximum ages are 22 and 74 years old, respectively, and the mean age is around 44 years.

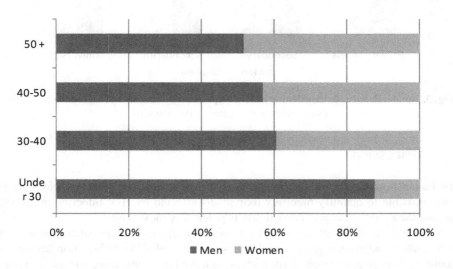

Fig. 2. Apulia. Age and sex structure of HIV infected population. 2010.
(Source: Own elaboration. HIV dataset.)

There were three options in the questionnaire to identify the origin of the respondent: born in Apulia, born in another region or foreign born. The correspondent values show that 85.8% were born in Apulia while 5.4% were born abroad. Considering only those born outside Italy, it is observed a feminization of the contingent with a proportion of women that reaches 75%. The lowest presence of foreigners in the sample might be a signal of differentials in the assistance for the medical consultation between Italians and foreigners, considering also that the majority of them were born in a non-EU member state.

The distribution of the educational level by sex shows a clear gender distinction. Regarding men, the distribution is concentrated on the two highest levels: 44% reached Secondary II while 16% got a university degree. On the contrary, most part of infected females declared the lowest levels: 32% primary or less and 36% Secondary I (Figure 3). It seems to be an interrelationship between educational level and high-risk behaviors that affects differently men and women.

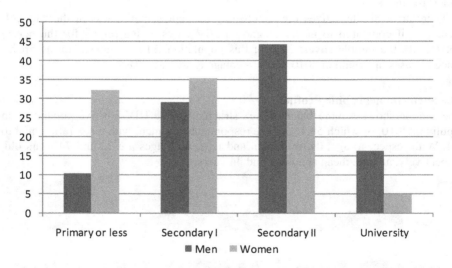

Fig. 3. Apulia. Percentage distribution of HIV population by educational level and sex. 2010. (Source: Own elaboration. HIV dataset).

5   Discussion

The Italian situation regarding HIV/AIDS is still surrounded by shadows. Not only it is not possible to quantify precisely individuals affected by HIV infection or AIDS, but also their public and private visibility is practically inexistent.

At the national level, attention has been focused exclusively on scientific research so all the public and especially the social issues related to HIV/AIDS infection have been secondary or even abandoned, despite having acquired over time more and more importance as recognized by all international institutions - including UNAIDS - and the Italian Ministry of Health. As a consequence, there is a clear lack of awareness-raising campaigns and activities. New actions have not been promoted, nor the specific prevention programs aimed at particular targets of HIV/AIDS population, and still public health institutions do not speak clearly on condom use.

In Apulia, the answers to an increasing trend of new infections, especially due to transmission of the virus through sexual intercourses, are still on standby. There have been only some prevention projects for specific targets: programs for the homosexual community, for foreign population or prostitutes, harm reduction interventions in favor of drug addicts are then supported according to the "sensitivity" of single administrations, without any organization and planning.

It is essential to invest resources and energy to block the rapid escalation of new infections and all its implications. It is of vital importance to support the improvement of the quality of life and the social inclusion of people with HIV/AIDS. The knowledge of such population gives significant clues about the interventions that are needed, and constitutes the basis for public health policy and planning at the local level.

References

1. ECDC/WHO. HIV/AIDS Surveillance in Europe, 2007. Stockholm and UNAIDS report (2008)
2. Fenton, K.A., Imrie, J.: Increasing rates of sexually transmitted diseases in homosexual men in Western Europe and the United States: why? Infectious Disease Clinics of North Am. 19(2), 311–331 (2005)
3. EURISPES. Sintesi dei risultati sul tema "italiani e la chiesa" (Synthesis of the results on "Italians and the Church"). Rapporto Italia (2006)
4. European Commission. The fight against HIV/AIDS in the European Union and neighbouring countries, 2009-2013. Brussels (2009)
5. ISS. Notiziario dell'Istituto Superiore di Sanità. Italia (2011)
6. ECDC/WHO. HIV/AIDS Surveillance in Europe, 2008. Stockholm and UNAIDS report (2009)
7. Decreto Ministeriale 31 marzo 2008. Istituzione del Sistema di Sorveglianza delle nuove diagnosi di infezione da HIV (Institution of Surveillance of newly diagnosis HIV infection). Gazzetta Ufficiale, Italy, 175 (luglio 28, 2008)
8. OER Puglia, Regione Puglia, ARES. Relazione sullo stato di salute della popolazione pugliese (Report on the State of Health of the Apulian Population) (2009)

How to Build Territorial Networks: Criteria for the Evaluation of the Networking Potential for Local Self-sustainable Development[*]

Pierangela Loconte and Francesco Selicato

[1] Polytechnic of Bari, via Orabona, 4, 70124, Bari, Italy
{pierangela.loconte,francesco.selicato}@poliba.it

Abstract. The processes of globalization have led to a substantial change in the urban and territorial systems and the role of each of their parts. In the age of technological change and knowledge society, the change in the structure of the city has started up a social and cultural reorganization, as a consequence of the response of the territories to the global dynamics. This is a multidimensional phenomenon, able to involve all the human dynamics related to free movement of goods, services, capital, knowledge and people. This paper seeks to understand whether the model of territorial networks may be the answer to the problems emerging from the territory. In particular the aim is to understand whether it is possible to evaluate the potentialities of the territory, i.e. its tangible and intangible resources, and build on them territorial networks able to activate processes of local self-sustainable development. The first part of the paper seeks to explore the theoretical link among reticular models, the territorial capital and local self-sustainable development. In the central part we propose a possible methodological approach for the construction of territorial networks based on existing resources and potentialities. In the final section, we apply the theoretical model to the case study of the district of Sibiu (Romania) in order to verify if the area of Valea Hartibaciului is really a territorial network whose identifying characteristics can be seen as a development opportunity. Finally, we try to define what are the limits of the proposed model and what are the future job prospects.

Keywords: Territorial networks, local self-sustainable development, indicators, cluster analysis.

1   Introduction

The need of comparison at the global scale has led to change the urban geography and to concentrate economic, cultural and social production in the major cities. This has led to an increase in the flow of information and people to the main centers, contributing to the impoverishment of the social parts of the territory not able to interface with

---

[*] The paper is the result of a joint work of the authors, also if chapter 1 has been written by Francesco Selicato, the others by Pierangela Loconte.

B. Murgante et al. (Eds.): ICCSA 2014, Part III, LNCS 8581, pp. 366–381, 2014.

global level. In addition, the increased levels of knowledge and the dissemination of information has ideally closed the gap making available what in the past wasn't, both in acquisition of goods and services, and in terms of knowledge and economies. From the urban point of view, the processes of globalization have led, to a radical change of urban and territorial systems.

The global phenomena were, therefore, able to change the balance between different historical geographic areas, creating new synergies and dissolving historic ties now non-functional to the economies of global relations. This means that the change taking place on our territories today is both morphological, i.e. on the structure of the system, and relational, i.e. between the parties that make it up. There is, therefore, the need to seek new forms of territorial organization able to respond to the requests made by global / local dichotomy. As a consequence, the concept of network is probably the one that can better explain the structure, dynamics and functions that characterize global cities, but at the same time, the local territorial systems.

The aim of this work is to study and develop a possible methodology for the construction of territorial identities, to understand what are the constituent parts, the peculiar characters, the relationships between them, and how they can form the basis for sustainable development. The objective is to intercept the natural propensity of the territory to the constitution of networks (information exchange - culture – synergy of economies) in order to enhance the common heritage and identify strategies that can open prospects of development.

Many experiences have shown that a situation characterized by strong local identities may emerge in a competitive way in the wider territorial area, if based on the enhancement of endogenous resources [1]. The winning strategy into the competition between territories, in fact, lies in its ability to deliver performances through local economic development [2] [3]. The concept of territorial network can also be reversed: it can be reversed i.e. the vision of the territorial network as an instrument that regulates the arrangement of settlements and social rights of its centers. In fact, also the action of the individual centers may influence the general structure of the network, and i.e. inducing knock-on effects on other centers, thus driving force to the wider territorial area through this particular form of pluralism [4]. In particular, forms of exchange of knowledge can give a great contribution to the establishment of system of relations of an innovative and pluralistic territorial network. The innovation is that the exchanged knowledge is the feature of the network and defines the characters of competitiveness in a global context in which it competes with other networks characterized by complex relationships. It will seek to understand, then, what would be the objectives the criteria and actions for the development of territorial networks, through the evaluation of the potential, often unspoken or hidden by the transformations that have occurred in recent times, not very sensitive to the peculiarities of local contexts.

2   The Territorial Networks and Local Self-sustainable Development

According to Amin et al. cities and regions are thereby interpreted as "sites within networks of varying geographical composition [...and as...] spaces of movement and circulation (of goods, technologies, knowledge, people, finance, information)"[5].

Given the inability of the urban and territorial to respond to the emerging questions, there is now the need to build a new city model, different from the past, which is able both to respond to global stress and market liberalization and to change the spaces and places of the economy, and to the local demand of development. We are facing to the change of the idea of territory and the evolution of its meaning, no more exclusively in physical terms but also relational ones, on different levels and scales and with different meanings and characters. The network concept is probably the best way to describe the complexity of relationships and spatial and relational patterns of development. A cross-disciplinary reading can be found in the definition of Boerzel, which describes "policy network as a set of relatively stable relationships which are of non-hierarchical and interdependent nature linking a variety of actors [...] who exchange resources to pursue these shared interests acknowledging that co-operation is the best way to achieve common goals"[6]. Looking at the territory, the concept of network is developed according to different theoretical references that clarify how these can intercept material and immaterial flows, taking shape as a tool for the definition of development strategies. "Network in general are made up of nodes (cities, firms, organizations, individuals), linkages between the nodes (infrastructure, relationships, ties), flows (people, goods, information, capital etc.) and meshes" [7].

As Dematteis says, the role of each center does not depend on its size but on its ability to create relationships by using their own economic, environmental, cultural and social resources: "the nodes, unlike what happens in the model equipotential, have a local endogenous component that is combined directly with an exogenous component of the network. So the issue of the networks is inside to the local development questions"[1] [8]. Then territory is a system constituted by endogenous and exogenous exchanges goods, values, information, economies that distinguish and characterize it in comparison to the others. These features are defined Local Milieu i.e. "a permanent set of socio-cultural characters have built up in a certain geographical area through the historical evolution of inter-relationships, in turn related to the mode of use of local natural ecosystems"[2] [9]. Inside the Local Territorial System, Local Milieu relates and interacts with the actors that compose it [9] [10] i.e. the settled communities and their cultures, configuring a dynamic system in evolution. The appropriation of the places from the community is based on "knowledge, acquired through a process of cultural transformation of the inhabitants, of the land value of the common assets (material and relational)"[3] [11]. In this context, it is clear that the development of a territory should necessarily begin from a local development based on the sustainable exploitation of tangible and intangible resources [11]. Therefore, the need to establish a new equilibrium between the parties involved in the development process exists. The key issue is to understand what conditions are at the base of the process, what should not be questioned or is not treatable, which invariants of the project can be taken into consideration to build specific targets and actions for

---

[1]   Our traslation.
[2]   Our traslation.
[3]   Our traslation.

development. The necessary choice is to refer to a model of sustainable local development that seeks to build virtuous relationships between environmental, social, territorial, economic, policy sustainability. So it makes sense to talk about local self-sustainable development as the creation of a system of relationships and the adoption of a unified strategy on the basis of local identity, founded "on the will to determine the balance between sustainable human settlements and the environment, reconnecting new uses, new knowledge and new technologies to the environmental knowledge of history"[4] [12]. To be self-sustaining, a local development process must be structured on a unified strategy firmly anchored in local identity that is able to stimulate the economic and social context, reducing the emerging gaps and that it is based on the relationships between the elements that form the context and the new relationships that can be built between the parties in order to activate the processes of exploitation and development. The territorial development policies should first and foremost help areas to develop existing assets, namely their Territorial Capital. It is defined as a collection of tangible and intangible assets including "geographical location, size, factor of production endowment, climate, traditions, natural resources, quality of life or the agglomeration economies provided by its cities [...] customs and informal rules that enable economic actors to work together under conditions of uncertainty, or the solidarity, mutual assistance and co-opting of ideas that often develop in small and medium-size enterprises working in the same sector (social capital). Lastly there is an intangible factor [...] which is the outcome of a combination of institutions, rules, practices, producers, researchers and policymakers, that make a certain creativity and innovation possible" [13]. This means that, although the general objectives and strategies of enhancement and local development may be common to most of the territories of Europe, in reality there is a need to establish specific objectives and actions which can affect with more efficacy on the positive features of the territory and activate latent opportunities. The territory, therefore, plays a central role in the processes of transformation and can be considered as a resource capable of generating opportunities of development: within it urban areas, networks and rural areas, historic landscapes and natural assets are intertwined with the historical background and common identity and intangible heritage. The Territorial Capital is a complex system that requires the adoption of a unified strategy, based on their identity, and it must be able to stimulate the economic and social context, reducing the emerging gaps. This strategy, therefore, must be based on the existing relations among the elements that form the context and the new relationships that can be built between the parties in order to activate the processes of exploitation and development of local self-sustainable. As claimed by Dematteis and Governa, "local sustainable development based on the complexity of the territory is configured as a process of the territorial development, based on the sustainable exploitation of tangible and intangible resources present in a given area, which also involves the social and cultural sphere and the ability of self-organization of the subjects" [11]. Then, talk about the Territorial Capital means also to refer to the Social and Cultural Capital in each territory. This shows the strong link that exists among

---

[4] Our traslation.

landscape, cultural heritage and society that have changed in a reciprocal manner within each context.

3   How to Build Territorial Networks for Local Self-sustainable Development: A Proposal for Methodology

"The metaphor of network emphasizes the complex and the strong relationships between the cities and such as the coherence and unit of the region [...] associated with economies of scale, critical mass and synergy" [7]. The construction of territorial networks can be seen as a possible strategy to be implemented for the enhancement of existing assets and for the activation of micro economies. We try to understand if there are suitable tools for the building of networks and, above all, how it is possible to evaluate the potential of the local territorial systems.

The general aim of the work is to propose an operational methodology by which, starting from the knowledge of the resources of the territory and, so by its territorial capital, it is possible to build network models on which private and public administrations can be able to promote strategies for local self-sustainable development through the enhancement of resources. The model should be able to:

- Refer to the general concept of Sustainable Development, trying to decline it at the local level;
- Aggregate various types of information and summarize ongoing phenomena translating them into strategies for local self-sustainable developing;
- Allow the construction and evaluation of different development scenarios;
- Be built on existing local resources, that we have defined the Territorial Capital, in its many forms.

The specific objectives of the work are:
- To identify the identity and relational characters of a given territorial area, i.e. its Territorial Capital;
- To understand what are the potentialities on which to build territorial networks and, consequently, on which to structure indications for the drafting of plans and projects on a local and regional level;
- To define criteria for the evaluation of the networking potential for local self-sustainable development;
- To identify methods of aggregation (territorial networks) based on the Territorial Capital.

Figure 1 shows the methodological approach that allows the construction of territorial networks and based on the evaluation of the territorial potentiality. Starting from the reflection on the Territorial Capital, it finds similar interesting references and applications in the international literature [14] [15] [16] [17] [18] [19] [20]. The methodological approach is divided into 3 main phases.

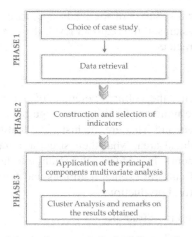

Fig. 1. Methodological approach for building territorial networks

Phase 1

It is that of defining the case study on which to apply the methodology and data retrieval for describing characters of the Territorial Capital. The difficulties are mainly related to the collection of data that will be processed in a GIS environment in the subsequent phase 2.

Phase 2

The objective is to identify what tool might be used for evaluating the potential for network able to describe the complexity and multidimensionality of the Territorial Capital. In particular, it seeks to define a possible core set of indicators able to respond to local self-sustainable development strategy, taking shape as an expression of the "best knowledge available " [21]. Indicators should be able to describe the territorial capital of a place on the basis of information obtained in phase 1. Moreover, they must possess homogeneous characteristics, must be uniquely defined for each geographical area in such a way as to make them correlated within the subsequent phase 3. The indicators are not only able to describe the states or environmental changes, but also to evaluate and set goals [22] [23] [24] and, then, they should serve to define the effective recognition of common relationships with the aim of building a territorial network and identify strengths and weaknesses, starting point for the project of local self-sustainable development.

Starting from the definition of Territorial Capital, it was possible to identify six key components that characterize closely related with each other, able to describe the tangible and intangible resources of the territory. Each of the six components, as follows, has been described through the identification of indicators that summarize the dominant characteristics:

1. Ecological Characters

The objective is to describe the natural potential of the scope of the study with reference to the basic concepts of landscape ecology [25] trying to understand what is the structure and the ecological richness of the landscape, as it is composed, what is the incidence of natural ecosystems at the regional level, the level of diversity and whether there are dominant characters.

2. Cultural Capital

The objective is to describe the cultural potential of a given territory, not only in terms of knowledge of the number of cultural heritage present within it but also, and more importantly, to understand what are the historical features of particular significance for a given geographical area, what their structural state (level of degradation, transformation, enhancement). In addition, the intangible heritage is also described, verifying the existence of cultural events, traditions and typical products both linked to local crafts and agricultural production.

3. Social Capital and Economic Structure

The objective is to describe the population and the existing economic structure, expressed both in the productive sectors and tourism and in connection with the territory (forms of associations, NGOs, etc.), the existence of specialized local productions, their inclusion into the context. It also describes what are the numbers of tourism, the attractiveness of the existing cultural heritage, the role that cultural events have on the population, the role of the associations in the local cultural context and in the dissemination of intangible heritage.

4. Infrastructural Character

The objective is to define the levels of accessibility of the territory and in relation to the road and pedestrian mobility, with particular reference to the cultural heritage and the sustainability of transport and pollution levels.

5. Urban System and Facilities

The objective is to describe what is the capacity of urban centers to meet the needs of the population and to define the level of quality of the urban areas.

6. Land Use

The objective is to define the main uses and systems of protection of suburban, rural and natural areas; groped to assess the levels of land consumption, environmental degradation, landscape value and the main agricultural vocations of the territory.

The descriptive indicators of Capital Regional are used to describe the territorial complexity, and to be used, shall be constructed and integrated into a Geographic Information System (GIS) a powerful tool designed for managing, transforming and representing georeferenced data [26] [27] [28] [18].

Phase 3
Finally, we will proceed to the correlation of the information available through the application of cluster analysis. First of all, however, it is necessary to deepen the study of the relationships between variables through factor analysis, particularly effective in solving territorial problems of a complex nature, for their description when it is necessary to use a large number of indicators. Then the method of Principal Component Multivariate Analysis analyze the relational structure of the set of quantitative variables through the derivation of a smaller number of variables, called *Principal Components*, able to explain a significant portion of the total variance of the data. On the base of the analysis of the principal components we'll define a cataloging system of entities from the observation of [...] "similar" or "resembling" [29] realities through the use of Cluster Analysis.

4   The Case Study: The Territorial Network of Valea Hartibaciului

4.1   Phase 1 – Definition of the Case Study and Data Retrieval

The choice of the case study fell on an area whose characters were examined and studied within the research project VIVA EASTPART for the enhancement of minor historical centers and their territories in the context of the study of Eastern Europe[5]. With the research project VIVA EASTPART we share the aim to enable the processes of local development through the definition of actions to put the focus of the environmental and cultural resources, and intangible assets exist, with particular reference to the minor historical centers through the establishment of pilot projects and action plans. The case study on which you have chosen to apply the methodology is that of Valea Hartibaciului. The area takes its name from the river called Hartibaciu, and around which a wide valley develops with 15 small municipalities, characterized by a strong polycentrism. All 15 municipalities are part of the district of Sibiu, in the heart of Transylvania and they are characterized by the homogeneity of its historic, cultural and environmental character. The region of Valea Hartibaciului is located in the eastern part of the district of Sibiu, and it is characterized by its mountains, hills and fortified churches of Saxon origin and by a great natural heritage preserved by Special Protection Areas (SPAs) and Site of Community Importance (SCI). In particular, through the proposed methodology, we try to understand if the 15 municipalities are consistent with the construction of local networks based on the identification of the Territorial Capital, with particular reference to the Cultural Capital. In order to do this, the analysis was extended to the whole district of Sibiu (64 municipalities), trying to understand what are the homogeneous characteristics and the strengths of the territory and what may be the relations to be enhanced.

With regard to the data retrieval, references is made by the Romanian National Geoportal[6] with regard to the territorial main characters (urban centers, infrastructure, etc.).

---

[5]   http://vivaeastpart.eu/ last access November 2013
[6]   http://geoportal.ancpi.ro/ last access July 2013

Fig. 2. District of Sibiu (source: http://pe-harta.ro/sibiu/ - modified by authors)

From the Romanian National Institute of Statistics[7] it was possible to find the descriptive data of the population for each municipality (total population, population by age, ethnicity, religion, employment levels) and data related to cultural and tourism, while from the website of the Ministry of Culture and National Heritage of Romania[8] it was possible to use the list of historical monuments of the district of Sibiu which are part of the national cultural heritage of Romania. For the retrieval of data about Ecological Characters and Land Use, we made reference to information disclosed by the European Environmental Agency[9] that provides all the data related to land cover (Corine Land Cover) and the perimeter of the protected areas at Community level (Natura 2000 Project). Through consultation of the UNESCO database were found the information relating to protected sites[10]. Additional information was obtained by consulting portal of the Local Action Group of Valea Hartibaciului[11] and OpenStreetMap[12].

4.2   Phase 2 – The Construction of the Indicators

Based on the information obtained, we have built the descriptive indicators of the Territorial Capital of Sibiu district: initially it was possible to build 80 indicators for

[7] http://www.insse.ro/ last access December 2013;
    http://www.sibiu.insse.ro last access December 2013
[8] http://www.cultura.ro/ last access September 2013
[9] http://www.eea.europa.eu/ last access July 2013
[10] http://whc.unesco.org/en/list/596/ last access July 2013
[11] http://www.gal-mh.eu/ last access June 2013
[12] http://www.openstreetmap.org/ last access July 2013

each of the 64 municipalities of the district of Sibiu, belonging to the 6 key components describing the Territorial Capital. Subsequently it was decided to evaluate them individually on the basis of specific characteristics in order to obtain structured data to be used in the factor analysis and cluster analysis. In particular, we have not taken into account all the data that:

- Had characteristics of collinearity that could produce homogeneity of variances;
- Were not present characters of independence with the other variables;
- Never had particular significance to the whole of the context from the point of view of information;
- Were present at a very low number of municipalities and weren't statistically significant.

Furthermore, in order to overcome some of the limitations just described and their data, in some cases, it was possible to combine several indicators. The analysis of the indicators constructed and the choices made by exclusion or merging resulted in a reduction of approximately 50% of initial information with the definition of 37 indicators. Some results are visible in the following figure 3.

Fig. 3. Example of same indicators build for the study of micro-region Valea Hartibaciului

In particular, we have identified 4 key indicators regarding the ecological characters, 3 indicators in relation to the Cultural Capital, 13 descriptive indicators of social capital and economic structure, 4 indicators related infrastructural, 4 indicators relating to the settlement system and facilities and, finally, 9 indicators related to the land use. The selected variables are able to describe what are the main features of the system, highlighting strengths and weaknesses, such as the presence of fortified churches or the lack of facilities for the population living in urban areas.

4.3   Phase 3 – Analysis by Principal Component and Cluster Analysis

Through the application of factor analysis, it was possible to correlate the variables defined in phase 2 and, in relation to levels of variance, define what are the main components of the system to reach the group using the Cluster Analysis. The construction of the correlations between the indicators and factor analysis was obtained using the software SPSS 17.0 statistical package. We started then the principal component analysis in order to extract the assumption of a number of factors equal to the number of input variables. Through the software, we have built the correlation matrix between the pairs of indicators, to verify the quality of the selected indicators (verifying the absence of linear correlations) and, consequently, simplify the initial model. In the present case, starting from the initial number of 37 variables, it has been possible to describe the phenomenon through the loss of a small amount of information, but retaining the same descriptive efficacy.

Table 1. Total variance explained

Component	Initial Eigenvalues			Weights of the not rotated factors			Weights of the rotated factors		
	Total	% of variance	% cumulative	Total	% of variance	% cumulative	Total	% of variance	% cumulative
1	8,299	23,054	23,054	8,299	23,054	23,054	5,222	14,506	14,506
2	6,228	17,301	40,355	6,228	17,301	40,355	4,47	12,417	26,923
3	3,994	11,094	51,449	3,994	11,094	51,449	4,1	11,388	38,312
4	2,272	6,311	57,76	2,272	6,311	57,76	3,081	8,558	46,869
5	1,951	5,419	63,18	1,951	5,419	63,18	2,729	7,582	54,451
6	1,557	4,324	67,504	1,557	4,324	67,504	2,644	7,344	61,795
7	1,378	3,827	71,331	1,378	3,827	71,331	2,02	5,611	67,406
8	1,316	3,656	74,987	1,316	3,656	74,987	1,971	5,476	72,883
9	1,079	2,997	77,984	1,079	2,997	77,984	1,837	5,102	77,984

Table 1 shows how in front of the definition of 9 principal components, in this case, there is a loss of information equal to about 22%. The table of explained variance (Table 2) shows the eigenvalues and the percentage of the variance for both the solution with a number of major components equal to the number of variables that the solution after the extraction of the nine main components.

The reading of the table shows that:

The first principal component describes the environmental and natural dominance of the characters in relation to the distribution of the population in the area.

The second principal component defines the character of the urban areas, able to offer development prospects to its inhabitants, within which there is also the greater part of the cultural heritage, both tangible and intangible.

The third principal component shows that cultural services are present in many small centers, characterized by elderly population with minimum levels of education while the active population is mainly concentrated in the larger towns, and with high density.

Table 2. Table of the principal components estracted

	Indicators	\multicolumn{9}{c}{Components}
		1	2	3	4	5	6	7	8	9
E	GRAIN	0,862	0,004	-0,111	-0,043	-0,022	0,032	-0,145	-0,018	0,177
E	MARGALEF	0,701	0,121	0,093	0,259	0,067	-0,151	0,178	-0,033	0,265
E	SHANNON	0,588	-0,135	-0,042	0,255	0,43	0,039	0,469	0,056	-0,147
E	EVENNESS	0,196	-0,214	-0,103	0,306	0,47	0,064	0,475	0,211	-0,31
C	FORTIFIED CHURCHES	-0,356	0,079	-0,329	-0,269	-0,001	-0,074	-0,384	-0,393	0,077
C	CULTURAL OFFER	0,072	-0,122	0,523	0,192	0,017	-0,038	-0,026	-0,49	0,057
C	CULTURAL HERITAGE	-0,051	0,805	-0,147	-0,106	0,029	0,3	-0,08	0,053	-0,047
I	PRIM + SEC ROADS	-0,119	0,728	0,092	0,12	0,179	0,105	0,047	-0,147	-0,028
I	ROADS (OTHERS)	-0,023	0,598	-0,22	0,063	0,175	0,592	0,191	0,099	0,047
I	RAILS	-0,071	0,374	-0,208	-0,018	0,05	0,444	0,005	0,22	0,651
I	RAILS/ROADS	0,206	0,014	-0,05	0,145	-0,114	-0,041	-0,207	0,062	0,819
LU	AGRIC. PERMANENT CROPS	-0,126	-0,125	-0,082	-0,012	0,028	-0,054	-0,817	0,17	0,082
LU	AGRIC. HETEROGENEOUS CROPS	-0,635	0,115	0,007	0,025	-0,304	-0,376	0,028	-0,253	-0,087
LU	AGRIC. ARABLE LAND	-0,475	0,072	0,043	0,067	-0,007	0,08	-0,15	0,717	0,219
LU	AGRICOLTURAL - TOTAL	-0,877	0,039	0,219	-0,054	-0,158	-0,076	-0,189	0,179	0,077
LU	NAT. SHRUB VEGETATION	0,414	-0,198	-0,14	0,174	0,537	-0,15	0,348	-0,058	-0,221
LU	NATURAL - TOTAL	0,855	-0,204	-0,203	0,036	0,16	0,087	0,186	-0,2	-0,111
LU	PROTECTED AREAS	0,166	-0,233	0,116	-0,323	-0,006	0,119	0,409	-0,537	-0,158
US	URBANIZED AREAS	-0,128	0,9	-0,063	0,078	-0,05	-0,018	-0,045	0,135	0,132
LU	ARTIFICIAL-NATURAL	-0,646	0,011	0,455	0,014	-0,104	0,074	-0,091	0,412	-0,041
LU	HYDROGRAPHY	0,055	0,14	-0,149	0,21	0,44	0,239	0,435	-0,023	0,43
US	INDUSTRY	0,026	0,777	-0,192	0,039	-0,081	-0,071	0,039	0,029	0,198
US	POPULATION DENSITY	0,405	0,578	-0,389	-0,012	-0,143	0,347	-0,006	0,137	-0,165
US	FACILITIES/1000ab	-0,007	-0,037	-0,133	-0,239	-0,159	-0,695	-0,013	0,344	-0,06
CS	ACCOMODATIONS	0,14	0,011	0,039	0,097	0,85	0,145	-0,036	0,009	0,023
CS	TURISTIC FLOWS	0,108	0,153	0,034	0,055	0,874	0,037	-0,009	-0,055	-0,014
CS	STRUCTURAL DEPENDENCE INDEX	-0,298	-0,201	0,83	-0,288	-0,028	-0,16	0,053	-0,066	-0,065
CS	DEPENDENCE ON OLDER INDEX	-0,137	-0,212	0,882	0,279	0,008	-0,069	0,034	0,056	-0,071
CS	OLDNESS INDEX	0,021	0,158	0,698	0,535	0,048	0,104	-0,023	0,155	-0,072
CS	HIGHER EDUCATION	-0,003	0,593	-0,323	0,212	0,051	0,543	0,028	0,16	0,071
CS	SECONDARY EDUCATION	0,14	-0,106	0,068	0,823	0,122	0,014	0,07	-0,011	-0,011
CS	PRIMARY PRIMARIA	-0,337	-0,26	0,396	-0,574	-0,142	-0,225	-0,048	-0,144	-0,103
CS	ILLITERATE	0,196	-0,185	-0,013	-0,66	-0,138	-0,105	-0,017	0,077	-0,213
CS	UNEMPLOYED	-0,256	-0,085	-0,084	-0,627	-0,012	-0,484	-0,098	-0,173	0,008
CS	WORKING POPULATION	0,276	0,242	-0,813	0,304	0,018	0,161	-0,073	0,087	0,059
CS	EMPLOYEES	0,101	0,301	-0,317	0,079	0,021	0,635	0,075	0,143	0,038

The fourth principal component describes the social structure of the study area.

The fifth principal component highlights the degree of attractiveness of the territory related to the naturalness and the capacity of areas to accommodate tourists.

The sixth principal component highlights the marginalization of smaller municipalities, due to the remoteness from major infrastructure networks though these possess a remarkable cultural heritage.

The seventh principal component describes the location of the assets and the dominance of environmental values compared to the historical and cultural heritage, mainly linked to a rural tradition.

The eighth principal component explains the shape of the local economy and the ability of urban centers to meet the needs of the population.

The ninth principal component has high correlation values between the railway and the distribution of the rivers in the area, confirming the substantial similarity of distribution.

Finally, the cluster analysis built on what has emerged from principal components clearly shows how the area of the district of Sibiu is characterized by a strong homogeneity and it is linked to the cultural and social environmental features that can be seen as in Figure 4. The cluster analysis confirms the fact that the microregion Valea

Hartibaciului is a territorial network structured on an homogeneous Territorial Capital. The presence of important natural areas protected by the Natura 2000 network and a huge tangible and intangible heritage, in some cases higher than what there is in the other towns of the district, which has its characteristic note the presence of a dense network of fortified churches, seem to be the distinctive and characteristic of the territory. Together with them, the system is structured on the presence of the river Hartibaciu, around which the valley extends and which is the backbone of the system as well as an ideal location to know and understand their peculiarities. The municipality of Sibiu and its conurbation Selimbar represent a sort of "gateway" to this territorial system, opening up perspectives and development opportunities and they are the heart of the tourist district. In social terms, the 49% of the total population live there and they are the center of main facilities. Finally, they have high accessibility thanks to the presence of the major road infrastructure, the railway and the only airport in the district. In particular, Sibiu is able to offer a wide cultural offer and has a strong appeal, both to the small towns of the district, and at national and international level.

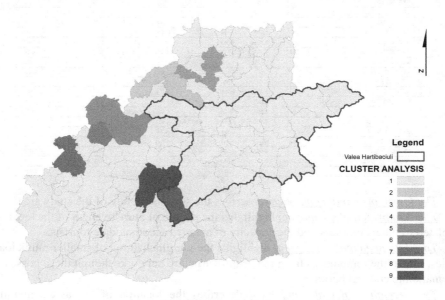

Fig. 4. Cluster Analysis

The application of the theoretical model to the case of Valea Hartibaciului, in the district of Sibiu in Romania has shown how this can be considered in all respects as a territorial network not only because among the existing municipalities there is a dense network of relationships, built over the years and now institutionalized through a Local Action Group, but also because it appears to be characterized by the same identity characters. With reference to the case study of Valea Hartibaciului, it is a homogeneous characters area, and it is aggregated within a single group. There is the presence of important natural areas have the upper hand on agriculture, there are small urban centers which tend to be depopulated or with a high percentage of elderly population, there is the presence of the river Hartibaciu that is configured as the backbone of the valley and which ideally connects to Sibiu area. The results obtained by cluster analysis and by indicators detect the presence of a huge tangible and intangible heritage, in some cases higher than what the heritage in the other towns of the district, with particular reference to the presence of fortified churches, which seem to be the distinguishing and connotative feature of the area. The correlation between this data, the urban areas, the socio-economic structure and tourist flows shows that this area has a potential that is not exploited today and that can be the starting point for the construction of development policies.

5   Conclusions

This paper seeks to explore the issues related to local development models through a methodological reflection on the assessment of the natural propensity of the territory to the establishment of networks in view of the enhancement of shared identity and identify strategies consistent with the objectives of local self-sustainable development. For the identification of territorial networks we have used operating procedures able to highlight network potentials through the creation of a core set of indicators subsequently correlated through the use of multivariate statistical methods. The methodological proposal allows to analyze and evaluate a huge amount of information, highlighting both the relational and identity existing characters which can be the base for the construction of a territorial network and the weaknesses of the territory and the characters that need to be enhanced or improved. These results can be used to structure indications for the drafting of plans and projects on a local and regional level and to define criteria for the evaluation of the networking potential for local self-sustainable development.

Some weaknesses in the methodology must be underlined: first, a basic problem preliminary to any study,   is the collection of the data or of their construction in order to create the system of indicators of territorial development. The available data must have a high degree of detail and, despite this, it is not said that the indicators built are able to translate effectively issues of territorial character. This is particularly true in the case of the development of indicators of intangible heritage. In the second instance, we must not lose sight of the fact that we are faced with an extremely complex system from the point of view of structure and information. The use of clustering techniques based on multivariate statistical analysis are surely a support, but must be related with a real knowledge of the territory. The identification of a strategy consistent with the morphology and territorial levels can afford to build relational development policies in order to protect the territories, landscapes and their resources and responding to the need of development and quality the life of the resident population.

This is a scientific approach that is relevant and necessary if compared to the usual arrangements, and administrative policies used to identify networks and territorial aggregations for participation in calls for funding or for the definition of territorial areas of wide area (i.e. in Italy the provinces or metropolitan areas, in Romania the provinces or the Local Action Groups). The results of the proposed methodology can be seen as an assessment tool on the basis of which to structure development paths based on local resources and territorial peculiarities, identifying the dominant characters or, conversely, elements which, although they may be considered as a resource, requiring policy development.

Future research may relate to the definition of further rules to study and data aggregation and for the involvement of the resident population within the process for the construction of territorial networks and the evaluation of their potential in order to enable local self-sustainable development paths.

References

1. Torre, C.M.: Attori e politiche della competizione urbana. In: Selicato, F. (ed.) Innovazione Procedurale e Progettuale Nella Pianificazione Attuativa. Urbanistica Dossier, vol. 23, pp. 3–5 (1999)
2. Bovaird, T.: Managing urban economic development: learning to change or the marketing of failure? Urban Studies 31, 573–603 (1994)
3. Bovaird, T., Löffler, E., Parrado-Díez, S.: Emerging practices in network management at local levels in Europe. Developing Local Governance Networks in Europe, Baden-Baden: Nomos 9-23 (2002)
4. Yates, D.: The ungovernable city: The politics of urban problems and policy making. MIT Press Cambridge, MA (1977)
5. Massey, D., Amin, A., Thrift, N.: Decentering the nation: A radical approach to regional inequality. Catalyst (2003)
6. Börzel, T.A.: Organizing Babylon-On the Different Conceptions of Policy Networks. Public Administration 76, 253–273 (1998)
7. Meijers, E.: Polycentric urban regions and the quest for synergy: Is a network of cities more than the sum of the parts? Urban Studies 42, 765–781 (2005)
8. Dematteis, G.: Progetto implicito. Il contributo della geografia umana alle scienze del territorio. Franco Angeli. Milano (1995)
9. Dematteis, G.: Possibilità e limiti dello sviluppo locale. Sviluppo Locale 1, 10–30 (1994)
10. Governa, F.: Il milieu urbano. L'identità territoriale nei processi di sviluppo. Franco Angeli. Milano (1997)
11. Dematteis, G., Governa, F.: Territorialità, sviluppo locale, sostenibilità: il modello SLoT.Franco ANgeli. Milano (2005)
12. Magnaghi, A.: Il progetto locale. Verso la coscienza di luogo. Nuova edizione accresciuta. Bollati Boringhieri (2010)
13. OECD: The Territorial State and Perspectives of the European Union. Paris (2011)
14. Roca, Z., Oliveira-Roca, M.D.N.: Affirmation of territorial identity: A development policy issue. Land Use Policy 24, 434–442 (2007)
15. Oliveira, J., Roca, Z., Leitão, N.: Territorial identity and development: From topophilia to terraphilia. Land Use Policy 27, 801–814 (2010)
16. Demšar, U., Harris, P., Brunsdon, C., Fotheringham, A.S., McLoone, S.: Principal component analysis on spatial data: An overview. Annals of the Association of American Geographers 103, 106–128 (2013)
17. Chuman, T., Romportl, D.: Multivariate classification analysis of cultural landscapes: An example from the Czech Republic. Landscape and Urban Planning 98, 200–209 (2010)

18. Vizzari, M.: Spatial modelling of potential landscape quality. Applied Geography 31, 108–118 (2011)

19. Owen, S., MacKenzie, A., Bunce, R., Stewart, H., Donovan, R., Stark, G., Hewitt, C.: Urban land classification and its uncertainties using principal component and cluster analyses: A case study for the UK West Midlands. Landscape and Urban Planning 78, 311–321 (2006)

20. Van Eetvelde, V., Antrop, M.: A stepwise multi-scaled landscape typology and characterisation for trans-regional integration, applied on the federal state of Belgium. Landscape and Urban Planning 91, 160–170 (2009)

21. OECD: Core set of indicators for environmental performance reviews. A synthesis report by the group of State of the Environment. Paris (1993)

22. Rempel, R.S., Andison, D.W., Hannon, S.J.: Guiding principles for developing an indicator and monitoring framework. The Forestry Chronicle 80, 82–90 (2004)

23. Dziock, F., Henle, K., Foeckler, F., Follner, K., Scholz, M.: Biological indicator systems in floodplains–a review. International Review of Hydrobiology 91, 271–291 (2006)

24. Heink, U., Kowarik, I.: What are indicators? On the definition of indicators in ecology and environmental planning. Ecological Indicators 10, 584–593 (2010)

25. Forman, R.T., Godron, M.: Landscape ecology. Wiley and sons New York etc. (1986)

26. Jones, C.B.: Geographical information systems and computer cartography. Longman Harlow (1997)

27. Murray, A.T., Tong, D.: GIS and spatial analysis in the media. Applied Geography 29, 250–259 (2009)

28. Smith, M.J., Goodchild, M.F., Longley, P.A.: Geospatial analysis. Matador (2009)

29. Fabbris, L.: Statistica multivariata: Analisi esplorativa dei dati. McGraw-Hill Libri Italia (1997)

Complex Values-Based Approach for Multidimensional Evaluation of Landscape

Maria Cerreta, Pasquale Inglese, Viviana Malangone, and Simona Panaro

Department Architecture (DiARC)
University of Naples Federico II, via Forno Vecchio 36, 80134 Naples, Italy
{maria.cerreta,viviana.malangone,simona.panaro}@unina.it,
pasqualeinglese@gmail.com

Abstract. The several meanings that landscape takes in all scientific studies and in the common speech highlight the complexity of a concept that finds in the richness of its dimensions the understanding key and the interpreting matrix for actions aimed at local sustainable development. A new concept of landscape identifies the relationships between the various points of view and different interpretive approaches, overcoming the consideration of territory as a physical-geometrical reality at the service of economic aspects. The paper, starting from the evolution of the landscape's concept, focuses on the management of its complexity in the transformation processes included in the dynamic context of landscape's cultural values and in development strategies designed to support and strengthen these values. It has been structured a multidimensional methodological framework oriented to the evaluation of landscape cultural values, tested in National Park of Cilento, Vallo di Diano and Alburni (Italy).

Keywords: Complex landscape, cultural values, multidimensional evaluation.

1 Introduction

Landscape is a place and a concept where insiders and disciplines meet, collide and, increasingly, interact. To improve interaction, and to assist those who care for and manage landscapes, it is important to find ways of achieving a more integrated and comprehensive approach to understanding landscape values.

Traditional landscape assessment methods, which focus on discipline-specific value typologies, may fall short of revealing the richness and diversity of cultural values in landscape held by insiders. Achieving a more integrated approach requires the establishment of a conceptual framework that is inclusive of perceptions founded in disciplinary methodologies and captures the rich and dynamic landscape experienced by insiders. While it is unnecessary for different forms of landscape knowledge to share a methodology or a theoretical foundation, the key is a common frame of reference that has a reasonable fit with the range of ways in which disciplines and communities perceive and value landscape [1].

According to the above perspective, the landscape framework provided by the Cultural Values Model [2] has attempted to offer a conceptual linkage between contemporary

B. Murgante et al. (Eds.): ICCSA 2014, Part III, LNCS 8581, pp. 382–397, 2014.

theory on landscape, space and time with the range of ways in which insiders and disciplines express what is important to them about landscape.

New transformations of landscape concern not just the physical landscape, but also the collective memories, meanings and identities that the landscape holds. Planning theory and practice currently offer relatively little guidance as to how to address meaning and value, particularly at a landscape scale. Recent literature from a variety of disciplines has stressed the need to develop holistic models of understanding landscape. The absence of integration between disciplinary approaches is relevant, and the need to involve communities in defining what is important and distinctive about their own landscapes. The Cultural Values Model sets out to develop a conceptual framework to assist in understanding multiple cultural values in landscape. Although the primary focus of the Model is to address the perceived shortcomings in planning theory and practice, its relevance to inter-disciplinary work also forms a major component of the approach [2]. According to the model, values in landscapes include those expressed by associated communities and those identified through a variety of disciplinary approaches. Using case studies, the nature and range of landscape values as expressed by those with special associations with particular landscapes, examining the nature of the meanings and values ascribed by disciplines with an interest in landscape, and how various disciplines model landscape to convey these values.

An analysis of these findings generates a landscape framework consisting of the Cultural Values Model [2] that offers a conceptual structure with which to consider the surface and embedded values of landscapes in terms of *forms*, *practices* and *relationships*. The landscape framework is found to be useful not only for generating a comprehensive picture of key landscape values, but also in offering an integrated evaluative approach useful both for planners and other landscape-related disciplines.

Much has been written about the significance of landscape to communities and their cultural identity [3,4]. Culture and identity are therefore not just about social relationships, but are also spatial. Inappropriate landscape development can change or obliterate locally distinctive characteristics and cultural meanings, creating a break between communities and their past [5].

The global groundswell of concern about such losses suggests that there may be shortcomings in the identification of landscapes' cultural significance, and that we should pay better attention to how to sustain landscape's contribution to cultural identity and diversity. A landscape's contribution to culture/s requires decision-makers to have a detailed knowledge of the particular values of that place, and how the values help support (or otherwise) cultural identity and diversity. Planning and management decisions would need to be taken in the context of the cultural dynamics of landscapes [6], and new development would need to be designed to support and enhance such values [7]. In order to support this, decision-makers would understand the nature and range of values that may be present in a given landscape, how these are spatially spread, and how they interact [8,9,10,11,12]. Yet current methods of landscape evaluation, as commonly incorporated into national legislation and institutionalized assessment mechanisms, may fail to do justice to the diverse, overlapping and irregularly spread values that are present in landscapes.

Formalized landscape assessments generally undertake to define set categories of value using predetermined criteria (aesthetic, historic, scientific, etc.) and are commonly set up to provide a series of parallel assessments by different disciplinary experts. What is perceived to be of value will depend on the particular interest of the discipline. The result can, firstly, be a static model of significance – a map of aesthetic, historic, and ecological values, for example – with no way of conceiving of the landscape's cultural dynamics as a whole [1].

Multi-disciplinary landscape assessments [13] offer a broader understanding of landscape values than a single discipline, but such collaborations can be hindered by the incompatibility of landscape-related theory and methodology.

The failure to understand landscape in a holistic sense requires an integrated, comprehensive theoretical and analytical framework that adequately address landscape study, assessment and planning. Ideally, such a framework would offer an effective unifying approach that enables the multiplicity of information (from whatever source) to be seen as an interlinked whole.

In relation to considering the cultural significance of landscape, a similarly holistic framework would be need to conceptualize landscape values as a whole, in a way that incorporates the very different assessments of value that might be made to from within different disciplines, as well as the values expressed by insiders for a given landscape [14]. According to this perspective the Cultural Values Model was developed in an attempt to respond to the above challenge by developing a holistic conceptual structure for considering the diversity of cultural values that might exist in any given landscape, and how these might relate to and reinforce one another. In order to avoid capture, a conscious choice was made to step aside from the lenses of predetermined value typologies, and instead to attempt to discover, from communities themselves, what it was about their landscapes that they particularly valued. The development of the model was informed by contemporary theories on the nature of landscape, and prevailing holistic models of landscape.

The paper, starting from the methodological framework proposed by the Cultural Values Model, proposes a multidimensional evaluation approach in order to identify cultural values of the National Park of Cilento, Vallo di Diano and Alburni, in Campania Region, Italy, through a testing conducted in the village of Castel Ruggero, municipality of Torre Orsaia. In the second section, we analyze the theoretical and methodological assumptions related to the concept of landscape cultural values. In the third section, we describe the evaluation process elaborated for Castel Ruggero case study. In the fourth section, we examine the results and express some reflections outlining future developments of the research.

2   Landscape Cultural Values: A Methodological Approach

Current interpretations propose that culture is a dynamic process whereby people are actively engaged in constructing group life and its products [15]. People are considered to live culturally rather than in cultures, with the generative source of culture being human practices rather than in representations of the world. These dynamic senses of culture are particularly relevant and can be a key interpretative concept of cultural values [15,16]. The concept of value is generally considered a social construction arising from the cultural contexts of a time and place. Brown et al. [17] suggest that people hold certain values but also express value for certain objects. In this sense, understanding how a landscape is valued involves understanding both the nature of the valued "object" (or aspect of landscape), and the nature of the expressed value/s for that object. These values do not speak for themselves: they can be identified when they are expressed by those who are part of the cultural context, or by those who are in a position to observe and understand.

Arising from the evolving meanings of "culture" and "values", cultural values are taken to be those values that are shared by a group or community, or are given legitimacy through a socially accepted way of assigning value. This suggests that there can be multiple ways of valuing landscapes: values shared by those within an associated group as well as those attributed by disciplinary experts [1].

At the same time, the perception generally differs between "insiders" and "outsiders": most experts, developers and policy-makers are outsiders to the area where they are to work, but an outsider can also be a person who does not belong to the local community, the same socio-economic group or have the same education and training.

Insiders and outsiders perceive, understand and create the landscape around them through the filter of their social and cultural background.

A relevant example of the Cultural Values Model implementation [2] is that of the two case studies of Bannockburn and Akaroa, in the South Island of New Zealand. The choice of case study areas was guided by a preference for landscapes that were distinctive, had recognized and varied cultural values, and had a resident community of which some people at least were likely to have developed strong connections with the landscape over time [18,19].

The Bannockburn area, a broad inland valley within rugged tussock-covered ranges, was extensively mined for gold the 19th century, and today is renowned for its quality vineyards. The Akaroa basin has at its heart a long narrow harbor, a shoreline is dotted with small settlements, and is encircled by rural and forested land rising to steep volcanic ridges. In both areas, Maori still retain close links with the land. Additionally, both landscapes are known to be undergoing relatively rapid modification from influxes of newcomers and land use changes.

The methodology applied consists of subjecting an in-depth semi-structured interview to a sample of people of different culture, profession, age and economic status, choices between permanent and temporary residents. The interviews were centered on the question: *what is important to you about this landscape?* What interviewees had to say about their landscape was used as the "way in" to understanding the meanings and values built up through their experience of the landscape.

The results were analyzed for statements that conveyed that the interviewees attributed some importance or significance to that matter, regardless of whether it fitted any preconceived notion of landscape held by the interviewer.

The data was further selected according to whether the expressed sentiments were shared or supported by others. From this broad picture of values as a whole, patterns and linkages were sought. Both case studies have revealed that the values are not limited to physical forms of landscape, but also to past and contemporary practices, and internal relations of the landscape itself. Although visual and experiential aspects of the landscape have emerged as important, members of the community have also given great importance to the values that had developed over time.

By the analysis result clearly many overlaps between the landscape interests of community members (insiders) and disciplines (outsiders). Insider perspectives were founded in personal experience and knowledge of place. It was also notable that insiders emphasized intangible values to a far greater extent that would usually be elicited through standard expert-based studies of landscape's material forms. As well, community members, did not generally confine themselves to landscape as defined through standard assessment typologies, but ranged freely across many topics. This is not to say that insider views are necessarily more right than those of outsiders: the crucial issue is that both forms of knowledge contribute to understanding landscape values as a whole. From the analysis and interpretation of the responses to the interviews the Cultural Values Model was developed and structured by three components of value. The first, denoted by the term *forms*, belong the physical, tangible and measurable elements of the landscape, both natural and artificial. The second, indicated by the term *relationships* is related to the second category to which belong the links generated by people-people interactions in the landscape, those generated by people-landscape interactions, and valued relationships within the landscape even where there is little or no direct human involvement. The third component, defined *practices*, is inclusive of both human practices and natural processes and include past and present actions, traditions and events; ecological and natural processes; and those practices/processes that incorporate both human and natural elements.

Indeed, human activity affects natural processes and, conversely, natural processes affect human activity. It thus appears how nature and culture are closely related and how natural processes are inseparable from the cultural ones [2].

These three fundamental components - *forms, relationships* and *practices* - offer the basis for an integrated understanding of landscape and its values and encompass the range of landscape values expressed by both disciplines and insiders. There is a clear call within contemporary thinking on landscape and space that it is necessary to move beyond static understandings, and to be inclusive of movement, social practice, and time. By considering the three model components in a dynamic sense, it can be seen that *practices, forms* and *relationships* are continually interacting to create landscape. Such interactions were implicit in many of the reported values from the case studies, and it was rare for interviewees to talk about one component (e.g. a *form*) without further elucidating its value in terms of *practices* or *relationships*, or both.

It is therefore proposed that these dynamic interactions help generate cultural values, and are also generated by them. Data analysis suggests that landscape has yet a further dimension: *temporality*. The time-thickness of landscapes was clearly evident in the case studies, where interviewees spoke of aspects of the past when referring to their landscapes.

Accordingly, a further variant on the model represents landscape as a *continuum*, bearing within it the forms, relationships and practices of the past that influence those of the present, and thereby shape landscape as it is perceived. It expresses the concept that landscape is created from the dynamic interactions of forms, practices and relationships, occurring over time, and that landscape values are contingent on elements from both the past and present. Landscape is thus always changing, carrying forward the threads of the past and weaving them into the future.

To describe the distinction between past and present value the terms *surface values* and *embedded values* are proposed [2]: surface values are the perceptual response to the directly perceived forms, relationships and practices, while embedded values arise out of an awareness of past forms, practices and relationships.

These concepts form the basis of the Cultural Values Model, offering a provisional framework for understanding multiple cultural values in landscapes: landscapes can be understood in an integrated way through consideration of forms, relationships and practices; the dynamic interactions amongst these; and the dimension of time. These components give rise to, and result from, cultural values in landscapes. Accordingly, it is necessary to take account of all of these landscape components to achieve an understanding of cultural values as a whole.

3 Cultural Values Model implementation: The Experience in Castel Ruggero

The Cultural Values Model approach has been the methodological framework considered for identifying and evaluating the landscape cultural values of Castel Ruggero, a village of Torre Orsaia municipality, in National Park of Cilento, Vallo di Diano and Alburni (here in after NPC). This experience is part of the Research Project "Cilento Labscape: an integrated model for the activation of a Living Lab in the National Park of Cilento, Vallo di Diano and Alburni", funded by FARO Program 2012-2014 "Funding for the Start of Original Research", University of Naples Federico II.

This research aims to develop a methodological framework that integrates the contribution of expert knowledge with context-aware knowledge to activate a Living Lab based on an approach of open innovation, in order to outline an innovative model of smart endogenous development and to enhance the local landscape resources. This proposal seeks to formulate an innovative approach that integrates the concept of Living Lab and the complex meaning of Smart Landscape by structuring a model of interpretation and evaluation of landscape cultural values, which can be implemented for the enhancement of the landscape of the NPC. This Park is enlisted as UNESCO World Heritage Site, MAB-UNESCO List of Biosphere Reserves, it is a Geopark and it is member of the UNESCO HELP-BASIN network.

The study area comprises 95 municipalities and Torre Orsaia, with Castel Ruggero village, is one of them. One of the aim of the research is testing a process of territorial co-design based on the interaction between landscape values and human economy ones. In Castel Ruggero a workshop was organized in order to address the issue of revitalization of abandoned (or in state of progressive abandonment) landscapes, helping to create a virtuous circle of introduction and management of innovation, by networking resources and planning. The village of Castel Ruggero (Figure 1) is the context where micro-actions of revitalization are co-designed, paying attention to local conditions and, at the same time, opening to innovation.

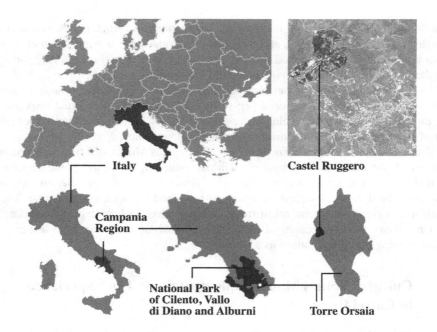

Fig. 1. Castel Ruggero localization

About fifty architecture students took part to the workshop of the University of Naples Federico II. With the support of tutors and experts, they participated to a three days activity in research, whose aim was to seek the conditions for village's transformations processes according to the Living Lab methodology: find out, co-design and test solutions with the users local community. This methodology is based on the involvement of specific groups of interest, in order to identify together particular needs and solutions to them.

In Castel Ruggero's workshop a group of insiders, mostly elderly inhabitants, has been compared to a group of outsiders, young architecture students, for letting the village's latent values come out and for making a map of values describing in an innovative way the context in new circuits.

The methodological approach has been articulated in three main phases (Figure 2):

1. Cognitive framework, aimed to surveying hard data and soft data;
2. Data processing, aimed to identifying values' meanings;
3. Evaluation, aimed to identifying relations between values and meanings.

In cognitive framework phase, the surveying of open spaces and buildings, elaborated in GIS maps was useful to carry out the knowledge of the village through specific analyses of physical features of buildings (details and levels of neglect of buildings, conditions of degradation, etc.) and of open space (use and characteristics of open spaces). At the same time, the cognitive framework was integrated with: a survey of in depth semi-structured interviews to the inhabitants aimed to identify the values and their meaning for them; a focus group with inhabitants and students for public sharing of visions and actions; a collection of storytelling of the architecture students, that give an external point of view, from those who spent time in the village has guests.

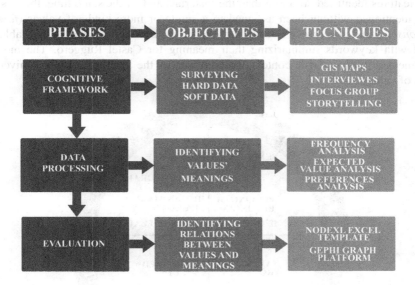

Fig. 2. The phases of the methodological approach

In order to identify the complex values for insiders and outsiders the Cultural Values Model approach has been used to decode soft data. This kind of information has been a very useful benchmark to understand in details not only different points of views, but also significant physical features of the village with different meanings assigned. This has helped the selection of a set of information relevant for triggering small co-designed actions with the local community.

More precisely the Cultural Values Model allowed to find out the potential of the range of values assigned to the Castel Ruggero's landscape, and to cope with the identification and the interpretation of values with an interdisciplinary attitude.

Starting from the interviews and the storytelling, the meanings and the values given to the landscape have been defined and classified in: *forms*, *relationships*, and *processes*. The concept of practices, just the original Model, has been replaced by that of processes, able to better explain the main dynamics that have characterized the changes and influenced the perception of the values.

In specific terms, *forms* include natural features, contemporary features, human-made features, and historic features; *relationships* express sense of community, stories, feeling of belonging and sense of place; *processes* identify natural processes and human processes; *temporality* recognizes embedded values.

This classification in its use reflects the values expressed, by testing their reliability and their applicability. It clearly shows, indeed, that the physical features of the land-scape are strictly linked to the immaterial ones, and, thus, to the values that assign them their meaning and importance. As it can be seen in Tables 1 and 2, the imma-terial features identified are more than the materials and, at the same time, there is no place mentioned without been assigned to a single or more kinds of values (*forms*, *relationships* and *processes*). Each of these latter has been represented in the Tables 1 and 2 with keywords summarizing their meaning for Castel Ruggero. The model, implemented in the specific context, aimed to survey the local meaning for universal forms of values.

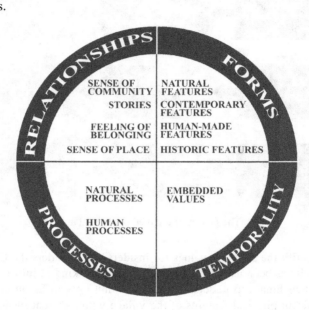

Fig. 3. Cultural Values Model interpretation: forms, relationships, processes and temporality

To analyze and identify the Castel Ruggero landscape values, the weight of the keywords and places mentioned in the interviews has been assigned according to their frequency and the expected values suggested by both the insiders and outsiders.

In details, the frequency regards the keywords and places referring to a particular value in the interviews and storytelling, while the expected value expresses the rank-ing of importance of keywords and places mentioned by the insiders and the outsid-ers. The preference is the synthesis of frequency and expected values. It allows to identifying a final ranking of keywords and places that define the specific values for Castel Ruggero village. Each keyword enriches the semantic domain of the observed issues, and knowledge is represented as a network of values linked by different asso-ciations. The main purpose of this form of interpretation is to find a finite set of basic semantic features defined without ambiguity that combined according to specific rules expresses landscape complex values of Castel Ruggero.

Table 1. Storytelling decoding: forms, relationships, and processes

	Values	Keywords	Places	Frequency	Expected value	Preference
Relation-ships	Stories		'Santa Maria'	2	1°	0,200
			'Pecorelli'	1	1°	0,100
			'Iannuzzi'	1	2°	0,050
			'Castello'	1	4°	0,025
	Sense of places	Rurality		1	1°	0,100
		Slowness		1	2°	0,050
		Nostalgia		3	1°	0,300
		Relation man-nature		1	1°	0,100
		Naivety		1	1°	0,100
		Serenity		2	2°	0,100
		Waiting for change		6	1°	0,600
		Value of time		1	1°	0,100
		Reliving childhood		1	2°	0,050
		Memory		1	1°	0,100
		Romantic		1	2°	0,050
	Sense of community	Friendly		2	1°	0,200
		Hospitality		2	1°	0,200
		Hostility		2	1°	0,200
		Isolation		1	2°	0,200
		Get together	'Puosti'	2	1°	0,050
			'Lavatoio'	1	2°	0,200
	Feeling of belonging	Home	'Torchio'	1	2°	0,200
			Luigia's home	5	1°	0,500
			Biagio's home	1	1°	0,100
			Mafalda's home	1	1°	0,100
			Corrado's home	1		0,100
		Roots	Carmine's home	1	2°	0,200
			'Castello'			

The graphic representation of the semantic networks identifies *nodes*, variably connected by arcs that indicate the *semantic relationship* between two values. Each value, considered as a node in a *network* of values, has a specific weight in the network depending on the quality and quantity of relationships that generates with the other nodes. The different values identify a *complex semantic network*, considered as large collection of interconnected nodes.

Networks are graphs that describe the structures of interacting systems and give substantial information about the patterns of connections between the nodes in a particular system. Knowing about the structure of networks and their arrangements enables one to make certain types of predictions about their behavior.

Referring to the graph theory and network analysis [20,21], to detect the weight of each value within its semantic domain, some analysis have been made that return the following indicators:

Table 2. Storytelling decoding: forms, relationships, and processes

	Values	Keywords	Places	Frequency	Expected value	Preference
Forms	Natural features	Sea		5	1°	0,500
	Contemporary	Mountains		3	2°	0,150
	features	View	'Castello'	8	1°	0,800
		Rural		3	1°	0,300
		Property speculation		2	2°	0,100
		Empty houses		2	1°	0,200
		Cars		1	3°	0,033
		Barriers		1	1°	0,100
		Watching		1	2°	0,050
		Uphill		2	1°	0,200
		Ruin		1	2°	0,050
		Mansions		2	2°	0,100
	Human-made	Alleys		3	1°	0,300
	structures	Mansions		1	1°	0,100
		Ruin		1	2°	0,050
	Historic features	Decay		2	1°	0,200
		Portal	'Imbriachi'	1	2°	0,050
			'Servi'	1	3°	0,033
			'Pecorelli'	7	1°	0,700
			'Mariosa'	6	1°	0,600
			'Iannuzzi'	1	2°	0,050
			'Santa Maria'	2	3°	0,066
			'Castello'	1	4°	0,025
Processes	Natural processes	Earthquake		1	3°	0,033
	Human processes					
		Sense of community		2	1°	0,200
		Depopulation		4	1°	0,400
		Property speculation		1	1°	0,100
		Physical neglect		2	1°	0,200
		Farming		1	1°	0,100

— *Betweenness* is a centrality measure of a vertex within a graph and quantifies the number of times a node acts as a bridge along the shortest path between two other nodes;

— *Closeness centrality* provides a measure of the distance of a node from all other nodes, indicating which points in the network minimize the average distance between the nodes.

— *Eccentricity* is a parameter associated with every conic section and it can be thought of as a measure of how much the conic section deviates from being circular.

— *Eigenvector centrality* is a measure of the influence of a node in a network. It assigns relative scores to all nodes in the network based on the concept that connections to high-scoring nodes contribute more to the score of the node in question than equal connections to low-scoring nodes.

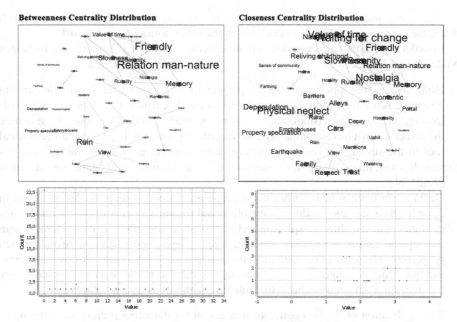

Fig. 4. Indicators: Betweenness and Centrality Distribution

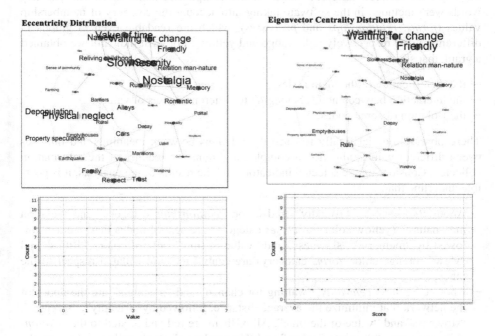

Fig. 5. Indicators: Eccentricity and Eigenvector Centrality

The above indicators take into account, in the calculation of the weight of each node, the relationships that it generates with the other nodes of the network.

In order to analyze the complex semantic network we used the software packages NodeXL and Gephi:

— NodeXL is a free, open-source SNA plug-in for use with Excel. It provides instant graphical representation of relationships of complex networked data, allowing collecting, analyze, and visualize a variety of networks. NodeXL is used to visualize the structure of conversations around specific topics. It is applied as an analytical tool in the social, information, and computer sciences as well as the focus of research in human computer interaction, data mining, and data visualization [22,23,24];
— Gephi is an open source software for graph and network analysis that works with complex data sets and produces valuable visual results. Gephi is a tool for Exploratory Data Analysis, able to exploring and understanding graphs: the user interacts with the representation, manipulate the structures, shapes and colors to reveal hidden properties. The goal is to support making hypothesis, intuitively discover patterns, isolates structure singularities or faults during data sourcing. It is a complementary tool to traditional statistics, as visual thinking with interactive interfaces [25].

The application of the two software was useful for detecting preferences coming to the relations between different values in the complex semantic network. The key words were included in the software taking into account the category of membership values (*forms*, *relationships* and *processes*), which are readable even in graphs in different colors (respectively red, orange and yellow). For each indicator we obtained a graph that shows:

— the identification of the main nodes;
— the interaction between nodes belonging to different classes of values;
— the isolation of some nodes.

These observations let finally to trace the relations between meanings and thus between different values, allowing a complex and dynamic reading of the information collected. Considering the selected indicators and the results of the analysis, it is possible underline that:

— About *Betweenness*, "friendly" (red color, *relationships* category) and "relation man-nature" (yellow color, *processes* category) are central nodes that interact, followed by "memory", "slowness" and "value of time" (all red color). "Ruins" and "view" (orange color, *forms* category) are central too, but isolated respect to the others.
— About *Closeness centrality*, "waiting for change" and "nostalgia" are the nodes of the network that minimize the average distance, followed by "friendly", "serenity", "slowness" and "value of the time". All of them are red and related to the *relationships* category. "Physical neglect" and "depopulation" (both yellow color) are the two other nodes related to the *processes* category.

— About *Eccentricity*, "nostalgia" and "slowness" are the main relevant nodes, followed by "waiting for change", "value of time", "friendly" and "serenity", with reference to the *relationships* category. All of them are of red color. In this case, too, "physical neglect" and "depopulation" (both yellow color) are the two other relevant nodes related to the *processes* category.

— About *Eigenvector centrality*, the nodes with more influence are "waiting for change", "friendly", followed by "value of the time". All of them are red and are part of the *relationships* category. "Relation man-nature" is the most relevant for the *processes* category (yellow color); "ruin", "empty houses" and "decay" (all orange color) are nodes of the *forms* category.

Therefore, we can identify the main relevant values that characterize the landscape of Castel Ruggero, considering the following groups:

1. "friendly", "waiting for change", "nostalgia" for the *relationships* category;
2. "ruins" for the *forms* category;
3. "relation man-nature", "physical neglect" and "depopulation" for the *processes* category.

According to the perceptions of insiders and outsiders, Castel Ruggero's landscape is expression of a network of values that make explicit the deep ties with the specificity of the context, but also the emotional dimension that derives from the cultural relationship with the places; they are not recognizable in measurable components and need an integrated approach to understanding intangible aspects.

4   Conclusions

The research of landscape values is, thus, a complex exercise that requires to investigate not only the meanings of the values, but also the relationships among them. This second stage of the study tests, therefore, the way the relations between meanings can increase the semantic domain of keywords, in order to better understanding the complex values and to express them by relating themselves to multiple categories of values.

The model adopted defines a systematic framework for understanding landscape values, and analyzing insiders and outsiders data and perceptions. The application of NodeXL and Gephi tools, according to the graph theory and network analysis, enables this information to be synthesized by capturing and locating key landscape-related values in a simulation of space and time.

The landscape framework provides a way of conceptualizing them by a structured approach of conceiving cultural values and linking them by a language based on interrelated key concepts. The model creates a basis for understanding, sharing and communicating landscape values. They also sketch out a theoretical structure that incorporates both qualitative and quantitative spatial significance.

In the further stages of the research, the complex values will be represented in the spatial dimension, reconsidering the places meanings. Indeed, the cultural values influence planning practice, it is necessary to be able to account for those values a spatial sense. Landscape's features mapping is an inappropriate model, since it offers little to the understanding of cultural values, while the support of cognitive maps may well be highly informative.

Landscape, as an inclusive concept, allows to overcome the fundamental division between nature and culture, incorporating the idea that assessments of natural values are a cultural construct.

References

1. Stephenson, J.: The Cultural Values Model: An Integrated Approach to Values in Landscapes. Landscape and Urban Planning 84, 127–139 (2008)
2. Stephenson, J.: Values in Space and Time: A Framework for Understanding and Linking Multiple Cultural Values in Landscapes. PhD, Geography. Otago University, Dunedin, NZ (2005)
3. Gray, J.: A Rural Sense of Place: Intimate Experience in Planning a Countryside for Life. Planning: Theory & Practice 4(1), 93–96 (2003)
4. Hay, R.: A Rooted Sense of Place in Cross-Cultural Perspective. Canadian Geographic 42(3), 245–266 (1998)
5. Antrop, M.: Why Landscapes of the Past are Important for the Future. Landscape and Urban Planning 70(1-2), 21–34 (2005)
6. Cerreta, M., De Toro, P.: Urbanization Suitability Maps: A Dynamics Spatial Decision Support System for Sustainable Land Use. Earth System Dynamics 3(2), 157–171 (2012), http://www.earth-syst-dynam.net
7. Fusco Girard, L., Torre, C.M.: The Use of Ahp in a Multiactor Evaluation for Urban Development Programs: A case Study. In: Murgante, B., Gervasi, O., Misra, S., Nedjah, N., Rocha, A.M.A.C., Taniar, D., Apduhan, B.O. (eds.) ICCSA 2012, Part II. LNCS, vol. 7334, pp. 157–167. Springer, Heidelberg (2012)
8. Cerreta, M.: Thinking through Complex Values. In: Cerreta, M., Concilio, G., Monno, V. (eds.) Making Strategies in Spatial Planning, Knowledge and Values, vol. 9, pp. 381–404. Springer, Dordrecht (2010)
9. Cerreta, M., Diappi, L.: Adaptive Evaluations in Complex Contexts. Introduction. Scienze Regionali – Italian Journal of Regional Science 13, 5–22 (2014)
10. Cerreta, M., Poli, G.: A Complex Values Map of Marginal Urban Landscapes: An Experiment in Naples (Italy). International Journal of Agricultural and Environmental Information Systems 4, 41–62 (2013), http://www.igi-global.com
11. Perchinunno, P., Rotondo, F., Torre, C.M.: The Evidence of Links between Landscape and Economy in a Rural Park. International Journal of Agricultural and Environmental Information Systems 3(2), 72–85 (2012), http://www.igi-global.com
12. Montrone, S., Perchinunno, P., Torre, C.M.: Analysis of Positional Aspects in the Variation of Real Estate Values in an Italian Southern Metropolitan Area. In: Taniar, D., Gervasi, O., Murgante, B., Pardede, E., Apduhan, B.O. (eds.) ICCSA 2010, Part I. LNCS, vol. 6016, pp. 17–31. Springer, Heidelberg (2010)
13. Hayden, D.: The Power of Place. The MIT Press, Cambridge (1995)
14. Williams, R.: The Country and the City. Oxford University Press, New York (1973)
15. Johnston, R., Gregory, D., Pratt, G., Watts, M.: The Dictionary of Human Geography. Blackwell, Oxford (2000)
16. Thrift, N., Whatmore, S.: Cultural Geography: Critical Concepts in the Social Sciences. Routledge, London (2004)

17. Brown, G., Reed, P., Harris, C.: Testing a Place-based Theory for Environmental Evaluation: An Alaska Case Study. Applied Geography 22, 49–76 (2002)
18. Stephenson, J.: Many Perceptions, One Landscape. Landscape Review 11(2), 9–30 (2007)
19. Stephenson, J., Bauchop, H., Petchey, P.: Bannockburn Heritage Landscape Study. Department of Conservation, Wellington (2004)
20. Aldous, J.M., Wilson, R.J.: Graphs and Applications: An Introductory Approach. Springer, Dordrech (2000)
21. Wasserman, S., Faust, K.: Social Network Analysis: Methods and Applications. Structural Analysis in the Social Sciences. Cambridge University Press, Cambridge (2008)
22. Bonsignore, E.M., Dunne, C., Rotman, D., Smith, M., Capone, T., Hansen, D.L., Schneiderman, B.: First Steps to Netviz Nirvana: Evaluating Social Network Analysis with NodeXL. In: Proceedings of the 1st IEEE International Conference on Computational Science and Engineering, CSE, pp. 332–339 (2009)
23. Hansen, D.L., Schneiderman, B., Smith, M.: Analyzing Social Media Networks with NodeXL: Insights from a Connected World. Morgan Kaufmann, Burlington (2010)
24. Mendes Rodrigues, E., Milic-Frayling, N., Smith, M., Shneiderman, B., Hansen, D.: Group-In-a-Box Layout for Multi-faceted Analysis of Communities. In: Proceedings of the 3rd IEEE International Conference on Social Computing, CSE, pp. 354–362 (2011)
25. Bastian, M., Heymann, S., Jacomy, M.: Gephi: An Open Source Software for Exploring and Manipulating Networks. In: International AAAI Conference on Weblogs and Social Media, pp. 1–2 (2009), http://gephi.org

Valuing Cultural Landscape Services: A Multidimensional and Multi-group SDSS for Scenario Simulations

Raffaele Attardi, Maria Cerreta, Alfredo Franciosa, and Antonia Gravagnuolo

Department of Architecture (DiARC), University of Naples Federico II, via Forno Vecchio 36, 80134 Naples, Italy
{raffaele.attardi,cerreta,alfredo.franciosa, antonia.gravagnuolo2}@unina.it

Abstract. The purpose of this paper is to define a methodological proposal towards a Spatial Decision Support System for strategic planning, based on the evaluation of Cultural Landscape Services (CLS). A combination of multidimensional evaluation techniques, multi-group analysis and Geographic Information Systems is applied to the simulation of landscape enhancement scenarios in the "National Park of Cilento, Vallo di Diano and Alburni", in order to explore the effectiveness and helpfulness of the evaluation of CLS in structuring both hierarchic and networking relationships among the municipalities comprised in the study area.

Keywords: Spatial Decision Support System, ecosystem services, landscape services, cultural services, multidimensional evaluation, multi-group analysis.

1 Introduction

In the last forty years, natural resources gained increasing attention in the global research agenda, drawing public attention to the issue of conservation of biodiversity [1]. Studies in this direction have been intensified in the last twenty years, in particular under the initiatives of the United Nations with regard to the assessment of the consequences of ecosystem changes on human well-being and the definition of a scientific basis for the implementation of actions aimed at the conservation and sustainable use of environmental resources [2]. The Millennium Ecosystem Assessment (MEA) [3] and The Economics of Ecosystems and Biodiversity [4] have been the first leading approaches focused on a global scale assessment of the direct and indirect benefits that people get from the ecosystem through the identification of ecosystem services, i.e. the various "utilities" that ecosystems provide to humans [3]. Ecosystem services have gained a key role in the scientific research, in order to investigate the close relationship between ecosystems and human well-being in an anthropocentric perspective [3]. The process that underlies the paradigm of ecosystem services is defined as a "cascade" process, which involves natural structures and environmental processes, human-induced ecological phenomena and individual or collective benefits [5].

B. Murgante et al. (Eds.): ICCSA 2014, Part III, LNCS 8581, pp. 398–413, 2014.

Several authors proposed systems of classification of ecosystem services [1,3, 6,7,8,9,10] starting from the theoretical model proposed by MEA, which identifies four major types of services:

— supply services of physical assets that produce direct benefits to people;
— services carried out by ecosystems in regulating environmental processes;
— services related to cultural and spiritual needs of the community;
— support services, which do not provide direct benefits to people but are required for the functioning of ecosystems.

In direct relation with them, human well-being and its socio-economic conditions may be affected in terms of safety, survival, enjoyment of basic materials for life and evolution, psychophysical health and opportunities of social relations.

In literature Landscape Services (LS), are examined as a further specification of ecosystem services considered at a regional scale [11,12], in which diverse and dynamic human and environmental forces need to be considered. According to De Groot et al. [1], on a global scale people receive only a part of the ecosystem services; while the landscape scale can reduces the distance between local actors and the environment, enhancing the services enjoyed. Landscape, as opposed to ecosystems, can be considered as an action context for not strictly ecological disciplines, and for a number of tangible and intangible services provided to humans [13,14], through which the conditions for the sustainable development of the territory are to be found.

LS are significant if they are interpreted as connections between the ecological knowledge of the landscape and the "cooperative" landscape planning (involving the knowledge and needs of local stakeholders) [15]. These theoretical premises call for integrated spatial assessment methodologies able to involve several fields of knowledge, in order to examine the benefits of services to local stakeholders and to identify the existing relationships among services in a specific geographic area.

The literature concerning the assessment of LS identifies different methodological frameworks for the identification, mapping and synthetic representation of LS. Nevertheless, three distinct phases can be recognized in most of the framework proposed so far: *knowledge*, *processing* and *selection*. The *knowledge* phase relates to the recognition of landscape services through the collection of *hard data* (maps, statistical datasets and other data based on conventional studies reported in the literature) and *soft data*, i.e. perspectives, verbal feedback or key concepts. The latter are generally gathered through the involvement of experts and stakeholders. The *processing* phase refers to the organization, standardization and cartographic representation of collected data in order to produce maps of services. Finally, the *selection* phase can be considered a synthesis of the *processing* phase and may be concluded with different outputs depending on the specific objectives: the evaluation of existing services, the evaluation of the transformations already in place and/or the simulation of alternative scenarios.

Among the different typologies of LS, this paper focuses on the study and assessment of Cultural Landscape Services (CLS).

Cultural services are defined by MEA as the intangible benefits that people receive from ecosystems through spiritual enrichment, cognitive development, reflection, recreation and aesthetic experience, including cultural systems, social relations and the aesthetic value [3]. A further specification is provided by Chan [16], which defines them as the contribution of the ecosystem to the intangible benefits (experience, skills) that people derive from human-ecological relationships. They are often dependent on intermediate services [10], and the cultural benefits are frequently combined with other forms of tangible and intangible capital [17]. In literature services defined as "cultural services" [18] are those which satisfy the needs of daily life, as a function of information [8], as comfort and gratification services [19], as comfort services[1,20], or as services for the satisfaction of socio-cultural needs [21]. CLS constitute an important category of services at a landscape scale, as they are able to express the "sense" of a place and the identity of a community interacting over time in a specific area. In fact, it is widely agreed that their main characteristic is the intangibility of values that they express [22,23]. Physical, emotional and psychological benefits of cultural products are often only implicitly expressed [24] through indirect manifestations. For this reason, although they are often mentioned, cultural services are treated as a residual category since they are difficult to assess [25] and, therefore, poorly integrated in landscape management plans [26]. Indeed, except from their recreational and aesthetic values, cultural heritage and educational values [20], CLS are rarely traced through economic based and negotiable indicators [27] and they rarely occur in policy-making processes since they are difficult to evaluate and to communicate [26], [17]. Therefore, the importance given to CLS is almost entirely associated with tangible services [25], [28] and it is closely related to local and personal value systems. As a result, many international studies focus on the mapping benefits rather than CLS, obtaining a quantification through monetary evaluation methodologies [18,29,27,30] applied in specific areas (e.g. protected areas in the case of recreational value), excluding the potential benefits, for which it is difficult to get reliable indicators [31,32]. However, monetary evaluation of CLS has been, however, largely contested in the literature [25]. Several authors argue that while the techniques of monetary evaluation can be successfully applied to the objects of cultural heritage, the evaluation of some aspects such as the identity and sense of place is still largely uncertain [33]. Other experiences has been conducted to quantify and map the cultural services on the basis of the aggregation of social interests within specific typologies of landscape [25,34]. In this case, only hard data have been used (land use, natural emergencies) to map services [34,35,36] through the Geographic Information System (GIS), limiting the understanding of all possible connections between ecological systems and social systems [25]. CLS, in fact, are not represented by purely ecological phenomena, but they are the result of complex and dynamic relationships between people and ecosystems within a specific landscape during large periods of time [37]. Starting from the approaches outlined so far, this paper proposes a methodology for the processing and evaluation of scenarios with the aim to enhance CLS through a multi-dimensional and multi-group approach [38,39]. The proposed methodology is applied to the "National Park of Cilento, Vallo di Diano and Alburni" (hereinafter NPC) in the Italian province of Salerno, Campania region. In Section 2, the methodological approach is introduced together with the description of the case study. In section 3, the stages of the methodological proposal is examined in relation to its application to the case study. Finally, in Section 4 we discuss the results and draw some methodological conclusions and possible future developments of the research.

2   Cultural Landscape Services Evaluation

This paper is part of the Research Project "Cilento Labscape: an integrated model for the activation of a Living Lab in the National Park of Cilento,Vallo di Diano and Alburni" funded by FARO Program 2012-2014 "Funding for the Start of Original Research", University of Naples Federico II. This research aims to develop a methodological framework that integrates the contribution of expert knowledge with context-aware knowledge to activate a Living Lab [40] based on an approach of open innovation, in order to outline an innovative model of smart endogenous development and to enhance the local landscape resources [12, 41]. This proposal seeks to formulate an innovative approach that integrates the concept of Living Lab and the complex meaning of Smart Landscape by structuring a model of interpretation and evaluation of LS, which can be implemented for the enhancement of the landscape of the NPC. This Park is enlisted as UNESCO World Heritage Site, MAB-UNESCO List of Biosphere Reserves, it is a Geopark and it is member of the UNESCO HELP-BASIN network. The study area comprise 95 municipalities and it is commonly known as "Cilento area". In this territorial context, a complex and multidimensional landscape services system can be found, in which multiple relationships interact and potentially regenerate themselves to be catalysts of sustainable development processes.

On this basis and for the stated purposes, the aim of this paper is to present a spatially explicit methodology for the evaluation of a broad set of cultural landscape services. In particular, the main goals are: the identification of appropriate indicators for the assessment of CLS; the processing of a complex map of the multi-functional landscape of Cilento; the definition of a multi-actor and multi-criteria decision problem for the scenario simulation, in order to create a network among the municipalities comprised in the study area. The overall methodological framework is based on data availability at a regional spatial scale, using the municipalities as service providing units. Several authors defined classification systems for the interpretation of CLS depending on the purpose of their research. For the purposes of our study, existing classifications of CLS [3,4], [8,9], [37], [42,43] have been appropriately aggregated into six categories and they have been investigated considering the 95 municipalities in the area as analysis units. In particular, CLS have been classified into six sub-services categories: aesthetic services, scientific and educational services, historical and cultural services, tourism services and recreational facilities, religious and spiritual services, identity services.

Based on to the phases described in Section 1, the methodological framework is structured into the following steps:

1. *knowledge* of cultural services;
2. *mapping* and *classification* of CLS spatial indicators;
3. *processing* and *evaluation* of CLS maps;
4. construction of CLS complexity maps and simulation (*selection* phase) of multifunctional landscape scenarios for the NPC (Figure 1).

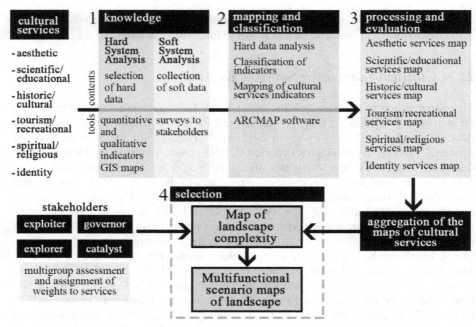

Fig. 1. The methodological framework: phases and contents

The *first phase* concerns the knowledge of the six categories of CLS in the cultural landscape of Cilento, which comprise two complementary approaches: Hard and Soft System Analysis [14], [39]. Through the application of Hard System Analysis, each service has been investigated on the basis of a selection of quantitative-qualitative indicators, which have been identified in order to structure a synthetic and objective picture of the study area through the rational choice of the statistical data and maps, which have been processed using a GIS. Soft System Analysis uses soft data which are expression of subjective perceptions and have been made explicit starting from the analysis of thirty interviews that involved a group of stakeholders on significant issues regarding the sustainable enhancement of landscape in Cilento area. The *second phase* focuses on the development of CLS spatial indicators, which have been mapped using appropriate GIS tools. The *third phase* is the evaluation and processing of synthetic maps of six CLS category, based on the spatial indicators, in order to define a hierarchy among the municipalities, based on their provision of CLS. The synthetic maps have then been aggregated in a single map of the complex multi-functional value of landscape. Finally, in the *fourth phase*, the point of view of the stakeholders has been considered, in order to define possible coalitions and develop a scenario simulation.

3   A Methodological Proposal

3.1   Knowledge and Perception of Cultural Services

The acquisition of knowledge about CLS has been performed through the Hard System Analysis. Required data have been gathered through available institutional databases and web information. In particular, the website of the NPC provides geo-referenced data in shape-file format, regarding its morphological, naturalistic and functional features, the architectural heritage and the transport facilities. Moreover, the NPC provided information of the one-hundred best innovative firms in Cilento area, in term of sustainable and ecological business leading. The database of the Chamber of Commerce, Industry, Handicraft and Agriculture of the province of Salerno, and the website of the Campania Region provided the number of firms by NACE classification. Through databases and websites the following data have been collected on a municipality base: the number of traditional, cultural, religious and wine-and-food events; the number of scientific national and international research programs in which the municipalities of the NPC are involved; the number of typical foods; the number of workshops activated and self-organized by local communities for the exploitation of local resources. Considering the information gathered, a set of qualitative indicators has been selected for each CLS. The choice of the indicators depends on their own significance in the comprehension and critical reading of the landscape of the NPC (Table 1).

Based on the Soft System Methodology [14], [39], soft data have been gathered through the analysis of an early sample of thirty structured interviews (to be increased) of four stakeholder groups with interests in the sub-region of NPC:

— Exploiters: representatives of common knowledge (18 respondents: locals and tourists);
— Explorers: representatives of scientific knowledge (5 respondents: scholars and researchers);
— Catalysts: representatives of technical knowledge (2 respondents: a hotelier and a retailer);
— Governors: representatives of the institutional knowledge (5 respondents: public administration managers and officers).

The interview is then structured to elicit the perception [44] of critical issues and potential future scenarios of transformation and their implementation strategies, as well as the interests of each social group.

Based on the interviews, a frequency analysis has been performed in order to identify a preference order of the CLS categories for each social group (Table 2). Starting from the preference frequency, each CLS category is evaluated on a semantic scale (perfect, very good, good, more or less good, moderate), in order to allow a multi-group analysis for the identification of possible coalitions of social groups in the construction of strategic scenarios (see par. 3.4).

Table 1. Cultural Landscape Services (CLS) and input data for the spatial indicators

CLS	ID	Input data for the processing of indicators
aesthetic	1	Municipal area occupied by pathways
	2	Panoramic points
	3	Scenic roads
scientific/ educational	4	Number of caves and resurgences
	5	Number of geosites
	6	Number of scientific researchs programs and number of R&D participatory workshops
	7	Number of educational farms
	8	Number associations and innovative firms
historic/ cultural	9	Surface area included in local archaeological areas
	10	Municipal area occupied by archaeological sites
	11	Number of historical monuments
	12	Municipal area occupied by historic routes
	13	Number of cultural events
tourism/ recreational	14	Municipal area occupied by tourist routes
	15	Number of transport nodes
	16	Municipal area occupied by provincial roads
	17	Municipal area occupied by local roads
	18	Number of accommodation services and restaurants
spiritual/ religious	19	Number of religious buildings
	20	Number of religious events
	21	Number of patronal feasts
identity	22	Number of local products
	23	Number of traditional events
	24	Number wine and food events
	25	Number of firms with the label "The 100 Friends of the Park"
	26	Number of manufacturing activities
	27	Number of farming, forestry and fishing activities

3.2   Mapping and Classification of CLS Spatial Indicators

On the basis of hard data on Table 1, appropriate indicators for CLS evaluation for each of the six categories of services have been selected. Consequently, a geodatabase with selected input data for each municipality has been structured. The data processing phase include the data standardization and classification into five classes valuated on a scale from 1 (the lowest performance class) to 5 (the highest performance class). Data relating to indicators 1,2,3,4,5,9,10,11,12,14,16,17,19 have been standardized with respect to the municipal area; data relating to the indicators 18,26,27 have been standardized with respect to the population of each municipality (population data gathered during the national Italian census 2011). After the standardization process, data have been classified into five classes through the Jenks Natural Breaks Algorithm. In order to model the indicators 7,8,13,20,21,23,24,25 (mostly related to the number of events per year), the municipalities have been clustered into five classes according to the number of inhabitants (less than 1000, between 1000 and 2000, between 2000 and 5000, between 5000 and 8000, more than 8000); on a second step, a set of *if…and…then* rules have been identified for the assignment of the indicator value to each municipality, as shown in Table 3. The problem that leads to choose the above-described approach is that when dealing with the number of events per year or similar data, the dimension of the municipality cannot be ignored and the standardization by the number of inhabitants would lead to unreliable assessments. On the other hand, a bonus should be assigned, for example, to small municipalities that exhibit particularly active in organizing events or bequeathing traditional events.

Table 2. Preferences expressed by stakeholder groups

CLS	Exploiter		Catalyst		Explorer		Governor	
	response rate	semantic scale	response rate	semantic scale	response rate	semantic scale	response rate	semantic scale
aesthetic	9	Very good	1	Moderate	3	Perfect	0	Moderate
scientific/ educational	13	Perfect	4	Very good	3	Perfect	1	Good
historic/ cultural	10	Very good	5	Perfect	3	Perfect	1	Good
tourism/ recreational	11	Perfect	1	Moderate	3	Perfect	2	Perfect
spiritual/ religious	3	More or Less good	2	More or Less good	1	More or Less good	0	Moderate
identity	7	Good	3	Good	2	Very good	0	Moderate

The indicator 15 (number of access nodes) has been modelled through a two steps process: at first, the different types of nodes (ports, docks, highway junctions, train stations, state highway junctions, intermodal stations) have been mapped. For each node, each municipality has been assigned a partial index "d", whose value is:

— '1' if the node is within the municipal boundaries;
— '0.5' if there is not a node within the considered municipal boundaries, but there is one within the boundaries of an adjoining municipality
— '0' to all other municipalities.

This operation has been repeated for each of the six types of nodes listed above. In this research, the above described approach has the limitation of not considering the nodes in the municipalities outside the area of investigation but bordering it. The nodes have been grouped into three transport modes (road, rail and water). Each type of node and each transport category has been assigned a weight (Table 4) for the aggregation into a single index (H) calculating the weighted average, as shown in the following equation (1):

$$H = 0,5(w_i d_i + w_a d_a + w_s d_s) + 0,3\, d_r + 0,2(w_p d_p + w_d d_d) \qquad (1)$$

The H indices of each municipalities have been classified into five classes with the Jenks Natural Breaks Algorithm.

The indicator 22 (number of typical products) has been derived from data without standardization and classified by the Jenks Natural Breaks Algorithm. The indicator 6 has been modelled with *if...and...then* rules (Table 5) combining the data on the number of scientific research and the number of research and development participated workshops activated. In this case, it is unnecessary to consider the standardization by the population as projects and workshops mostly involve several municipalities with particular characteristics or included in area of scientific environmental/cultural interest.

Table 3. Rules adopted for the classification of indicators 7,8,13,20,21,23,24,25

IF (population)	AND (n. of events)	THEN (indicator value)
Any	0	1
<1000	1	4
<1000	>1	5
[1000;2000]	1	3
[1000;2000]	2	4
[1000;2000]	>2	5
[2000;5000]	1	2
[2000;5000]	2	3
[2000;5000]	3	4
[2000;5000]	>3	5
[5000;10000]	1	2
[5000;10000]	2	3
[5000;10000]	[3;5]	4
[5000;10000]	>5	5
>10000	1	1
>10000	2	2
>10000	[3;4]	3
>10000	[5;7]	4
>10000	>7	5

Table 4. Classification of transport nodes typology

Transport mode	Typology of nodes
Road transport $w_h=0.5$	Intermodal stations (d_i); $w_i=0.4$
	Highway junctions (d_a); $w_a=0.35$
	State road junctions (d_s); $w_s=0.25$
Rail transport $w_r=0.3$	Railway station (d_r)
Water transport $w_w=0.2$	Ports (d_p); $w_p=0.7$
	Docks (d_d); $w_d=0.3$

Table 5. Rules adopted for classification of indicator 6

IF (number of research projects)	AND (number of participatory labs)	THEN (indicator value)
0	0	1
0	1	2
1	0	3
0	2	3
1	1	4
0	3	4
2	0	5
2	2	5

3.3   Processing and Evaluation of Cultural Services Maps

After obtaining the values of all the indicators examined in each municipality, a map related to each of the six categories of CLS has been processed by calculating the average value of the indicators describing each category.

We obtained six carriers v_i ($i = 1, ... 6$) with 95 components a_{ij} ($j=1,...95$), which represent the value of each aggregated indicator for each municipality with respect to the CLS category j. The maps of the six categories of CLS have been aggregated by calculating the vector V, that is the arithmetic average value of the vectors v_i, thus obtaining a map of the complexity of landscape values of the NPC (Figure 2).

Tyrrhenian Sea

Cultural Landscape Services complex map

■ low   medium   ■ high

Fig. 2. Complex multi-functional landscape map

3.4   CLS Complexity Maps and Simulation of Multifunctional Landscape Perceived Scenarios

The introduction of the preferences of social groups allows the definition of a multi-group decision problem for the simulation of scenarios for landscape enhancement. In this sense, the scenario maps are functional to the creation of a network of municipalities, identifying *leader municipalities* (high ranking), *bridge municipalities* (medium ran-king) and *isolate municipalities* (low ranking).

From the map of the complex multi-functional value of landscape it is possible to develop feasible scenarios if the preferences expressed by the four stakeholder groups are introduced in the process of aggregation of the CLS category maps, By analysing the preferences of the stakeholders, we identified the possible coalitions. This resulted in the construction of a shared scenario. Therefore, a multi-group analysis of the preferences has been structured with the NAIADE (Novel Approach to Imprecise Assessment and Decision Environments) [45,46].

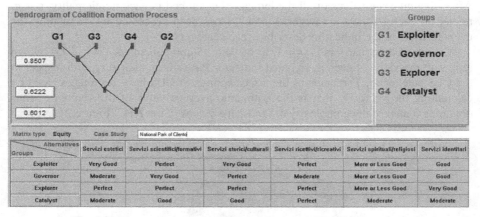

Fig. 3. Equity analysis results obtained through the NAIADE method

The NAIADE method allows to perform the analysis of equity with a fuzzy clustering algorithm. Equity analysis starts with the creation of the equity matrix which gives a linguistic indication of the interest group judgment ruling for each of the alternatives (i.e. CLS categories). Semantic distance is used to calculate the similarity indexes among interest groups. A similarity matrix is then computed starting from the equity matrix. The similarity matrix gives an index, for each pair of interest groups, of the similarity of judgment over the proposed alternatives. Through a sequence of mathematical reductions the dendrogram of coalition formation is built which shows possible coalition formation for decreasing values of the similarity index and the degree of conflict among social groups. The results of the multi-group analysis carried out by the NAIADE method are shown in Figure 3. Furthermore, for each coalition the indices of conflict relating to each CLS category are computed. These indices with values between 0 and 1 describe the level of conflict related to each category within the coalition: the higher the index, the lower the consensus over a specific category.

In order to calculate the weight that each coalition gives to each category, the indices of conflict have been used to calculate the complement to unity and normalizing in respect to the sum. The weights of each coalition (Table 6) are useful for the processing of maps of landscape complexity for each coalition, which are simulations of scenarios for the creation of a network among municipalities maintaining the hierarchy of *leaders*, *bridges* and *isolates* (Fig. 4). The final preference order of the CLS categories is significant not only because it expresses the achievement of a higher level of consensus among stakeholders, but also because it allows the rational allocation of weights to each CLS for the construction of a shared scenario of landscape services.

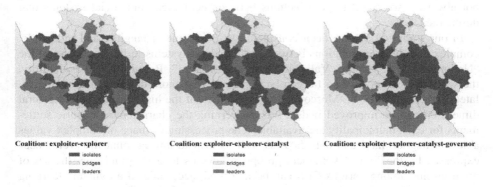

Fig. 4. Scenario simulation for three possible coalitions of stakeholders

Table 6. Coalitions for CLS categories

CLS Categories	Coalition 1: Exploiter-Explorer	Coalition 2: Exploiter-Explorer-Catalyst	Coalition 3: Exploiter-Explorer-Catalyst- Governor
aesthetic	0,18	0,13	0,09
scientific/educational	0,19	0,20	0,26
historic/cultural	0,18	0,20	0,26
tourism/recreational	0,19	0,26	0,16
spiritual/religious	0,11	0,09	0,09
identity	0,15	0,12	0,14

4   Discussion and Conclusions

The assessment of Cultural Landscape Services (CLS) is generally associated with tangible services [25,28] and it is closely related to the local and personal value systems.

CLS are often dependent on intermediate ecosystem services [10]; consequently the cultural benefits arise from the combination of different forms of capital [17], whose spatial representation is still an open question.

Many studies have focused on benefits rather than mapping services, resulting in a quantification through monetary evaluation methodologies [18,29,27,30].

Furthermore, the methodologies presented in the literature have been applied in specific areas, excluding from the maps those areas of potential benefits for which it is difficult to get reliable indicators [31,32]. Other attempts have been made to quantify and map the CLS based on the aggregation of social interests for specific services within specific types of landscape [34,25,47], using specific classes of objects (land use, natural emergencies, etc.). Using GIS to map the cultural values [34,35,36], does not always describe all the connections between ecological and social systems that define the CLS.

In our study, CLS have been considered as the result of tangible and intangible, complex and dynamic relations between man and ecosystems in the landscape of the "National Park of Cilento, Vallo di Diano and Alburni". Based on data availability, the set of indicators used so far can be improved with more detailed information related to the hard data set. Moreover, the assessment of the historical and architectural dimension can be improved if data sets concerning the characters of historic settlements for each municipality are available. The processing of maps of complex values able to integrate hard and soft data, as well as to take into account the preferences expressed by the four stakeholders groups, allows us to understand how the set of relationships among various CLS can be modified, according to a dynamic learning process that determines different network synergies and complementarities between the different municipalities.

Performing multi-group analysis with NAIADE is very helpful for the identification of possible coalitions among stakeholders groups, even with an unstructured data set of social preferences. Comparing the map of the objective complexity of the landscape values, where the relations between the different services are represented, with maps of the possible scenarios that take into account the point of view of the three coalitions (Exploiter-Explorer, Exploiter-Explorer-Catalyst; Exploiter-Explorer-Catalyst-Governor), we point out how the aggregation of categories of CLS changes, making explicit how the role of certain municipalities changes, moving from *bridges* to *leaders*, or from isolates to *bridges*. Thus, the methodology allows to outline a network among municipalities, taking into account the role played by CLS, enabling internal municipalities to become driving forces of the inland areas, in the same way of those located along the coast. Therefore, this multidimensional assessment allows to include both tangible and intangible values, making explicit the different components that characterize landscape in a dynamic and incremental sense.

References

1. De Groot, R., Alkemade, R., Braat, L., Hein, L., Willemen, L.: Challenges in integrating the concept of ecosystem services and values in landscape planning, management and decision making. Ecological Complexity 7, 260–272 (2010)

2. Cerreta, M.: Thinking through complex values. In: Making Strategies in Spatial Planning, vol. 9, pp. 381–404. Springer, Dordrecht (2010)
3. MA (Millennium Ecosystem Assessment): Ecosystems and Human Well-being: The Assessment Series (Four Volumes and Summary). Island Press, Washington, DC (2005)
4. TEEB: The Economics of Ecosystems and Biodiversity: Ecological and Economic Foundations. Earthscan, London (2010)
5. Haines-Young, R., Potschin, M.: The links between biodiversity, ecosystem services and human well-being. In: Ecosystem Ecology: A New Synthesis, pp. 110–139. University Press, Cambridge (2010)
6. Costanza, R., Folke, C.: Valuing Ecosystem Services with Efficiency, Fairness, and Sustainability as Goals. In: Nature's Services: Societal Dependence on Natural Ecosystems, pp. 49–68. Island Press, Washington, DC (1997)
7. Heal, G.: Valuing ecosystem services. In: Ecosystems, vol. 3(1), pp. 24–30. Springer, Dordrecht (2000)
8. De Groot, R., Wilson, M.A., Boumans, R.M.J.: A typology for the classification, description and valuation of ecosystem functions, goods and services. Ecological Economics 41, 393–408 (2002)
9. De Groot, R.: Function-analysis and valuation as a tool to assess land use conflicts in planning for sustainable, multi-funcional landscapes. Landscape and Urban Planning 75, 175–186 (2006)
10. Fisher, B., Turner, R.K., Morling, P.: Defining and classifying ecosystem services for decision making. Ecological Economics 68, 643–653 (2009)
11. Limburg, K.E., O'Neill, R.V., Costanza, R., Farber, S.: Complex systems and valuation. Ecological Economics 41, 409–420 (2002)
12. Cerreta, M., De Toro, P.: Integrated Spatial Assessment for a Creative Decision-Making Process: A Combined Methodological Approach to Strategic Environmental Assessment. International Journal of Sustainable Development 13(1-2), 17–30 (2010)
13. Musacchio, L., Wu, J.: Collaborative landscape-scale ecological research: Emerging trends in urban and regional ecology. Urban Ecosystem 7, 175–178 (2004)
14. Cerreta, M., Poli, G.: A Complex Values Map of Marginal Urban Landscapes: An Experiment in Naples (Italy). International Journal of Agricultural and Environmental Information Systems 4, 41–62 (2013)
15. Termorshuizen, J.W., Opdam, P.: Landscape services as a bridge between landscape ecology and sustainable development. Landscape Ecology 24, 1037–1052 (2009)
16. Chan, K.M.A., Guerry, A.D., Balvanera, P., Klain, S., Satterfield, T., Basurto, X., Bostrom, A., Chuenpagdee, R., Gould, R., Halpern, B.S., Hannahs, N., Levine, J., Norton, B., Ruckelshaus, M., Russell, R., Tam, J., Woodside, U.: Where are cultural and social in ecosystem services? A framework for constructive engagement. BioScience 62(8), 744–756 (2012)
17. Chan, K.M.A., Goldstein, J., Satterfield, T., Hannahs, N., Kikiloi, K., Naidoo, R., Vadeboncoeur, N., Woodside, U.: Cultural services and non-use values. In: Natural capital: Theory and practice of mapping ecosystem services. Oxford University Press, Oxford (2011)
18. Costanza, R., D'Arge, R., de Groot, R., Farber, S., Grasso, M., Hannon, B., Limbyrg, K., Naeem, S., O'Neill, R., Paruelo, J., Raskin, R.G., Sutton, P., van del Belt, M.: The value of the world's ecosystem services and natural capital. Nature 387, 253–260 (1997)
19. Boyd, J., Banzhaf, S.: What are ecosystem services? The need for standardized environmental accounting units. Ecological Economics 63, 616–626 (2007)
20. Kumar, P.: The Economics of Ecosystems and Biodiversity (TEEB) Ecological and Economic Foundations. Earthscan, London and Washington (2010)

21. Wallace, K.J.: Classification of ecosystem services: problems and solutions. Biological Conservation 39, 235–246 (2007)

22. Adekola, O., Mitchell, G.: The Niger Delta Wetlands: threaths to ecosystem services, their importance to dependent communities and possible management measures. International Journal of Biodiversity Science, Ecosystem Services & Management 7, 50–68 (2011)

23. Daw, T., Brown, K., Rosendo, S., Pomeroy, R.: Applying the ecosystem services concept to poverty alleviation: The need to disaggregate human well-being. Environmental Conservation 38, 370–379 (2011)

24. Kenter, J.O., Hyde, T., Christie, M., Fazey, I.: The importance of deliberation in valuing ecosystem services in developing countries — evidence from the Solomon Islands. Global Environmental Change 21, 505–521 (2011)

25. Daniel, T.C., Muhar, A., Arnberger, A., Aznar, O., Boyd, J.W., Chan, K.M.A., Costanza, R., Elmqvist, T., Flint, C.G., Gobster, P.H., Gret-Regamey, A., Rebecca, L., Muhar, S., Penker, M., Ribe, R.G., Schauppenleher, T., Sikor, T., Soloviy, I., Spiernburg, M., Taczanowska, K., Tam, J., von der Dunk, A.: Contributions of cultural services to the ecosystem services agenda. PNAS 109(23), 8812–8819 (2012)

26. De Groot, R., Ramakrishnan, P.S., Berg, A.V.D., Kulenthran, T., Muller, S., Pitt, D., Wascher, D.: Cultural and amenity services. In: Findings of the Condition and Trends Working Group of the Millennium Ecosystem Assessment, Ecosystems and Human well-being: Current State and Trends. Millennium Ecosystem Assessment Series, vol. 1, pp. 455–476. Island Press, Washington, D.C (2005)

27. Martín-López, B., Gómez-Baggethun, E., Lomas, P.L., Montes, C.: Effects of spatial and temporal scales on cultural services valuation. Journal of Environmental Management 90, 1050–1059 (2009)

28. Milcu, A., Hanspach, J., Abson, D., Fischer, J.: Cultural ecosystem services: A literature review and prospects for future research. Ecology and Society 18(3), 44 (2013)

29. Angulo-Valdes, J.A., Hatcher, B.G.: A new typology of benefits derived from marine protected areas. Marine Policy 34, 635–644 (2009)

30. Zhang, Y., Singh, S., Bakshi, B.R.: Accounting for ecosystem services in life cycle assessment. Part I: A critical review. Environmental Science & Technology 44, 2232–2242 (2010)

31. Anderson, B.J., Armsworth, P.R., Eigenbrod, F., Thomas, C.D., Gillings, S., Heinemeyer, A.: Spatial covariance between biodiversity and other ecosystem service priorities ecosystem service priorities. Journal of Applied Ecology 46, 888–896 (2009)

32. Eigenbrod, F., Armsworth, P.R., Anderson, B.J., Heinemeyer, A., Gillings, S., Roy, D.B., Thomas, C.D., Gaston, K.J.: The impact of proxy-based methods on mapping the distribution of ecosystem services. Journal of Applied Ecology 47, 377–385 (2010)

33. Butler, C.D., Oluoch-Kosura, W.: Linking future ecosystem services and future human well-being. Ecology and Society 11(1), 30 (2006)

34. PEER (Partnership for European Environmental Research): A spatial assessment of ecosystem services in Europe: methods, case studies and policy analysis – phase 1 (2011), http://www.peer.eu

35. Plieninger, T., Dijks, S., Oteros-Rozas, L., Bieling, C.: Assessing, mapping, and quantifying cultural ecosystem services at community level. Land Use Policy 33, 118–129 (2013)

36. van Berkel, D.B., Verburg, P.H.: Spatial quantification and valuation of cultural ecosystem services in an agricultural landscape. Ecological Indicators 37, 163–164 (2014)

37. Fagerholm, N., Käyhkö, N., Ndumbaro, F., Khamis, M.: Community stakeholders' knowledge in landscape assessment-Mapping indicators for landscape services. Ecological Indicators 18, 421–433 (2012)

38. Attardi, R., De Rosa, F., Di Palma, M., Piscitelli, C.: A Multi-criteria and Multi-group Analysis for Historic District Quality Assessment. In: Murgante, B., Misra, S., Carlini, M., Torre, C.M., Nguyen, H.-Q., Taniar, D., Apduhan, B.O., Gervasi, O. (eds.) ICCSA 2013, Part IV. LNCS, vol. 7974, pp. 541–555. Springer, Heidelberg (2013)

39. Cerreta, M., Panaro, S., Cannatella, D.: Multidimensional Spatial Decision-Making Process: Local Shared Values in Action. In: Murgante, B., Gervasi, O., Misra, S., Nedjah, N., Rocha, A.M.A.C., Taniar, D., Apduhan, B.O. (eds.) ICCSA 2012, Part II. LNCS, vol. 7334, pp. 54–70. Springer, Heidelberg (2012)

40. Eriksson, M., Niitamo, V.-P., Kulkki, S.: State of the art in utilizing Living Labs approach ti user-centric ICT innovation – a European approach (2005), http://www.vinnova.se

41. Cerreta, M., De Toro, P.: Urbanization Suitability Maps: A Dynamics Spatial Decision Support System for Sustainable Land Use. Earth System Dynamics 3(2), 157–171 (2012)

42. Frank, S., Fürst, C., Koschke, L., Makeschin, F.: A contribution towards a transfer of the ecosystem service concept to landscape planning using landscape metrics. Ecological Indicators 21, 30–38 (2012)

43. Luesink, E.: Cultural heritage as specific landscape service Stimulus of cultural heritage in the Netherlands. Wageningen University, Wageningen (2013)

44. Franciosa, A.: La valutazione della qualità percepita del paesaggio: il caso studio della Regione di Valencia. BDC. Bollettino del Centro Calza Bini 13(1), 119–144 (2013)

45. Munda, G.: Multicriteria evaluation in a fuzzy environment. Theory and applications in ecological economics. Contributions to Economics Series. Physica-Verlag, Heidelberg (1995)

46. Montrone, S., Perchinunno, P., Di Giuro, A., Rotondo, F., Torre, C.: Identification of "Hot Spots" of Social and Housing Difficulty in Urban Areas: Scan Statistics for Housing Market and Urban Planning Policies. In: Murgante, B., Borruso, G., Lapucci, A. (eds.) Geocomputation and Urban Planning. SCI, vol. 176, pp. 57–78. Springer, Heidelberg (2009)

47. Fusco Girard, L., Cerreta, M., De Toro, P.: Integrated Assessment for Sustainable Choiches. Scienze Regionali Italian Journal of Regional Science 13(1), 111–142 (2014)

Methods and Techniques for Integrated Mesoscale and Microscale Analysis of Urban Thermal Behavior: The Case of Bari (Italy)

Claudia Ceppi[1], Mariella De Fino[1], Giovanna Mangialardi[2], Simona Erario[3], and Francesco Selicato[2]

[1] DICATECh-Technical University of Bari, Italy
{mariella.defino,claudia.ceppi}@poliba.it
[2] DICAR- Technical University of Bari, Italy
ing.mangialardi@gmail.com, francesco.selicato@poliba.it
[3] Independent Researcher
erariosimona@libero.it

Abstract. Within the general study of urban heat islands, some specific aspects were investigated in the city of Bari, South Italy, where the analyzes were carried out at two different scales, in order to explore the thermal patterns of all the areas including public housing complexes. In fact, it was considered that those are among the most vulnerable to urban and building decay, also due to poor original environmental and constructional quality, so that improvement actions at different decision-making levels are highly desirable. Specifically, starting from a GIS-based analysis of the overall urban area at the mesoscale, through some representative parameters and their relationships with the land surface temperatures, the assessment was further carried out at the microscale, which suggested retrofitting strategies on the buildings, in order to address both district heat mitigation and envelope energy efficiency.

Keywords: urban climate alteration, surface temperature, multi-scale approach, building energy saving.

1 Introduction

The urban heat island (UHI) and the climate anomalies in urban areas result from climate modification, climate change and urban growth. In particular, the UHI describes the influence of urban surfaces on temperature patterns in urban areas as opposed to surrounding areas [1, 2]. In general, the main effects of climate alteration are: damaging for the health of the inhabitants, worsening of outdoor and indoor comfort conditions and increase of global energy consumptions [3].

In literature, two main layers have been recognized to address the study of the atmospheric urban heat island and, thus, of the thermal behaviour: the Urban Canopy Layer (UCL) and the Urban Boundary Layer (UBL). The UCL concerns the urban atmosphere extending upwards from the surface to approximately the mean building height, while the UBL is the layer above the UCL that is influenced by the underlying urban surface [4]. Similarly, the thermal behaviour under canopy can and should be

B. Murgante et al. (Eds.): ICCSA 2014, Part III, LNCS 8581, pp. 414–429, 2014.

broken down into different scales of analysis, that, according to Oke (2004), have been taking into account herein: the mesoscale and the microscale.

The mesoscale is commonly opposed to regional scales and it concerns the study of the whole urban area. Differently, the microscale, namely the district scale, is referred to specific objects, i.e. a building, a road, with dimension ranging from one meter to hundreds of meters, which can influence the microclimate due to their proper thermal properties (emissivity, reflectivity, thermal inertia, ...).

The choice of the multi-scale approach is motivated by the strict interrelation between the different levels of analysis, although they require different approaches and different methods. In fact, the environmental conditions can affect the thermal behaviour of the buildings, whereas building characteristics, namely envelope surfaces rather than shape and structure, can affect the local thermal behaviour (the terms mesoscale, local and microscale scale are used according to[5]).

It is acknowledged that there are close relationships among global climate, local climate, urban planning, urban morphology and building typology. Both global and local climates affect urban planning and morphology and, thus, outdoor and indoor comfort and building energy consumption, since climatic variables such as solar radiation, air temperature and wind, combined with surface/volume ratio, height/distance ratio, natural shading distribution, are vital psychological and functional components of a place. Similarly, the interaction of the building thermal behaviour, due to surface characteristics, construction components and technological facilities with the local climate is relevant. For instance, it is recognized that a major threat to urban heat island and climate anomalies within the urban area is constituted by heat waves. Besides, the cooling loads of the buildings, which are strictly connected to the optical properties of the finishing materials and to the thermal performances of the envelope, is considered one of the primary causes of higher energy consumption and, thus, one of the factors that most influence the climate behavior at the microscale.

The urban environment is therefore faced with a challenge that requires two levels of sensitivity to understand the individual parts and the whole system [6]. On the one hand, the urban planning at the mesoscale can determine the morphology and, on the other hand, the building design of technical and construction characteristics at the microscale can affect the urban climate. Both levels interact and combine to determine the regional climate and relative global consequences, besides affecting the fulfilment of residents' needs. Therefore, in order to analyze those phenomena, the residential complexes of public housing have been taken into account, within the city. Those are considered the most vulnerable to the phenomenon of heat waves, also due to poor original environmental and constructional quality. Moreover, since they are generally owned and managed by public bodies, they might more easily undergo coordinated and integrated improvement measures.

1.1   Methodology

To assess the thermal behaviour at the urban scale, remote sensing data is used derived by ASTER VNIR (Advanced Spaceborne Thermal Emission and Reflection Radiometer- Visible and Near Infrared) images [7]. In particular, the land surface temperature (LST) is considered as derived data. Despite the surface temperature and the air temperature are distinct, in terms of investigated areas and characteristics, these are related. The term surface UHI is used to distinguish the measurements obtained by the remote sensor in alternative to air temperature.

The advantage of using such data is the ability to study large areas, which allows to analyse the emission characteristics of the surfaces diversified by materials, namely colours and finishing, and thus to identify heterogeneous thermal behavioural patterns in the urban area [2].

Based on the preliminary assessment, a representative neighbourhood was further analysed at the urban and building scales. Specifically, urban materials/surfaces were mapped, in terms of solar reflectivity and thermal emissivity, and investigated by infrared thermography, in order to detect the temperature distribution and identify the most critical situations with regard to the heat island effect. Then, building envelope components were assessed, in terms of thermal transmittance compared with normative thresholds, in order to define their influence on the heating and cooling energy demand.

Based on the overall results, some retrofitting strategies were proposed for those cases, where both surface properties and component performances were found poor, in order to merge heat island mitigation strategies and energy saving solutions, according to a multi-scale approach.

1.2   The Case Study

The case study is the city of Bari, Apulia Region, located in Southern Italy.

The city has been thoroughly analyzed within a three-year strategic research project, funded by Apulia Region and European Union and concluded in 2013. The ECOURB project could achieve some interesting results about the assessment of thermal patterns of the city. The present work moves from those results in order to integrate them according to the abovementioned different scales and approaches.

2   The Meso-Scale (Urban) Framework

The meso-scale analysis allows to investigate the relationship between the land surface temperature and the features of the urban settlement. The use of the LST is widely recognize as indicator of the overall thermal pattern and used in conjunction with indicators, such as land cover, land use, vegetation index, urban morphology and others [2, 6, 8-12]. In fact, the LST is an indicator to assess the effects of surface radiative proprieties, energy exchange, internal climate of buildings and human comfort in the cities [11, 13].

Thirty-two areas within the metropolitan area of Bari have been examined in order to recognize their thermal pattern. These areas consist of residential complexes of social housing with public ownership and variously distributed within the city.

Those complexes are hubs, with variable extension, that feature most Italian cities and that might be integrated, on the boundaries or isolated from the consolidated urban area. They show very peculiar morphological configuration with clear settlement connotation, particularly in the applications of the early XX century, compared with the 50's and 60's and, even more, with the recent cases. They give evidence of specific construction and technical traditions throughout the years. They have social and political relevance, as they were designed and built as whole neighborhoods ("167 areas", as designated by the law n.167/1962 before, and "PEEP areas", so defined by the law n. 865/72 afterwards) to meet the housing needs – not without contradictions and failures. Finally, the debate about public housing is challenging within the current policies of urban development, for the fulfillment of the diffuse housing demand of a wide social class and the current use and management requirement of the built heritage.

In the light of the abovementioned aspects, social housing complexes were selected for the present study, also considering their vulnerability to the heat island phenomenon [23]and the possibility to implement their improvement policies directly by the government bodies, owning either/or managing the heritage, with relative positive impacts on the urban areas.

The areas mainly vary for year of construction, ranging from 1903 and 2007, and urban morphology of the districts, in which they are included.

The urban morphology represents an indicator of some key features of the urban space, such as geometry, density, building typology and functions. Moreover, its relation to the spatial pattern of surface temperature within the urban area has been proven [4, 9, 14]. According to the classification of the city of Bari in nine morphology classes [7], the relationship between the district and the urban morphology has been assessed in a GIS environment.

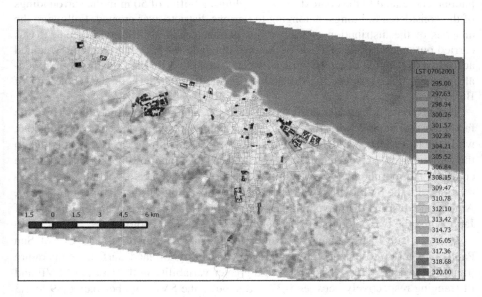

Fig. 1. Thermal image of the city of Bari acquired on 07.06.2001, which has been superimposed on the distribution of public housing complex

The urban morphology that characterizes most of the areas can be identified as "open tissue", as the various complexes of buildings are mostly localized in the peripheral areas of the city centre. It is worth mention that, for the above mentioned purposes of investigation, the in-depth analysis of those areas is particularly interesting, since they represent the border areas with respect to the phenomenon of heat islands, forming a boundary between heavily urbanized areas and suburban areas. Through spatial representation in GIS, four indicators, that are considered particularly descriptive of the districts in relation to their thermal behaviour, are determined. In detail, the thermal data, processed by the ASTER level 1B through the ENVI software, is related to satellite in summer period acquired on 07.06.2001.The other parameters are the Normalized Difference Vegetation Index (NDVI), processed from the same thermal image with a resolution of 15 m, the morphometric protection index and the Sky View Factor (SVF).

The morphometric protection index is a dimensionless parameter that expresses the level of protection from the surrounding areas. With reference to urban areas, it defines the most protected areas inside the buildings and then synthesizes information such as the distance between the buildings and their relative height. It is therefore used, in the present study, as a proxy to determine the effects of sunshine on structures.

The SVF is considered a measure of the radiation geometry of a given site. It ranges from zero to one, representing totally obstructed and free spaces, respectively [15]. The different data were resampled at a resolution of 10 m and were intersected with the data related to digital mapping of the city. In particular, in order to study the parameters related to the context of the buildings, a buffer of 50 m in the surroundings of the considered buildings, belonging to various districts, was created. Following the analysis of the distribution of the various parameters in a GIS environment, it was carried out a descriptive analysis of the data obtained for each building and weighted according to the area of intersection between grid and vector data relating to individual buildings. Fig. 1 shows the distribution of the thermal data for the whole city of Bari.

The bivariate correlation analysis between the factors shows that the SVF is the factor best correlated with the surface temperature, reaching a value of about 0.6, while the NDVI shows an almost negligible correlation with the LST, probably due to the general lack of vegetated surfaces, replaced by impermeable surfaces. On the contrary, the morphometric protection index shows, as expected, a good inverse correlation with the SVF, with a well explained relatively low surface temperature report. Among the districts that have a higher variability of this parameter are the districts of Japigia and the two belonging to the neighbourhood of San Paolo

Both the sub district (named "Comparto A" and "Comparto B and C") of San Paolo, are also those that show a greater variability of land surface temperature (ranges between 307.32-316.61 K) and a greater variability in the values of SVF and PI (ranging respectively between 0,37 and about 1 the SVF, and between 0,39 to 1,3 the PI). [Fig.2][Fig.3].This variability linked to the wider range of surface temperatures within the same district, and the morphological type (open tissue) has lead to the choice of the case study for microscale analysis on San Paolo area, as will be illustrated in the following paragraphs.

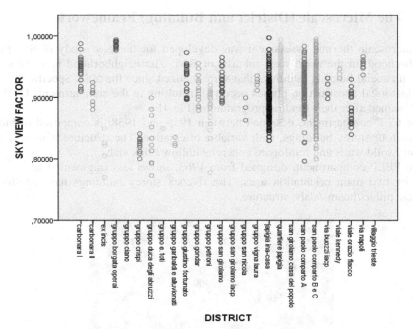

Fig. 2. Scatter plot of the distribution of SVF for the various districts

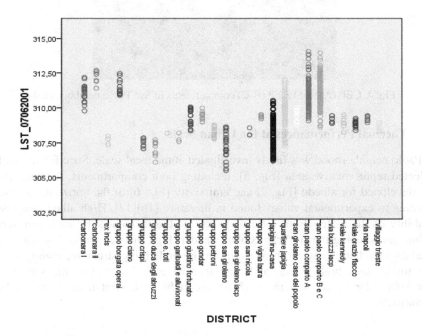

Fig. 3. Scatter plot of the distribution of LST for the various districts

3   The Microscale (District and Building) Framework

The microscale thermal assessment was developed for the case study of San Paolo neighborhood, in the north-west suburbs of Bari. The neighborhood is a residential area with social housing buildings that were realized since the 50's. Specifically, two main historical construction phases occurred, resulting in the compartments CEP and PEEP, named after their funding programmes [Fig.4].

The CEP compartment, designed between 1955 and 1958, is composed of blocks of three/four storey buildings, with variable orientation. The structure is mixed, with masonry solid walls and reinforced concrete/ hollow brick slabs.

The PEEP compartment, designed from 1963, shows less fragmented layout, featured by two main orientation axes. The five/six storey buildings have reinforced concrete pillars/beams/slabs structure.

Fig. 4. CEP (A) and PEEP (B-C) compartments in San Paolo neighborhood

3.1   Thermal Performances at the Urban Scale

San Paolo neighborhood was firstly investigated at the local scale. Specifically, within a selected representative area [Fig. 5], including both compartments, thematic maps were developed for albedo [Fig. 7] and emissivity [Fig. 6] of the horizontal surfaces, according to experimental values found in literature [16][17]. High albedo expresses the ability of the materials of reflecting most of the incident solar radiation during daytime and keeping the surfaces cooler, whereas high thermal emissivity allows the materials to radiate away the heat stored in the underlying structure, mainly during night time. Thus, high albedo and high thermal emissivity allow the surfaces to show high solar reflectance index (SRI) and reduce the heat transfer to the built environment.

Within the case study, the emissivity is quite high, ranging from 0.9 to 0.98, whereas albedo varies from 0.08 to 0.3. Particularly, lowest albedo values feature the roofs that are generally finished by asphalt waterproofing membrane.

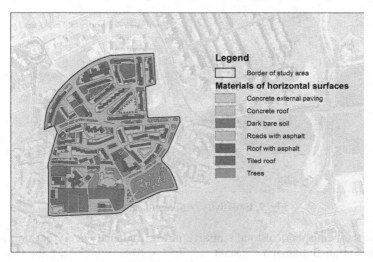

Fig. 5. Materials of horizontal surfaces

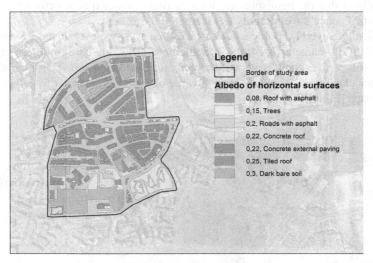

Fig. 6. Albedo of horizontal surfaces

Furthermore, IR thermography mapping of the surfaces was carried out in order to assess their thermal behaviour. IR thermography is a non destructive technique to detect the temperature distribution on a surface, based on the emitted infrared radiation, according to the well known equation:

$$q = \sigma \, \varepsilon \, T^4 \qquad (1)$$

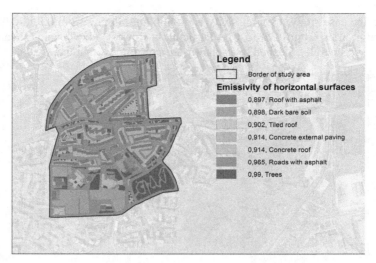

Fig. 7. Emissivity of horizontal surfaces

where q is the hemispherical total emissive power (radiated energy per unit area, W/m2), σ the Stefan-Boltzmann constant (5.67051 x 10-8 W/m2K4), ε the total hemispherical emissivity of the surface (0 < ε < 1) and T the surface absolute temperature (K). It follows that the hemispherical total emissive power q is directly related to the temperature T, once the emissivity ε is known.

Specifically, several typologies of surfaces/materials were analyzed by Avio Neothermo TVS-700 camera (8-14μm) in September 2012, including plastered building facades, roofs with asphalt, roads with asphalt, trees and bare soil. Thermograms were preliminary elaborated, using an average value of ε = 0,95, in order to get qualitative thermal maps [Fig.8][Fig.9].

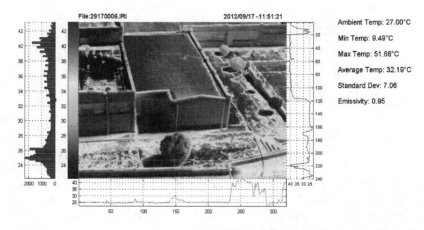

Fig. 8. Thermal map, including plastered building facades and roofs with asphalt

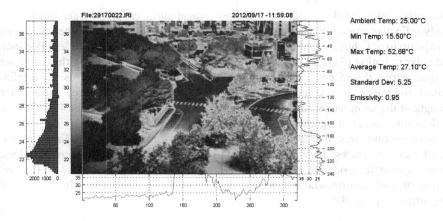

Fig. 9. Thermal map including roads with asphalt, trees and bare soil

Then, quantitative assessment of the surface temperatures was achieved by iteratively correcting the emissivity for all the different materials [Tab.1]. It is worth mention that the correction of emissivity would generally result in a slight temperature variation of 1-2°C on average.

Table 1. Surface/Material versus Emissivity [1]

Surface/Material	Emissivity (8-13 μm)
Trees	0,990
Dark bare soil	0,898
Roads with asphalt	0,965
Roof with asphalt	0,897
Plastered facades	0,950

As an example, for the thermograms showed above, a grid of 10 measurement points was used for each surface/material [Tab. 2].

Table 2. Temperature values

	T_{max} (°C)	T_{min} (°C)	$T_{average}$ (°C)
Fig. 2			
Plastered facade	26,40	25,42	25,81
Roof with asphalt	41,54	40,37	41,02
Fig. 3			
Road with asphalt	38,28	35,10	36,90
Tree	28,28	24,57	26,38
Bare soil	29,50	26,28	27,76

Results confirmed that roofs are the most critical surfaces, as they are finished by poorly reflective materials. At about 12 a.m. in a summer day, their temperature is on average up to 4°C higher that the roads and 16°C higher than close facades. The incident solar radiation plays a key role in the process, as well, considering that roofs are the most exposed surfaces, typically not shaded by surrounding elements. As a reference, a simulation by Ecotect of a group of 25 buildings in the area shows that, throughout the summer season, horizontal surfaces are exposed to about 475 kWh/m^2 incident solar radiation against 143,203 kWh/m2 of vertical surfaces.

Thermography was also used to assess energy performances of buildings. As expected, most structures show missing thermal insulation, resulting in thermal bridges, due to unprotected concrete framework mainly in PEEP compartment, as well as reduction of wall section where heating radiators are located, both in CEP and PEEP compartments [Fig. 10].

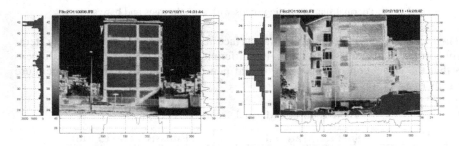

Fig. 10. Thermal bridges due to unprotected concrete framework and heating radiators

3.2   Thermal Performances at the Building Scale

The CEP compartment was built according to the common criteria featuring the social housing complexes by INA Casa Institute in Italy, as thoroughly documented in bibliographic and archivist records, including the four "Handbooks of INA Casa". Particularly, the handbooks were released to support design and construction of buildings belonging to the Institute and comprise guidelines and recommendations from urban planning to technical detailing.

As a consequence, with specific reference to that compartment and based on the available information and the direct investigation of the buildings, the materials and construction techniques were defined and the energy performances of building components were assessed. In detail:

Solid walls were generally made out of plastered calcarenite stone, locally called "tufo". The thickness varied: 60cm at the ground floor, 55cm at the first floor, 50cm at the second floor, 45cm for the third floor. Sometimes, they could be composed of two outer leaves and an inner air cavity, with 43cm total thickness. Plaster was usually made out of cement or lime based mortar with stone powder added.

Basement floors were separated from the ground by a 15-20 cm thick layer of unshaped stone blocks, with cross ventilation by ducts, and finished by concrete/gravel slab, cement mortar and cement tiles.

Roofs were made out of 20-25cm slab with reinforced concrete beams and hollow clay bricks, light concrete 15% slope layer, 7-10 mm asphalt waterproof membrane and eventually cement mortar and cement tiles.

Windows were generally single glazed, with wooden frame and wooden shutters. Most frames are currently made out of aluminum, steel and PVC.

A summary of construction details for the most representative components is shown in Tab.3, along with the calculated thermal transmittance - and relative threshold values according to Italian codes - as well as the attenuation ratio and delay time due to the thermal inertia. The information for all the buildings of the compartment, including photographic documentation, technical drawings and envelope/plant performance details, were stored and managed by GIS platform [Fig.12].

Table 3. Thermal performances of representative envelope components

Envelope opaque components					
Component	Thickness (cm)	$U_{calculated}$ (W/m²K)	$U_{threshold}$ (W/m²K)	Delay time (h)	Attenuation ratio
Solid wall	50	0,99	0,4	8,89	0,1
Cavity walls	43	0,84	0,4	9,06	0,41
Roof	43	1,45	0,38	8,31	0,34
Envelope transparent components					
Component				$U_{calculated}$ (W/m²K)	$U_{threshold}$ (W/m²K)
Single glazed window with wooden frame and outer wooden shutters				2,33	2,6
Single glazed window with aluminum frame and outer PVC shutters				5,8	2,6

Results show that the replacement of wooden frames with aluminum ones doubled the thermal transmittance of the original windows. Moreover, all the traditional envelope components do not accomplish the normative thresholds. The roofs show the poorest performances, requiring a transmittance reduction of 74%, against 60% and 52% of solid and cavity walls respectively.

4   Energy Retrofitting Strategies

Based on the abovementioned assessment, some retrofitting strategies have been evaluated, with specific focus on roof components, that showed the most critical behaviour, in terms of both surface solar reflection index and thermal component transmittance. Specifically, green roofs and cool roofs were assessed by Design Builder and Energy Plus simulations on a building-type, based on the most recurring construction envelope characteristics.

The solutions are widely acknowledged by the scientific community as highly performing, both for mitigation of the urban island effect [17,18,19] and for reduction of building energy consumptions and peak cooling demand [20,21,22]. Specifically, green roofs exploit the capability of the vegetation to absorb the solar radiation for supporting its life-cycle, including photosynthesis, evapotranspiration and respiration. Moreover, the soil layer gives an added insulation to the building roof and the water content increases the thermal inertia of the structure. Differently, cool roofs limit the heat transfer to the underlying structure, due to high reflective/ low absorbing materials, whose high thermal emissivity also allows heat dissipation when the environmental temperature drops.

The building-type is north-south oriented. It is a four storey housing complex, including 3 staircases and 24 apartments, with 70 m^2 net surface. The load bearing structure is composed of 50 cm thick solid masonry walls and 25 cm thick reinforced concrete/ hollow brick slabs. Windows are single glazed and wooden framed with outer wooden shutters. HVACs include gas heating and electricity cooling. Simulations compared the annual heating and cooling demand before and after the interventions, particularly for the apartments at the upper floor, that are supposed to undergo the most significant retrofitting benefit.

Tab. 4 summarizes the simulated scenarios and relative thermal transmittance values. It is worth mention that cool roof and green roof were designed to equally accomplish the normative threshold.

Table 4. Simulation roof scenarios

Component	Layers (from inside to outside)	Thickness (mm)	U (W/m^{2}K)
Original roof	Cement plaster	15	1,45
	Cast concrete (beam + slab)	300	
	Lightweight sloped cast concrete	100	
	Asphalt waterproof membrane	10	
Cool roof	Cement plaster	15	0,27
	Cast concrete (beam + slab)	300	
	Aerogel insulating board	40	
	Lightweight sloped cast concrete	100	
	Asphalt waterproof membrane	10	
	Lime sand render	20	
	White cement tiles (ρ=0,8)	30	
Green roof	Cement plaster	15	0,28
	Cast concrete (beam + slab)	300	
	EPS insulating board	20	
	Lightweight sloped cast concrete	100	
	Asphalt waterproof membrane	10	
	Daku FSD	80	
	Vegetation soil	180	

Table 5. summarizes the annual heating and cooling demand

Component	Heating demand (kWh/m^2)	Cooling demand (kWh/m^2)
Original roof	114,24	32,78
Cool roof	101,45	21,96
Green roof	98,00	21,62

Results show that cool roof and green roof similarly address cooling energy decrease, about 33% saving, whereas the green roof perform slightly better, in terms of heating energy reduction, about 14% against of the cool roof. Although both solutions were designed with same thermal transmittance, it is reasonable that the green roof exploits its thermal inertia during the winter time, too.

5   Conclusion

The paper meant to highlight how meso-scale and micro-scale should interact to address mitigation strategies for climate anomalies within the city and their relative impacts on the comfort of the citizens.

Specifically, the assessment at the district level, based on the outcome of the urban relationship analyses and further specified through the qualification of the buildings, in terms of optical and thermal properties of external surfaces and energy performances of envelope components, enabled the identification of some relevant critical issues for the heat island phenomenon, namely the solar gain of the urban areas and the consumptions/emissions due to the construction and technological systems. As a result, retrofitting strategies and solutions were proposed that were validated at the building level, but can be easily replicated for a variety of cases at the compartment level, toward the sustainable requalification of the built heritage as urban planning target, with positive environmental and energy impacts both at the micro-scale and the meso-scale.

In all, along with building retrofitting, urban requalification should be based on energy quality. Since that requalification mainly concerns the government bodies, public housing complexes, which were mainly examined in the present work, can offer a valuable chance for integrated and coordinated intervention. Nevertheless, such an approach is certainly consistent with the current legislation, e.g. the Landscape Plan of Apulia Region, where it is higly recommended and awarded by Green Certificates the development of green zones in peri-urban areas, as compensation for the consumption of soil and the contribution to the pollution increase.

Acknowledgments. The authors would like to acknowledge Rocco Rubino, Francesco Mancini and Pasquale Balena, Technical University of Bari. Their contribution for supporting data acquisition and elaboration is much appreciated.

References

1. Oke, T.R.: The energetic basis of the urban heat island. Quarterly Journal of the Royal Meteorological Society 108(455), 1–24 (1982)
2. Schwarz, N., Lautenbach, S., Seppelt, R.: Exploring indicators for quantifying surface urban heat islands of European cities with MODIS land surface temperatures. Remote Sensing of Environment 115(12), 3175–3186 (2011)
3. Bouyer, J., Inard, C., Musy, M.: Microclimatic coupling as a solution to improve building energy simulation in an urban context. Energy and Buildings 43(7), 1549–1559 (2011)
4. Voogt, J.A., Oke, T.R.: Thermal remote sensing of urban climates. Remote Sensing of Environment 86(3), 370–384 (2003)
5. Oke, T.R.: Initial Guidance to Obtain Representative Meteorological Observations at Urban Sites, IOM Editor. World Meteorological Organization, Geneva (2004)
6. Zhao, C., et al.: Urban planning indicators, morphology and climate indicators: A case study for a north-south transect of Beijing, China. Building and Environment 46(5), 1174–1183 (2011)
7. Loconte, P., Ceppi, C., Lubisco, G., Mancini, F., Piscitelli, C., Selicato, F.: Climate Alteration in the Metropolitan Area of Bari: Temperatures and Relationship with Characters of Urban Context. In: Murgante, B., Gervasi, O., Misra, S., Nedjah, N., Rocha, A.M.A.C., Taniar, D., Apduhan, B.O. (eds.) ICCSA 2012, Part II. LNCS, vol. 7334, pp. 517–531. Springer, Heidelberg (2012)
8. Gluch, R., Quattrochi, D.A., Luvall, J.C.: A multi-scale approach to urban thermal analysis. Remote Sensing of Environment 104(2), 123–132 (2006)
9. Guo, Z., et al.: Assess the effect of different degrees of urbanization on land surface temperature using remote sensing images. Procedia Environmental Sciences 13(0), 935–942 (2012)
10. Jiang, J., Tian, G.: Analysis of the impact of Land use/Land cover change on Land Surface Temperature with Remote Sensing. Procedia Environmental Sciences 2(0), 571–575 (2010)
11. Weng, Q.: Thermal infrared remote sensing for urban climate and environmental studies: Methods, applications, and trends. ISPRS Journal of Photogrammetry and Remote Sensing 64(4), 335–344 (2009)
12. Yuan, F., Bauer, M.E.: Comparison of impervious surface area and normalized difference vegetation index as indicators of surface urban heat island effects in Landsat imagery. Remote Sensing of Environment 106(3), 375–386 (2007)
13. Voogt, J.A., Oke, T.R.: Effects of urban surface geometry on remotely-sensed surface temperature. International Journal of Remote Sensing 19(5), 895–920 (1998)
14. Weng, Q., Lu, D., Schubring, J.: Estimation of land surface temperature–vegetation abundance relationship for urban heat island studies. Remote Sensing of Environment 89(4), 467–483 (2004)
15. Unger, J.: Connection between urban heat island and sky view factor approximated by a software tool on a 3D urban database. Int. J. of Environment and Pollution 36(1/2/3), 59–80 (2009)
16. Sobrino, J.A., Oltra-Carrió, R., Jiménez-Munoz, J.C., Julien, Y., Sòria, G., Franch, B., Mattar, C.: Emissivity mapping over urban areas using a classification-based approach: Application to the Dual-use European Security IR Experiment (DESIREX). International Journal of Applied Earth Observation and Geoinformation 18(2012), 141–147 (2013)
17. Santamouris, M.: Cooling the cities – A review of reflective and green roof mitigation technologies to fight heat island and improve comfort in urban environments (July 2012)

18. Coutts, A.M., Daly, E., Beringer, J., Tapper, N.J.: Assessing practical measures to reduce urban heat: Green and cool roofs. Building and Environment 70, 266–276 (2013)
19. Zinzi, M., Agnoli, S.: Cool and green roofs: An energy and comfort comparison between passive cooling and mitigation urban heat island techniques for residential buildings in the Mediterranean region. Energy and Buildings 55, 66–76 (2012)
20. Castleton, H.F., Stovin, V., Beck, S.B.M., Davison, J.B.: Green roofs; building energy savings and the potential for retrofit. Energy and Buildings 42, 1582–1591 (2010)
21. Jaffal, I., Ouldboukhitine, S.-E., Belarbi, R.: A comprehensive study of the impact of green roofs on building energy performance. Renewable Energy 43, 157–164 (2012)
22. Akbari, H., Konopacki, S.: Calculating energy-saving potentials of heat-island reduction strategies. Energy Policy 33, 721–756 (2005)
23. Mirzaei, P.A., Haghighat, F., Nakhaie, A.A., Yagouti, A., Giguère, M., Keusseyan, R., et al.: Indoor thermal condition in urban heat island e development of a predictivetool. Building and Environment 57, 7–17 (2012)

"Scrapping" of Quarters and Urban Renewal: A Geostatistic-Based Evaluation

Raffaele Attardi[1], Emanuele Pastore[2], and Carmelo M. Torre[2]

[1] Department of Architecture, University Federico II of Naples
Via Roma 402, Naples, Italy
[2] Department of Civil Engineering and Architectural Science, Polytechnic of Bari
Via Orabona 4, Bari Italy
`raffaele.attardi@unina.it, emanuele.pastore@live.it,`
`carmelomaria.torre@poliba.it`

Abstract. The environmental and social costs of the extensive and intensive development of new areas not yet built are increasingly relevant . For this reason land use decisions are increasingly oriented to consider the development of the city from its existing perimeter , giving up models that generate urban sprawl.

It is also being used increasingly in urban redevelopment - but it requires substantial financial resources that in times of economic hardship are difficult to find by the individual municipal administrations .

Therefore, the public-private partnership becomes an important way forward, not for ideological choice, by certain administrations and researchers, to achieve program objectives and design of all involved subjects.

With new knowledge provided by Geo-analysis, more than 37% of Italian housing property was carried out between 1946 and 1971 and that about all of this amount is currently in poor or very poor condition . And ' possible to speak of " the city to be scrapped ".

The papers show a geo-statistic analysis aiming at finding best for urban renewal by the plus-value recapture of property, deriving from a plan for urban densification

Keywords: Scrapping, Hot spots, Urban densification, land value recapture.

1 The Concept of "Urban Scrapping"

In urban redevelopment we must assess the effectiveness of new instruments based on market rules, such as incentives and rewarding , whether or volumetric tax .
So far these tools have been widely used as an alternative to the expropriation , while the objective is to verify their effectiveness in demolition and reconstruction.
Speaking of urban demolition and reconstruction means intervene heavily in the urban context with radical actions to amend entire urban mesh, coming to warping and changing their composition and volumetric spatial general.

It is also being used increasingly in urban redevelopment but require substantial financial resources in times of economic hardship are difficult to find from the

B. Murgante et al. (Eds.): ICCSA 2014, Part III, LNCS 8581, pp. 430–445, 2014.

individual municipal administrations. Therefore, the public-private partnership becomes an important way forward, not so much for ideological choice by certain administrations and research resources and intelligence to achieve program objectives and design of all stakeholders. By the new data provided by Geographical Information Systems shows that more than 37% of Italian households was carried out between 1946 and 1971 and that about all of this amount is currently in poor or very poor condition. And 'possible to speak of "urban scrapping".

You have to analyze the conditions under which the transfer of value by individuals by means of additional volumes and other reward is enough to make it economically viable for the private project of demolition and reconstruction [1].

2   The Feasibility of the Urban Redevelopment of Degraded Properties

Below is spoken about the argument introduced by Micelli [], by a study aimed at identifying the cost-effectiveness of these operations. An important assumption made by the author consists in the total absence of transaction costs due to the shift register of the lots. In fact, these procedures due to the split of the property can be complex and costly , and decidedly uncertain.

The basic relationship is:

$$Vp > Ve$$

where:
• Vp is the value of the final volume (volume of the most rewarding plan) ;
• Ve represents the capital value of the existing building .
And ' possible to rewrite the inequality analyzing in detail the individual values:

$$Vmt\ ip \times A \times > \times S\ Vme$$

In particular :
• A represents the area of the abutments subject to transformation ;
• Ip is the Index sized in the Buildable Volume deriving from the densification;
• Vmt is the economic value of the potential development in the area (expressed in square meters or cubic meters).

It can also be defined as the product of the unit market value for newly built properties " Vn " multiplied by the coefficient of impact area "a" (which takes into account the quality of the positional subject to change) ;
• S is the gross leasable area of the property;
• Vme is the average market value for existing housing units. It can be defined as the product of the unit value of the market for newly built properties " Vn " multiplied by an eventual coefficient of depreciation due to obsolescence "b".

Therefore we can rewrite the expression by simplifying Vn , introducing the existing density factor "it" being the ratio S / A and then deriving the coefficient " ip" :

$$ip > en \times b / a$$

So the ratio b / a is basically a multiplier of the stocking density exists. It is clear that the condition of feasibility of the intervention in contiguous areas is to assign a building index higher than it was multiplied by the ratio b / a.

If you're planning urban development is divided into non-contiguous areas , the formula is modified as follows :

$$ip > en \times n \times b / to$$

where:

• b / ai is basically a multiplier of existing density "it" obtained from the ratio between the depreciation coefficient "b" and the index of incidence area of effective use of rights (and not of the generation of rights) ;

• n is a correction factor value and expresses the ratio between the market values of the areas of generation of rights, and the market values of the areas of application volumes. In practice, if the areas on which will rest the rights have lower value than the areas that have generated the right n will be greater than 1, and vice-versa.

From the formulas above it shows that, for the assessment of the condition of convenience are a few parameters :

• a "redevelopment index" attributed to the areas of intervention , ie the premium volume ;

• the existing building density ;

• coefficient of incidence of buildable area ;

• coefficient of reduction of value due to the obsolescence of the property;

The formulas shows that with increasing volumes granted by the administration, increasing the quality of the receiving area and the decrease in the quality of sending area , it increases the possibility of obtaining a radical transformation of the building fabric obsolete.

Conversely, the increase of building density and quality of the goods of sending area decreases the viability of the project of demolition and reconstruction.

In fact, in the general case the value of real estate never drops below 50-60% of the market value of the properties of new construction, therefore, the destruction of the city will never be realized only with the increase in volumes but only by introducing new tools and incentives for economic and financial nature .

In general, the redevelopment index to be assigned to a fund must have two fundamental characteristics:

• The index must be enough high to produce a property plus-value larger than that produced by the procedure of expropriation, with the aim of recaptuirn a miece of the value for public facilities;

• the existing index should be enough low that it does not disadvantage the acquisition of areas by the Public Administration .

In general there are three basic methods :

1 the seek for a priori incentive redevelopment index based on the characteristics of the actual and legal areas .

In this case, the axis of the negotiation is strongly shifted towards the Public Administration , which does not really have such a strong bargaining power and that in any case goes to meet accountability and independence are too high in a field that so far has not a legislative reference national ;

2 The bargaining between parties between public and private. In practice, it leaves the field open to game - force between the parties involved that will set their partisan interests to the main objective of equalization , that is, the urban transformation .
The solution would be to start from a base rate and then add the values of rewards , however this can only be done in the case of fair exchange [2] [3];

3 The design of a new urban redevelopment, where the planning choices define the amount of the index. In this case the choice of the index, which always falls on the responsibility of Directors, is certainly justified as being made in retrospect. In addition, the variation of yield ΔR take into account the marginal social costs of the interventions to be performed on the sub-fund [4].

3   Urban Plus Value Creation and Recapture

How to evaluate the benefit [1]? You have to compare the change in income with the old mode implemented with the annuity due to equalization . In general the plus value of rent ΔR is :

$$\Delta R = [\, Ve + Vv \,] - Va$$

In which:
• Ve is the value of the new realized property;
• Vv is the amount of the compensation for expropriation ;
• Va should be the value of the first choices of the plan;
    This variation of rent therefore depends on the new building index , the market value of real estate and realized the value of incidence of the area.
Instead , obtained by equalizing the annuity is:

$$\Delta Rp = [\, Vep - Va \,]$$

In which:
• Ve is the value of the new realized property;
• Va should be the value of the first choices of the plan;
    The difference between the two reports is:

$$\Delta R - \Delta Rp = Vv$$

    Comparing the two expressions can be seen that the variation of annuity is higher if obtained with the old methods.
    All this in accordance with the logic of the equalization fund , which sees all parties lose part of the capital gain , ie the gain due to urban transformation . This loss is mainly due to the fact that individuals lose some of the volumes being subject to a single rate equalization [5].

4   The Case of Study

Having dealt with at a theoretical level the process of "urban destruction", an attempt was made through a simulation to identify the potential outcome of the application of the equalization of urban densification and exposed in the previous chapters in different scenarios [6].

These scenarios have been identified within the urban area of Andria, an Italian town of 100 244 inhabitants (ISTAT data on 30/06/2013), which is the fourth largest city in the Apulia region by population. The choice was not random but motivated by the presence in the urban area of a number of scenarios with different characteristics. The simulation was divided into several stages follow-up.

A very important step in preparation for the spatial analysis and provides for the setting of the database containing all the information necessary to obtain the final ranking of areas where you can achieve urban regeneration interventions in the form of demolition and reconstruction. Please remember that was activated about an internship associated with the thesis aims to develop skills in the use of data infrastructure to support spatial statistical analysis .

Using data obtained from the Sit Puglia (an acronym for Geographic Information Service) and data from the 2001 Istat census and 2011 , were analyzed with the aid of cartographic software ArcMap 10.1. This software does have the ability to query large amounts of data , principally based on vector polygons included in shapefiles which in turn contain a variety of information , called " attributes " , tabulated and editable .

Leveraging the "Join and RelatesTables ", " Select" and "Clip" , you can generate and export shapefiles defined and contain all the information taken from different sources (in different formats) . This operation generates a single " input" is essential in order to exploit the full potential of subsequent spatial analysis and statistics.

In general, the input data can be classified and processed in various ways , through the objective functions or through the nature of the data . In our case we used the data already presented in the previous chapter leveraging the query (QUERY), which provide :

- Use of base maps that meet the selected criteria in the identification stage of the case study ;
- Use of vector data and raster data ;
- Using geometric properties and attributes ;

In this phase , are introduced cartography base , ie the orthophoto and shapefiles , of which shows the jpeg images .

In the preliminary steps have been withdrawn from regional chart polygons related to the residential fabric, then separate into its main components, namely in the historic core, woven fabric and consolidated residential suburb, recognizable by the increased dispersion and the different types of buildings. Below is the elaborate such information.

Below are selection operations carried out on the residential fabric in a "skim" all that fabric that will never be subject to any kind of intervention.

- Excerpt from chart constraints architectural, archaeological and environmental issues present within the urbanized area; In fact, in these areas should be undertaken to more safeguard actions, recovery and restoration of existing buildings;

- Excerpt from the maps of the areas of recent construction, especially in the areas that fall under the "167" areas (popular housing areas) north of the city, the neighborhood Valentine's recently regenerated and outside the perimeter of the buildings erected;
- Identification of the availability of the standards and community facilities, in order to identify the areas most served and deficit. In fact, the attention of the public is particularly targeted at areas where there is the need to overcome these shortcomings of services;
- From this preliminary operation has been possible to identify the actual work area on which to operate for the simulation.

In particular, the polygons have been cataloged individually identified and each of them have been added to the "Join Tables" table values divided by Istat census sections.

4.1   Analyses

In the second phase we have gone to the realization of a spatial statistical analysis that allowed us to draw , considering all the variables weighed, a ranking in which you read clearly what are the possible scenarios for the implementation of the simulation.

As a reminder theory, it should be noted that for spatial analysis refers to a wide range of processes whose results depend on the geographical location of the objects, that is, depend on the shape , size, position data input.

At the urban level , this type of analysis becomes a very useful tool for decision support since it can handle and synthesize large amounts of data . In particular, the output data can then be used both to analyze the state of the fact of a territory is to predict future scenarios.

In particular, the spatial analysis was carried out through the following phases:

1 . Defining the objective, in this case the proper choice of suitable areas to support processes of demolition and reconstruction ;
2 . Set criteria , namely the choice of data to be analyzed. In our case, have been used :

- Data on the time of construction of the buildings;
- Data on heights and densities of buildings;
- Data relating to landslide risk in urban areas ;
- Data relating to the market value of immovable property for residential use ;
- Data on the presence of services and standards in the urban area ;
- Data relating to environmental, landscaping and architectural features in the urban area ;

The data used were sourced from two different sources :

- cartographic bases obtained by SIT Apulia ;
- Shapefiles ISTAT analysis with aggregated by census sections ;

3 . Preparation and execution of space operations , with the help of matrices and tabular data ;
4 . Evaluation and interpretation of the outputs;
5 . Production of the final ranking ;

Defining the objective, setting and prioritization criteria . Having chosen as its main objective the optimal location of the areas that can accommodate process of demolition and reconstruction , the definition of the matrix multi-criteria will be oriented to give more weight to those criteria that the developer (and all the more reason for the government) deems certainly useful for this purpose. In this case, the more weight will comply with the following list , from highest to lowest:

- Hydrogeological risk (greater weight) ;
- Presence of constraints of the plan;
- Density of buildings;
- Market value ;
- Availability of services;
- Age of construction (less weight) .

The decision to give less weight to the criterion based on the age of the construction of buildings might seem at first sight a mistake, however, the consideration to make is that while it is true that the obsolescence of artifacts is an important determinant for demolition processes , on the other hand it is also true that the type of analysis carried out , based on census sections , does not allow for a detailed analysis of the actual state of the artefacts, with the risk of going into the operational phase to invalidate the whole ' architecture analysis.

Therefore it was decided to weigh less than the criterion linked obsolescence of the artifacts and then return in the second phase, to analyze the section on time (or groups of sections) of choice in the census lists to confirm the location of the same on the list.

4.2   Spatial Autocorrelation

Once spatial autocorrelation analysis described all done, we moved on to the next phase which includes analysis of the top level or analysis that allow the transition from localized and timely input in the individual sections of a census of the output raster (grid) [7].

And ' appropriate, in order to better interpret the reports of the various analyzes , recall some basic concepts related to spatial geo-referenced statistical analysis . The first key concept is that of "clusters" . Objective analysis of clustering is to group objects into groups with a certain degree of homogeneity . It ' also known as unsupervisedclassification , whose purpose is to segment the data without assigning class labels , but only to group classes of similar objects. A good method of clusteringprodurrà clusters of high quality with :

- High intra -class similarity , where distances are minimized
- Lower similarity in which the inter-class distances are maximized ;

It ' also important to define the goodness of clustering, and it depends on the similarity measure used , or the specific algorithm used . The requirements necessary to identify a good clustering method are:
- Scalability ;

- Ability to use different types of attributes ;
- To discover clusters with arbitrary shapes ;
- Not feeling the ordering of input record ;
- Interpretabilitàe usability of the results.

Finally , please note that the spatial autocorrelation in general can have three outcomes:

- Positive spatial autocorrelation in the case , stating the presence of clusters in the study area ;
- Spatial autocorrelation nothing in the case where it is not detectable any spatial effect at both positional both at the level of the attributes of individual events ;
- Negative spatial autocorrelation , or repulsion , while in the presence of events which, although close in space exhibit considerable differences regarding the properties that characterize them.

For the ultimate goal , namely to identify the area of the city where it is most possible to operate with demolition and reconstruction, the following analyzes were performed :

- NearestNeighborDistance , statistical analysis , which describes the spatial distribution of events by measuring the distance . This technique analyzes the distance of an event than most others and the result is described in a simplified way by the law of Tobler that " everything is related to everything, but near things are more related things away " [7].
- Index of Moran, which allows you to turn a simple auto-correlation in the spatial correlation. This index takes into account the number of events that occur in a given area and their intensity. This index is between -1 and 1. If the term is greater than 1 the autocorrelation is positive, if it is less than 0 is negative and if it is close to 0 is nothing.
- G function of Getis and Ord, takes into account measures disaggregated autocorrelation, considering the similarity or difference in some areas. This index measures the number of events with similar characteristics within a certain distance D.

4.3   Analysis of Hotspots and Overlay

Once checked all the parameters of spatial autocorrelation, we have moved to carry out the analysis Hotspots.

This tool, based on the index Getis-OrdGi * allows for the output raster in which there is a homogenization of the attributes according to the logic of the cluster. Below are all the hotspots analysis on the criteria adopted for the selection of the scenario.

The final step in preparation for the identification of the best graphics consists of a procedure, that is an overlay that allows you to identify areas of particular interest thematic maps. The result is a raster image that will serve as a basis to identify using orthophotos scenario on which to operate the simulation [8] [9] [10] [11] [12].

The analysis phase of the urban context in which you will be working with interventions of urban regeneration is essential. The quality and the objectivity of the analysis deteminante has an impact on the choices that the developer , be it a government or a company Urban Transformation , urban transformations apply to physical, economic and social issues.

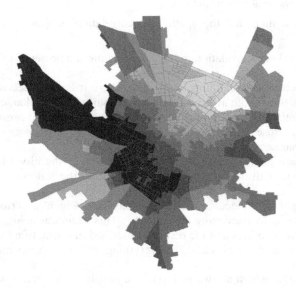

Fig. 1. Relationship between Plus Value and one level buldings – low real estate value

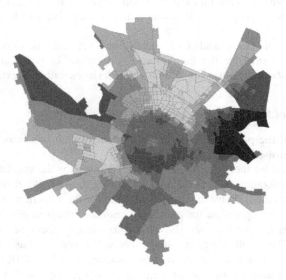

Fig. 2. Relationship between Plus Value and two level buildings– low real estate value

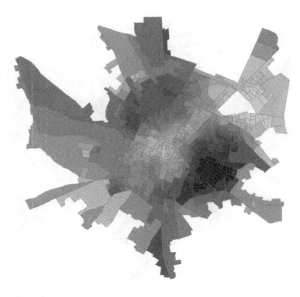

Fig. 3. Relationship between Plus Value and three level buildings high real estate value

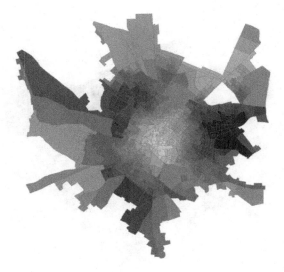

Fig. 4. Relationship between Plus Value and after second world war buildings - low real estate value

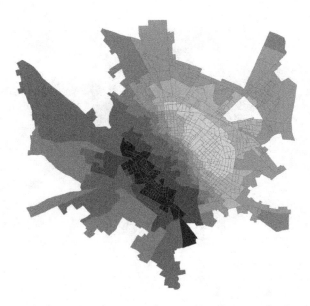

Fig. 5. Relationship between Plus Value and After Sixties Buildings low real estate value

Fig. 6. Relationship between Plus Value and and hydrogeolocal Constrains – Low real estate values

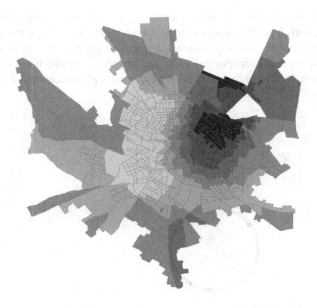

Fig. 7. Relationship between Plus Value and Facilities – high real estate value

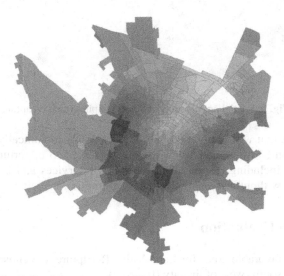

Fig. 8. Spatial aggregation of favourable plus value recapture

The analyses were divided into two branches: physical and spatial statistics . It 's on the latter that has focused most attention , with the aid of data infrastructure to support analysis , using programs such as ArcMap 10.1 and CAD data to extrapolate both vector and statistics.

The use by developers of these technologies allows to objectify qualitative information , both from the point of view of both calculations from the graphical point of view . the final result is obtained through the overlay mapping data based on analysis Hotspots , was represented by a satellite photo of which are evident in urban areas on which will be more convenient to work with processes of demolition and reconstruction .

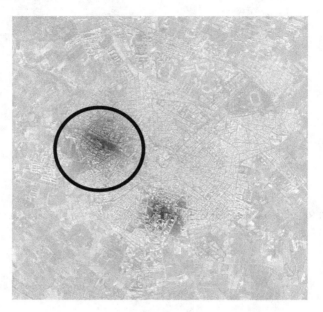

Fig. 9. Spatial aggregation of favorable plus value recapture

A convenience that will not be tied only to occasions economically favorable (low cost of demolition , low market value), but also linked to opportunities for public administrations , including resolution of the lack of services and traffic problems , increased density without increasing the land consumption) .

5   Scenario Evaluation

Among the most favorable areas for Land Value Recapture, the chosen one is named Camaggio, in the south -west of the city (figure 9).

The next part of the paper explains simulations on the economic feasibility of the process of demolition and reconstruction.

Once you find the scenario, in this chapter we enter the planning part. At this stage, within the area identified through the physical and geostatistical analysis, it will be performed a simulation of a process of" destruction" Urban or the ability to implement through a process (governed by the agency of the Society for Urban Transformation) of demolition and reconstruction within the built settlements.

The scenario analysis involves the following steps :

1) analysis of the area;

2) identification of the volumes and their current market value ;

3) simulations based on the variation of the reward and the new volume indices derived from the theoretical formulas analyzed in the first chapters of the thesis ;

4) general reflections on the impact of individual simulations , through the analysis of the ecological footprint of the various interventions ;

5) calculation of the new market value to derive the total economic benefit of the project;

6) analysis of the economic benefit of the intervention and its distribution among the various investors;

7) master plan and preliminary project area.

Fig. 10. The area for scenario analysis

Looking at it from a strictly economic point of view , financial , it is clear , given the obvious profit margins , that the operations of demolition and reconstruction if properly coordinated and encouraged by rewarding volume are always cost-effective . By analyzing the four simulations , it is possible to make some observations related to the increase in volume of premiums multiplier [13]:

- The higher the reward , increase the volumes of Land granted to the owners of the goods , so you can just say that the choice of multiplier has a strong influence on the ability of negotiation and consultation with the individual owners ;

- It reduces the cost of construction of buildings , in view of economy of scale, thereby increasing the total gross benefit ;

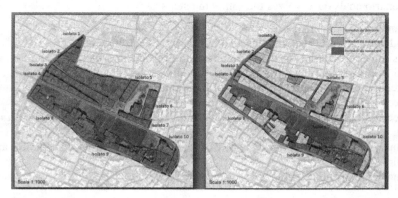

Fig. 11. Hypotesis of most probable (Red) and probable (Orange) demolition

- Simulations with multiplier too low (1.50 to 2.0) in the face of a clear equitable redistribution of the total gross benefit , which coincides exactly with the benefit of a private concession fees and gross of management fees , have a very low palatability economic . This is due to the fact that the total gross benefit in these cases can cover only the capital gains of the individual owners involved , not allowing individuals outside of or benefit from investing in the project through the participation of the Society for Urban Transformation;

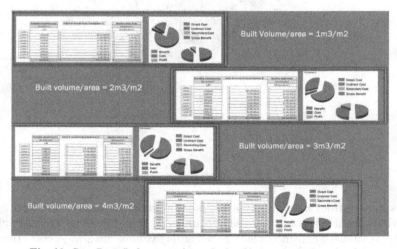

Fig. 12. Cost-Benefit for scenario analysis of increasing urban density

In summary can be seen as a practical level , in agreement with the theory of Micelli [1], there is a lower limit below which fail all the interventions of urban regeneration based on private investment and new planning instruments related to the market .

References

1. Micelli, E.: Development Rights Markets to Manage Urban Plans in Italy. Urban Studies 39, 141–154 (2002)
2. Conrad Jon, M., LeBlanc, D.: The Supply of Development Rights: Results from a Survey in Hadley, Massachusetts. Land Economics 55, 269–276 (1979)
3. Daniels, T.L.: The Purchase of Development Rights: Preserving Agricultural Land and Open Space. Journal of the American Planning Association 57, 421–431 (1991)
4. Levinson, A.: Why Oppose TDRs: Transferable Development Rights Can Increase Overall Development. Regional Science and Urban Economics 27, 283–296 (1997)
5. Machemer, P.L., Kaplowitz, M.D.: A Framework for Evaluating Transferable Development Rights Programmes. Journal of Environmental Planning and Management 45(6), 773–795 (2002)
6. Morano, P., Tajani, F.: The transfer of development rights for the regeneration of brownfield sites. Applied Mechanics and Materials 409(2013), 971–978 (2013)
7. Murgante, B., Las Casas, G., Sansone, A.: A spatial rough set for locating the periurban fringe. Revue des nouvelles technologies de l'information 857, 101–125 (2008)
8. Torre, C.M., Balena, P., Zito, R.: An Automatic Procedure to Select Areas for Transfer Development Rights in the Urban Market. In: Murgante, B., Gervasi, O., Misra, S., Nedjah, N., Rocha, A.M.A.C., Taniar, D., Apduhan, B.O. (eds.) ICCSA 2012, Part IV. LNCS, vol. 7333, pp. 583–598. Springer, Heidelberg (2012)
9. Cerreta, M., De Toro, P.: Assessing urban transformations: A SDSS for the master plan of Castel Capuano, Naples. In: Murgante, B., Gervasi, O., Misra, S., Nedjah, N., Rocha, A.M.A.C., Taniar, D., Apduhan, B.O. (eds.) ICCSA 2012, Part II. LNCS, vol. 7334, pp. 168–180. Springer, Heidelberg (2012)
10. Fusco Girard, L.: Analytic Hierarchy Process (AHP) and Geographical Information Systems (GIS): An Integrated Spatial Assessment for Planning Strategic Choices. International Journal of the Analytic Hierarchy Process 4(1), 6736–6744 (2012)
11. Cerreta, M., Mele, R.: A landscape complex value map: Integration among soft values and hard values in a spatial decision support. In: Murgante, B., Gervasi, O., Misra, S., Nedjah, N., Rocha, A.M.A.C., Taniar, D., Apduhan, B.O. (eds.) ICCSA 2012, Part II. LNCS, vol. 7334, pp. 653–669. Springer, Heidelberg (2012)
12. Lake, I.R., Lovett, A.A., Bateman, I.J., Day, B.H.: Improving Land Compensation procedures via GIS and hedonic pricing. Environment and Planning C: Government and Policy 18, 669–681 (2000)
13. Morano, P., Tajani, F.: Break Even Analysis for the financial verification of urban regeneration projects. Applied Mechanics and Materials 439, 1830–1835 (2013)

Spatial Multicrierial Evaluation of Soil Consumption as a Tool for SEA

Pasquale Balena, Valentina Sannicandro, and Carmelo Maria Torre

Department of Civil Engineering and Architecture
Polytechnic of Bari, Via Orabona 4, Bari Italy
p.balena@poliba.it, sanni.vale@gmail.com, cartorre@yahoo.com

Abstract. The paper represent a check of the use of multicriteria evaluation in order to add a qualitative evaluation to the traditional quantitative measure of the sustainability of soil consumption. The experiment starts analysing all deriving measure from measures of different typology of soil consumption and land use as criteria to evaluate which part of urbanised land is more expendable for land transformation.

This application results quite interesting when utilised to create a set of indicators useful for Strategic Environmental Assessment (SEA), as instrument of measuring impact and of monitoring future urban development

The work is subdivided at the urban scale in three different stages. The setting of measures, the creation of complex indicators as basis for evaluation criteria, the classification of priority n soil consumption. Results show the opportunity of densification inside existing settlements, but they extend the utility of the evaluation in profiling measures for tegional policies of containment of soil consumption.

Keywords: Smoothing, TDR, Densification, Electre, SMCA.

1 Soil Consumption and Planning

The land consumption can be broadly defined as a process that generates the progressive anthropogenic transformation of natural or agricultural areas through the construction of buildings and infrastructure, when it is assumed that the restoration of the pre-existing state of the environment is very difficult, or impossible, to because of the nature of the distortion of the earth matrix. This definition is characterized in a negative way, because it is perceived negatively the problem of the diversion of agricultural or natural surfaces considering the finiteness of the terrestrial surface, and would therefore be more correct to refer to the concept of transformations of soils.

After the ecological footprint, the soil consumption is a spatial measure widely used to assess the sustainability of territorial transformations. The ecological footprint is a theoretical measure, which synthesizes many aspects, while the soil consumption is real: it represents an element of environmental pressure and natural consumption of a resource, that is the earth surface, covered by a layer of soil generated by natural processes [1].

B. Murgante et al. (Eds.): ICCSA 2014, Part III, LNCS 8581, pp. 446–458, 2014.

The attention to the issue has become relevant in fairly recent times; however, especially the causes and the consequences on environmental status and quality of urban life, delineate an alarming scenario in terms of waterproofed areas. Just in 2012 the Italian Institute for protection and environmental Research (ISPRA) estimates an occupation of 8 sqm / second of soil by new constructions.

Despite to many open debates on, the subject remains unresolved about the extent of the measures. In fact, you should create shared criteria and consistent estimation starting from a unique and real definition of "land use", taking in account, when planning, a balance between urban and rural areas, and between the occupied ground and the real need of consuming.

2   Problem Definition

The objective of this study is therefore to:
• Quantify and qualify soil consumption for planning strategies aimed at reducing the phenomenon
• Assess the quality of the urban landscape transferring development rights and weighting the dynamic of urban expansion according to different land use, and consequently
• Develop a methodology for mapping and monitoring of the state of the territory;
• Identify the actions to be pursued in the communal area to stop the use of land through the exposure of possible scenarios after the planned expansion
• Provide a new tool for environmental assessment and monitoring the effects of town planning

The control and limitation of the transformations of the soil is institutionally anchored to the procedures for strategic environmental assessment.

The consumption of soil then becomes a synthetic indicator of the ongoing reformation. The assessment is used, however, several "steps". Referring to the classification ISPRA, some measures are context indicators, other than change of state and other processes. In reference to the issue of land use, the effectiveness of any policy of containment of the transformations of use determining degradation and/or loss of soil can be evaluated and monitored only if it is based on the availability of data for use and land cover that are up to date, comparable and scalable to different levels within which decisions are made by the territorial government [2].

The overall knowledge of the phenomena that affect the dynamics of soil and land use are of paramount importance to make a determination on the changes taking place and to intervene in the planning process.

The control and the limitation of the transformations of the soil is institutionally anchored to the procedures for strategic environmental assessment. The consumption of soil then becomes a synthetic indicator of the ongoing transformation . However, the assessment needs several "measures". Referring to the classification of the Italian "High Institute for Environmental Protection and Research" (ISPRA), about indicators some measures are referable to the context, others are of state-change indicators and further are indicators of process [3].

As regards the issue of land use:

- Context indicators represent the status of the soil as it is today;
- Indicators of state-change represent the extent of a differential land use in a given time interval;
- Process indicators represent measures of actions with the purpose of regulating, containing or generating the consumption of soil, that can be represented in a plane of procedures for decisions on the expansion and recovery, for the definition of the limits of expansion and sustainability of the recovery, for the economic values created in the policies of expansion and those of containment.

More complex is the review of the policies and instruments used by municipalities. The first question, in this regard, concerns the possibility to coordinate the planning and the guidelines of the over-local level with the activation of effective measures for the detection and control of land use at the scale of detail.

It seems more appropriate to pass a logic of control of the transformation of land use that is based on exclusive and objective method of quantitative detection, to develop ways to government and direction that improve a responsible and integrated approach of the same uses, starting from a fair knowledge and the sharing of quality of soils affected by planning policies [4] [5].

3   Research

To investigate the issue and to develop a methodology that can overcome the problems above introduced, it were conducted analyzes aimed at specific needs (controls in agriculture, land use planning, environmental assessment, basic statistics, etc..), having the need to define some classification systems, as the relationship Land Cover vs. Land Uses, which have proved unsuitable were to provide a complete evaluation of land use, characterized by: consistency, thematic accuracy, units of inquiry.

Therefore, the inventory of this data base is insufficient, because it takes very detailed information for the classification of soils, not always available only with advanced uses of satellite observation and digital interpretation, but also with specific methods of sampling and census of the territory that presuppose a specific knowledge/experience technique .

Above all, the European context calls substantial and consistent information's on Member States, constant updates, integration with EU and international directives, especially in terms of classifications adopted.

In this regard, follows a methodological proposal aims to study and implement a system that contains a structured legend in a hierarchical manner, depending on the specific characteristics of each area of application. In this way we obtain a general mapping of the entire municipality and it is possible to assign a "tool" to every single part of the land.

Therefore, the search starts from the insertion of a series of digital spatial information layers (layer) related to land cover and physical characteristics of the territory, found by SIT Puglia c(the Geographic Regional Apulian System).

As a base map, so in-put data to be entered into the system, it is important to support the Regional Technical Map (CTR) and the Regional Map of Using Soil (UDS), both because of their easy accessibility from the site of the Regional Cartographic Service and, above all, because they are tools for self georeferenced and marked by a series of attributes.

The other data were extrapolated from these maps and municipal plans and approved by sites such as the Italian Population Census (ISTAT).

The analysis of the City Sample, chosen as a case study, has been carried out through the use of ArcGIS software, with the ability to give at each item it's real space coordinates and the ability to perform topological analysis of the data, or to manage and analyze all kinds of spatial relationship between all the elements that are part of the data base.

The research was developed in two phases: the first is a cognitive and analytical phase, the second is a operational phase.

In the first phase of investigation, in broad terms, the methodology is divided as follows:

-    in-depth analysis of the City;
-    identification of the urbanized area;
-    characterization of the urbanized area;
-    construction of some representative indicators of the state of the soil;
-    elaboration of a coded map of land-use;
-    analysis of the estimated needs valued from the Master Plan and quantification of area as still developable;
-    diachronic survey between the built detected by Regional GIS (CTR), dated to 2006, and the map extrapolated from the digitization of the PRG of 1990;
-    interpretation of results.

The second phase starts from the estimate of the size of spatial development plan intended as a "land use", but above all it assesses if the predictions of the planning are legitimated with respect to zoning and to vocations and characteristics of the sub-municipal areas examined.

4   Case of Study

The introduction to the case of study begin with the choice of the City "Sample" to observe and analyze. It was chosen the City of Bitonto as a sample to test the analysis above.

It is characterized by a rather large municipal area if it is compared to that of the towns of Salento and it has a relevant territorial structure, consisting of three compact and homogeneous urban cores, in relation to the urban settlement: the core that is developed in the crown of the historic center and two peripheral cores represented by the suburbs of Palombaio and Mariotto.

The following table shows what are in-put data, its source and what is the calculation of the simple index that uses the corresponding data.

Afterward, were analyzed for each sample the two main types of land use, one related to its size, understood as the total area of the urbanized surface, and one related to its configuration, connected to the physical form of the same surface and hence at the type of territorial development.

Through the interpretation and processing of the Puglia CTR and UDS it is possible to identify all the technical, physical and environmental characteristics of the municipality in question, such as:

Urbanized surface: given by the sum of the built area and the infrastructured area, any populated area not classified as agricultural land or natural. This surface corresponds to the following categories of the Regional Map of Real Use of Soil: urbanized residential areas and related appurtenances, production areas, services, networks and infrastructure, areas affected by mining and landfills, construction sites and reworked soils, green urban areas, parks, cemeteries and sports activities;

Permeable surface: any surface, free from buildings above or below the ground, able to guarantee the absorption of water and able to promote the productivity of the soil;

Waterproof Surface: every cement-covered surface, used and covered with any type of structure;

Protected surface: a public or private place of great natural, historical or artistic value that the State, or any other association, protects in order to prevent it from being damaged or destroyed;

Bound Surface: every area in which the insertion of building works and infrastructure is tied to a higher-level view, making the most compatible human activities with the beauty and value of these places.

DATA	SOURCE	CALCULATION - INDICES	VALORE
municipal area	Municipal limits	Land survey (Area Tot.)	Area Tot. = 172.388.094,788 mq
population	Italian census	Growth rate of population (T)	$T = (Pop\ t1 - Pop\ t0)\ /\ Pop\ t0 = -0,008$
Quote Altitude	Italian census	Altimetrical (EA)	$EA = Qmax - Qmin = 454$ m
Living green spaces	Regional GIS	% Green space (S.V.U.)	S. V. U. = 93 %
built-up area	Regional GIS	Index1 of urbanization(I URB1)	$I\ URB1 = Built\ Area/Tot. = 0,012$
Roads	Regional GIS	Index2 of urbanization(I URB2)	$I\ URB2 = Area\ Infrastr.\ /\ Area\ Tot. = 0,048$
Impervious surface	Regional GIS	Index of impermeability(I IMP)	$I\ IMP = Area\ Imp.\ /\ Area.\ Tot. = 0,06$

Permeable surface	Regional GIS	Index of permeability (I P)	I P = Area Perm. / Area. Tot. = 0,94
Use of the land	Regional GIS	Area I Perimeter class	
Intended Use	Town Planning and land use maps	Estimation of develop-able areas	
System of protection	PPTR	Estimation of sensitive areas	Surface of sensitive areas = 61.996.497,38 mq
Conservation areas	Map of Cultural Heritage	Estimation of conserva-tion areas	Surface of Conservation Areas= 6.824.913,71819 mq

The spatial data, correlated with those of the resident population, lead to the creation of complex indicators:

-   Comparison between trends of urban expansion and urban population = Growth population rate= population growth on the initial one = - 8 %
-   Growth rate of urban areas = Built Area in a given time interval on the initial one = 48 %
-   Index of urbanization = urbanized surface (Built Area+ Infrastructured Area) on the municipal area = 0,06
-   Difference between permeation rate and of green surface = 1,005
-   Index of sensitive areas = sensitive areas on municipal areas = 0,35
-   Index of conservation areas = conservation areas on municipal areas = 0,39

The construction of the "mapping of soil" consists in dividing of municipal territory into a number of particles using a special grid (50x50 m); the matrix is used to analyze in detail each part of the soil, in such a way to highlight also the percentage and with which mode it is used.

For each particle will calculate:

-   Percentage of impermeable soil;
-   Type of water proofing;
-   Activities carried out;
-   Quantity of activities that carried out.

The purpose of the fragmentation is to simplify the monitoring in the future, because the systematic and uniqueness of the procedure makes it possible to update the database annually, so that the data can also be used in other contexts and is useful to compare the different situations of the various municipalities.

To continue the analysis compared to the state of the of the "consumption of the soil", the study has specifically investigated the spatial density, which is considered in the literature the main cause of land consumption.

The smoothing used for this phase, has the objective to provide an idea of the variation of density between two adjacent areas, through a compensation between the exact value and the average value, then to make less definite the transition from a very dense area to little dense one.

However, it is noted that this type of model does not provide an estimate of the error and applies to flat surfaces; does not consider the volume of the built, which produces a consumption of land depending to several induced factors such as human pressure and the area devoted to compliance with planning standards; the state of the vegetation is not thorough or evaluated.

The size of the mesh is 10 x 10 (m); while the size of the buffer is 200m.

Following , was made a random division of density values into three classes: the darker color indicates a denser surface, the color with media saturation indicates the transition area and the lighter color identifies the free area.

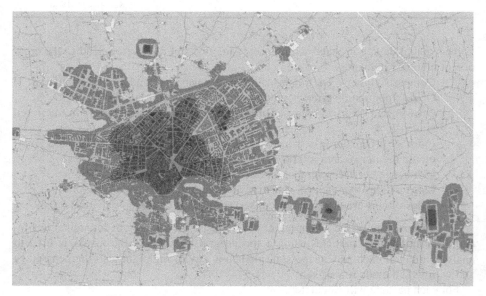

Fig. 1. Areas of potential TDR, identified by smoothing

The limitations manifested by the application of smoothing, leading to consider this item incomplete, so it can only be used as an additional diagnostic criterion and not as a parameter for discriminating choices of future plans [7] [8].

5   Identification of Areas That are Object of the Future Transformation

The third phase is instead aimed at identifying the areas that are the object of future transformation, and classifying them with respect to the sustainability of the implementation with the indications of plan applies to them [8] [9].

To this purpose, several measures are used to produce indicators of Pressure, Land Use, and Fragility [10] [11].

The use of the soil and the fragility represent the set of context and process indicators, on which are grafted indicators of changes dictated by environmental pressure, generated by the activities already located in the settlements.

In order to perform this analysis we have selected the areas for the future expansion of the plan.

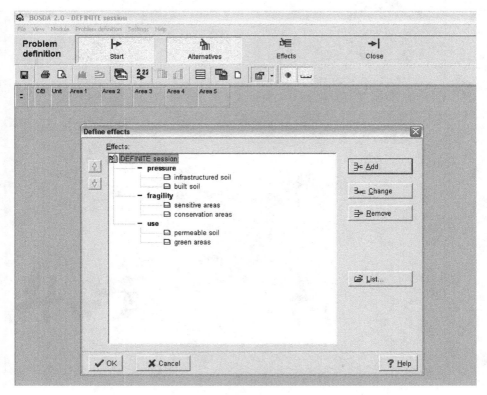

Fig. 2. Criterias and group of criterias in Multidimensional evaluation

The processing areas are characterized by the presence of manufacture and residential buildings. In addition, some areas are partially built, while others are completely free and intended for future transformations.

These elements of differentiation between areas develop different problems; there are partially transformed areas (areas 3 and 5 in figure) that lead to increased environmental pressure and totally free areas that need of increased urbanization and involve a greater consumption of land, still free (areas 4 and 5 in figure).

Pressure Factors considered are the infrastructure index and the index of urbanization. The index of infrastructure is a element key in favor of limiting the use of the soil, because it allows you to use areas that in case of similar levels of urbanization have a pro-vided networks; therefore they can be used for new settlements in compact form , that unpack the environmental pressure. Thus, at a greater infrastructural facilities in an area already built, corresponds a greater chance to act in contexts already built with the least expenditure.

Fig. 3. Areas of potential TDR, identified by smoothing

The index of consumption of soil, instead, is proportional to the environmental pressure, and generates a damage, which is not compared to the soil from re-builds, but to the existing environmental heritage that provides the amount of natural resources and energy necessary for the existing settlement. So, to increased amounts of land urbanization correspond more requests for resources, already present.

The consumption of the Soil is qualified in two respects: the permeable soil and the green surface. These two aspects are seen as resources and as a potential compensation of the additional consumption of soil.

Finally, the fragility of the environment are represented by the presence of sensitive areas (under protection) with a convertibility regulated and by restricted areas for which it is impossible the transformation.

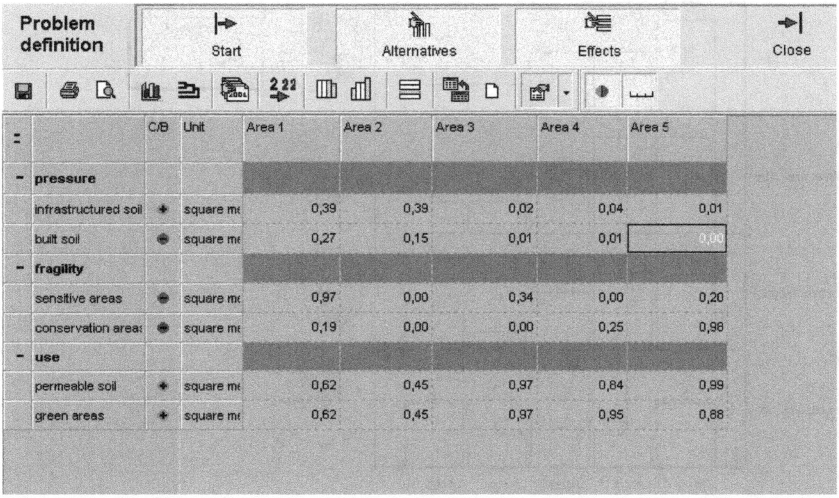

Problem definition
Start
Alternatives
Effects
Close
C/B
Unit
Area 1
Area 2
Area 3
Area 4
Area 5
pressure
infrastructured soil + square me 0,39 0,39 0,02 0,04 0,01
built soil square me 0,27 0,15 0,01 0,01 0,00
fragility
sensitive areas square me 0,97 0,00 0,34 0,00 0,20
conservation area: square me 0,19 0,00 0,00 0,25 0,98
use
permeable soil + square me 0,62 0,45 0,97 0,84 0,99
green areas + square me 0,62 0,45 0,97 0,95 0,88

Fig. 4. Effects of soil consumption as criteria measures

For the purposes of ELECTRE method 2 [12], the criteria were weighted with the AHP method [13], and the standardized criteria in terms of the incidence rate of the soil.

Multicriteria analysis
Start
Standardize
Weights
Rank
Close
Method: Electre 2
Analysis Description: MCA 1: Electre 2 (goal; Direct (fragility: 0,4))
pressure
infrastructured soil
built soil
fragility
sensitive areas
conservation areas
use
permeable soil
green areas
MCA 1: Electre 2 (goal; Direct (fragility: 0,4)): Discordance table
Alternatives
Area 1
Area 2
Area 3
Area 4
Area 5
Thresholds for discordance
Strong graph: 0,489
Weak graph: 0,588
OK
Cancel
? Help

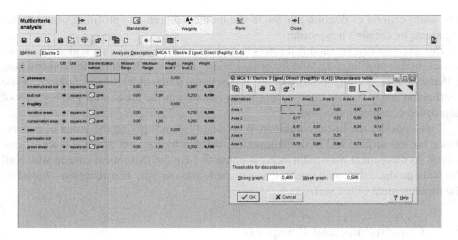

Fig. 5. Concordance and weights

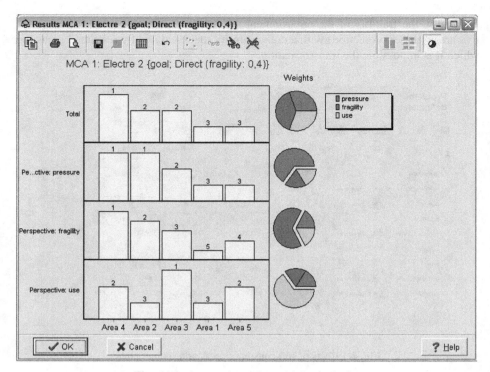

Fig. 6. Final pre-order and sensitivity analysis

6   Final Reflections on the Case of Study

The multi-criteria evaluation allows, in the municipal area, the profiling of the building areas, by creating a pre-order.

This exercise can be valid if it refers to a communal area, both in a regional context. In fact, in this profiling, the land for urbanization, in areas with priority for construction, is that to associate a high priority of use proportional to the index of concordance, while coefficients away from a good concordance can weight the areas with greatest risk of unjustified consumption of soil.

If we consider the linear combination obtained by multiplying the index of concordance for the surface soil of transformation, divided by the sum of the surfaces themselves, its theoretical maximum value is equal to 1 (in the theoretical case in which all the soils produce the same qualitative-weighed- and quantity consumption - based on the measurement of the surface).

7   Conclusions and Prospects

The multidimensional approach and the sensitivity analysis show that in the valuations made by the method ELECTRE 2, preference is given to areas where the presence of infrastructure and the presence of fewer constraints, make it more easily transformable areas, already affected by edification (areas 4 and 2 in the Figure), as some studies often put on evidence [14].

For priorities determined by the multi-criteria evaluation are weighed also constrained areas and the sensitive areas, often more present in completely undeveloped areas (areas 5 and 1 in the Figure).

The result of the assessment is a different weight that can be given to building soils in the various areas [15].

The assessment, at this point, becomes a support to the valuation of the environmental transformation of a settlement plan, and the measure of the ability of load performance at the regional level as a consequence of possible transformations [16].

References

1. Funtowicz, S.O., Martinez-Alier, J., Munda, G., Ravetz, J.R.: Information tools for environmental policy under conditions of complexity, Bruxelles. EC Environmental Issues Series, vol. 9 (1999)
2. Murgante, B., Las Casas, G., Sansone, A.: A spatial rough set for locating the periurban fringe. Revue des nouvelles technologies de l'information 857, 101–125 (2008)
3. Balena, P., Sannicandro, V., Torre, C.M.: Spatial Analysis of Soil Consumption and as Support to Transfer Development Rights Mechanisms. In: Murgante, B., Misra, S., Carlini, M., Torre, C.M., Nguyen, H.-Q., Taniar, D., Apduhan, B.O., Gervasi, O. (eds.) ICCSA 2013, Part IV. LNCS, vol. 7974, pp. 587–599. Springer, Heidelberg (2013)
4. Kim, D., Batty, M.: Modeling Urban Growth: An Agent Based Microeconomic Approach to Urban Dynamics and Spatial Policy Simulation, CASA Working Paper 165. UCL Press, London (2010)
5. Torre, C.M., Balena, P., Zito, R.: An Automatic Procedure to Select Areas for Transfer Development Rights in the Urban Market. In: Murgante, B., Gervasi, O., Misra, S., Nedjah, N., Rocha, A.M.A.C., Taniar, D., Apduhan, B.O. (eds.) ICCSA 2012, Part I. LNCS, vol. 7333, pp. 583–598. Springer, Heidelberg (2012)
6. Levinson, A.: Why Oppose TDRs: Transferable Development Rights Can Increase Overall Development. Regional Science and Urban Economics 27, 283–296 (1997)
7. Machemer, P.L., Kaplowitz, M.D.: A Framework for Evaluating Transferable Development Rights Programmes. Journal of Environmental Planning and Management 45(6), 773–795 (2002)
8. Conrad, J.M., LeBlanc, D.: The Supply of Development Rights: Results from a Survey in Hadley, Massachusetts. Land Economics 55, 269–276 (1979)
9. Daniels, T.L.: The Purchase of Development Rights: Preserving Agricultural Land and Open Space. Journal of the American Planning Association 57, 421–431 (1991)
10. Cerreta, M., De Toro, P.: Assessing urban transformations: A SDSS for the master plan of Castel Capuano, Naples. In: Murgante, B., Gervasi, O., Misra, S., Nedjah, N., Rocha, A.M.A.C., Taniar, D., Apduhan, B.O. (eds.) ICCSA 2012, Part II. LNCS, vol. 7334, pp. 168–180. Springer, Heidelberg (2012)
11. Cerreta, M., Mele, R.: A landscape complex value map: integration among soft values and hard values in a spatial decision support. In: Murgante, B., Gervasi, O., Misra, S., Nedjah, N., Rocha, A.M.A.C., Taniar, D., Apduhan, B.O. (eds.) ICCSA 2012, Part II. LNCS, vol. 7334, pp. 653–669. Springer, Heidelberg (2012)

12. Roy, B.: The outranking approach and the foundations of ELECTRE Methods. Theory and Decision 31, 49–73 (1991)
13. Saaty, T.: The Analytic Hierarchy Process: Planning, Priority Setting, Resource Allocation. McGraw-Hill, New York (1980)
14. Attardi, R., De Rosa, F., Di Palma, M., Piscitelli, C.: A Multi-criteria and Multi-group Analysis for Historic District Quality Assessment. In: Murgante, B., Misra, S., Carlini, M., Torre, C.M., Nguyen, H.-Q., Taniar, D., Apduhan, B.O., Gervasi, O. (eds.) ICCSA 2013, Part IV. LNCS, vol. 7974, pp. 541–555. Springer, Heidelberg (2013)
15. Torre, C.M., Selicato, M.: The support of multidimensional approaches in integrate monitoring for SEA: A case of study. Earth Syst. Dynam. 4, 51–61 (2013)
16. Fusco, G.L.: Sustainability, creativity, resilience: toward new development strategies of port areas through evaluation processes. International Journal of Sustainable Development 13(1-2:), 161–184 (2010)

The Economic Effect of Sale of Italian Public Property: A Relevant Question of Real Estate Appraisal

Sebastiano Carbonara[1], Davide Stefano[1], and Carmelo Maria Torre[2]

[1] Department of Architecture, University "G. d'Annunzio" of Chieti-Pescara
Viale Pindaro, 42, 65127 Pescara, Italy
[2] Department of Civil Engineering and Architecture, Polytecnhnic of Bari
Via Orabona 4, Bar, Italy
s.carbonara@unich.it, davide_stefano@hotmail.com,
carmelomaria.torre@poliba.it

Abstract. In Italy , for 20 years , governments that have taken place , they tried to reduce the public deficit through the sale of part of the public property .

The results , to date, have been modest, for a set of reasons, including the difficulty of organizing appropriate sales plans. In the case of public residential properties , sales programs are easier, more complex is the case of other public goods, such as the military bases disused. In these cases , for the purpose of sale, a number of factors come into play, from the conditions in which they are places and buildings, and finally to the future transformation that may receive. A frequent mistake to place on the market these goods, evaluating the price based on the prices of goods only in theory similar compared: permeable soil sand devoid of buildings assessed as farm land, the buildings as urban property. It is not a correct approach, because the land, buildings and any other building structure present in military bases, require extensive work of transformation to be allocated to new uses.

C. Torre writes the introduction, S. Carbonara developes the explaination of the criticality of the case of study, and D. Stefano traced the assessment.

Keywords: Sale of public property, real estate appraisal.

1 Introduction. The Relevance of Italian Public Property as Economical Asset

The issue of valuation of property in public ownership has become especially important in Italy in recent years. The main reasons for this importance are at least as follows.

The first is the need to provide real estate appraisals for tax purposes of the assets of the public and private sectors to match the level of taxation on the income of real estate.

The second is related to the fact that the Italian public property is relevant , and often it is difficult to identify appropriate market values. In fact not only building services , and social housing belong to the public real estate heritage, but also historic buildings, monuments , and ordinary buildings but of considerable size.

B. Murgante et al. (Eds.): ICCSA 2014, Part III, LNCS 8581, pp. 459–470, 2014.

In this kind of heritage acquired assets also are associated with unordinary characters, due to the way in which they have become public property , such as real property owned by the criminal mafia that the State confiscated and took away.

These aspects are often invoked in the political debate every time that coming and going governments have attempted to reduce some of the debt through the sale of real estate.

Any minister of the economy and the tax authorities that try to make a budget estimate of what can be the great contribution by the sale of public property made to the economy of the state and to the reduction of the debt deal really difficult to make predictions .

Generally makers produce most optimistic predictions of expert evaluators , creating additional risks and difficulties to the already difficult economic maneuvers government forecasts .

Another problematic aspect is related to the fact that not only is it difficult to give a value to the public , but it is also difficult to identify all properties that are publicly owned .

A study just prior to the last update of the laws on the land register of a sample of public ownership of the largest Italian municipalities (Fig. 1) showed that the number of buildings intended for housing and commercial offices supposedly publicly owned uncertain were more than twice those of certain properties .

Metropolitan City	Public buildings	Supposed Public buildings	Total	Metropolitan City	Public buildings	Supposed Public buildings	Total
BARI				NAPOLI			
housing	19		19	housing		364	554
offices	12		12	offices		82	82
commercial		64	64	commercial		195	195
BOLOGNA				PADOVA			
housing	100		100	housing	33		33
offices	103		103	offices	-		-
commercial	149		149	commercial	-		-
BRESCIA				PALERMO			
housing		71	71	housing	275		278
offices		16	16	offices	55		55
commercial	4		4	commercial		131	131
CATANIA				TRIESTE			
housing		122	122	housing	200		200
offices		26	26	offices		18	18
commercial		65	65	commercial	100		100
FIRENZE							
housing		136	136				
offices		31	31	Total			
commercial		72	72	Metropolitan City	Public buildings	Supposed Public buildings	Total
GENOVA							
housing		230	230	housing	630	1022	1.652
offices	14		14	offices	184	174	358
commercial	54		54	commercial	307	524	831

Fig. 1. A Survey of hypothetical public property in some Italian metropolitan cities (Survey made by Valori Urbani, 2004)

Finally, the properties of public property to sell are placed in the so-called "public plan for sale and yield", which decides on the intended use they may have for individuals who buy . The public sale plan is a not flexible tool inside the market, and therefore does devalue the value of real estate for sale .

This phenomenon has been investigated in several literature cases in the past in more advanced real estate market, where those consideration rise up to the fore:

-    the influence of land-use plans [1] [2]
-    the difficulty of estimating prices and rents [3] [4]
-    several "bubble effect" [5] [6]
-    the peculiar character of Italian market [7] [8]

2   The Peculiarity of Military Property

The results , to date, have been modest, for a set of reasons, including the difficulty of organizing appropriate sales plans. In the case of public residential properties , sales programs are easier, more complex is the case of other public goods, such as disused military bases [9] [10].

The Ministry of Defence has granted to the Italian Government several times the selling of many military areas.

The so-called "Decree of the military barracks" of 2011, for example, permitted the sale of 76 military complexes, with a surface area ranging from 2,500 to nearly 400,000 square meters (see Tab.1 Tab.2 and Tab.3)

This decree was following after a similar law in 2004, which hypothesized the possible sale of 560 military complexes.

The current Premier Matteo Renzi in 2013, involves the reuse of the military stations for the construction of social housing.[1]

In these cases , for the purpose of sale, a number of factors come into play, from the conditions in which they are places and buildings, and finally to the future transformation that may receive.

A frequent mistake to place on the market these goods, evaluating the price based on the prices of goods only in similar compared theory: permeable soils and devoid of buildings assessed as farm land, the buildings as urban property. It is not a correct approach, because the land, buildings and any other building structure presenting military bases, require extensive work of transformation to be allocated to new uses; but also because it does not consider the highest and best use of land that the market can be attributed.

Another and no less important aspect concerns the urban destination of these areas, which are characterized in the plans of the municipal level as a "military zone". In Italy, the change of use planning is an obligatory act and formal power to establish a new feature compared to the previous year. Therefore, this problem also arises because the soils and buildings of the former military bases, are placed on the market with the original destination, with no guarantee that potential buyers can then operate freely subsequent changes of use. Only in the sale occurred, it will open a negotiation

---

[1]   All these news were deeply reported in the newspaper of the National Confederation of Industry "Il Sole 24 Ore.

with the local authority to agree the change of use. All this creates a very big risk for investors.

The instrument to decide wich public areas or real estate could be sale is the "Piano di Alienazione" (Plan for sale of public Property), where it is decided the new destination of use of the property, before of any kind of assessment or Hypothesis of best destination from the economic point of view. The instrument of the "Piano di Alienazione" is considered an instrument for urgency, and therefore is free from many form of control. For instance the Piano di Alienazione is one of the few instruments that are not submitted to the Strategic environmental assessment provided by the National and the European legislation. Therefore it is considered so urgent that it is not necessary to assess the environmental effect of the reuse of military properties. Such exemption is relevant since military properties are often characterized by the need of a de-polluting action, due to the presence of Asbestum and other chemical toxic substances in the land.

Most of these areas are concentrated in Northern Italy . It is a heritage A huge , covering an area of 6,026,249 square meters , often in the interior of the city, with about 734 500 copertti buildings with a total volume of 4,341,693 cubic meters.

Most of these areas, especially those aimed at the barracks and military housing , are dated at the end of World War II , and today are incorporated into the urban fabric.

A striking example is the "Rossani Barracks" (Fig. 1), a military area of about 100,000 square meters, located in the city center of Bari This area has been the subject of barter between the municipality of Bari and the Italian Government.

Fig. 2. The Area of Rossani Barracks in the core of the city of Bari

The Italian government has transferred the ownership of the Barracks to the Municipality of Bari , who sold the property in exchange for the Palace of the Prefecture and the Russian Church of St. Nicholas of Bari.

The ownership of the Russian Orthodox Church of St. Nicholas of Bari, was ceded to the Russian nation in 2008 by Prime Minister Romano Prodi to President Putin .

The value of the exchange for the abandoned area of Barracks Rossani was estimated at about 32 million. The value of Rossani Barrack, according to the estimate of the National Agency of Public Property, corresponds to the sum of value of Palace of the Prefecture and the Russian Church of St. Nicholas of Bari.

Today, the debate on the reuse of the area which is the largest property in the context of the city is built fairly open and is at the center of the municipal administration of the programs , which would make a great park , converting the buildings included in a Centre for Art and Culture [11].

3   The Case of Study

The case in the paper concerns a ex ammunition depot, long abandoned, located in a region of central Italy.

Fig. 3. The military base Area

The military base consists of a wide extension of the hilly land and various buildings (command headquarters building cafeteria, dormitories, sheds, small barracks, etc.), partly underground, destined in large part to the ammunition depot. Area also includes other structures, such as tanks and underground corridors, made several meters above the ground level.

The area has a total surface of approximately 200 hectares and is only a few kilometres from a small town. It benefits from excellent access to important road and rail infrastructure. It is very near a new industrial plant.

There are about 10,000 cubic meters of buildings, and about 80,000 cubic meters of building structures for the storage of ammunition.

As brief and largely incomplete or secret, description, available today in the military area, will allow you to say that - the state of things - they can not be sold on the market. Considered in its entirety as it appears today, the military compendium cannot have a destination other than the military.

This account covers both the soils free from buildings that the same buildings. These last, in fact, could be used for civilian purposes as they appear, because of the shape of the volumes, total or partial undergrounding of some of them, the quality of construction materials used (not excluding asbestos).

In addition, the same location of the buildings on the area, the viability of the service, the organization of space, make it completely incompatible area for other functions, other than the military.

4   The Institutional Assessments

At first, the Ministry devoted to the sale of public assets, has given the military base with a market value of approximately € 11 million , to carry out the public auction. Do you know the origin of this value , as it has been determined.

Under these conditions, the auction is not successful .

A potential buyer has asked for a re-evaluation. In this second case, the Public Agency who made the estimate , proceeded according to the approach set out in the introduction. They were considered the prices quoted in public databases for agricultural land and residential real estate , industrial warehouses and craft of the village near the military base . They have been developed by applying the reduction coefficients to take account of the level of deterioration of the buildings.

The market value has been defined slightly more than ten millions of euros , reduced to little more than seven millions of euros due to the costs that will be needed for the remediation of polluted wont substances released from stored ammunition.

5   Alternative Assumptions for Estimating

Considering the current characteristics of the military base, such as alternative hypotheses could be used?

The mistake so far has been to tie the estimation procedures used in the original function of the area, without considering the new features that could be attributed to it. In other words , just as it appears today, the military base could have no other use other than that due to the Italian Army, being completely incompatible to accommodate other functions than that for which it has been used so far .

That said, it will be clear that the search for a likely market price cannot be separated from assumptions redevelopment of the locatable property, related to new activities in the area, but also considering the costs that will be incurred to restore adequate conditions for civilian use .

And ' common sense , even before the market, suggesting the evaluation of those volumes is not as marketable artifacts , but for the potential related to their rebuilding

Compared to the costs of restoration , the process of regeneration must start from the restoration of the pre-existing conditions to transformations carried out by the Armed Forces .

This means to make four preliminary works:
-    trimming
-    military remediation
-    environmental remediation
-    demolition of existing buildings.

At that point might confront alternative hypotheses of new zoning laws, in order to "hook " the urban land market to determine the value of the estimate.

The trimming is necessary as a preliminary treatment of the soil to be subjected to remediation of ordnance . Is to remove the trees and shrubs that may hinder the use of the apparatus on the detector surface to be the clean-up operation . Should be performed with caution , according to established procedures (for "fields" and "strips" of land reclamation) and appropriate method of collection of plant material removed . As a rule, is carried out simultaneously with the surface reclamation operations .

The surface reclamation is performed to search for, identify and locate explosive devices or ferrous objects buried at depths not exceeding cm . 100 , through the use of appropriate metal detector. This operation is preliminary remediation that occurs through deep drilling (at depths greater than 100 cm) , generally run in the middle of squares of side m . 2.80.

Subsequent phases of removal and defusing of the bombs remain the sole responsibility of the Army Corps of Engineers .

Concerning the environmental remediation , it represents the voice of cost more problematic to estimate . In order to proceed with the correct quantification , it would be necessary to have the results of the Study of characterization of the soil , so as to know the pollutants present and hence to define the procedures and costs of remediation . The state of current information is certainly the asbestos (derived from the shells of some buildings) and possibly hydrocarbons (arising from the storage tanks on the site). For the rest, the functions hosted in the past , one could hypothesize the presence of lead and even residues of depleted uranium. It would , in any event, mere assumptions that, given the information currently available, may not be excluded nor confirmed with certainty. In other words, the soil characterization procedure is necessary to reach a definitive estimate of the costs necessary to complete redevelopment of the site.

All these operations associated with the demolition of the buildings , generating a cost that can be estimated to exceed seven million of Euros. This is a significant figure that should be subtracted from the value defined for the military base.

6   Hypothesis of Agricultural Use

The premise for this hypothesis lies in the consideration that the entire area of the base, given the current configuration planning (military area) could have used as the only alternative agriculture. For the reasons stated above about the nature and state of advanced deterioration of the existing buildings , they cannot have another assessment if not the one linked to the cost of demolition.

The simulation carried out made it possible to define a total value of estimated related to the agricultural use of only slightly more than three million of Euros .

Of course, the estimation value so defined should be compared with the total cost of restoration of the area .

The conclusion that can be drawn is that the assessment that stems from this assumption - as completely adherent to the local market of agricultural soils and consistent with the current planning regime - does not have the requirement of feasibility on the financial plan for the buyer, less than a negotiation cost recovery , especially in relation to war and environmental remediation and costs of demolition.

7   Hypothesis of Partial Industrial Use

Another hypothesis considers the possibility of allocating part of the compendium of the military part of the agricultural activity and industrial production activities .

In the absence of any reference to urban planning (target , building index , size of lots, etc.), a possible approach for understanding the meaning of the enhancement of the well, is to consider the conversion of land for industrial purposes. This hypothesis is standing in close proximity to military compendium of recent industrial settlements and favorable with respect to the accessibility of major transport infrastructure , road and rail .

This assumption becomes realistic provided that the City Council recognizes and incorporates within its planning instrument volumes that currently exist , providing the opportunity to rebuilt with concentrating on a part of the former ammunition depot.

Continuing in this perspective you could think of for industrial reconstruction of about 90000 cubic meters exist today.

Considering a height of 6.00 ml of warehouses would lead to hypothesize a retail area of approximately 15 thousand square meters. At this point , based on the market prices of industrial buildings , could be defined as a percentage of the value of the impact on the value of built.

Following this principle , this would lead to a total value on the building potential of just under 2.7 million euro .

Considering the estimated value of the surface that would express building potential and the estimated value of its undeveloped assessed as agricultural , this would lead to a total value associated to this hypothesis reuse of just under six millions of euros. Also on the gross amount of seven million of restoration costs .

So it is also in this case the hypothesis punt financially unsustainable.

Table 1. The list of Military Properties free for sale in 2011 in Northern Italy (Source Ministry of Defense)

	Region	City	Area (square meters)	Built Surface (square meters)	Volume (cubic meters)
1.	Friuli	Aquileia (Ud)	47.818	3.540	23.821
2.	Friuli	Chiusaforte (Ud)	156.046	26.423	165.353
3.	Friuli	Lucinico (Go)	23.450	5.142	24.060
4.	Friuli	Palmanova (Ud)	10.505	3.318	19.506
5.	Friuli	S.Lorenzo Isontino (Go)	111.649	10.855	82.373
6.	Friuli	Trieste	54.985	3.838	20.139
7.	Friuli	Udine	91.254	16.918	202.214
8.	Friuli	Udine	20.256	6.495	36.595
9.	Friuli	Venzone (Ud)	1.987	820	16.210
10.	Friuli	Villa Vicentina (Ud)	19.144	5.499	32.700
11.	Veneto	Bagnoli Di Sopra (Pd)	59.826	7.486	28.711
12.	Veneto	Castelnuovo Del Garda (Vr)	23.000	4.200	36.000
13.	Veneto	Chioggia (Ve)	36.639	13.406	74.814
14.	Veneto	Dosson Di Casier (Tv)	6.236	1.600	18.780
15.	Veneto	Feltre (Bl)	131.280	31.930	164.300
16.	Veneto	Isola Rizza (Vr)	113.080	19.480	77.920
17.	Veneto	Padova	104.877	104.877	121.877
18.	Veneto	Padova	31.265	5.976	35.836
19.	Veneto	Padova	657.128	15.092	45.276
20.	Veneto	Sappada (Bl)	413.087	7.950	24.000
21.	Veneto	Valeggio Sul Mincio (Vr)	163.366	2.202	8.600
22.	Veneto	Valeggio Sul Mincio (Vr)	92.231	10.052	35.182
23.	Veneto	Venezia	156.400	4.000	20.000
24.	Veneto	Venezia	21.450	3.780	24.800
25.	Veneto	Venezia			
26.	Veneto	Verona			
27.	Veneto	Vigodarzere (Pd)	993	570	5.601
28.	Lombardy	Brescia	29.750	10.650	73.300
29.	Lombardy	Cremona			
30.	Lombardy	Legnano (Mi)	69.000	12.170	73.020
31.	Lombardy	Milano	10.993	3.335	16.147
32.	Lombardy	Milano	71.718	23.459	143.550
33.	Lombardy	Milano	16.066	4.336	52.016
34.	Lombardy	Milano	101.501	26.224	117.160
35.	Lombardy	Milano	28.500	10.000	90.000
36.	Lombardy	Milano	58.000	22.000	264.000

Table 1. (*Continued*)

37.	Lombardy	Milano	36.000	0	0
38.	Piedmont	Monteu Da Po' (To)	3.820	1.820	12.590
39.	Piedmont	Prazzo (Cn)	7.410	4.084	50.482
40.	Piedmont	Torino	19.445	7.960	122.161
41.	Piedmont	Torino	383.445	8.988	15.317
42.	Piedmont	Torino	43.461	24.500	122.225
43.	Liguria	La Spezia	600	600	3.000
44.	Liguria	La Spezia			
45.	Liguria	La Spezia	46.693	1.800	8.300
46.	Liguria	La Spezia			
47.	Liguria	La Spezia			
48.	Liguria	La Spezia			
49.	Liguria	La Spezia	9.913	3.844	14.424
50.	Liguria	La Spezia			

Table 2. The list of Military Properties free for sale in 2011 in Central Italy (Source Ministry of Defense)

	Region	City	Area (square meters)	Built Surface (square meters)	Volume (cubic meters)
1.	Emilia Romagna	Bologna	23.300	7.000	54.113
2.	Emilia Romagna	Parma	64.900	8.443	40.975
3.	Emilia Romagna	Ravenna	135.000	43.624	364.414
4.	Tuscany	Scandicci (Fi)	25.564	20.308	104.899
5.	Marche	Ascoli Piceno	8.825	3.233	44.114
6.	Lazio	Civitavecchia (Rm)	64.250	13.363	48.454
7.	Lazio	Manziana (Rm)	8.030	970	3.547
8.	Lazio	Roma	497.000	12.275	67.814
9.	Lazio	Roma	5.248	4.034	37.000
10.	Lazio	Roma	4.846	4.240	74.240
11.	Abruzzo	Sulmona (Aq)	2.421	1.391	11.292

Table 3. The list of Military Properties free for sale in 2011 in Southern Italy (Source Ministry of Defense)

	Region	City	Area (square meters)	Built Surface (square meters)	Volume (cubic meters)
1.	Campania	Capua (Ce)	74.500	2.650	32.853
2.	Campania	Napoli	2.500	0	0
3.	Apulia	Bari	53.536	24.060	181.372
4.	Apulia	Lecce	74.742	17.166	88.639
5.	Apulia	Lecce	20.949	8.245	60.000

Table 3. (*Continued*)

6.	Apulia	Montemarano (Ta)	2.448	200	3.360
7.	Apulia	Taranto	21.085	411	1.233
8.	Apulia	Taranto	533.644	45.867	400.514
9.	Calabria	Campo Calabro (Rc)	144.950	10.600	58.300
10.	Sicily	Catania	961	880	4.543
11.	Sicily	Isola Delle Femmine (Pa)	457.593	1.320	5.752
12.	Sicily	Marsala	108.000	4.430	32.985
13.	Sicily	Messina	125.830	7.760	38.350
14.	Sicily	Patti (Me)	81.860	10.811	60.570

8   Conclusions

In the case of the sale of military areas is required :

- estimate the costs of restoration (decontamination , demolition , etc.).
- value assets such as land builders , who ascribe appropriate destinations and building volumes in order to cover the costs of rehabilitation and return to allow the public administration ;
- the best choice for the new use must begin with an analysis of the characteristics of the local context and economic environment in which these goods are located.

The seek for the best and highest use should be make free, despite to the indication of any form of pre-constituted sale plan.

References

1. Alterman, R.: Takings International: A Comparative Perspective on Land Use Regulations and Compensation Rights. American Bar Association Publications, Chicago (2010)
2. Cerreta, M., De Toro, P.: Assessing urban transformations: A SDSS for the master plan of Castel Capuano, Naples. In: Murgante, B., Gervasi, O., Misra, S., Nedjah, N., Rocha, A.M.A.C., Taniar, D., Apduhan, B.O. (eds.) ICCSA 2012, Part II. LNCS, vol. 7334, pp. 168–180. Springer, Heidelberg (2012)
3. Tse, R.Y.C.: Housing Price, Land Supply and Revenue from Land Sales. Urban Studies 35(8), 1377–1392 (1998)
4. Ong, S.E., Sing, T.F.: Price discovery between private and public housing markets. Urban Studies 39(1), 57–67 (2002)
5. Rose, C.: The Comedy of the Commons: Custom, Commerce, and Inherently Public Property. The University of Chicago Law Review 53(3), 711–781 (1986)
6. Meen, G.: Regional house prices and the ripple effect: A new interpretation. Housing Studies 14(6), 733–753 (1999)
7. Manganelli, B., Morano, P., Tajani, F.: House Prices and Rents. The Italian experience. Transactions on Business and Economics 11 (2014)

8. Torre, C.M., Perchinunno, P., Rotondo, F.: Estimates of housing costs and housing difficulties: An application on Italian metropolitan areas. In: Housing, Housing Costs and Mortgages: Trends, Impact and Prediction, pp. 93–108. Nova Science Publ., New York (2013)

9. Van Driesche, J., Lane, M.: Conservation through Conversation: Collaborative Planning for Reuse of a Former Military Property in Sauk County, Wisconsin, USA. Planning Theory & Practice 3(2), 133–153 (2002)

10. Morrison, P.D.: State Property Tax Implications for Military Privatized Family Housing Program. AFL Rev. 56, 261–268 (2005)

11. Signorile, N., Rossani, D.: La difesa dello spazio pubblico e la privatizzazione della città Caratteri mobili, Bari (2014)

Web Potential to Promote Equity Planning Practices

Maria Giovanna Altieri, Francesco Rotondo, and Carmelo Maria Torre

Department of Civil Engineering and Architecture, Polytechnic of Bari,
Via Orabona 4, 70125 Bari, Italy
{carmelomaria.torre,f.rotondo}@poliba.it
mariagiovannaaltieri@gmail.com

Abstract. The concept of Urban Services, like the product between supply and demand in a city, dominates, although in an indirect way, urban planning. Nowadays we need to organize planning processes based on a study of actual needs, from which come down services to promote in urban center. It's like a "matriosca" scheme: the first part, the bigger, is the city, and then we have a lot of subsystems connected together, who are Government, citizens, needs and services. This process comes down from the city's change, that before the Modern Era was just a geographical place, where the walls were the limits. Today the city is the union between the geographical place and citizens, where social and urban polices are the basis for the economic, social and environmental growth. In the case study, concerning an urban regeneration plan in the city of Bari in Southern Italy, these aspects have been analyzed and practiced. It shows the integration between public participation and use of ICT, for its promotion.

Keywords: Community Impact Evaluation, Web Participation, Social Learning.

1 Introduction

In recent years planning activities are trying to change their aspect becoming more competitive, because urban politics are more focused on reduction of landscape use and promotion of sustainable development. In this case, the need to grow up the information and the communication between stakeholders and decision making is one of the most important things.

The active participation of citizens, with others fildes like environmental protection and the cost-effectiveness of project, directs the urban planning in a sustainable way. In this way, city planning is like a multisectoral subject , where the social and environmental evaluations are connected to economic analysis.

Nowadays, also, the society is more digital (digital city) and computerized, so an important share is given by the use of Information and Communication Technolgies (called ICT), that create the base for a development of informative flux and database, used for the supervision of plan/program, during processing and management stage. In the examined case of study, that is about a requalification plan in Torre a Mare, Bari's district, there are sum up this aspects, and it shows the integration between public participation and use of ICT, for their promotion.

B. Murgante et al. (Eds.): ICCSA 2014, Part III, LNCS 8581, pp. 471–482, 2014.

The first chapter talks about the concepts (already well known in the literature) exercised in the case study on the use of ICT for assessment of urban development plans and to encourage public participation. In the next chapter, the paper recalls the basic elements of Community Impact Evaluation method developed by Lichfield (2005). In the third chapter, it deals with the fundamental issue of the relationship between participation and contemporary urbanism trying to overcome the rhetoric of community that still characterize it. In the following chapter it discusses the process of upgrading the district of Torre a Mare in the capital of the Apulia Region , and the results and prospects of the use of ICT for the development of these practices. In the fifth chapter we evaluate the choices of the urban regeneration plan on the basis of expert and common knowledge, emerged on the basis of the results of social surveys. In the conclusions the paper discusses the results and prospects of the work done.

2   Information and Communication Technologies in Urban Planning Evaluation and Public Participation

Town planning evaluation process are necessary tools to integrate environmental, social and economic observation, to improve the entire decision making quality. In fact they should follow the plan from the identification of choices and strategies until their complete approval, within a sustainable framework. Therefore, it is an indispensable tool to design the plan, to implement it and to improve it (ex-ante, ongoing and ex-post [1]. But evaluation tools are not sufficient if they are not included in a dialectic context of citizen participation, where the agreement, built by dialogue and communication between citizens, becomes very important for the growing up of a sustainable development [1]. From this point of view, there is the need to integrate, in a more effective way, the social part, composed by inhabitants, with innovation technologies, and both of them are basic pillars of Smart City.

In this way, the planners give up the traditional model of planning that dominated during the postwar years, due to the emergence of environmental issues, which has led to the emergence of new urban policies focused on landscape reduction. In fact in city centers there is the need to activate participatory processes of regeneration and redevelopment of underutilized or brownfield, that provide to increase services and urban greenery, rather than pursue new developments in agricultural areas, expanding the city to outlying areas not equipped with the necessary urbanization. The need to pursue different models of planning in uncertain decision-making environments, has led to a strategic planning model, which is an interactive and retrospective process , where there is a constant integration between urban and social policies to promote sustainable local development.

Within this scheme, participation should be preferably located upstream of the urban planning process and decision processes, during the definition of the ideas, the guidelines, the response to the needs and strategies to be adopted [2], then keeping constant attention to the citizens and users needs of the sites through a consultation and a collaborative process in the early stages of implementation and management of planning decisions [3].

Thanks to the increasing contribution of Information and Communication Technologies (after called ICT), it is possible to promote effective strategies in plans evaluation, linked to participation and supervision with information raising.

This has been possible from the electronic digital revolution, which has increasingly blurred the boundaries between the Information and the Communication Technologies, the first mainly linked to the processes of storage and management of complex information and the second dedicated to interactive forms of communication in real time [4]. Sometimes, however, it is not enough only to introduce these technologies within society, but it is necessary to establish organizational processes [3], which cannot be reduced to the technical and procedural size, where participation is seen as a means to the expected product, rather than as a goal. In this way, the Information Society, has a lot of benefits, as facilitation in public participation practice of urban planning processes, where database organization and information raising becomes easier than before, and also it's possible to supervise the participation. In this sense, the Information Society, finds a lot of benefits, as facilitation in the practice of public participation in urban planning processes, where is increased the organization of data and information collection and it is possible to monitor participation, as well as the ability to monitor the concrete acts of urban planning through the official websites of public authorities involved. Through web accesses analysis affected by the project, it's possible to understand causes and strengths of connections trends. So with this information we are able to underline the most interesting issues. In fact, the Information gives us the basic and necessary elements useful for the start of a participation process and plan evaluation in a critical, conscious and exhaustive way. Moreover it encourages the development of a public space, not only as a physical space for meeting, but also as an intellectual one.

A, not very recent case study [5] underlines that the majority of citizens would like to take part to decision making processes in a concrete way, but it don't believe that its experiences and advices could be really taken. Moreover this research underlines that the use of word may be adapt to the discussion with citizens, rather than requiring citizens to the adoption of this language, promoting more direct and simple communication shape.

The exploitation of information and telematics technologies gave birth to new forms of democracy, called e-democracy, that determine the social inclusion within the information society, in contrast to the digital divide phenomenon.

The goal of ICT is not to replace to traditional processes, but to strengthen and support them in particular situation, where there is a difficult supervision. This may be the case of strategic environmental assessments, where is required an appropriate supervision of initial parameters, with the use of environmental indicators of the impact that the program or plan may have on the environment during the operating phase. Moreover, thanks to technological advances, it is possible to achieve better and transparent information , that allows citizens to be informed and to inform themselves quickly and completely. This kind of progress is at the base of spatial information development, called SIT, not yet available on most Italian Municipalities websites [6]. These systems provide a picture of interest local situation and they settle information, communication and participation method in planning practice [7]. This leads not only designers, but all those who are interested in urban and environmental issues, to take on a spatial awareness that serves as a guide in urban planning choices.

The geographic information system, known as, today represents one of the most important instruments of territorial and environmental knowledge [8], that associated with remote sensing techniques, help to outline a framework supporting information to the urban transformation. The correct use of these technologies allows the system does not deviate from the curve of urban sustainability, for the benefit of smart city development, to a lower level of urban entropy [8].

Area's knowledge through georeferenced systems lets us to develop more accurate planning skills, that really take into place peculiarities and services. So, in this way, it's possible to create bijective relationship between life quality/well-being and planning one. In addition, EU policies are increasingly incorporating forms of communication and consultation through the use of ICT; for example the site Your voice[1], the European portal for public consultation in the field of European law, or even the Demos project (Delphi Mediation On line System) which was attended by the Municipality of Bologna.

As planners we refer to DSS Technology, that is Decision Support Systems, to increase effectiveness of decisions processes. In particular, in the case study, referred to the regeneration plan of Torre a Mare, Bari's district, we use the ELECTRE methods, very useful to compare the alternatives to one problem. The ELECTRE method is a multi-criteria analysis, composed by two matrices, the first called Impact Matrix and the second correlation Matrix, that transform the decision making process in analytical way.

The relationship between use of ICT, urban planning and public participation is examined in the case of study, where we are a little place, but quiet active in communication and information field.

3   Impact Community Evaluation

The evaluation processes accompany the plan / program from the preliminary stage until the operation phase of the same, in order to constantly monitor the impacts that it causes in environmental economic and social terms. The application of the Community Impact Evaluation [9] has as main objective the identification of the effects and its expected impacts on the different sections of the community that the planning action involves. This identification allows us to understand how the proposed intervention and later realized increases the welfare of society or otherwise causes disadvantages. This valuation approach is interposed between the cost-benefit analysis and multi-objective one, trying to identify the objectives of the different areas that divides the community, the distribution of costs and benefits for each of them, the possible conflicts that could arise, and then arrive at a classification of design alternatives, depending on the degree of appreciation received from various community groups.

---

[1] To have more information about this themes, it is possible to consult
http://ec.europa.eu/yourvoice/index_it.htm

Such procedure has got great efficacy in the evaluation of regeneration plans or recovery of brownfield sites, where are analyzed in the different proposed alternatives, and the level of compliance [10].

The union between the community impact evaluation and ICT, is a strong point in the practice of urban planning, because it allows the spread of information and the raising of the same.

4   A Different Relationship between Urban Planning and Participation

The term "design-action", referring equally to both public and private spaces, refers to the idea of dynamism typical of the relationship between living and building, that allows us to understand what are the spaces, situations, conditions where people can exert a truly planning shared [2]. In this way, the citizens become the real protagonists in the construction of dwelling places, in 'ownership' of the space, reviving the concept of membership. Then, the project is not a tool for few experts, characterized by widespread timelessness, but it should have the characters of sharing, of dynamic planning between designers and inhabitants.

The authority, that is typical of the modern era, now seems to come to the decline, mainly because it has created city unrelated to the needs of the population, characterized by a widespread lack of services, that now feel the need to be re-thought. It is positioned inside the stadium of the "reconfiguration" mentioned by Ricoeur [11],

The dwell here is understood as a response and reaction the act of build, takes care of plural action, and therefore, the 'man who dwells, now can starting from his state to revise the act of building [11]. Cities are not only geographical locations, where the walls set out the end [12], but they have to represent the union between space and citizen, where social and urban policies are the basis for the economic, social and environmental growth.

The discipline of urban planning should be reviewed in a perspective of increasing participation, and urban project needs to demonstrate as a result of the interaction between human society and environment. Therefore, the management of urban transformations should point to the exploitation of environmental resources, the emergence and proliferation of social policies and the economic competitiveness of the interventions. These aspects are not covered in traditional planning, characterized by lack of interdisciplinary and a hierarchical- pyramidal model. We need to focus on the integration of bottom-up and top-down models, where the local community, the citizens, interacts with public administration, in the living spaces definition and the planning processes promotion.

In the urban planning another important tool is Reticular model, that is based on the relationship's network between public and private subjects that operate in the area, not on the consent of a decision already taken. It's necessary to practice a 'third generation strategic planning', where the plot is characterized by a deep knowledge about environmental, social, economic and urban context and constant relationship between the actors of the process.

5   An Urban Regeneration Plan

The experience carried out in Torre a Mare, district of Bari, the capital of Apulia Region, as part of a redevelopment plan, has seen the promotion of practices of participatory planning, through the promulgation of questionnaires, that are published on the web or delivered in the most interesting places of the district. The analysis shows that most of the recipients are residents of the district, who have a common interest to increase the life quality and the urban services. Thanks to their contribution, there was an improvement of planning choices. This project also marked the importance of accessibility and advertising of urban policies, because the citizen is not uninterested in these issues, but he/she is poorly informed or sometimes the procedures constitute a barrier in the expression of their own judgment.

The main goal of redevelopment plan is to create a new relationship between sea and city, through the recovery and re-definition of coastal areas and most interest places of Torre a Mare.

The area affected by the Redevelopment Plan and by the Detailed one (A2 zona), that are in process of drafting by Division of Urbanism and Private Construction, is composed by:

1. Environmental interest area, A2
2. Completion area, B3
3. Expansion area C2
4. Primary activities B, that are near the riverbed of Lama Giotta. Lama Giotta runs through the West size of Torre a Mare, and its mouth is stopped by Trulli street.
5. neighborhood green, that are both in the east site of Environmental area A2 (near Vittorio Veneto square) in central position and in the west site near Grotta Regina Area.
6. Residence services
7. "Urban green, type A", that are located along the coastal area, in the north site of Redeveloped Plan, near Punta la Penna Archaeological zone

The area, interested by the regeneration plan, is included into the Thematic Territorial Development Plan, known as PUTT/p, approved by resolution of G.R. n.1748 del 15/12/2000, and so it is necessary to obtain landscape opinion for the art.5.03 of N.T.A. PUTT/P[2].

In particular, the redevelopment plan includes the realization of four actions, as written:

1. Realization of covered weekly market with basement parking
2. Renovation of fish market, located near port area
3. Realization of Urban Park
4. Renovation of waterfront

---

[2] References to the redevelopment plan of the district of Torre a Mare are extrapolated from the technical report of the plan drawn up by the technicians of the city of Bari, Section of Urbanism and private construction.

These actions had proposed into the questionnaire, that we talk about it later, to evaluate citizens point of view. Among the main plan objectives there are: protection and valorization of the city center, creation of new squares and open space for social and fun activities, new commercial, professional and craft activities, improvement of road conditions and of coastal areas.

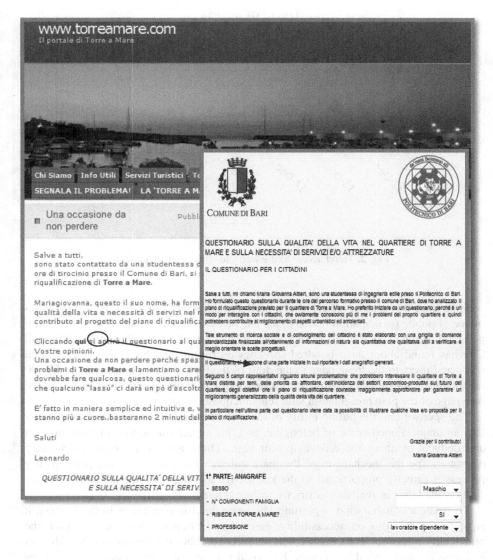

Fig. 1. Web Page

This work was possible thanks to the integration of participation practices and information technologies; arrived properly completed forty-five questionnaires, in particular thirty-two from the web page of Torre a mare and the others thirteen from paper. It is possible to connect the obtained results to the period of presence on the web, not very long (about 10 days) and the small size of the district, mainly characterized by residences non-continuous, but concentrated in the summer.

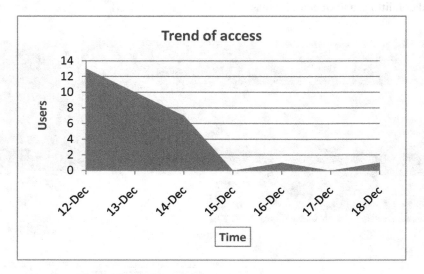

Fig. 2. Trend access to the web site www.torreamare.com

Start from this case of study, it's possible to talk about others two important experience of public participation linked to the use of web site, that relate the Pug of Monopoli City and the PTCP of Foggia Province. The project PartecipaPug of the Monopoli City has found about three hundred speeches on forums and blogs and two hundred registered users to the login page to the draft of the PUG out of a total of about 50,000 inhabitants. Less subsidiary was the experience of PTCP of Foggia, where access to the web page of the plan presentation with relevant questionnaire were quite content.

The sponsor and the public utility of the questionnaire, created in the citizens district an unquestioned sense of belonging, and this aspect was leaked in their answers and deductions about the redevelopment plan. These are social processes, so they cannot escape the mechanism of the main culture. The success of the participatory process is directly proportional to the knowledge and depth inside the subject, especially if the latter is available in strictly technical terms.

To create a collaboration opportunity between the different actors in the process, it is necessary to focus on accessibility, ease of understanding and use of a tool, the practicality response. The citizens were questioned about the services of their district and possible and desirable changes that could lead to processes of economic and social growth. In this sense, the term "design" becomes wider and incorporates the rules of the participation process, which if it doesn't show up correctly formulated, it leads to inconsistencies and unreliable results.

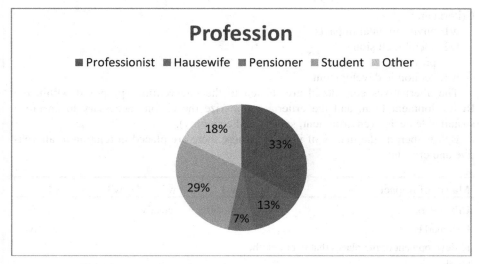

Fig. 3. Job and Social Profile of users

6   Alternatives Evaluation: ELECTRE Method

The multi-criteria analysis can address complex problems, considering individually but in an integrated way all the variables, giving each of them their relative importance. The multi-criteria evaluation allows you to examine a problem from several points of view, like economical, social and environmental (Boggia, 2007).

The evaluation through ELECTRE method, called ELiminationEt Choix Traduisant the Realité, developed during the mid-60s by Bernard Roy, and renewed during the years with the versions ELECTRE 2 and 3 [13], allows to manage in an integrated way the alternatives related to planning problem, and inserts the technique of correlation, that provides a measure of the prevalence of each alternative on the others, criterion by criterion .

The implementation of this methodology involves the construction of two matrices, the first call of the impact matrix, where is possible to evaluate the relationship between alternatives and criteria, according to qualitative and quantitative evaluations. It is then formulated the matrix of concordance, where for each criterion is assigned a weight, which can vary depending on the point of view that you decide to take. In particular is possible to underlines five alternatives and four criteria [14].
Alternatives are:

1. Urban park,
2. Fish market,
3. Redevelopment of the places that overlook the coast,
4. Weekly market with basement parking,
5. Redevelopment of the old city center.

Criteria are:
  w1: environmental impacts
  w2: social inclusion
  w3: project costs
  w4: economic development
The alternatives considered are related to the interventions proposed within the Redevelopment Plan, and the criteria summarize the canons necessary to ensure a sustainable design (environment, society and economy).

Below there is the matrix of impacts, where there are placed in relation to alternative and criteria:

Matrix of impact	w1	w2	w3	w4
Urban Park	positive	positive		
Fish market				positive
Redevelopment of the places that overlook the coast			negative	
Weekly market with basement parking	negative		negative	positive
Redevelopment of old city center	positive	positive		positive

In order to process the correlation matrix and the development of the index, we have divided the analysis into four points of view, which led to the obtaining of four matrices:
  1. Environmentalist point of view
  2. Political-administrative point of view
  3. Business point of view
  4. Social point of view

Depending on the scenario examined, the relative weights of the criteria have undergone some changes. In particular:

Weights	Environmentalist	Political-administrative	Business	Social
w1	0,4	0,15	0,20	0,30
w2	0,3	0,15	0,20	0,40
w3	0,15	0,40	0,20	0,15
w4	0,15	0,30	0,40	0,15

The analysis of the concordance matrixes shows that the intervention more suited to the district of Torre a Mare is the creation of an Urban Park, followed by the development and renovation of the old city center.

Concordance: environment	P.U.	M.P.	R.C.	M.S.	N.A.	Index
Urban Park (PU)		0,775	0,925	0,85	0,425	0,74375
Fish Market (MP)	0,225		0,65	0,775	0,15	0,39375
Redevelopment of the places that overlook the coast (RC)	0,075	0,35		0,625	0	0,24375
Weekly market with basement parking (MS)	0,15	0,225	0,375		0,075	0,16875
Redevelopment of old city center (NA)	0,575	0,85	1	0,925		0,69375

Concordance: political-adm	P.U.	M.P.	R.C.	M.S.	N.A.	Index
Urban Park (PU)		0,5	0,85	0,7	0,35	0,6
Fish Market (MP)	0,5		0,85	0,775	0,35	0,49375
Redevelopment of the places that overlook the coast (RC)	0,15	0,15		0,425	0	0,14375
Weekly market with basement parking (MS)	0,3	0,225	0,575		0,15	0,2375
Redevelopment of old city center (NA)	0,65	0,65	1	0,85		0,625

Concordance: business	P.U.	M.P.	R.C.	M.S.	N.A.	Index
Urban Park (PU)		0,5	0,8	0,6	0,3	0,55
Fish Market (MP)	0,5		0,8	0,7	0,3	0,45
Redevelopment of the places that overlook the coast (RC)	0,2	0,2		0,4	0	0,15
Weekly market with basement parking (MS)	0,4	0,3	0,6		0,2	0,275
Redevelopment of old city center (NA)	0,7	0,7	1	0,8		0,625

Concordance: social	P.U.	M.P.	R.C.	M.S.	N.A.	Index
Urban Park (PU)		0,775	0,925	0,85	0,425	0,74375
Fish Market (MP)	0,225		0,65	0,725	0,15	0,38125
Redevelopment of the places that overlook the coast (RC)	0,075	0,35		0,575	0	0,23125
Weekly market with basement parking (MS)	0,15	0,275	0,425		0,075	0,19375
Redevelopment of old city center (NA)	0,575	0,85	1	0,925		0,69375

7   Conclusions

The support provided by the net increase in the case of study the ability to collect community's point of views. Especially as regards time and cost of survey, the facilities provided by the net are considerable.

The same result can be easily submitted, by the use of the technology to different stakeholders and community sector, and they can be used to re-develop a more shared vision of the world.

Furthermore the democracy is enforced by the approach. The quality of methodological approach in some way is not so relevant as in other more structured, cost benefit evaluation, because it is not so important to have a monetary feedback, more than to support the social debate.

References

1. Marchi, G., Lenti, L.: La valutazione nei processi di piano – Strumenti di trasformazione urbana. Franco Angeli editore, Milan (2003)
2. Cellamare, C.: Progettualità dell'agire urbano. Processi e pratiche urbane. Carocci, Rome (2011)
3. Caperna, A.: Elementi di ICT nella pianificazione e progettazione urbana. DIPSU, University of Rome 3 (2008)
4. Rotondo, F., Selicato, F.: ICT to evaluate participation in urban planning: remarks from a case study. In: Murgante, B., Gervasi, O., Misra, S., Nedjah, N., Rocha, A.M.A.C., Taniar, D., Apduhan, B.O. (eds.) ICCSA 2012, Part I. LNCS, vol. 7333, pp. 545–560. Springer, Heidelberg (2012)
5. Coleman, S., Gøtze, J.: Bowling Together: Online Public Engagement in Policy Deliberation. Hansard Society, London (2001)
6. De Cindio, D.: Guidelines for Designing Deliberative Digital Habitats: Learning from e-Participation for Open Data Initiatives. The Journal of community informatics 8(2) (2012)
7. Orlando, M.: Il ruolo dei sistemi informativi territoriali nel processo di recupero dei centri storici. Franco Angeli Editore Milan (2008)
8. Fistola, R., La Rocca, R.A.: New technologies for Sustainable Energy in Smart City, the WET Theory. TeMa Journal of Land Use Mobility and Environment, 29–42 (2014)
9. Nathaniel Lichfield, The Community Impact Evaluation (2005)
10. Fusco, G.L.: Sustainability, creativity, resilience: Toward new development strategies of port areas through evaluation processes. International Journal of Sustainable Development 13(1-2), 161–184 (2010)
11. Ricoeur, P.: Lezioni. La persona. Morcellina, Brescia (1997)
12. Bill Cooke, Uma Kothari Partecipation: The new Tiranny (2001)
13. Roy, B.: Classement et choix en présence de points de vue multiples (la méthode ELECTRE). La Revue d'Informatique et de Recherche Opérationelle (RIRO) 8, 57–75 (1968)
14. Munda, G.: Multicriteria Evaluation in a Fuzzy Environment. Physica, Heidelberg (1995)

Classical and Bayesian Goodness-of-fit Tests for the Exponential Model: A Comparative Study

Maria J. Polidoro[1,4], Fernando J. Magalhães[2,4],
and Maria A. Amaral Turkman[3,4]

[1] CIICESI, ESTGF-Polytechnic Institute of Porto, Felgueiras, Portugal
mariapolidoro@eu.ipp.pt
[2] ISCAP-Polytechnic Institute of Porto, Porto, Portugal
fjmm@iscap.ipp.pt
[3] Faculty of Sciences of University of Lisboa, Lisboa, Portugal
maturkman@fc.ul.pt
[4] CEAUL-Center of Statistics and Aplications of University of Lisboa, Portugal

Abstract. Most common statistical methodologies assume a parametric model for the data and inference is made based on that assumption. If the model does not fit the data, the resulting inference will be mislead. Thus, evaluation of the fitting of a proposed parametric statistical model to a given dataset becomes an important issue.

In several practical situations, namely in reliability and life sciences problems, the exponential model has been widely used and several classical tests were already proposed for its fitting evaluation. In this work we suggest two Bayesian tests when an exponential model is proposed to describe the data, and using a simulation study, we compare their power with the classical ones.

Keywords: Goodness-of-fit test, Bayesian nonparametric model, Bayes factor, mixture of finite Polya trees, power of test.

1 Introduction

In statistical methodology it is often necessary to find a suitable model for the data under study. An important issue in modeling is to evaluate the fitting of the proposed parametric statistical model to the given dataset.

In many problems, namely in reliability engineering and life sciences, the exponential model is widely used; however, a wrong choice of the parametric model may mislead the statistical inference and hence it is important to find the best available test to evaluate the fit.

In a classical approach, formal methods, often called goodness-of-fit tests, involve the test of a null hypothesis where the parametric model is defined without the specification of an alternative hypothesis or alternative model. There are many kinds of goodness-of-fit tests in literature (see, e.g. D'Agostino and Stephens [1]). Some of them are very specific, but others are quite broad tests which are applicable to general cases. The most commonly used are Pearson's chi-squared test and those based on the empirical distribution function.

B. Murgante et al. (Eds.): ICCSA 2014, Part III, LNCS 8581, pp. 483–497, 2014.

In the last decades, the interest in the problem of goodness-of-fit test for the exponential case has increased and new test statistics emerged (see, e.g. [2–6] and references therein).

The Bayesian foundation for fit evaluation is conveyed by the predictive distribution through exploratory methods or by using formal posterior predictive model checks, like Bayesian p-values (see, e.g. [7–10]). More recently, Johnson [11, 12] proposed a Bayesian chi-squared goodness-of-fit test of a parametric model by generalizing the classical Pearson's chi-squared statistic and discussed the use of pivotal test statistics.

An enhanced Bayesian nonparametric alternative consists on embedding the proposed parametric model (H_0) in an alternative nonparametric model (H_1). To validate the proposed model the parametric fit is compared with the nonparametric one, using the Bayes factor as a measure of evidence against H_0, based on the observed values.

These testing problems require the formulation of Bayesian nonparametric models (see, for example, the book by Hjort et al. [13], for some discussion on the subject). Bayesian literature on nonparametric goodness-of-fit tests is still scarce. For a continuous density function, particularly for the normal density, Verdinelli and Wasserman [14], Berger and Guglielmi [15] and Tokdar and Martin [16] proposed a Bayesian nonparametric goodness-of-fit test, which assigns a mixture of Gaussian processes, a mixture of Polya trees and a Dirichlet process mixture, respectively, for the alternative model. All the three tests allow calculation of the Bayes factor, however only the alternative model based on a mixture of Polya trees is computationally more accessible and intuitively simpler.

In this work we suggest the Bayesian nonparametric goodness-of-fit test of Berger and Guglielmi [15] and the Bayesian chi-squared test of Johnson [11], to evaluate the goodness-of-fit in the exponential case and we compare the power of these tests with some classical test statistics.

The paper is organized as follows: in Section 2 we review some definitions on nonparametric Bayesian statistical models as well as the Bayesian nonparametric test and the Bayesian chi-squared goodness-of-fit test. In Section 3 we describe some of the classical test statistics and the Bayesian tests for the problem of testing the adequacy of an exponential model. In Section 4 we carry out a simulation study to compare the power of the different proposed tests. Finally, in Section 5, conclusions are summarized based on the obtained simulation results.

2   Bayesian Approach

Let $(X_1, X_2, \ldots, X_n)$ be a vector of continuous, identically distributed, conditionally independent observations drawn from a probability density function $f(x|\theta)$ defined with respect to Lebesgue measure and indexed by an s-dimensional parameter vector $\theta \in \Theta \subset \mathbb{R}^s$, which is unknown. Our goal is to test the adequacy of the assumed model $f(x|\theta)$ based on the observed data $(x_1, x_2, \ldots, x_n)$. The assumed model $f(x|\theta)$ is the "null" model or "null" hypothesis, represented by H_0. For the Bayesian formulation, if an "alternative" hypothesis, H_1 is to be specified, it will be associated to a nonparametric Bayesian model named by G.

2.1   Nonparametric Bayesian Model

In a nonparametric Bayesian context, the random sample is defined by an unknown random probability measure G, and the goal is to place a prior directly on the class of random probability measures. Lavine [17, 18] proposed Polya trees as an useful nonparametric prior distribution for random probability measures G on the sample space Ω of the random variable X. Reference papers on Polya trees are also Hanson and Johnson [19] and Hanson [20]. The Polya tree can be easily constructed as follows.

A finite Polya tree for a distribution G is built by dividing the sample space Ω, into a sequence of ever finer partitions. Let $\{B_0, B_1\}$ be a measurable partition of Ω at the first level; then, let $\{B_{00}, B_{01}\}$ and $\{B_{10}, B_{11}\}$ be measurable partitions of B_0 and B_1, respectively, at the second level; continue in this way until M levels (i.e. $m = 1, 2, \ldots, M$) are achieved and call the set of measurable partitions a finite binary tree partition of Ω. Let, at the m-th level, $\varepsilon_{1:m} = \varepsilon_1 \varepsilon_2 \cdots \varepsilon_m$, with each $\varepsilon_j \in \{0, 1\}$, $j = 1, 2, \ldots, m$, so that each $\varepsilon_{1:m}$ defines a unique subset $B_{\varepsilon_{1:m}}$. It is clear that the number of partitions at the m-th level is 2^m and that $B_{\varepsilon_{1:m}}$ will split into $B_{\varepsilon_{1:m}0}$ and $B_{\varepsilon_{1:m}1}$ at level $(m+1)$. The finite binary tree partition structure of the Polya tree can be denoted by $\Pi = \{B_{\varepsilon_{1:m}}, m = 1, 2, \ldots, M\}$.

In order to define random measures on Ω we construct random measures on the sets $B_{\varepsilon_{1:m}}$ for $m = 1, 2, \ldots, M$. Starting at Ω, we can move into B_0, with probability Y_0, or into B_1, with probability $Y_1 = 1 - Y_0$. Generally, on entering $B_{\varepsilon_{1:m}}$, we can either move into $B_{\varepsilon_{1:m}0}$, with conditional probability $Y_{\varepsilon_{1:m}0}$ or into $B_{\varepsilon_{1:m}1}$ with conditional probability $Y_{\varepsilon_{1:m}1} = 1 - Y_{\varepsilon_{1:m}0}$. The marginal probability of a subset in the m-th partition is

$$G(B_{\varepsilon_{1:m}}) = \left(\prod_{j=1, \varepsilon_j=0}^{m} Y_{\varepsilon_1 \cdots \varepsilon_{j-1}0} \right) \left(\prod_{j=1, \varepsilon_j=1}^{m} (1 - Y_{\varepsilon_1 \cdots \varepsilon_{j-1}0}) \right)$$

with the marginal probability for the first level, i.e. for $j = 1$, being given by Y_0 or $1 - Y_0$.

For instance, for $m = 2$, $G(B_{00}) = Y_0 Y_{00}$, $G(B_{01}) = Y_0(1 - Y_{00})$, $G(B_{10}) = (1 - Y_0)Y_{10}$ and $G(B_{11}) = (1 - Y_0)(1 - Y_{10})$. In Polya trees, these probabilities are random and independent Beta variables, i.e., $Y_0 \sim \mathrm{Beta}(\alpha_0, \alpha_1)$ and for every $\varepsilon_{1:m}$, $Y_{\varepsilon_{1:m}0} \overset{\mathrm{ind}}{\sim} \mathrm{Beta}(\alpha_{\varepsilon_{1:m}0}, \alpha_{\varepsilon_{1:m}1})$, with nonnegative parameters $\alpha_{\varepsilon_{1:m}0}$ and $\alpha_{\varepsilon_{1:m}1}$. Denoting the collection of parameters α's by $\mathcal{A} = \{\alpha_{\varepsilon_{1:m}}, m = 1, 2, \ldots, M\}$, the particular finite Polya tree distribution with M levels is defined by the partitions in Π and the Beta parameters in $\mathcal{A}$, and is denoted by $G \sim \mathrm{FPT}_M(\Pi, \mathcal{A})$.

The parameters of the Polya tree can be chosen such that G is absolutely continuous with probability one. In particular, any $\alpha_{\varepsilon_{1:m}} = \rho(m)$ such that $\sum_{m=1}^{\infty} \rho(m)^{-1} < \infty$ guarantees G to be absolutely continuous as referred by Schervish [21]. For example, Berger and Guglielmi [15] considered $\alpha_{\varepsilon_{1:m}} = c\rho(m)$, $c > 0$ and $\rho(m) = m^2, m^3, 2^m, 4^m$ and 8^m.

By defining Π and $\mathcal{A}$ the Polya tree can be centered on some particular parametric distribution $f(x|\theta)$ on Ω, so that $E[G(B_{\varepsilon_{1:m}})|\theta] = F_\theta(B_{\varepsilon_{1:m}}) = \Pr(X_i \in B_{\varepsilon_{1:m}}|\theta)$, where θ is the parameter vector of the parametric distribution. One way of doing it is to fix a nested partition sequence Π, not depending on θ, and then choose the parameters $\alpha_{\varepsilon_{1:m}}$'s depending on θ as mentioned in [15].

For instance, if $X_i \in \mathbb{R}^+$, define $B_0 = (0, F_{\hat{\theta}}^{-1}(0.5)]$, $B_1 = (F_{\hat{\theta}}^{-1}(0.5), +\infty)$ and more generally, at level m,

$$B_{\varepsilon_{1:m}} = \left\{ \left(F_{\hat{\theta}}^{-1} \left(\frac{k-1}{2^m} \right), F_{\hat{\theta}}^{-1} \left(\frac{k}{2^m} \right) \right], \ m = 1, 2, \ldots, M, \ k = 1, 2, \ldots, 2^m \right\} \ ,$$

where $F_{\hat{\theta}}^{-1}(\cdot)$ are quantiles of X_i substituting θ by its m.l.e. vector, $\hat{\theta}$. Then, for $h > 0$, define

$$\alpha_{\varepsilon_{1:m-1}0}(\theta) = h^{-1}\rho(m) \left(\frac{F_\theta(B_{\varepsilon_{1:m-1}0})}{F_\theta(B_{\varepsilon_{1:m-1}1})} \right)^{1/2} \tag{1}$$

and

$$\alpha_{\varepsilon_{1:m-1}1}(\theta) = h^{-1}\rho(m) \left(\frac{F_\theta(B_{\varepsilon_{1:m-1}1})}{F_\theta(B_{\varepsilon_{1:m-1}0})} \right)^{1/2} . \tag{2}$$

For example, since $G(B_0|\theta) = Y_0 \sim \text{Beta}(\alpha_0(\theta), \alpha_1(\theta))$, then

$$E[G(B_0)|\theta] = E[Y_0|\theta] = \frac{\alpha_0(\theta)}{\alpha_0(\theta) + \alpha_1(\theta)} = F_\theta(B_0) \ ,$$

and for any $B_{\varepsilon_{1:m}} \in \Pi$, $E[G(B_{\varepsilon_{1:m}})|\theta] = F_\theta(B_{\varepsilon_{1:m}})$.

The function $\rho(\cdot)$ controls the smoothness of the distribution upon which the Polya tree distribution concentrates its mass and h refers to an overall scale factor which, in some sense, controls the overall variance of the Polya tree about its mean which is the parametric distribution. Further details on these two measures will be referred to later.

Finally, uncertainty about θ can also be modeled as $\pi(\theta)$, generating a mixture of finite Polya trees distribution for G.

The notation $G|\Pi, \mathcal{A}_\theta \sim \text{MFPT}_M(\Pi, \mathcal{A}_\theta)$ is used to denote that G has a mixture of finite Polya trees prior distribution with M levels, fixed partition Π and the remaining Polya tree parameters, $\mathcal{A}_\theta$, are updated throughout the procedure.

2.2 Berger and Guglielmi's Bayesian Nonparametric Goodness-of-fit Test

Berger and Guglielmi's Bayesian nonparametric goodness-of-fit test is defined by $H_0 : X \sim f(x|\theta), \theta \in \Theta$, versus $H_1 : X \sim G|\Pi, \mathcal{A}_\theta, \theta \in \Theta$, with a prior density, $\pi(\theta)$, usually noninformative. The test is performed based on the Bayes factor for H_0 against H_1, given by

$$\text{BF}_{01}(\boldsymbol{x}) = \frac{p(\boldsymbol{x}|H_0)}{p(\boldsymbol{x}|H_1)} = \frac{\int_\Theta f(\boldsymbol{x}|\theta)\pi(\theta)d\theta}{\int_\Theta p(\boldsymbol{x}|\theta)\pi(\theta)d\theta} \ , \tag{3}$$

where $\boldsymbol{x} = (x_1, x_2, \ldots, x_n)$, $f(\boldsymbol{x}|\theta) = \prod_{i=1}^{n} f(x_i|\theta)$, and $p(\boldsymbol{x}|\theta)$ is the marginal density of the sample under a Polya tree scheme, which following Lavine [17], is given by

$$p(\boldsymbol{x}|\theta) = f(\boldsymbol{x}|\theta)\psi(\theta) \ , \tag{4}$$

with

$$\psi(\theta) = \prod_{j=2}^{n} \prod_{m=1}^{m^*(x_j)} \frac{\alpha'_{\varepsilon_{1:m}(x_j)}(\theta) \left(\alpha_{\varepsilon_{1:m-1}0(x_j)}(\theta) + \alpha_{\varepsilon_{1:m-1}1(x_j)}(\theta) \right)}{\alpha_{\varepsilon_{1:m}}(x_j)(\theta) \left(\alpha'_{\varepsilon_{m-1}0(x_j)}(\theta) + \alpha'_{\varepsilon_{m-1}1(x_j)}(\theta) \right)} \ , \tag{5}$$

where $\varepsilon_{1:m}(x_j)$ is the index $\varepsilon_1\varepsilon_2 \cdots \varepsilon_m$ such that x_j belongs to $B_{\varepsilon_1 \cdots \varepsilon_m}$, for each level m, and $\alpha'_{\varepsilon_{1:m}(x_j)}(\theta)$ is equal to $\alpha_{\varepsilon_{1:m}(x_j)}(\theta)$ plus the number of observations among $\{x_1, x_2, \ldots, x_{j-1}\}$ which belong to $B_{\varepsilon_1 \cdots \varepsilon_m}(x_j)$. For each x_j, the product in (5) is up to the smallest level $m^*(x_j)$, such that no x_i, $i < j$, belongs to $B_{\varepsilon_{1:m}(x_j)}$.

The Bayes factor measures the evidence in favor of the null model against the alternative, based on observed values $\boldsymbol{x}$, i.e., values of $\mathrm{BF}_{01}(\boldsymbol{x})$ larger or smaller than one are interpreted as evidence given by the data, respectively, in favor or against H_0. More formally, let c_{BF} be a fixed threshold that controls type I error rate; then for any given data set $\boldsymbol{x}$, if $\mathrm{BF}_{01}(\boldsymbol{x}) > c_{BF}$ ($<$) then we accept (reject) the null model.

The computation of the Bayes factor is simplified because by (4), (3) can be written as

$$\mathrm{BF}_{01}(\boldsymbol{x}) = \left(\int_{\Theta} \psi(\theta)\pi(\theta|\boldsymbol{x})d\theta \right)^{-1} \ ,$$

where $\pi(\theta|\boldsymbol{x}) = f(\boldsymbol{x}|\theta)\pi(\theta)/p(\boldsymbol{x}|H_0)$, i.e., the Bayes factor can be written as the inverse of a posterior expectation of $\psi(\theta)$ under H_0. Thus, generating a sample $(\theta_1, \theta_2, \ldots, \theta_L)$ from $\pi(\theta|\boldsymbol{x})$, an approximation of the Bayes factor is given by

$$\widehat{\mathrm{BF}}_{01}(\boldsymbol{x}) = \frac{L}{\sum_{l=1}^{L} \psi(\theta_l)} \ . \tag{6}$$

Berger and Guglielmi [15] proposed examining a plot of the Bayes factor as a function of the scale factor h, because it determines how concentrated the defined mixture of finite Polya trees prior distribution is about its mean, F_θ. As $h \to 0$, the mixture of finite Polya trees prior distribution concentrates both in terms of similarity in shape and distance from the fixed F_θ, and the Bayes factor will converge to one. As $h \to \infty$ the mixture of finite Polya trees prior distribution will, usually, be dispersed from the fixed F_θ, and the Bayes factor will be quite large. Between these two extremes, the Bayes factor will sometimes increase with h, but also will first decrease and then increase. Thus, they minimize over h to make a conservative choice of the Bayes factor as mentioned in [16].

2.3   Johnson's Bayesian Chi-squared Test

Denote by $F(x|\theta)$ the cumulative distribution corresponding to the density $f(x|\theta)$. Let $\breve{\theta}$ be a random observation from the posterior distribution of θ, and let $0 \equiv a_0 < a_1 < \ldots < a_k \equiv 1$ with $p_k = a_k - a_{k-1}$, $k = 1, 2, \ldots, K$ be a partition of $[0, 1]$. Usually, a noninformative prior distribution is specified on θ and is recommended to use $K \simeq n^{0.4}$ bins. Then Johnson's Bayesian chi-squared test statistic is given by

$$R_n^B(\breve{\theta}) = \sum_{k=1}^{K} \frac{(m_k(\breve{\theta}) - np_k)^2}{np_k} \ , \tag{7}$$

where $m_k(\breve{\theta})$ represents the number of observations that fall into the k-th bin, i.e., the number of x_i's satisfying $F(x_i|\breve{\theta}) \in (a_{k-1}, a_k]$, $i = 1, 2, \ldots, n$. Johnson [11] proves that under the null hypothesis, $R_n^B(\breve{\theta})$ is asymptotically distributed as $\chi^2_{(K-1)}$, independently of the dimension of the underlying parameter vector θ.

3   Tests for Exponentiality

In this Section, preliminary notation is provided and classical and Bayesian tests for exponentiality are introduced.

A non-negative random variable X denoting the time to failure (or lifetime) of some item of interest has an exponential distribution, $X \sim \text{Exp}(\lambda)$, if its density is given by

$$f(x|\lambda) = \lambda \exp(-\lambda x), \quad x \geq 0$$

where $\lambda > 0$ defines the unknown failure rate.

The problem consists in testing the null hypothesis

$$H_0 : X \sim \text{Exp}(\lambda) \text{ for some } \lambda > 0 \ ,$$

against the general alternative that X is not exponentially distributed, based on a sample $(x_1, x_2, \ldots, x_n)$ of identical distributed and conditional independent observations of X.

3.1   Classical Tests

A large number of classical tests for exponentiality have been proposed in the literature. The tests are based on different characteristics of the exponential distribution, and can be classified into several categories. According to Henze and Meintanis [5] there are: tests based on the empirical distribution function; tests based on the integrated empirical distribution function; tests based on spacings and the Gini index; tests based on the entropy characterization; tests based on the statistic of Cox and Oakes; tests based on a characterization via the mean residual life function; tests statistics derived from the empirical Laplace transform; test statistics derived from the empirical characteristic function, among others. Henze and Meintanis [5] compared twenty one test statistics for the exponentiality against eighteen alternative distributions. This exhaustive study concludes that there is no test statistic which is better than the others in terms of power. However, the study indicates that the Cox and Oakes [22] statistic CO_n, the Epps and Pulley [23] statistic EP_n, the modified Cramér-von Mises type statistic $\overline{CM}_n$ of Baringhaus and Henze [3], based on a characterization of the mean residual life function, the Baringhaus and Henze [2] classical test statistic $BH_{n,a}$, based on the empirical Laplace transform, and the new test statistic $T_{n,a}$, proposed in Henze and Meintanis [5], based on the empirical characteristic function, are among the most powerful test statistics and they are quite easy to evaluate. Note that the two latter test statistics depend on an arbitrary constant a which affects drastically the power of the test. According to the studies of Henze and Meintanis [5], is now taken as $a = 1$, $a = 1.5$ and $a = 2.5$ for $BH_{n,a}$ and $a = 1.5$ and $a = 2.5$ for $T_{n,a}$. The statistic $T_{n,a}$ is considered for small samples only. The Anderson and Darling [24] test statistic AD_n was not considered in the study by Henze and Meintanis [5] but it is included in our study because of its wide use in literature.

In this work, the above mentioned six classical test statistics will be used and compared in terms of power, with the Bayesian tests referred to in the previous Section. With that purpose in mind, each one of these classical test statistics is defined following closely the notation in [5]. We just present the expressions of the corresponding tests statistics used to evaluate the adequacy of the exponential distribution; all theoretical details can be found in the literature.

Let $Y_i = X_i/\bar{X}$ and $\bar{X} = n^{-1} \sum_{i=1}^{n} X_i$. Then,

1. The test statistic of Cox and Oakes [22] is given by

$$CO_n = n + \sum_{i=1}^{n} (1 - Y_i) \log(Y_i) \ .$$

2. The normalized Epps and Pulley [23] test statistics is defined by

$$EP_n = (48n)^{1/2} + \left[\frac{1}{n} \sum_{i=1}^{n} \exp(-Y_i) - \frac{1}{2} \right] \ .$$

3. The Cramér-von Mises type statistic $\overline{CM}_C$ of Baringhaus and Henze [3] is computed as

$$\overline{CM}_n = \frac{1}{n} \sum_{i,k=1}^{n} \left(2 - 3e^{-\min(Y_i,Y_k)} - 2\min(Y_i, Y_k)(e^{-Y_i} + e^{-Y_k}) \right.$$
$$\left. + 2e^{-\max(Y_i,Y_k)} \right) \ .$$

4. The Baringhaus and Henze [2] test statistic is defined by

$$
\mathrm{BH}_{n,a} = \frac{1}{n} \sum_{i,k=1}^{n} \left[\frac{(1 - Y_i)(1 - Y_k)}{Y_i + Y_k + a} - \frac{Y_i + Y_k}{(Y_i + Y_k + a)^2} \right.
$$
$$
\left. + \frac{2Y_i Y_k}{(Y_i + Y_k + a)^2} + \frac{2Y_i Y_k}{(Y_i + Y_k + a)^3} \right] .
$$

5. The new test statistic proposed in Henze and Meintanis [5] is computed as

$$
\mathrm{T}_{n,a} = \frac{a}{n} \sum_{i,k=1}^{n} \left[\frac{1}{a^2 + (Y_i - Y_k)^2} + \frac{1}{a^2 + (Y_i + Y_k)^2} \right]
$$
$$
- \frac{2a}{n^2} \sum_{i,k=1}^{n} \sum_{l=1}^{n} \left[\frac{1}{a^2 + (Y_i - Y_k - Y_l)^2} + \frac{1}{a^2 + (Y_i + Y_k + Y_l)^2} \right]
$$
$$
+ \frac{a}{n^3} \sum_{i,k=1}^{n} \sum_{l,m=1}^{n} \left[\frac{1}{a^2 + (Y_i - Y_k - (Y_l - Y_m))^2} \right.
$$
$$
\left. + \frac{1}{a^2 + (Y_i + Y_k + (Y_l - Y_m))^2} \right] .
$$

6. The Anderson and Darling [24] test statistic is given by

$$
\mathrm{AD}_n = -n - \frac{1}{n} \sum_{i=1}^{n} (2i - 1) \left[\log(W_{(i)}) + \log(1 - W_{(n-i+1)}) \right] ,
$$

where $W_{(i)} = 1 - \exp(-Y_{(i)})$, $1 \le i \le n$, and $Y_{(1)} \le Y_{(2)} \le \cdots \le Y_{(n)}$ are the order statistics of $(Y_1, Y_2, \ldots, Y_n)$.

For the first two test statistics, their authors proved that under the null hypothesis, EP_n and $\mathrm{CO}_n^* = \left(\frac{6}{n} \right)^{1/2} \left(\frac{\mathrm{CO}_n}{\pi} \right)$ are assymptotically standard normal distributed; therefore we reject the null hypothesis (that the observations are from an exponential distribution) for large values of $|\mathrm{EP}_n|$ and $|\mathrm{CO}_n^*|$. For the other four test statistics, the distribution under the null hypothesis has not been obtained analytically. To determine the critical values (empirical quantiles) of the respective test statistic distribution, Monte Carlo simulations are employed. The null hypothesis is rejected if the observed value of each one of these test statistics exceeds the corresponding critical value.

In order to obtain the empirical quantiles of each test statistic, 100,000 random samples of size n are generated from the standard exponential distribution and the value of the test statistic is calculated. The sample sizes considered are $n = 25$, $n = 50$ and $n = 100$, and the chosen significance levels are $\alpha = 0.1$, $\alpha = 0.05$ and $\alpha = 0.025$, for the statistics $\overline{\mathrm{CM}}_n$, AD_n and $\mathrm{BH}_{n,a}$. For the test statistic $T_{n,a}$, the sample sizes considered are only $n = 25$ and $n = 50$, and the chosen significance levels are only for $\alpha = 0.1$ and $\alpha = 0.05$. The empirical critical value for each classical statistic, presented in Tables 1, 2 and 3, for each n,

Table 1. Empirical critical values of test statistics $\overline{\mathrm{CM}}_n$ and AD_n

	$\overline{\mathrm{CM}}_n$			AD_n		
α	0.1	0.05	0.025	0.1	0.05	0.025
$n = 25$	0.341	0.451	0.564	1.042	1.292	1.562
$n = 50$	0.344	0.455	0.567	1.053	1.312	1.573
$n = 100$	0.348	0.464	0.583	1.058	1.325	1.595

Table 2. Empirical critical values of test statistic $\mathrm{T}_{n,a}$, for $a = 1.5$ and $a = 2.5$

	$\mathrm{T}_{n,a=1.5}$		$\mathrm{T}_{n,a=2.5}$	
α	0.1	0.05	0.1	0.05
$n = 25$	0.275	0.411	0.077	0.109
$n = 50$	0.256	0.359	0.075	0.104

Table 3. Empirical critical values of test statistic $\mathrm{BH}_{n,a}$, for $a = 1$, $a = 1.5$ and $a = 2.5$

	$\mathrm{BH}_{n,a=1}$			$\mathrm{BH}_{n,a=1.5}$			$\mathrm{BH}_{n,a=2.5}$		
α	0.1	0.05	0.025	0.1	0.05	0.025	0.1	0.05	0.025
$n = 25$	0.219	0.304	0.385	0.140	0.194	0.251	0.071	0.098	0.127
$n = 50$	0.220	0.305	0.396	0.142	0.199	0.258	0.073	0.102	0.131
$n = 100$	0.221	0.311	0.401	0.143	0.199	0.260	0.073	0.104	0.135

is determined with the quantile $(1 - \alpha) \times 100\%$ from the corresponding empirical distribution (for $\mathrm{BH}_{n,a}$ and $\mathrm{T}_{n,a}$, the empirical critical values are very close to the ones obtained in [3] and in [5]).

3.2   Bayesian Tests

Let us begin by defining the test based on Berger and Guglielmi [15] which assumes two models, a parametric one and a nonparametric one, such that the former is embedded in the latter. The parametric Bayesian model is given by

$$X_i | \lambda \overset{\mathrm{iid}}{\sim} \mathrm{Exp}(\lambda), \text{ for } i = 1, 2, \ldots, n,$$
$$\lambda \sim \pi(\lambda)$$

and the nonparametric Bayesian model is

$$X_1, X_2, \ldots, X_n | G \overset{\mathrm{iid}}{\sim} G$$
$$G | \Pi, \mathcal{A}_\lambda \sim \mathrm{MFPT}_M(\Pi, \mathcal{A}_\lambda)$$
$$\lambda \sim \pi(\lambda) ,$$

where $\mathrm{MFPT}_M(\Pi, \mathcal{A}_\lambda)$ defines a mixture of finite Polya trees prior distribution, with parameters $(\Pi, \mathcal{A}_\lambda)$ and M pre-specified levels. We consider a conjugate noninformative prior distribution for the unknown parameter λ, i.e., $\lambda \sim \mathrm{Gamma}(a, b)$, with $a, b \to 0$. In practice, we took $a = b = 0.0001$.

The binary partitions, which are fixed not depending on λ, are given by

$$B_{\varepsilon_{1:m}} = \left\{ \left(F_{\hat{\lambda}}^{-1} \left(\frac{k-1}{2^m} \right), F_{\hat{\lambda}}^{-1} \left(\frac{k}{2^m} \right) \right], \ m = 1, 2, \ldots, M, \ k = 1, 2, \ldots, 2^m \right\} ,$$

where $F_{\hat{\lambda}}^{-1}(\cdot)$ defines the quantiles of the exponential distribution taking as parameter value its m.l.e., $\hat{\lambda} = 1/\bar{x}$, with $\bar{x} = \frac{\sum_{i=1}^n x_i}{n}$, as suggested by Berger and Guglielmi [15].

The values for the parameters $\alpha_{\varepsilon_{1:m}}(\lambda)$ are obtained using (1) and (2) with $\rho(m) = 4^m$. As h must take a range of values near to and far from zero, a suggestion by Tokdar and Martin [16] is considered. This suggestion proposes evaluating Bayes factor for 13 values of h, defining $h = 2^s$, where s takes the values of all the integers in the interval $[-6, 6]$.

For the Johnson's Bayesian chi-squared test we use the test statistic defined in (7) and a fixed significance level $\alpha = 0.05$. We substitute $\breve{\theta}$ by $\breve{\lambda}$, a random observation from the posterior distribution considering a conjugate noninformative prior distribution, $\mathrm{Ga}(a, b)$, $a, b \to 0$. This test is applied to binned data, and the author recommended to use $K \simeq n^{0.4}$ bins. We also tried different number of bins, but the results are very similar. In the simulation study it is considered $K = 4$, 5 and 6 for $n = 25$, 50 and 100, respectively.

4   Simulation Study

In order to investigate the power of the classical and Bayesian tests, a simulation study is carried out. The goal is to test $H_0 : X \sim \mathrm{Exp}(\lambda)$, where $\lambda > 0$ is unknown.

Several different alternative distributions are considered in this work and they are summarized in Table 4. These distributions were chosen such that the most common time to failure distributions are included and a wide variety of distributions with different failure rates and other characteristics is covered. For example, the Gamma and Weibull distributions have increasing failure rate for $a > 1$ and decreasing failure rate for $0 < a < 1$. For $a = 1$, they both reduce to standard exponential distribution with constant failure rate. The failure rate for the LogNormal distribution initially increases over time and then decreases (non-monotonous). Half-Normal distribution has increasing failure rate and $\chi^2(1)$ distribution has decreasing failure rate. Half-Cauchy is an heavy tailed distribution.

The choice of the parameters of the different alternative distributions is done in such a way that the density gradually differs from a standard exponential distribution $(\mathrm{Exp}(\lambda = 1))$. All these distributions are depicted in Fig. 1 as well as the standard exponential distribution, in order to allow visualization of the density curve.

Bayes factor is approximated via direct Monte Carlo method as defined in (6), for each one of the h values, based on $L = 20,000$ simulated values from the posterior distribution and by defining a mixture of finite Polya trees prior distribution with $M = 8$ levels. This choice of M was made after a study was carried

Table 4. Alternative distributions

Distribution	Notation	Density
Gamma	$Ga(a,1)$	$\Gamma(a)^{-1}x^{a-1}\exp(-x)$
Weibull	$Wei(a,1)$	$ax^{a-1}\exp(-x^a)$
LogNormal	$LN(0,1)$	$x^{-1}(2\pi)^{-1/2}\exp(-\log^2(x)/2)$
Half-Normal	$HN(0,1)$	$(2/\pi)^{1/2}\exp(-x^2/2)$
Chi-squared	$\chi^2(a)$	$1/(2^{a/2}\Gamma(a/2))x^{a/2-1}\exp(-x/2)$
Half-Cauchy	$HCa(0,1)$	$(2/\pi)(x^2+1)$

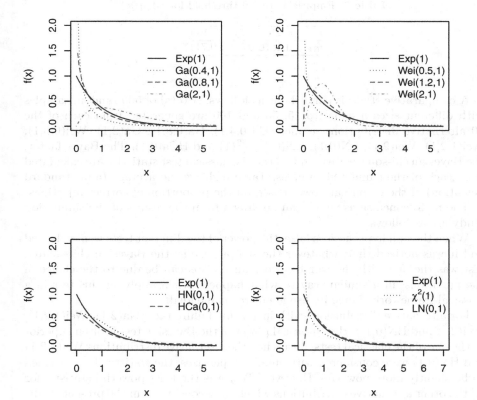

Fig. 1. Density plots

out analysing the effect of M values on the results. We tried $M = 5, 6, 7, 8, 9$ and 10 concluding that for $M \geq 8$ there were no significant differences on the Bayes factor results. The final estimate for the Bayes factor is obtained by evaluating the minimum value of the thirteen estimates for all values of h. We decided to reject the exponential model whenever the final estimate for the Bayes factor is less than the critical threshold c_{BF}.

To determine the critical threshold, we use the same idea behind the frequentist type I error. As the conditional type I error cannot be evaluated for fixed partitions, as mentioned in Berger and Guglielmi [15], we considered a frequentist type I error. We generate 500 samples of sizes n from the standard exponential distribution and compute the Bayes factor $BF_{01}(x)$ for each sample size n. The empirical critical threshold for the Bayes factor and for every n ($n = 25$, $n = 50$ and $n = 100$) is determined by the quantile $\alpha \times 100\%$ from the empirical distribution of $\widehat{BF}_{01}(x)$. In Table 5 we present the empirical critical threshold, obtained by simulation, necessary to achieve a type I error rate of 0.05.

Table 5. Empirical critical threshold for $\widehat{BF}_{01}(x)$

n	25	50	100
c_{BF}	0.4876	0.5697	0.7117

A comparative study is carried out as follows. A total of 500 random samples with different sizes n equal to 25, 50 and 100 are generated from each of the 10 alternative distributions, namely Ga(0.4,1), Ga(0.8,1), Ga(2,1), Wei(0.5,1), Wei(1.2,1), Wei(2,1), LN(0,1), HN(0,1), $\chi^2(1)$ and HCa(0,1). The Bayes factor, the Bayesian chi-squared test and all the six classical test statistics are calculated from each of the simulated samples. Table 6 shows the average (and standard deviation) of the empirical power based on the proportion of correct rejections.

The main conclusions which can be drawn from the results of the simulation study are as follows.

When the null hypothesis is false, the power of the different tests was evaluated and it was noticed that, whatever the sample size is, the Bayesian chi-squared test was the one with the worst performance. This can be due to the choice of the partitions, in a similar way to what happens when applying this test in a classical framework. Thus, its use is not recommended.

For alternative distributions with increasing failure rate (Ga(2,1), Wei(1.2,1), Wei(2,1) and HN(0,1)), the empirical power of the Bayesian test is often superior to the one of the classical tests. Just when the alternative distributions Wei(1.2,1) and HN(0,1) were considered with small sample sizes, the statistic $T_{n,a}$ revealed to be slightly more powerful. However, $T_{n,a}$ was the least powerful statistic for all the other alternative distributions which were considered in the present study.

When samples are generated from alternative distributions with decreasing failure rate like Ga(0.4,1) and Wei(0.5,1), it was noticed that the Bayesian test was often, at least, as potent as the classical tests. However it was slightly less potent than the classical tests for distributions with decreasing failure rate like Ga(0.8,1) and $\chi^2(1)$. This may be due to the fact that those distributions are quite close to the standard exponential distribution.

For the Half-Cauchy distribution, the empirical power of the Bayesian test is similar to the power of the classical tests. When the LogNormal distribution was considered, particularly for small sample sizes, the Bayesian test was the one with the best performance.

Table 6. Average (and standard deviations) of the empirical proportion of H_0 rejections out of 500 samples, for 3 different samples sizes n and for 8 different tests. For the first distribution, Exp(1), this corresponds to empirical type I error rate, and for the alternative distributions it represents the empirical power.

Dist.	n	BF_{01}	R_n^B	EP_n	CO_n	CM_n	AD_n	$BH_{n,a}$ $a=1$	$BH_{n,a}$ $a=1.5$	$BH_{n,a}$ $a=2.5$	$T_{n,a}$ $a=1.5$	$T_{n,a}$ $a=2.5$
Exp(1)	25	0.050	0.052	0.038	0.034	0.048	0.046	0.042	0.042	0.046	0.048	0.046
		(0.028)	(0.028)	(0.022)	(0.021)	(0.030)	(0.023)	(0.024)	(0.027)	(0.025)	(0.028)	(0.036)
	50	0.050	0.052	0.034	0.042	0.044	0.052	0.044	0.040	0.042	0.05	0.052
		(0.030)	(0.041)	(0.017)	(0.031)	(0.023)	(0.027)	(0.026)	(0.021)	(0.022)	(0.017)	(0.023)
	100	0.050	0.052	0.056	0.050	0.050	0.048	0.054	0.056	0.054	–	–
		(0.031)	(0.043)	(0.032)	(0.033)	(0.027)	(0.037)	(0.031)	(0.039)	(0.033)		
Ga(0.4, 1)	25	0.898	0.404	0.820	0.956	0.824	0.942	0.886	0.868	0.844	0.302	0.498
		(0.040)	(0.084)	(0.042)	(0.025)	(0.042)	(0.026)	(0.041)	(0.045)	(0.046)	(0.061)	(0.061)
	50	0.990	0.856	0.984	0.998	0.984	0.998	0.996	0.992	0.986	0.882	0.918
		(0.011)	(0.037)	(0.021)	(0.006)	(0.021)	(0.006)	(0.008)	(0.014)	(0.019)	(0.022)	(0.017)
	100	1	1	1	1	1	1	1	1	1	–	–
Ga(0.8, 1)	25	0.110	0.042	0.114	0.148	0.110	0.146	0.136	0.130	0.126	0.004	0.026
		(0.046)	(0.024)	(0.049)	(0.052)	(0.044)	(0.052)	(0.047)	(0.052)	(0.046)	(0.008)	(0.019)
	50	0.214	0.080	0.212	0.266	0.202	0.248	0.232	0.216	0.222	0.05	0.092
		(0.071)	(0.048)	(0.067)	(0.068)	(0.068)	(0.078)	(0.077)	(0.076)	(0.068)	(0.029)	(0.037)
	100	0.318	0.310	0.348	0.452	0.328	0.392	0.380	0.372	0.348	–	–
		(0.083)	(0.054)	(0.063)	(0.095)	(0.079)	(0.093)	(0.074)	(0.074)	(0.058)		
Ga(2, 1)	25	0.696	0.310	0.588	0.578	0.592	0.554	0.650	0.638	0.612	0.674	0.676
		(0.060)	(0.054)	(0.088)	(0.066)	(0.088)	(0.076)	(0.077)	(0.076)	(0.082)	(0.088)	(0.092)
	50	0.989	0.630	0.924	0.984	0.918	0.928	0.956	0.942	0.928	0.936	0.912
		(0.019)	(0.064)	(0.025)	(0.025)	(0.030)	(0.021)	(0.033)	(0.022)	(0.023)	(0.036)	(0.045)
	100	1	0.932	0.996	0.998	0.996	0.998	0.998	0.998	0.996	–	–
			(0.035)	(0.008)	(0.006)	(0.008)	(0.006)	(0.006)	(0.006)	(0.008)		
$\chi^2(1)$	25	0.576	0.25	0.626	0.810	0.614	0.782	0.704	0.672	0.656	0.130	0.250
		(0.068)	(0.057)	(0.067)	(0.043)	(0.075)	(0.049)	(0.079)	(0.077)	(0.078)	(0.045)	(0.054)
	50	0.880	0.592	0.882	0.982	0.872	0.962	0.926	0.910	0.890	0.674	0.738
		(0.034)	(0.074)	(0.033)	(0.020)	(0.025)	(0.028)	(0.028)	(0.025)	(0.027)	(0.061)	(0.031)
	100	1	0.928	0.992	1	0.988	1	1	0.998	0.992	–	–
			(0.030)	(0.014)		(0.017)			(0.006)	(0.014)		
Wei(0.5, 1)	25	0.990	0.664	0.938	0.976	0.940	0.974	0.966	0.954	0.944	0.452	0.706
		(0.024)	(0.071)	(0.037)	(0.025)	(0.035)	(0.019)	(0.023)	(0.027)	(0.036)	(0.103)	(0.068)
	50	1	0.986	0.998	1	1	1	1	1	1	0.974	0.998
			(0.019)	(0.006)				(0.019)	(0.010)			
	100	1	1	1	1	1	1	1	1	1	–	–
Wei(1.2, 1)	25	0.208	0.084	0.150	0.124	0.164	0.144	0.164	0.160	0.158	0.226	0.214
		(0.040)	(0.049)	(0.041)	(0.048)	(0.041)	(0.047)	(0.042)	(0.036)	(0.045)	(0.070)	(0.072)
	50	0.294	0.132	0.270	0.260	0.276	0.224	0.242	0.284	0.288	0.312	0.318
		(0.038)	(0.065)	(0.039)	(0.057)	(0.056)	(0.053)	(0.071)	(0.070)	(0.047)	(0.061)	(0.058)
	100	0.635	0.242	0.598	0.594	0.574	0.530	0.612	0.606	0.594	–	–
		(0.029)	(0.038)	(0.033)	(0.057)	(0.042)	(0.024)	(0.044)	(0.040)	(0.033)		
Wei(2, 1)	25	0.990	0.742	0.982	0.972	0.980	0.978	0.970	0.984	0.984	0.988	0.988
		(0.019)	(0.045)	(0.020)	(0.021)	(0.019)	(0.017)	(0.017)	(0.016)	(0.016)	(0.017)	(0.017)
	50	1	0.994	1	1	1	1	1	1	1	1	1
			(0.009)									
	100	1	1	1	1	1	1	1	1	1	–	–
HCa(0, 1)	25	0.766	0.514	0.764	0.742	0.770	0.754	0.758	0.768	0.782	0.038	0.250
		(0.030)	(0.059)	(0.039)	(0.033)	(0.036)	(0.038)	(0.035)	(0.036)	(0.036)	(0.037)	(0.058)
	50	0.968	0.752	0.950	0.928	0.956	0.936	0.946	0.948	0.958	0.346	0.564
		(0.020)	(0.039)	(0.025)	(0.030)	(0.023)	(0.025)	(0.021)	(0.025)	(0.024)	(0.041)	(0.059)
	100	0.998	0.958	0.998	0.998	0.998	0.998	0.998	1	1	–	–
		(0.004)	(0.017)	(0.006)	(0.006)	(0.006)	(0.006)	(0.006)				
LN(0, 1)	25	0.270	0.114	0.134	0.090	0.170	0.170	0.122	0.144	0.168	0.076	0.056
		(0.038)	(0.031)	(0.049)	(0.037)	(0.040)	(0.052)	(0.050)	(0.061)	(0.053)	(0.028)	(0.021)
	50	0.352	0.210	0.168	0.140	0.266	0.342	0.206	0.204	0.216	0.094	0.072
		(0.042)	(0.057)	(0.067)	(0.057)	(0.075)	(0.078)	(0.057)	(0.065)	(0.063)	(0.034)	(0.039)
	100	0.740	0.446	0.234	0.184	0.446	0.704	0.296	0.294	0.302	–	–
		(0.038)	(0.064)	(0.048)	(0.063)	(0.074)	(0.047)	(0.062)	(0.046)	(0.061)		
HN(0, 1)	25	0.267	0.118	0.232	0.160	0.246	0.198	0.246	0.254	0.256	0.314	0.322
		(0.036)	(0.054)	(0.042)	(0.056)	(0.041)	(0.044)	(0.037)	(0.031)	(0.035)	(0.036)	(0.046)
	50	0.586	0.192	0.500	0.372	0.514	0.408	0.462	0.480	0.506	0.576	0.586
		(0.040)	(0.065)	(0.085)	(0.079)	(0.071)	(0.073)	(0.087)	(0.077)	(0.082)	(0.069)	(0.060)
	100	0.866	0.398	0.848	0.722	0.864	0.792	0.810	0.834	0.850	–	–
		(0.038)	(0.043)	(0.048)	(0.068)	(0.044)	(0.057)	(0.061)	(0.057)	(0.044)		

When the samples were generated from distributions quite different from the exponential, the empirical power values are very high (> 0.8) and for large samples ($n = 100$) the empirical power often equals to one.

The classical tests present some instability, in terms of power, as it was already stated by Henze and Meintanis [5], that is, there is not a classical test which is better than all the others. The same happens with the Anderson and Darling test which is, in some cases, superior, in terms of power, than the other classical tests but very similar when compared with the results obtained when using the Cox and Oakes statistic.

It was also noticed that, for the considered alternative distributions, the number of correct rejections increases, as expected, as the sample size increases, independently of the test.

5 Conclusion

In this work, we applied two Bayesian goodness-of-fit tests to the specific problem of testing for an exponential model. A Monte Carlo simulation study was carried out for power comparisons with some of the most powerful classical tests proposed in literature and for some close alternative distributions.

The simulation results showed a poor performance of the Johnson's Bayesian chi-squared test. The Bayesian nonparametric test showed good discriminatory power for most of the alternative distributions under consideration, particulary for those with increasing failure rate, and it showed to be at least as powerful as the classical tests for most of the remaining alternative distributions with decreasing failure rate.

Acknowledgments. The research was partially sponsored by national funds through the FCT, Fundação Nacional para a Ciência e Tecnologia, under the project PEst-OE/MAT/UI0006/2014 and by Polytechnic Institute of Porto.

References

1. D'Agostino, R.B., Stephens, M.A.: Goodness-of-fit Techniques. Marcel Dekker, New York (1986)
2. Baringhaus, L., Henze, N.: A class of consistent tests for exponentiality based on the empirical Laplace transform. Ann. Inst. Statist. Math. 43, 551–564 (1991)
3. Baringhaus, L., Henze, N.: Tests of fit for exponentiality based on a characterization via the mean residual life function. Statist. Paper 41, 225–236 (2000)
4. Choi, B., Kim, K., Song, S.H.: Goodness of fit test for exponentiality based on Kullback-Leibler information. Comm. Statist. Simulation Comput. 33(2), 525–536 (2004)
5. Henze, N., Meintanis, S.: Recent and classical tests for exponentiality: A partial review with comparisons. Metrika 61, 29–45 (2005)
6. Grané, A., Fortiana, J.: A directional test of exponentiality based on maximum correlations. Metrika 73, 255–274 (2011)

7. Gelman, A., Meng, X.L., Stern, H.: Posterior predictive assesssment of model fitness via realized discrepancies (with discussion). Statist. Sinica 6, 733–807 (1996)
8. Robins, J.M., Vaart, A., van der Ventura, V.: Asymptotic distribution of p-values in composite null models. J. Amer. Statist. Assoc. 95, 1143–1159 (2000)
9. Bayarri, M.J., Berger, J.O.: P-values for composite null models. J. Amer. Statist. Assoc. 95, 1127–1142 (2000)
10. Hjort, N.L., Dahl, F.A., Steinbakk, G.H.: Post-processing posterior predictive p-values. J. Amer. Statist. Assoc. 101, 1157–1174 (2006)
11. Johnson, V.E.: A Bayesian chi-squared test for goodness-of-fit. Ann. Statist. 32(6), 2361–2384 (2004)
12. Johnson, V.E.: Bayesian model assessment using pivotal quantities. Bayesian Analysis 2, 719–734 (2007)
13. Hjort, N.L., Holmes, C., Müller, P., Walker, S.G.: Bayesian Nonparametrics. Cambridge University Press, New York (2010)
14. Verdinelli, I., Wasserman, L.: Bayesian goodness-of-fit testing using infinite-dimensional exponential families. Ann. Statist. 26, 1215–1241 (1998)
15. Berger, J.O., Guglielmi, A.: Bayesian and conditional frequentist testing of a parametric model versus nonparametric alternatives. J. Amer. Statist. Assoc. 96, 174–184 (2001)
16. Tokdar, S.T., Martin, R.: Bayesian test of normality versus a Dirichlet process mixture alternative. ArXiv e-prints (2011)
17. Lavine, M.: Some aspects of Polya tree distributions for statistical modeling. Ann. Statist. 20, 1222–1235 (1992)
18. Lavine, M.: More aspects of Polya tree distributions for statistical modeling. Ann. Statist. 22, 1161–1176 (1994)
19. Hanson, T., Johnson, W.: Modeling regression errors with a mixture of Polya trees. J. Amer. Statist. Assoc. 97, 1020–1033 (2002)
20. Hanson, T.: Inference for mixture of finite Polya tree models. J. Amer. Statist. Assoc. 101, 1548–1565 (2006)
21. Schervish, M.J.: Theory os Statistic. Springer, New York (1995)
22. Cox, D., Oakes, D.: Analysis of Survival Data. Chapman and Hall, New York (1984)
23. Epps, T., Pulley, L.: A test for exponentiality vs. monotone hazard alternatives derived from the empirical characteristic function. J. R. Stat. Soc. Ser. B 48(2), 206–213 (1986)
24. Anderson, T.W., Darling, D.A.: A test of goodness-of-fit. J. Amer. Statist. Assoc. 49, 765–769 (1954)

Predictive Models for Mutation Carriers in Brugada Syndrome Screening

Carla Henriques[1,2], Ana Matos[1,3], and Luís Ferreira Santos[4]

[1] School of Technology and Management, Polytechnic Institute of Viseu, Portugal
[2] Centre for Mathematics of the University of Coimbra (CMUC), Portugal
[3] Centre for the Study of Education, Technologies and Health (CSETH), Portugal
{carlahenriq,amatos}@estgv.ipv.pt
[4] Department of Cardiology, Tondela-Viseu Hospital Center, Portugal
luisferreirasantos@gmail.com

Abstract. In this study we consider logistic regression models to predict mutation carriers in family members affected by Brugada Syndrome. This Syndrome is an inherited cardiopathy that predisposes individuals without structural heart disease to sudden cardiac death. We focused on five electrocardiographic markers, which have been explored as good discriminators between carriers and non-carriers of the genetic mutation responsible for this disease. Logistic regression models which combine some of the five markers were investigated. Our objective was to assess the predictive ability of these models through internal validation procedures. We also applied shrinkage methods to improve calibration of the models and future predictive accuracy. Validation of these models, using bootstrapping, point to some superiority of two models, for which fairly good measures of predictive accuracy were obtained. This study provides confidence in these models, which offer greater sensitivity than the usual screening by detecting a characteristic pattern in an electrocardiogram.

Keywords: Bootstrapping, Logistic Regression, Ridge Regression, Calibration, Discrimination.

1 Introduction

Logistic regression models may be constructed for prognostic proposes; however, predictive ability may be compromised if samples are too small, compared to the number of prognostic variables, or if some of these variables are fairly correlated [3], [8], [14]. In fact, these two difficulties are common in life science research studies and have been widely addressed in the literature. Some papers worth mentioning, handling with prognostic models for clinical outcomes, are [14], [21], [22], [25], just to cite a few. Small samples may cause overfitting, since estimated models may be overly adjusted to the sample, being influenced by special features (noise) that may exist in the data set [3], [14], [21]. Consequently, these models may not generalize well to future data. High correlations between prognostic variables lead to large standard error of parameter estimates. Both issues, of course,

B. Murgante et al. (Eds.): ICCSA 2014, Part III, LNCS 8581, pp. 498–511, 2014.

lead to unstable coefficient regression estimates [8], [14]. The reliability of a prognostic model constructed in such circumstances should then be evaluated, using adequate procedures that can yield reliable measures of predictive accuracy [6], [23]. It is well known that the apparent accuracy of a model is optimistic, that is, the accuracy obtained when evaluating a model with the same data set that was used to estimate it, will yield an optimistic evaluation [3], [5], [8]. To overcome this weakness, several procedures for internal validation have been proposed and investigated in the literature [2], [6], [8], [23]. In Steyerberg et al. [23], we can find a comparison of some internal validation methods for logistic regression models, using random samples of different sizes drawn from a large data set of clinical records. The authors recommend bootstrapping, as it gives stable and nearly unbiased estimates of model performance. Bootstrapping has been frequently applied to validate logistic regression predictive models in life sciences (e.g. [2], [6], [13], [17], [25]).

In this work we are interested in models that can predict carriers of the genetic mutation responsible for Brugada Syndrome (BS). This syndrome has a very rare diagnostic - the worldwide prevalence is estimated at 1 to 5 per 10000 habitants [1]. Nevertheless, it is estimated that it is responsible for at least 20% of the cases of sudden cardiac deaths (SCD) in individuals with normal hearts and at least 4% of all cases of SCD [1]. Moreover, being an inherited disorder, family member screenning is imperative, once a brugada diagnosis is made [19].

For a definitive diagnosis of BS, a specific pattern in an electrocardiogram (ECG), called Type 1 pattern or Brugada pattern, must be identified [18], [20]. However, it is well known that Brugada patients' ECGs often fluctuate between normal and non-normal patterns, which makes diagnosis challenging [10], [19].

This work is based on the records of 64 members of two Portuguese families, 37 of whom carry the genetic mutation. In earlier studies, [11], [18], [20], five ECG markers were identified to have good ability to discriminate between carriers and non-carriers of the genetic mutation. Henriques et al. [11] explored different ways of combining these five markers in order to enhance the ability of each one to discriminate between the two groups. The studied models exhibited excellent discriminative ability, opening the possibilities for diagnostic tools in Brugada Syndrome. Yet, these models have the weaknesses mentioned above: they are based on a small sample data set and some of the five markers are relatively high correlated. In this study, we use bootstrapping to obtain estimates of prediction accuracy for the logistic regression models studied in [11]. Furthermore, we use shrinkage methods to improve calibration and future predictive accuracy of the models. Shrinkage techniques are well known tools to improve predictive models in situations like the two referred above: small sample size and correlated predictive variables [6], [8], [21], [22], [25]. Shrinkage is related to overfitting. Because models are estimated to fit to a sample, the predictive performance of a model will generally be better in the data set used to estimate it than in a new data set. This is called shrinkage. Of course, we expect minimum shrinkage for a good model, indicating that it can produce good predictions outside the sample used to construct it. We can estimate the amount of shrinkage by a shrinkage factor and then use this factor to re-calibrate the model, multiplying it by the estimated regression coefficients (excluding the intercept) [8]. This will be referred to as linear shrinkage. Penalized maximum likelihood (PML), also known as ridge regression, and Lasso (least absolute shrinkage and selection operator) are other methods of shrinkage widely applied, yet these methods allow shrinkage of some regression coefficients more than others [8], [14], [22], [24]. PML adds a penalty factor to the likelihood function in order to shrink the coefficients according to the variance of each covariable [22]. In Lasso, shrinkage is combined with selection of predictors, since some coefficients are shrunk to zero [22], [24]. Generally speaking, shrinkage methods shrink the coefficients bringing the model to a better calibration [8].

Building on clinical data, Steyerberg et al. [22] compared the performance of three shrinkage techniques in small data sets: linear shrinkage, PML and Lasso. They found that shrinkage of regression coefficients may improve substantially the performance of a prognostic model. Also, they reported no major differences between the three shrinkage techniques. Furthermore, with clinical data as well, Vágó and Kemény [25] investigated the effect of sample size on the estimation of logistic regression models, comparing maximum likelihood (ML) estimation and ridge estimation, and found that, for small samples, the ridge method was much more effective because it produced estimates with much lower variability, offsetting the bias.

In this study, we apply penalized maximum likelihood estimation of regression coefficients, which consists of adding a penalty term to the likelihood function. For internal validation of these models we again use bootstrapping. The obtained performance measures revealed great potential of these models as predictors of mutation carriers. For all the analysis we used the package rms for R [9].

2 Materials and Methods

Having as a starting point a BS diagnosis in a Portuguese Hospital Center (Tondela Viseu Hospital Center), records from family members were collected [18], [20] and a database began to be built. For this study, we considered the records of 64 members of two Portuguese families, from which we had information related to the five ECG markers being studied. These markers are: P-wave duration (PR), transmural dispersion of repolarization (dQT) between V1 and V3, filtered QRS duration (QRSf), where QRS stands for the combination of three of the graphical deflections seen on a typical electrocardiogram (typically an ECG has five deflections, arbitrarily named P, Q, R, S and T waves), low-amplitude signal duration (LAS) and root-mean-square of the voltage in the last 40 ms of the fQRS (RMS40), the last three taken in a signal average ECG (see e.g. [18]). It was possible to offer genetic counseling to patients, by which 37 mutation carriers were identified within the 64 individuals. As mentioned before, the diagnosis of BS is definitive only when the Brugada pattern is identified in an ECG, but this entails the problem of the constant fluctuations of the Brugada pattern [10], [19]. Earlier studies [11], [18], [20] have opened up new options in the diagnosis of BS, identifying five ECG markers with good potential to distinguish mutation from non-mutation carriers [18], and exploring ways of linearly combining these markers in order to enhance the ability of each one to discriminate between the two groups [11]. Exploring the logistic regression models of Henriques et al. [11] further, in this study we assess the predictive ability of those models and also we apply shrinkage techniques to improve calibration and future predictions. First, the stability of coefficient estimates will be evaluated through 95% bootstrap percentile confidence intervals, which will be presented in Section 3.

To evaluate the predictive accuracy of the models, we use calibration, discrimination and overall performance measures [8]. The measures considered in this study are commonly applied in logistic regression model evaluation (e.g., [6], [14], [21], [22], [23]). Calibration refers to whether the predicted probabilities agree with the observed probabilities [21] and can be assessed through the calibration slope [4]. If there is an independent data set, the model under evaluation may be applied to obtain predicted values of the log odds of the outcome for the independent data, which are known as prognostic indexes - PI_i. Then PI_i may be used to explain the observed values of the independent data set through a simple logistic regression model, whose slope coefficient is taken as the calibration slope [4]. Bootstrapping may well be used to quantify calibration slope, as we will explain below [6], [21], [22], [23]. Discrimination concerns the ability to distinguish individuals of one group from individuals of the other. For logistic regression models, discrimination is often measured by the area under the ROC curve (AUC) [6], [8], [21], [22], [23]. Other measures commonly used to quantify the overall accuracy of predictions are the Brier Score (or average prediction error), D statistic and Nagelkerke's R^2 [7], [14], [15], [23]. The Brier score is given by

$$\frac{1}{n} \sum (y_i - \hat{p}_i)^2 \ ,$$

where y represents the outcome, $\hat{p}$ the predicted probabilities and n the number of subjects in the sample. This measure quantifies agreement between predicted and observed values, in a quadratic scale [23]. To quantify this agreement in a log-likelihood scale, we consider the chi-square model, χ_m^2, also referred as the model likelihood ratio, which is the $-2 \log$ of the ratio between two likelihoods, the likelihood of the model with only the intercept, L_0, and the likelihood of the actual (full) model, L_m:

$$\chi_m^2 = -2 \log \left(\frac{L_0}{L_m} \right) = -2 \log L_0 + 2 \log L_m \ .$$

The model likelihood ratio, χ_m^2, is the test statistic for the know likelihood ratio test, but in this study, as in Steyerberg et al. [23], we use it as an accuracy measure, through its scaled version D which is given by

$$D = \left(\chi_m^2 - 1 \right) / n \ .$$

Nagelkerke's R^2 is another widely used measure in the assessment of a predictive model, which also compares those two likelihoods L_0 and L_m,

$$R^2 = \frac{1 - \exp\left(-\chi_m^2/n\right)}{1 - \exp\left(2 \log L_0/n\right)} \, .$$

The five measures described above were considered in evaluating the logistic regression models under study. As mentioned before, the apparent measures of performance are generally optimistic, but this optimism may be corrected through internal validation procedures. In this study we use bootstrapping, as recommended by Steyerberg et al. [23]. Let M denote one of the measures presented above to assess the model's performance. The bootstrap procedure is accomplished in the following steps [5]:

1. Determine the apparent measure of performance, M_{app}, which is the measure obtained when evaluating the model with the same data set used to construct it.
2. Estimate the logistic regression model with a bootstrap sample (a sample of the same size drawn with replacement from the sample of patient records). Compute the measure M in the bootstrap sample, M_{boot}, which represents the apparent performance measure. Evaluate the measure applying the bootstrap model to the original sample and computing the measure M, M_{orig}. The optimism of the apparent performance can be estimated discounting M_{orig} to M_{boot}.
3. Repeat the last step B times (B=100 is enough as indicated in Steyerberg et al. [23]) and average the optimism measures to obtain a stable estimate of the optimism, O.
4. Correct the apparent measure M_{app}, discounting the optimism O.

The procedure just described was applied to correct the optimism of the apparent performance measures. Furthermore, for each performance measure described above, the last procedure was applied 100 times, thus obtaining 100 optimism-corrected measures. The distribution of these values will be exhibited through boxplots presented in Section 4.

As mentioned in the Introduction, when sample data sets are small or when some of the covariates are highly correlated, shrinkage techniques may substantially improve the predictive accuracy of logistic regression models. These techniques shrink the regression coefficients in order to better calibrate the model. In this study we applied penalized maximum likelihood estimation, which adds a penalty factor to the likelihood function in order to shrink the coefficients according to the variance of each covariable [22]. Strictly speaking, the ordinary maximum likelihood estimates are obtained maximizing the log of the likelihood function, $\log L$; still, a penalty factor may be included, so that the function to maximize becomes $\log L - 1/2\lambda\beta'P\beta$, where β' represents the transpose of the coefficients vector (excluding the intercept) and P is a penalty matrix. We took P as a diagonal matrix, whose diagonal elements are the variances of the covariables, which makes the penalty to the log likelihood unitless (this is the default option for the lrm function of package rms by Harrel [9]). For the choice of the penalty factor, we varied it over a grid of values. The penalty factor was chosen as the optimal value according to the modified Akaike Information Criterion (AICc), as in studies [21] and [22].

Moreover, to assess overfitting (model fitting to noise, which results in unstable coefficients) we also used van Houwelingen and le Cessie heuristic shrinkage estimator [26], $\widehat{\gamma} = \left(\chi_m^2 - p\right)/\chi_m^2$, where p is the number of parameters (excluding the intercept), which gives how much of the model fit is "noise".

3   Logistic Regression Models to Predict Mutation Carriers

In Henriques et al. [11], several methodologies were applied to combine the five ECG markers presented above, in order to obtain a prediction model for mutation carriers. Three combination alternatives were highlighted: combination of PR and QRSf, combination of these two and dQT, and combination of the five markers. These will be henceforth designated by Model 1, Model 2 and Model 3, respectively. The logistic regression models for these combinations are:

$$\text{Model 1:} \ln(odds) = -23.642 + 0.066\,PR + 0.13\,QRSf\,,$$
$$\text{Model 2:} \ln(odds) = -26.459 + 0.069\,PR + 0.13\,QRSf + 0.068\,dQT\,,$$
$$\text{Model 3:} \ln(odds) = -22.1 + 0.127\,PR + 0.172\,QRSf + 0.109\,dQT$$
$$-0.373\,LAS - 0.252\,RMS40\,.$$

The apparent performance measures for these models are gathered in Table 1.

Table 1. Apparent performance of the models

	Model 1	Model 2	Model 3
AUC	0.936	0.959	0.973
Calibration slope	1	1	1
D	0.7	0.824	0.984
R^2	0.687	0.764	0.85
Brier Score	0.093	0.068	0.056

The apparent measures presented in Table 1 indicate good model performance. The AUC values, being greater than 0.9, suggest an excellent ability of all three models to discriminate carriers from non-carriers of the genetic mutation ([12], p. 162). The Brier score values are indicative of a low average prediction error. Furthermore, the values for the D statistic and for the Nagelkerke's R^2 are also fairly good (Hosmer and Lemeshow [12], p. 167, refer that low R^2 values in logistic regression are the norm, yet the three studied models have relatively high R^2 values).

Moreover, using van Houwelingen and le Cessie's shrinkage estimator [26], we obtain, for models 1, 2 and 3, 0.9563, 0.9441 and 0.9375, respectively. Note that shrinkage values below 0.85 are reason for concern, because this would mean that more than 0.15 of the model fit would be "noise" [8]. For the three models under consideration, we can expect overfitting of only 0.0437, 0.0559 and 0.0625 (i.e., non-replicable noise). Hence, the apparent measures are visibly indicative of very good performance for all three models. Nevertheless, given the well-known apparent optimism of the model performance, when evaluation of the model is made by the data set used to construct it, care must be taken when considering these measures as good indicators of model performance. In fact, when these models are estimated with one hundred bootstrap samples, the 95% percentile confidence intervals, displayed in Table 2, show that only Model 1 does not suffer from a fair amount of instability in estimated coefficients. This was, in fact, already expected, as studies had demonstrated that we should have at least 10 events for each covariable in logistic regression estimation, to avoid major problems (see e.g [16]).

Table 2. 95% percentile confidence intervals for logistic regression coefficients

	Model 1	Model 2	Model 3
PR	0.039 - 0.193	0.041 - 2.282	0.094 - 3.182
QRSf	0.058 - 0.369	0.032 - 9.134	0.049 - 10.266
dQT		0.032 - 2.217	0.057 - 3.863
LAS			-11.568 - 0.244
RMS40			-6.964 - 0.46

In order to correct the optimism of the apparent measures, we conducted an internal validation of the predictive abilities of the models. The results are included in the next section.

4 Internal Validation of the Logistic Regression Models

The bootstrap methodology described in Section 2 was used to internally validate the models under study. Five hundred bootstrap samples were considered, but one hundred would suffice (see e.g. [23], p. 776). The estimated measures of predictive accuracy are presented in Table 3. Table 3 displays the apparent measures, the measures corrected for their optimism and the percentage of the apparent measures' overestimation. For Model 1, corrected measures of performance are quite good, having changed just slightly when compared with the apparent measures. This means that the apparent measures were not too optimistic. Hence, this model is expected to have good predictive ability. Model 2 exhibits better performance measures than Model 1, except that it has poor calibration. For models 2 and 3 the changes from the apparent measures are bigger, which was already expected, as these models have more coefficients to estimate than Model 1. Also, Model 3 does not seem to have better performance than Model 2: it has poor calibration and all the other measures are similar to Model 2.

Table 3. Performance of the three models: apparent performance, performance corrected from apparent optimism and percentage of overestimation of the apparent measure

	Apparent	Corrected	% of overestimation
Model 1			
AUC	0.936	0.931	0.515
Calibration slope	1	0.922	8.495
D	0.7	0.654	7.083
R^2	0.687	0.662	3.855
Brier Score	0.093	0.105	11.292
Model 2			
AUC	0.96	0.955	0.555
Calibration slope	1	0.891	12.196
D	0.824	0.766	7.549
R^2	0.764	0.734	4.047
Brier Score	0.068	0.08	14.662
Model 3			
AUC	0.973	0.955	1.906
Calibration slope	1	0.623	60.488
D	0.984	0.849	15.928
R^2	0.85	0.777	9.451
Brier Score	0.056	0.08	29.825

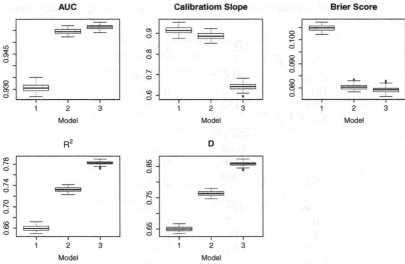

Fig. 1. Distribution of the internally validated measures, obtained repeating 100 times the bootstrap procedure

This is in agreement with the conclusions of Henriques and al. [11], who found a clear indication that the combination of PR, QRSf and dQT provide a good model to discriminate the two groups, but could not establish the advantage of including LAS and RMS40 in the models.

Repeating the internal validation procedure 100 times, we obtained 100 values for each performance measure, whose distribution is illustrated by the boxplots shown in Figure 1. All boxplots exhibit a low variability of the validated performance measures. Indeed, bootstrap validation yields stable and nearly unbiased estimates of model performance [23].

The boxplots again point to a certain superiority of Model 2 with respect to Model 1: better discrimination (high AUC), lower prediction error (low Brier score) and better overall performance measured by the D statistic and Nagelkerke's R^2. It is also patent that Model 2, compared with Model 3, has better calibration (slope) and similar performance quantified by the other measures.

5   Penalized Maximum Likelihood Estimation Results

Shrinkage methods are often applied to recalibrate and improve predictive logistic regression models [6], [8], [21], [22], [25]. We applied penalized Maximum likelihood (PML) estimation, choosing the penalty factor according to the modified AIC (AICc), as explained in Section 2. Table 4 shows the estimated coefficients, as well as the coefficients estimated by ordinary ML, for easy comparison.

Table 4. Estimates of logistic regression coefficients by ordinary ML and penalized ML (PML)

	ML	PML
Model 1		
Const.	-23.642	-21.490
PR	0.066	0.060
QRSf	0.130	0.119
Model 2		
Const.	-26.459	-18.778
PR	0.069	0.049
QRSf	0.130	0.093
dQT	0.069	0.047
Model 3		
Const.	-22.091	-19.967
PR	0.127	0.106
QRSf	0.172	0.151
dQT	0.109	0.092
LAS	-0.373	-0.297
RMS40	-0.252	-0.201

The stability of Model 1 is reinforced, as estimated coefficients from PML where similar to the ones estimated by ML. Similarly, for Model 2, the estimated coefficients obtained by the two methodologies are very close. However,

Model 3 has some changes in estimated coefficients. That is, Model 3 was further calibrated than the other two.

Validating the PML models via bootstrapping, the performance measures obtained are summarized in Table 5, were we added the ones obtained by the ML models, for easy comparison.

Table 5. Performance of the models, corrected of the apparent optimism

	ML	PML
Model 1		
AUC	0.931	0.929
Calibration slope	0.922	1.039
D	0.654	0.680
R^2	0.662	0.660
Brier Score	0.105	0.106
Model 2		
AUC	0.955	0.954
Calibration slope	0.891	1.080
D	0.766	0.805
R^2	0.734	0.737
Brier Score	0.080	0.082
Model 3		
AUC	0.955	0.953
Calibration slope	0.623	0.779
D	0.849	0.856
R^2	0.777	0.771
Brier Score	0.080	0.082

In general, as expected, the PML models are better calibrated. This is more noticeable in models 2 and 3 (for Model 2 slope rose from 0.89 to 1.08, and for Model 3 from 0.62 to 0.78).

Shrinkage methodologies are not appropriate tools to correct discrimination problems [8],[22], so the AUC hardly changed by applying the PML method. Note also that, in general, Model 1 performance practically does not improve by applying PML. In fact, the sample is not too small to estimate this model, since we have at least ten events for each covariable in the regression model [16].

The stability of these models against variations in the data set is analyzed via bootstrapping, by obtaining 95% percentile confidence intervals for coefficient estimates. These are included in Table 6. It is clear that PML models are far more stable in the estimation of the coefficients for models 2 and 3. We note also that, even for the PML estimation method, Model 3 coefficients for LAS and RMS fluctuate between negative and positive values. This may be due to the relatively high correlations between these and some of the other three variables (for example, the correlation coefficient between LAS and QRSf is 0.725 and between the latter and RMS40 is −0.68), which suggests that these two markers may be dispensable in a multivariate model for the identification of mutation carriers.

Table 6. 95% bootstrap percentile confidence intervals for regression coefficients by ML and PML models (obtained with 100 bootstrap samples)

	ML	PML
Model 1		
PR	0.039 - 0.193	0.040 - 0.112
QRSf	0.058 - 0.369	0.048 - 0.238
Model 2		
PR	0.041 - 2.282	0.034 - 0.105
QRSf	0.032 - 9.134	0.033 - 0.266
dQT	0.032 - 2.217	0.033 - 0.106
Model 3		
PR	0.094 - 3.182	0.078 - 0.236
QRSf	0.049 - 10.266	0.041 - 0.536
dQT	0.057 - 3.863	0.044 - 0.246
LAS	-11.568 - 0.244	-0.593 - 0.082
RMS40	-6.964 - 0.46	-0.401 - 0.014

Shrinkage could also be obtained by multiplying the regression estimated coefficients by a shrinkage factor - linear shrinkage. We again note that Steyerberg et al. [22] reported no big differences between linear shrinkage and PML. Linear shrinkage may be done multiplying ML model coefficients by the calibration slope as done in Steyerberg et al. [22]. This produces similar coefficient estimates, as compared with those of the PML, especially for models 1 and 2 (see Table 7).

Table 7. Coefficient estimates by linear shrinkage and penalized ML (PML)

	Linear Shrinkage	PML
Model 1		
const.	-23.642	-21.490
PR	0.061	0.060
QRSf	0.12	0.119
Model 2		
const.	-23.459	-18.778
PR	0.061	0.049
QRSf	0.116	0.093
dQT	0.061	0.047
Model 3		
const.	-22.091	-19.967
PR	0.079	0.106
QRSf	0.107	0.151
dQT	0.068	0.092
LAS	-0.232	-0.297
RMS40	-0.157	-0.201

6   Discussion

The internal model validation study revealed good predictive performance of the models in general, but especially of models 1 and 2. We consider these two the preferred models. Model 1 combines the information of PR and QRSf and Model 2 considers the dQT in addition to these two markers. The latter model seems to be the best option, having higher discriminative ability than Model 1, and also a lower mean prediction error. Model 3, which adds information from LAS and RMS40, did not prove to be a better option than Model 2, because it has poor calibration and one of the estimated coefficients as a counterintuitive sign (LAS). The high correlations between the two latter markers and some of the other three, suggests that the latter do not add much information to predictive models and are, therefore, expendable. Models 1 and 2 estimated by PML exhibited more stability against variations in the data set, when compared to the ML versions. Additionally, they showed good performance measures in the internal validation study. This suggests that these models have good predictive ability. We note that, the actual diagnosis is made by detecting a Brugada pattern in an ECG. But, given the known fluctuations between diagnostic and non-diagnostic ECGs, the sensitivity for a given ECG is very low. For example, Santos el al. [18] reported a sensitivity of 12.5% and a specificity of 100% in the screening of 122 individuals with a single ECG, and a sensitivity of 30%, maintaining the specificity, with several ECGs performed during the follow-up (see Table 1 of [18]). Using the studied models, the sensitivity of one ECG is greatly enhanced. In fact, the ROC curves of the models yield several cutoff values with higher sensibility and specificity near 100%. For instance, using Model 1 we can obtain a sensitivity of 67.6% and a specificity of 96.3%, and using Model 2 it is possible to attain a sensitivity of 92% with the same specificity. In short, this study shows that the combination of the markers PR, QRSf and dQT, building a regression model by PML, has good predictive accuracy and should be considered in the diagnosis of Brugada Syndrome.

As previously stated, Brugada Syndrome is a rare diagnostic, so, data collection for a study of this nature is always very difficult. The size of the sample is a limitation of this study. We intended to overcome this limitation through the internal validation procedure, which yielded good results for the preferred models. This means that, if these models are intended for internal use (patients from a similar population as where the sample came from), we expect them to perform well. However, this study is based merely on records of Portuguese families, which is a serious limitation with regards to generalizing to other countries and ethnicities. Yet, the good performance exhibited by the studied models should be taken into account in future research on Brugada Syndrome and, ideally, these models ought to be investigated with records of families of various nationalities.

Acknowledgments. This work was partially supported by the Centro de Matemática da Universidade de Coimbra (CMUC), funded by the European Regional Development Fund through the program COMPETE and by the Portuguese Government through the FCT, under the project PEst-C/MAT/UI0324/2013.

References

1. Antzelevitch, C.: Brugada Syndrome. Pacing Clin Electrophysiol 29, 1130–59 (2006)
2. Brunelli, A., Rocco, G.: Internal Validation of Risk Models in Lung Resection Surgery: Bootstrap versus Training-and-Test Sampling. The Journal of Thoracic and Cardiovascular Surgery 131, 1243-1247 (2006)
3. Copas, J. B.: Regression, Prediction and Shrinkage. Journal of the Royal Statistical Society, Series B, Methodological 45, 311–354 (1983)
4. Cox, D. R.: Two further applications of a model for binary regression. Biometrika 45, 562–565 (1958)
5. Efron, B., Tibshirani, R. J.: An Introduction to the Bootstrap. Monographs on Statistics and Applied Probability 57, Chapman & Hall, New York (1993)
6. Gude, J. A., Mitchell, M. S., Ausband, D. E., Sime, C. A., Bangs, E. E.: Internal Validation of Predictive Logistic Regression Models for Decision-Making in Wildlife Management. Wildlife Biology 15, 352–369 (2009)
7. Harrel, F. E., Lee, K. L., Califf, R. M., Pryor, D. B., Rosati, R. A.: Regression modelling strategies for improved prognostic prediction. Statistics in Medicine 3, 143-152 (1984)
8. Harrell, F., Lee, K. L., Mark, D. B.: Tutorial in Biostatistics; Multivariable Prognostic Models: Issues in Developing Models, Evaluating Assumptions and Adequacy, and Measuring and Reducing errors. Statistics in Medicine 15, 361–387 (1996)
9. Harrel, F. E. Jr.: Package 'rms' - R Package Version 3.6-3. Available at: http://biostat.mc.vanderbilt.edu/rms (2013)
10. Henriques, C., Matos, A., Santos, L. F.: Study of the Electrocardiographic Fluctuations on Brugada Syndrome Screening. In: Lita da Silva, J., Caeiro, F., Natário, I., Braumann, C.A. (eds.) Advances in Regression, Survival Analysis, Extreme Values, Markov Processes and Other Statistical Applications, Studies in Theoretical and Applied Statistics, pp 231-238. Springer-Verlag, Berlin Heidelberg (2013)
11. Henriques, C., Matos, A., Santos, L. F.: Brugada Syndrome Diagnosis - Three Approaches to Combining Diagnostic Markers. To appear in: Pacheco, A., Santos, R., Oliveira, M.R., Paulino, C.D. (eds.) New Advances in Statistical Modeling and Applications, Studies in Theoretical and Applied Statistics, Springer
12. Hosmer, D. W., Lemeshow, S.: Applied Logistic Regression (2nd edition). John Wiley & Sons, New York (2000)
13. Knudby, A., Brenning, A., LeDrew, E.: New Approaches to Modelling Fish-Habitat Relationships. Ecological Modelling 221, 503-511 (2010)
14. Le Cessie, S., Van Houwelingen, J.: Ridge Estimators in Logistic Regression. Applied Statistics 41, 191–201 (1992)
15. Nagelkerke, N. J. D.: A Note on the General Definition of the Coefficient of Determination. Biometrika 78, 691-692 (1991)
16. Peduzzi, P., Concato, J., Kemper E., Holford, T.R., Feinstein, A.R.: A Simulation Study of the Number of Events per Variable in Logistic Regression Analysis. Journal of Clinical Epidemiology 49, 1372–1379 (1996)
17. Sánchez-Nieto, B., Goset, K. C., Caviedes, I., Delgado, I. O., Córdova, A.: Predictive Models for Pulmonary Function Changes After Radiotherapy for Breast Cancer and Lymphoma. International Journal of Radiation Oncology*Biology*Physics 82, 257-264 (2012)

18. Santos, L.F., Rodrigues, B., Moreira, D., Correia, E., Nunes, L., Costa, A., Elvas, L., Pereira, T., Machado, J. C., Castedo, S., Henriques, C., Matos, A., Santos, O.: Criteria to Predict Carriers of a Novel SCN5A Mutation in a Large Portuguese Family Affected by the Brugada Syndrome. Europace 14 (6), 882–888 (2012)
19. Santos, L. F., Correia, E., Rodrigues, B., Nunes, L., Costa, A., Carvalho, J. L., Elvas, L., Henriques, C., Matos, A., Santos, J. O.: Spontaneous Fluctuations Between Diagnostic and Nondiagnostic ECGs in Brugada Syndrome Screening: Portuguese Family with Brugada Syndrome. Ann Noninvasive Electrocardiol 15, 337–343 (2010)
20. Santos, L.F., Pereira, T., Rodrigues, B., Correia, E., Moreira, D., Nunes, L., Costa, A., Elvas, L., Machado, J. C., Castedo, S., Henriques, C., Matos, A., Santos, O.: Critérios de Diagnóstico da Síndrome Brugada. Podemos Melhorar? Portuguese Journal of Cardiology 31, 355–362 (2012)
21. Steyerberg, E. W., Eijkemans, M. J. C., Harrel, F. E. Jr., Habbema, J. D. F.: Prognostic Modelling with Logistic Regression Analysis: a Comparison of Selection and Estimation Methods in Small Data Sets. Statistics in Medicine 19, 1059–1079 (2000)
22. Steyerberg, E. W., Eijkemans, M. J. C., Habbema, J. D. F.: Application of Shrinkage Techniques in Logistic Regression Analysis: a Case Study. Statistica Neerlandica 55, 76–88 (2001)
23. Steyerbreg, E. W., Harrel, F. E., Borsboom, G. J. J. M., Eijkemans, M. J. C., Vergouwe, Y., Habbema, J. D. F.: Internal Validation of Predictive Models: Efficiency of some Procedures for Logistic Regression Analysis. Journal of Clinical Epidemiology 54, 774–781 (2001)
24. Tibshirani, R.: Regression Shrinkage and Selection via the Lasso. Journal of the Royal Statistical Society, Series B 58, 267–288 (1996)
25. Vágó, E., Kemény, S.: Logistic Ridge Regression for Clinical Data Analysis (a Case Study). Applied Ecology and Environmental Research 4(2), 171–179 (2006)
26. Van Houwelingen, J. C., Le Cessie, S.: Predictive Value of Statistical Models. Statistics in Medicine 9, 1303–1325 (1990)

Outlier Detection and Robust Variable Selection for Least Angle Regression

Shirin Shahriari[1], Susana Faria[1], A. Manuela Gonçalves[1], and
Stefan Van Aelst[2,3]

[1] DMA-Department of Mathematics and Applications,
CMAT-Centre of Mathematics, University of Minho, Guimarães, Portugal
[2] Department of Mathematics, K.U. Leuven, Leuven, Belgium
[3] Department of Applied Mathematics, Computer Science and Statistics,
Ghent University, Ghent, Belgium
shirin.shahriari22@gmail.com, {sfaria,mneves}@math.uminho.pt,
stefan.vanaelst@wis.kuleuven.be

Abstract. The problem of selecting a parsimonious subset of variables from a large number of predictors in a regression model is a topic of high importance. When the data contains vertical outliers and/or leverage points, outlier detection and variable selection are inseparable problems. Therefore a robust method that can simultaneously detect outliers and select variables is needed. An outlier detection and robust variable selection method is introduced that combines robust least angle regression with least trimmed squares regression on jack-knife subsets. In a second stage the detected outliers are removed and standard least angle regression is applied on the cleaned data to robustly sequence the predictor variables in order of importance. The performance of this method is evaluated by simulations that contain vertical outliers and high leverage points. The results of the simulation study show the good performance of this method in both outlier detection and robust variable selection.

Keywords: Outlier Detection, Robust Variable Selection, Least Angle Regression.

1 Introduction

Regression models are used in many different areas. The ability to collect data has grown dramatically, yielding larger numbers of potentially relevant predictor variables. Usually, there are some correlated predictor variables, but including all of them in a statistical model will not necessarily improve the model's prediction performance. On the other hand, the interpretation of models with fewer predictors is easier than for models with a large number of predictors. Model selection focuses on finding a model with good prediction properties, but puts less emphasis on sequencing the variables in order of importance. Variable selection refers to the selection of the best variables to enter the model. Therefore, finding the best variables among all candidate explanatory variables is an important problem to study. When the data contains atypical observations, we need

B. Murgante et al. (Eds.): ICCSA 2014, Part III, LNCS 8581, pp. 512–522, 2014.

a robust variable selection method that is resistant to outliers in order to select variables reliably. A small proportion of outliers in the data may largely influence likelihood-type model selection methods such as AIC [1], Mallows' C_p [2], and BIC [3]. Recently, robust variable selection methods have received more attention in the literature. There are various robust variable selection approaches that are based on robustifying classical selection criteria, namely robust AIC [4], robust C_p [5], robust final prediction error [6], and robust selection criteria based on stratified bootstrap [7] or fast and robust bootstrap [8]. An added variable t-test in the context of regression based on the forward search procedure for variable selection has been proposed by Atkinson and Riani [9]. Also, for generalized linear models, a robust selection criteria has been proposed [10].Salibian-Barrera and Van Aelst [8] used the fast and robust bootstrap to achieve a faster model selection method based on bootstrap, which makes it feasible to consider larger numbers of predictors. Most of the robust model selection methods need to fit a large number of submodels robustly. When p, the number of predictors is large, then it is computationally more efficient to use variable selection methods that sequence the predictors according to their importance, such as forward selection and backward elimination [11].The Least Angle Regression (LARS) algorithm proposed by Efron et al. [12], is a modified version of the forward stagewise procedure. It is a powerful and computationally efficient technique to sequence the predictor variables for least squares regression. LARS is not robust to the presence of a small amount of anomalies in data since it is based on the pairwise correlation between the predictors and the response variable. Khan et al. [13] showed that LARS is only based on the means, variances and correlations of the variables, and replaced them with robust counterparts. They proposed robust LARS (RLARS) which is a robust linear model selection method based on LARS by using medians, median absolute deviations (MAD) and robust pairwise correlation estimates. In this paper, we propose an algorithm which combines RLARS with least trimmed squares (LTS) regression, which is a highly robust regression method, and apply it on jack-knife (JK) subsets [14] to perform outlier detection. JK subsets are used to find the regression model with optimal predictive ability as measured by the MAD of the prediction errors obtained by cross-validation. This optimal model is used to detect outliers in the data. In the second stage the detected outliers are removed and LARS is applied on the cleaned data to find a robust version of LARS sequenced predictor variables. This paper is organized as follows. In Section 2, we briefly describe the LARS algorithm and the RLARS method. In Section 3, our strategy for outlier detection and robust variable selection are explained as well as our proposed algorithm, jack-knife robust least angle regression (JKRLARS). The design of simulation studies that have been conducted to compare the performance of JKRLARS with LARS and RLARS are described in Section 4. Section 5 contains the results and discussion of the methods on the simulation scenarios. Finally, in Section 6, the conclusions are presented.

2   Least Angle Regression (LARS)

The least angle regression algorithm which has been proposed by Efron et al. [12] enters variables in the model in order of their importance. LARS is related to forward stagewise selection (FSS), which sequentially adds variables to the model in small steps. FSS uses in each step the variable that is most correlated with the residuals, and updates the fit by adding only a small fraction of the least squares contribution of this variable to the model. Usually, FSS takes a large number of steps using the same variable before another variable yields a higher correlation with the remaining residual. Therefore, FSS is not computationally efficient. By deriving a simple mathematical formula for the optimal step size of a selected variable, LARS largely reduces the number of steps and speeds up the computation time.Consider a data set $(y_i, \mathbf{x}_i), i = 1, \ldots, n$, where $y_i \in \mathbb{R}$ and $\mathbf{x}_i = (x_{i1}, x_{i2}, \ldots, x_{ip})^T \in \mathbb{R}^p$ are the response and predictor values, respectively. We are interested in fitting the following linear model

$$y_i = \beta_0 + \beta^T \mathbf{x}_i + e_i, \tag{1}$$

where $\beta \in \mathbb{R}^p$. The errors e_i are assumed to have $E(e_i) = 0$ and $\mathrm{Var}(e_i) = \sigma^2$. Without loss of generality, we assume that $\beta_0 = 0$. This can be achieved by centering both the predictor variables and the response variable. Then, the linear model does not contain an intercept anymore.

The steps of the LARS algorithm can be summarized as follows:

1. Start with all coefficients equal to zero and set the initial residual vector equal to the response vector $\mathbf{y}$;
2. Find the predictor that is most correlated with the response variable, say $\mathbf{x}_1$;
3. Take the largest step possible in the direction of this predictor until some other predictor, say $\mathbf{x}_2$, has the same maximal correlation (in an absolute value) with the current residual;
4. Proceed in the equiangular direction of the two predictors until a third variable $\mathbf{x}_3$ earns its way into the "most correlated" set; The equiangular direction is the direction in which the correlation of the predictors decreases at the same pace so that these correlations remain equal at all times;
5. Repeat until all predictors have been entered.

LARS is only based on the means, variances and correlations of the data and so it is sensitive to the presence of contamination in the data. Therefore, Khan et al. [13] proposed a robust LARS by replacing the mean, variances and correlations of the data with robust counterparts. As robust center and scale measures they used the computationally fast median and MAD, respectively. They introduced a fast robust pairwise correlation estimator based on bivariate winsorization (a generalization of univariate winsorization as introduced in Huber (1981) [15]).

Bivariate winsorization of robustly standardized bivariate data is based on an initial robust bivariate correlation matrix R_0 and a corresponding tolerance ellipse. For example, for $\mathbf{x} = (x_1, x_2)^t \in \mathbb{R}^2$ consider the Mahalonobis distance $D(\mathbf{x})$ based on initial bivariate correlation matrix R_0 and set the tuning constant $c = 5.99$, which is the 95 % quantile of the χ_2^2 distribution. By using the bivariate transformation $u = \min(\sqrt{c/D(\mathbf{x})}, 1)\mathbf{x}$ with $\mathbf{x} = (x_1, x_2)^t$ outliers are then shrunken to the boundary of the 95% tolerance ellipse and thus will less influence the resulting correlation estimate. Hence, a more robust correlation estimate is obtained.

Choosing an appropriate initial correlation matrix R_0 is a crucial part of the bivariate winsorization procedure. Khan et al. proposed a new method called adjusted winsorization as initial estimator (see details in [13]).

3   Our Outlier Detection and Robust Variable Selection Strategy

We propose an algorithm that can perform outlier detection and robust variable selection simultaneously. We aim to detect outliers in the data and at the same time determine a robust sequence of the predictor variables in order of their importance. To produce jack-knife subsets, first the data is divided into l folds, that is l randomized non-overlapping equally-sized subsets. Then, each subset is left out once and we sequence the variables in order of importance using RLARS on the remaining subsets to find l sequence predictor variables. The most relevant predictor variables according to these l RLARS sequences are used to fit a robust regression model. We use the highly robust and computationally efficient LTS regression which is an intuitively appealing and well-known robust regression method. Denote the vector of squared residuals by $\mathbf{r}^2(\beta) = (r_1^2, \ldots, r_n^2)^T$ with $r_i^2 = (y_i - \beta^T \mathbf{x}_i)^2, i = 1, \ldots, n$. Then the LTS estimator is defined as

$$\hat{\beta}_{LTS} = \arg\min_{\beta} \sum_{i=1}^{h} (\mathbf{r}^2(\beta))_{i:n}, \tag{2}$$

where $(\mathbf{r}^2(\beta))_{1:n} \leq \cdots \leq (\mathbf{r}^2(\beta))_{n:n}$ are the order statistics of the squared residuals and $h \leq n$, for $h = \lfloor (n + p + 1)/2 \rfloor$ the LTS breakdown point equals 50% whereas for greater h its break down point is $(n - h)/n$. The usual choice $h \approx \lfloor (n + 1)0.75 \rfloor$ yields the LTS break down point of 25%. Therefore, LTS regression corresponds to finding the subset of h observations whose least squares fit produces the smallest sum of squared residuals. Based on the LTS regression model, predicted values of the observations are computed by cross-validation. The median absolute deviation (MAD) values of the prediction errors for each LTS regression model are calculated. Then, the optimal LTS regression model which yields to minimum MAD value, is selected. To identify outliers, the standardized predictions of this optimal model are computed. The detected outliers are left out, and LARS is applied on the cleaned data to find an improved sequence of predictor variables. The goals of detecting outliers in data and robustly sequencing predictor variables simultaneously, reflect the specifications of our algorithm.

3.1 Description of the Algorithm

Step 1. The observations $(y_i, \mathbf{x}_i), i = 1, \ldots, n$ are partitioned into l randomized non-overlapping equally-sized JK subsets $I_f \subset I = \{1, \ldots, n\}$, with $f = 1, \ldots, l$. Clearly, I_f contains the indices of the observations in f-th subset, with $|I_f| \approx \frac{n}{l}$.

Step 2. With the f-th subset left out, RLARS is applied to the set $(y_i, \mathbf{x}_i), i \notin I_f$, of $l - 1$ subsets to find a sequence of predictor variables $\mathbf{x}_j^{(f)}, j \in J_1$ with $J_1 \subset J = \{1, \ldots, p\}$ such that $|J_1| \ll p$, where p is the number of predictor variables. Clearly, J_1 contains the indices of the $|J_1|$ most important predictor variables returned by RLARS.

Step 3. The covariates $\mathbf{x}_j^{(f)}, j \in J_1$, are used as predictor variables for LTS regression. We thus consider the regression model

$$y_i = \beta^{(f)T} \mathbf{x}_i^{(f)} + e_i, \tag{3}$$

where $\mathbf{x}_i^{(f)}$ contains the values of the predictors $\mathbf{x}_j^{(f)}, j \in J_1$ for the ith observation and $\beta^{(f)}$ is estimated by using LTS.

Step 4. To evaluate the prediction performance of each LTS regression fit, we perform l-fold cross-validation. In order to detect outliers, we compute the predicted values $\hat{y}_i$ which are defined as

$$\hat{y}_i = \hat{\beta}^{(f)T} \mathbf{x}_i^{(f)}, i \in I_f, \tag{4}$$

where $\hat{\beta}^{(f)}$ denotes the LTS estimates of the regression coefficients with the f-th subset left out as obtained in the previous step. Thus, predicted values for all the observations are obtained. For each LTS regression the prediction errors $\text{PE}_i = y_i - \hat{y}_i$, and the corresponding MAD value of the prediction errors as a measure of the model's predictive ability are calculated. The minimum MAD value over all LTS models is obtained to find the LTS model with optimal predictive ability. Then, the corr- esponding standardized prediction errors of this optimal model are used to detect outliers. The standardized prediction errors are defined by $\frac{\text{PE}_i}{\hat{\sigma}}, i = 1, \ldots, n$ where $\hat{\sigma} = c_{h,n} \sqrt{\frac{1}{h} \sum_{i=1}^{h} (r^2)_{i:n}}$, and $c_{h,n}$ makes $\hat{\sigma}$ consistent and unbiased at Gaussian error distribution [16]. It should be mentioned that the LTS scale estimate $\hat{\sigma}$ is itself highly robust and therefore, can be used to identify vertical outliers by $\frac{\text{PE}_i}{\hat{\sigma}}$. As in [17] we define the set of the indices of outlying observations as $I_{Out} = \left\{ i \in I : |\frac{\text{PE}_i}{\hat{\sigma}}| > \sqrt{\chi_{1,0.975}^2} \right\}$.

Step 5. The detected outliers $(y_i, \mathbf{x}_i), i \in I_{Out}$, are removed (or given weight zero) and LARS is applied on the cleaned data $(y_i, \mathbf{x}_i), i \in I_{Out}^c$, with I_{Out}^c the complement of I_{Out}. The covariates $\mathbf{x}_j, j \in J_2$ with $J_2 \subset J$ at the beginning of this LARS sequence yield a robust version of LARS sequenced predictor variables.

3.2 Summary of the Jack-knife Robust LARS (JKRLARS) Algorithm

Consider a data set $(y_i, \mathbf{x}_i), i = 1, \ldots, n$. Let $J = \{1, \ldots, p\}$ and $I = \{1, \ldots, n\}$ be the set of indices for the candidate predictors and observations respectively, and $q \ll p$ the length of the RLARS sequenced predictor variables that will be used in the first step. Then the basic steps of JKRLARS algorithm are as follows:

1. Partition data $(y_i, \mathbf{x}_i)$ into l randomized non-overlapping equally-sized JK subsets I_f with $|I_f| \approx \frac{n}{l}$;
2. With the f-th subset left out, perform RLARS on $(y_i, \mathbf{x}_i), i \notin I_f$, to find a sequence $\mathbf{x}_j{}^{(f)}, j \in J_1$ with $|J_1| = q$. This leads to l sets of sequenced variables;
3. For each sequence use $\mathbf{x}_j{}^{(f)}, j \in J_1$ as predictors in LTS regression and compute the LTS estimates $\{\hat{\beta}_j^{(f)} : j \in J_1\}$;
4. For $i = 1, \ldots, n$, compute the predicted values $\hat{y}_i = \hat{\beta}^{(f)^T} \mathbf{x}_i^{(f)}$ by applying LTS regression using cross-validation and the corresponding MAD value of the prediction errors. For $f = 1, \ldots, l$ obtain the minimum MAD value over all LTS models to find the optimal LTS model. Calculate the optimal LTS model's standardized prediction errors $\frac{\mathrm{PE}_i}{\hat{\sigma}}, i = 1, \ldots, n$. Flag the i-th observation as an outlier if $|\frac{\mathrm{PE}_i}{\hat{\sigma}}| > \sqrt{\chi_{1,0.975}^2}$;
5. Find the covariates $\mathbf{x}_j, j \in J_2$ with $J_2 \subset J$, from candidate covariates as the robust version of LARS sequenced predictor variables by applying LARS on cleaned data by removing the detected outliers.

4 Simulation Study

This section presents a simulation study to investigate the behavior of the JKR-LARS algorithm and compare its performance to other methods. LARS and RLARS are performed in R [19] with packages *lars* [20] and *robustHD* [21], respectively. To calculate LTS inside JKRLARS we prefer to take a relatively larger trimming proportion subset size than $h \approx \lfloor (n + 1)0.75 \rfloor$ to guarantee a break down point of 20%. The number of $l = 10$ subsets is considered for this algorithm.We consider the simulation setting of Khan et al. [13], which is based on the design of Frank and Friedman [18]. The linear model is created as follows:

$$\mathbf{y} = \mathbf{l}_1 + \ldots + \mathbf{l}_k + \sigma\epsilon, \tag{5}$$

with $k = 6$ latent variables, where $\mathbf{l}_1, \mathbf{l}_2, \ldots, \mathbf{l}_k$ and ϵ are independent standard normal variables. The value of σ is chosen such that the signal to noise ratio is equal to 3. Let $\mathbf{f}_1, \ldots, \mathbf{f}_p$ be independent standard normal variables, then the set of p predictors is created as follows:

$$\mathbf{x}_i = \mathbf{l}_i + \tau\mathbf{f}_i, \quad i = 1, ..., k$$
$$\mathbf{x}_{k+1} = \mathbf{l}_1 + \delta\mathbf{f}_{k+1}$$
$$\mathbf{x}_{k+2} = \mathbf{l}_1 + \delta\mathbf{f}_{k+2}$$
$$\mathbf{x}_{k+3} = \mathbf{l}_2 + \delta\mathbf{f}_{k+3}$$
$$\mathbf{x}_{k+4} = \mathbf{l}_2 + \delta\mathbf{f}_{k+4}$$
$$\cdot$$
$$\cdot$$
$$\cdot$$
$$\mathbf{x}_{3k-1} = \mathbf{l}_k + \delta\mathbf{f}_{3k-1}$$
$$\mathbf{x}_{3k} = \mathbf{l}_k + \delta\mathbf{f}_{3k}$$
$$\text{and } \mathbf{x}_i = \mathbf{f}_i, \quad i = 3k + 1, ..., p$$

where $\delta = 5$ and $\tau = 0.4$ so that target predictor variables $\mathbf{x}_1, ..., \mathbf{x}_k$ are formed by low noise perturbations of the latent variables. Variables $\mathbf{x}_{k+1}, ..., \mathbf{x}_{3k}$ are noise predictor variables that are correlated with the target predictor variables, and variables $\mathbf{x}_{3k+1}, ..., \mathbf{x}_p$ are independent noise predictor variables. We consider a fraction of contamination $a = 0.1$. Five error distributions are considered to generate contamination:

(a) $\epsilon \sim (1 - a)N(0, 1) + aN(0, 1)/U(0, 1)$, symmetric, slash contamination;
(b) $\epsilon \sim \text{Cauchy}(0, 1)$, heavy-tailed cauchy contamination;
(c) $\epsilon \sim (1 - a)N(0, 1) + aN(20, 1)$, asymmetric, shifted normal contamination;
(d) same as (a), with high leverage X values, $X \sim N(50, 1)$;
(e) same as (b), with high leverage X values, $X \sim N(50, 1)$;

We generate 100 independent data sets of size $n = 150$ from the above five simulation scenarios with $p = 50$ predictors, and each time we perform all the aforementioned methods on the same data set.

4.1 Performance Measures

The performance of the JKRLARS concerning outlier detection is evaluated by the true positive rate (TPR) and false negative rate (FNR). A true positive is an observation that is contaminated in the data and is also detected as outlier. Analogously, a false negative is an observation that is contaminated in the data, but is flagged as a regular observation. Denote by $I_R \subset I = \{1, ..., n\}$ the set of the indices of the regular observations in the data and by $I_O = I_R^c \subset I = \{1, ..., n\}$ the set of the indices of the contaminated observations in the data. The TPR and FNR can then be defined as

$$TPR = \frac{|\{i : i \in I_{Out} \land i \in I_O\}|}{|I_O|}; \tag{6}$$

$$FNR = \frac{|\{i : i \in I_{Out}^c \land i \in I_O\}|}{|I_R|}. \tag{7}$$

Higher TPR and smaller FNR show that JKRLARS performs well concerning outlier detection.

In order to compare the performance of JKRLARS with RLARS, we plot the a- verage number of target covariates t_m included in the first m sequenced predictor variables entering the model as a function of varying m. With k number of target covariates, good performance is achieved when the method can find the k target covariates in the first t_k sequenced predictor variables, with t_k equal or close to k.

5   Results and Discussion

The simulation results for the different data configurations are presented and discussed in this section. We perform LARS, RLARS and JKRLARS on each of the 100 independent data sets with $n = 150$, $p = 50$ and $k = 6$ number of target covariates.

First, we examine the performance of JKRLARS to detect the outliers in the data sets. The results for the TPR and FNR averaged over the 100 data sets are reported in Table 1 for the 5 types of contamination considered. From Table 1 we can see that the TPR and FNR of the outlier detection procedure is almost perfect in scenarios c,d, and e. High leverage points and clear vertical outliers can thus be detected with high accuracy. At first sight the TPR and FNR for scenarios a and b look much worse. However, in these cases the observations are contaminated by generating errors from the long-tailed slash and Cauchy distributions, respectively. Not all errors generated from these distributions will lie in the tails of the distribution. Hence, in these scenarios not all of the 10% contaminated observations will be actual outliers. Considering the cut-off value $\sqrt{\chi^2_{1,0.975}}$ for detecting outliers, it can easily be checked that only 35.2% of the generated slash errors and 26.7% of the generated Cauchy errors are expected to produce outlying observations. Comparing the TPR with these fractions, we see that the outlier detection procedure still performs reasonably well in these difficult scenarios. Thus, JKRLARS does a good job of outlier detection considering both TPR and FNR in all scenarios.

We performed LARS, RLARS and JKRLARS on all 100 data sets in each scenario to sort the predictor variables in order of importance. The curves for comparing LARS, RLARS and JKRLARS in terms of sequencing the predictor variables in each simulation scenario are displayed in Figure 1.

For each sequence of predictor variables we determine the number t_m of target covariates included in the first m sequenced predictor variables entering the model with m ranging from 1 to 25.As expected, Figures 1 (a)-(e) show that the performance of LARS considerably diminishes in presence of contamination, while the robustified LARS methods are much less influenced by the contamination in the data. In scenarios a, b and c JKRLARS shows the same excellent performance as RLARS in sequencing the $k = 6$ target predictors at the top of the sequence (Figures 1 (a)-(c)). In the high leverage scenarios d and e, RLARS has much more problems to pick up the target variables in the beginning. Figures 1 (d) and (e) show that in these scenarios JKRLARS succeeds much better to pick up most of the important predictors in the beginning of the sequence.

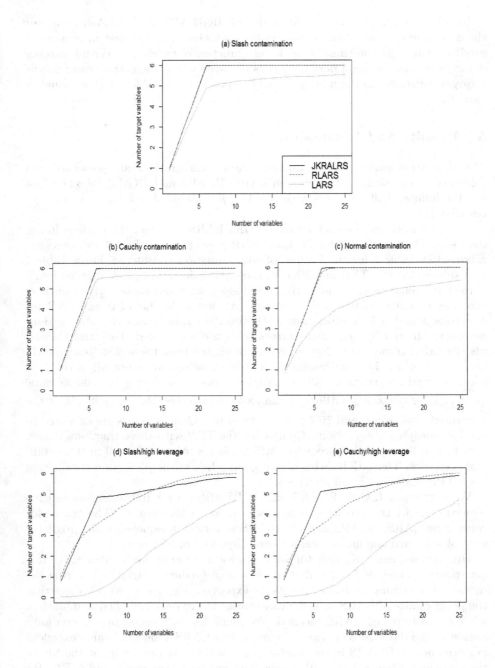

Fig. 1. Average number of target variables t_m versus m for LARS, RLARS and JKR-LARS for scenarios (a)-(e). The lines shown in all plots follow the legend of figure (a).

In particular, in models with (less than) 10 predictors JKRLARS captures 5 of the 6 important covariates, while RLARS needs models with up to 15 predictors to include at least 5 of the 6 important covariates.

Table 1. The true positive rate (TPR) and the false negative rate (FNR), averaged over 100 simulation runs are reported for JKRLARS

Case	a	b	c	d	e
TPR	0.22	0.17	1	0.96	0.97
FNR	0.09	0.09	0	0.004	0.003

6 Conclusion

Outlier detection and variable selection are inseparable problems. Therefore robust methods that can detect outliers and select relevant variables simultaneously are needed. We proposed a new robust algorithm to detect outliers and robustly select variables at the same time. Our JKRLARS procedure combines RLARS with LTS regression as a highly robust regression method. Jack-knife subsets are used to find the regression model with optimal predictive ability as measured by the MAD of the prediction errors obtained by cross-validation. This optimal model is used to detect outliers in the data set. Then, the detected outliers are removed and LARS is applied on cleaned data to find the robust sequenced predictor variables. The results of a simulation study showed that JKRLARS does a good job in outlier detection for the presented scenarios. It also succeeds well in robustly sequencing the predictor variables for the different data configurations containing vertical outliers and leverage points. JKRLARS thus yields a robust version of LARS sequenced predictor variables. JKRLARS can outperform RLARS in robustly sequencing the predictor variables in situations with high leverage points.

Acknowledgments. The author Shirin Shahriari has a Ph.D. scholarship by FCT SFRH/BD/51164/2010. This research was financed by FEDER Funds through "Programa Operacional Factores de Competitividade-COMPETE" and by Portuguese Funds through FCT "Fundação para a Ciência e a Tecnologia", within the PEst-OE/MAT/UI0013/2014.

References

1. Akaike, H.: Statistical predictor identification. Annals of the Institute of Statistical Mathematics 22(1), 203–217 (1970)
2. Mallows, C.L.: Some comments on c_p. Technometrics 15(4), 661–675 (1973)
3. Schwarz, G.: Estimating the dimension of a model. The Annals of Statistics 6(2), 461–464 (1978)

4. Ronchetti, E.: Robust model selection in regression. Statistics & Probability Letters 3(1), 21–23 (1985)
5. Ronchetti, E., Staudte, R.G.: A robust version of mallows' c_p. Journal of the American Statistical Association 89(426), 550–559 (1994)
6. Maronna, R.A., Martin, R.D., Yohai, V.J.: Robust Statistics: Theory and Methods. J. Wiley & Sons (2006)
7. Müller, S., Welsh, A.: Outlier robust model selection in linear regression. Journal of the American Statistical Association 100(472), 1297–1310 (2005)
8. Salibian-Barrera, M., Van Aelst, S.: Robust model selection using fast and robust bootstrap. Computational Statistics & Data Analysis 52(12), 5121–5135 (2008)
9. Atkinson, A.C., Riani, M.: Forward search added-variable t-tests and the effect of masked outliers on model selection. Biometrika 89(4), 939–946 (2002)
10. Cantoni, E., Ronchetti, E.: Robust inference for generalized linear models. Journal of the American Statistical Association 96(455), 1022–1030 (2001)
11. Weisberg, S.: Applied Linear Regression. J. Wiley & Sons, New York (2005)
12. Efron, B., Hastie, T., Johnstone, I., Tibshirani, R.: Least angle regression. The Annals of Statistics 32(2), 407–499 (2004)
13. Khan, J.A., Van Aelst, S., Zamar, R.H.: Robust linear model selection based on least angle regression. Journal of the American Statistical Association 102(480), 1289–1299 (2007)
14. Efron, B.: The jackknife, the bootstrap and other resampling plans, vol. 38. SIAM NSF-CBMS (1982)
15. Huber, P.J., Ronchetti, E.M.: Robust Statistics. Wiley, New York (2009)
16. Pison, G., Van Aelst, S., Willems, G.: Small sample corrections for lts and mcd. Metrika 55(1-2), 111–123 (2002)
17. Hubert, M., Rousseeuw, P.J., Van Aelst, S.: High-breakdown robust multivariate methods. Statistical Science, 92–119 (2008)
18. Frank, L.E., Friedman, J.H.: A statistical view of some chemometrics regression tools. Technometrics 35(2), 109–135 (1993)
19. Core Team, R.: R: A Language and Environment for Statistical Computing. R Foundation for Statistical Computing, Vienna (2012)
20. Hastie, T., Efron, B.: lars: Least Angle Regression, Lasso and Forward Stagewise, R package version 1.2 (2013)
21. Alfons, A.: robustHD: Robust methods for high-dimensional data, R package version 0.4.0 (2013)

Forecasting Household Packaging Waste Generation: A Case Study

João A. Ferreira, Manuel C. Figueiredo, and José A. Oliveira

University of Minho, Centre Algoritmi,
Campus de Azur ém, 4800-058, Guimarães, Portugal
joao.aoferreira@gmail.com, {mcf,zan}@dps.uminho.pt

Abstract. Nowadays, house packaging waste (HPW) materials acquired a great deal of importance, due to environmental and economic reasons, and therefore waste collection companies place thousands of collection points (ecopontos) for people to deposit their HPW.

In order to optimize HPW collection process, accurate forecasts of the waste generation rates are needed.

Our objective is to develop forecasting models to predict the number of collections per year required for each ecoponto by evaluating the relevance of ten proposed explanatory factors for HPW generation.

We developed models based on two approaches: multiple linear regression and artificial neural networks (ANN). The results obtained show that the best ANN model, which achieved an R^2 of 0.672 and MAD of 9.1, slightly outperforms the best regression model (R^2 of 0.636, MAD of 10.44).

The most important factors to estimate HPW generation rates are related to ecoponto characteristics and to the population and economic activities around each ecoponto location.

Keywords: Forecasting, Municipal Solid Waste Generation, House Packaging Waste, Waste Collection, Recycling, Multiple Linear Regression, Artificial Neural Network.

1 Introduction

Over the last decades, the recycling of waste materials became a very important subject for society, as the environment benefits greatly from any advances made in direction of a cleaner future. In fact, as the process of recycling became really important, it also turned into an interesting resource management problem, in particular when referring to collection of waste for recycling, which involves teams of workers and vehicles. Therefore, the collection process is crucial in order for recycling to go on, and so, fleet management may frequently deal with various issues. One main problem lies in finding optimal collection routes, where a set of collection points is targeted, and each point is given a priority level. This problem can be described as a Vehicle Routing Problem (VRP), yet more flexibility is needed when it comes to choose only part of the collection points

B. Murgante et al. (Eds.): ICCSA 2014, Part III, LNCS 8581, pp. 523–538, 2014.

to be visited, instead of the whole set. Thus, a more fitting description of the selective waste collection process may be the Team Orienteering Problem (TOP). In this context, the TOP can be described as the problem of designing optimized collection routes to be assigned to a fleet of vehicles that perform the collection of different types of waste stored along a network of collection points. Each one of these collection points contains a certain amount of waste that directly quantifies the respective priority level. The collection routes have maximum durations or distances, and consequently, the selection of collection points to be visited by the vehicles is made by balancing their priorities and their contributions for the route duration or distance. The objective is to maximize the total amount of waste collected by all routes while respecting the time or distance constraints.

Aside from the routing problem, there are other issues related to the process of waste collection for recycling, especially when dealing with real scenarios and the activity of real waste collection companies. One of these issues is the prediction of waste material quantities generated over time at each collection point in a given collection network, which enables the determining of a waste generation rate (WGR) for each collection point. Determining the WGR values is usually helpful during the designing phase of the collection network. Collection points are assigned to certain places based on probable WGR values that were previously assessed (predicted). Based on forecasts for waste generated along the network, the principles of the TOP can be employed to design the collection routes, as each collection point is assigned with a certain priority level according to their WGR, which translates the need for each collection point to be emptied.

In Portugal, household packaging waste (HPW) is separated by citizens at the local recycling site, named ecoponto ("ecological point"). Here, the waste is divided into three main containers, typically identified with different colours to help people to separate the waste: 1) glass (green) 2) paper/cardboard (blue), and 3) plastic/metal (yellow). These ecoponto sites are provided by the municipalities for only household waste to be recycled in these containers. Given the goals Portugal has to fulfil for the recycling and recovery of HPW, there is a permanent need for increased efficiency in waste collection.

The work presented in this paper is integrated in a R&D project named Genetic Algorithm for Team Orienteering Problem (GATOP), which is financed by the Portuguese Foundation for Science and Technology (FCT). The major goal of GATOP is the development of a more complete and efficient solutions for several real-life multi-level Vehicle Routing Problems, with emphasis on the waste collection management. Within the project scope, one important task is the development of forecasting models to predict the quantities of waste generated at collection points. In this work we focused on developing models to predict the generation of recyclable waste along a network of ecopontos. We aimed to achieve the best approximation possible of the predicted values to the real values by using regression models. We also explored other forecasting methods based on Artificial Neural Network models.

This paper is structured in 6 sections. Section 2 presents the real problem. In section 3, several forecasting methods and models found in the literature that are applied in the context of waste management are discussed. The followed methodologies and the developed models are presented and analysed in the fourth section. Computational experiments are described in the fifth section and the results are discussed. Finally, on section 6, the main conclusions of this study are presented.

2   The Real Problem

In this paper we intend to solve a real-world problem faced by Braval, an intermunicipal waste management company in the Cávado sub region of northern Portugal. Braval takes action across six municipalities: Braga, Vieira do Minho, Vila Verde, Póvoa do Lanhoso, Amares and Terras de Bouro. Braval currently operates a network of more than 1,200 ecopontos where residents can start the recycling process of their HPW. These sites are located across the municipalities in a variety of easily accessible areas. Within the six municipalities where Braval operates, there is a mix of urban and rural areas, which prompts the demand of different strategies for waste management.

Braval's fleet does not visit all the ecoponto sites every workday, and so it is necessary to select a subset of ecopontos to visit each time route planning is done. Furthermore, given a planning horizon of, for example, a week, or a month, Braval must decide which ecoponto sites must be visited, which ecoponto sites can be visited, and which sites can be skipped during the collection routes, and then design effective routes to perform the selective collection of HPW. The priority level of an ecoponto to be visited is highly related to the amounts of waste it holds during the route planning phase, as well as their own WGR. Taking into account its priority level, an ecoponto is either selected or not to be visited during the established planning horizon.

In order to help Braval performing better route planning, reliable predictions of the amount of waste generated at each ecoponto are necessary. There are several forecasting methods and models presented in the literature that feature real-world waste management problems, and good results were achieved with those strategies on those situations. Before deciding on forecasting methodologies, we should keep in mind the kind of information resources put at our disposal by Braval. These informations consist of time series with waste collection data, more specifically records of previous waste collections performed during a period of one year, where a count was kept of how much times each ecoponto was visited and emptied during that time interval. Therefore, our main goals with this study are: 1) Determining significant factors of HPW generation as well as their relevance once applied in forecasting models; 2) Predict the number of times each ecoponto should be visited during a certain period of time, that being a week, a month, a trimester or a year. These predictions are highly related to the WGR factor, and once its value is determined for each ecoponto, all of them can be categorized into different groups based on WGR value and overall collection priority. Once the categorization is done, the obtained information can be used by the vehicle routing optimization models previously developed for the GATOP project.

3   Literature Review

The subject addressed by this study is predicting or explaining the generation of HPW for recycling purposes, but in fact, this subject is related to a larger research field which is forecasting the generation of municipal solid waste (MSW). Forecasting is a necessity for the development of waste management infrastructures. It is also important for the improvement and optimization of logistics associated to waste management. Therefore, reliable data and precise forecasts are needed in order to avoid cases of insufficient or excessive waste disposal and high or low usage of infrastructures (transportation, processing, incineration or landfilling).

A review of previously published approaches [2] revealed a great amount of methods applied to forecast MSW generation. The methods referred in the review range from purely application-oriented models to very sophisticated tools, and all of them can be identified in seven different categories: group comparison, correlation analysis, multiple regression analysis, single regression analysis, input–output analysis, time series analysis and system dynamics.

In their review, Beigl et al. [2] concluded that MSW generation is best predicted by time series analysis (when assessment of seasonal impacts is necessary), alongside correlation and regression methods. In respect to the waste generation contributing factors, Wang and Nie [12] stated that a rapid growth of the urban population and gross domestic product (GDP) were the most important ones. In an attempt to forecast MSW generation based on more factors besides the previously mentioned ones, linear regression models have been employed since the 1950's. Grossman et al. [5] enhanced forecasting methods by including in the linear regression model other factors such as: increase of population, income level and housing type. Later studies pointed out that waste generation can be related to predicted production level and consumption [3, 7], and also to private consumption [4]. More detailed analysis showed the growth of the urban population to have a greater impact than GDP on the total amount of MSW produced. Also, with factors like the increasing income and the quality of life, MSW seems to change more in composition rather than increasing in total amount generated. Other factors that may influence the generation and composition of waste are the average living standards or the average people's income, climate, living habits, level of education, religious and cultural beliefs, and social and public attitudes [1, 6].

Usually, time series forecasting models may be employed to predict MSW generation when there is access to significant amount of historical data. This method does not rely on the estimation of the social and economic factors, which can be a not so accurate procedure. Based on the comparison of several analysed forecasting methods, Beigl et al. [2] imply that a forecasting tool based on the relationship between social-economic conditions and the amount of waste generated was more suitable than a single time series analyses. In most cases, the application of modelling methods such as correlation, regression analyses, and group comparisons, seems to be the better option when the goal is to test the relationship between the level of affluence and the generation of total MSW or a material-related fraction, and to identify significant effects of waste management activities on recycling quotas. The application of time series analyses and input–output analyses is advantageous when there is a need for special information (i.e., assessment of seasonal effects for short-term forecasts). Sorting analyses are indispensable, if impacts on the quantity of separately collected waste streams (i.e., of recyclables) are to be quantified.

After reviewing several studies from other authors within the subject of MSW or HPW generation forecasting, it was clear that all of them focused on analysing at a different level that did not match our intent, which was predicting waste generation for each collection point in a collection network, so the emphasis is on the operational level of waste collection. Our study surely pursues a different depth or degree of analysis and a more problem-specific approach, but some studies found in the literature will certainly be helpful in the process of solving the forecasting problem we presented in the previous chapter.

4   Methodology

The data we accessed consists of records with a registry of all waste collections performed by Braval in all six municipalities they operate. In more detail, these records show how much times each ecoponto in Braval's network was emptied each month during the year of 2013. Our aim is to predict, with the smallest error possible, the number of times each ecoponto needs to be emptied each year. Therefore, this number of collections per year (and per ecoponto), hereafter referred as CPY, is the dependent variable to be considered in the forecasting models to be developed. In the next subsections, a brief classification of the developed models is given, followed by a more detailed description of each one.

4.1   Model Classification

The forecasting models yet to be presented in this study can be categorized on different aspects according to "(...) four characteristic classification criteria: regional scale, type of modelled waste streams, type of independent variables and modelling method." as stated by [2]. In terms of regional scale, our modelling approach is certainly between household scale and settlement area scale. In respect to the type of modelled waste streams, and following the concepts referred by [2], the waste streams to be modelled are collection streams. More specifically, the aim is to model source separated waste streams related to recyclable materials such as paper/cardboard, plastics/metals and glass.

Regarding the data sources for the dependent variable, CPY, these are solely based on waste collection statistics with information extracted from Braval's reports. As for the independent variables, which are factors for the prediction of waste generation, we intend to rely only on easily accessible information that is manly included in waste management related infrastructure data sources, as well as on simple socio-economic and demographic data. Other information may be collected using some Geographic Information System (GIS). The factors we believe to be of relevance are listed in table 1. We expect these factors might help explain the behaviour of the dependent variable. The factors will be included in the two modelling methods presented in the following subsections: Multiple Regression and Artificial Neural Networks.

Table 1. List of considered contributing factors for HPW generation

Factor	Description	Acronym
1	Number of *Ecopontos* per civil parish	NE
2	Population Density (residents per square-kilometre) in the civil parish	PD
3	Number of Residents per *Ecoponto*	NRE
4	*Ecoponto* Density (number of ecopontos per square-kilometre)	ED
5	Ecoponto Type (containers/bins can be at street level (Type 1) or underground (Type 2)	ET
6	Ecoponto Position (containers/bins placed within an enclosed area (i.e. a school) or placed in open area	EP
7	Ecoponto Capacity (capacity of the containers/bins)	EC
8	Number of Ecopontos within a **300** metres radius around considered ecoponto	NE 300
9	Demographic Factor (household density around each ecoponto in a 300 metres radius around it)	DF
10	Socio-Economical Factor (based on the number of schools, local management infrastructures, quantity of commercial activities, local attractions and relevant monuments, tourism and lodging infrastructures, leisure and sports infrastructures, restaurants, cafés and bars)	SEF

4.2   Multiple Regression Model

Regression models are widely used for predicting purposes, and have proven to be a very efficient method in many studies. The first modelling method we decided to explore was a linear regression model. Since we selected several independent variables (or factors) to explain the outcome of the dependent variable, we shall employ a multiple regression model (MRM).

In order to design the model, we started by assembling all available data in order to obtain values for all previously mentioned factors (table 1). As stated earlier, waste collection data and records were made available by Braval. Simple demographics were also found on Braval's data sources. Most of socio-economic information was obtained using GIS software tools such as Google Earth and Google Maps. The point was to use easily accessible informations to apply in forecasting models.

4.3   Artificial Neural Network Model

Even though our first choice of a modelling method relied on linear regression models, we also intended to explore other forecasting methods. We opted to develop a method based on an Artificial Neural Network (ANN), which is a machine learning technique that includes algorithms able to create some kind of artificial intelligence with learning capabilities, and has been widely applied in forecasting models and with great success and acceptance by the research community. ANNs are networks of artificial nodes called artificial neurons, and their functioning is inspired on how the human brain works. ANNs can be used to model non-linear relationships between input and output data, and also to find patterns within extensive data.

For our ANN models, we opted to use the Multilayer Perceptron structure with only one hidden layer at first, since it is the most applied structure in similar studies [8, 9, 10, 13], and also because the complexity of our problem might not require the use of an extra hidden layer. Concerning the kind of ANN topology, a feed-forward network was chosen. The employed training methods were Back-Propagation and Levenberg-Marquardt.

5   Experiments and Results

An experimentation phase took place to explore two methodologies: Regressions models and Artificial Neural Network models. Our intent was two develop the best performing models for each adopted methodology, and thus compare the results achieved, so that we can determine the overall best model based on error performance, stability and robustness. Since we also want to determine the most contributing factors for the WGR, we tested several combinations between the factors presented in table 1. Therefore, we obtained several different models that allowed the assessment of each factor's relevance.

5.1   Data Processing

Before engaging on experiments, data processing needed to be done. We used a list of all ecopontos placed in two municipalities, Amares and Vila Verde, as well as specific characteristics of those ecopontos, such as their GPS coordinates, containers type and capacity, placement date, civil parish to where each ecoponto belongs, among others. Then, we combined those informations with the available collection statistics which referred to the number of performed collections for each recyclable waste type (paper/cardboard, plastic/metal and glass) during the year of 2013 (CPY). For this study we opted to consider only one waste stream, which was paper/cardboard, or hereafter simply referred to as cardboard stream. A distribution of the variable CPY for cardboard stream is represented in figure 1, with CPY ranging between 6 and 130, and the ecopontos are sorted in ten categories. For demographics, we used census informations to determine population density values for each civil parish. We also took record of how many ecopontos were placed in each parish.

Data treatment for socio-economic factors involved the use of GIS software such as Google Earth and Google Maps. These tools also supplied some extra demographic information referring to each ecoponto individually, by analysing their surroundings in a 300 metres radius. We used this specific length by indication of the Braval's Manager, since people tend to not travel longer distances in order to place their HPW in ecopontos. So, considering a 300 metres radius area of effect for each ecoponto, and with assistance from the GIS software, we where able to count the number of schools, local management infrastructures, commercial activities and establishments, local attractions and relevant monuments, tourism and lodging infrastructures, leisure and sports infrastructures, restaurants, cafés and bars. These informations translated to a specific measurement that we called Socio-Economic Factor (or SEF), with values ranging on a scale from 1 to 3. The SEF values were calculated based of six sub-factors (table 2), and following equations 1 and 2.

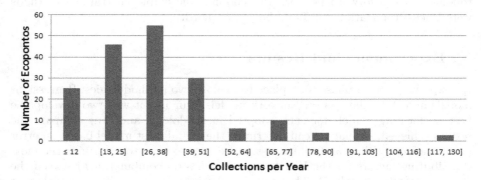

Fig. 1. Distribution of CPY for cardboard stream in 2013

Table 2. Socio-Economic sub-factors used to determine the SEF value

Factor	Description
SEF_1	Number of schools and education-related infrastructures
SEF_2	Number of local government infrastructures
SEF_3	Number of restaurants, cafés, bars, discos, bakery and cake houses
SEF_4	Number of shops, supermarkets, other trading-goods facilities
SEF_5	Number of hotels and other lodging facilities
SEF_6	Number of local attractions, monuments, parks, leisure and sports infrastructures

$$x = SEF_1 + SEF_2 + SEF_3 + SEF_4 + SEF_5 + SEF_6 \qquad (1)$$

$$SEF = \begin{cases} 1, & \text{if } x \leq 2 \\ 2, & \text{if } 2 > x \leq 4 \\ 3, & \text{if } x > 4 \end{cases} \qquad (2)$$

During the socio-economic analysis, we kept track of how many other ecopontos were placed within each ecoponto area of effect, which directly translates to a certain competition factor between ecopontos. We assumed that a higher competition level could mean a decrease in the affluence to each ecoponto, since there are more disposal options, and the nearest option tends to be the preferred one. This situation lead to an important factor related to demographics, and we named it Demographic Factor (DF). To measure the DF for each ecoponto, its area of effect was rated on a scale from 1 to 3 (low, medium and high), in terms of households concentration. Since the values for DF depend on human-eye perception and direct observation, this factor might be susceptible to the personal arbitration of the observer. Nonetheless, an observation rule was set, so that the DF level corresponds to value 1 when up to a third of the area of effect is filled with households; the value 2 is when then household concentration ranges from one third to about two thirds of that area; finally, the DF value 3 is given when more than two thirds of the effect area has households. The DF factor is not directly related to population density, because in this case, only the populated area around the ecopontos is considered. In addition, the considered households on the map, in a top-down perspective, can either be a sole family house or a whole building with many apartments.

Not all ecopontos from Amares and Vila Verde municipalities were used in the forecasting models. Only ecopontos with a minimum amount of 6 yearly collections were considered, resulting in a total of 185 eligible ecopontos. Although some ecopontos were removed, the NE value was not updated, since these ecopontos still existed in their respective civil parish. In the following subsections, a detailed description of the experiments done with multiple regression models and ANN models are presented. Other assumptions and hypotheses are exposed and explained, and experiment results with various models are compared. While testing the models, we focused on predicting the CPY for each ecoponto considering only one waste stream: cardboard.

5.2   Experiments with Multiple Regression Model

The experiments based on a multiple regression model, hereafter MRM, were performed using the software ForecastPro. The previously treated data was transferred to the software after the ecopontos, the lines in the table, were scrambled in such a way that a balanced distribution of values was achieved, which means the first half of ecopontos was equivalent to the second half in terms of average values for waste collections, population density (PD), ecopontos in civil parish (NE), DF and SEF. For the forecasting experiments, the MRM was constructed with all ecopontos except the least 50 ones on the list, and so, the CPY values were predicted for those remaining ecopontos. Various combinations of factors resulted in different models. At first, each factor was assessed independently, resulting in ten simple regression models (just one independent variable). Later, the factors that seemed to have greater coefficient of determination (R^2) were selected and combined aiming to produce a stronger model. We also employed an inverse process of model construction, by starting with all the factors combined, and then, taking out the ones with the lowest significance to the model, from one combination to the next one. We grouped all the developed models under the name MR1. In table 3, the most important results achieved in this first phase of experiments are presented. The considered performance measurements were R^2 and Mean Absolute Error (MAD)

Table 3. Forecast results with MR1

MODEL	FACTORS	R^2	MAD
MR1-1	DF	0,290	15,26
MR1-2	SEF	0,219	16,25
MR1-3	DF, SEF	0,403	14,3
MR1-4	DF, EC	0,318	15,02
MR1-5	DF, ED	0,325	14,55
MR1-6	DF, EP	0,290	15,28
MR1-7	DF, ET	0,397	14,27
MR1-8	DF, NE	0,362	13,8
MR1-9	DF, NE300	0,321	14,66
MR1-10	DF, NRE	0,293	15,25
MR1-11	DF, PD	0,332	14,35
MR1-12	DF, SEF, ED	0,425	14,06
MR1-13	DF, SEF, ET	0,449	13,75
MR1-14	DF, SEF, NE	0,438	13,79
MR1-15	DF, SEF, NE300	0,417	14,22
MR1-16	DF, SEF, NRE	0,409	14,24
MR1-17	DF, SEF, PD	0,430	14
MR1-18	DF, SEF, ET, NE, PD	0,488	13,26
MR1-19	DF, SEF, ET, NE	0,487	13,23
MR1-20	DF, SEF3	0,331	14,77
MR1-21	DF, SEF4	0,367	15,2
MR1-22	DF, SEF6	0,302	15,04
MR1-23	DF, ET, NE	0,460	13,39
MR1-24	DF, ET, SEF3	0,420	14
MR1-25	DF, ET, SEF4	0,584	12,14

After analysing the results, it seemed clear that some factors stood out more than others, namely DF, SEF, NE and ET. The combination of these factors resulted in a model that achieved an R^2 value of 0.487 (model MR1-19). Regarding SEF, it seemed unclear if the way it was calculated could be masking the influence of its sub-factors (SEF1, SEF2, etc.), and so we tested each individual contribution to the model. Once tested, this hypothesis revealed a better performing model which combined DF, ET and SEF_4, with an R^2 value of 0.584 (MR1-25) and slightly lower errors, but with SEF_4 having low significance level. This result hinted us into modifying the SEF formula (equation 2). We decided to analyse the data once again, looking into the table listing all the ecopontos and their values. We paid particular attention to the SEF values and came to a conclusion: if an ecoponto had all SEF sub-factors equal to zero, it was still being awarded a SEF level equal to one, therefore, the SEF formula was modified so that when that situation happens, the SEF value would be set to zero. This modification lead to a new set of experiments but it did not translated into a superior performance of the models, considering the most relevant factors and combinations, especially when including the SEF values. Also, the significance level of SEF in a single-variable MRM dropped. We concluded that in the MR-1 group, the socio-economic factor was not as important to the model as we thought in the beginning of the experimental phase.

Table 4. Forecast results with MR2

MODEL	FACTORS	R^2	MAD
MR2-1	DF	0,526	16,62
MR2-2	DF, ED	0,535	16,53
MR2-3	DF, ET	0,571	15,88
MR2-4	DF, PD	0,534	16,55
MR2-5	DF, SEF	0,536	16,51
MR2-6	DF, EC, ED, ET,NE, NE300, NRE, PD, SEF	0,643	14,89
MR2-7	DF, EC, ED, ET, NE300, NRE, PD, SEF	0,642	14,84
MR2-8	DF, EC, ED, ET, NE300, NRE, PD	0,638	14,86
MR2-9	DF, EC, ED, ET, NRE, PD, SEF	0,636	14,92
MR2-10	DF, EC, ED, ET, NRE, PD	0,636	10,44
MR2-11	DF, ET, SEF4	0,584	15,69

Although SEF showed to have less importance compared to other factors, there were other ones that seemed to have even lower effect on the outcome of the MRM predicting performance. One of them is the ecoponto positioning (EP), and it caught our attention after another data review and analysis. Although the EP values turned to be irrelevant, the fact that an ecoponto is positioned within an enclosed area (i.e. a restaurant) or inside a facility (i.e. a school) tends to shadow the influence of other factors, since that ecoponto is mainly accessible to the users of that particular area or facility. A new hypothesis consisted on evaluating the performance of the MRM after removing the ecopontos with EP value of 1 (placed inwards), and so, a new series of tests took place, but this time only considering the remaining 185 ecopontos, and keeping the least 50 in the list for prediction of CPY. The performed tests generated a new group of forecasting models named MR2, and the most relevant results achieved are presented in table 4. This last hypothesis turned out to have a positive impact on the predicting performance, and revealed other factors once considered less relevant in previous experiments, to have great influence when combined all together with DF and ET. Also, the MR2-3 model showed that with only these two factors, a strong performance was achieved in terms of R^2, with a value of 0.571, and in terms of error analysis, with lesser average error and mean absolute deviation once compared to the best performing models in the MR1 group (MR1-19 and MR1-25). This was exactly one of our goals for this study, since the aim was to develop a simple forecasting model using as few factors as possible without compromising too much the predicting performance when compared to models with several factors. Nonetheless, the overall best performing regression model is MR2-10, with a R^2 value of 0.636 and MAD equal to 10.44. All the variables used for this model presented a high level of significance, and so we can conclude the best factors to use when predicting CPY using regression methods are: DF, EC, ED, ET, NRE and PD. The MR2-10 model is represented in equation 4. The parameter estimates for this model are presented in table 5.

Table 5. Parameter estimates for the best performing regression model - MR-10

Factor	Coefficient	Std. Error	t-Statistic	Significance
DF	18.208770	1.889130	9.638708	1.000000
EC	-0.006169	0.001458	-4.232523	0.999977
ED	8.313962	2.404589	3.457540	0.999455
ET	21.141376	3.716969	5.687800	1.000000
NIE	0.034961	0.011332	3.085258	0.997966
PD	-0.026852	0.009635	-2.786869	0.994678

$$CPY_i = 18.209 \cdot DF - 0.006 \cdot EC + 8.314 \cdot ED + 21.141 \cdot ET + 0.035 \cdot NRE - 0.027 \cdot PD \qquad (3)$$

5.3   Experiments with Artificial Neural Network Model

The development and testing of ANN models was achieved using a software tool called Neuro Solutions 6. This software offers several options in terms of ANN model construction and structuring, as well as a wide variety of training methods. For the experiments, we chose the main model characteristics based on previous studies on forecasting MSW using ANN methods [8, 9, 10, 13]. Our ANN models have the following base characteristics:

- Network Type: MLP
- Network Structure: 3 layers (input, output and one hidden layer)
- Network Topology: Feed-forward
- Training samples: 125
- Cross-validation samples: 10
- Testing samples (forecast): 50

There are several parameters that need to be tuned such as the number of neurons or processing units per layer in the ANN. Other parameters are related to the training process, and there are different options in the software for the training algorithm (learning rule) and activation functions for the neurons in the hidden and output layers. For ANN training, we opted to experiment two algorithms: Levenberg-Marquardt (L-M) and Back-propagation (BP). As for the activation functions, we opted to explore the tangent hyperbolic and sigmoid functions. The number of epochs, or training periods was set to 1000, which determines the stopping criterium for the training process. We also tested a value of 2000 to see if with a longer training could improve ANN performance. There are many possible combinations for these parameters. The most interesting ANN models and their respective results are showed in table 5. Although we also tested ANN models using the BP algorithm, the results performed obtained were worse than the models using M-L, and so we did not include them in table 6.

An analysis over the presented results showed interesting outcomes. Considering the performance indicators R^2 and MAD, the overall best model is ANN-7, yet similar results were achieved by other models such as ANN-2, ANN-3

Table 6. Forecast results with ANN models

MODEL	ANN Config.	Activ. Function	Epochs	FACTORS	R^2	MAD	Max. Error
ANN-1	2-4-1	Tanh	1000	DF,SEF	0.624	9.80	60.34
ANN-2	2-8-1	Tanh	1000	DF,SEF	0.669	9.59	50.88
ANN-3	2-4-1	Sigmoid	1000	DF,SEF	0.672	9.55	55.28
ANN-4	2-8-1	Sigmoid	1000	DF,SEF	0.649	9.74	63.78
ANN-5	4-4-1	Tanh	1000	DF,ET,NE,SEF	0.577	11.29	50.56
ANN-6	4-4-1	Tanh	2000	DF,ET,NE,SEF	0.467	12.57	49.21
ANN-7	4-4-1	Sigmoid	1000	DF,ET,NE,SEF	0.672	9.14	64.46
ANN-8	4-4-1	Sigmoid	2000	DF,ET,NE,SEF	0.547	11.78	47.71
ANN-9	4-8-1	Sigmoid	2000	DF,ET,NE,SEF	0.682	9.19	65.06
ANN-10	6-12-1	Tanh	1000	DF,EC,ED,ET,NRE,PD	0.588	10.68	47.68
ANN-11	6-12-1	Tanh	2000	DF,EC,ED,ET,NRE,PD	0.662	11.07	42.04

and ANN-9. Evaluating other error measurements such as maximum absolute error, the ANN-11 model has the lowest score. So, to determine the overall best ANN model we opted to select ANN-7 by giving priority to the MAD value. The factors used for this model were only four: DF, ET, NE and SEF. We believe to have reached our goals with this ANN model, with simplicity in terms of factors used, and with a promising performance.

5.4   Multiple Regression and ANN Model Comparison

The experiments conducted with the two chosen methodologies allow the comparison of regression and ANN models at predicting HPW generation, or more specifically at predicting the number of waste collections needed for each collection point in a certain network. The best regression model we developed, MR2-10, performed quite well. On the other hand, the best ANN model, ANN-7, achieved better performance than MR2-10 by obtaining higher R^2 lower MAD values. In terms of maximum absolute error in the testing sample, MR2-10 got 47.83, which is better than ANN-7 with 64.46. In figures 2 and 3, the distribution of error values within the sample (135 ecopontos) used to designed both the regression model and ANN model. In figure 4, a graphical representation of the predicted values with both ANN-7 and MR2-10 models is presented.

We considered ANN-7 the best performing model, but in ANN-3, where only two factors were applied (DF and SEF), the superior capacity of ANNs to understand relationships between variables is very clear ($R^2 = 0.672$), once compared to regression models with the same factors (i.e. MR2-5 with R^2 of 0.536). Although ANN models can achieve better performance than linear regression models, the modelling set-up was more difficult to manage, since there are many parameters involved that can greatly influence the outcome and overall success of the prediction process. Designing a neural network, deciding on the training algorithms and correctly tuning the parameters can be time consuming.

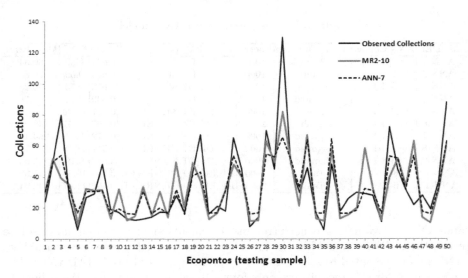

Fig. 2. Forecast results with the best ANN and regression models

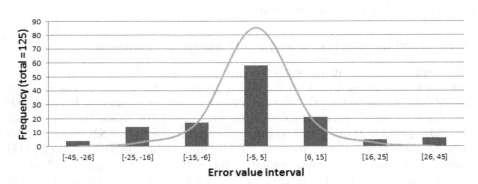

Fig. 3. Error distribution with MR2-10 model

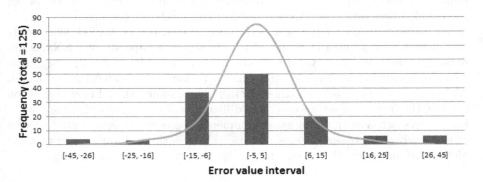

Fig. 4. Error distribution with ANN-7 model

6   Conclusions and Future Work

In this study we presented a real-world problem faced by a company that collects House Packaging Waste (HPW) for recycling stored along a network of waste collection points (ecopontos). Our main goal was to develop a forecasting model in order to predict the number of waste collections per year for each ecoponto in the network. Other purpose of this study was to evaluate which are the most important factors for HPW generation. To accomplish that, we developed Multiple Linear Regression (MLR) and Artificial Neural Network (ANN) models. The two methodologies were tested and compared against each other. The best models developed with each methodology achieved similar performances, with a slight advantage for the ANN model, with R^2 equal to 0.672 and MAD of 9.14.

Both ANN and regression models failed to explain around 35% of the dependent variable CPY (32.8% ANN and 36.4% MLR), and that may be due to a random component that we were not able to unveil, probably because the available data did not included the quantities of waste collected at each ecoponto, and since an ecoponto can be collected when it is only half-full, the number of yearly collections we predicted may not be accurately related to waste generation rate (considering completely filled ecoponto). This missing data could help at determining a more accurate waste generation rate, based on the exact waste quantities collected instead of the number of collections made. In addition, some of the factors, such as DF and SEF, can be affected by human-error.

For future work, since MLR and ANN models have similar performances but are very different from each other in terms of methodology, we intend to combine the predictions from both models trying to improve prediction accuracy. Other hypothesis could also be tested to achieve better results for both MLR and ANN models, for example:

- Add more detail to the SEF sub-factors (check incompleteness).
- Test different ways of determining SEF.
- To correct the estimates of CPY for seasonality.

Regarding the future applications, it is our intention to label each ecoponto with a collection priority level, based on their HPW generation rate, which can be determined using the predictions of yearly waste disposals. Other relevant use for the forecasting models would be to analyse possible new locations for ecopontos along an existent network, or when designing a new collection network.

Acknowledgments. This work has been supported by FCT – Fundação para a Ciência e Tecnologia within the Project Scope: PEst-OE/EEI/UI0319/2014.

References

1. Bandara, N.J.G.J., Hettiaratchi, J.P.A., Wirasinghe, S.C., Pilapiiya, S.: Relation of waste generation and composition to socio-economic factors: A case study. Environmental Monitoring and Assessment 135(1-3), 31–39 (2007)
2. Beigl, P., Lebersorger, S., Salhofer, S.: Modelling municipal solid waste generation: A review. Waste Management 28(1), 200–214 (2008)

3. Bruvoll, A., Spurkland, G.: Waste in Norway up to 2010, reports 95/8. Statistics Norway (1995)
4. Coopers and Lybrand: Cost-Benefit Analysis of the Different Municipal Solid Waste Management Systems: Objectives and Instruments for the Year 2000. Brussels, Belgium: European Commission DG XI, final report (1996)
5. Grossman, D., Hudson, J.F., Mark, D.H.: Waste generation methods for solid waste collection. Journal of Environmental Engineering-ASCE 6, 1219–1230 (1974)
6. Marquez, M.Y., Ojeda, S., Hidalgo, H.: Identification of behavior patterns in household solid waste generation in Mexicalis city: Study case. Resources, Conservation and Recyclin 52(11), 1299–1306 (2008)
7. Nagelhout, D., Joosten, M., Wierenga, K.: Future waste disposal in the Netherlands. Resources, Conservation and Recycling 4(4), 283–295 (1990)
8. Noori, R., Abdoli, M.A., Jalili GhaziZade, M., Samieifard, R.: Comparison of Neural Network and Principal Component Regression Analysis to Predict the Solid Waste Generation in Tehran. Iranian J. Publ. Health 38(1), 74–84 (2009)
9. Pham, D., Liu, X.: Neural Networks for Identification, Prediction and Control. Springer, Berlin (1995)
10. Shahabi, H., Khezri, S., Ahmad, B.B., Zabihi, H.: Application of Artificial Neural Network in Prediction of Municipal Solid Waste Generation (Case Study: Saqqez City in Kurdistan Province). World Applied Sciences Journal 20(2), 336–343 (2012)
11. Skovgaard, M., Hedal, N., Villanueva, A.: Municipal Waste Management and Greenhouse Gases. European Topic Centre on Resource and Waste Management, working paper 2008/1. In: European Topic Centre on Sustainable Consumption and Production. Publications (2008)
12. Wang, H.T., Nie, Y.F.: Municipal solid waste characteristics and management in China. Journal of the Air and Waste Management Association 51(2), 250–263 (2001)
13. Zade, M., Noori, R.: Prediction of Municipal Solid Waste Generation by Use of Artificial Neural Network: A Case Study of Mashhad. Int. J. Environ. Res. 2(1), 13–22 (2008)

Distributions Families in Counting Bacteria for Compound Sampling

Miguel Felgueiras, João Paulo Martins, and Rui Santos

School of Technology and Management, Polytechnic Institute of Leiria and
Center of Statistics and Applications of University of Lisbon, Leiria, Portugal
{mfelg,jpmartins,rui.santos}@ipleiria.pt

Abstract. The sensitivity and the specificity of a compound test depend on the distribution underlying the phenomenon. In this paper we consider count distributions unified under Panjer recursive formula and belonging to the Morris family, which verifies useful properties. The influence of the tail weight of the count distributions (that varies among infected and uninfected elements), evaluated in terms of the dispersion index, is investigated in the sensitivity and the specificity of the compound test.

Keywords: Compound tests, Panjer system, NEF–QVF.

1 Introduction

Dorfman seminal work [2] in the identification of the American soldiers infected with syphilis during World War II was the first work published in compound analysis. Dorfman pioneer idea was minimizing the costs of performing a large number of blood tests by mixing the blood of n individuals. A negative result of a compound test implies that no one is infected and therefore $n-1$ tests are spared. By the other hand, a positive result means that at least one of the members is infected and therefore individual tests should be performed, increasing in one the total number of performed tests for the group ($n + 1$ tests are performed). The optimal batch size minimizes the total number of tests, and therefore the total cost since the mixing cost is usually irrelevant. However, Dorfman methodology does not take into account classification errors (measured by the test sensitivity and specificity) and it is restricted to a binary classification of the individuals (presence or absence of some substance in the analysed fluid). To overcome the above limitations, different methodologies were proposed by several authors. Boswell *et al* [1] presents an extensive work on this subject, including different applications of these methodologies. Later, Kim *et al* [6] also tackles the group testing application with a more modern approach. Finally, Hwang [5] and Santos *et al* [13] deal with the dilution problem. The dilution problem is quite relevant since, when measuring a quantitative variable, the concentration of some type of bacteria in blood (for instance) will be diminished by the mixing process (cf. [15] for the HIV problem).

Accordingly, and after a brief revision of some families of distributions (section 2), the connection to the dispersion index concept is explained (section 3).

B. Murgante et al. (Eds.): ICCSA 2014, Part III, LNCS 8581, pp. 539–551, 2014.

Afterwards, the appropriate models for counting bacteria are described (section 4), and the application of tests for over and under dispersion data is discussed in section 5. Section 6 provides a revision of the most applied measures of misclassification in the compound analysis context. The performance of the proposed score test is analysed for different count distributions considering healthy and non healthy populations (section 7). Finally, the main conclusions are displayed in section 8.

2 Panjer and Natural Exponential with Quadratic Variance Functions Families of Distributions

In this section we present an overview of some families of distribution which unify standard counting distributions.

2.1 Panjer Family

After some preliminary work on the aggregate claims distribution, the Panjer recursive formula was simultaneously presented in Panjer [11] and in Sundt and Jewell [14], again in the aggregate claims context and when the recursion formula holds for $n \geq 0$. Later, Willmot et al [16] extended the Panjer family for distributions where the formula can only be applied when $n \geq 1$, and finally Hess et al [4] identify all distributions from the Panjer class of order k with arbitrary k, $n \geq k$. The Panjer family is therefore characterized by the Panjer recursion formula,

$$\mathrm{P}(N = n + 1) = \left(a + \frac{b}{n+1} \right) \mathrm{P}(N = n), \ n = k, \ k+1, \dots,$$

where N is a counting variable, $k \in \mathbb{N}_0$ and $a + b > 0$. When $k = 0$, we obtain the binomial (B), the negative binomial (NB) and the Poisson distributions (Po). For these distributions, the mean value is

$$\mathrm{E}(N) = \sum_{n=0}^{\infty} n\mathrm{P}(N = n) = \sum_{n=1}^{\infty} n \left(a + \frac{b}{n} \right) \mathrm{P}(N = n - 1) =$$

$$= a \sum_{n=1}^{\infty} \mathrm{P}(N = n - 1)(n - 1 + 1) + b = a\mathrm{E}(N) + a + b.$$

Note that $a = 1$ leads to $b = -1$, which isn't an interesting combination of parameters (see Table 1). In order to simplify we will consider from now on $a \neq 1$. Therefore,

$$\mathrm{E}(N) = \frac{a+b}{1-a}.$$

A similar approach can be applied in order to obtain the variance, leading to

$$\mathrm{V}(N) = \frac{a+b}{(1-a)^2}.$$

2.2   Natural Exponential Families with Quadratic Variance Functions

The natural exponential family with quadratic variance functions (NEF–QVF) was introduced by Morris [8] and several of its properties derived in [9]. Later, Morris and Lock [10] provided an elegant way to represent the NEF–QVF family within the NEF family. Under the exponential family, the density or probability mass function has the form $f(x) = \exp\left[A(x)B(\theta) + C(x) + D(\theta)\right]$, where θ is the parameter of interest. The natural exponential family has the restriction of $A\,(.)$ linear and, finally, the NEF–QVF is a subset of the natural exponential family where

$$\sigma^2 = v_0 + v_1\mu + v_2\mu^2\,,$$

that is, the variance σ^2 is at most a quadratic function of the mean μ.

There are only six NEF–QVF distributions, and among them the discrete ones are the binomial, the negative binomial and the Poisson distributions. Thus, discrete NEF–QVF distributions are the only ones that fulfil Panjer recursion formula with $k = 0$. However, the NEF–QVF distributions can be extended under linear transformation and convolution. If X_i, $i = 1, ..., n$, are independent and identically distributed (i.i.d.) random variables (r.v.s) with NEF–QVF distribution, then

$$Y = \sum_{i=1}^{n} \frac{(X_i - \theta)}{\beta}, \; \beta > 0$$

is also a NEF–QVF with variance given by

$$\sigma_Y^2 = v_0^* + v_1^*\mu_Y + v_2^*\mu_Y^2\,,$$

where v_0^*, v_1^*, and v_2^* are directly connected with the original v_0, v_1, and v_2 (cf. [9]).

3   Connecting Panjer Family, NEF–QVF and Dispersion Index

A tail weight measure for finite variance distributions, known as the dispersion index, is introduced in this section. Moreover, this index is investigated for some specific distributions belonging to the families of distributions stated in the previous section.

3.1   Dispersion Index

The dispersion index is defined as

$$CD = \frac{\sigma^2}{\mu}\,.$$

Furthermore, if $CD > 1$ the distribution is called over-dispersed and if $CD < 1$ the distribution is called under-dispersed. For the Panjer family with $k = 0$, and parameters a and b, hereafter denoted as Panjer$(a, b, 0)$, the CD is given by

$$CD = \frac{1}{1 - a}\,.$$

3.2   Connecting the Dots

For the Panjer$(a, b, 0)$ family, the parameters a and b can be computed using the standard parameters of each distributions (see Table 1), and therefore $\text{CD}_\text{B} = 1 - p < 1$, $\text{CD}_\text{Po} = 1$ and $\text{CD}_\text{NB} = \frac{1}{p} > 1$. Thence, the binomial distribution is under-dispersed and the negative binomial distribution is over-dispersed. The Poisson distribution represents the turning point between under and over dispersed distributions. Furthermore, $a < 0$ implies under-dispersion and $a > 0$ implies over-dispersion. Moreover, CD is an increasing and monotonous function of a.

Table 1. Relationship between the parameters of the Panjer family and the usual count distributions parameters

	$\text{B}(n, p)$	$\text{Po}(\lambda)$	$\text{NB}(r, p)$
a	$-p(1-p)^{-1}$	0	$1-p$
b	$(n+1)p(1-p)^{-1}$	λ	$(r-1)(1-p)$
$a+b$	$np(1-p)^{-1}$	λ	$r(1-p)$

The NEF–QVF parameter v_2 is also connected with the distribution CD (see Table 2). Besides, the variance can be expressed in a simple formula as

$$\sigma^2 = \mu + v_2\mu^2. \tag{1}$$

Clearly, $v_2 < 0$ entails under-dispersion and $v_2 > 0$ implies over-dispersion. Further, CD is an increasing and monotonous function of v_2. In conclusion, a and v_2 hold the same type of information about the dispersion index (under or over dispersion).

Table 2. The NEF–QVF parameters

	$\text{B}(n, p)$	$\text{Po}(\lambda)$	$\text{NB}(r, p)$
v_0	0	0	0
v_1	1	1	1
v_2	$-n^{-1}$	0	r^{-1}

4   Counting the Number of Bacteria

Let us consider that the number of bacteria[1] Y_i in $1\,ml$ of blood from an uninfected individual is described by $Y_i \sim \text{Panjer}(a, b, 0)$. Thus, the total number of bacteria in $n\,ml$ of an uninfected pool sample will be given by the r.v.

---

[1]   Note that this type of problem is not restricted to the count of bacteria. Any numeric characteristic measurable in a blood or urine analysis, or even in another mixable fluid or gas, can be analysed.

$B_n = \sum_{i=1}^{n} Y_i$, where Y_i are i.i.d. and n is the sample size. Clearly, B_n distribution belongs to the NEF–QVF family and hence to the Panjer system. For a homogeneous mixture, if we randomly retrieve $1\,ml$ of this amalgamated sample then its number of bacteria is given by the r.v. $B_1 \sim B(B_n, 1/n)$. Moreover, this hierarchical model (cf. [13]) remains at the Panjer family as well (see Table 3).

Table 3. Compound sample distribution

Y_i distribution	$B(m, p_1)$	$Po(\lambda)$	$NB(r, p_1)$
B_n distribution	$B(nm, p_1)$	$Po(n\lambda)$	$NB(nr, p_1)$
B_1 distribution	$B\left(nm, \frac{p_1}{n}\right)$	$Po(\lambda)$	$NB\left(nr, \frac{np_1}{np_1+1-p_1}\right)$
a for B_1 distribution	$-p_1(n-p_1)^{-1}$	0	$(1-p_1)(np_1+1-p_1)^{-1}$
b for B_1 distribution	$(nm+1)p_1(n-p_1)^{-1}$	λ	$nr(1-p_1)(np_1+1-p_1)^{-1}$

The main problem arises whenever there are infected individuals in the group. In these cases the number of bacteria X_i of $1\,ml$ of blood from an infected individual is characterized by a different distribution. Otherwise, it would not be possible to distinguish between an infected individual and an uninfected individual by the number of bacteria. Thus, the number of bacteria B_n in the pool sample (with at least one infected individual and one uninfected individual) does not belong to the Panjer family. For instance, if the number of bacteria in $1\,ml$ of blood from an uninfected individual is given by a Poisson distribution and by a negative binomial (over dispersed) distribution in an infected individual, then the distribution of the total number of bacteria in the pool sample no longer belong to the Panjer $(a, b, 0)$ family. When a sample of size n is divided in $I^{[n]}$ infected and $n - I^{[n]}$ uninfected elements, with $1 \leq I^{[n]} \leq n-1$, then

$$B_n = \sum_{i=1}^{I^{[n]}} X_i + \sum_{i=1}^{n-I^{[n]}} Y_i \nsim \text{Panjer}(a, b, 0),$$

whenever X_i and Y_i distributions differ in more than parametrization.

5   Testing the Poisson Dispersion

In this section is presented and analysed a hypothesis tests for the dispersion, wherein the null hypothesis corresponds to the equi-dispersion. This is a constraint to the problem, but it is a logical option to assume equi-dispersion as the standard one, given that the Poisson model is the simplest one concerning the bacteria counting. Thus, the null hypothesis is

$$H_0 : a = 0 \quad \text{(Poisson Dispersion)}.$$

As we previously noted, $a = 0$ is equivalent to $v_2 = 0$ in the NEF–QVF family, thus it is indistinguishable to consider $a = 0$ or $v_2 = 0$ for the hypothesis test performance. Hence, by equation (1), $v_2 > 0$ leads to over-dispersion and $v_2 < 0$ implies under-dispersion.

5.1 Testing the Slope of a Regression Line

As stated before, within Panjer $(a, b, 0)$ family,

$$P(N = x) = \left(a + \frac{b}{x} \right) P(N = x - 1)$$

and therefore

$$\frac{P(N = x)}{P(N = x - 1)} x = ax + b. \tag{2}$$

Let us assume, as a consequence of the Law of Large Numbers, that $P(N = x) \approx \frac{n_x}{n}$ where n_x is the absolute frequency of x in a set of pooled samples. Applying this approximation in equation (2),

$$\frac{n_x}{n_{x-1}} x = ax + b,$$

and thence a regression model may be applied to the points $\left(x, \frac{n_x}{n_{x-1}} x \right)$ and then, according with the confidence interval for the parameter a:

- if the confidence interval for a contains 0, a Poisson model is the more suitable to the data;
- if the confidence interval for a is above 0, a negative binomial model is the more suitable to the data;
- if the confidence interval for a is below 0, a binomial model is the more suitable to the data.

Note that, for instance, we can consider that a is above 0 but even though the negative binomial model can reveal lack of fit. This can happen when the bacteria distributions for individuals are not *i.i.d.* random variables.

5.2 Using a Score Test

Ghahfarokhi *et al* [3] and Molla and Muniswamy [7] claim that under H_0 and using a score test it is possible to obtain an assintotically distributed variable, since

$$Z = \sum_{i=1}^{k} \frac{\left(B_{1i} - \overline{B_1} \right)^2 - B_{1i}}{\sqrt{2} \ \overline{B_1}} \overset{\circ}{\sim} N(0, 1) \tag{3}$$

leads to the locally most powerful test, considering k as the number of pooled subsets where each subset has n^* individuals, B_{1i} the number of bacteria obtained in $1\,ml$ of each amalgamated sample and $\overline{B_1}$ the B_1 data mean. The pooled sample of size $n = n^* \times k$ will be called as the big sample. For the alternative hypothesis, it can be considered

$$H_1^1 : a > 0 \quad \text{or} \quad H_1^2 : a < 0.$$

6   Sensitivity and Specificity

Consider the problem of estimating the prevalence rate of some disease. Let $W_i = 1$ denote an infected individual (with a number of bacteria model by X_i) and $W_i = 0$ denote a uninfected individual (with a number of bacteria model by Y_i). In addition, let W_i^+ stand for a positive test result and W_i^- stand for a negative test result. In order to assess the sources of error two measures will be considered: sensitivity and specificity.

Let W_i be individually tested. As it is usual, the probability

$$\varphi_s = \mathrm{P}\left(W_i^+|W_i = 1\right)$$

defines the test sensitivity and

$$\varphi_e = \mathrm{P}\left(W_i^-|W_i = 0\right)$$

the test specificity. Nevertheless, if a pooled or a group test is performed then the probability of having a positive result from an infected group (a group with at least one infected individual) may be lower than φ_s, due to the dilution effect (see [5]). Thus, Kim *et al* [6] defines the pooling sensitivity $\varphi_s^{[n]}$ as the probability of getting a positive result from an infected group, i.e.

$$\varphi_s^{[n]} = \mathrm{P}\left(W^{[+,n]}|\ \sum_{i=1}^{n} W_i \geq 1\right) = \mathrm{P}(W^{[+,n]}|\ \mathrm{I}^{[n]} \geq 1),$$

where $W^{[+,n]}$ denotes a positive compound result from a group with size n. Analogously, the pooling specificity $\varphi_e^{[n]}$ is defined as the probability of getting a negative compound result $W^{[-,n]}$ from a uninfected group (no infected element within the group), i.e.

$$\varphi_e^{[n]} = \mathrm{P}\left(W^{[-,n]}|\ \sum_{i=1}^{n} W_i = 0\right) = \mathrm{P}(W^{[-,n]}|\ \mathrm{I}^{[n]} = 0).$$

Thereby, the probability of getting a negative outcome in a uninfected group sample is equal to φ_e as there is no dilution problems, because the amount of bacteria in each individual within the group is described by i.i.d. r.v.s. Nevertheless, $\varphi_s^{[n]}$ depends on the group size n and on the number of infected members in the group $\mathrm{I}^{[n]}$ due to the dilution and consequent rarefaction of the number of bacteria. To simplify, let us consider that the quantity of bacteria found in a ml of blood from an infected individual is expected to be higher than from an uninfected individual. Moreover, to perform the compound test $1\,ml$ of blood from each of the n individuals within the group are pooled and $1\,ml$ of this pooled blood is analysed in order to perform the compound test. Consequently, if the blood of n (with n large) is pooled and also $\mathrm{I}^{[n]} = 1$, then the higher number of bacteria in the infected blood will be diluted in the blood of the others uninfected individuals and consequently the probability of the test correctly reveal the presence of an infected individual may be quite low. In order to model these compound probabilities Santos *et al* [13] define the sensitivity conditional to j infected individuals in the group as $\varphi_s^{[j,n]} = \mathrm{P}(W^{[+,n]}|\ I^{[n]} = j)$. This rarefaction factor can be added in the $\varphi_s^{[n]}$ computation, as a consequence of the Total Probability Theorem, obtaining

$$\varphi_s^{[n]} = \sum_{j=1}^{n} \varphi_s^{[j,n]}\ \mathrm{P}(I^{[n]} = j|\ I^{[n]} \geq 1)\,.$$

Moreover, as for low prevalence rates $\mathrm{P}(I^{[n]} = 1|\ I^{[n]} \geq 1) \approx 1$, then $\varphi_s^{[n]} \approx \varphi_s^{[1,n]}$ in most applications of compound test analysis. Although it may be difficult to screen a infected pool sample as positive, Santos *et al* [12] showed, under the application of continuous r.v.s., that these measures depend on the prevalence rate p of the infection, the group dimension n and the right tail weight of the continuous distribution that describes the phenomenon. Moreover, it concluded that compound test can be applied with quite low probability of misclassification whenever heavy-tailed distributions and low prevalence rates occur. However, no analysis was performed for counting distributions.

Furthermore, Santos *et al* [13] define the concepts of specificity and sensitivity of some specific methodology of classification or estimation $\mathcal{M}$ (based in hierarchical algorithms or in array-based group testing), which depend on the pooling sensitivity and pooling specificity measures previously defined. These measures assess the quality of an outcome provided by some methodology $\mathcal{M}$. Thus, the methodology sensitivity (or procedure sensitivity) is the probability of an infected individual being correctly identified by the methodology $\mathcal{M}$, that is,

$$\varphi_s^{\mathcal{M}} = P_{\mathcal{M}}\left(W_i^+|W_i = 1\right)\,.$$

Analogously, the methodology specificity (or procedure specificity) stands for the probability of an uninfected individual being correctly classified by the methodology $\mathcal{M}$, that is,

$$\varphi_e^{\mathcal{M}} = P_{\mathcal{M}}\left(W_i^-|W_i = 0\right)\,.$$

7   Score Test Application

In the previous two subsections of section 5, dispersion tests which can be performed to select the underlying type of dispersion using the sample data were presented. Now, we present a small simulation study to check out the performance of the score test. In order to assess its usefulness, the obtained results are compared to Dorfman methodology. Both classification and estimation problems are considered.

7.1   Estimation Problem

The usual criteria to decide if a sample pool of size n is classified as infected or not is to use the threshold of the individual test. Assume that an individual is classified as infected if the number of bacteria exceeds some threshold

$t = F_Y^{-1}(1 - \alpha)$ where $F_Y^{-1}(1 - \alpha)$ stands for the generalized inverse of the distribution function Y in $1 - \alpha$ (a similar reasoning is applied if the rule is to declare a positive sample when the number of bacteria is lower than some threshold t).

For a perfect mixture of the individual samples, the observed number of bacteria is described by a hierarchical model as previously discussed. A pooled sample is infected if any of the individuals $W_1, \cdots, W_n$ is infected, that is, if $\sum_{i=1}^n W_i \geq 1$. As $P(\max(Y_1, \cdots, Y_n) > t) = F_Y^n(t) = 1 - \alpha$ the threshold of the pooled sample test is given by $t^{[n]} = F_Y^{-1}\left((1 - \alpha)^{\frac{1}{n}}\right)$.

For low prevalence rates and small values of k and n^*, it is likely to have only one infected element in only one sample. Thence, a positive results means (almost surely) that only one of the $k \times n^*$ individuals is infected and all the others are not infected. Thus, when the study goal is estimation, it is only required to perform an experimental test to the big samples. Thence, the individuals are gathered in the big samples for a batched testing and no further tests are performed. The innovative idea is to use the score test to decide whether the big sample is infected or not. The relative cost (expected number of tests for the classification of each individual) is given by $\frac{1}{n} = \frac{1}{kn^*}$.

7.2 Classification Problem

When the study purpose is the identification of the infected elements, and if the null hypothesis is rejected, it is required the performance of individual tests. However, as previously stated, for low prevalence rates there is (almost surely) just one infected individual in the big pooled sample. For avoiding the performance of a great number of individual experimental tests ($n^* \times k$ individual tests) we suggest restraining our attention only to the pooled sample with the highest observed value of B_1 (n individual tests) if $E(X) > E(Y)$. Or, otherwise, to the pooled sample with the lowest observed value of B_1. Hence, only the n^* individuals in that pooled sample should be individually tested. This restriction of the number of individuals tests that are performed allows attaining a lower relative cost than the cost obtained by Dorfman methodology (D_c). However, as the identification of the infected individuals depends on the performance of individual tests it is expected a sensitivity lower than Dorfman procedure sensitivity.

7.3 A Case Study by Simulation

In this section a case study is analysed by simulation, using ℝ software, in order to check out the performance of the score test for both problems: estimation and classification.

Let us consider that the number of bacteria in a ml of blood from an uninfected element is given by $Y \sim \text{Po}(\lambda)$, whereas in an infected element is given by $X \sim \text{NB}(r, p_1)$. The number of infected elements in a pooled sample with n^* elements is, as previously, denoted by the r.v. $I^{[n^*]} \sim B(n^*, p)$, where p is the prevalence rate of the infection. Thus, we began by simulating the $I^{[n^*]}$ values, and then the total number of bacteria in a pooled sample, $B_{n^*} = \sum_{i=1}^{n^*-I^{[n^*]}} Y_i + \sum_{i=1}^{I^{[n^*]}} X_i$, and thereafter the number of bacteria in $1\,ml$ of the amalgamated sample, $B_1 \sim B\left(B_{n^*}, 1/n^*\right)$. This procedure was repeated k times in order to obtain the data set $(B_{11}, B_{12}, ..., B_{1k})$. Finally, the entire data set was tested using (3).

The next results were obtained using 10^4 replicates of big samples. The relative cost (RC), i.e., the ratio between the number of expected experimental tests and the total number of individuals, the methodology sensitivity $\varphi_s^{\mathcal{M}}$, the methodology specificity $\varphi_e^{\mathcal{M}}$, the obtained estimate (Estim.) and the estimate bias (Bias) are presented for the estimation (S_e) and classification problem (S_c). For Dorfman methodology (D_c) (using the optimal pool sample size) and for a procedure that divides all the individuals in groups of size n and performs a single pooled sample test to each group (D_e), a similar simulation was performed. In this last type of procedure the estimate computed for the prevalence rate is equal to the ratio between the number of positively classified pooled samples over the number of pooled samples.

Note that when the study goal is the estimation of the prevalence rate, the sensitivity $\varphi_s^{\mathcal{M}}$ of the process can be estimated by the number of detected data sets $(B_{11}, B_{12}, ..., B_{1k})$ with infected elements among the total number of data sets with infected elements, and the specificity $\varphi_e^{\mathcal{M}}$ can be estimated by the number of detected data sets $(B_{11}, B_{12}, ..., B_{1k})$ with no infected elements among the total number of data sets without infected elements in N replicates. Table 4 presents the results for a prevalence rate $p = 0.01$.

Table 4. Simulation results for $p = 0.01$

	$\alpha = 0.1; k = 8; \lambda = 100; p_1 = 0.01$									
	$r = 5$					$r = 8$				
	S_e	S_c	D_c	D_e	D_e	S_e	S_c	D_c	D_e	D_e
	$n^* = 5$	$n^* = 5$	$n = 11$	$n = 11$	$n = 40$	$n^* = 5$	$n^* = 5$	$n = 11$	$n = 11$	$n = 40$
RC	0.0250	0.0801	0.2651	0.0909	0.0250	0.0250	0.0827	0.2678	0.0909	0.0250
$\varphi_s^{\mathcal{M}}$	0.7661	0.7821	0.8628	0.8222	0.3480	0.8229	0.8427	0.9793	0.9276	0.5494
$\varphi_e^{\mathcal{M}}$	1	0.9934	0.9782	0.9003	0.8595	1	0.9931	0.9790	0.8933	0.7726
Estim.	0.0144	0.0110	0.0302	0.0158	0.0050	0.0115	0.0152	0.0303	0.0161	0.0071
Bias	1.9×10^{-5}	1.0×10^{-6}	4.1×10^{-4}	3.2×10^{-5}	2.4×10^{-5}	2.3×10^{-6}	2.7×10^{-5}	4.3×10^{-4}	4.1×10^{-5}	9.5×10^{-6}

	$\alpha = 0.05; k = 8; \lambda = 100; p_1 = 0.01$									
	$r = 5$					$r = 8$				
	S_e	S_c	D_c	D_e	D_e	S_e	S_c	D_c	D_e	D_e
	$n^* = 5$	$n^* = 5$	$n = 11$	$n = 11$	$n = 40$	$n^* = 5$	$n^* = 5$	$n = 11$	$n = 11$	$n = 40$
RC	0.0250	0.0778	0.2107	0.0909	0.0250	0.0250	0.0827	0.2321	0.0909	0.0250
$\varphi_s^{\mathcal{M}}$	0.7580	0.7755	0.8009	0.7507	0.2549	0.8211	0.8418	0.9713	0.9158	0.4555
$\varphi_e^{\mathcal{M}}$	1	0.9967	0.9934	0.9135	0.8963	0.9967	1	0.9920	0.8932	0.8159
Estim.	0.0106	0.0110	0.0145	0.0109	0.0033	0.0111	0.0117	0.0178	0.0128	0.0052
Bias	2.9×10^{-7}	9.4×10^{-7}	2.1×10^{-5}	8.8×10^{-7}	4.5×10^{-5}	1.1×10^{-6}	2.9×10^{-6}	5.8×10^{-5}	7.2×10^{-6}	2.3×10^{-5}

	$\alpha = 0.01; k = 8; \lambda = 100; p_1 = 0.01$									
	$r = 5$					$r = 8$				
	S_e	S_c	D_c	D_e	D_e	S_e	S_c	D_c	D_e	D_e
	$n^* = 5$	$n^* = 5$	$n = 11$	$n = 11$	$n = 40$	$n^* = 5$	$n^* = 5$	$n = 11$	$n = 11$	$n = 40$
RC	0.0250	0.0778	0.2107	0.0909	0.0250	0.0250	0.0739	0.1713	0.0909	0.0250
$\varphi_s^{\mathcal{M}}$	0.7580	0.7755	0.8009	0.7507	0.2549	0.7594	0.7767	0.6820	0.6351	0.1421
$\varphi_e^{\mathcal{M}}$	1	0.9967	0.9934	0.9135	0.8963	1	0.9994	0.9990	0.9268	0.9420
Estim.	0.0106	0.0110	0.0145	0.0109	0.0033	0.0098	0.0084	0.0080	0.0073	0.0016
Bias	2.9×10^{-7}	9.4×10^{-7}	2.1×10^{-5}	8.8×10^{-7}	4.5×10^{-5}	2.5×10^{-8}	2.5×10^{-6}	5.1×10^{-6}	8.7×10^{-6}	7.2×10^{-5}

Concerning to the estimation problem, the application of the score test produces, in general, better estimates than the other procedures. The reason for this good performance of the score test lies on the fact that the procedure specificity is in practice equal to one or, at least, very near to one. This means that all non infected individuals (which are the majority of the individuals) are correctly classified. Besides, one can verify that the best results are associated to a very high methodology specificity and a moderate methodology sensitivity. On the other hand, when the classification problem is considered it is quite interesting to observe that although the ratio between the relative cost of the methodology D_c and S_c is quite high, the differences between the sensitivity and specificity methodologies are very low. This shows that the score test can be an alternative to take into account even when classification is the aim. Note that the relative cost of the big pooled sample methodology is equal to $1/40$ as only one test is performed to each big pooled sample. Table 5 provides the results for a low prevalence rate, $p = 0.005$.

Table 5. Simulation results for $p = 0.005$

	\multicolumn{10}{c}{$\alpha = 0.1; k = 8; \lambda = 100; p_1 = 0.01$}
	\multicolumn{5}{c}{$r = 5$}	\multicolumn{5}{c}{$r = 8$}
	S_e	S_c	D_c	D_e	D_e	S_e	S_c	D_c	D_e	D_e
	$n^* = 8$	$n^* = 8$	$n = 32$	$n = 32$	$n = 64$	$n^* = 8$	$n^* = 8$	$n = 32$	$n = 32$	$n = 64$
RC	0.0156	0.0629	0.1990	0.0313	0.0156	0.0156	0.0676	0.2204	0.0313	0.0156
φ_s^M	0.7056	0.7268	0.7360	0.7100	0.2384	0.8552	0.8418	0.9547	0.9240	0.3852
φ_e^M	1	0.9943	0.9838	0.9425	0.9206	0.9939	1	0.9819	0.9247	0.8758
Estim.	0.0059	0.0092	0.0197	0.0088	0.0023	0.0065	0.0103	0.0228	0.0102	0.0030
SqBias	$9.4{\times}10^{-7}$	$1.8{\times}10^{-5}$	$2.2{\times}10^{-4}$	$1.6{\times}10^{-5}$	$8.1{\times}10^{-6}$	$2.5{\times}10^{-6}$	$2.9{\times}10^{-5}$	$3.2{\times}10^{-4}$	$2.8{\times}10^{-5}$	$4.05{\times}10^{-6}$
	\multicolumn{10}{c}{$\alpha = 0.05; k = 8; \lambda = 100; p_1 = 0.01$}
	\multicolumn{5}{c}{$r = 5$}	\multicolumn{5}{c}{$r = 8$}
	S_e	S_c	D_c	D_e	D_e	S_e	S_c	D_c	D_e	D_e
	$n^* = 8$	$n^* = 8$	$n = 32$	$n = 32$	$n = 64$	$n^* = 8$	$n^* = 8$	$n = 32$	$n = 32$	$n = 64$
RC	0.0156	0.0591	0.1537	0.0313	0.0156	0.0156	0.0639	0.1729	0.0313	0.0156
φ_s^M	0.6929	0.7047	0.6724	0.6326	0.1376	0.8279	0.8437	0.9261	0.8905	0.2863
φ_e^M	1	0.9973	0.9948	0.9494	0.9559	1	0.9973	0.9940	0.9296	0.9064
Estim.	0.0054	0.0062	0.0085	0.0058	0.0011	0.0060	0.0069	0.0106	0.0071	0.0019
Bias	$2.6{\times}10^{-7}$	$1.5{\times}10^{-6}$	$1.2{\times}10^{-5}$	$6.0{\times}10^{-7}$	$1.4{\times}10^{-5}$	$1.1{\times}10^{-6}$	$3.8{\times}10^{-6}$	$3.1{\times}10^{-5}$	$4.1{\times}10^{-6}$	$1.0{\times}10^{-5}$
	\multicolumn{10}{c}{$\alpha = 0.01; k = 8; \lambda = 100; p_1 = 0.01$}
	\multicolumn{5}{c}{$r = 5$}	\multicolumn{5}{c}{$r = 8$}
	S_e	S_c	D_c	D_e	D_e	S_e	S_c	D_c	D_e	D_e
	$n^* = 8$	$n^* = 8$	$n = 32$	$n = 32$	$n = 64$	$n^* = 8$	$n^* = 8$	$n = 32$	$n = 32$	$n = 64$
RC	0.0156	0.0542	0.1114	0.0313	0.0156	0.0156	0.0607	0.1348	0.0313	0.0156
φ_s^M	0.6787	0.6987	0.4973	0.4784	0.0544	0.8451	0.8265	0.8520	0.8183	0.1423
φ_e^M	1	0.9995	0.9995	0.9638	0.9823	1	0.9995	0.9991	0.9384	0.9547
Estim.	0.0048	0.0040	0.0030	0.0030	$3.8{\times}10^{-4}$	0.0056	0.0048	0.0051	0.0045	$7.9{\times}10^{-4}$
Bias	$6.3{\times}10^{-9}$	$8.9{\times}10^{-7}$	$3.8{\times}10^{-6}$	$3.8{\times}10^{-6}$	$2.2{\times}10^{-5}$	$3.5{\times}10^{-7}$	$7.8{\times}10^{-8}$	$1.6{\times}10^{-8}$	$1.7{\times}10^{-7}$	$1.7{\times}10^{-5}$

Table 5 shows that when the prevalence rate is very low it is even more important to have a high methodology specificity in order to produce estimates with a reduced bias. Considering the estimation problem it is surprising to observe that the less costly S_c methodology can outperform the methodology D_c sensitivity in spite of the use of the optimal pooled sample size of D_c.

8   Conclusions

Compound analysis can be applied in order to save resources whenever low prevalence rates exist, although its decreasing accuracy. In quantitative compound analysis the rarefaction factor is crucial in the evaluation of its performance (cf.[12]), mainly the characterization of the sensitivity $\varphi_s^{[1,n]}$. General results concerning sensitivity and specificity, simultaneously with results regarding dispersion and its influence in pooled samples tests, can be applied making use of distributions families that include the main applied count distributions.

The score test procedure is designed to NEF–QVF distributions and uses their intrinsic characteristics, namely the dispersion index, to provide a more efficient method of estimating and classifying with a similar quality to other standard methods.

Acknowledgments. Research partially funded by FCT, Portugal, through the project Pest-OE/MAT/UI0006/2014.

References

1. Boswell, M.T., Gore, S.D., Lovison, G., Patil, G.P.: Annotated bibliography of composite sampling, Part A: 1936–1992. Environ. Ecol. Stat. 3, 1–50 (1996)
2. Dorfman, R.: The detection of defective members in large populations. Ann. Math. Statistics 14, 436–440 (1943)
3. Ghahfarokhi, M.A., Iravani, H., Sepehri, M.R.: Application of Katz Family of Distributions for Detecting and Testing Overdispersion in Poisson Regression Models. Proceedings of World Academy of Science: Engineering and Technology 44, 544–550 (2008)
4. Hess, K.T., Liewald, A., Schmidt, K.D.: An extension of Panjer's recursion. Astin Bull. 32(2), 283–297 (2002)
5. Hwang, F.K.: Group testing with a dilution effect. Biometrika 63, 671–673 (1976)
6. Kim, H., Hudgens, M., Dreyfuss, J., Westreich, D., Pilcher, C.: Comparison of group testing algorithms for case identification in the presence of testing errors. Biometrics 63, 1152–1163 (2007)
7. Molla, D.T., Muniswamy, B.: Power of Tests for Overdispersion Parameter in Negative Binomial Regression Model. IOSRJM 1(4), 29–36 (2012)
8. Morris, C.: Natural Exponential Families With Quadratic Variance Functions. Annals Stat. 10(1), 65–80 (1982)
9. Morris, C.: Natural Exponential Families With Quadratic Variance Functions: Statistical Theory. Annals Stat. 11(2), 515–529 (1983)
10. Morris, C., Lock, K.: Unifying the Named Natural Exponential Families and Their Relatives. Am. Stat. 63(3), 253–260 (2009)
11. Panjer, H.: Recursive Evaluation of a Family of Compound Distributions. ASTIN Bull. 12(1), 22–26 (1981)
12. Santos, R., Felgueiras, M., Martins, J.P.: Known mean, unknown maxima? Testing the maximum knowing only the mean. Commun. Stat. Simulat (Published online: January 23, 2014) doi:10.1080/03610918.2013.773345

13. Santos, R., Pestana, D., Martins, J.P.: Extensions of Dorfman's Theory. In: Oliveira, P.E., et al. (eds.) Recent Developments in Modeling and Applications in Statistics. Studies in Theoretical and Applied Statistics, Selected Papers of the Statistical Societies, pp. 179–189. Springer (2013)
14. Sundt, B., Jewell, W.S.: Further results on recursive evaluation of compound distributions. Astin Bull. 12(1), 27–39 (1981)
15. Tu, X.M., Litvak, E., Pagano, M.: On the informativeness and accuracy of pooled testing in estimating prevalence of a rare disease: Application to HIV screening. Biometrika 82(2), 287–297 (1995)
16. Willmot, G.E.: Sundt and Jewell's family of discrete distributions. Astin Bull. 18, 17–29 (1988)

Item Response Models in Computerized Adaptive Testing: A Simulation Study

Maria Eugénia Ferrão[1,2] and Paula Prata[1,3]

[1] University of Beira Interior, Portugal
[2] Centre for Applied Mathematics and Economics (CEMAPRE), Portugal
[3] Instituto de Telecomunicações (IT), Portugal
meferrao@ubi.pt, pprata@di.ubi.pt

Abstract . In the digital world, any conceptual assessment framework faces two main challenges: (a) the complexity of knowledge, capacities and skills to be assessed; (b) the increasing usability of web-based assessments, which requires innovative approaches to the development, delivery and scoring of tests. Statistical methods play a central role in such framework. Item response models have been the most common statistical methods used to address such kind of measurement challenges, and they have been used in computer-based adaptive tests, which allow the item selection adaptively, from an item pool, according to the person ability during test administration. The test is tailored to each student. In this paper we conduct a simulation study based on the minimum error-variance criterion method varying the item exposure rate (0.1, 0.3, 0.5) and the test maximum length (18, 27, 36). The comparison is done by examining the absolute bias, the root mean square-error, and the correlation. Hypotheses tests are applied to compare the true and estimated distributions. The results suggest the considerable reduction of bias as the number of item administered increases, the occurrence of ceiling effect in very small size tests, the full agreement between true and empirical distributions for computerized tests of length smaller than the paper-and-pencil tests.

Keywords: Item response model, computerized adaptive testing, measurement error.

1 Introduction

The importance of the educational assessment in Europe is recognized as: "assessments which are well-devised can enhance learning, whereas those that are poorly designed or unthinkingly applied can have a negative impact on students' learning progress" (Association for Educational Assessment - Europe, 2010; p.7), and also that, so far, "relatively little attention at a European level has been given to how (the quality of) tests, examinations, assessments and assessment programs used across Europe compare to one another" (op.cit. p. 27). Then, it is expected to pay more attention on the conceptual and operational framework of educational assessments in Europe. The challenges posed by the digital world, such as (a) the complexity of knowledge, capacities and skills to be assessed; and (b) the increasing usability of

B. Murgante et al. (Eds.): ICCSA 2014, Part III, LNCS 8581, pp. 552–565, 2014.

web-based assessments, requires innovative approaches to the development, delivery and scoring of tests. Statistical methods play a central role in such framework. Classical test theory [2, 3] and item response models (IRM) [4] have been the most common statistical methods used to address such kind of measurement challenges [5, 6]. IRMs are used in computer-based adaptive tests (CAT), which allow the item selection adaptively, from an item pool, according to the student's ability during test administration. Thus, the test is tailored to each student. Conceptually, because the CAT platform may capitalize on the power of web to deliver a more efficient test to a wider population (the size and composition of the population may vary), it is expected to obtain improvements on the reliability and comparability of assessment results. Despite the obvious advantages of adaptive testing, there are still some challenges such as the optimization of the test size conditional on a given level of measurement error. In addition, according to Yao (2012) [7], multidimensional CAT "can reduce the test length and increase the precision of ability estimates compared to unidimensional CAT".

In this paper we apply the minimum error-variance criterion method [8] implemented in SimuMCAT [9] to the study of the items usage and to the comparison of bias and true ability, of correlation between true and estimated ability, and also to compare their frequency distributions. The calibrated item pool refers to the Portuguese grade 6 Mathematics, extended with replications. A sample of 300 students, three test lengths of size 18, 27, and 36, and item exposure rates of 0.1, 0.3, 0.5 are considered. The remaining sections of this paper are organized as follows. Section 2 describes in detail data and methods used in this simulation study, and also used for the item pool calibration. Section 3 reports the results and complementary statistical analysis, followed by the discussion in section 4. Finally, section 5 pinpoints some directions for future work.

2   Data and Methods

The class of statistical models used in the simulation study was also used five years ago for item pool calibration. The methodological approaches, definitions and model specifications are described in the following subsections.

2.1   Data

The item pool is composed by items calibrated in the scope of the research project 3EM [10, 11] and in the project for the improvement of the instruments and scales in the Portuguese educational assessment [12], with replication of items. The calibration was conducted by applying the two-parameter logistic model for a dichotomously response. Items assess students' Mathematics ability and skills at grades from 1st to 9th of the Portuguese core curriculum. The items development was based on a reference matrix that includes the purposes of the assessment, the contents and descriptors, and some operational specifications [13], and the format of administration was paper-and-pencil. The content and descriptors derive from the national curriculum organization programme [14, 15]. For the purpose of this paper, the grade 6 is considered.

The domains are geometry, numbers and calculus, functions, and statistics. The number of items per domain is not enough to allow a domain specific modelling. Thus, for the purpose of this work the overall ability score is considered. In the academic year 2007/8, two paper-and-pencil tests, comprising the total number of 56 items, were administered to each of the 300 students, and his/her 5-level classification was given by the Ministry of Education, based on the official assessment test, as the outcome of the Portuguese educational assessment.

2.2   Methods

Simulation is used when CAT intends to reduce the length of the test that has been administered conventionally. The objective of applying the simulations is to determine how much reduction in test length is achieved by "re-administering" the items adaptively, without significant changes in the properties of the test scores. The simulation proceeds as described by L.Yao [7], [9], [16], applying the method multidimensional adaptive testing with a minimum error-variance criterion [8], and the restriction set to the number of dimensions. The simulation stops if either the required precision has been achieved or the selected number of items has reached the maximum limit. Following the Yao's notation [17], J is the number of items of known parameters in a test, $I_J(\vec{\theta})$ is the test information for a given ability score point $\vec{\theta}$. The weight for selecting item j is denoted as $\vec{w}_j$. The multidimensional three-parameter logistic model for a dichotomously scored item j is given by eq. (1). The probability of a correct response to item j for the examinee i with the ability $\vec{\theta}_i = (\theta_{i1}, \dots, \theta_{iD})$, where D is the number of domains, is P_{ij1}

$$P_{ij1} = P(x_{ij} = 1 | \vec{\theta}_i, \vec{\beta}_j) = \beta_{3j} + \frac{1-\beta_{3j}}{1+e^{(-\vec{\beta}_{2j}\circ\vec{\theta}_i^T+\beta_{1j})}}, \tag{1}$$

where $x_{ij} = 0$ or 1 is the response of examinee i to item j, $\vec{\beta}_{2j} = (\beta_{2j1}, \dots, \beta_{2jD})$ is a vector of dimension D for the item discrimination parameters, β_{1j} is the difficulty parameter, β_{3j} is the lower asymptote or the guessing parameter, and $\vec{\beta}_{2j}\circ\vec{\theta}_i^T = \sum_{l=1}^{D}\beta_{2jl}\theta_{il}$. For the purpose of this work, D=1 and $\beta_{3j} = 0$.

Algorithm for Item Selection

1. For (j=1,..J),
    Compute weight $\vec{w}_{j-1} = (w_1, \dots, w_D), w_l = \frac{1}{D}, l = 1, \dots, D$
2. Select item j=m such that $(\vec{w}_{j-1})[I_{j-1}^m(\vec{\theta}^{j-1})]^{-1}(\vec{w}_{j-1})^T$ has minimum value
3. Update ability $\vec{\theta}^j$ and information $I_j(\vec{\theta}^j)$ based on the selected j items.

Item Exposure Control

For the jth item, let r_j denote its exposure rate. For each selection step, n_j is the number of examinees to whom the item j had already been administered. The index for the item exposure control is[17]

$$f_{jl} = max\left\{\left(r_j - \frac{n_j}{N}\right)/r_j, 0\right\}, \tag{2}$$

where N is the total number of examinees.

Evaluation Criteria

For each combination of design values $r_j=0.1, 0.3, 0.5$ and $J=18, 27, 36$ the absolute bias (ABIAS) for the overall score, the root mean square error (RMSE), and the correlations between the estimates and the true scores are computed.   ABIAS and RMSE are given by eq. (3) and (4), respectively,

$$ABIAS = \frac{1}{N}\sum_{i=1}^{N}|\hat{\theta}_i - \bar{\theta}_i|, \tag{3}$$

$$RMSE = \sqrt{\frac{1}{N}\sum_{i=1}^{N}\left(\hat{\theta}_i - \bar{\theta}_i\right)^2}. \tag{4}$$

The comparison of true and estimated ability score distributions is conducted by applying the chi-square test with the statistic,

$$X^2 = \sum_{i=1}^{N}\frac{\left(\hat{\theta}_i - \bar{\theta}_i\right)^2}{\bar{\theta}_i}. \tag{5}$$

The true population distribution is the 5-level classification attributed to each 6th year student in the context of the Portuguese educational assessment. Nine 5-level classification scales are derived from the percentiles of the estimate score distributions and compared to the true population distribution.

3   Results

Table 1 presents the mean of the absolute bias for test length of 18, 27, 36 and for item exposure rates of 0.1, 0.3, 0.5. Considering the same design, the table 2 presents the root mean square error.

Table 1. Absolute BIAS for test length values of 18, 27 and 36, considering item exposure rates of 0.1, 0.3 and 0.5

	Test length		
Exposure rate	18	27	36
0.1	0.4601	0.4001	0.3616
0.3	0.4764	0.4070	0.3620
0.5	0.4623	0.4002	0.3391

Table 2. RMSE for test length values of 18, 27 and 36, considering item exposure rates of 0.1, 0.3 and 0.5

Exposure rate	Test length		
	18	27	36
0.1	0.5980	0.5311	0.4793
0.3	0.6097	0.5268	0.4568
0.5	0.6112	0.5356	0.4444

The absolute bias varies from 0.46 to 0.48 for J=18, it is approximately 0.40 for J=27, and it varies from 0.34 to 0.36 to J=36. The RMSE is approximately 0.61 for J=18, 0.53 for J=27, and it ranges from 0.44 to 0.48 to J=36. Therefore, the results show a considerable reduction of the absolute bias and of the RMSE as the test length increases.

The correlations between the ability estimates and true theta are presented in Table 3. The correlation is approximately 0.81 for J=18, 0.86 for J=27, and 0.90 for J=36. As expected, the correlation between the ability score estimates and the true ability increases as the increasing of the number of items in the test.

Table 3. Correlations for test length values of 18, 27 and 36, considering item exposure rates of 0.1, 0.3 and 0.5

Exposure rate	Test length		
	18	27	36
0.1	0.8172	0.8582	0.8854
0.3	0.8064	0.8583	0.8978
0.5	0.8106	0.8546	0.9027

To analyze the differences between the nine design sets, the bias and the ability score estimates were plotted against the true values (true theta). The pattern is similar for all item exposure rates. The graphs presented in Figures 1, 2 and 3, concern the exposure rate of 0.5 as an example. In particular the Figures 1a, 2a and 3a show the relationship between the bias and true theta and Figures 1b, 2b, and 3b show the relationship between the estimate and true theta for varying test lengths.

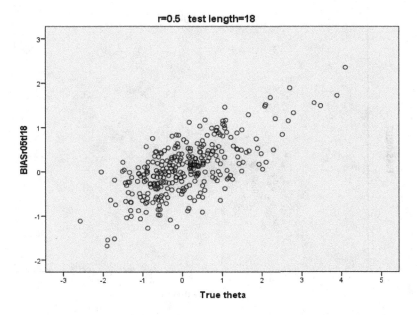

Fig. 1a. BIAS-True Theta for test length =18 and exposure rate 0.5

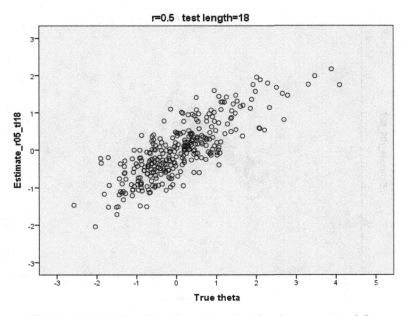

Fig. 1b. Estimate-True Theta for test length =18 and exposure rate 0.5

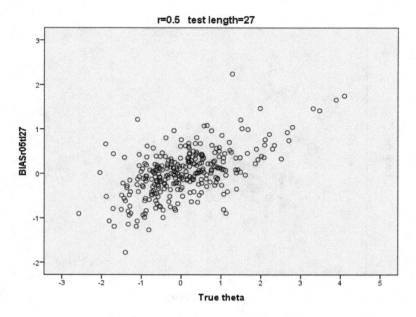

Fig. 2a. BIAS-True Theta for test length =27 and exposure rate 0.5

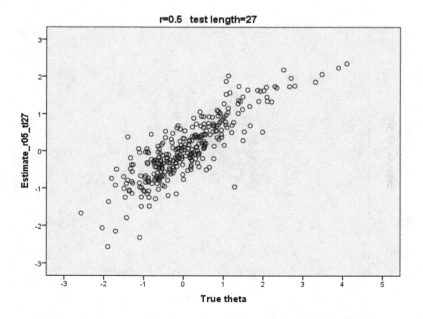

Fig. 2b. Estimate-True Theta for test length =27 and exposure rate 0.5

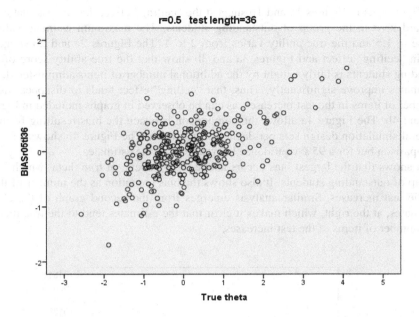

Fig. 3a. BIAS-True Theta for test length =36 and exposure rate 0.5

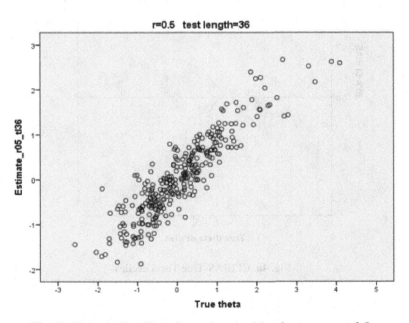

Fig. 3b. Estimate-True Theta for test length =36 and exposure rate 0.5

The graphs in Figures 1a and 1b suggest the "ceiling" effect, i.e., a systematic bias toward zero in the group of outstanding students. The maximum estimate value is close to 1.5 and the true ability varies from 2 to 4. The Figures 2a and 2b suggest a slight "ceiling" effect and Figures 3a and 3b show that the true ability score of outstanding students is fairly caught by the additional number of items administered. The estimates improve significantly. Thus, that "ceiling" effect tends to disappear as the number of items in the test increases, as can be observed in graphs included in Figures 4a and 4b. The Figure 4a allows the comparison between the bias resulting from the several simulation design sets per deciles of true theta. The Figure 4b shows a similar comparison but for a 95% confidence interval (CI) of the estimates.

It shows that the largest bias occurs above the 9th decile of true theta, which is the group of outstanding students. It also shows the bias reduction as the number of items in the test increases. Similar analysis emerges from the second graph of the ability estimates, at the right, which makes it clear that the estimates tend to the true theta as the number of items of the test increases.

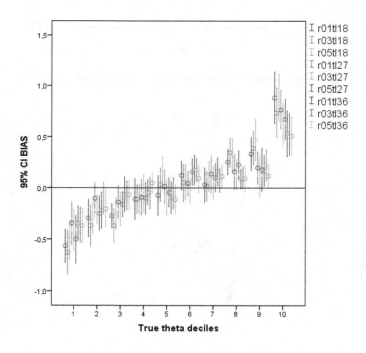

Fig. 4a. CI BIAS-True Theta deciles

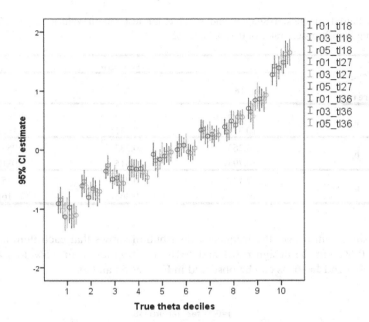

Fig. 4b. CI Estimate-True Theta deciles

The chi-square hypotheses test was applied to the comparison of empirical and population distributions. As can be observed in Table 4, for all of the nine derived score distributions, the null hypothesis is rejected at the level of significance of 5%. It suggests the agreement between any one of the 5-level classification of the ability estimates and the population distribution.

Table 4. Chi-square statistics (p-value<0.01) for test length values of 18, 27 and 36, considering item exposure rates of 0.1, 0.3 and 0.5

Exposure rate	Test length		
	18	27	36
0.1	142.544	159.095	113.556
0.3	102.163	109.447	146.956
0.5	97.953	145.154	145.703

In addition, the hypotheses test for paired samples was used for the comparison of examinees' true scores and CAT scores. Table 5 includes the t-Student statistic and its p-value for each set of values. It can be observed that, with the exception of two out of nine sets (r=0.5 and J=18; r=0.1 and J=27), the null hypothesis should not be rejected for any other set of values. That is, at the level of significance of 5%, there is no statistically significant difference between the examinees' true scores and ability score estimates obtained by CAT.

Table 5. Paired samples: t-Student and p-value for test length values of 18, 27 and 36, considering item exposure rates of 0.1, 0.3 and 0.5

Exposure rate	Test length		
	18	27	36
0.1	0.896 (0.371)	2.098 (0.037)	1.307 (0.192)
0.3	0.263 (0.793)	0.235 (0.815)	1.753 (0.081)
0.5	2.431 (0.016)	0.910 (0.363)	0.815 (0.416)

Regarding item usage, the frequency distribution shows that each item is selected 0.5% or 0.6% in the design r=0.1 and J=18 and it varies from 0.5% to 1% in the design r=0.5 and J=36, as can be observed in Figures 5a and 5b.

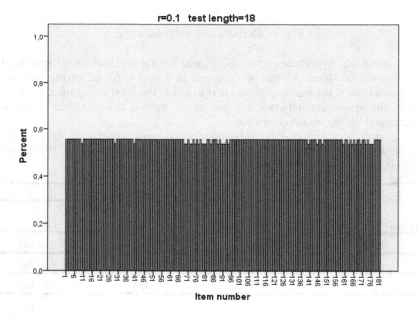

Fig. 5a. Frequency distribution of item usage for item exposure 0.1 and test length 18

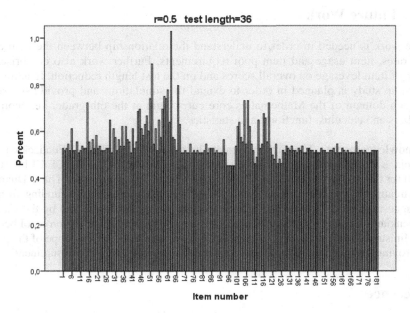

Fig. 5b. Frequency distribution of item usage for item exposure 0.5 and test length 36

4   Discussion

Reporting the overall scores is the ultimate goal of the Portuguese educational assessment, as any other educational assessment system in the world. In the Portuguese system, a 5-level classification scale is adopted based on the delivery and scoring of paper-and-pencil tests administered to the population of students at the 4th, 6th, and 9th grades. Reported scores should be valid, comparable, reliable, and tests developed according to protocols, such as those published by [18], and taking into account the posterior use of the reported scores. Considering the benefit for educational improvement, the next generation of educational assessment will capitalize on the power of the web to deliver more efficient tests. Thus, it is expected to obtain improvements on the reliability and comparability of assessment results. The prior optimization of the test size conditional on a given level of measurement error makes the simulation of adaptive tests as the crucial step.

The simulation study presented above shows the successfull reduction of test length from 56 items of paper-and-pencil administered tests to 36 items CAT (36%). The extreme reduction from 56 to 18 items (68%) provide overall scores poorly estimated and with the ceiling effect. This may be due to the insufficiency of the item pool in providing items properly mentored for outstanding students. Even in this situation, the comparison between the true and estimated score distributions suggests their agreement at the level of significance of 5%.

5   Future Work

More work is needed in order to understand the relationship between the item exposure rates, item usage and item pool requirements. Further work also comprises the study of item leverage on overall scores and on the test length reduction. In addition, a bootstrap study is planned in order to extend the simulations and provide sub-scores for each domain of the Mathematics core curriculum at the 6th grade, i.e. geometry, numbers and calculus, functions, and statistics.

Acknowledgements. The authors gratefully acknowledge FCT – Fundação para a Ciência e Tecnologia (references PEst-OE/EGE/UI0491/2013, PEst-OE/EEI/LA0008/2013) for financial support, Lihua Yao for the SimuMCAT and support, Diana Duarte for files treatment. The longitudinal survey 3EM was made possible by funding from the Portuguese Ministry of Science Technology and Higher Education and by the Calouste Gulbenkian Foundation. The data subset was made available under the protocol between the Ministry of Education and UBI, signed in November 2007, in the scope of the project 'Improving the quality of educational assessment tests and scales of measurement'.

References

1. Association for Educational Assessment - Europe: European Framework of Standards for Educational Assessment 1.0 (2010)
2. Lord, F.: A Theory of Test Scores. Psychometric Monograph, vol. (7). Richmond, VA (1952)
3. Lord, F.M., Novick, M.R., Birnbaum, A.: Statistical Theories of Mental Test Scores. Addison-Wesley, Oxford (1968)
4. Lord, F.M.: Applications of Item Response Theory to Practical Testing Problems. Erlbaum, Hillsdale (1980)
5. Lin, C.: Comparisons between Classical Test Theory and Item Response Theory in Automated Assembly of Parallel Test Forms in Automated Assembly of Parallel Test Forms. J. Technol. Learn. Assess. 6, 1–43 (2008)
6. Eignor, D.R.: Future Challenges in Psychometrics: Linking Scores Across Computer and Paper-Based Modes of Test Administration. In: Handbook of Statistics (Psychometrics), pp. 1099–1102 (2007)
7. Yao, L.: Multidimensional CAT Item Selection Methods for Domain Scores and Composite Scores: Theory and Applications. Psychometrika 77, 495–523 (2012)
8. Van der Linden, W.J.: Multidimensional adaptive testing with a minimum error-variance criterion. J. Educ. Behav. Stat. 24, 398–412 (1999)
9. Yao, L.: simuMCAT: Simulation of multidimensional computer adaptive testing. Monterey (2011)
10. Ferrão, M.E., Costa, P., Navio, V.M., Dias, V.M.: Medição da competência dos alunos do ensino básico em Matemática, 3EMat: uma proposta. In: Machado, C., Almeida, L., Guisande, M.A., Gonçalves, M., Ramalho, V. (eds.) XI Conferência Internacional Avaliação Psicológica: Formas e Contextos, pp. 905–915. Psiquilíbrios, Braga (2006)

11. Costa, P., Ferrão, M.E., Fernandes, N., Soares, T.: Uma aplicação da análise factorial na detecção das dimensões cognitivas em testes de avaliação em larga escala em Portugal (An application of Factorial Analysis for the Detection of Cognitive Dimensions in Large Scale Assessment in Portugal). In: Anais do XLI Simpósio Brasileiro de Pesquisa Operacional – Pesquisa Operacional na Gestão do Conhecimento, pp. 391–402. SOBRAPO, Rio de Janeiro (2009)

12. Ferrão, M.E., Costa, P.: Melhoria da Qualidade dos Instrumentos e Escalas de Aferição dos Resultados Escolares: Ligação entre Escalas das Provas de Aferição de Matemática e 3EMat. Covilhã (2009)

13. Ferrão, M.E., Loureiro, M.J., Navio, V.M., Coelho, I.: Aferição das Aprendizagens em Matemática no Ensino Básico: A proposta 3EMat. Universidade da Beira Interior, Covilhã (2009)

14. Ministério da Educação: Currículo Nacional do Ensino Básico: Competências Essenciais (National Curriculum in Primary, Elementary and Lower Secondary Education: essential competencies). Ministério da Educação, Lisboa (2001)

15. Ministério da Educação: Currículo Nacional do Ensino Básico: Organização Curricular e Programas do Ensino Básico – 1° ciclo (National Curriculum in Primary, Elementary and Lower Secondary Education: curricular organization). Ministério da Educação, Lisboa (2001)

16. Yao, L.: The BMIRT Toolkit. Monterey (2013)

17. Yao, L.: Comparing the performance of five multidimensional CAT selection procedures with different stopping rules. Appl. Psychol. Meas. 37, 3–23 (2012)

18. AERA, APA, NCME: The Standards for Educational and Psychological Testing. Washington DC (1999)

Accommodating Maternal Age in CRIB Scale: Quantifying the Effect on the Classification

Maria Filipa Mourão[1], Ana C. Braga[2], and Pedro Nuno Oliveira[3]

[1] School of Technology and Management-Polytechnic Institute of Viana do Castelo,
4900-348 Viana do Castelo, Portugal
fmourao@estg.ipvc.pt

[2] Department of Production and Systems Engineering, Algoritmi Research Centre,
University of Minho,
4710-057 Braga, Portugal
acb@dps.uminho.pt

[3] Biomedical Sciences Abel Salazar Institute, University of Porto,
4050-313 Porto, Portugal
pnoliveira@icbas.up.pt

Abstract. Receiver operating characteristic (ROC) curves are a well-accepted measure of accuracy of diagnostic tests using in continuous or ordinal markers. Based on the notion of using a threshold to classify subjects as positive (diseased) or negative (no diseased), a ROC curve is a plot of the true positive fraction (TPF) versus the false positive fraction (FPF)for all possible cut points. Thus, it describes the whole range of possible operating characteristic for the test and hence its inherent capacity for distinguish between two status. The clinical severity scale CRIB - Clinical Risk Index for Babies, emerged in 1993 to predict the mortality of newborn at less than 32 weeks of gestation and very low birth weight ($< 1500gr$) [4]. In previous work of Braga [3] this index was reported as showing a good performance in assessing risk of death for babies with very low birth weight (less than 1500 g weight). However, in some situations, the performance of the diagnostic test, the ROC curve itself and the Area Under the Curve(AUC) can be strongly influenced by the presence of covariates, whether continuous or categorical [5], [32], [33]. The World Health Organization and the Ministry of Health, defined as "late pregnancy" that thus occurs in women over 35 years. In this work, using the conditional ROC curve, we analyze the effect of one covariate, maternal age, on the ROC curve that representing the diagnostic test performance. We chose two age status, less than 35 years old and equal or greater than 35 years old, to verify the effects on the discriminating power of CRIB scale, in the process classification using R and STATA software.

Keywords: Conditional ROC (Receiver Operating Characteristic) curve, CRIB (Clinical Risk Index for Babies), maternal age, AUC (Area Under the Curve).

B. Murgante et al. (Eds.): ICCSA 2014, Part III, LNCS 8581, pp. 566–579, 2014.

1   Introduction

The discriminatory capacity of a continuous marker or diagnostic test X, is usually measured by means of the receiver operating characteristic (ROC) curve [1] [2]. Under the conventional assumption that high marker values are indicative of disease, classification on the basis of X of an individual as healthy ($\bar{D}$) or diseased (D) can be made by the choice of a cut-off value c, such that, if $X \geq c$, the individual is classified as diseased, and if $X < c$, the individual is classified as healthy. Hence, for each cut-off value, c we define the true positive fraction,

$$TPF(c) = P[X \geq c|D] \tag{1}$$

and the false positive fraction,

$$FPF(c) = P[X \geq c|\bar{D}] \tag{2}$$

In such a situation, the ROC curve is defined as the set of all pairs for these fractions that can be obtained on the variation of cut-off value, c, $(TPF(c), FPF(c))$, $c \in (-\infty, +\infty)$, or, equivalently, as a function:

$$ROC(p) = S_D(S_{\bar{D}}^{-1}(p)), \text{for } p \in (0, 1) \tag{3}$$

where S_D and $S_{\bar{D}}$ denotes the survival functions of X in diseased and healthy subjects, respectively [29]. The Area Under the ROC Curve (AUC) is considered as an effective measure of inherent validity of a diagnostic test. This index is useful in evaluating the discriminatory ability of a test to correctly pick up diseased and non-diseased subjects and finding optimal cut-off point to least misclassify diseased and non-diseased subjects.

In many practical situations, however, a marker's discriminatory capacity may be affected by a set of continuous and/or categorical covariates.

1.1   Maternal Age as Covariate

Nowadays, fortunately, few women have a risky pregnancy. However, in cases where there is such a quiet evolution, because there is a chronic illness or because medical or pregnancy related problems arise during the nine months, it is essential a proper and timely monitoring. By definition, there is a high-risk pregnancy when the probability, of an adverse outcome for the mother and/or infant during or following pregnancy and delivery, is increased above the mean baseline risk in the general population by the presence of one or more risk factors. These factors can be divided into maternal and fetal. The maternal factors include maternal age (less than 15 years and upper 35 years). Women over 35 years old are at greater risk of having a baby with chromosomal abnormalities. The existence of hypertension, diabetes and complications during labor is also more frequent. Below 15 years old, low birth weight and preeclampsia, may occur. There are many works in which the mother's age is the subject of study, either in association with low birth weight, either because it can be the cause of preterm birth [8], [30], [31]. Most research indicates that very young mothers imply an increased risk of low birth weight and premature births. Few studies have been conducted to older mothers, although this is the current trend in most developed societies. The few studies associate old maternal age to a potential decrease in babys development during pregnancy. Aras [9] presents several studies in which it is concluded that low birth weight depends not only on the age of the mother but also of health care during pregnancy, biological characteristics, race, socio-economic factors, weight gain during pregnancy, among others. Friede et al. [7], in a study to examine the effect of maternal age on low birth weight and infant mortality, used data from young mothers and found a strong association between young maternal age and infant mortality and also with a high prevalence of infants with low birth weight.

2   Empirical Estimators of the ROC Curve

There are many estimators proposed for the ROC curve when it is defined by the expression (3). Assuming that two independent samples, one from diseased population (D) and another from healthy ($\bar{D}$) are avaiable, we may obtain the empirical estimators of survival functions of X in diseased and healthy subjects, $\widehat{S}_D$ and $\widehat{S}_{\bar{D}}$, respectively. Thus we can write the empirical ROC curve estimator by the following expression:

$$\widehat{ROC}(p) = \widehat{S}_D(\widehat{S}_{\bar{D}}^{-1}(p)), \text{for } p \in (0,1) \tag{4}$$

For each possible cut-off value c, the fractions of true positives and false positives are obtained by:

$$\widehat{TPF}(c) = \frac{1}{n_D} \sum_{i=1}^{n_D} I\left(X_{Di} \geq c\right) \tag{5}$$

and

$$\widehat{FPF}(c) = \frac{1}{n_{\bar{D}}} \sum_{i=1}^{n_{\bar{D}}} I\left(X_{\bar{D}i} \geq c\right) \tag{6}$$

Thus, the estimated ROC curve is obtained by representing pairs of values

$$(\widehat{TPF}(c), \widehat{FPF}(c)), c \in (-\infty, +\infty) \tag{7}$$

Similarly, the empirical estimators of survival function of X in diseased and healthy subjects presents in the expression for the ROC curve estimator (4) can be obtained by

$$\widehat{S}_D(c) = \frac{1}{n_D} \sum_{i=1}^{n_D} I\left(X_{Di} \geq c\right) \tag{8}$$

and

$$\widehat{S}_{\bar{D}}(c) = \frac{1}{n_{\bar{D}}} \sum_{i=1}^{n_{\bar{D}}} I\left(X_{\bar{D}i} \geq c\right) \tag{9}$$

One of the biggest advantages of using ROC curves is the possibility of comparing diagnostic tests, such in medical diagnosis, trough the AUCs obtained from these curves. In this study, we use the statistical approach to the Wilcoxon-Mann-Whitney test to calculate the estimate of AUC index, which summarize each ROC curve in terms of area bellow it. One possibility to test weather the difference between two ROC curves is statistically significant, involves the AUC index. Consider AUC_1 and AUC_2 the areas obtained from two ROC curves. The relevant hypothesis to test, H_0 is that the two data sets come from ROC curves with the same AUC:

$$H_0: \text{AUC}_2 - \text{AUC}_1 = 0 \quad \text{vs} \quad H_1: \text{AUC}_2 - \text{AUC}_1 \neq 0$$

A method for testing the difference between the two areas for independent samples is based on critical ratio Z [25]:

$$Z = \frac{AUC_2 - AUC_1}{\sqrt{SE_{AUC_1}^2 + SE_{AUC_2}^2}} \sim N(0,1) \tag{10}$$

3   ROC Curve with Covariates

It is well known that a diagnostic test performance may be strongly influenced by covariates. In such situation, the ROC curve (and its summary indices, such as the area under the curve) may be underestimated if important covariates are neglected. In most studies, paralell to the diagnostic test used to classify individuals in D and $\bar{D}$ class, one or more covariates may be associated with the classification variable and can provide extra information about the individuals classification and increase the discriminating power of the curve. In such situations, the interest must be focused on assessing the discriminatory capacity of marker Xs regarding to the values assumed by the covariate that we represent by Y. The ROC curve in this case can be considered for each value of the covariate, y. Changes that occurs in curve, due to these values, might mean that the covariate has effect on the discriminating power of the diagnostic test. If the conditional survival functions of X_D and $X_{\bar{D}}$, given Y, are denoted by S_{DY} and $S_{\bar{D}Y}$ respectively, the conditional ROC curve is defined as

$$ROC_y(p) = S_{DY}(S_{\bar{D}Y}^{-1}(p)), \text{for } p \in (0,1) \tag{11}$$

The first works developed in the area of adjustment of covariates on the discrimination or classification of individuals into classes, are assigned to Cochran and Bliss [10], Cochran [11] and Rao [13]. These publications were focused on the selection of discriminating variables. The rules to adjust covariates requires the formulation of appropriate probability models and are developed later in works of Lachenbrush [12] and McLachlan [14]. However, these works focus on classification rules, providing only background of interest to the study. The first works on adjustment of covariates were from Guttman et al. [15] and Tolsteson and Begg [16]. The first authors, in the context of stress strength models, have obtained the ROC curve as a function of covariates by fitting linear regression models to X (diseased subjects) and Y (healthy subjects), assuming normality with different variances. The adjustment of ROC curves from continuous scales was studied by Smith and Thompson [17], Pepe [18], [19], [20], Faraggi [21], Janes and Pepe [22], [23] among others. The summary measures from ROC curves with adjustment of covariates was studied by Faraggi [21] and by Dodd and Pepe [24].

4   Covariate ROC Curve Adjustment

To assess the possible covariate effects on the ROC curve, two different approaches have been suggested in the statistical literature. The **Induced** methodology proposed by Tolsteson and Begg [16]; Zheng and Heagerty [26]; Faraggi [21], in which the covariates effect on diagnostic test is modeled in the two populations (D and $\bar{D}$) separately. The covariate effects on the associated ROC curve can then be computed by deriving the *induced* form of the ROC curve.

Otherwise the **Direct** method proposed by Pepe [20]; Alonzo and Pepe [28]; Cai [27], assumes a ROC curve with *direct* effect of the covariates on it.

4.1   Induced Adjustment

Suppose there is a set of covariates Y_D associated with the diseased population and a set of covariates $Y_{\bar{D}}$, associated with the healthy population. In many applications, some if not all covariates are common to both populations, but there is no need for two sets are identical. If α_D and $\alpha_{\bar{D}}$ are scalars, β_D and $\beta_{\bar{D}}$ are vectors of unknown parameters with Y_D and $Y_{\bar{D}}$ elements, then the average values associated with the diagnostic test for the diseased and healthy populations, for certain values of the covariates, can be modeled as [29]

$$\mu_D(Y_D) = \alpha_D + \beta_D^T Y_D \tag{12}$$

and

$$\mu_{\bar{D}}(Y_{\bar{D}}) = \alpha_{\bar{D}} + \beta_{\bar{D}}^T Y_{\bar{D}} \tag{13}$$

The model specification is complete assuming that these average values are normally distributed with standard deviation σ_D and $\sigma_{\bar{D}}$, respectively. This model is essentially the binormal by specifying the population average. Then it follows that the equation of the ROC curve is given by

$$ROC_y(p) = \phi \left(\frac{\mu_D(Y_D) - \mu_{\bar{D}}(Y_{\bar{D}}) + \sigma_{\bar{D}} \phi^{-1}(p)}{\sigma_D} \right), p \in (0,1) \tag{14}$$

The least squares regression can be used to obtain point estimates of α_D, $\alpha_{\bar{D}}$, β_D and $\beta_{\bar{D}}$. Substituting these estimates in the above expression for certain y_D, $y_{\bar{D}}$ values of Y_D and $Y_{\bar{D}}$, we obtain the ROC curve for covariates.

4.2   Direct Adjustment

This approach, contrasts with the approach set before. Instead of separately modeling the effect of covariate in the two distributions results of the diagnostic test and then obtains the ROC curve from the modified conditional distributions, direct adjustment model the covariate effects directly on the ROC curve. There are several advantages to the direct modeling approach. Foremost amongst them is that of any parameter associated with the covariate has a direct interpretation in terms of the curve. The heart of this approach is the specification of a suitable model that captures the effect of the covariate on the ROC curve, which allows a flexible and easy interpretation. The flexibility of the model is in terms of how this relationship works in the ROC curve, to preserve the condition that either the domain and the curve lie in the interval $(0, 1)$ and that the curve continues to be monotone increasing in this range. The knowledge of least squares method to fit measures of variables, such as diagnostic test, and to evaluate the relationship between his results and the subject diseased status, suggested that the relationship between the ROC curve and the covariate effect was measured. This relationship is evaluated by using Generalized Linear Models (GLM). This methodology, leads the general form of the conditional ROC curve given by:

$$ROC_Y(p) = g\left(Y'\beta + h(p)\right), p \in (0,1) \tag{15}$$

where Y is a p-dimensional vector of covariates, β is a p-dimensional vector of unknown parameters, h is an unknown monotone increasing function of the $FPFs$, and g is a known link function, describing the functional relationship between the ROC curve and the covariates. Models such as (15) define the so-called class of ROC-GLMs [29].

The most common link functions are probit, logistic and logarithmic. Together with the advantage of directly evaluating the effect of the covariate on the ROC curve, direct methodology has some other appealing features:

- the ROC curve property of being invariant to monotonic transformation of the test result is preserved,
- any possible interaction between covariates and FPFs is easy to incorporate into the regression model,
- allows modeling the performance parameters and allows the comparison of different classifiers.

5   Application and Results

5.1   Study Description

In this work, to check if the age of the mother affects the discriminating ability of the CRIB index, we used a sample of 187 infants of very low birth weight (< 1500gr and/or gestational age < 32 weeks) from a hospital in North of Portugal. Thus, the variable X will correspond to the CRIB, and the covariate of interest will be the age of the mother. We will consider the age of the mother as a binary variable, with reference to the age of 35 years old. So, we will have two categories: age greater or equal than 35 years and younger than 35 years. Of the 187 infants, 152 are newborns from mothers with less than 35 years old and the remaining 35 are newborns of mothers aged 35 years or older. The rating assigned according to the CRIB scale regarding their clinical status was, for the first newborn mothers, 8.55% of the babies classified as "dead" and 14.29% of the babies on the second group with the same classification. In the sample, 9.63% of babies are classified as "dead".

To asses the objective of our study, we choose to apply the direct method, because the advantages listed above.

5.2   Results

We begin by characterizing the distribution of babies, according to the classification assigned by the CRIB scale, when we consider that the mother's age has no effect on this classification and then considering this possible effect (Fig. 1).

According to this characterization, we obtained the respective empirical ROC curves (Global and separately by mother's age) with R software, which are shown in Fig. 2.

We compute the AUC, the standard deviation and the confidence intervals for each of the ROC curves plotted above. These values are summarized in Table 1.

Obtained the intervals, with 95% of confidence, for the ROC curves, we proceed to its graphical representation to visualize possible dispersion of the pairs of values $(1 - specificity, sensitivity)$. Figure 3 show the ROC curves with the corresponding confidence interval, obtained by the application of R software.

To verify if maternal age has effect on the CRIB scale when used to classify clinical babies status, we used the critical reason Z. The value computed by this statistic was $Z = 2.28$ ($p - value = 0.0226$).

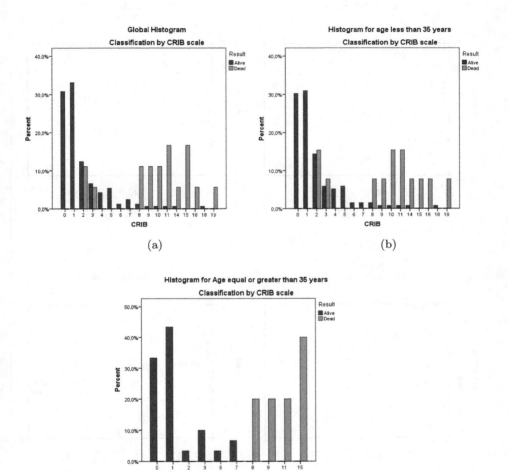

Fig. 1. Histograms - classification assigned by the CRIB scale

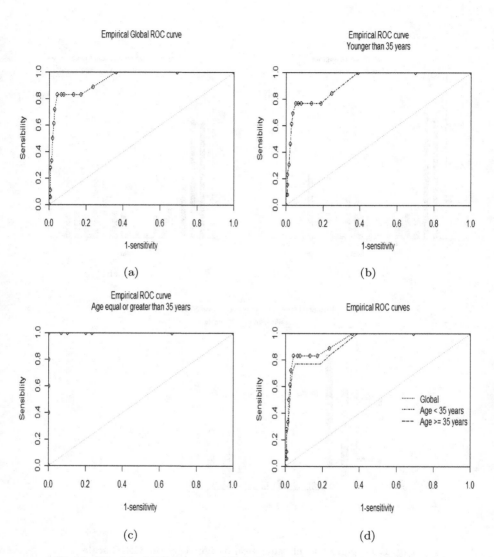

Fig. 2. Empirical ROC curves (Global and separately by mother's age)

Table 1. ROC curves Results

	AUC	SE	95% IC Lower Bound	95% IC Upper Bound
Global	0.942	0.025	0.8929	0.991
Maternal age less than 35 years	0.9203	0.035	0.8515	0.9891
Maternal age equal or greater than 35 years	1.000	6.8E-9	1.000	1.000

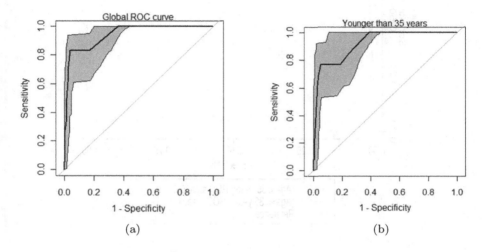

Fig. 3. ROC curves and Confidence Intervals

5.3   Direct Adjustment of Covariate with Stata Software

The results above suggest that the impact of maternal age on CRIB scale should be statistically adjusted in the ROC analysis. Stata software, using ROC regression, models the ROC curve (CRIB ROC curve) as a function of covariates (maternal age), via GLM. In statistic, the GLM is a flexible generalization of ordinary linear regression that allows for response variables that have error distribution models other than normal distribution. The GLM generalizes linear regression by allowing the linear model to be related to the response variable using a link function and by allowing the magnitude of the variance of each measurement to be function of its predicted value. We start by computing the graph of the ROC curve conditioned to maternal age (Fig. 4) with correspondent AUCs (Fig. 5) and test of equality of those areas (Fig. 6). The AUC for the conditioned CRIB ROC curve and the correspondent confidence intervals ((N)- normal confidence interval; (P) - percentile confidence interval; (BC) - bias-corrected confidence interval) is shown in (Fig. 7).

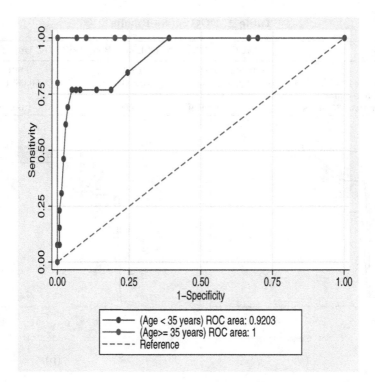

Fig. 4. ROC curve conditioned by Maternal Age (from STATA)

Obs	ROC Area	Std. Err.	—Asymptotic Normal — [95% Conf. Interval]	
152	0.9203	0.0351	0.85153	0.98909
35	1.0000	0.0000	1.00000	1.00000

Fig. 5. AUCs for ROC curve conditioned by Maternal Age (from STATA)

Regarding the ROC curve for the CRIB, conditioned on maternal age, we found that the behavior is identical to that observed in the study described above with R software. The CRIB for age less than 35 year old mother discriminates 92.03% while for maternal age equal or greater than 35 years, discriminates in 100% of cases. We also found that a greater variability in the results is observed for the curve generated from the first data set. When performed the test of equality of the two areas, we found the following results:

```
Ho: area( 0) = area( 1)
    chi2( 1) =      5.16        Prob>chi2 =      0.0232
```

Fig. 6. ROC curve conditioned by Maternal Age - comparison test (STATA output)

that leads to the rejection of the null hypothesis.

AUC	Observed Coef.	Bias	Bootstrap Std. Err.	[95% Conf. Interval]	
	.931255	.0009726	.0309972	.8705016	.9920084 (N)
				.8638773	.9858114 (P)
				.8551341	.9808154 (BC)

Fig. 7. AUC for the CRIB ROC curve conditioned by Maternal Age (STATA output)

By the results presented in (Fig. 7), we found a discrimination power for the CRIB ROC curve in 93.13% of the babies (AUC = 0.9313) and with a standard deviation equal to 0.0309. The confidence intervals to evaluate the dispersion of the pairs of values $(1 - specificity, sensitivity)$ was also computed.

6   Conclusions

From the estimated values of AUCs for the global ROC curve we may observe that, without considering the effect of maternal age, the CRIB scale discriminates between babies "alive" and "dead" in 94.2% of cases. This power, when considering maternal age as covariate, is 92.03% for age less than 35 years old and equal to 100% when maternal age is equal or greater than 35 years. We can see too, that the ROC curve for maternal age less than 35 year old shows greater variability (Table 1) and (Fig. 3). Applied the statistical test previously set out in Section 2 to determine whether maternal age has an effect on the classification of baby's, when CRIB scale is used, we conclude that there is a significant difference in the performance of CRIB scale when conditioned on maternal age and maternal age equal or greater than 35 years old has effect on the discrimination power of CRIB scale. From the results obtained above, it is apparent that mother's age appears to affect the discriminatory power of CRIB scale when used to classify babies. For this reason, we consider useful to verify the performance of this scale when conditioned on maternal age by using the ROC-GLM regression models (CRIB ROC analysis with correspondent CRIB ROC curve). The results presented in (Fig. 7) suggest that the performance of CRIB scale decreases when one incorporates the age of the mother in her review (AUC = 0.9313). The standard deviation, in turn, is greater than the standard deviation of the ROC curves obtained without the covariate. Consequently, it also increases the dispersion of the pairs of values $(1 - specificity, sensitivity)$.

Fig. 6 shown the results for comparison test by the STATA software that indicated also that maternal age has effect on the classification of babies by the CRIB scale. CRIB scale perform better for maternal age equal or greater than 35 years old.

Motivated by these results we propose as future work, to check if the sex of the babies along with the mother's age, also have an effect on the classification of infants when using the CRIB scale.

References

1. Metz, C.E.: Basic principles of ROC analysis. Seminars in Nuclear Medicine 8, 183–298
2. Swets, J.A., Pickett, R.M.: Evaluating of Diagnostic System: Methods from Signal Detetion Theory. Academic Press, New York
3. Braga, A.C., Oliveira, P., Gomes, A.: Avaliação do risco de morte em recém-nascidos de muito baixo peso: uma comparação de índices de risco baseada em curvas ROC. IV Congresso Anual da Sociedade Portuguesa de Estatística. Editores: Luísa Canto e Castro, Dinis Pestana, Rita Vasconcelos, Isabel Fraga Alves. Edições Salamandra
4. Dorling, J.S., Field, D.J., Manketelow, B.: Neonatal disease severity scoring systems. Arch. Dis. Child. Fetal Neonatal 90, F11–F16
5. López-de-Ullibarri, I., Cao, R., Cardaso-Suárez, C., Lado, M.J.: Nonparametric estimation of conditional ROC curves: application to discrimination tasks in computerized detection of early breast cancer. Computational Statistics & Data Analysis 52(5), 2623–2631
6. Metz, C.E.: Statistical Analysis of ROC Data in Evaluating Diagnostic Performance. Multiple Regression Analysis: Application in Health Sciences. American Institute of Physics 13, 365–384
7. Friede, A., Baldwin, W., Rhodes, P.H., Buehler, J.W., Strauss, L.T., Smith, J.C., Hogue, C.J.R.: Young Maternal Age and Infant Mortality: The role of low birth weight. Public Health Report 102(2) (March-April)
8. Friede, A., Baldwin, W., Rhodes, P.H., et al.: Older maternal age and infant mortality in the United States. Obstet. Gynecol. 72, 1527
9. Aras, R.Y.: Is maternal age risk factor for low birth weight? Archives of Medicine and Health Sciences 1(1) (January-June)
10. Cochran, W.G., Bliss, C.I.: Discriminant functions with covariance. Ann. Math. Statist. 19(2), 151–291
11. Cochran, W.G.: Comparison of two methods of handling covariates in discriminant analysis. Annals of the Institute of Statistical Mathematics 16, 43–53
12. Lachenbrush, P.A.: Covariance adjusted discriminant functions. Annals of the Institute of Statistical Mathematics 29, 247–257
13. Rao, C.R.: On some problems arising out of discrimination of multiple characters. The Indian Journal of Statistics 9, 343–366
14. McLachlan, G.J.: Discriminant analysis and pattern recognition. Wiley, New York
15. Guttman, I., Johnson, R.A., Bhattacharayya, G.K., Reiser, B.: Confidence limits for stress-strenght models with explanatory variables. Technometrics 30(2), 161–168
16. Tolsteson, A.N., Begg, C.B.: A general regression methodology for ROC curve estimation. Med. Decision Making 8(3), 204–215

17. Smith, D.J., Tompson, T.J.: Correcting for confounding in analising receiver operating characteristic curves. Biometrical Journal 38, 357–863
18. Pepe, M.S.: A regression modelling framework for receiver operating characteristic curve in medical diagnostic testing. Biometrika 84, 595–608
19. Pepe, M.S.: Three approaches to regression analysis of receiver operating characteristic curves for continuous tests results. Biometrics 54, 124–135
20. Pepe, M.S.: An interpretation for the ROC curve and inference using GLM procedures. Biometrics 56, 352–359
21. Faraggi, D.: Adjusting receiver operating characteristic curves and related indices for covariates. Journal of the Royal Statistical Society: Series D (The Statistician) 52(2), 179–192
22. Janes, H., Pepe, M.S.: Adjusting for covariates effects on classification accuracy using the covariate-adjusted ROC curve. UW Bioestatistics Workin Paper Series. Working paper 283, http://biostats.bepress.com/uwbiostat/paper283
23. Janes, H., Pepe, M.S.: Adjusting for covariates in studies of diagnostic, screening, or prognostic markers: An old concept in a new setting. UW Bioestatistics Working Paper Series. Working paper 310, http://biostats.bepress.com/uwbiostat/paper310
24. Dodd, L.E., Pepe, M.S.: Partial AUC estimation and regression. Biometrics 59(3), 614–623
25. Hanley, J.A., McNeil, B.J.: A method of comparing the areas under receiver operating characteristic curves derived from the same cases. Radiology 148, 839–843
26. Zheng, Y., Heagerty, P.J.: Semiparametric estimation of time-dependent ROC curves for longitudinal marker data. Biostatistics 4, 615–632
27. Cai, T.: Semiparametric ROC regression analysis with placement values. Biostatistics 5, 45–60
28. Alonzo, T.A., Pepe, M.S.: Distribution free ROC analysis using binary regression techniques. Biostatistics 3, 421–432
29. Pepe, M.S.: The Statistical Evaluation of Medical Tests for Classification and Prediction. Oxford University Press, New York
30. Carolan, M., Frankowska, D.: Advanced maternal age and adverse perinatal outcome: A review of the evidence. Midwifery, doi:10.1016/j.midw.2010.07.006
31. Linda, J., Heffner, M.D.: Advanced Maternal Age How Old Is Too Old? The New England Journal of Medicine 351, 19
32. Rodriguez-Álvarez, M.X., Roca-Pardiñas, J., Cardaso-Suárez, C.: ROC curve and covariates: Extending induced methodology to the non-parametric framework. Statistical and Computing 21, 483–499
33. González-Manteiga, W., Pardo-Fernández, J.C., Keilegom, I.: ROC curves in Non-Parametric Location-Scale Regression Models. Scandinavian Journal of Statistics 38(1), 169–184

Joint Modelling for Longitudinal and Time-to-Event Data: Application to Liver Transplantation Data

Ipek Guler[1], Laura Calaza-Díaz[1], Christel Faes[2], Carmen Cadarso-Suárez[1],
Elena Giraldez[3], and Francisco Gude[4]

[1] Unit of Biostatistics, Department of Statistics and Operations Research,
University of Santiago de Compostela, Spain
{ipek.guler,laura.calaza,carmen.cadarso}@usc.es

[2] Interuniversity Institute for Biostatistics and Statistical Bioinformatics,
Universiteit Hasselt, Belgium
christel.faes@uhasselt.be

[3] Intensive Care Unit, Hospital Clínico, Santiago de Compostela, Spain
elegiva@hotmail.com

[4] Clinical Epidemiology Unit, Hospital Clínico, Santiago de Compostela, Spain
francisco.gude.sampedro@sergas.es

Abstract. The joint modelling approaches are often used when an association exists between time-to-event and longitudinal processes. They are recognized for their efficiency involving the association structure between these two processes. Recently, [17] and [14] suggested alternative joint modelling approaches. In this paper, we will focus our attention on the Rizopoulos' approach. This methodology was applied to Orthotopic Liver Transplantation data (OLT) with a flexible environment for both longitudinal and survival sub-models. Different regression models were fitted to the OLT data and their predictive performances were compared by using time-dependent ROC curves, also, dynamic predictions were obtained for the survival process. Computational aspects (including software) related to the use of the joint modelling approach in practice, were also discussed. The application of joint modelling revealed a hitherto unreported effect: for non-diabetic patients, the longitudinal Glucose levels have a significant effect on survival. In addition the discrimination ability improves over time. However for diabetic patients the association between these two processes is not significant.

Keywords: Joint Modelling, longitudinal, survival data, time-dependent ROC curves, Area Under Curve (AUC), dynamic predictions; transplantation.

1 Introduction

The longitudinal studies are becoming increasingly popular, especially in biomedical research. Many of these studies are aimed to characterise the relationship

B. Murgante et al. (Eds.): ICCSA 2014, Part III, LNCS 8581, pp. 580–593, 2014.

between the longitudinal outcomes and time-to-event. To study the association between longitudinal responses and particularly time-to-event survival process there exist several methods in the literature. The extended Cox models [1] and the two stage approaches [20], were proposed to handle this association.

Although these methods have an advantage of fast computing, they don't take into account longitudinal data measured with error or informative censoring in longitudinal process. Joint modelling approaches [18], [14] became popular for their several advantages [16], [10], [21].

In this paper we will use the joint modelling approach to be applied in Liver Transplantation data. The motivation behind the joint modelling proposal was a liver transplantation database. Orthotopic Liver Transplantation (OLT) is the established treatment for end-stage liver disease and acute fulminant hepatic failure, and more than 80,000 OLT have been performed in Europe. Advances in both medical management and surgical technique have led to an increase in the number of long-term survivors [6]. Because liver transplant recipients live longer, it is necessary to understand and anticipate causes of morbidity and mortality. Several investigators have consistently reported a significant association between increased glycemic variability and worse outcome in critically ill patients. [7], [5], [11], [13] In their analysis, blood glucose variability is measured by using standard deviation, percentile values, successive changes in blood glucose, and by calculating the coefficient of variation. However, it is recognized that compared with the use of only single-moment biomarker values, serial biomarker evaluations may carry important additional information regarding the prognosis of the disease under study [22].

In the present study we aimed to investigate the abilities of postoperative glucose profiles to predict the death of patients who underwent OLT, distinguishing between patients with and without a previous diagnosis of diabetes mellitus. The relationships between glucose profiles and the risk of death were modelled by using flexible joint modelling of longitudinal data and time-to-event analysis [17], and the predictive capacity is analyzed by time dependent ROC curves [8].

The outline of the paper is as follows. The Liver Transplantation Data is described in Section 2. It is common in longitudinal studies to miss repeated measures during the follow-up. To handle the missingness issues of the dataset, an imputation was realised for baseline covariates and Glucose levels. This procedure was described in detail in subsection 2.1. In Section 3, the longitudinal and survival submodels of the joint regression model and the description of dynamic predictions were introduced. Following-up with an analysis of the results obtained in Section 4. Finally, the computational aspects and conclusions are discussed in Sections 5 and 6, respectively.

2   Liver Transplantation Data

From the institutional clinical database, adult patients who underwent OLT in the Hospital Clínico Universitario de Santiago, between July 1994 and July 2011, were identified. Patients who were lost to follow-up and those who died in the first 72 hours were excluded. Registry data that did not conform within a range of expected results were rejected and reevaluated. A total of 644 patients were available for study. The participants were observed until either the primary endpoint (death) was reached or 31 July 2012 (5.6 [0.1, 17.5] years). Research use of the Registry was approved by the Institutional Review Board (Comité Ético de Investigación Clínica de Galicia, Santiago, Spain).

Patients were classified as known diabetic patients if they had been informed of this diagnosis by a physician before admission or were on oral antihyperglycemic agents, insulin, or diet therapy.

The outcome studied was death from any cause before July 2011. Patients were followed up by the study team throughout their hospital stay. After discharge, vital status information was acquired by reviewing the Galician Health Registry, by contacting patients or their families individually and, if the patient had been rehospitalized, by reviewing the hospital records of major clinical events.

```
> library(JM)
> trasplante<-read.spss("trasplante.sav")
> attach(trasplante)
> trasplante.long<-reshape(trasplante,
 +varying = list(names(trasplante)[c(24:31)]),direction = "long",
 +v.names = c("Glu"),idvar = "id", times = c(0:7))
```

The postoperative glucose profiles for individuals with and without previous diabetes, and for those who died in the first year post-transplantation, can be seen in Figures 1 and 2, respectively.

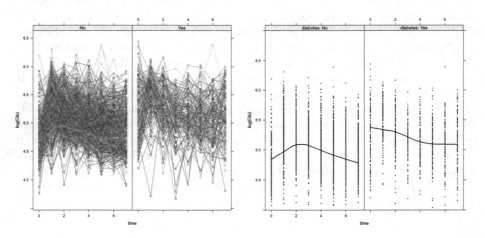

Fig. 1. Subject specific trajectories of Glucose levels for the diabetic and non-diabetic patients and the overall trajectories of Glucose with a p-spline method

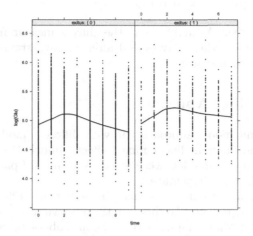

Fig. 2. Overall trajectories of Glucose levels for patients with exitus versus nonexitus

Multivariate Cox models of all-time risk of death were constructed that, together with glucose profiles, included variables of known prognostic value, and
variables of unproven prognostic value that nevertheless emerged as significant
predictors of mortality. In view of the results obtained in the two-covariate analysis with diabetes, independent multivariate Cox models were constructed for
diabetic and non-diabetic patients. In both cases, the variables included because
they proved significant in the two-covariate analyses were *age*, body mass index (*bmi*), red blood cells or platelets transfusion (*TH*), hepatitis C virus (*vhc*),
carcinoma (*carc*), *meld* (model for end-stage liver disease), time in days of total
parenteral nutrition, as well as daily glucose concentrations. In all the analyses described in this paragraph, natural splines were employed to model glucose
dependence.

2.1   Imputation

Because of missing data in the database's variables (% of missing values): *TH*
(0.4%); *bmi* (0%); *age* (0%); *meld* (57%); *vhc* (0.1%); *carc* (0%); apart from
missing values in glucose levels, multiple imputation was used to estimate the
missing values. Chained equations imputation performs a separate imputation
model for each incomplete variable and updates the missing data for each variable
in turn.

The Multivariate Imputation by Chained Equations approach was made using
the R package *mice* [3], with the number of imputations by default ($m = 5$):

```
> library(mice)
> imputation=mice(transplante)
```

and choosing the first complete data set:

```
> transplante=complete(imputation)
```

We had to define as "Not Available, NA" the glucose measurements imputed in patients who died before the 7-day period after surgery, staying only with the measurement before their death.

3   Joint Modelling Approach

We introduced a joint model approach [19] with different modelisations of survival sub-models and different linear mixed effects models [15] for the longitudinal sub-model. Different models were fitted by using *JM* package [17], the R project joint modelling implementation.

From now on, a parallel study was made involving two different groups: diabetic and non-diabetic patients. To this aim, the data was divided in two different databases: "dm1" and "dm0", for diabetic and non-diabetic patients respectively. We will show simultaneously both situations with their corresponding models. In addition, the R code for diabetic patients below each section was added to illustrate the process.

3.1   Survival Sub-model

To study the survival process, significant covariates were selected by using a backward stepwise procedure. The final models considered were the following:

$$h_{i,diab}(t) = h_0(t) \exp(\lambda_1 vhc_i + \lambda_2 meld_i + \alpha \log(Glucose)_i(t)),$$
$$h_{i,nodiab}(t) = h_0(t) \exp(\lambda_1 age_i + \lambda_2 carc_i + \lambda_3 meld_i$$
$$+ \lambda_4 bmi_i + \lambda_5 TH_i + \alpha \log(Glucose)_i(t)),$$

where $h_0(t)$ is the baseline risk function, t is the time-to-event and $log(Glucose)$ is the true (unobserved) value of the longitudinal outcome.

```
> fitSurv<-coxph(Surv(timeExitus, exitus)
+ ~ vhc+meld,data=dm1,x=TRUE)
```

The *JM* package permits to specify different types of survival sub-models to be fitted and the type of numerical integration method such as a Weibull model with a relative risk function and a spline-approximated baseline risk function. We will show in Section 4 the results of these approaches and the final model choosed comparing their Akaike Information Criterion, AIC [2].

3.2   Longitudinal Sub-model

We introduced three different longitudinal sub-models to study the longitudinal outcome, namely,

a) *Intercept Model*

$$\log(Glucose)_{i,diab} = \beta_0 + \beta_1 time_i + \beta_2 vhc_i + \beta_3 meld_i + U_{0i} + \epsilon(t_{ij}),$$
$$\log(Glucose)_{i,nodiab} = \beta_0 + \beta_1 time_i + \beta_2 age_i + \beta_3 carc_i + \beta_4 meld_i +$$
$$+ \beta_5 bmi_i + \beta_6 TH_i + U_{0i} + \epsilon(t_{ij}),$$

where *time* is the time that repeated measurements are taken and U_{0i} is the random intercept effect for each patient.

```
> fitLME.int<-lme(log(Glu)~time+vhc+meld,random=~1|id,
+data=subset(trasplante.long,trasplante.long$dm=="Yes"),
+na.action=na.omit)
```

b) *Slope Model*

$$\log(Glucose)_{i,diab} = \beta_0 + \beta_1 time_i + \beta_2 vhc_i + \beta_3 meld_i + U_{0i} +$$
$$+ U_{1i} t_{ij} + \epsilon_i(t_{ij}),$$
$$\log(Glucose)_{i,nodiab} = \beta_0 + \beta_1 time_i + \beta_2 age_i + \beta_3 carc_i + \beta_4 meld_i +$$
$$+ \beta_5 bmi_i + \beta_6 TH_i + U_{0i} + U_{1i} t_{ij} + \epsilon_i(t_{ij}),$$

in this model we additionally have $U_{1i}(t_{ij})$ which represents the random slope effect of the different Glucose trajectories of each patient.

```
> ctrl <- lmeControl(opt='optim')
> fitLME.slope<-lme(log(Glu)~time+vhc+meld,
+random=~time|id,data=subset(trasplante.long,
+trasplante.long$dm=="Yes"),na.action=na.omit,control=ctrl)
```

c) *Spline Model*

$$log(Glucose)_{i,diab} = (\beta_0 + b_{i0}) + (\beta_1 + b_{i1})B_n(t,d1) + (\beta_2 + b_{12})B_n(t,d_2)$$
$$+ (\beta_3 + b_{i3})B_n(t,d_3) + \beta_4 vhc_i + \beta_5 meld_i + \epsilon_i(t),$$
$$log(Glucose)_{i,nodiab} = (\beta_0 + b_{i0}) + (\beta_1 + b_{i1})B_n(t,d1) + (\beta_2 + b_{12})B_n(t,d_2)$$
$$+ (\beta_3 + b_{i3})B_n(t,d_3) + \beta_4 age_i + \beta_5 carc_i +$$
$$+ \beta_6 meld_i + \beta_7 bmi_i + \beta_8 TH_i + \epsilon_i(t),$$

where $\{B_n(t,d_k); k = 1,2,3\}$ denotes a B-spline basis matrix for a natural cubic spline (de Boor, 1978).

```
> fitLME.spline<-lme(log(Glu)~ns(time,3)+vhc
++meld,random=list(id=pdDiag(form=~ns(time,3))),data =
+subset(trasplante.long,trasplante.long$dm=="Yes"),
+na.action=na.omit)
```

3.3 Dynamic Predictions

In joint modelling approaches the objective is to study the association between the survival process and longitudinal outcomes. Such models can be used to provide predictions for the survival and longitudinal outcomes. The predictions of a joint model have a dynamic nature, coming from the effect of repeated measures taken in time t to the survival up to time t. Permitting us to update the prediction when we have new information recorded for the patient. Thus, the conditional probability is of primary interest, described as,

$$\pi_i(u/t) = P(T_i^* \geq u/T_i^* > t, Y_i(t), \omega_i, D_n),$$

where u is the followed-up time $(u > t)$, D_n denotes the sample on which joint model was fitted. Rizopoulos (2011) uses a Bayesian formulation of the problem and Monte Carlo estimates of $\pi_i(u/t)$.

4   Results

In this Section we applied the joint modelling approach described in Section 3, to assess the effect of glucose profiles in survival after Liver transplantation.

Different longitudinal sub-models are analysed with an only intercept, intercept and slope analysis and a non-linear subject specific evolutions for the Glucose levels. Using Akaike Information Criterion (AIC), we chose the longitudinal sub-model for each group with the less AIC value as shown in Table 1.

Table 1. AIC values of different Longitudinal submodels

	AIC	
Model	Diabetes	No Diabetes
Intercept	1042.675	3305.060
Slope	1044.719	3255.652
Spline	1024.077	2244.641

Comparing different sub-models for the survival process of transplant data such as mentioned above: I) a time-dependent relative risk model with Weibull baseline risk function and II) a time-dependent relative risk model with the log baseline risk function is approximated using B-splines:

```
>fit.JM.weibull<-jointModel(fitLME.spline,
fitSurv,timeVar="time")
>fit.JM.spline<-jointModel(fitLME.spline,fitSurv,
+timeVar="time",method="spline-PH-aGH")
```

Following the same procedure, we obtained the final model which the less AIC value as shown in Table 2.

The results indicate that final models for diabetic and non-diabetic patients take a relative risk model with Weibull baseline risk function and with a spline longitudinal sub-model. In Table 3 and 4 we synthesized all the information of both joint models. We observed that non-diabetic patients with higher Glucose

Table 2. AIC values of different survival submodels

| | AIC | |
Model	Diabetes	No Diabetes
fit.JM.weibull	1759.246	5377.097
fit.JM.spline	-	8475.035

level have a worse survival through the coefficients of association 0.0032 (95% CI: 0.002 − 0.004). However, for diabetic patients the association between the Glucose levels and survival process is not statistically significant, 0.0002 (95% CI: −0.0033 − 0.0004).

Table 3. Fitted values of the final model for the joint model approaches for diabetic patients

Joint Models (*JM*) - Diabetic patients		Coef	Std. error
	Intercept (β_0)	5.3261	0.0591
	β_1	-0.4126	0.0487
	β_2	-0.1523	0.0760
Longitudinal	β_3	-0.2286	0.0376
Process	β_4	-0.0696	0.0508
	β_5	-0.0006	0.0039
	Glucose	0.0002	0.0020
Survival	vhc	0.8412	0.3349
Process	meld	0.0464	0.0300
LogLikelihood		−869.1573	

4.1   Predictive Accuracy

Distinguishing the predictions between patients with and without diabetes, we compared the joint modelling approach, described in Section 3, with the following survival models, to notice the advantage and disadvantages of each:

Survival model (1)
   In these models we introduced the baseline covariates: *vhc* and *meld* for diabetic patients and *age*, *carc*, *meld*, *bmi* and *TH* for non-diabetic patients.

$$h_{i,diab}(t) = h_0(t) \exp(\lambda_1 vhc_i + \lambda_2 meld_i),$$
$$h_{i,nodiab}(t) = h_0(t) \exp(\lambda_1 age_i + \lambda_2 carc_i + \lambda_3 meld_i$$
$$+ \lambda_4 bmi_i + \lambda_5 TH_i).$$

Survival model (2)
   In the second model we added a baseline Glucose level (*glu0*) as another covariate.

Table 4. Fitted values of the final model for the joint model approaches for non-diabetic patients

Joint Models (*JM*) - Non-Diabetic patients		Coef	Std. error
	Intercept (β_0)	4.4489	0.0602
	β_1	-0.1849	0.0183
	β_2	0.6881	0.0332
	β_3	-0.3461	0.0136
Longitudinal	β_4	0.0032	0.0007
Process	β_5	-0.0130	0.0195
	β_6	0.0026	0.0012
	β_7	0.0030	0.0021
	β_8	0.0053	0.0009
	Glucose	0.0032	0.0010
	age	0.0293	0.0079
Survival	carc	0.6604	0.1899
Process	meld	0.0755	0.0116
	bmi	-0.0517	0.0224
	TH	0.0216	0.0073
LogLikelihood		-2659.553	

$$h_{i,diab}(t) = h_0(t)\exp(\lambda_1 vhc_i + \lambda_2 meld_i + \lambda_3 glu0),$$
$$h_{i,nodiab}(t) = h_0(t)\exp(\lambda_1 age_i + \lambda_2 carc_i + \lambda_3 meld_i$$
$$+ \lambda_4 bmi_i + \lambda_5 TH_i + \lambda_6 glu0).$$

In order to carry out such comparison, we used the linear predictors at time t to compute the ROC curves and the Area Under Curve for each time point [8]. This calculation is implemented in R package *risksetROC* [9]. As we can observe in Figure 3, the behaviour is completely different depending on a positive or negative diabetes diagnosis.

Dynamic Predictions. Figure 4 shows the dynamic predictions of particular cases: a diabetic patient (subject 51) and a non-diabetic patient (subject 40), with their longitudinal observations to observe the effect of the longitudinal outcome to the survival probability of these patients.

5   Computational Aspects

Joint modelling approach for longitudinal and time-to-event data requires a combination of a double numerical integration and optimization. The function `jointmodel()` implements a hybrid optimization procedure to locate the maximum likelihood estimates, starting with EM algorithm and if not converge switches to a quasi-Newton algorithm until it converges. These requirements of both double optimization and numerical integration may lead us to experience

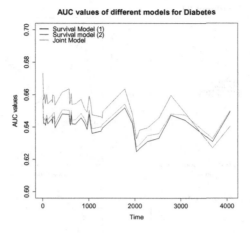

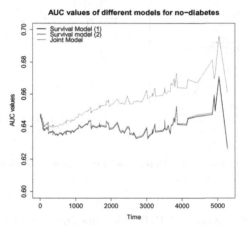

Fig. 3. Time dependent AUCs for each model separately for diabetic and non-diabetic patients

convergence problems. Although the function `jointmodel()` permits choices for default control arguments such as number of quadratic points, number of iterations, convergence tolerances, it is not guaranteed to work in all datasets. These aspects are described and discussed in [19] . The author also mentions that in the majority of the cases, the converge problems can be avoided by changing the stating values, increasing EM iterations or choosing other locations for the knots if the piecewise constant of spline-based baseline hazard functions are used. In this study we experienced converge problems while fitting a spline-based baseline hazard function for the survival sub-model.

```
>fit.JM.spline1<-jointModel(fitLME.spline,fitSurv,timeVar="time",
 +method="spline-PH-aGH",verbose=T)
```

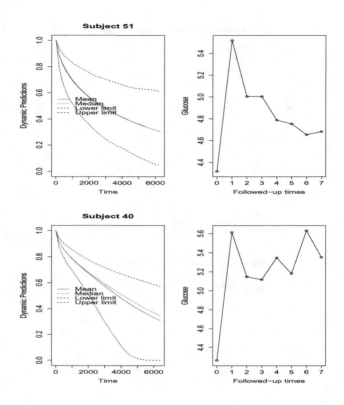

Fig. 4. Dynamic predictions for the final model for a diabetic patient (subject 51) and a non-diabetic patient (subject 40)

Setting the "verbose" option to TRUE, the function gives us the optimization path:

```
iter: 1
log-likelihood: -2302.596
betas: 5.2325 -0.412 -0.1326 -0.2248 -0.0766 0.0059
sigma: 0.3613
gammas: 1.2437 8e-04
alpha: 0.6122
gammas.bs: -12.7378 -11.4164 -11.4444 -11.027 -10.9222 -10.7309 -10.6004
         -10.4899 -10.4587
D: 0.0393 0 0 0.0318


...

iter: 119
```

```
log-likelihood: 0
betas: 5.2326 -0.4117 -0.1327 -0.2248 -0.0763 0.0059
sigma: 0.4144
gammas: 1.2437 8e-04
alpha: 0.6122
gammas.bs: -12.7378 -11.4164 -11.4444 -11.027 -10.9222 -10.7309 -10.6004
            -10.4899 -10.4587
D: NaN NaN NaN NaN

iter: 120
log-likelihood: 0
betas: 5.2326 -0.4117 -0.1327 -0.2248 -0.0763 0.0059
sigma: 0.4144
gammas: 1.2437 8e-04
alpha: 0.6122
gammas.bs: -12.7378 -11.4164 -11.4444 -11.027 -10.9222 -10.7309 -10.6004
            -10.4899 -10.4587
D: NaN NaN NaN NaN


quasi-Newton iterations start.

Error en optim(thetas, LogLik.splineGH, Score.splineGH, method = "BFGS",   :
  valor no finito provisto por optim
```

We can observe in the output the first iteration has a reasonable log-likelihood value but from the iteration 3 the log-likelihood values are 0 and then higher estimations for the longitudinal sub-model coefficients have obtained. As mentions [19], this problem usually comes from a failure of the numerical integration rule or from parameter scale problem.

Besides its several advantages in estimations, the joint modelling doesn't have a fast computation in which some further studies required. For this particular study, the timing of the joint model with different survival sub-models and a cubic spline model for the longitudinal process are as follows, 4.50 for B-spline survival model and 7.20 minutes for the Weibull model. The timings above-mentioned are based on Intel Core(TM)2.30 GHz 8GB RAM, Windows Vista.

6   Conclusions

In this paper two different joint regression models for diabetic and non-diabetic patients are obtained. From these models, for non-diabetic patients we have obtained a significant effect post Glucose levels on survival and this effect improves the predictive performance of the model according to AUC values. As we can observe in Figure 4, the dynamic predictions of the survival process of patient 40, which has no diabetes, decrease after a certain time point having a higher post-transplant Glucose levels in the first week.

The behaviours of AUC values are different for diabetic and non-diabetic patients. It can be observed in Figure 3 that for non-diabetic patients, the discrimination capacity is low for a model with a single Glucose measure. The model improves by adding longitudinal observations of the Glucose levels.

In joint modelling of longitudinal data and time-to-event analysis and time-dependent ROC curves are useful tools to show the abilities of postoperative glucose profiles in predicting death in patients who underwent liver transplantation.

Acknowledgments. This work was supported by grants PI11/002219 from the Instituto de Salud Carlos III and the MTM2011-28285-C02-01 from the Ministry of Economy and Competitiveness.

References

1. Anderson, P.K., Gill, R.D.: Cox's Regression Model for Counting Process: A Large Sample Study. Annals of Statistics 10, 1100–1120 (1982)
2. Akaike, H.: A new look at the statistical model identification. IEEE Transactions on Automatic Control 19(6), 716–723 (1974)
3. van Buuren, S., Groothuis-Oudshoorn, K.: MICE: Multivariate Imputation by Chained Equations in R. Journal of Statistical Software 45(3) (2011)
4. de Boor, C.: A Practical Guide to Splines. Springer, Berlin (1978)
5. Dossett, L.A., Cao, H., Mowery, N.T., Dortch, M.J., Morris Jr., J.M., May, A.K.: Blood glucose variability is associated with mortality in the surgical intensive care unit. Am. Surg. 74(8), 679–685 (2008)
6. Dutkowski, P., De Rougemont, O., Clavien, P.A.: Current and Future Trends in Liver Transplantation in Europe. Gastroenterology 138(3), 802–809 (2010)
7. Egi, M., Bellomo, R., Stachowski, E., French, C.J., Hart, G.: Variability of blood glucose concentration and short-term mortality in critically ill patients. Anesthesiology 105(2), 244–252 (2006)
8. Heagerty, P.J., Zheng, Y.: Survival model predictive accuracy and ROC curves. Biometrics 61(1), 92–105 (2005)
9. Heagerty, P.J., Saha-Chaudhuri, P.: risksetROC: Riskset ROC estimation from censored survival data. Biometrics 61(1), 92–105 (2012)
10. Ibrahim, J.G., Chu, H., Chen, L.M.: Basic concepts and methods for joint models of longitudinal and survival data. J. Clin. Oncol. 28(16), 2796–2801 (2010)
11. Krinsley, J.S.: Glycemic variability: A strong independent predictor of mortality in critically ill patients. Crit. Care Med. 36(11), 3008–3013 (2008)
12. Laryea, M., Watt, K.D., Molinari, M., Walsh, M.J., McAlister, V.C., Marotta, P.J., et al.: Metabolic syndrome in liver transplant recipients: prevalence and association with major vascular events. Liver Transpl. 13, 1109–1114 (2007)
13. Meyfroidt, G., Keenan, D.M., Wang, X., Wouters, P.J., Veldhuis, J.D., Van den Berghe, G.: Dynamic characteristics of blood glucose time series during the course of critical illness: effects of intensive insulin therapy and relative association with mortality. Crit. Care Med. 38(4), 1021–1029 (2010)
14. Philipson, P., Sousa, I., Diggle, P., Williamson, P., Kolamunnage-Dona, R., Henderson, R.: joineR: Joint modelling of repeated measurements and time-to-event data (2012)

15. Pinheiro, J., Bates, D.: Mixed-Effects Models in S and S-PLUS. Springer, New York (2000)
16. Ratcliffe, S.J., Guo, W., Have, T.R.T.: Joint modeling of longitudinal and survival data via a common frailty. Biometrics 60(4), 892–899 (2004)
17. Rizopoulos, D.J.M.: An R package for the joint modelling of longitudinal and time-to-event-data. Journal of Statistical Software 35(9), 1–33 (2010)
18. Rizopoulos, D.: Dynamic predictions and properspective accuracy in joint models for longitudinal and time-to-event data. Biometrics 67, 819–829 (2011)
19. Rizopoulos, D.: Joint Models for Longitudinal and Time-to-Event Data. With Applications in R. Chapman & Hall/CRC Biostatistics Series (2012)
20. Self, S., Pawitan, Y.: Modelling a marker of disease progression and ofset of disease. In: Jewell, N.P., Dietz, K., Farewell, V.T. (eds.) AIDS Epidemiology: Methodological Issues, Birkhauser, Boston (1992)
21. Sweeting, M.J., Thompson, S.G.: Joint modelling of longitudinal and time-to-event data with application to predicting abdominal aortic aneurysm growth and rupture. Giome. J. 53(5), 750–763 (2011)
22. Wolbers, M., Babiker, A., Sabin, C., et al.: Pretreatment CD4 cell slope and progression to AIDS or death in HIV-infected patients initiating antiretroviral therapy the CASCADE collaboration: A collaboration of 23 cohort studies. PLoS Med 7, e1000239 (2010)

Performance Comparison of Screening Tests for POAG for T2DM Using Comp2ROC Package

Ana C. Braga[1], Lígia Figueiredo[2], and Dália Meira[2]

[1] Department of Production and Systems Engineering,
Algoritmi Research Centre, University of Minho,
4710-057 Braga, Portugal
acb@dps.uminho.pt
[2] Department of Ophthalmology,
Centro Hospitalar de Vila Nova de Gaia/Espinho EPE
4434-502 Vila Nova de Gaia, Portugal
ligia_figueiredo@hotmail.com, daliamartinsmeira@gmail.com

Abstract. Primary open angle glaucoma (POAG) is a leading cause of blindness [8]. The commonly accepted risk factors for POAG include older age, high intraocular pressure, positive family history of glaucoma, and race [3].

Individuals with type 2 *diabetes mellitus* (T2DM) are at greater risk of having open-angle glaucoma (POAG) than those without T2DM. Thus, screening programs for diabetic retinopathy may be an excellent opportunity for screening for POAG and implementation of additional tests.

In this work we intend to evaluate the performance of GDx measures for detecting POAG through the methodology of ROC curves.

One retrospective cross-sectional study was carried out on individuals with T2DM with evaluation of diabetic retinopathy from 2008 to 2010 in Hospital Centre of Vila Nova de Gaia Northern Portugal. Individuals who had a positive screen were referred for scanning the nerve fiber layer of the retina with GDx[TM]device. In this study, the diagnosis of POAG was based on a computerized ophthalmoscopy and static analysis and based on this, the eyes were classified as healthy ($n_N = 85$) and with definite glaucoma ($n_A = 37$).

The comparison of diagnostic indicators was performed by *Comp2ROC* package [2].

Keywords: ROC (Receiver Operating Characteristic), AUC (Area Under the Curve), *Comp2ROC*.

1 Introduction

The index area under the ROC (Receiver Operating Characteristic) curve is a widely used measure of accuracy for evaluating the performance of diagnostic systems. A comparison of two imaging systems through this index was developed by authors such as Metz [10], considering the binormal model, and by Hanley and McNeil [6], using an approach based on nonparametric statistic of Wilcoxon-Mann-Whitney. As suggested by Rockette [15], this nonparametric procedure

B. Murgante et al. (Eds.): ICCSA 2014, Part III, LNCS 8581, pp. 594–605, 2014.
© Springer International Publishing Switzerland 2014

should be considered as a viable alternative to maximum likelihood (ML) used by Metz, when there are practical problems with convergence, or when it is believed that the conditions necessary for correct application of ML (i.e. binormal distribution) are not satisfied.

When one intends to compare two diagnostic systems based on the index area under the ROC curve (AUC), if the ROC curves associated with these two diagnostic systems do not intersect, the system whose ROC curve is closer to the upper left corner provides greater discriminatory power. However, it may happen that the ROC curves intersect and the test based on the index area under the ROC curve becomes compromising for evaluation [1].

Thus, the need to compare the two systems in terms of performance evaluation by ROC methodology has led Frade and Braga [2] to create a package in R [14], *Comp2ROC*, that would allow make this comparison. The choice of the R language relates to the fact that this is a language with an integrated development environment for statistical calculations and graphics.

Using ROC curves methodology for diagnostic test comparison, we illustrate this application with a screening method for POAG, running along with the diabetic retinopathy screening program, measured by GDx VCC.

2   Problem Description

According Weinreb [9] the primary open-angle glaucoma (POAG) is the most common form of glaucoma. This disease is a progressive optic neuropathy and will be treatable. Without adequate treatment, glaucoma can progress to visual disability and eventual blindness.

Because the visual impairment caused by glaucoma is irreversible, early detection is essential. The early diagnosis of POAG depends on examination of the optic disc, retinal nerve fibre layer, and visual field. New imaging and psychophysical tests can improve both detection and monitoring of the progression of the disease.

By another way, people, predominantly white, with Type 2 Diabetes Mellitus (T2DM) have an independent higher risk of having open angle glaucoma than those without T2DM, so screening programs for diabetic retinopathy are an excellent opportunity for screening POAG, implementing additional testing.

The GDx VCC is an instrument that uses laser to determine the thickness of the nerve fiber layer. Older glaucoma tests have centered around measuring eye pressure or measuring the effect that glaucoma has on your overall visual field. Although these tests are extremely important in the treatment and management of glaucoma, it would be helpful to measure or test early what damage glaucoma can cause to the nerve fiber layer in the back of the eye [20]. This procedure is noninvasive and not painful.

The nerve fiber layer consists of millions of individual fibers called axons involving the optic nerve. Patients with glaucoma has a decreased nerve fiber layer, and in most cases, when they feel the loss of vision, there are already a significant loss of this fiber layer. Hence the GDx could be a good test to detect patients with asymptomatic glaucoma.

The nerve fiber layer measured by GDx is compared to the nerve fiber layer of normal eyes which are in a database. The decrease of the fibers indicates glaucoma. This information is then made available in the form of pictures, graphs and statistics that indicate the probability for glaucoma (Fig. 1).

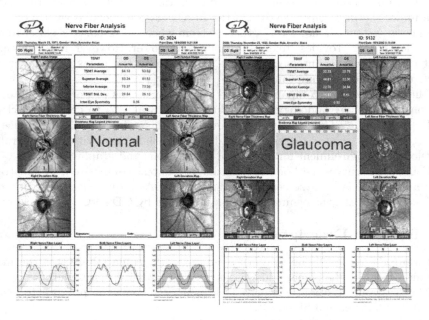

Fig. 1. Example of output of GDx VCC test [21]

In addition to measure the nerve fiber layer, in a global way, the optic nerve is divided into four quadrants and the fibers of the upper and lower quadrants are also evaluated separately. The evaluated measures are:

- **TSNIT** (Temporal-Superior Nasal Inferior-Temporal) Average: indicates the average of retinal nerve fiber layer thickness around the entire calculation circle(lower value indicates a diagnosis of glaucoma);
- **GDxS**, Superior Average: average of retinal nerve fiber layer of the upper quadrant average (lower value indicates a diagnosis of glaucoma);
- **GDxI**, Inferior Average: average of retinal nerve fiber layer of the lower quadrant (lower value indicates a diagnosis of glaucoma);
- **NFI**, Nerve Fiber Indicator: it is based on global measure of retinal nerve fiber layer (higher value indicates a diagnosis of glaucoma).

3   Methodology

3.1   Comparison Using Two ROC Curves

In ROC analysis, for a given test, the compromise between the False Positive Fraction (FPF) and the True Positive Fraction (TPF) can be presented through a graph called ROC curve. Consider a diagnostic system for disease detection, then we have the non disease (ND) and disease (D) samples that can be represented by X_i^{ND} and X_j^D, with $i = 1, ..., n_{ND}$ and $j = 1, ..., n_D$, respectively. If we assume that the diagnostic system is based on a continuous measurement T and that a patient is classified as non diseased if $T < c$ and diseased otherwise, the distributions of T for each group can be denoted by $FPF(c) = P(T \geq c|ND)$ and $TPF(c) = P(T \geq c|D)$. Thus, $TPF(c)$ is the sensitivity and $1 - FPF(c)$ is the specificity of the diagnostic test. The ROC curve can be defined as $ROC(t) = TPF(FPF^{-1}(t))$ where $0 \leq t \leq 1$, i.e., for all different thresholds c [1]:

$$ROC(.) = (FPF(c), TPF(c)), c \in (-\infty, \infty).$$

The usual measures for testing ROC curves are based on the comparison of the areas below it. The area under the ROC curve (AUC) is a global measure of accuracy across all possible threshold values (c).

$$AUC = \int_0^1 ROC(t)dt = \int_{-\infty}^{+\infty} TPF(c)dFPF(c) \tag{1}$$

When comparing two diagnostic systems based on the AUC index, if the two curves do not cross each other, it is well-known that the ROC curve with the greatest value of the index corresponds to the system which presents a better performance, i.e., a greater discriminant power. In ROC unitary space this corresponds to the curve closer to the left upper corner.

The comparison of two diagnostic systems was primarily developed by [11]. These authors use the approximation to a normal model to calculate and test the value of the index as a function of the two parameters of the straight line on the binormal plane.

Hanley and McNeil [6] suggested a nonparametric approximation to the Wilcoxon-Mann-Whitney statistic to calculate the AUC index and developed a Z test to compare two ROC curves. In these approaches they assume that the two curves not cross each other.

However, the AUC index can produce the same value for quite different curves that cross each other. Furthermore, it can be argued that the area under the ROC curve includes regions of little interest in terms of false positive fractions(FPF). For this reason, several works have addressed the problem of comparing ROC curves in specific regions of interest with respect to a fixed value of FPF [5, 7, 12, 17–19].

The nonparametric methodology proposed by Braga et al [1] considers the entire ROC curve and, while providing a global comparison, specifies the ranges of FPF over which the two ROC curves differ. Thus, this methodology can deal with ROC curves that cross each other at more than one point. Moreover, it is also possible to identify and compare particular portions of the ROC curves based on measures used to assess the performance of multi-objective algorithms [1].

3.2   *Comp2ROC* Package Description

The version 1.0 of *Comp2ROC* is the result of compiling functions that implement a methodology for evaluating two diagnostic systems based on measures of ROC curves performance developed by Braga et al [1].

The package allows drawing empirical ROC curves in the unitary ROC plan, distances graphs and exhibits the curves comparison results. The application is developed in R (requires version 2.15.1 or higher and **boot** and **ROCR** packages)[2].

In this section it will describe the functions used on *Comp2ROC*.

curvesegslope function. This function calculates the slopes of the lines that together build the empirical ROC curve. It takes as parameters the TPF (true positive fraction) and FPF (false positive fraction) of the ROC curve.

curvesegsloperef function. Calculates the slopes of the lines that join the points of the ROC curve to the reference point (1,0) in unitary ROC plan. The function parameters are the TPF, the FPF and also the reference point.

lineslope function. This is used to calculate the slopes of the sampling lines, depending on the number desired by the user (usually $K = 100$). It returns a vector with these values.

linedistance function. This function calculates the intersection points of the sampling lines with the ROC curves to compare. It also calculates the distance between these points and the reference point. The parameters of this function are the TPF and FPF points, the slopes calculated by *curvesegslope*, *curvesegsloperef*, *lineslope* and also the reference point.

areatriangles function. This function is built to calculates the triangles areas formed by two sequential points and the reference point. In addition it also calculates the total area, based on previous triangles. The parameters of this function are the results obtained in *lineslope* and *linedistance*.

diffareatriangles function. Calculates the difference between triangles areas of the two curves. This also allows calculating the difference between the total areas, that define de TS (test statistic). The parameters of this function are the triangles areas of the two curves obtained in *areatriangles*.

rocsampling function. Creates the two curves and compare them by extension and location. All previous functions are used here. It takes the TPF and FPF points of each ROC curve in ROC space, and also the number of sampling lines, K. As output value, returns a list of the overall areas, the proportions and the locations measures for each curve, the slopes of the sampling lines, the differences between areas and finally the distances calculated by *linedistance*.

rocsampling.summary function. Receives and presents the results obtained through the function *rocsampling*.

comp.roc.curves function. Using bootstrapping, this function estimates the real distribution for the entire set for TS. It takes as parameters the result obtained by *rocsampling*, two flags and the name to the plot. One flag indicates if the user wishes calculate the confidence interval and the other if the user wants to make the plot. As output values,it returns the test statistic (TS), two p-values (one-sided and two-sided) and the confidence interval.

comp.roc.delong function. This function is related to the areas calculation and some statistical measures. Firstly, it divides the data into two categories, negative and positive. Then it calculates the Wilcoxon-Mann-Whitney matrix for each test. Next there are calculated some values such as areas, standard errors and global correlations, which are the output of the function. It takes as parameters the data of each curve and their status.

roc.curves.plot function. This function draws the graph of the two ROC curves. Curve 1 is shown using solid line and curve 2 is shown using dotted line. The random line is shown using dashed line.

read.manually.introduced function. This function reads the data that will be used in comparison. Only works with data already introduce in the R workspace. It needs the name of two modalities and their dependency. If they are independent the user must give second test status. This variable must be ordered by state, first negatives values followed by the positives. At the end the user must identify the test direction for each test to compare.

read.file function. Reads the data that will be used in comparison, like *read. manually.introduced*, but in this case only works with data saved in txt and csv formats, so the user must say if there's any header in the file. In addition to the parameters already described in *read.manually.introduced* is also needed to indicate the column and decimal separator.

roc.curves.boot function. This is the core function of the package. The user must give the data, an α for the confidence level, the graphics display name and the number of permutations. Apart this, is also needed the name of the two modalities that will be compared. As return value, it give a list with the areas of each test standard errors, confidence intervals, the areas through the trapezoidal rule, the correlation coefficient, the sum of the differences of areas and corresponding confidence interval, the Z-stats and its p-value and the number of existing crossings.

rocboot.summary function This function displays the data obtained in *roc. curves.boot* in the R command line. These are structured according to Fig. 3.

save.file.summary function This function saves the results as the structure shown in *rocboot.summary function* in a `txt` file. To save this data the users must provide the results obtained, the file name and also the names of the two modalities as well a parameter that indicates how the file is saved (overwrite or attach).

3.3   Using Comp2ROC Package

In this section is explained how to use Comp2ROC to compare related samples. We test the package using data set created in `.csv` format.

The comparison test should be done using the following commands in the order they are presented:

1. `nameE="nome"` - graphs and output file name
2. `moda1="measure1"` - test one column name
3. `moda2="measure2"` - test two column name
4. `data=read.file("data.csv",TRUE,";",",",moda1,TRUE,moda2,TRUE,"status",` `TRUE)` - the file is read. According to the details, this file has header, ";" indicates the column separator and "," is the decimal separator. The names of the modalities are followed by the direction of the test. At the end the logical value TRUE, indicates that the modalities are related.
5. `results=roc.curves.boot(data,name=nameE,measure1=moda1,measure2=moda2)` - calculates the comparison parameters
6. `rocboot.summary(results,measure1,measure2)` - the results are shown in command line
7. `save.file.summary(results,nameE,app=TRUE,measure1,measure2)` - results are saved in file, and in mode attachment

If the user wishes to introduce manually the data vectors instead of the command `read.file` the user must use the command `read.manually.introduced` and proceed according.

4   Results

The used data set was obtained from a retrospective cross-sectional study carry out at a hospital at North of Portugal. People with T2DM screened for diabetic retinopathy between 2008 and 2010, were submitted to a two-stage screening method for POAG.

Per subject, the data were obtained from one eye randomly selected. After excluding eyes with missing data or with poor quality scans, the data of 185 eyes was analyzed. From these eyes 85 were classified as healthy ($n = 85, 45.9\%$), 63 as glaucoma suspect ($n = 63, 34.1\%$) and 37 as definitive glaucoma ($n = 37, 20\%$). For this study we do not consider the eyes with diagnostic of glaucoma suspect. The final sample consist of 122 eyes with GDx VCC results.

4.1   R Script for 2 Related Samples

In the following code is an example of a script to compare two paired measures to the data set of measures mentioned above.

Example of a script in R

```
####################################################################
#Comparison NFI vs GDxS
nameE="glaucoma_comp1"
NFI="NFI"
GDxS="GDxS"
data=read.file("data_glaucoma.csv",TRUE,";",",",NFI,TRUE,GDxS,
FALSE,"status",TRUE)
results=roc.curves.boot(data,name=nameE,mod1= NFI,mod2=GDxS)
rocboot.summary(results,NFI,GDxS)
save.file.summary(results,nameE,app=TRUE,NFI,GDxS)
####################################################################

####################################################################
#Comparison GDxS vs GDxTSNIT
nameE="glaucoma_comp6"
GDxS="GDxS"
GDxTSNIT="GDxTSNIT"
data=read.file("data_glaucoma.csv",TRUE,";",",",GDxS,FALSE,GDxTSNIT,
FALSE,"status",TRUE)
results=roc.curves.boot(data,name=nameE,mod1= GDxS,mod2=GDxTSNIT)
rocboot.summary(results,GDxS,GDxTSNIT)
save.file.summary(results,nameE,app=TRUE,GDxS,GDxTSNIT)
####################################################################
```

The R results are presented while the steps are performed, as illustrated in Fig. 2 for the first comparison (NFI vs GDxS).

Through analysis of Fig. 2(a) when NFI index is compared with GDxS, NFI is superior to GDxS in 35.6% of the ROC space. On the other hand, GDxS is superior to NFI in 25.7% of ROC space. In the remaining 38.6% of the space, the indexes present identical performance These results could be illustrated by the graph in Fig. 2(b).

Fig. 2(c) represents the difference between areas (solid line) with their confidence intervals (dotted and dashed line). The analysis shows that all values above 0 represent a win to NFI, in the opposite GDxS wins. We can see if both limits of confidence intervals are above or below 0, there is a significant difference between indexes. Fig. 2(d) gives $t*$ (test statistic) distribution and also a Quantile-Quantile chart as result of permutation test.

ROC SAMPLING RESULTS
Number of sampling lines: 100
Proportion NFI :   0.3564356
Proportion GDxS :   0.2574257
Proportion ties: 0.3861386

(a) First results obtained

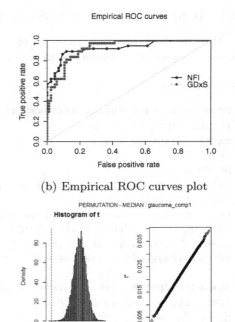

(b) Empirical ROC curves plot

(c) Area Plot

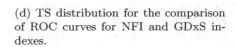

(d) TS distribution for the comparison of ROC curves for NFI and GDxS indexes.

Fig. 2. Comp2ROC output images

For this example, final results are shown in Fig. 3. This results can be saved in a .txt file and represents a summary of individual results for each index to be compared, in terms of AUC values, their standard errors, and confidence intervals for each corresponding area.

These results show that in terms of AUC value, the NFI and GdxS exhibit equal performance. The value of correlation coefficient between areas reveals that two indexes are related.

In comparison tests results, the value of Z-stats (based on Z proposed by Hanley and McNeil [6]) and respective p-value, indicate that on the whole there are no statistically significant differences between the two indexes. The values obtained for the two limits on confidence interval for the difference between areas (proposed method) also indicate that on the whole there are no statistically significant differences between the two modalities. However the information in the graph of Fig. 2(c) allows detect one distinct region were one curve performs better than other.

```
-------------------------------------------------
NFI
-------------------------------------------------
Area:                                   0.9259141
Standard Error:                         0.02837218
Area through Trapezoidal Method:        0.9258187
CI Upper bound (Percentil Method):      0.9745787
CI Lower bound (Percentil Method):      0.8659658
-------------------------------------------------


-------------------------------------------------
GDXS
-------------------------------------------------
Area:                                   0.9259141
Standard Error:                         0.02249312
Area through Trapezoidal Method:        0.9257377
CI Upper bound (Percentil Method):      0.9653418
CI Lower bound (Percentil Method):      0.8744038
-------------------------------------------------

Correlation Coefficient between areas:  0.5807365

TEST OF DIFFERENCES
Z stats:   0
p-value:   1

Sum of Global Areas Differences (TS):   0
CI Upper bound (Percentil Method):      0.0704372
CI Lower bound (Percentil Method):     -0.07408983

Number of Crossings:   1
```

Fig. 3. Final results saved in .txt file

Table 1. Summary results for all comparisons 2 by 2

	NFI	GDxS	GDxI	TSNIT
AUC	0.9259	0.9259	0.8184	0.8251
SE	0.02837	0.02249	0.05072	0.04947
NFI	-	No Sig.	Sig.	Sig.
		1	0	0
GDxS	-	-	Sig.	No Sig.
			1	3
GDxI	-	-	-	No Sig.
				2

In Table 1 are presented the results for all comparison tests to the measures obtained in GDx VCC in terms of area under curve (AUC), standard error for area(SE), result of statistical test and number of crossing points in ROC space.

5   Conclusions and Future Work

The package *Comp2ROC* solve the problem of comparing two ROC curves that cross each other according the methodology proposed by Braga et al [1]. This package also allows the user to visualize graphically the region on ROC space where one curve is superior to another.

All code and associated functions have resulted in one package which is designated *Comp2ROC* published in R CRAN. The notes in help file and examples were created from an example in the literature [19].

Regarding to evaluated screening tests this study has shown that, in terms of the value of the area under the ROC curve, the NFI and GDxS exhibit good discrimination capacity for the existence of primary open-angle glaucoma.

In the assessments using the comparative tests performed in *Comp2ROC* it was found that the NFI has better capacity compared with other indicators.

Regarding the package *Comp2ROC* there is some future work that will be implemented in further versions. At this point, the user just needs to use seven steps to perform the functionality of this. One of the changes to do is divide the function *roc.curves.boot* into smaller functions for intermediate values could be saved.

As future work we intended to display the intersection points as well to allow the user to identify the coordinates of these points in ROC space.

Acknowledgments. This work has been supported by FCT - Fundação para a Ciência e Tecnologia within the Project Scope: PEst-OE/EEI/UI0319/2014.

References

1. Braga, A.C., Costa, L., Oliveira, P.: An alternative method for global and partial comparison of two diagnostic system based on ROC curves. Journal of Statistical Computation and Simulation 83(2), 307–325 (2013)
2. Braga, A.C., Frade, H.: Comp2ROC: R Package to Compare Two ROC Curves. In: Mohamad, M.S., Nanni, L., Rocha, M.P., Fdez-Riverola, F. (eds.) 7th International Conference on Practical Applications of Computational Biology & Bioinformatics (PACBB 2013). AISC, vol. 222, pp. 127–136. Springer, Switzerland (2013)
3. Chopra, V., Varma, R., Francis, B.A., Wu, J., Torres, M., Azen, S.P.: Type 2 diabetes mellitus and the risk of open-angle glaucoma the Los Angeles Latino Eye Study. Ophthalmology 115(2), 227–232 (2008)
4. DeLong, E., DeLong, D., Clarkepearson, D.: Comparing the areas under 2 or more correlated receiver operating characteristic curves - a non parametrical approach. Biometrics 44(3), 837–845 (1988)
5. Dodd, L.E., Pepe, M.S.: Partial AUC estimation and regression. Biometrics 59(3), 614–623 (2003)
6. Hanley, J., McNeil, B.: A method of comparing the areas under receiver operating characteristic curves derived from the same cases. Radiology 148(3), 839–843 (1983)
7. Jiang, Y.L., Metz, C.E., Nishikawa, R.M.: A receiver operating: Characteristic partial area index for highly sensitive diagnostic tests. Radiology 201(3), 745–750 (1996)
8. Leske, M.C.: The epidemiology of open-angle glaucoma: A review. Am. J. Epidemiol. 118(2), 166–191 (1983)
9. Weinreb, R.N., Khaw, P.T.: Primary open-angle glaucoma. Lancet 363, 1711–1720 (2004)
10. Metz, C.E.: Statistical Analysis of ROC Dara in Evaluating Diagnostic Performance. In: Proceeding of First Midyear Topical Symposium: Multiple Regression Analysis Application in the Health Science, vol. 13, pp. 365–384 (1986)

11. Metz, C.E., Wang, P.-L., Kronman, H.: A new approach for testing the significance of differences between ROC curves measured from correlated data. In: Proceedings of the 8th Conference in Information Processing in Medical Imaging, Brussels, pp. 432–445 (1983)
12. McClish, D.K.: Analyzing a portion of the ROC curve. Med. Decis. Making 9(3), 190–195 (1989)
13. Pepe, M.S.: The Statistical Evaluation of Medical Tests for Classification an Prediction. Oxford Statistical Science Series. Oxford University Press, New York (2003)
14. R Development Core Team. R: A language and environment for statistical computing. Vienna: R Foundation for Statistical Computing (2006), http://www.R-project.org
15. Rockette, H.E., Obuchowski, N.A., Gur, D.: Nonparametric estimation of degenerate ROC data sets used for comparison of imaging-systems. Invest. Radiol. 25(7), 835–837 (1990)
16. Swets, J.A., Pickett, R.M.: Evaluation of Diagnostic Systems Methods from Signal Detection Theory. Academic Press, London (1982)
17. Thompson, M.L., Zucchini, W.: On the statistical analysis of ROC curves. Stat. Med. 8(10), 1277–1290 (1989)
18. Wieand, S., Gail, M.H., James, B.R., James, K.L.: A family of nonparametric statistics for comparing diagnostic markers with paired or unpaired data. Biometrika 76(3), 585–592 (1989)
19. Zhang, D., Zhou, X., Freeman, D., Freeman, J.: A nonparametric method for the comparison of partial areas under ROC curves and its application to large health care data sets. Stat. Med. 21(5), 701–715 (2002)
20. Glaucoma tests, http://vision.about.com/od/eyeexamination1/f/GDx.htm
21. Sharma, A., Sobti, A., Wadhwani, M., Panda, A.: Evaluation of Retinal Nerve Fiber Layer using Scanning Laser Polarimetry, www.jaypeejournals.com/eJournals/

Degradation Prediction Model for Friction in Highways

Adriana Santos[1], Elisabete Freitas[1], Susana Faria[2], Joel R.M. Oliveira[1],
and Ana Maria A.C. Rocha[3]

[1] Department of Civil Engineering, C-TAC – Territory,
Environment and Construction Centre, University of Minho, Portugal
asantos@ascendi.pt, {efreitas,joliveira}@civil.uminho.pt
[2] Department of Mathematics and Applications, CMAT-Centre of Mathematics,
University of Minho, Guimarães, Portugal
sfaria@math.uminho.pt
[3] Department of Production and Systems, Algoritmi Research Centre,
University of Minho, Portugal
arocha@dps.uminho.pt

Abstract. The purpose of this paper is to develop a multiple linear regression
model that describes the pavement's friction behaviour using a degradation evolution law that also considers the effects of weather, vertical alignment and traffic factors.

This study is based on real data obtained from two different highways with
an approximate total length of 43 km. These sections present different alignment features (plan/profile), different Annual Average Daily Traffic and are
subjected to different weather conditions. Nevertheless, both comprise the same
type of upper layer.

The efficiency of the linear regression model in approaching and explaining
data was demonstrated. The most relevant factors involved in the degradation
process of pavements' friction were identified.

Keywords: Prediction, regression model, friction, highways.

1 Introduction

Pavements can become degraded in several ways and their structural and functional
status can be described – according to the European project COST 324 – by seven
indicators: Longitudinal Profile, Transversal Profile, Surface Cracking, Structural
Cracking, Structural Adequacy, Surface Defects and Skid Resistance (friction) [2].

Pavements maintenance conditions are strongly conditioned by a significant number of variables, with several pondering factors. All these variables are inter-related
and affect the pavement's entire life cycle, both in terms of functionality and structure. Those variables can be used in pavements degradation models, which are a fundamental component of any pavement management system. These models are based
on the study of performance data and establish the evolution law concerning the
pavements degradation, by determining the relationships between the causes and
effects of pavement degradation [11].

B. Murgante et al. (Eds.): ICCSA 2014, Part III, LNCS 8581, pp. 606–614, 2014.

The main goal of a degradation model is to characterize the response-variable throughout time, as well as to determine whether this is related with a set of previously selected explanatory variables, such as the age, the traffic, the structure of the pavement and the weather conditions, among others.

In a modelling study of pavements degradation, in each sample unit, both the response-variable and the explanatory variables are constantly monitored. Some of these variables represent fixed characteristics of the sample unit, such as the traffic directions, the kilometric point and the type of lane (left, right and additional); therefore these are time independent explanatory variables, whereas others, such as the temperature and the traffic are time-dependent explanatory variables.

There are several pavement degradation models for the different status indicators, individual or global (that consider the individuals with pondering). Markov's prevision model is an example of a frequently used model of predicting the global conditions of a pavement, due to its ability to include pavement's rehabilitations and degradation rates in one single probability transition matrix [1].

On the other hand, when historical data are insufficient to develop numerical algorithms, non-formulated methods can be used, such as the Intelligent Systems, among which we highlight the artificial neural networks [6] and the adaptive systems (fuzzy) which are another category of modelling techniques [7]. With the combination of both (artificial neural networks and fuzzy systems), a hybrid system arose, named Neuro-fuzzy System [8].

Nevertheless, in the last decades many progresses have been made in this field, leading to the birth of other prediction models, such as the Nonlinear mixed-effects models that, due the flexibility they provide in handling imperfect data, such as those which are incorrect and/or inconsistent, provide more reliable previsions [10, 14].

Many degradation models have been developed for the longitudinal and transversal profile indicators, as well as for the cracking indicators. However, regarding the friction, few models are presented and even fewer are discussed in terms of modelling, as can be seen in COST 324 [3].

Friction has been acknowledged as one of the main factors contributing to the number of traffic accidents and is therefore essential to the assessment of the pavements quality, integrated within management systems. This indicator is influenced by a large number of parameters, such as the traffic volume and respective driving speed, the pavements characteristics and the grading composition, the precipitation and drainage capacity, as well as the temperature. In a study in the United Kingdom, it was recognized that the level of friction demanded for roads under operation would have to vary along the road, according to its geometry and other factors [13].

The main objective of this study is to describe the pavement's friction behaviour through a multiple linear regression model taking into account the effects of weather, vertical alignment and traffic factors.

The remainder of the paper is organized as follows. Section 2 describes the data, the variables and multiple linear regression models. The statistical analysis of the variable is done in Section 3. The results of applying multiple regression linear models to describe the pavement's friction behaviour are presented in Section 4. Finally, we make some concluding remarks in Section 5.

2   Materials and Methodology

2.1   Sample Design

This study is based on real data obtained from 100 m pavement sections of two different highway lanes with an approximate total length of 43 Km. The first stretch
(Section 1) is located in the Northern and the second (Section 2) in the Centre of Portugal. These sections present different alignment features (plan/profile), different
Annual Average Daily Traffic (AADT) and are subject to different weather conditions. Nevertheless, both comprise the same type of upper layer.

Each section was monitored for a period of approximately 8 years. The data comprises a total of 3366 observations.

2.2   The Data

For the development of this model we considered the variables that mostly influence
the pavement's friction degradation, such as the traffic, the weather conditions, the
texture and the alignment features, especially in terms of longitudinal profile.

The FRICTION is taken, for modelling purposes, as the dependent variable. This
variable represents the component of adherence and friction between the tyre and the
pavement surface and is responsible for a significant part of the highway accidents.
The value of friction can be expressed in several units, but in this study it was expressed in GN (Grip Number).

The explanatory variables are:

(a) TEXTURE (mm): designated by macro-texture, this corresponds to the texture
    with a wavelength between 0.5 and 50 mm. This is important since it enables the
    dispersion of surface water, providing drainage channels. Macro-texture is ob-
    tained by the adequate proportion between the grading and the surface mortar or
    by surface finishing techniques;
(b) Ac.AADT (vehicles): is the accumulated annual average daily traffic (light and
    heavy vehicles) up to each data acquisition moment;
(c) MAAT (°C): is the mean of the average air temperature obtained in the month and
    period referring to the Climatological Normal;
(d) VA: is the vertical alignment (1- Slopes, i; 2- Concave Curves, Ccv; 3- Convex
    Curves, Ccx), as shown in Fig. 1.

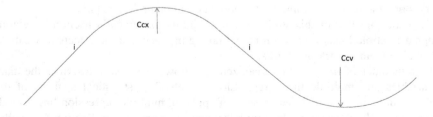

Fig. 1. Vertical Alignment in the highway

2.3   Statistical Modelling

Regression analysis is one of the most widely used statistical tools because it provides simple methods for establishing a functional relationship between a response and a set of explanatory variables. It has extensive applications in many subject areas, like medicine, economics, biology, chemistry, engineering, sociology, etc.

Consider the multiple linear regression model

$$Y = X\beta + \varepsilon$$

Where Y is an $n \times 1$ vector of response, X is an $n \times k = (p + 1)$ full rank matrix of explanatory variables including one constant column, β is a $k \times 1$ vector of unknown finite parameters and ε is an $n \times 1$ vector of i.i.d. random disturbances each of which follows $N(0, \sigma^2)$. We predict the output Y using the least squares method via the model $\hat{Y} = X\hat{\beta}$, where $\hat{\beta} = (X^T X)^{-1} X^T Y$.

Considering that the behaviour of friction is closer to the linear when compared with other parameters, such as alligator cracking, in this study we developed a multiple linear regression model that enabled the analysis of the relation between a dependent variable (FRICTION) and a set of explanatory variables, previously described.

The selection of an appropriate subset of explanatory variables to be used in a multiple linear regression model is an important aspect of a statistical analysis. In this study, we applied Stepwise regression which chooses a subset of explanatory variables exploring only a few possible sub-models [9].

These models were estimated using data collected from Section 1, Section 2 and from both sections (Section 1+2). In the model constructed with all data, a binary variable for the section (SECT) was also included.

Since the major role of regression modelling is to predict unknown future samples [5], it may be of interest to estimate their prediction performance. The root mean-squared error of prediction (RMSEP) of a model represents its prediction ability. The most common approaches to measure prediction accuracy use resampling. The standard resampling method in this context is cross-validation while bootstrap can be used as an alternative [4]. In this work, we used K-fold cross-validation to estimate the root mean-squared error of prediction.

Data were statistically analyzed using the freeware R, version 2.14.1 [12].

3   Statistical Analysis

For a better understanding of the data under study, a statistical analysis of the continuous variables was initially made. The results are shown in Table 1.

Section 1 showed friction values significantly higher than Section 2 (t-test = −40.72, p-value < 0.001). This result may be explained by the fact that Section 2 has a higher accumulated annual average daily traffic (Ac.AADT) than Section 1 (t-test = 21.28, p-value < 0.001). Furthermore, there is a greater variability of friction in Section 2.

Table 1. Descriptive Statistics of continuous variables

Section	Descriptive Statistics	FRICTION (GN)	TEXTURE (mm)	Ac.AADT (vehicles)	MAAT (°C)
1	Minimum	0.45	0.79	48 945	12.0
(N=1580)	Maximum	0.85	1.69	6.00E+06	20.9
	Mean	0.69	1.11	3.04E+06	16.5
	StdDev	0.07	0.14	2.98E+06	4.5
2	Minimum	0.35	0.79	246 691	8.0
(N=1786)	Maximum	0.84	1.85	11.88E+06	21.0
	Mean	0.57	1.25	6.52E+06	15.7
	StdDev	0.11	0.22	4.74E+06	5.1
1+2	Minimum	0.35	0.79	48 945	8.0
(N=3366)	Maximum	0.85	1.85	11.88E+06	21.0
	Mean	0.62	1.19	4.89E+06	16.1
	StdDev	0.11	0.20	4.37E+06	4.8

We may also observe that both Sections present a similar maximum value of the mean average air temperature (21.0°C). However, Section 1 displayed a mean of the average air temperature (MAAT) significantly higher than Section 2 (t- test = −4.55, p-value < 0.001).

Concerning the texture, Section 1 displayed texture values significantly lower than Section 2 (t- test = 25.82, p-value < 0.001).

We studied the presence or absence of an association between each explanatory variable and the dependent variable FRICTION, aiming to reveal the individual predictive capacity of the explanatory variable. For continuous explanatory variables, the correlation coefficient of Pearson was employed and for categorical explanatory variables that of the ANOVA test.

In Table 2, the coefficients of correlation between the variable friction and each of the continuous explanatory variables are presented along with the respective p-value. The variables Ac.AADT and TEXTURE reveal the most significant negative correlation, whereas the variable Mean Average Air Temperature (MAAT) reveals different correlations in Section 1 and Section 2.

Table 2. The correlation coefficient for continuous covariates and variable Friction

Section	TEXTURE	Ac.AADT	MAAT
1	−0.490[***]	−0.831[***]	−0.831[***]
2	−0.557[***]	−0.871[***]	0.369[***]
1+2	−0.609[***]	−0.876[***]	0.021[***]

[***]Correlation is significant at the 0.001 level (2-tailed).

The friction distribution per vertical alignment characteristics is shown in Table 3.

Table 3. Friction distribution per vertical alignment characteristics

Section	Descriptive Statistics	Ccv	Ccx	i
1	N	88	348	1144
		(5.57%)	(22.03%)	(72.41%)
	Minimum	0.55	0.53	0.45
	Maximum	0.79	0.83	0.85
	Mean	0.69	0.69	0.69
	Std. Dev	0.06	0.06	0.07
2	N	359	586	841
		(20.00%)	(32.43%)	(47.57%)
	Minimum	0.35	0.36	0.35
	Maximum	0.84	0.82	0.82
	Mean	0.56	0.56	0.56
	Std. Dev	0.10	0.11	0.10
1+2	N	447	934	1985
		(13.28%)	(27.75%)	(58.97%)
	Minimum	0.35	0.36	0.35
	Maximum	0.84	0.83	0.85
	Mean	0.59	0.61	0.64
	Std. Dev	0.11	0.12	0.11

The effect of alignment characteristics and sections on FRICTION and their interaction was analyzed by a two-way analysis of variance (ANOVA). Results of the two-way ANOVA showed that the interaction was not significant (F = 0.117, p-value = 0.890).

In the situation of non-significant interaction, a common practice is proceeding by fitting an additive model, whose results are shown in Table 4. We found no significant differences between the alignment characteristics in terms of FRICTION (F=0.405; p-value = 0.667) and, as shown, the mean friction in Section 1 was significantly higher than in Section 2.

Table 4. ANOVA table

Model	Df	Mean Square	F-value	P-value
Section	1	12.080	1457.629	<0.001
Alignment	2	0.003	0.405	0.667
Error	3362	0.008	0.53	0.45

4   Results

Table 4 indicates the estimates of the parameters for the three multiple linear regression models. For each model, the adjusted coefficient of determination R_a^2 is also computed. To measure prediction error in these models, we used the average of 2000 random 5-fold cross-validation runs.

In all models, the included explanatory variables are highly significant. We also found that when the texture and accumulated annual average daily traffic increase, in average the friction value decreases. As expected, we observed that the vertical alignment does not influence FRICTION.

As can be seen in Table 5, the quality of fit of the models is good ($0.693 < R_a^2 < 0.826$). We also can observe that the prediction performance of three models is similar.

Table 5. Estimated parameters

	Section 1		Section 2		Section 1+2	
Variable	Estimate	StdError	Estimate	StdError	Estimate	StdError
Constant	0.784***	0.008	0.771***	0.010	0.742***	0.007
TEXTURE	-0.034***	0.008	-0.039***	0.007	-0.035***	0.006
Ac.AADT	-1.83E-08***	3.79E-10	-1.96E-08***	3.42E-10	-9.30E-09***	2.46E-10
MAAT			-0.002***	2.78E-04	-1.00E-03**	1.87E-04
SECT(ref.:1)						
2					-0.056***	0.002
Summary Statistics						
R_a^2	0.693		0.767		0.826	
RMSEP	0.048		0.051		0.053	

** Denotes the significance at the 0.01 level

*** Denotes the significance at the 0.001 level

Plots of the residuals are widely used to assess the adequacy of a regression model. Figure 2 provides the QQ-plot of the standardized residuals of the three models, which indicate that the standardized residuals follow normal distribution. The normality test computed for each model (Shapiro–Wilks) also does not lead to reject the null Hypothesis.

Figure 3 shows the standardized residuals versus fitted values. This panel does not show any pattern in the spread of the residuals and we concluded that homogeneity assumption is verified.

We also constructed standardized residuals versus each explanatory variable to check independence assumption and we found no residual patterns.

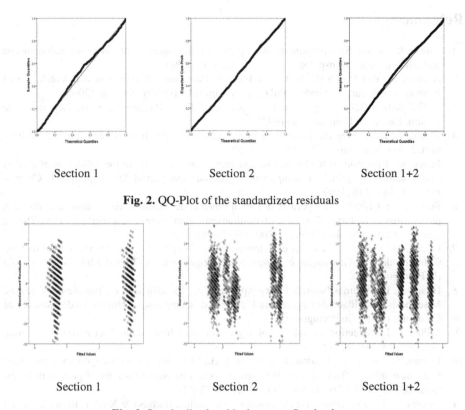

Section 1   Section 2   Section 1+2

Fig. 2. QQ-Plot of the standardized residuals

Section 1   Section 2   Section 1+2

Fig. 3. Standardized residuals versus fitted values

5   Conclusions

For the real data available, this study developed a multiple regression model that describes the prevision of the pavements' friction degradation and identifies the effects of various factors upon its behaviour.

This study also revealed the efficiency of the linear regression model in approaching and explaining that data, especially with regard to the identification of the most relevant factors involved in the evolution process of pavements' friction.

In the future, mixed-effects models will be applied to the data used in this paper taking into account that the data comprise a series of test results measured throughout the time.

Acknowledgements. This research was financed by FEDER Funds through "Programa Operacional Factores de Competitividade - COMPETE" and by Portuguese Funds through FCT -"Fundação para a Ciência e a Tecnologia", within the Projects PEst-OE/MAT/UI0013/2014, PEst-OE/ECI/UI4047/2014 and PEst-OE/EEI/UI0319/2014. The authors would also like to thank AdI – Innovation Agency, for the financial support awarded through POFC program, for the R&D project SustIMS – Sustainable Infrastructure Management Systems (FCOMP-01-0202-FEDER-023113).

References

1. Abaza, K., Ashur, S.: Optimum decision policy for management of pavement maintenancae and rehabilitation. Transp. Res. Record. 1655, 8–15 (1999)
2. Agardh, S.: Rut Depth Prediction on Flexible Pavements, Calibration and Validation of Incremental-Recursive Models. PhD Thesis, Lunds University, Sweden (2005)
3. COST Action 324: Long Term Performance of Road Pavements, Final Report of the Action. European Commission (1997)
4. Efron, B.: Estimating the Error Rate of a Prediction Rule: Improvement on Crossvalidation. J. Am. Stat. Assoc. 78, 316–331 (1983)
5. Faber, K., Kowalski, B.R.: Propagation of measurement errors for the validation of prediction obtained by principal component regression and partial least squares. J. Chemometr. 11, 181–238 (1997)
6. Freitas, E.: Contribuição para o desenvolvimento de modelos de comportamentos dos pavimentos rodoviários flexíveis – fendilhamento com origem na superfície. PhD Thesis, Universidade do Minho, Portugal (2004) (in Portuguese)
7. Falcão, D.: Conjuntos, Lógica e Sistemas Fuzzy, Technicalreport, Instituto Alberto Luiz Coimbra de Pós-Graduação e Pesquisa de Engenharia, COPPE/UFRJ, Brazil (2002) (in Portuguese)
8. Huamaní, I.: Sistemas Neurais Fuzzy Aplicados em Identificação e Controle de Sistemas, MScThesis, Faculdade de Engenharia Elétrica e de Computação, Universidade Estadual de Campinas (2003) (in Portuguese)
9. Miller, A.J.: Selection of subsets of regression variables (with discussion). J. R. Stat. Soc. 147(A), 389–425 (1984)
10. Lorino, T., Lepert, P., Marion, J., Khraibani, H.: Modeling the road degradation processs: non-linear mixed effects models for correlation and heteroscedasticity of pavement longitunal data. Procedia Soc. Behav. Sci. 48, 21–29 (2012)
11. Pereira, P., Picado-Santos, L.: Pavimentos rodoviários, Barbosa & Xavier, Braga, Portugal (2002) (in Portuguese)
12. R Development Core Team: R: A Language and Environment for Statistical Computing. R Foundation for Statistical Computing, Vienna, Austria (2014)
13. Sinhal, R.: The Implementation of a Skid Policy to Provide the Required Friction Demand on the Main Road Network in the United Kingdom. Highways Agency, UK (2004)
14. Yuan, X., Pandey, M.: A nonlinear mixed-effects model for degradation data obtained from in-service inspections. Reliab. Eng. Syst. Safe. 94, 509–519 (2009)

Design and Deployment of a Dynamic-Coupling Tool for EFDC

Vladimir J. Alarcon[1,*], Donald Johnson[2], William H. Mcanally[2], John van der Zwaag[2], Derek Irby[2], and John Cartwright[2]

[1] Civil Engineering School, Universidad Diego Portales, 441 Ejercito Ave., Santiago, Chile
vladimir.alarcon@udp.cl
[2] Geosystems Research Institute, Mississippi State University, 2 Research Blvd.,
Starkville, MS 39759, USA
{donaldj,mcanally,vanderz,derek,johnc}@gri.msstate.edu

Abstract. A dynamic-coupling tool designed to link several hydrodynamic models is presented. The tool is able to dynamically transfer time-series data among models that are geographically adjacent. Dynamic data transfer is implemented at the models' common boundaries. The Message Passing Interface (MPI) and a coupling code were used for implementing the dynamic link. Several issues that had to be overcome during the development of the tool (such as porting of the code to a Linux environment, MPI implementation, and compiler flags used for optimum performance) are discussed. The tool is applied to a test case in which three hydrodynamic models built with the Environmental and Fluid Dynamics Code (EFDC) are run with the dynamic-coupling tool in a Linux cluster. Run times were compared to a sequential run of the three models in a Windows environment. A speed up of 8.53 was achieved by exploring and finding an optimal combination of Intel Fortran compiler flags.

Keywords: Hydrodynamic modeling, EFDC, dynamic-coupling.

1   Introduction

Loose-coupling of computational models is a technique that involves the sequenced execution of individual programs in which the programs (or models) exchange data via files. The exchange is usually performed manually or via scripts requiring a large number of invocations, often of several different programs [1].

Loose coupling can be used for coupling different computer codes that model different processes, or for coupling different models that use the same code. For example, [2] used the technique for coupling watershed hydrology models and hydraulic models; [3] and [4] coupled hydrodynamic and water quality models; [5] linked wave models with hydrodynamic models; and [6] coupled one computer model with itself for nested hydrodynamic modeling.

In nested hydrodynamic modeling several geographically adjacent hydrodynamic models are loosely-linked at the common geographical boundaries. This sequential

---

* Corresponding author.

B. Murgante et al. (Eds.): ICCSA 2014, Part III, LNCS 8581, pp. 615–624, 2014.

linking of models is done due to the different spatial scale of computational domains: processes occurring in oceans often use coarse computational meshes, and estuary or river models usually require fine computational grids. Once the models are generated, they are run independently and then they are loosely-coupled by using the results of each model as boundary conditions for the spatially neighboring model. This approach is very effective, but the sequential independent execution of "smaller" models plus the additional time required for loose-linking (through the results) is very time-consuming.

Modern hydrodynamic models are two or three-dimensional codes that solve the mass and momentum transport equations numerically, using either finite-difference or finite-element schemes. Those equations are the widely known momentum and continuity equations for fluids [7] that are usually simplified to the more commonly known Navier-Stokes equations (as in: [7], [8], [9], and [10]). An example of a widely used hydrodynamic code is the Environmental and Fluid Dynamics Code (EFDC). EFDC [11] uses a finite-difference numerical scheme, and as many other hydrodynamic codes, uses extensive input data and produce even more extensive output data. In this context, loose-linking is the limiting factor for achieving results expeditiously when using EFDC as the primary modeling tool.

This paper presents a novel dynamic-coupling tool designed to link several hydrodynamic models that were built using the Environmental Fluid Dynamics Code. The tool is able to transfer time-series data dynamically from one model to other models that are geographically adjacent. The dynamic data transfer is implemented at the models' common boundaries, and is completely transparent for the user, without requiring user´s intervention or oversight. The dynamic-coupling tool was tested with three hydrodynamic models of water bodies located in the Northern Gulf of Mexico. The paper discusses several issues that had to be overcome during the development of the tool: such as porting of the code to a Linux environment, MPI implementation, and compiler flags used for optimum performance.

2   Methods

2.1   Hydrodynamic Code

We used the Environmental Fluid Dynamic Code (EFDC) as the modeling tool. EFDC [11] is a 3-D model that is used extensively in the United States [12]. The model belongs to the suite of water modeling tools of the US Environmental Protection Agency. EFDC was developed by John Hamrick at the Virginia Institute of Marine Science [11] with primary support from the State of Virginia. It is presently maintained by Tetra Tech, Inc. [6]. The EFDC code is currently used by federal, state and local agencies, consultants, and universities. The EFDC version used throughout this research was the Dynamics Solutions LLC's EFDC_DSI Version 2009_11_01 [13]. This EFDC version is a Windows application that is coded in Fortran. Throughout this paper, the EFDC_DSI version will be denominated EFDC for brevity. The proprietors of this code facilitated the full Fortran code and it was modified for it to be able to run under a Linux environment, and also for using the Message Passing Interface (MPI) in the implementation of the dynamic-coupling tool.

2.2 Dynamic-Coupling and Nested Hydrodynamic Approach

The dynamic-coupling tool used in this research is able to run (in parallel) several EFDC models, each corresponding to a different geographical part or segment of a water body. Results of each model are passed to the neighboring model (or models) using the Message Passing Interface (MPI). Figure 1 illustrates the tool implementation.

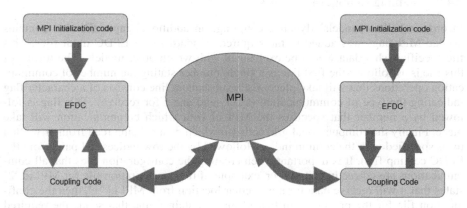

Fig. 1. Dynamic-coupling of EFDC models. The Message Passing Interface (MPI) and a coupling code are used to dynamically pass results from one EFDC model to the other.

2.3 Porting EFDC

The Windows source code for EFDC was successfully compiled on a Linux environment. The compiler used was the Intel Fortran Compiler (ifort) to minimize changes in the code. Several details, however, had to be taken into account for the compiling. The source code of EFDC makes extensive use of the non-standard variable attribute 'static'; this attribute is only a valid input to ifort. The effect of this attribute is exactly the same as the standard attribute 'save'. Therefore another possible solution would have been to change all occurrences of 'static' to 'save' but this would have required extensive modification of the source and was deemed unnecessary, as ifort was available.

The second issue was that the code references to three C functions in the Fotran source. Specifically, the functions 'getch', 'kbhit', and 'atexit' are referenced using their symbol names of '__getch', '__kbhit', and '__atexit" respectively. The problem with this type of referencing is that symbol names generated will vary even between different versions of the same compiler. This results in link errors, as the requested symbol names cannot be found. The error for the function 'atexit' can be resolved by changing its symbol name to 'atexit' instead of '__atexit'. The Linux compilers do not add underscores to symbol names. Fixing the error for the function 'getch' is more complicated, because the function in Linux is a macro that calls the function 'wgetch' (Windows getch). This function uses a window definition structure that is not portable to Fortran, hence it should not be called directly. The solution is to create a C wrapping function that calls 'getch' and returns the result of that call. This wrapper function was named 'getch2'. After this function is created, the Fortran source that interfaced with 'getch' needs to be changed to link to the symbol 'getch2'.

With regards to the referenced function 'kbhit', it is not a standard Linux function and requires to be implemented. Working Linux source code for this function can easily be located using Internet search engines. Once the "kbhit" function is implemented, the Fortran reference should be changed from '__kbhit' to 'kbhit'. After the changes detailed above are made, EFDC compiles in a Linux environment.

2.4   Coupling Configuration Files

In order to allow models' dynamic-coupling, an additional input file that contains needed MPI flags was added to the required standard set of EFDC input files. This file specifies what data would be exchanged with which other models. The format of this file is as follows: the first line is a single number stating the number of communication operations that will take place. Each subsequent line consists of a character flag indicating the type of communication (s for send and r for receive). This flag is followed by a number that specifies the MPI id with which communication will take place. Finally the computational grid cells (to which data is being read from or written to) is specified with the column indices followed by the row indices (I,J pairs from the EFDC cell.inp file). It is important when creating the configuration files that all communications are correctly paired. For 'example, if the configuration file for MPI id '2' states that it will receive data for a particular location from MPI id '1', then the configuration file for the process with that id must contain a line that sends the required location. In addition, if there are multiple "receives" expected from a process, then the corresponding "send" lines must be in the same order. Failure to have a "send" that pairs with a "receive" will cause the coupled models to deadlock. Failure to correctly order "send" and "receive" pairs will cause errors in the flow of data.

2.5   Enabling Coupling

In order to allow coupling of different instances of EFDC, running separate models in overlapping special domains, the follow changes were made to the EFDC source:

- The creation of a coupling initialization function, and
- The insertion of a coupling-code portion at the end of EFDC's main loop, so that it would be called as the final operation in each time step of the simulation.

The coupling initializations code has four tasks:

- Initialize the MPI Environment (this allows communication between model instances),
- Change the working directory of the model based on its MPI id. All programs launched by MPI start in same directory. However, EFDC expects its input files to be in the working directory. To solve this issue, the initialization code determines its MPI id and then changes the working directory. Specifically the id '0' will change to the subdirectory 'm1', the id '1' will change to the subdirectory 'm2' and so on.
- Read the coupling configuration file 'mpi-cfg.txt'. Since each model is now in a different directory this file will be different for each model. The information stored in this file is used to create a communication list that is used in the coupling code.
- Pass control to the original EFDC code by calling the subroutine "EFDC"

The coupling code is inserted at the end of EFDC main computation loop. As there are two possible loop (one in the file 'HMDT.for' and the other in the file 'HMDT2.for') it is inserted in both files. When called coupling code makes the send and receive operations indicated in the coupling configuration file. Receive operations update the value for water elevation at cell center 'HP' and send operation read from this value to determine the value to send.

2.6   Testing of the Dynamic Coupling Tool

The testing of the dynamic coupling tool was performed by linking tree hydrodynamic models of different spatial resolutions. The purpose was to produce a (nested) composite hydrodynamic model of Mobile Bay, Weeks Bay and the Fish River (water bodies located in coastal Alabama, USA). More details on the models are found in [14]. The different scale of the models is reflected in grid cell dimensions and total number of cells: the Mobile Bay model cells have average dimension of 400 m x 400 m (total number of cells: $\sim 10^3$ cells), the Weeks Bay grid cells are 89 m by 89 m (total number of cells: $\sim 3*10^4$ cells), and the Fish River grid cells are 26 by 26 meters (total number of cells: $\sim 10^5$ cells).

The testing process was as follows:

- The three hydrodynamic models were run consecutively and independently, linking them after each model run by inputting one model's results as boundary condition of the neighboring model at the common geographical boundaries. The runs were done in dual-core Windows box, running at 3.10 Hz, with 4GB RAM.
- The EFDC code and dynamic-coupling tool were compiled in a Linux cluster using several different combinations of Fortran optimization flags.
- Then, the models were run using the compiled EFDC and dynamic-coupling tool in the linux cluster for each compiled binary resulting from each combination of Fortran optimization flags. The cluster was a 3072 core cluster composed of 256 IBM iDataPlex nodes, each with two six-core Intel Westmere processors (2.8GHz) and 24 GB of memory (for a total of 6TB).

The total run time for a 10-day simulation, and using loose-coupling of the three models took 30.631 hours. This time accounts for: running each model consecutively and independently, and then linking them by inputting one model's results as input of the neighboring model at the common geographical boundaries.

2.7   Compiler Flags

In order to optimize run-times of the coupled model, several combinations of Fortran compiler flags were used to find the combination that provided the best runtime. The following table provides details on the compiler flags used, and a summary of the properties of each one.

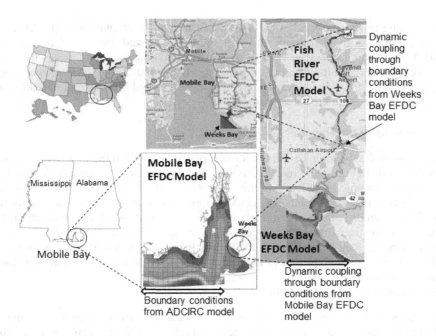

Fig. 2. Coupling of EFDC models for Mobile Bay, Weeks Bay, and Fish River. Grid cells at the Mobile Bay-Weeks Bay boundary, and Weeks Bay-Fish River boundary were used for dynamic transfer of water surface elevations values (modified after [14]).

Table 1. Intel Fortran compiler flags description used for performance optimization, for IA-32 and/or IA-64 architectures. (After [15]).

Flag	Effect
-O0	Disables all -O<n> optimizations. Sets the -fp option
-O1	Enables optimizations for speed, while being aware of code size. Also disables intrinsic recognition and the –fp option. Enables optimizations for server applications.
-O2	Default for optimizations. Enables optimizations for speed, including global code scheduling, software pipelining, predication, and speculation.
-fp-model precise	Enables value-safe optimizations on floating-point data and rounds intermediate results to source-defined precision. Disables optimizations that can change the result of floating-point calculations. Ensure the accuracy of floating-point computations, but may slow performance.
-fp-model strict	Enables precise and except. This is the strictest floating-point model.
-fp-model fast	Enables more aggressive optimizations when implementing floating-point calculations. Increase speed, but may alter the accuracy of floating-point computations.
-ip	Enables additional inter-procedural optimizations for single file compilation. Allows the compiler to perform inline function expansion for calls to functions defined within the current source file.
-ipo	Enables multi-file inter-procedural (IP) optimizations (between files). Performs inline function expansion for calls to functions defined in separate files.

Table 1. (*Continued*)

-xS	Can generate SSE4 Vectorizing Compiler and Media Accelerators instructions for future Intel processors that support the instructions.
-xT	Can generate SSSE3, SSE3, SSE2, and SSE instructions for Intel processors, and it can optimize for the Intel(R) Core(TM)2 Duo processor family.
-xP	Expands –XT to Core(TM) microarchitecture and Intel NetBurst(R) micro-architecture.
-xO	Can generate SSE3, SSE2, and SSE instructions, and it can optimize for Intel processors based on Intel Core microarchitecture and Intel Netburst microarchitecture.
-xHost	Tells the compiler to generate instructions for the highest instruction set available on the compilation host processor.

3   Results

Table 2 shows results of the different runs performed with the newly compiled EFDC code and the dynamic-code code. As shown, run times without optimization (O0 flag) already provide a speed up of 2.05 with respect to sequential execution of the loosely-coupled hydrodynamic models. However, enabling optimization (O1 and O2 flags) produce speed ups within the range of 5.58 to 7.51. Enabling value-safe optimizations on floating-point data (fp-model precise, and fp-model strict) do not optimize run times further than the optimization achieved by using O1 and O2 flags.

Table 2. Run time speed up after optimization flags

Fortran flags	Run time (hours)	Speed up with respect to sequential execution
-O0	14.932	2.051
-O1	5.490	5.579
-O2	4.077	7.513
-fp-model precise –O2	4.255	7.199
-fp-model strict –O2	4.616	6.636
-fp-model fast –O2 –ip	4.064	7.537
-fp-model fast -O2 –ip –ipo	3.646	8.401
-fp-model fast -O2 –ip –ipo –xS	3.590	8.532
-fp-model fast -O2 –ip –ipo –xT	3.792	8.078
-fp-model fast -O2 –ip –ipo –xP	4.291	7.138
-fp-model fast -O2 –ip –ipo –xO	4.563	6.713
-fp-model fast -O2 –ip –ipo -xHost	4.672	6.556

With the '-fp-model fast –O2 –ip' flag combination that enables more aggressive optimizations (with regards to floating-point calculations) and additional inter-procedural optimizations, run time was improved to provide a speed-up of 7.53. However, enabling multi-file inter-procedural optimizations with -ipo, and SSE4 Vectorizing Compiler and Media Accelerators with -xS, the optimum run time was

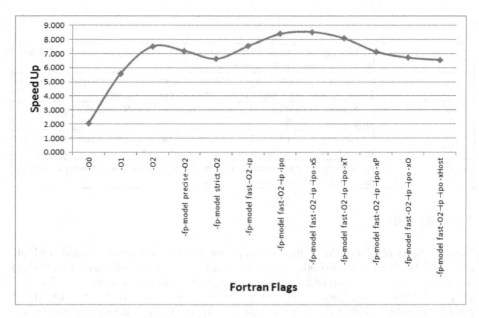

Fig. 3. Speed up achieved with the dynamic-linking tool

achieved: 3.59 hours. Other flag combinations were explored but none of them provided better run-time than 3.59 hours. This optimum run time corresponds to a speed-up of 8.53, meaning that the optimum run time is 800% faster than the one corresponding to sequential execution. Figure 6 illustrates speed up values for all combinations of Intel Fortran compiler flags.

The flag combination '-fp-model fast -O2 –ip –ipo –xS', which provided the fastest run time, can be interpreted as a set of flags that enables aggressive optimizations for floating-point calculations, multi-file inter-procedural optimizations for single and multiple files compilation (that allow the compiler to perform inline function expansion for calls to functions defined within the current source file, as well as for calls to functions defined in separate files). Moreover, the flags also enable to detect operations in the program that can be done in parallel (such as simple loops), and then converts the sequential operations like one SIMD instruction that processes 2, 4, 8 or up to 16 elements in parallel.

4   Conclusions

A novel dynamic-coupling tool designed to link several hydrodynamic models is presented. The tool is able to transfer time-series data dynamically from one model to a geographically adjacent model or models. The dynamic data transfer is implemented at the models' common geographical boundaries, and its operation is completely transparent for the user.

The dynamic-coupling tool was tested with three hydrodynamic models of water bodies located in the Northern Gulf of Mexico that were built using the Environmental

Fluid Dynamics Code (EFDC). Performance of the dynamically-linked model was measured by comparing run times of the Windows EFDC code to processing times of the dynamically-linked model in a cluster.

Run times at the cluster without optimization flags produced a speed up of 2.05 with respect to sequential execution of the three hydrodynamic models in a Windows environment. Enablement of basic optimization (O1 and O2 flags) produced speed ups within the range of 5.58 to 7.51. Enabling multi-file inter-procedural optimizations with -ipo, and SSE4 Vectorizing Compiler and Media Accelerators with -xS, an optimum run time of 3.59 hours was achieved. Other flag combinations were explored but none of them provided better run-times. The speed up corresponding to the optimum time was calculated as 8.53 (approximately a 600% improvement to sequential execution run time).

References

1. Zhang, Z., Espinosa, A., Iskra, K., Raicu, I., Foster, I., Wilde, M.: Design and evaluation of a collective I/O model for loosely-coupled petascale programming. In: Proc. MTAGS Workshop and SC 2008 (2008),
   http://arxiv.org/ftp/arxiv/papers/0901/0901.0134.pdf
2. He, L., Wang, G.Q., Zhang, C.: Application of Loosely Coupled Watershed Model and Channel Model in Yellow River, China. Journal of Environmental Informatics 19(1), 30–37 (2012)
3. Wool, T., Davie, S., Rodriguez, H.: Development of Three-Dimensional Hydro-dynamic and Water Quality Models to Support Total Maximum Daily Load Decision Process for the Neuse River Estuary, North Carolina. Journal of Water Resources Planning and Management 129(4) (July 1, 2003)
4. Liu, Z., Hashim, N.B., Kingery, W.L., Huddleston, D.H., Xia, M.: Hydrodynamic modeling of St. Louis Bay estuary and watershed using EFDC and HSPF. Journal of Coastal Research 52, 107–116 (2008)
5. Dietrich, J.C., Zijlema, M., Westerink, J.J., Holthuijsen, L.H., Dawson, C., Luettich Jr., R.A., Jensen, R.E., Smith, S.G.S., Stone, G.W.: Modeling hurricane waves and storm surge using integrally-coupled, scalable computations. Coastal Engineering 58(2011), 45–65 (2011)
6. Tetratech, Inc., 2012. Draft user's manual for environmental fluid dynamics code Hydro Version (EFDC-Hydro) Release 1.00. Tetra Tech., Inc., Fairfax, Virginia (2002),
   http://snl-efdc.sourceforge.net/EFDC-Hydro_Manual.pdf
   (accessed May 2013)
7. Tabuenca, P., Cardona, J., Samartín, A.: Numerical model for the study of hydrody-namics on bays and estuaries. Applied Mathematical Modelling 16(2), 78–85 (1992)
8. Berger, R.C., Tate, J.N., Brown, G.L., Savant, G.: Guidelines for Solving Two-Dimensional Shallow Water Problems with the Adaptive Hydraulics Modeling System AdH Version 4.3 (2013),
   http://chl.erdc.usace.army.mil/Media/1/2/9/9/
   AdH_Manual_4.3.pdf

9. North Carolina Department of Environment and Natural Resources, NCDECNR, Falls Lake Nutrient Response Model: Final Report. Modeling & TMDL Unit, Division of Water Quality, North Carolina Department of Environment and Natural Resources (2009), http://portal.ncdenr.org/c/document_library/get_file?uuid=33 debbba-5160-4928-9570-55496539f667&groupId=38364

10. Cook, C.B., Rakowski, C.L., Richmond, M.C., Titzler, S.P., Coleman, A.M., Bleich, M.D.: Numerically Simulating the Hydrodynamic and Water Quality Environment for Migrating Salmon in the Lower Snake. Pacific Northwest Laboratory (Prepared for the U.S. Department of Energy) (2003)

11. Hamrick, J.M.: A three-dimensional environmental fluid dynamics computer code: theoretical and computational aspects. Special Report 317 in Applied Marine Science and Ocean Engineering. The College of William and Mary, Virginia Institute of Marine Science, 63 p. (1992)

12. Environmental Protection Agency, EPA, Environmental Fluid Dynamics Code, EFDC (2013), http://www2.epa.gov/sites/production/files/ documents/EFDC_Brochure.pdf

13. Craig, P.M.: User's Manual for EFDC_Explorer: A Pre/Post Processor for the Environmental Fluid Dynamics Code (Rev 00). Dynamic Solutions Intl, LLC Knoxville, TN (2009),
http://efdc-explorer.com/documents/
EFDC_Explorer6%20Users%20Manual%20(2011_06_10%20Rev00).htm

14. Alarcon, V.J., Johnson, D., McAnally, W.H., van der Zwaag, J., Irby, D., Cartwright, J.: Nested hydrodynamic modeling of flooding events for a coastal river applying dynamic-coupling (2013) (submitted)

15. Intel Corporation, Intel(R) Fortran Compiler Options (2007),
http://www.rcac.purdue.edu/userinfo/resources/common/
compile/compilers/intel/man/ifort.txt

Taking Account of Uncertain, Imprecise and Incomplete Data in Sustainability Assessments in Agriculture

Jérôme Dantan[1,2], Yann Pollet[2], and Salima Taibi[1]

[1] Esitpa, Agri'terr, Mont-Saint-Aignan, France
{jdantan,staibi}@esitpa.fr
[2] CNAM, CEDRIC, Paris, France
yann.pollet@cnam.fr

Abstract. To manipulate incomplete, uncertain and hypothetic information about farms, the future evaluation software in the agriculture domain will have to make computations and combine fuzzy numerical values. The approach we propose relies on combination of incomplete, uncertain and hypothetic object values based on particular application of the Dempster-Shafer evidence theory. We then propose the integration of this approach in classic object oriented programming via the implementation of a C++ library.

Keywords: Dempster-Shafer model, belief, plausibility, conjunctive combination.

1 Introduction

One of the main challenges of today's society is the management of large amounts of data that are becoming more and more often uncertain, imprecise or incomplete. Current researches in the domain of Information and Communication Technologies, especially in the field of artificial intelligence, describe and extend the existing formalisms to develop intelligent and autonomous systems that compute uncertain, imprecise or incomplete data. Many actors of several domains have to cope with such data, especially to assist humans in their decisions by merging data from many sources of information (measurements, sensors, observations) to model the behaviour of complex systems.

This is the case in the field of agriculture: both crop and livestock systems depend on many parameters that may be: uncertain (e.g. the weather for the next year or decisions of the agricultural stakeholders), or imprecise/incomplete (e.g. several possible values for discrete data, area allocated to different crops, especially in developing countries where they are not always accurately known), or random (e.g. crop forecasts and prices). In addition, farms have to aim more and more goals; in today's context, the objective of farms is not only to be profitable, but also to preserve ecosystems, to reduce waste, etc. In summary, farmers have to produce better and to contribute to sustainable development. They also have to optimize the resources and costs of farms, while limiting the impact of agricultural activities on the environment.

That's why the field of decision support in agricultural information systems is promising; its potential contributions are going to deeply impact the future. Decision

B. Murgante et al. (Eds.): ICCSA 2014, Part III, LNCS 8581, pp. 625–639, 2014.

support tools aim to assist the experts of agriculture in their decisions on specific issues: what crops should be used? How much input should be used? Etc. For this, the stakeholders of the agriculture domain have to cope with increasing amounts of information, to solve issues from basic production evaluations to complex sustainability indicators.

For this, the stakeholders of the agriculture domain have to cope with increasing amounts of information, to solve issues from basic production evaluations to complex sustainability indicators. Data experts must deal with data that are based on both heterogeneous and increasing sources and formats (such as agricultural field samples, production cost and income, soil heavy metal rates, etc). Indeed, there is an exponential development of means of recording and data storage ("big data" buzz word) from multiple sensors, from intelligent soil analyzing augers from GPS coordinates linked with weight measurements of crops at harvest, all in real time. These data cannot therefore be systematically validated and formatted for classical statistical analysis. Furthermore, it seems unrealistic to consider absolute accuracy about measuring means. Rather than ignore the problem (by considering only the median of a sample, for example), it is instead interesting to integrate the knowledge we have about the imprecision of some data in addition to data suitable for classical statistics analysis.

We propose an approach for modeling the combination of uncertain, imprecise and random knowledge in a common formalism through the Dempster-Shafer theory and the transferable Belief Model. The article includes the following sections: (1) a state of the art, (2) the presentation of our approach, (3) some examples to illustrate the approach, (4) the application on an object-oriented language, and (5) our conclusions and a brief opening to our future work.

2   Background

2.1   Probability Theory

For a long time, uncertainty modeling remained addressed by the probability theory. Let Ω be a finite set. In the probabilistic approaches, events of Ω are associated with a reference probability measure, i.e. a P application from Ω to $[0; 1]$, satisfying $P(\Omega) = 1$ and $P(\cup_i A_i) = \sum_i P(A_i)$ (additive constraint) where $\{A_i\}_{i \in \{1,...,n\}}$ are a finite family of disjoint events of Ω. These axioms imply in particular that the probability of an event completely determines that of its opposite, i.e. $P(A) = 1 - P(\neg A)$. However, this approach is little suitable for total ignorance representation, and objective interpretations of probabilities, assigned to such events remain difficult when handled knowledge are no longer linked to random and repeatable phenomena [2]. As against, it is possible to model uncertainty thanks to the possibility theory.

2.2   Possibility Theory

Possibility theory [2], [8] removes the strong probability additive constraint and associates the events of Ω to a possibility measure denoted Π and a necessity measure denoted N, that are both applications from Ω to $[0; 1]$, respectively satisfying: $\Pi(A \cup B) = \max\left(\Pi(A), \Pi(B)\right)$ and $N(A \cap B) = \min\left(N(A), N(B)\right)$. The relationship between the possibility of an event and its opposite is given by $\Pi(A) = 1 - N(\neg A)$ and total ignorance is then given by $\Pi(A) = \Pi(\neg A) = 1$, which implies: $N(A) = N(\neg A) = 0$. This approach also allows representing imprecision using notions of fuzzy sets and distributions of possibilities. Thus, a fuzzy set [7] F of a set E is defined by a membership function μ_F from E to $[0,1]$, which associates each element x of E its membership degree $\mu_F(x)$, to the subset F (i.e.: x belongs "more or less" to F). When this membership function is normalized (a x value from E such as $\mu_F(x) = 1$ exists), $\mu_F(x)$ can then interpreted as the chance that F takes the value x ($\mu_F(x)$ is then a possibility distribution).

2.3   Belief Functions

A state of the art on the belief functions applied to the processing of information is provided in [6]. The results on upper and lower limits of a family of probability distributions of Dempster introduced the theory of evidence [1], which were then used to develop the Shafer theory of belief functions and show their interest for modeling uncertain knowledge [4]. Then Smets [5] introduced the TBM or Transferable Belief Model, a non-probabilistic model of uncertain reasoning based on belief functions completing the Dempster-Shafer theory.

Let A be a subset of Ω ; $\Omega = \{\omega_1, \omega_2, \dots, \omega_n\}_{n \in N}$ is a finite set, called universe or frame of discernment. A belief function is defined by a basic belief assignment (bba) m defined in 2^Ω (i.e. all parts of Ω) in $[0,1]$, which satisfies $\sum_{A \subseteq \Omega} m(A) = 1$. The $m(A)$ values are called basic belief masses (bbm).

Each subset A of Ω with m(A) > 0 is called a focal element of m. $m(A)$ is the degree of belief assigned to the A subset which has not been assigned to a specific subset of A, given the current status of knowledge.

The degree of belief Bel of A quantifies the total amount of potential specific support given to A: it is obtained by adding all the masses of B that are included in A:

$$\forall A \subseteq \Omega, Bel(A) = \sum_{B \subseteq A, B \neq \phi} m(B) \tag{1}$$

The degree of plausibility Pl of A quantifies the maximum of evidence that could potentially be given to A: it obtained by adding all the masses of B that have a non-empty intersection (and are therefore compatible) with A:

$$\forall A \subseteq \Omega, Pl(A) = \sum_{B \cap A \neq \phi} m(B) \tag{2}$$

Belief and plausibility check the following relationship: $Pl(A) + Bel(\neg A) = 1 - m(\phi)$.

In the Transferable Belief Model, Smets [5], two levels of modeling are described:

- The credal level where beliefs are modeled and revised and
- The pignistic level in which belief functions are transformed into probability function for decision making.

In the TBM, the weight allocated to the empty set is not assumed to be zero, and therefore the condition $\sum_{A \subseteq \Omega, A \neq \phi} m(A) = 1$ is not necessarily true. This specificity allows TBM to introduce the concept of open world, where the empty set is considered as the not clearly defined hypothesis in the frame of discernment, instead of a closed world where the set of assumptions is considered exhaustive.

An attenuation coefficient a may be introduced if the information source is not completely reliable. This coefficient is used to transfer a part of the belief to Ω. A weakened mass function denoted m_a may be expressed with m: $\forall A \subset \Omega, A \neq \Omega$, $m_a(A) = am(A)$, and $m_a(\Omega) = 1 - a + am(\Omega)$.

2.4   Combination of Belief Functions

If we have two sources of information, with two sets of masses m_1 and m_2 defined on Ω. m_1 and m_2 may be aggregated by the combination operator of Dempster (also called orthogonal sum), that corresponds to the conjunctive rule of combination of two masses (called also conjunctive sum) denoted $\cap$, normalized through a factor that depends on the conflict between the two sets of masses: $\left(\frac{1}{1-m_\cap(\phi)}\right)$. $m_\cap(\phi)$ is the degree of conflict between m_1 and m_2 masses. The resulting bbm is expressed as:

$$[\text{conjunctive sum}] \; \forall A \subseteq \Omega, m_\cap(A) = (m_1 \cap m_2)(A) = \sum_{B \cap C = A} m_1(B)m_2(C) \quad (3)$$

$$[\text{orthogonal sum}] \; \forall \phi \neq A \subseteq \Omega, m_\oplus(A) = (m_1 \oplus m_2)(A) = \frac{(m_1 \cap m_2)(A)}{1 - m_\cap(\phi)} \quad (4)$$

$$[\text{degree of conflict}] \; m_\cap(\phi) = (m_1 \cap m_2)(\phi) = \sum_{B \cap C = \phi} m_1(B)m_2(C) \quad (5)$$

The orthogonal sum is possible only if the conflict between the sets of masses is not total, i.e. if at least two focal elements of m_1 and m_2 have their intersection which is not empty.

The pignistic level is used for decision-making: the mass function m is transformed into a so-called pignistic probability function defined on Ω. For any singleton $\omega \in \Omega$:

$$\text{BetP}(\omega) = \frac{1}{1 - m_\cap(\phi)} \sum_{\omega \in A} \frac{m(A)}{|A|} \quad (6)$$

with $|A|$ the cardinal of the A subset. If $\text{BetP}(\omega)$ is maximum, ω is decided.

3   Approach

The goal of this article is to combine any uncertain, imprecise/incomplete, and random data in a common formalism, as specified in the introduction. Indeed, the domain of agriculture is appropriated to combine such data: the actors of this domain have to make decisions in an uncertain context with many constraints due to external factors such as climate, the economic context, environmental standards, etc.

3.1 Possibility / Probability Distributions and Focal Elements

Let $\Omega = \{\omega_1, \omega_2, ..., \omega_n\}_{n \in N}$ be a finite set, called frame of discernment. In the remainder of the article, we consider the world is closed (i.e. $m(\phi) = 0$). The Transferable Belief Model is usually related to neither a probability model nor a possibility model. But, in some particular cases, the belief or plausibility deduced from the bbm may follow either a probability distribution or a possibility distribution, as stated in the following theorems (proofs in [3]):

Theorem 1: the focal elements are totally ordered by inclusion iff Bel and Pl are respectively a measure of necessity and possibility.

Nota bene: A possibility distribution is not a belief distribution because the sum of Π is generally NOT 1. Possibility distributions need to be converted into bbm distributions in this way: if fuzzy intervals $I_1, I_2, ..., I_i, ..., I_n$ are discrete, there is a simple relationship between the masses $m_1, ..., m_i, ..., m_n$ and alpha-cuts $\alpha_1 \geq ... \geq \alpha_i \geq \cdots \geq \alpha_n$ (with $\alpha_1 = 1$):

$$m_1 = \alpha_1 - \alpha_2 \; ; \; m_2 = \alpha_2 - \alpha_3 \; ; \; ... \; ; \; m_i = \alpha_i - \alpha_{i+1} \; ; \; ... \; ; \; m_n = \alpha_n \qquad (7)$$

This is an illustration of a possibility distribution with n = 4:

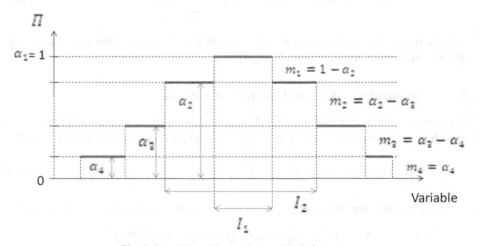

Fig. 1. Possibility distribution and belief masses

Theorem 2: A belief on a finite set is a probability iff the focal elements are singletons.

In this case, P is a probability, i.e. an application from Ω to $[0; 1]$, satisfying $P(\Omega) = 1$ and $P(\cup_i A_i) = \sum_i P(Ai)$ where $\{A_i\}_{i\epsilon\{1,...,n\}}$ are a finite family of disjoint events of Ω, and $P(A) = 1 - P(\neg A)$.

3.2   Data Base Application / "Where" Clause

In a given database, the actual value of an attribute in a given tuple might be not exactly known. Furthermore, we can admit that some values of a given attribute are more suitable than others. Indeed some values may be either more probable – they can be assigned to a probability distribution – or more possible – they can be assigned to a possibility distribution – or more believable – they can be assigned to a belief distribution.

Suppose the value of an attribute A is made of the disjunction of several subsets of the domain of believable values which is supposed to be finite. Suppose the database stores attributes that are characterized by a tuple in a given relation. Suppose the information about the attribute A of a record T is described by the bbm $m(I_i)_{i\epsilon\{1,...,n\}}$, with $(I_i)_{i\epsilon\{1,...,n\}}$ focal elements. To query the tuples $WHERE$ the attribute A belongs to I, you need to create a new relation $R = "A \epsilon I"$. The belief and plausibility that T should be selected are:

$$Bel(T \text{ belongs to } R) = Bel(I) = \sum_{I_i \subseteq I, I_i \neq \phi} m(I_i) \qquad (8)$$

$$Pl(T \text{ belongs to } R) = Pl(I) = \sum_{I_i \cap I \neq \phi} m(I_i) \qquad (9)$$

Each tuple has a pair of weights that quantify the degree of belief and plausibility that a tuple belongs to the relation R.

3.3   Combination of More or Less Reliable Sources

Suppose that the p distributions of belief masses come from p sources $(S_j)_{j\epsilon\{1,...,p\}}$, with their respective reliability $(a_j)_{j\epsilon\{1,...,p\}}$, the weakened mass function are denoted:

$$(m(I_i))_{i\epsilon\{1,...,n\}} * (a_j)_{j\epsilon\{1,...,p\}} = \left(m_{a_j}(I_i)\right)_{i\epsilon\{1,...,n\}; j\epsilon\{1,...,p\}} \qquad (10)$$

The results that are obtained by combining the bbm of the $(S_j)_{j\epsilon\{1,...,p\}}$ sources are:

$$\forall A \subseteq \Omega, \forall k, l\epsilon\{1,...,p\}, k \neq l, m_\cap(I) = \left(m_{a_k} \cap m_{a_l}\right)(A)$$

$$= \sum_{I_i \cap I_j = A} m_{a_k}(I_i) m_{a_l}(I_j) \qquad (11)$$

Then we can deduce the degree of belief and plausibility that the tuple belongs to the relation R:

$$Bel(T \text{ belongs to } R) = Bel(I) = \sum_{I_i \subseteq I, I_i \neq \phi} m_{\cap}(I_i) \qquad (12)$$

$$Pl(T \text{ belongs to } R) = Pl(I) = \sum_{I_i \cap I \neq \phi} m_{\cap}(I_i) \qquad (13)$$

3.4   Combination of Possibility and Probability Distributions

Let us suppose the following hypothesizes:

- Let Bel and Pl respectively be measures of necessity and possibility on the frame of discernment Ω_{Π}. Then, according to theorem 1, the focal elements are totally ordered by inclusion.
- Let P be a probability measure on the frame of discernment Ω_P, i.e. an application from Ω_P to $[0; 1]$, satisfying $P(\Omega_P) = 1$ and $P(\cup_i A_i) = \sum_i P(A_i)$ (additive constraint) where $\{A_i\}_{i \in \{1,...,n\}}$ are a finite family of disjoint events of Ω_P, characterized by a discrete probability mass function i.e. the set of possible values for the random variable is countable. According to theorem 2, the focal elements are singletons.
- The considered probability and possibility distributions are independent.

The processing of any computations between measures from Ω_{Π} and Ω_P need the combination of a density of possibility with a density of probability; for this, we can use the Demspter rule of mass combination, with the hypothesis of independence of the combined distributions.

The computations of the X random and Y possible values have many potential values that are neither probable nor possible, but check a plausibility measure that results from the Dempster-Shafer rule of combination. In the general case, the resulting mass of a combination of Dempster is a sum of products of masses: $\forall A \subseteq \Omega$, $m_{\cap}(A) = \sum_{B \cap C = A} m_1(B) m_2(C)$. As the focal elements are disjoint in the probability distribution case, the resulting bbm is simply a product of masses:

$$\forall (X, Y) \subseteq (\Omega_P X \, \Omega_{\Pi}), m_{\cap}(X, Y) = (P \cap m_{\Pi})(X, Y) = P(X) * m_{\Pi}(Y) \qquad (14)$$

or

$$\forall (X, Y) \subseteq (\Omega_P X \, \Omega_{\Pi}), m_{\cap}(X, Y) = P(X) * (\alpha_i - \alpha_{i+1})(Y) \qquad (15)$$

with alpha-cuts $\alpha_1 \geq ... \geq \alpha_i \geq \cdots \geq \alpha_n \, / \, m_{\Pi}(Y) = (\alpha_i - \alpha_{i+1})(Y)$

Trivially, the overall sum on $(\Omega_P X \, \Omega_{\Pi})$ is equal to 1:

$$\sum_{A \subseteq (\Omega_P X \Omega_{\Pi})} m_{\cap}(A) = 1 \qquad (16)$$

The resulting masses $m_\cap(X, Y)$ check a Plausibility measure.

The combined masses check the following equations:

$$Pl(X, Y) = \sum_{(A,B) \cap (X,Y) \neq \phi} m_\cap(A, B) \tag{17}$$

$$Bel(X, Y) = \sum_{(A,B) \subseteq (X,Y), (A,B) \neq \phi} m_\cap(A, B) \tag{18}$$

The potential worlds match with the whole combinations of potential values. Indeed, the combination of possible values (respectively necessary) and probable values results in plausible values (respectively credible), i.e. in $card(\Omega_P) * card(\Omega_\Pi)$ potential worlds; each potential combination of values is related to both a probability and a possibility, that are obtained by projection of the resulting bbm respectively on a probability axis and a possibility axis. In the remainder of this article, some examples will be provided to illustrate the given approach.

3.5   The Pignistic Level

Finally, at the pignistic level, it is possible to associate probabilities to decisions by computing pignistic probalbilities from random, imprecise or incomplete data distributions for decision support, as described in the state of art. An example is given just below.

4   Combination of Imperfect Data: Use Cases

In general, we consider that he first step of the approach, i.e. the attribution of bbm to the focal elements of Ω, is already done.

4.1   Database Application

Context.
Suppose the value of the "cultivated area of a given farm" attribute is made of the disjunction of several subsets of the domain of believable values which is supposed to be $[0; 200]$. Suppose that (1) the attribute is the cultivated area (e.g. where maize is cultivated), characterized by a tuple in a "farm" relation and (2) that the actual values of the cultivated area of the farm F may be described by the farmer's belief about it. The farmer of the farm F (in a developing country) describes the cultivated area by the following bbm (with hectare unit of area) on the domain of the value:

$$m([24; 26]) = 0.7$$
$$m([15; 40]) = 0.2$$
$$m([0; 200]) = 0.1$$

To query the tuples $WHERE$ the pasture surface belongs to $[15; 30]$, a new relation R = "cultivatedArea $\in[15; 30]$" is created. The belief and plausibility that farm F should be selected are respectively:

$$Bel(farm\ F\ belongs\ to\ R) = 0.7$$
$$Pl(farm\ F\ belongs\ to\ R) = 1$$

Combination of More or Less Reliable Sources.
Each tuple is associated to both degrees of belief and plausibility that the tuple belongs to the relation R. The data given in the relation FARM (please see the following table) means that, for example, FAO (Food and Agriculture Organization of the United Nations) believes that the cultivated area of farm F is from 24 to 26 ha with belief 0.7, from 15 to 40 ha with belief 0.2 (thanks to precise but incomplete studies with inhabitants of the region of the farm). Furthermore, a given artificial geography satellite believes that it is between 20 and 30 (due to photos imprecision).

Table 1. Data about farm F

Farm	cultivated area (ha)	Source
Farm F	$m([24; 26]) = 0.7$ $m([15; 40]) = 0.2$ $m([0; 200]) = 0.1$	FAO
Farm F	$m([20; 30]) = 1$	Geography satellite

The users, who are also experts in the agriculture domain, have opinions about the reliability of the sources (FAO and satellite), which values are:

Table 2. Sources reliability

Source	Reliability
FAO (Food and Agriculture Organisation)	$m_{a_1} = 0.7$
Geography satellite	$m_{a_2} = 0.6$

To assess the agent's belief about the cultivated area of farm F, we have to apply the weakening factors of reliability of sources, for each given mass. The resulting masses are provided in the following table:

Table 3. Computed bbm with sources reliability

Farm	cultivated area (ha)	Source
Farm F	$m_{a_1}([24; 26]) = 0.7 * 0.7 = 0.49$ $m_{a_1}([15; 40]) = 0.2 * 0.7 = 0.14$ $m_{a_1}([0; 200]) = 1 - 0.63 = 0.37$	FAO
Farm F	$m_{a_2}([20; 30]) = 0.6$ $m_{a_2}([0; 200]) = 0.4$	Geography satellite

The results that are obtained by combining the two bbm given by the preceding table are:

$$m_{a_1 a_2}([24; 26]) = 0.49 * 0.6 + 0.49 * 0.4 = 0.49$$
$$m_{a_1 a_2}([20; 30]) = 0.6 * 0.14 + 0.6 * 0.37 = 0.51 * 0.6 = 0.306$$
$$m_{a_1 a_2}([15; 40]) = 0.14 * 0.4 = 0.056$$
$$m_{a_1 a_2}([0; 200]) = 0.37 * 0.4 = 0.148$$

We can then deduce the degree of belief and plausibility that the tuple belongs to the relation $R = "cultivatedArea \in [15; 30]"$.

$$Bel(farm\ F\ belongs\ to\ R) = 0.796$$
$$Pl(farm\ F\ belongs\ to\ R) = 1$$

Nota bene: we have merged the sets of masses from two independent sources through the combination rule of Dempster. Let us approximate the results just obtained as follows:

$$m([24; 26]) = 0.5$$
$$m([20; 30]) = 0.3$$
$$m([15; 40]) = 0.05$$
$$m([0; 200]) = 0.15$$

In fact, this distribution of masses is a special case of possibility. Indeed, focal events are totally ordered by the inclusion, i.e. $[24; 26] \subset [20,30] \subset [15, 40] \subset [0,200]$. According to theorem 1, the belief and plausibility functions associated with the masses that describe the cultivated area of farm F match respectively with measures of necessity and possibility.

Fuzzy intervals being discrete, there is a simple relationship between the masses and alpha-cuts (please refer to equation (7)). We deduce $\Pi([24; 26]) = 1$; $\Pi([20; 30]) = 0.5$; $\Pi([15; 40]) = 0.2$; $\Pi([0; 200]) = 0.15$

This second example has shown how to apply beliefs to imprecise/incomplete data.

4.2   Combination with Probabilistic Measures

In the later paragraphs, we consider that crop yield refers to the measure of the yield of a given crop per unit of cultivation area (in kilograms per hectare; kg/ha). According to theorem 2, "a belief on a finite set is a probability if and only if the focal elements are singletons".

In the following case, we suppose that the probability distribution of crop yields for a given type of culture (e.g. maize) in farm F is characterized by a discrete probability mass function i.e. the set of possible values for the random variable is countable. We assign a probability to each possible value. Let "CY" be the discrete random variable that describes the probability distribution of maize crop yields of maize in farm F.

$$P(1000 \leq CY < 3000) = 0.2$$
$$P(3000 \leq CY < 4000) = 0.6$$
$$P(4000 \leq CY < 8000) = 0.2$$

With the discrete probabilities:

$$If\ CY \in [1000; 3000[,\ P(X = CY) = \frac{0.2}{2000} = 10^{-4}$$

$$If\ CY \in [3000; 4000[,\ P(X = CY) = \frac{0.6}{1000} = 6.\,10^{-4}$$

$$If\ CY \in [4000; 8000[,\ P(X = CY) = \frac{0.2}{4000} = 5.\,10^{-5}$$

Then: Average(CY) $= 2000 * 0.2 + 3500 * 0.6 + 5000 * 0.2 = 400 + 2100 + 1200 = 3700$

The distribution of probability is illustrated in the following schema:

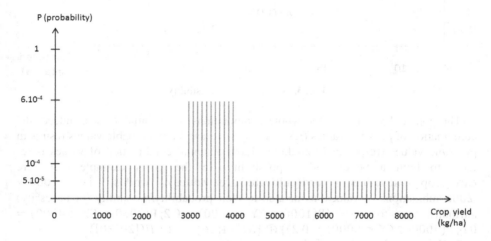

Fig. 2. Distribution of probability

We suppose that cultivated area and crop yield distributions are independent. The processing of the total farm F maize yields (in kg) needs the combination of a density of possibility, and a density of probability; for this, we apply the Demspter rule of mass combination, with the hypothesis of independence of cultivated area and crop yields: $\Pi([24; 26]) = 1$; $\Pi([20; 30]) = 0.5$; $\Pi([15; 40]) = 0.2$; $\Pi([0; 200]) = 0.15$; which implies: $m([24; 26]) = 0.5$; $m([20; 30]) = 0.3$; $m([15; 40]) = 0.05$; $m([0; 200]) = 0.15$. The following schema illustrates the distribution of possibility about cultivated area values:

The computation of cultivated area CA and crop yield CY of farm F (CA*CY) has many potential values that are neither probable nor possible, and result from the Dempster-Shafer rule of combination. The combined masses check the following equations:

$$m_\cap(CY, CA) = (P \cap m_\Pi)(CY, CA) = P(CY) * m_\Pi(CA) \qquad (19)$$

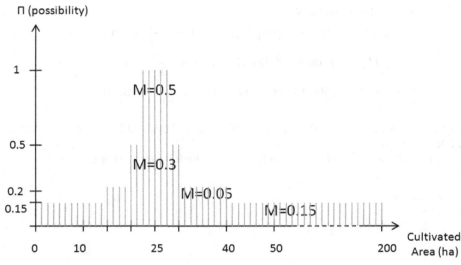

Fig. 3. Distribution of possibility

The potential worlds are the whole combinations of potential values. Indeed, the combination of possible values (respectively necessary) and probable values results in plausible values (respectively credible). Each potential combination of values is related to both a belief and a possibility. In the given example there are card(crop yield values) ∗ card(cultivated area values)   worlds,   i.e.   7000 ∗ 200combinations of values, associated to their probability/possibility (or necessity) attributes   as   follows: {P(1000 ≤ CY < 3000) = 0.2; P(3000 ≤ CY < 4000) = 0.6; P(4000 ≤ CY < 8000) = 0.2)} ⊕ {Π([24; 26]) = 1 ; Π([20; 30]) = 0.5; Π([15; 40]) = 0.2 ; Π([0; 200]) = 0.15}

In the following illustration, the combination of a cultivated area between 20 and 30 ha (possibility 0.5) and of crop between 3000 and 4000 kg/ha (probability 0.6) yield, results in a total production value between 60000 kg (60 tons) and 120000 kg (120 tons), with plausibility 0.3. The resulting measurement is the conjunctive combination of a probability distribution (projection on the axis of probabilities) and its dual (projection on the axis of possibilities).

The two by two intersection of each of focal masses, conjunctive combination of Dempster-Shafer, is distributed on smaller subsets, so that the orthogonal sum of all the masses allocated to each world is equal to 1. Here is a table containing the results of the combination of the crop yield and the cultivated area of the farm F:

m([24; 26]) = 0.5 ; m([20; 30]) = 0.3 ; m([15; 40]) = 0.05 ; m([0; 200]) = 0.15

4.3   Combination and Conflict with Experts' Advices

Suppose that a second data source, the farmer who is familiar with its farm, believes that the total production of the year is going to be either from 5×10^4 and 6×10^4kg (with confidence 0.8), or from 4×10^4 et 7×10^4kg (with confidence 0.2).

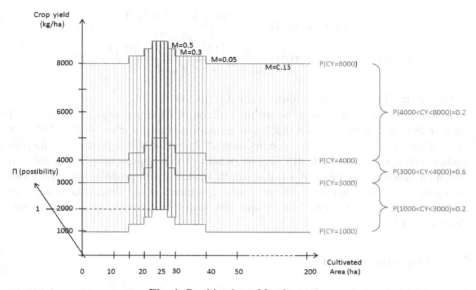

Fig. 4. Combination of focal masses

Table 4. Dempster combination of P and m

	$m([24;26])$ $= 0.5$	$m([20;30])$ $= 0.3$	$m([15;40])$ $= 0.05$	$m([0;200])$ $= 0.15$
P([1000,3000[) = 0.2	0.1	0.06	0.01	0.03
P([3000,4000[) = 0.6	0.3	0.18	0.03	0.09
P([4000,8000[) = 0.2	0.1	0.06	0.01	0.03

This example has shown how to combine imprecise/incomplete data with probabilistic measures.

The results of combined masses show that there is a conflict between the two data sources (one source from a combination of probabilistic and possibility data and the other from the expert's advice). The obtained computation of the degree of conflict between the sources is: $m_\cap(\phi) = 1 - 0.74 = 0.26$

Then, let us compute the masses allocated to the intersected intervals through the combination rule of Dempster, multiplied by coefficient of conflict $\frac{1}{1-m_\cap(\phi)}$:

$$m_\oplus([5 \times 10^4; 6 \times 10^4[) = \frac{m_\cap([5 \times 10^4; 6 \times 10^4[)}{1 - m_\cap(\phi)} = 0.75$$

$$m_\oplus([4 \times 10^4; 7 \times 10^4[) = \frac{m_\cap([4 \times 10^4; 7 \times 10^4[)}{1 - m_\cap(\phi)} = 0.19$$

$$m_{\oplus}([6 \times 10^4; 7 \times 10^4[) = \frac{m_{\cap}([6 \times 10^4; 7 \times 10^4[)}{1 - m_{\cap}(\phi)} = 0.05$$

$$m_{\oplus}([4.5 \times 10^4; 7 \times 10^4[) = \frac{m_{\cap}([4.5 \times 10^4; 7 \times 10^4[)}{1 - m_{\cap}(\phi)} = 0.01$$

Nota Bene: The results have been rounded up to 10^{-2}.

In the domain of agriculture, where the opinion of experts (usually farmers) is of great importance, it is observed that the mass allocated to the interval $[5 \times 10^4; 6 \times 10^4[$, although weakened by the combination of the result of the product of the production acreage, remain high (0.75). The benefit of this approach is to mix data on which uncertainties are random, i.e. that can be modeled with probabilistic tools, with imprecise or incomplete data, i.e. due to either ignorance or too limited knowledge of these data, which can be remedied by opinions from experts, limiting the wrong estimates of results.

5   Implementation

The concepts described above have been implemented as a C++ library. This library consists of a class provided with the four classic overloaded operators +, -, *, /. Agents and web services calling such C++ objects take into account the private internal attributes such as the masses of the beliefs functions of each operand, probabilistic or possible, and their Dempster Shafer combination. The methods such as getProba, getPossibility, getbelief used determine rigorously a projection on the possible and / or probable worlds.

6   Conclusion

The approach we have introduced in this article provides a method to manipulate and combine measures that are random, incomplete and uncertain. With the API that we have implemented, it is possible to compute and automatically combine the belief distributions for given values.

The results we achieved seem to confirm the benefits of our approach. However, they should be completed in many ways. Several directions that have been only drafted here will be provided for the continuation of this research. Firstly, the development of programming concepts has been quickly investigated and should be, in our opinion explored. On the other hand, approaches for combining both conflicting and continuous quantities have to be continued.

References

1. Dempster, A.: Upper and lower probabilities induced by multivalued mapping. Annals of Mathematical Statistics AMS-38, 325–339 (1967)
2. Dubois, D., Prade, H.: Théorie des possibilités, Application à la représentation des connaissances en informatique. Masson (1988)

3. Gacôgne, L.: 1997. Eléments de logique floue. CNAM, Institut d'informatique d'Entreprise, p. 47 (Mai 1997)
4. Shafer, G.: A mathematical theory of evidence. Princeton University Press (1976)
5. Smets, P., Kennes, R.: The transferable belief model. Artificial Intelligence 66(1994), 191–234 (1994)
6. Vannoorenberghe, P.: Un état de l'art sur les fonctions de croyance appliquées au traitement de l'information. Revue I3 3(3(2)), 945 (2003)
7. Zadeh, L.A.: Fuzzy Sets. Information and Control, vol. 8. Academic Press (1965)
8. Zadeh, L.A.: Fuzzy Sets as a basis for a Theory of Possibility. Fuzzy Sets and Systems 1 (1978)

Variables Selection for Ecotoxicity and Human Toxicity Characterization Using Gamma Test

Antonino Marvuglia[1], Mikhail Kanevski[2], Michael Leuenberger[2], and Enrico Benetto[1]

[1] Public Research Centre Henri Tudor, Resource Centre for Environmental Technologies (CRTE), 6A, avenue des Hauts-Fourneaux, L-4362 Esch-sur-Alzette, Luxembourg
[2] University of Lausanne, Faculty of Geosciences and Environment,
Institute of Earth Surface Dynamics, Geopolis building - 1015 Lausanne, Switzerland
{antonino.marvuglia,enrico.benetto}@tudor.lu,
{mikhail.kanevski,michael.leuenberger}@unil.ch

Abstract. Toxicity characterization of chemicals' emissions is a complex task which proceeds via multimedia fate and exposure models attached to models of dose–response relationships. Several different environmental multimedia models exist, but in any case a vast amount of data on the properties of the chemical compounds being assessed is required. This paper deals with the selection of informative variables in the problem of deriving characterization factors for eco-toxicology and human toxicology of chemical compounds starting from molecular-based properties. The Gamma Test algorithm has been applied to single out the most informative variables. The set of variables retained varies with the subset of the original dataset used to carry out the analysis. In particular, 16 different subsets have been used. They have been created selecting each time only those entries in the data set where one chosen input variable was available only from measurements/estimations, respectively.

Keywords: Toxicity characterization, Life Cycle Assessment, USEtox®, Gamma Test, features selection.

1 Introduction

Toxicity characterization of chemical emissions is currently performed via multimedia fate and exposure models attached to models of dose–response relationships. Several environmental multimedia models exist. They have different scope and modelling principles and hence producing different characterization factors (CF), i.e. factors that, once multiplied by the corresponding inventory inputs (chemical substances), translate them into directly comparable impact indicators.

In life cycle impact assessment (LCIA) the CF is built with a number of separate elements (Rosenbaum et al. 2008):

- The fate factor (FF), which takes into account the successive steps of fate. It represents the persistence of the chemical into the environment;
- The aspect of exposure or intake, symbolised by the exposure factor (XF). It expresses the bioavailability of a chemical, represented by the fraction of it dissolved in the environmental compartment at stake;

B. Murgante et al. (Eds.): ICCSA 2014, Part III, LNCS 8581, pp. 640–652, 2014.

- The aspect of effect, symbolised by the effect factor (EF);
- The combined aspect of fate and human exposure, represented by the intake fraction (iF).

This results in a set of scale-specific CFs, given as the product:

$$CF = EF \times \underbrace{XF \times FF}_{iF} = EF \times iF \tag{1}$$

The model USEtox® (Rosenbaum et al. 2008) is the result of a scientific consensus process involving comparison of and harmonization among existing environmental multimedia fate models.

The compound specific fate factors consist of 66 parameters for each chemical (11 fate factors for each emission compartment: continental urban air, continental rural air, continental freshwater, continental sea water, continental natural soil, continental agricultural soil), of which all are needed to perform a full characterization (i.e. calculation of CFs for emission from all possible emission compartments to all final compartments) in USEtox®. The model allows the calculation of 42 exposure factors, 1 eco-exposure factor, 1 eco-effect factor on freshwater and 4 human health effect factors, for a total of 114 factors for each substance.

Several substance specific input variables are required by USEtox® to calculate the full set of factors. However, not all of them have the same explanatory power with respect to each of the output factors (FF, XF, EF, iF). In this connection, the aim of the *feature selection* we perform in this paper is finding the configuration of inputs that allows the best prediction of the output factors with the available data. Based on the USEtox® database of CFs, consisting of 3,073 data records, Birkved and Heijungs (2011) derived multidimensional bilinear models for emission compartment specific fate characterisation of chemical emissions applying Partial Least Squares Regression (PLSR). However, the authors did not perform a model-based features selection, but "decided to let data availability determine the grouping of the variables" (Birkved and Heijungs, 2011). To our knowledge, no other attempt has been performed in this direction in the literature.

The aim of this paper is analysing the input and output manifolds given by the USEtox® database and performing a ranking of the input variables based on their contribution to explain the variance of the output variables using a particular nonlinear technique known as Gamma Test (Stefánsson et al., 1997).

2   Data Sets Preparation

The data used for the analysis are issued from the database made available by Huijbregts et al. (2010). For the sake of simplicity, only the organic compounds have been taken into account in this work. This simplification facilitates the development and testing of the approach pursued in this paper, while not affecting the consistency of the outcomes and their applicability to the other compounds.

The predictors describing our manifold are: **1)** Molecular weight (MW); **2)** Partitioning coefficient between octanol and water (K_{OW}); **3)** Partitioning coefficient between organic carbon and water (K_{OC}); **4)** Henry law coefficient at 25°C (K_H25C); **5)** Vapour pressure at 25°C (Pvap25); **6)** Solubility at 25°C (Sol25); **7)** Degradation rate in air (Kdeg$_A$); **8)** Degradation rate in water (K$_{degW}$); **9)** Bioaccumulation factor in fish/biota (BAF$_{fish}$).

Two additional predictors, the degradation rate in sediment (Kdeg$_{Sd}$) and the degradation rate in soil (Kdeg$_{Sl}$) were present in the original database. These variables were originally included in the input data set and subsequently excluded from the rest of the analysis as a scatterplot of all the possible pairwise combinations pointed out a perfectly linear dependency from variable #8 (K$_{degW}$).

In order to simplify calculations and results presentation only the following five factors have been chosen and used as output variables in this study:

- Output 1= the fate factor from urban air to urban air (FF_airU_airU);
- Output 2= the fate factor from urban air to continental air (FF_airU_airC);
- Output 3= the fate factor from urban air to continental freshwater (FF_airU_fr.waterC);
- Output 4= the fate factor from continental freshwater to continental freshwater (FF_fr.waterC_fr.waterC)
- Output 5= the intake fraction from continental freshwater to drinking water (iF_fr.waterC_dr.waterC).

A choice of some representative output variables out of the full set to study was necessary since, as it will be explained in section 3, we performed a *full embedding* search of the features space; therefore we explored all the possible combinations of inputs. This means that for *m* inputs and *n* outputs, we explored $n(2^m - 1)$ models. In our case the number of inputs is quite small, but exploring the whole set of 114 USEtox® factors would not have allowed reasonable calculation times.

It is important to remark that some of the substance-related data contained in the USEtox® database come from experimental measurements, others are estimated. In particular, when a piece of data is not available from measurements, a set of inter-relationships (see Huijbregts et al., 2010, page 12) have been used in USEtox® to derive them.

As already done in (Birkved and Heijungs, 2011), in order to reduce the differences in scale (thus performing a normalization of the data set), before further processing the data have been logarithmically transformed (using Log$_{10}$).

Since in the database there were 22 substances with a null value of K$_{degA}$, they were excluded from the logarithmic transformation, thus reducing the total number of observations used for the analysis from 3,073 to 3,051.

In order to determine if estimated data can affect the results of the variable selection, we generated several data sets starting from the original one, according to the following rationale: for each variable (except for the molecular weight, which is always exactly known) we split the original database in two subsets; one contains only those substances for which the variable at stake is known from measurements, and the other contains only those substances for which it was estimated. We repeated the procedure for all the 8 variables, thus obtaining 16 different subsets of data, of which, however, the one containing only measured values for the variable K$_{degW}$ was not statistically representative, since it contained only 6 samples. It was not possible to select a database containing for each and every variable only measured values, since the resulting set would have contained only a handful of substances. Table 1 shows the number of data contained in each subset.

Table 1. Number of observations contained in each data subset. The row header "EST" means that the database is obtained considering those substances for which the variable named in the column header is estimated from other variables using the equations contained in (Huijbregts et al., 2010); the row header "EXP" refers to the database which contains measured data.

	MW	K_{OW}	K_{OC}	K_H25C	Pvap25	Sol25	K_{degA}	K_{degW}	BAF_{fish}
EST	0	1313	2521	1905	1648	1218	2692	3045	2607
EXP	3051	1738	530	1146	1403	1833	359	6	444

3   The Gamma Test

The knowledge of the noise level in the input space provides an important metric in a feature selection task. Given a database of m possible inputs for a corresponding output y, *feature selection* is the selection of the configuration of inputs that allows the best prediction of y with the available data.

The Gamma Test (Stefánsson et al., 1997) computes an estimate of the noise (or error variance) present in a dataset directly from the data itself. Moreover, it does not make any assumption regarding the parametric form of the equations governing the system under investigation. The only requirement is that the system is smooth (i.e. the transformation from input to output is continuous and has bounded partial derivatives over the input space) and can be modelled by a smooth function f of the following form:

$$y = f\left(x_1 \ldots x_m\right) + r \tag{2}$$

where x_m is the generic observation vector (point in the modelling space) and r is a random variable representing the noise with a distribution variance σ^2 and sample variance $\hat{\sigma}_M^2$.

Given sufficient data, the Gamma Test can efficiently estimate the sample variance of the noise variable $\hat{\sigma}_M^2$. Since $\hat{\sigma}_M^2 \to \sigma^2$ in probability as $M \to \infty$, it follows that the Gamma Test asymptotically estimates σ^2.

Let $x_N\left[i,k\right]$ denote the k-th nearest neighbour, in terms of Euclidean distance, of the point $x_i \left(1 \le i \le M\right), \left(1 \le k \le p\right)$. The equations required to compute Γ are:

$$\delta_M\left(k\right) = \frac{1}{M} \sum_{i=1}^{M} \left|x_{N[i,k]} - x_i\right|^2 \quad \left(1 \le k \le p\right) \tag{3}$$

where M is the total number of points, p is the maximum number of neighbours of point i (which has to be set in advance) and $\left|\cdot\right|$ denotes Euclidean distance, and:

$$\gamma_M\left(k\right) = \frac{1}{2M} \sum_{i=1}^{M} \left|y_{N[i,k]} - y_i\right|^2 \quad \left(1 \le k \le p\right) \tag{4}$$

In order to compute Γ a least squares fit regression line is constructed for the p points $\left(\delta_M(k), \gamma_M(k) \right)$. The intercept on the vertical axis is the Γ value. If Γ is large (compared to the variance of the output) then it is not likely that the outputs can be determined by the inputs from a smooth model, whereas if the Γ is small or close to zero this becomes more likely (Jones, 2004).

As shown by Evans and Jones (2002a):

$$\gamma_M(k) \to Var(r) \text{ in probability as } \delta_M(k) \to 0 \tag{5}$$

A formal mathematical justification of the method can be found in (Evans and Jones, 2002a, Evans and Jones, 2002b).

Calculating the gradient of the regression line provides an estimation of the complexity of the unknown surface f one is trying to model. A good regression line with points $\left(\delta_M(p), \gamma_M(p) \right)$ approaching $(\delta, \gamma) = (0,0)$ indicates that the scalar output values of input-near-neighbours are close. If the regression line has a steep slope this indicates that the modelling function f that one seeks to approximate is likely to be quite difficult to construct and a large number of data points M will be required. If the line is almost horizontal (*pure nugget effect*) the process can be modelled with a smooth function.

The number of near neighbours p has to be selected in advance, but one has to bear in mind that if the underlying function f is non-linear, then setting a "large" value for p will make *non-linearity errors* arise in the linear regression of the $\left(\delta_M(p), \gamma_M(p) \right) (1 \le k \le p)$ points. In fact, as M decreases the local density of points decreases and the non-linearities of f become an issue. In particular, the higher the average local curvature of the function f, the higher M needs to be in order to ensure that the local density of points is sufficient to get a good $\left(\delta_M(p), \gamma_M(p) \right)$ regression line.

However, we often cannot increase M when we have a non-linear function with high local gradients, therefore the best alternative is to decrease p so that the neighbours $\left(\mathbf{x}', \mathbf{x}'', ..., \mathbf{x}^p \right)$ of the generic point $\mathbf{x}$ are not too far away from it and the function f does not change too much at the local scale.

In order to choose a suitable number of neighbours one can compute the Gamma error defined by Eq. (6) and observe its evolution as a function of p:

$$\Gamma_{error} = \Gamma(p,M) - \hat{\sigma}_M^2 \tag{6}$$

where $\Gamma(p,M)$ is the Gamma statistic for a specified p and $\hat{\sigma}_M^2$ is the sample noise variance.

To "track" the Γ_{error} the standard error (*SE*) of the regression given by Eq. (7) can be used:

$$SE = \sqrt{\frac{1}{p-2} \sum_{k=1}^{p} \left(\Gamma(k,M) - \overline{\Gamma} \right)^2} \tag{7}$$

where $\Gamma(k,M)$ is the generic k-th Gamma regression point value and $\bar{\Gamma}$ is their mean.

The value of p for which the SE begins to raise provides an upper bound for p; using a value of p higher than this will result in a bigger influence of non-linearity errors in the (δ, γ) regression fit, in turn leading to a poor estimate of σ^2.

A further measure that allows a judgement on how well the output can be modelled by a smooth function (independently on the output range) is the V-Ratio, which can be defined as follows:

$$V = \frac{\Gamma}{\sigma^2(y)} \tag{8}$$

where $\sigma^2(y)$ is the variance of outputs y. A V-Ratio close to zero means that there is a high degree of predictability of the given output y. If the V-Ratio is close to one the output is equivalent to a random walk (not a smooth underlying model) (Monte, 1999). The benefit of this statistics is that it is scale-independent.

As mentioned above, the goal of the application of Gamma Test in this paper is performing a variable selection. In particular, for any specific output, we chose the set of inputs that has the minimum asymptotic value of the Gamma statistic. It has to be taken into account that, because of sampling error, if the variance of the noise level on an output is very small the Gamma statistic may sometimes be negative, even though a variance can never be negative. If this occurs we use the absolute value of the Gamma statistic.

The minimum value of Γ is calculated as the smallest over a list of values obtained with a *full embedding*, i.e. trying every possible combination of inputs for every possible output. The software Wingamma (Jones, 1998) has been used to carry out the experiments. In order to choose the number of neighbours p for each output, the Gamma Test was run for increasing number of neighbours. The chosen value of p was the one for which the SE started to have a rapid increase. This value coincided or was always very close to the minimum of the absolute value of the Gamma Test. Obviously, the minimum possible value for p is always 2, since this is the minimum required number of points to calculate a regression line.

4   Results and Discussion

As an example, Table 2 shows the minimum value of Γ, and the corresponding number of neighbours used, gradient, standard error and V-Ratio obtained for the two data sets K_{degA}_EST and K_{degA}_EXP and for all the models obtained using one of the five outputs at the time. For each model, if the generic variable $\#X$ was used as input, then the 9-elements vector presented in the rightmost column of Table 2 has the digit "1" in its X-th position. The graphs of the Gamma statistics and the SE for increasing number of neighbours are shown in Fig. 1 and Fig. 2 for the data of the K_{degA}_EST and K_{degA}_EXP sets, respectively.

Table 2. Results of the Gamma Test for the two data sets Kdeg$_A$_EST and Kdeg$_A$_EXP with data logarithmically transformed

Data set	Output variable used	p	min(Γ)	Gradient	Standard error	V-Ratio	Selected inputs
	Output 1	2	1.85E-06	1.75E-03	0.00E+00	3.34E-04	000101101
	Output 2	2	2.10E-04	4.42E-02	0.00E+00	6.20E-04	100111111
K$_{degA}$_EST	Output 3	2	1.42E-04	4.68E-02	0.00E+00	1.34E-04	111100110
	Output 4	4	2.15E-03	3.00E-02	1.35E-04	2.01E-02	101100010
	Output 5	5	4.26E-05	6.44E-02	5.45E-03	1.01E-04	011100010
	Output 1	5	1.21E-05	2.64E-03	6.75E-05	3.99E-03	001000110
	Output 2	4	8.75E-06	1.07E-01	1.00E-03	2.08E-05	110000101
K$_{degA}$_EXP	Output 3	10	6.31E-06	6.94E-02	4.33E-03	2.83E-06	011110110
	Output 4	2	2.78E-05	1.68E-02	0.00E+00	3.01E-04	100111010
	Output 5	30	2.48E-05	4.78E-03	3.07E-03	2.81E-04	111111111

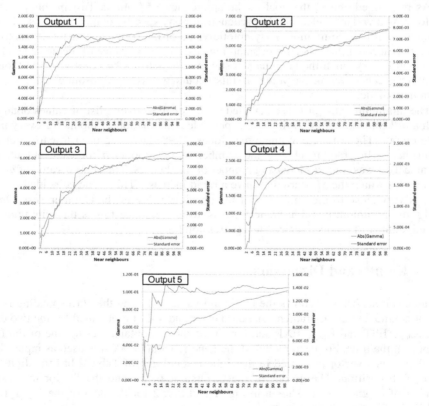

Fig. 1. Graph of the absolute value of the Gamma statistic and the standard error for increasing number of neighbours, using the logarithmically transformed data of the K$_{degA}$_EST data set and the different outputs

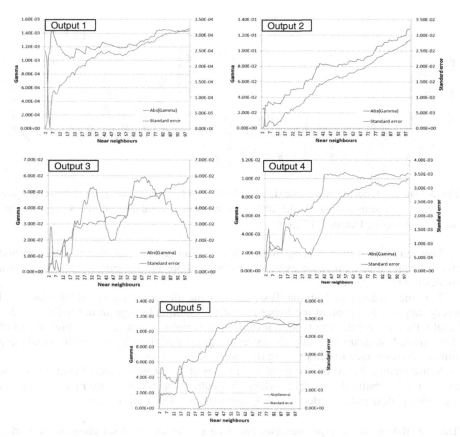

Fig. 2. Graph of the absolute value of the Gamma statistic and the standard error for increasing number of neighbours, using the logarithmically transformed data of the K_{degA}_EXP data set and the different outputs

By plotting the frequency histogram of the Gamma statistic it is possible to get further information. As an example Fig. 3 (left) shows the Gamma frequency histogram obtained with the K_{degA}_EST data set and considering Output 1.

The histogram is bimodal: the first peak contains the feature sets that produced results having $\Gamma < 3.5 \cdot 10^{-4}$ (approximately). A frequency graph of these features is shown in Fig. 3 (right). It shows that the 7-th input variable (K_{degA}) is the most significant, and it is in fact one of the features selected for Output 1 (see Table 2).

If we plot the frequency graph of the features for values of Γ corresponding to other intervals (in particular higher values of Γ), the other features shown in Table 2 will emerge (i.e. they will have higher frequency than K_{degA} in regions of higher Γ) and K_{degA} will instead have a low frequency. Therefore K_{degA} represents the most favourable feature (i.e. the one which is most directly related to the analysed output) because it has higher frequency than average in regions of low Γ.

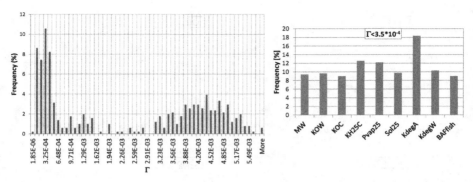

Fig. 3. Gamma frequency histogram (left hand side) and frequency histogram of the feature sets that produced results having $\Gamma < 3.5 \cdot 10^{-4}$ (right hand side) obtained with the K_{degA}_EST data set and considering Output 1 (FF_airU_airU)

This may look as a foregone finding, but it is remarkable the fact that the method did not require any previous expert knowledge about the importance of K_{degA} when modelling FF_airU_airU.

This nice feature of Gamma Test can be generalized for every set of inputs and every output, i.e. a peak at the lower end of the Gamma histogram normally contains results that use all the available relevant input variables; a peak at the higher end of the gamma histogram normally shows results generated from input variables that have little or no relevance in determining the output (Durrant, 2001).

Subsequently, based on the number of times an input was selected (regardless the output used to build the model in which that output was retained) we ranked the input variables in descending order. Table 3 shows the results of this ranking.

Table 3. Raking of the input variables based on the frequency of selection in the models according to the Gamma statistic (experiment run only on K_{degA}_EST and K_{degA}_EXP).

Order in the ranking	Input variable (using both data sets)	Input variable (using K_{degA}_EST)	Input variable (using K_{degA}_EXP)
I	K_H25C	K_H25C	K_{degA}
II	K_{degW}	K_{degW}	K_{degW}
III	K_{degA}	MW	MW
IV	MW	K_{OC}	K_{OW}
V	K_{OC}	K_{degA}	K_{OC}
VI	K_{OW}	K_{OW}	K_H25C
VII	Pvap25	Sol25	Pvap25
VIII	Sol25	BAF_{fish}	Sol25
IX	BAF_{fish}	Pvap25	BAF_{fish}

Even though only five outputs probably represent a quite small sample to build a ranking based on the frequency of selection, some observations can be formulated. One can observe that when basing the ranking on both data sets, the variable K_H25C is the one resulting in the highest selection rate (8 times out of 10) and this holds true also when using only the K_{degA}_EST dataset (in this case it is always selected), but not when using the data of the K_{degA}_EXP dataset. Furthermore, the list of the five most selected variables, even with some differences in the ranking position, is the same when using the results of both datasets together or only the K_{degA}_EST dataset, but not when using only the K_{degA}_EXP dataset. The experiment should be in principle repeated using all the 114 outputs of USEtox® to have a more consistent estimation of the selection rate of each variable, but from the observation of this result and of the graphs related to the data of the K_{degA}_EXP dataset (not shown in this paper due to space constraints) we can certainly assert that the two datasets show different patterns. A similar effect has also been observed for the other output variables (not presented in this paper).

Finally, we also run the M test. The M test is used to check how quickly the estimate returned by the Gamma Test will stabilise to a close approximation of the noise variance. It consists in computing the Γ statistic for an increasing number of data points M. By plotting the Γ values over M it can be seen whether the graph appears to be approaching a stable asymptote. Figures 4 and 5 show the M test graphs obtained with the logarithmically transformed data of the K_{degA}_EST and K_{degA}_EXP data sets, respectively.

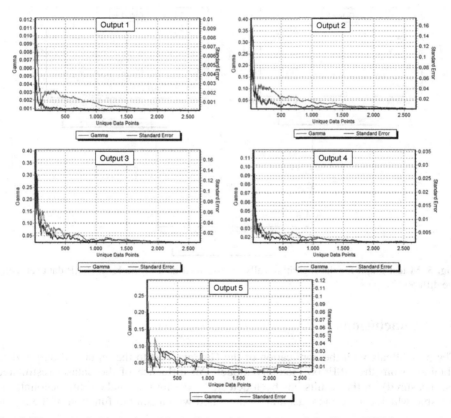

Fig. 4. M test graph for the logarithmically transformed data of the K_{degA}_EST data set using the different outputs

It is apparent that the graphs obtained with the data of the K_{degA}_EST set exhibit an exponential-like decrease and reach a quite clear asymptote. In the graphs obtained with the data of the K_{degA}_EXP set the values of Γ did not stabilize to an asymptote (except for Output 2, see Fig. 5). This indicates that these latter data are not sufficient (M is not sufficiently large) so the error in estimating the Gamma statistic for any particular embedding (i.e. selection of input variables chosen from all the possible inputs) is sufficiently high to make the outcome of a full embedding search not reliable.

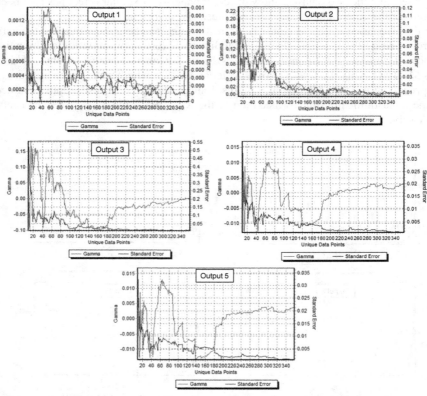

Fig. 5. M test graph for the logarithmically transformed data of the K_{degA}_EXP data set using the different outputs

5   Conclusions

The paper deals with the application of the Gamma Test to the input and output data obtained from the USEtox® toxicity model. An influence of the dataset (estimated vs. measured) on the results has been detected. Among those taken into account, the variable which can be more accurately modelled with a smooth function of the inputs is the intake fraction from continental freshwater to drinking water and only if in the input data the variable K_{degA} comes from estimates and not from experimental measurements. This is due to the fact that the measurements of K_{degA} in the USEtox® database are too few to build a smooth function. This latter statement holds true also for the other variables.

The application of the Gamma Test allowed performing an informative variables selection (embedding) for each case, according to which output was modelled (see Table 2), with K_H25C being the variable selected the most of the times (regardless the output modelled) when using the K_{degA}_EST data set, but not when using the K_{degA}_EXP data set.

It is worth noting that the Gamma Test has the nice capability of detecting the "most favourable" feature(s), i.e. the input variable(s) which are more directly linked to the modelled output variable. This information is inferred by coupling the Gamma frequency histogram with the frequency histogram of the feature sets that produce values of Γ lying at the lower end of the interval of variation of Γ. An example is shown in the paper for the model having the fate factor from urban air to urban air as the output (Fig. 2). In this case K_{degA} evidently appears as the most relevant feature.

Repeating the experiment for the other output variables we following features were singled out as the most relevant: K_{degA} in the case of Output 2 (fate factor from urban air to continental air); K_H25C and K_{degA} in the case of Output 3 (fate factor from urban air to continental freshwater); K_{degW} in the case of Output 4 (fate factor from continental freshwater to continental fresh-water) and K_{OC} in the case of Output 5 (intake fraction from continental freshwater to drinking water).

These results are not surprising, but it is worthy highlighting here that they were obtained by the model without any a priori knowledge of the physicochemical mechanisms governing substances fate.

References

1. Rosenbaum, R.K., et al.: USEtox®—the UNEP-SETAC toxicity model: recommended characterisation factors for human toxicity and freshwater ecotoxicity in life cycle impact assessment. Int. J. Life Cycle Assess. 13(7), 532–546 (2008)
2. Birkved, M., Heijungs, R.: Simplified fate modelling in respect to ecotoxicological and human toxicological characterisation of emissions of chemical compounds. Int. J. Life Cycle Assess. 16(8), 739–747 (2011)
3. Stefánsson, A., Končar, N., Jones, A.J.: A note on the Gamma Test. Neural Comput. Appl. 5, 131–133 (1997)
4. Huijbregts, M., Hauschild, M., Jolliet, O., Margni, M., McKone, T., Rosenbaum, R.K., et al.: USEtox® User manual (2010)
5. Jones, A.J.: New tools in non-linear modelling and prediction. Comput. Manag. Science 1(2), 109–149 (2004)
6. Evans, D., Jones, A.J.: Asymptotic moments of near neighbour distance distributions. Proc. of R. Soc. Ser. A 458(2028), 2839–2849 (2002a)
7. Evans, D., Jones, A.J.: A proof of the Gamma Test. Proc. of R. Soc. Ser. A 458(2027), 2759 (2002b)
8. Monte, R.A.: A random walk for dummies. MIT Undergrad. J. Math. 1, 143–148 (1999)

9. Jones, A.: The winGamma™ User Guide (1998),
   http://users.cs.cf.ac.uk/O.F.Rana/Antonia.J.Jones/
   GammaArchive/IndexPage.htm
10. Durrant, P.: winGamma™: A non-linear data analysis and modelling tool with applications
    to flood prediction system. Ph.D. thesis, University of Wales, Cardiff, Wales U.K (2001),
    http://www.cs.cf.ac.uk/user/Antonia.J.Jones

A Multiobjective Model for Biodiesel Blends Minimizing Cost and Greenhouse Gas Emissions

Carla Caldeira[1,2], Ece Gülsen[3], Elsa A. Olivetti[3], Randolph Kirchain[3], Luís Dias[1], and Fausto Freire[2]

[1] INESC-Coimbra and Faculty of Economics, University of Coimbra, Coimbra, Portugal
`carla.caldeira@dem.uc.pt, ldias@inescc.pt`
[2] ADAI-LAETA, Dept. of Mechanical Engineering, University of Coimbra, Coimbra, Portugal
`fausto.freire@dem.uc.pt`
[3] Materials Systems Laboratory, Engineering Systems Division, MIT, Massachusetts, USA
`{elsao,kirchain}@MIT.EDU`

Abstract. The goal of this paper is to present a multiobjective model to optimize the blend of virgin oils for biodiesel production, minimizing costs and life-cycle Greenhouse Gas (GHG) emissions. Prediction models for biodiesel properties based on the chemical composition of the oils were used to establish technical constraints of the model. Biodiesel produced in Portugal from palm, rapeseed and/or soya was used as a case study. The model was solved using the ε-constraint method and the resulting Pareto curve reveals the trade-off between costs and GHG emissions, from which it was possible to calculate GHG abatement costs. Illustrative results are presented: GHG emissions (not accounting for direct and indirect Land Use Change -LUC) and biodiesel production costs (focused on oil feedstock). Analyzing the blends along the Pareto curve, a reduction in GHG emissions is obtained by progressively replacing rapeseed by soya and reducing the palm share in the blend used for biodiesel production.

Keywords: Biodiesel, blend, Life-Cycle, Greenhouse Gas (GHG), Multiobjective model.

1 Introduction

According the OECD-FAO outlook 2011-2020, biodiesel use in the European Union (EU) will increase by almost 85% over the projection period [1]. The selection of vegetable oil feedstocks for biodiesel production depends on factors such as supply, cost, storage properties and engine performance [2]. Nowadays, rapeseed and sunflower oils are mainly cultivated and used in Europe for biodiesel production, palm oil has the biggest share in tropical countries and soybean oil is the major feedstock in the United States [2]. According to information provided by the Portuguese energy agency (DGEG), in 2012 the main feedstocks used for biodiesel production in Portu-gal were soya (49%), rapeseed (34%) and palm (14%).

The usual process for biodiesel production is transesterification - reaction of a triglyceride (main component of vegetable oils) and an alcohol in the presence of a

B. Murgante et al. (Eds.): ICCSA 2014, Part III, LNCS 8581, pp. 653–666, 2014.

catalyst. A triglyceride is an ester derived from glycerol and three fatty acids (FA). A FA is a carboxylic acid with a long aliphatic tail (chain), which is either saturated or unsaturated. The most common used alcohol is methanol and the products of the reaction are glycerol and Fatty Acid Methyl Ester (FAME) – the biodiesel. Each feedstock presents a typical FA profile that will influence the final properties of the biodiesel. It is generally assumed that the FA compositional profiles remain unchanged during conversion of the feedstocks to fuels via transesterification. For this reason, the fatty esters properties are directly related to the FA profile. Structural features such as chain length, degree of unsaturation and branching of the chain determine the fuel properties [4].

Although several factors can influence the final properties of the biodiesel, those that are directly related with the FAME profile are: viscosity, density, cetane number (CN), cloud point (CP), pour point (PP), cold filter plugging point (CFPP), flash point (FP), iodine value (IV) and heating value. Oxidative stability (OS) besides being related to the FAME composition it is also influenced significantly by the FAME clean-up and storage practices employed [3, 4]. Moreover, the anti-oxidant additives that some biodiesel samples contain modify stability of the FAME [4, 5].

Recently, beyond chemical properties, sustainability of biodiesel production emerged as an important aspect to be addressed. For example, the European Directive 2009/28/EC on the promotion of the use of energy from renewable sources (Renewable Energy Directive - RED) established sustainability criteria that prevent production of biofuels on recently deforested land or ecosystems with high biodiversity, and also requires Life-Cycle (LC) GHG savings of 35% when biodiesel replaces fossil diesel [6]. This target will become more restricted after 1 January 2017, when the GHG savings shall be at least 50 % relative to fossil diesel. Biodiesel that does not comply with the sustainability criteria will not be taken into account to reach the mandatory targets established in the RED of including at least 10% of all energy in road transport fuels, to be produced from renewable sources by 2020. Thus, constraints regarding biodiesel LC GHG emissions will have a significant influence on the selection of the vegetable oil feedstocks, available in the market, for biodiesel production.

Gülsen et al. [3] developed a blend optimization model for biodiesel production, considering the uncertainty and variation of feedstock chemical composition and GHG emissions. The blending algorithm determines the "recipe" that minimizes cost having as constraints of the fuel physical properties and GHG emissions. The main goal was to evaluate the performance of three different emission control policy scenarios, inspired by current regulations, in the context of blending and the implications of those policies on the property distributions as well as costs of biodiesel. Although not applied to biodiesel production, other blending models have been developed: Fröhling et al. [7] derive linear input-output functions for the relevant material and energy flows by using multiple linear regression having as a support the thermodynamic simulation of the processes; Li et al. [8] developed a nonlinear time-varying model for cement blending process; and, Lingshuang et al. [9] presented a stochastic optimization method based on Hammersley Sequence Sampling (HSS) technique and expert knowledge for aluminium blending process with parametric uncertainty.

The goal of this paper is to present a multiobjective (MO) model to optimize the blend of virgin oils for biodiesel production, minimizing costs and life-cycle GHG emissions. Prediction models for biodiesel properties (density, CN, CFPP, IV and OS), based on the FA composition of the oils, were used to establish constraints of the model, allowing the compliance with biodiesel technical specifications (established in the European [10] and American standards [11]). GHG abatement costs were calculated by comparing the costs of the efficient solutions with highest and lowest GHG emissions. Biodiesel produced in Portugal from palm, rapeseed and/or soya is used as a case study.

2 Methodology

A MO blending model was developed taking into account: i) the relationship between the chemical composition of vegetable oils and biodiesel properties; ii) LC GHG emissions; and iii) biodiesel production costs. The model was applied considering the production of biodiesel in Portugal, on the basis of blending palm, rapeseed and/or soya oils. In the following paragraphs a description of the methodology applied to assess technical aspects (sub-section 2.1), GHG emissions (sub-section 2.2) and production costs (sub-section 2.3) is presented. The mathematical formulation of the model and the approach used to obtain the Pareto curve is explained in sub-section 2.4. Sub-section 2.5 describes the methodology applied to calculate GHG abatement costs.

2.1 Chemical Composition of the Vegetable Oil and Biodiesel Properties

Each virgin oil feedstock presents a typical FA profile that will influence the final properties of the biodiesel. It is generally assumed that FA compositional profiles remain unchanged during conversion of the feedstocks to fuels via transesterification. For this reason, the fatty esters properties are directly related to the FA profile. Structural features such as chain length, degree of unsaturation and branching of the chain determine the fuel properties [4]. Table 1 shows the compositional profile for biodiesel (FAME) from palm, rapeseed and soya. To each FA is associated a nomenclature CX:Y, where X is the number of carbon atoms and Y the number of carbon–carbon double bonds in the FA chain. There are five FA that dominate the composition of FAME: palmitic acid (C16:0), stearic acid (C18:0), oleic acid (C18:1), linoleic acid (C18:2), and linolenic acid (C18:3) [4].

Although several factors can influence the final properties of the biodiesel, those that are directly related with the compositional profile are: viscosity, density, cetane number (CN), cold flow properties (cloud point (CP), pour point (PP), and cold filter plugging point (CFPP)), flash point (FP), iodine value (IV) and heating value [3, 4]. Oxidative stability (OS) is related to the FAME composition but also significantly influenced by the FAME clean-up and storage practices employed. Moreover, the anti-oxidant additives that some biodiesel samples contain modify the OS of FAME [4, 5]. In the literature [5, 10, 12, 13, 14], prediction models based on the FA composition are used for the following biodiesel proprieties: density, CN, CFPP, IV and OS. These prediction models were used as technical constraints.

Density - Giakoumis [5] demonstrated a notable correlation (R2 = 0.86) between density and the degree of unsaturation (DU) given by Eq. (1).

$$\text{Density (kg/m3)} = 9.17 * DU + 869.25 \tag{1}$$

CN - Bamgboye & Hansen [12] proposed an equation to predict the CN of biodiesel (Eq. (2)) based on the composition of specific FAME (x_i - composition of FAME; 3- Lauric; 4-Myristic; 5- Palmitic; 8- Stearic; 6- Palmitoleic; 9- Oleic; 10- Linoleic; 11- Linolenic).

$$CN = 61.1 + 0.088x_4 + 0.133x_5 + 0.152x_8 - 0.101x_6 - 0.039x_9 - 0.243x_{10} - 0.395x_{11} \tag{2}$$

Table 1. FA compositional profile (%) for palm, rapeseed and soya adapted from [4]

FA Common name	j (FA index)	FA nom	Palm	Rapeseed	Soya
Caprylic	1	C8:0	0.8		
Capric	2	C10:0	0.5	0.6	
Lauric	3	C12:0	0.3	0.1	0.1
Myristic	4	C14:0	1.1		0.1
Palmitic	5	C16:0	42.5	4.2	11.6
Palmitoleic	6	C16:1	0.2	0.1	0.2
Heptadecenoic	7	C17:1	0.1	0.1	0.1
Stearic	8	C18:0	4.2	1.6	3.9
Oleic	9	C18:1	41.3	59.5	23.7
Linoleic	10	C18:2	9.5	21.5	53.8
Linolenic	11	C18:3	0.3	8.4	5.9
Arachidic	12	C20:0	0.3	0.4	0.3
Gondoic	13	C20:1	0.1	2.1	0.3
Eicosatrienoic	14	C20:2		0.1	
Behenic	15	C22:0		0.3	0.3
Erucic	16	C22:1		0.5	0.1
Lignocric	17	C24:0			0.1
Nervonic	18	C24:1	0.9	4.3	4.1

IV - The European standard [10] provides a model to calculate IV, based on the sum of the individual contributions of each methyl ester, obtained by multiplying the methyl ester percentage (MEP) by the respective factor (Eq. 3 and Table 2).

$$IV = \sum MEP * ME\ Factor \qquad (3)$$

Table 2. Methyl ester (ME) factors for IV calculation

Methyl ester (ME)	ME Factor
C X:0	0
C16:1	0.950
C18:1	0.860
C18:2	1.732
C18:3	2.616
C20:1	0.785
C22:1	0.723

CFPP - Ramos *et al.* [13] analyzed the correlation between CFPP and the parameter Long Chain Saturated Factor (LCSF) obtaining a very strong correlation factor (R^2 = 0.97). LCSF is an empiric parameter determined taking into account the composition of saturated FA and giving more weight to the composition of FA with long chain. The relation between LCSF and CFPP is given by Eq. (4).

$$CFPP = 3.1417*LCSF - 16.477 \qquad (4)$$

OS - Park *et al.* [14] studied the relation between the content of unsaturated FAMES, linoleic (C18:2) and linolenic (C18:3), and the stability of the biodiesel and developed a predictive equation for biofuel stability (Eq. (5)) based on the content of these two FAMEs in wt% (X).

$$OS\ (h) = 117.9295/X + 2.5905 \qquad (5)$$

Table 3. Results of the prediction models (M) and literature reference (R) values for the properties included in the model

Property			Palm	Rapeseed	Soybean
Density (kg/m³)	M	Giakoumis [5]	876	880	881
	R	Hoekman *et al* [4]	[857, 889]	[859, 899]	[869, 899]
Cetane Number	M	Bamgboye & Hansen [12]	63.4	51.0	46.9
	R	Hoekman et al [4]	[54.7, 69.1]	[47.9, 59.5]	[42.1, 60.5]
Iodine Value (g iodine/100 g)	M	EN 14 214 [10]	49.5	111.7	126.8
	R	Hoekman *et al.* [4]	[41.8, 66.2]	[102.7, 129.5]	[114.7, 136.3]
CFPP (°C)	M	Ramos *et al* [13]	6	-9	-4
	R	Hoekman *et al.*[4]	[-1, 19]	[-25, -1]	[-8, 0]
OS (hours)	M	Park *et al.* [14]	14.6	6.5	4.6
	R	Giakoumis [5]	[6.6, 16.2]	[3.8, 11.0]	[-0.2, 10.2]

Table 3 presents the results of the prediction models using the compositional pro-files of FA presented in Table 1. Table 3 shows for each property the value obtained by the model (M) and an interval of values (average ± 2*standard deviation) obtained from the literature (R). It can be observed that there is agreement between the prediction models and the values available in the literature.

2.2   Life-Cycle GHG Emissions

LC GHG emissions for biodiesel produced from palm, rapeseed and soya were considered in the model, based on data available in the literature [15, 16, 17, 18] and research currently carried on at the University of Coimbra. The LCA includes the following stages: cultivation, extraction, refining, biodiesel production and transportation. The functional unit selected is 1 MJ of biodiesel. Allocation based on the energy content (lower heating value) of biodiesel was performed. For the sake of simplicity, GHG emissions due to Land Use Change (LUC) were not considered since there is a large range of values and scenarios [15]. This has no implication on the formulation of the MO optimization problem, although adding LUC scenarios would entail adding more variables to the problem to separate different origins of the same feedstock.

Castanheira *et al.* [15] presented an LC modeling and inventory for biodiesel pro-duced in Portugal from palm oil imported from Colombia with a comprehensive as-sessment of alternative fertilization schemes and LUC scenarios. A detailed LC inventory based on primary data was performed for a specific plantation, equipped with its own mill, in the Orinoquía Region of Colombia. In the cultivation stage, the GHG emissions (excluding LUC) arise from: field application of fertilizers and mill byproducts (empty fruit bunches, treated PO mill effluent, ashes); fossil fuel used in agricultural operations; palm oil mill effluent treatment and production of fertilizers, fuels, and electricity. The authors accessed different fertilization schemes in order to evaluate the influence of using different N-fertilizers: ammonium sulfate (#AS), calcium ammonium nitrate (#CAN), urea (#U) and poultry manure (#Poultry). Transportation (to Portugal), refining and biodiesel production were also considered in the model, based on data collected in mills in Portugal [16]. An average value for the GHG emissions (32 g CO_2 eq /MJ) considering alternative different fertilization schemes (and assuming no LUC emissions) was used in the optimization model.

The LC GHG emissions of rapeseed-based biodiesel used in the model (average value of 48 g CO_2 eq /MJ) were obtained from Malça *et al.* [17], which studied alternative cultivation practices and locations (Spain, France, Germany and Canada). Data collection from references was carried out to build LCI tables of the biodiesel production system for each alternative. Rapeseed cultivation includes: soil preparation, fertilization, sowing, weed control, and harvesting. Seeds are separated from the rest of the plant during harvesting. The straw (consisting of stalks, pods and leaves) is usually plowed back into the field. Following harvesting, oilseeds are cleaned and dried.

The typical moisture content of oilseeds is reduced, as required by oil extraction facilities and to ensure stability in storage. Oilseeds are transported to Portugal, where extraction, refining and biodiesel production take place. The inventory of these stages was developed based on data collected in Portuguese mills [16].

An LC model and inventory of biodiesel from soybean produced in four states in Brazil (Mato Grosso- MT, Goiás- GO, Paraná –PR and Rio Grande do Sul-RS) was implemented in research carried out in the University of Coimbra [18, 19]. In this article, a GHG emission average of 39 g CO2 eq /MJ was used. Oil extraction data was collected in a Brazilian mill [18]. LCI of crude vegetable oil refining, which are the weighted average values calculated on the basis of 2009 to 2010 data, collected in four Portuguese plants. The LCI of soybean oil extraction and biodiesel production was developed based on 2009 to 2010 data, collected in Portuguese mills [16].

Table 4 presents the values obtained from the mentioned references for each LC stage and the averages used in the MO model.averages used in the MO model.

Table 4. GHG Emissions (gCO_2 eq/MJ) for biodiesel from Soya, Palm and Rapeseed

| | GHG Emissions g CO_2 eq /MJ | | | | | | | | | | | |
| | Soya | | | | Palm | | | | Rapeseed | | | |
	MT	GO	PR	RS	#AS	#CAN	#U	#Poultry	Fr	Can	Sp	Ger
Cultivation	13	14	12	14	13	17	15	13	34	30	40	30
Extraction	1	1	1	1	2	2	2	2	3	3	3	3
Refining	1	1	1	1	1	1	1	1				
Production	5	5	5	5	5	5	5	5	5	5	5	5
Transport	22	18	17	19	9	9	9	9	5	8	4	9
Total	42	39	36	40	30	34	32	30	48	46	52	48
Average Value	39				32				48			

2.3   Biodiesel Production Costs

As discussed by Gülsen [3], 77 - 88 % of the biodiesel production cost is associated with feedstock costs. In this context, the biodiesel cost addressed in the model focused on oil feedstock prices. Figure 1 shows the price evolution (in euros) in the past five years [20]. Rapeseed always presented the highest price and palm the lowest. To tackle the relative oil price influence in the blending, a preliminary analysis (for dates with significantly different relative prices between the various feedstock) was performed showing that when the difference between the soya and palm prices is not significant or when there is a balanced difference between the price of the three feedstocks, the model gives a single Pareto optimal solution. The model is illustrated using the price values for July 2013 (Palm 558 €/t; Rapeseed 767€/t; Soya 765€/t) where multiple Pareto optimal blend solutions are obtained. In practice, the model should be solved by the biodiesel producer on a regular basis using the latest available prices.

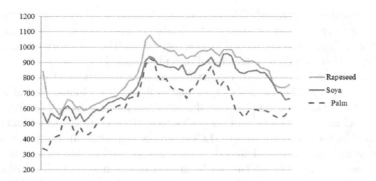

Fig. 1. Prices (in euros per ton) for rapeseed, soya and palm, from Nov -08 to Nov- 13 [20]

2.4   Multiobjective Blending Model

The model includes the minimizing of two objective functions: biodiesel production-feedstock costs and LC GHG emissions. Costs are given by Eq. (6), multiplying the quantity of each feedstock by its market price. GHG emissions are given by Eq. (7), multiplying the quantity of each feedstock by its life-cycle emissions. The model is subject to demand and supply constraints - Eq. (8) and Eq. (9) and mass balance constraints - Eq. (10). Eq. (11) to Eq. (16) represents the technical specifications constraints. Since the goal is to analyze the proportions of each feedstock in the blend, the demand was set equal to 1. The constraints were defined in accordance with the European Standard EN 14214 [10] that establish the FAME requirements for diesel engines. This standard limits biodiesel density between 860 and 900 kg/m3 – Eq. (11) and Eq. (12); the CN has a minimum of 51 – Eq. (13); the IV has a maximum of 120 g iodine/100 g biodiesel – Eq. (14); the limit for CFPP was defined as 0° C (grade B for temperate climates) - Eq. (15); and OS to a minimum of 6 hours – Eq. (16). In a second phase, the model was also applied to restrict the OS constraint according to the value established in the American Standard [11] – minimum of 3 hours.

The mathematical formulation is the following:

Minimize,

$$\sum_{i \in F} P_i Q_i \tag{6}$$

$$\sum_{i \in F} GHG_i Q_i, \tag{7}$$

Subject to:

$$\sum_{i \in F} Q_i = D , \tag{8}$$

$$Q_i \leq S_i , \quad \forall_i \in F \tag{9}$$

$$q_j = \sum_{i \in F} Q_i * q_{ij} \quad , \forall_j \in A \tag{10}$$

$$\text{Den}_B \leq 900 \tag{11}$$

$$\text{Den}_B \geq 860 \tag{12}$$

$$\text{CN}_B \geq 51 \tag{13}$$

$$\text{IV}_B \leq 120 \tag{14}$$

$$\text{CFPP}_B \leq 0 \tag{15}$$

$$\text{OS}_B \geq 6 \tag{16}$$

Where:

F = {soya, palm, rapeseed}

A = {1,2,...,18} , FA index (third column of Table 1)

P_i , $\forall_i \in F$: Price of feedstock i

Q_i , $\forall_i \in F$: Quantity of feedstock i

GHG_i , $\forall_i \in F$: GHG emissions of feedstock i

D: Demand

S_i , $\forall_i \in F$: Supply of feedstock i

q_j , $\forall_j \in A$: Quantity of FA j in the blend

q_{ij}, $\forall_i \in F$; $\forall_j \in A$: Quantity of FA j in feedstock i

Den_B : Density of the blend

CN_B: Cetane Number of the blend

IV_B : Iodine Value of the blend

CFPP_B : Cold Filter Plugging Point of the blend

OS_B : Oxidative Stability of the blend

The physical parameters that were established by the models presented in 2.1 are given by Eq. (17) to Eq. (21):

$$\text{Den}_B = 9.17 \left(\sum_{k \in K} q_k + \sum_{l \in L} 2 * q_l + 3*q_{C11} \right) + 869.25 \tag{17}$$
$$K = \{6, 9, 13, 16, 18\} \quad ; \quad L = \{10, 14\}$$

$$\text{CN}_B = 61.1 + 0.088 * q_4 + 0.133 * q_5 + 0.152 * q_8$$
$$- 0.101 * q_6 - 0.039 * q_9 - 0.243 * q_{10} - 0.395 * q_{11} \tag{18}$$

$$\text{IV}_B = 0.950 * q_6 + 0.860 * q_9 + 1.732 * q_{10} +$$
$$2.616 * q_{11} + 0.785 * q_{13} + 0.723 * q_{16} \tag{19}$$

$$CFPP_B = 3.1417(0.1 * q_5 + 0.5 * q_8 + 1 * q_{12}$$

$$+1.5 * q_{15} + 2 * q_{17}) - 16.477 \tag{20}$$

$$OS_B = \frac{117.9295}{q_{10} + q_{11}} + 2.5905 \tag{21}$$

The MO problem was modelled in GAMS 23.7.3 [21] and solved using Linear Programming, minimizing costs and GHG emissions. The Pareto-optimal curve was obtained using the $\mathcal{E}$-constraint method, where one of the two conflicting objectives is treated as a constraint (which changes from run to run), while the other is optimized taking into the account the constraints.

2.5 GHG Abatement Costs

The GHG abatement costs (C_A) were calculated dividing net reductions in GHG emissions by net difference in costs, as summarized by Eq. (22) [22, 23]:

$$C_A = (C_i - C_{REF}) / (E_{REF} - E_i) \tag{22}$$

where:

C_A - GHG abatement costs (€/t CO_2 eq)
C_i - costs associated with the blend with lower GHG emissions (€/t)
C_{REF} - costs associated with the blend with higher GHG emissions (€/t)
E_i - GHG emissions of the blend with lower GHG emissions (tCO_2 eq/MJ)
E_{REF} - GHG emissions of the blend with higher GHG emissions (t CO_2eq/MJ)

3 Results and Discussion

The resulting Pareto curve shown in Figure 2 reveals the trade-off that exists between biodiesel production costs (represented in the x-axis) and GHG emissions (represented in the y-axis).The Pareto curve represents the set of all the efficient solutions considering both objectives: for these solutions, it is not possible to improve one of the objectives without worsening the other objective.

The analysis (performed for several dates with different relative prices between the various feedstock) shows that when the difference between the soya and palm prices is not significant or when there is a balanced difference between the price of the three feedstocks, the model yields the (Pareto optimal) solution: blending of 0.454 palm: 0.083 rapeseed: 0.463 soya, corresponding to GHG emissions of 37 gCO_2eq/MJ.

The results of the model, constrained according to the European standard [10], for the feedstock prices of July 2013 (for instance) are shown in Figure 2.

The lower cost / higher GHG emissions combination is obtained with blend 1 and the higher cost / lower GHG emissions combination occurs with blend 7. In terms of GHG emissions, no significant difference is obtained between blend 1 and 7 (about 1 g CO_2eq/MJ). This is due to similar values of the GHG emissions of each feedstock and to the requirements of the technical constraints.

The solution that represents lower costs is obtained by blending palm and rapeseed (blend 1). To reach the lower GHG emissions (blend 7) requires a trade-off of about 35 euros. In this particular case, since the difference between the LC GHG emissions of each feedstock considered is not very high, the GHG emissions are varying over small ranges.

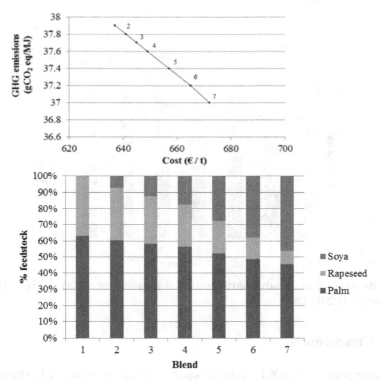

Fig. 2. Pareto curve and blends. Model constrained according the EN 14214 [10], feedstock prices July 2013 [20].

Analyzing the blends along the Pareto curve, a reduction of GHG emissions is obtained by progressively replacing rapeseed by soya (that presents lower GHG emissions: 39 against the 48 gCO_2eq/MJ of rapeseed) and reducing the palm quantity in the blend. Since palm provides the lowest price, this reduction implies an increase in the blend cost. In this particular context, the abatement cost for using a blend with a better environmental performance is very high: 1000 €/tCO_2 eq.

Figure 3 shows the Pareto curve for the model constrained according with the American standard for FAME [11] (assuming that biodiesel is sold in the US), which as mentioned in section 2.4 is more flexible for OS. A new blend (without rapeseed) can be observed: blend 6 (more expensive and with lower GHG emissions).

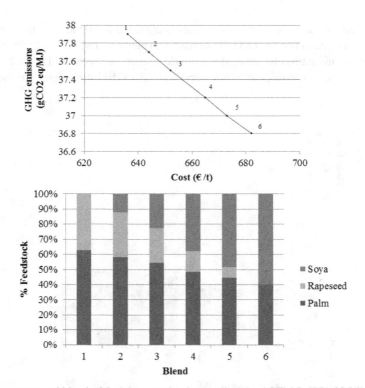

Fig. 3. Pareto curve and blends. Model constrained according the ASTM D6751-08 [11], feedstock prices July 2013 [20].

4   Conclusions

This paper presents a MOLP model for optimal virgin oils blend for biodiesel production minimizing biodiesel production costs and LC GHG emissions, subject to mass balance and technical constraints. The MOLP was solved using the ε-constraint method and the resulting Pareto curve reveals the trade-off between biodiesel production costs and the LC GHG emissions. It was observed that in many time periods, when the difference between the soya and palm prices is not significant or when there is a balanced difference between the price of the three feedstocks, the model gives a single Pareto optimal solution – blending of 0.454 palm: 0.083 rapeseed: 0.463 soya. When the prices of soya and rapeseed are closer, multiple Pareto optimal blend solutions are obtained. The analysis of the blends along the Pareto curve shows that the GHG emission reduction is obtained by progressively replacing rapeseed by soya and reducing the palm quantity in the blend. Since palm presents the lowest price, this reduction implies an increase in the blend cost. In this particular context, the abatement cost for using a blend with a better environmental performance is very high – 1000 €/tCO$_2$ eq. Results comparing the European and American biodiesel standard specifications, show that different optimal blends can be obtained. The lower OS requirement in the American Standard (3 hours instead of the 6 hours established in the European Standard) allows a new blend made without rapeseed.

This study presents limitations that future research will address. In particular, the objective function for costs does not include other costs (energy, process, capital, transportation) besides the feedstock price. Although these other costs are less important and with a low negotiation margin to biodiesel producers, a more comprehensive model will include these other costs, including transportation costs by considering multiple variables for each feedstock (one per origin). Another limitation is considering only average GHG emissions concerning LUC. That will be addressed by considering multiple variables for each feedstock (one for each LUC). These modifications will make the problem to be solved larger in size, and will increase the data detail required, but will not change the structure of the model. Another aspect that will also be considered in the development of this research is the integration in the model of other relevant environmental impacts such as acidification, eutrophication and water depletion. However, this inclusion will change the structure of the problem to be solved and modelling challenges are expected.

Acknowledgments. The authors gratefully acknowledge to Fundação para a Ciência e a Tecnologia (FCT) for support under the projects /SET/0014/2009: Capturing uncertainty in biofuels for transportation: resolving environmental performance and enabling improved use, PTDC/SEN-TRA/117251/2010: Extended "well - to- wheels" assessment of biodiesel for heavy transport vehicles and PTDC/EMS-ENE/1839/2012: Sustainable mobility: Perspectives for the future of biofuel production.. Carla Caldeira acknowledges the financial support from FCT, through grant SFRH/BD/51952/2012. This work has been framed under the Energy for Sustainability initiative of the University of Coimbra and supported by the R&D project EMSURE (CENTRO-07-0224-FEDER-002004).

References

1. OECD-FAO: Agricultural Outlook 2011-2020 OECD/FAO (2011)
2. Refaat, A.A.: Correlation between the chemical structure of biodiesel and its physical properties. International Journal of Environmental Science & Technology 6(4), 677–694 (2009)
3. Gülsen, E., Olivetti, E., Freire, F., Dias, L., Kirchain, R.: Impact of feedstock diversification on the cost-effectiveness of biodiesel. Applied Energy 126(1), 281–296 (2014)
4. Hoekman, S.K., Broch, A., Robbins, C., Ceniceros, E., Natarajan, M.: Review of biodiesel composition, properties, and specifications. Renewable and Sustainable Energy Reviews 16(1), 143–169 (2012)
5. Giakoumis, E.G.: A statistical investigation of biodiesel physical and chemical properties, and their correlation with the degree of unsaturation. Renewable Energy 50, 858–878 (2013)
6. European Comission: Directive 2009/28/EC of the European Parliament and of the Council on the promotion of the use of energy from renewable sources, 16–62 (2009)
7. Fröhling, M., Rentz, O.: A case study on raw material blending for the recycling of ferrous wastes in a blast furnace. Journal of Cleaner Production 18(2), 161–173 (2010)

8. Li, X., Yu, H., Yuan, M.: Modeling and Optimization of Cement Raw Materials Blending Process. Mathematical Problems in Engineering 2012, 1–30 (2012)
9. Lingshuang, K., Chunhua, Y., Shenping, X., Gang, C.: Stochastic Optimization Method Based on HSS Technique and Expert Knowledge for a Metallurgical Blending Process. In: 2013 Third International Conference on Intelligent System Design and Engineering Applications, pp. 1290–1293 (2013)
10. European Committee for Standardization. EN 14214: automotive fuels – fatty acid methyl esters (FAME) for diesel engines – requirements and test methods.Report no. EN 14214:2008. Management Centre (2008)
11. Standard specification for biodiesel fuel blend stock (B100) for middle distillate fuels. Report no. D6751-08. ASTM (2008)
12. Bamgboye, A.I., Hansen, A.C.: Prediction of cetane number of biodiesel fuel from the fatty acid methyl ester (FAME) composition, 21–29 (2008)
13. Ramos, M.J., Fernández, C.M., Casas, A., Rodríguez, L., Pérez, A.: Influence of fatty acid composition of raw materials on biodiesel properties. Bioresource Technology 100(1), 261–268 (2009)
14. Park, J.-Y., Kim, D.-K., Lee, J.-P., Park, S.-C., Kim, Y.-J., Lee, J.-S.: Blending effects of biodiesels on oxidation stability and low temperature flow properties. Bioresource Technology 99(5), 1196–1203 (2008)
15. Castanheira, É.G., Acevedo, H., Freire, F.: Greenhouse gas intensity of palm oil produced in Colombia addressing alternative land use change and fertilization scenarios. Applied Energy 114, 958–967 (2014)
16. Castanheira, É.G., Freire, F.: Avaliação de Ciclo de Vida das emissões de Gases com Efeito de Estufa da Produção de Biodiesel de Soja em Portugal. Relatório Final elaborado para a APPB, no âmbito do Protocolo de colaboração entre a APPB e a ADAI. Coimbra (2011)
17. Malça, J., Coelho, A., Freire, F.: Environmental Life-Cycle Assessment of rapeseed-based biodiesel: alternative cultivation systems and locations. Applied Energy 114, 837–844 (2014)
18. Castanheira, É.G.: Environmental sustainability assessment of soybean and palm biodiesel systems: A life-cycle approach, University of Coimbra (2014)
19. Castanheira, É.G., Grisoli, R., Freire, F., Coelho, S.: Environmental sustainability of biodiesel in Brazil. Energy Policy 65, 680–691 (2014)
20. Index Mundi, http://www.indexmundi.com/
21. GAMS Development Corporation. General Algebraic Modeling System (GAMS) Release 23.7.3 Washington, DC, USA (2011)
22. Diekmann, J., Jochem, E.: Methodological Guideline for Assessing the Impact of Measures for Emission Mitigation, Jülich (1998)
23. UNEP: Mitigation and Adaptation Assessment. Concepts, Methods and Appropriate Use. UNEP Collaborating Centre on Energy and Environment, Roskilde, Denmark, p. 169 (1998)

An Integrated Approach for Exploring Opportunities and Vulnerabilities of Complex Territorial Systems

Valentina Ferretti, Marta Bottero, and Giulio Mondini

Politecnico di Torino, Department of Regional and Urban Studies and Planning
{valentina.ferretti,marta.bottero,giulio.mondini}@polito.it

Abstract. Dealing with territorial transformations assessment means addressing the challenge represented by the inherent complexity and multidimensionality of these systems. This requires an integrated approach for the evaluation in order to obtain concise final judgments. Moreover, when dealing with territorial systems, the analysis of the geographical patterns of the elements under investigation plays a fundamental role. The paper thus proposes an innovative approach for the analysis of a complex territorial system based on the integration of Geographic Information Systems (GIS) and a specific Multicriteria Analysis technique, named Analytic Network Process (ANP). In particular, starting from a real case related to a mountain area in Northern Italy, the present paper explores the potentialities of spatial Multicriteria analysis for supporting strategic planning and sustainability assessment procedures.

Keywords: Scenario analysis, Multicriteria Analysis, Spatial Decision Support Systems, Analytic Network Process.

1 Introduction

Territorial transformation projects are affected by high levels of uncertainty and refer to long-term perspectives. In this context, a very useful role is played by scenario analysis which supports the decision-making process for the definition and assessment of future development strategies for a certain area [1]. In particular, scenario analysis attempts to develop and judge a set of hypothetical policy or development alternatives for a complex decision-making system, in order to generate a rational frame of reference for evaluating different options [2]. Scenarios studies have usually an experimental nature and play an important role in the field of spatial planning and analysis [3].

The use of scenario analysis in decision-making processes related to spatial planning is based on the assumption that the future is not predetermined, but it consists in the product of causal chains of events that are determined from exogenous or endogenous elements of the spatial system [1]. Planning actions aim to guide these events in order to achieve political objectives.

An important problem arising when assessing territorial systems refers to their complexity which requires an integrated and systemic approach for the evaluation.

B. Murgante et al. (Eds.): ICCSA 2014, Part III, LNCS 8581, pp. 667–681, 2014.

Following this reasoning, there is a need for quantitative methods that are able to synthesize the full range of aspects involved in the transformation, from the impacts on the environmental system to the effects in terms of mobility and accessibility, from the social and economic consequences of a certain strategy to the outcomes with reference to landscape and cultural heritage.

In the current debate regarding sustainability assessment and integrated approaches, spatial Multicriteria Analysis [4] plays a fundamental role by solving semi-structured spatial problems through the integration of Geographic Information Systems (GIS) and Multicriteria Decision Aiding (MCDA) techniques.

From the methodological point of view, the present application proposes the integration between GIS and a specific MCDA technique named Analytic Network Process (ANP; Saaty 2005), which represents the evolution of the Analytic Hierarchy Process (AHP; Saaty 1980). Since the incorporation of the AHP calculation block in the IDRISI 3.2 software package, it has become much easier to apply this technique to solve spatial problems. Applications of the ANP, which is particularly suitable for dealing with complex decision problems that are characterized by interrelationships among the elements at stake, are instead scarce.

The purpose of the paper is thus to investigate the potentialities of the ANP-GIS integration and to present the innovative methodological framework with reference to a case study dealing with the identification of future opportunities and vulnerabilities in a mountain area in Northern Italy. Moreover, the research also explores the potentialities of a decision support process which makes use of a panel of experts for the implementation of the evaluation model.

2   Scenario Analysis and Spatial Decision Support Systems

2.1   State of the Art

Scenario analysis has been developed as a scientific tool for supporting policy-making processes under uncertainty conditions. According to Kahn and Wiener [5], a scenario can be defined as a possible, often hypothetical, sequence of events constructed in an internally consistent way for the purpose of focusing attention on casual processes and decision points.

Following this first definition, several attempts were made in the scientific literature for better clarifying the concept of scenario. Warfield [6] defined the scenario as "a narrative description of a possible state of affairs or development over time. It can be very useful to communicate speculative thoughts about future developments to elicit discussion and feedback, and to stimulate the imagination. Scenarios generally are based on quantitative expert information, but may include qualitative information as well".

Ratcliffe [7] states that "the principal objective of scenario analysis is to enable decision-makers to detect and explore the full range of alternative futures so as to clarify present actions and subsequent consequences".

According to Godet [8], "scenarios should aim to detect the key variables that emerge from the relationship between the many different factors describing a particular system, especially those relating to the particular actors and their strategies". Moreover, Schwartz [9] highlights that scenarios "provide a context for thinking clearly about the otherwise impossible complex array of factors that affect any decision; give a common language to decision-makers for talking about these factors and encourage them to think about a series of 'what-if' stories; help lift the 'blinkers' that limit creativity and resourcefulness; and lead to organisations thinking strategically and continuously learning about key decisions and priorities".

The purpose of scenario analysis is not just about constructing scenarios, but it is about informing decision-makers and influencing and enhancing, decision-making, thus creating a learning process.

The methodological base of scenario building, as with all future studies, is broad, diverse and comprises a wide range of approaches and techniques. It has been noticed that the integrated use of scenario analysis and Multicriteria Analysis can efficiently support decision-making process [10, 11].

With specific reference to the context of spatial planning, many applications exist in the literature related the use of Spatial Decision Support Systems (SDSS) in the domain of scenario analysis. In order to better contextualize our study and highlight its innovation with respect to the state of the art, Table 1 syntheses the main scientific papers, highlight the field of application, the objective of the analysis and the methodology applied. As it is possible to see from Table 1, the sphere of the researches is very vast, including application for environmental risk analysis and energy planning. The principal aim the studies is the creation of a suitability map with the projection of effects produced by the considered scenarios.

Table 1. Examples of applications of (SDSS) for scenario analysis in land use planning and management

Author	Field of application	Objective of the evaluation	Methodology
Volk et al. [12]	Management of water resources	Analysis of different land use scenarios from the point of view of the ecological and socio-economic effects.	FLUMAGIS (GIS-based integrated ecological-economic model)
Duzgun et al. [13]	Evaluation of seismic vulnerability	Map of the seismic vulnerability index of an urban area with 3D visualizations	MC-SDSS (Multicriteria-Spatial Decision Support Systems) where the MCA module is based on a set of indicators obtained by means of a series of questionnaires to DM and key actors
Zerger e Wealands [14]	Hydrogeological risk management	Map of the effects of different potential risk scenarios	Integration between GIS and hydrodynamic models

Table 1. (*Continued*)

Scholten et al. [15]	Hydrogeological risk management	Ranking of alternative management scenarios	SDSS developed through the IDRISI software
Brody et al. [16]	Management of energy resources	Suitability map for the identification of the best sites for the production of oil and gas in Texas (USA)	MC-SDSS based on a set of statistical indicators
Volk et al. [17]	Management of landscape and river basins	Critical review of different SDSS approaches	FLUMAGIS Elbe-DSS CatchMODS MedAction
Ballas et al. [18]	Evaluation of public policies	Analysis of the effects of different scenarios for the city of Leeds s	SDSS based on micro-simulations for the generation of predictions of census data (Micro-MaPPAS)
Grueau e Rodriguez, [19]	Environmental planning and management	Evaluation of the environmental effects of different land use scenarios	Integration of multi-agents models with GIS
Rutledge et al. [20]	Regional planning	Evaluation of different land use scenarios in New Zealand	MC-SDSS based on a set of spatial indicators developed through the software GEONAMICA
Danese et al [21]	Geomorphological and geostatistical analysis	Macroseismic damage effects in urban areas	Integrated Geological, Geomorphological and Geostatistical Analysis

2.2   Spatial Multicriteria Analysis: Methodological Background

Scenario analysis plays a crucial role particularly in the field of sustainability assessments and territorial transformation processes, both at the urban and rural scale. These contexts give rise to complex decision problems due to the presence of different and often conflicting objectives to be pursued, the public/private nature of the goods under investigation, the existence of several values (historical, naturalistic, cultural, economic, etc.) and the presence of different actors (public government representatives, architects, architectural historians, citizens, developers and owners).

Moreover, when dealing with territorial transformation processes, an undeniable important role is played by the spatial distribution of the characteristics and consequences of each option under analysis.

The availability of analytical frameworks able to support the process is thus getting more and more important.

Within this context, a fundamental support may be provided by spatial Multicriteria Analysis [4] which combines Geographic Information Systems and Multicriteria Decision Aiding in order to provide a collection of methods and tools for transforming and integrating geographic data (map criteria) and Decision Maker's preferences and uncertainties (value judgments) to obtain information for decision-making and an overall assessment of the decision alternatives.

Spatial Multicriteria Analysis is an increasingly popular tool in decision-making processes and in policy making, thanks to its significant new capabilities in the use of spatial or geospatial information. In recent years there has thus been a growing interest towards the development and application of spatial Multicriteria Analysis across many scientific fields for solving different decision problem typologies [22], thanks to the ability of this integrated approach to both generate alternatives during the strategic planning phase and to compare them during the evaluation phase.

In particular, spatial Multicriteria Analysis is most commonly applied to land suitability analysis in the urban/regional planning, hydrology and water management and environment/ecology fields and is usually based on a loose coupling approach and on a value focused thinking framework [22].

Within these fields an emerging trend seems to focus on the application of spatial Multicriteria Analysis for scenarios generation and evaluation, thanks to the ability of combining both qualitative and quantitative data representing the spatial consequences of different future courses of actions.

From the methodological point of view, the steps needed for the development of a spatial Multicriteria Analysis to specifically support planning and decision-making processes are summarized in Figure 1.

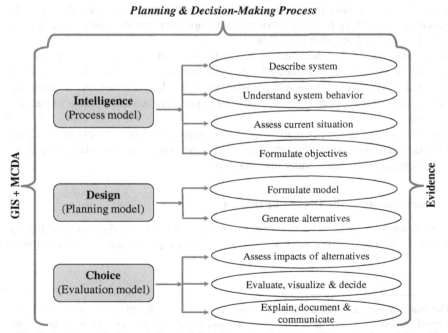

Fig. 1. Framework for planning and decision-making process (Source: adapted from [23])

In this model, there is a flow of activities from intelligence to design to choice phase, as well as steps in each phase.

In particular, the intelligence phase refers to the examination of the environment in order to identify problems or opportunity situations and includes the structuring of the problem, during which the system under consideration is defined and the objectives to pursue are explored. One or more criteria, or attributes, are then selected to describe the degree of achievement of each objective.

The design phase involves the development and analysis of possible courses of action. During the choice phase alternatives are evaluated and a selection of specific courses of action is performed; furthermore, detailed analyses, such as the sensitivity analysis, are deemed appropriate in order to obtain useful recommendations.

Finally, evidence is defined as the total set of data, information, and knowledge at disposal of the planner, Decision Makers and analysts.

3   Application

3.1   Presentation of the Case

The case study considered in the application refers to a small town in Northern Italy named Ormea[1]. The town has a population of 1750 inhabitants and is located in the Alpine territory of the Piedmont Region, on the border with the Liguria Region and with France (Figure 2).

In the past, the city used to be very important from the point of view of the industrial activities concerning the production of wool and paper. Moreover, thanks to the presence of the railway line, the town was an important tourism centre with tourists coming from many European countries.

Nowadays, due to the phenomenon of the abandon of mountain areas, many economic activities have been relocated; also the tourism sector is suffering and the trends of the presences has been decreasing since the last decades. As a result, the town is experiencing a deep crisis and new strategies for the development of the area are needed.

The objective of the paper is to support the creation of future development scenarios for the area, with particular reference to the analysis of the opportunities and the risk characterizing the town under investigation.

3.2   Structuring of the Decision Problem

Starting from the overall objective of the analysis, which is the definition of the opportunities and risks of the territory of Ormea, a comprehensive set of evaluation

---

[1]   The material used to illustrate the case study application is based on the thesis work performed by Elisa Piolatto at the Architectural and Landscape Heritage post-graduates specialization school of the Polytechnic University of Turin (Italy).

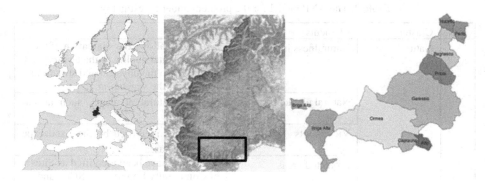

Fig. 2. The location of the area under investigation

criteria that reflect all the concerns relevant to the decision problem has been identified according to a value focused thinking approach, which assumes the values as
fundamental elements in the decision analysis and, based on the values and criteria
structure, develops and evaluates feasible options [24].

Due to the presence of different interrelated factors and to the intrinsic spatial nature of the problem, the method of the Analytic Network Process (ANP, [25]) has
been coupled with Geographic Information Systems (GIS). The ANP represents the
evolution of the Analytic Hierarchy Process (AHP, [26]) in order to take into account
interactions and feedbacks among the decision elements. According to the ANP, the
problem structuring phase involves identifying groups (or clusters) constituted by
various elements (nodes) that influence the decision. All the elements in the network
can be related in different ways since the network can incur feedbacks and complex
inter-relationships within and between clusters, thus providing a more accurate modeling of complex settings.

In the present application the model has been developed according to the complex
network structure [25]. The problem has thus been divided four five clusters (namely,
natural system, historical and cultural heritage, economic aspects, territorial system)
that have been organized according to the categories of Opportunities and Risks. In
this case, the opportunities and the risks have been considered, respectively, as positive and negative aspects of the transformation in the long time period, for which it is
difficult to make any prevision

According to the ANP methodology, once the network has been identified, it is necessary to represent the influences among the elements [25].

Moreover, a raster map was linked to each criterion, within which each pixel has a
suitability value. These maps were derived from basic raster GIS operations (map
overlay, buffering, distance mapping, spatial queries, etc.).

Table 2 presents the criteria identified for the analysis while Figure 3 represents as
an example the Opportunities subnetwork of the ANP model.

Table 2. The ANP model for the problem under investigation

O/R	Cluster	Elements	Description
OPPORTUNITIES	Natural system	Naturalness index	This index allows the evaluation of the environmental quality of the territory, according to the physical and structural features of the vegetation
		Natural elements	Specific natural elements, such as caverns
		Viewpoints	Viewpoints identified by the Landscape Plan of the area
		Protected areas	Areas that have been identified as Sites of Community Importance (SCIs) and Special Protection Areas (SPAs)
	Historical and cultural heritage	Historical monuments	Archaeological sites and historical monuments such as castles, towers, churches, industrial archaeology buildings, etc.
		Historical settlements	Important settlements identified in historical sources
		Cultural events	Cultural events such as feasts, religious events, etc.
	Economic aspects	Accommodation structures	Hotels, bed & breakfast, mountain dews, etc.
		Sport pathways	The elements is related to the paths destined to the practice of trekking and other sport activities
		Sport facilities	Facilities for climbing, sport fishing, skiing, hand gliding, etc.
		Picnic areas	Areas for the temporal stop of tourists
	Territorial systems	Accessibility	Infrastructural roads for arriving at Ormea from Piedmont, Liguria and France
		Local roads	Local roads for reaching the different parts of the town of Ormea
RISKS	Natural system	Hydrogeological risk	Areas which are characterized by an high level of hydrogeological risk (water bodies, areas subjected to avalanches and landslides)
	Historical and cultural heritage	Abandoned historical pathways	This elements is related to the presence of pathways that used to be employed for the transhumance or for reaching seasonal settlements
		Tracks of ancient cultivations and productive activities	The element concerns the presence of ancient rural activities that nowadays are disappearing (for example, terraces, mill runs, etc.)

Economic aspects	Power lines	Network infrastructure for the transportation of the electric energy
	Quarries	Presence of quarries for the extraction of the marble
	Abandoned industrial areas	Former productive areas which now are abandoned (for example, the building of the paper mill)
	Distribution of the population	This elements concerns the different distribution of the population in the centre of Ormea and in the small outlying suburb hamlet
Territorial systems	Slope	This element is related to the slope of the ground which constitutes an obstacle for the accessibility
	Soil consumption	The element concerns the progressive soil consumption, comparing the actual situation with the situation registered in the historical maps

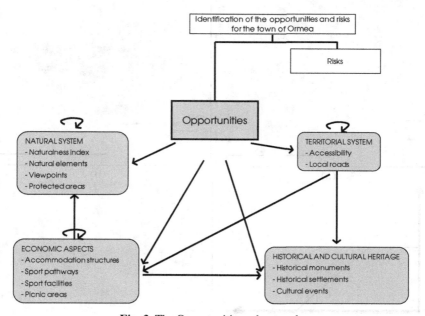

Fig. 3. The Opportunities subnetwork

3.3   Standardizing the Criteria Maps

For decision analysis the values and classes of all the maps associated with each considered criteria should be converted into a common scale. Such a transformation is commonly referred to as standardization [27].

Through standardization the original factor scores (each expressed in its own unit of measurement) are converted into dimensionless scores in the 0 (worst situation) or 1 (best situation) range.

In the present study standardization was performed by means of a focus group of both experts in different fields (economic evaluation, environmental engineering, and landscape ecology) and real stakeholders coming from the Ormea municipality. The training of a panel of experts allows to overcome some difficulties and biases which characterize the decision processes based on a single expert. In the present application, a close attention was devoted to the formation of a group of experts having a balanced background composition.

Through the active participation of all the experts and stakeholders the control points used for the standardization of each criterion have thus been discussed and decided during the aforementioned focus group.

With the aim of providing an illustrative example, Figure 4 shows the initial raw map (Fig.4a), the intermediary source map (Fig.4b), the standardization function (Fig.4c) and the standardized map (Fig.4d) for the factor "panoramic viewpoints" under the Opportunities sub-network.

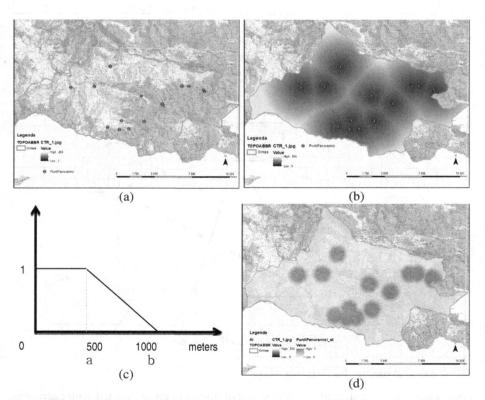

Fig. 4. The initial raw map (Fig.4a), the intermediary source map (Fig.4b), the standardization function (Fig.4c) and the standardized map (Fig.4d) for the factor "panoramic viewpoints"

In particular, control point *a* and control point *b* in Figure 4c are the points that govern the shape of the standardization function. In this case, the first control point (*a*) indicates the value where the membership function starts to decrease. This is due to the fact that during the focus group the experts and the stakeholders agreed that after 500 meters the perceived utility of being near to a panoramic viewpoint starts to decrease. The second control point (*b*) in this case indicates the point at which the function reaches membership 0, since the focus group result was that a distance greater than 1000 meters is not worth being covered to reach a panoramic viewpoint.

The same kind of reasoning has been repeated for all the considered factors inside the Opportunities and Vulnerabilities sub-networks.

3.4   Weighing and Aggregation

Once all the maps have been standardized in the 0-1 range, the next step of the decision process consisted in weighing all the factors according to the pair-wise comparison approach underpinning the Analytic Network Process methodology. The different experts thus worked together in order to achieve a consensus with reference not only to the standardization of each factor map but also to the weighing of the elements involved in the decision. According to the ANP methodology, the comparison and evaluation phase is based on the pair-wise comparison of the elements under consideration which can be divided into two levels: the comparison between clusters which is more general and strategic and the comparison between nodes which is more specific and detailed.

In paired comparisons, the smaller element is used as the unit, and the larger element becomes a multiple of that unit with respect to the common property or criterion for which the comparisons are made. A ratio scale of 1–9, that is, the Saaty's fundamental scale, is used to compare any two elements. The main eigenvector of each pair-wise comparison matrix represents the synthesis of the numerical judgements established at each level of the network [26].

As an example, Figure 5 shows the graphical representation of the following question asked to the panel of experts and stakeholders during the focus group: *with reference to the valorisation of the area under analysis, which of the following aspects can better enhance the opportunities of the territory? And how much more?*

Fig. 5. Graphical representation of one of the pair-wise questions asked during the focus group. In case of disagreement, the dimension of the circles in the picture is proportional to the number of votes obtained by each value.

The results of the collaborative procedure for weighing all the elements are summarised in Table 3. In particular, the final priorities showed in Table 3 are those obtained from the progressive formation of the unweighted supermatrix, the weighted supermatrix and the limit supermatrix on which the ANP development is based [26].

Table 3. Final priorities for the elements under analysis

O/V	Cluster	Elements	Final priorities
OPPORTUNITIES	Natural system (0,39)	Naturalness index	0,02
		Natural elements	0,09
		Viewpoints	0,11
		Protected areas	0,17
	Historical and cultural heritage (0,15)	Historical monuments	0,06
		Historical settlements	0,06
		Cultural events	0,02
	Economic aspects (0,39)	Accommodation structures	0,21
		Sport pathways	0,02
		Sport facilities	0,05
		Picnic areas	0,10
	Territorial systems (0,07)	Accessibility	0,06
		Local roads	0,01
VULNERABILITIES	Natural system (0,15)	Hydrogeological risk	0,15
	Historical and cultural heritage (0,21)	Abandoned historical pathways	0,10
		Tracks of ancient cultivations and productive activities	0,11
	Economic aspects (0,56)	Power lines	0,07
		Quarries	0,02
		Abandoned industrial areas	0,35
		Distribution of the population	0,11
	Territorial systems (0,08)	Slope	0,07
		Soil consumption	0,01

In order to obtain the final opportunities and vulnerabilities' maps, a weighted linear combination was used, combining the respective factor maps according to the following formula:

$$S_j = \sum W_I X_I$$
(1)

where S_j represents the overall value of pixel j, W_i represents the weight of factor i, and X_i represents the standardized criterion score of factor i.

The results of the proposed study are thus represented by two maps highlighting the spatial distribution of opportunities and vulnerabilities within the area under examination. These maps represent a first synthesis of negative and positive aspects for the region under analysis (Figure 6a and Figure 6b, respectively) and allow to derive

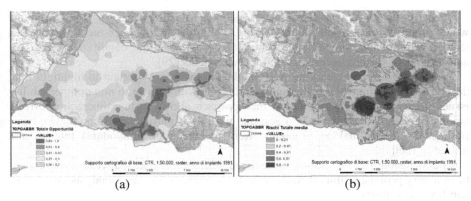

Fig. 6. Overall distribution of the Opportunities (5a) and Vulnerabilities (5b) for the area under analysis

useful indications with reference to warning spots needing specific mitigation or monitoring measures.

As it is possible to notice from the results of the analysis, the Opportunities and Vulnerabilities seem to concentrate in the South Eastern portion of the area under investigation, where the city centre is located.

The subsequent steps of the study will allow to draw policy recommendations and to support the strategic planning phase in order to foster the opportunities and minimize the vulnerabilities for the region. These results are of crucial importance for the subsequent generation and comparison of valorisation scenarios based on enhancement strategies of the strengths of the region.

4   Conclusions

This paper describes the development of a spatial Multicriteria Analysis to identify future opportunities and vulnerabilities for a specific region. The proposed methodology was illustrated with reference to a mountain area in the North of Italy which represents an environmental system characterized by multiple values.

The obtained results show that spatial Multicriteria Analysis can handle heterogeneous information and provide a significant contribution in the strategic decision-making phase. Moreover, one of the most significant strengths of the adopted methodological approach is represented by the fact that the evaluation is organized in a learning perspective. The decision maker thus gains more awareness with reference to the elements at stake while structuring the model (by means of standardization functions and trade-offs elicitation) and thus learns about the problems throughout the decision process [28].

By identifying opportunities and vulnerabilties for the area under analysis, the adopted approach also allows to foresee different future strategies (scenarios) for the management and valorization of the entire area. Consequently, different policy strategies could then be studied and evaluated in order to select the most sustainable one.

Scenarios can thus assist decision makers in the selection of proper policy solutions which produce robust results under varying conditions, in the assessment of strategies to cope with threats from particular natural and socio-economic conditions, and in risk assessments of various uncertain future developments [29].

With specific reference to the ANP methodology, it is important to highlight that, despite the limitations of a linear aggregation rule with respect to non-compensability issues in sustainability assessments, the approach allows to take interaction effects among the decision elements into account and this is particularly important in environmental decision-making problems, were the different components interact and influence each other.

In conclusion, the integration of decision aiding tools and spatial analysis constitutes a very promising line of research in the context of scenario analysis and sustainability assessments.

Acknowledgments. The paper is the result of the joint work of the three authors. Despite the overall responsibility being equally shared, Marta Bottero is responsible for paragraphs 2.1, 3.1 and 3.2 while Valentina Ferretti is responsible for paragraphs 2.2, 3.3 and 3.4. The abstract, paragraph 1 and the conclusions are the result of the joint work of the three authors. Finally, the authors would like to express a special thanks to Dr. Elisa Piolatto for the work done during her thesis at the Architectural and Landscape Heritage post-graduates specialization school of the Polytechnic University of Turin (Italy).

References

1. Torrieri, F., Nijkmap, P.: Scenario Analysis in Spatial Impact Assessment: a methodological approach. In: Vreeker, R., Deakin, M., Curwell, S. (eds.) Sustainable Urban Development. Volume 3: The Toolkit for Assessment. Routledge, Oxon (2009)
2. Nijkamp, P., Rienstra, S., Vleugel, J.: Transportation and the Future. Wiley, New York (1997)
3. Nijkamp, P., Rienstra, S.A.: The pivotal role of the private sector in achieving transport policy objectives. European Spatial Research and Policy, 93–113 (1998) (Spring)
4. Malczewski, J.: GIS and Multicriteria Decision Analysis. John Wiley and Sons, New York (1999)
5. Kahn, H., Weiner, A.: The Year 2000: A Framework for Speculation on the Next Thirty-Three Years. McMillan, New York (1967)
6. Warfield, J.: An Overview of Futures Methods. In: Slaughter, R. (ed.) The Foresight Principle. Praeger, Connecticut (1995)
7. Ratcliffe, R.: Scenario Building: A Suitable Method for Strategic Property Planning? Property Management 18(2), 127–144 (2000)
8. Godet, M.: Scenarios and Strategic Management. Butterworths, London (1987)
9. Schwartz, P.: The Art of the Long View. Currency Doubleday, New York (1996)
10. Stewart, T., Franch, S., Rios, J.: Integrating multicriteria decision analysis and scenario planning—Review and extension. Omega 41, 679–688 (2013)
11. Montibeller, G., Franco, A.: Decision and Risk Analysis for the Evaluation of Strategic Options. In: O'brien, F.A., Dyson, R.G. (eds.) Supporting Strategy: Frameworks. Methods and Models. John Wiley & Sons, Chichester (2007)

12. Volk, M., Hirschfeld, J., Schmidt, G., Bohn, C., Dehnhardt, A., Liersch, S., Lym-burner, L.: A SDSS-based Ecological-economic Modelling Approach for Integrated River Basin Management on Different Scale Levels – The Project FLUMAGIS. Water Resour. Manag. 21, 2049–2061 (2007)

13. Duzgun, H.S.B., Yucemen, M.S., Kalaycioglu, H.S., Celik, K., Kemec, S., Ertugay, K., Deniz, A.: An integrated earthquake vulnerability assessment framework for urban areas. Nat. Hazards 59, 917–947 (2011)

14. Zerger, A., Wealands, S.: Beyond Modelling: Linking Models with GIS for Flood Risk Management. Nat.l Hazards 33, 191–208 (2004)

15. Scholten, H.J., LoCashio, A., Overduin, T.: Towards a spatial information infra-structure for flood management in The Netherlands. Journal of Coastal Conservation 4, 151–160 (1998)

16. Brody, S.D., Grover, H., Bernhardt, S., Tang, Z., Whitaker, B., Spence, C.: Identifying Potential Conflict Associated with Oil and Gas Exploration in Texas State Coastal Waters: A Multicriteria Spatial Analysis. J. Environ. Manage. 38, 597–617 (2006)

17. Volk, M., Lautenbach, S., van Delden, H., Newham, L.T.H., Seppelt, R.: How Can We Make Progress with Decision Support Systems in Landscape and River Basin Management? Lessons Learned from a Comparative Analysis of Four Different Decision Support Systems. J. Environ. Manage. 46, 834–849 (2010)

18. Ballas, D., Kingston, R., Stillwell, J.: Using a Spatial Microsimulation Decision Support System for Policy Scenario Analysis. In: van Leeuwen, J.P., Timmermans, H.J.P. (eds.) Recent Advances in Design and Decision Support Systems in Architecture and Urban Planning. Kluwer Academic Publishers, Dordrecht (2004)

19. Grueau, C., Rodrigues, A.: Simulation Tools for Transparent Decision Making in Environmental Planning. Paper presented at the Second Annual Conference of GeoComputation 1997 & SIRC 1997, August 26-29. University of Otago, New Zealand (1997)

20. Rutledge, D.T., Cameron, M., Elliott, S., et al.: Choosing Regional Futures: Challenges and choices in building integrated models to support long-term regional planning in New Zealand. Regional Science Policy & Practice 1, 85–108 (2008)

21. Danese, M., Lazzari, M., Murgante, B.: Integrated geological, geomorphological and geostatistical analysis to study macroseismic effects of 1980 Irpinian earth-quake in urban areas (southern Italy). In: Gervasi, O., Taniar, D., Murgante, B., Laganà, A., Mun, Y., Gavrilova, M.L., et al. (eds.) ICCSA 2009, Part I. LNCS, vol. 5592, pp. 50–65. Springer, Heidelberg (2009)

22. Ferretti, V.: Le Analisi Multicriteri spaziali a supporto delle procedure di pianificazione e valutazione: analisi e classificazione della letteratura scientifica. Geoingegneria Ambientale e Mineraria L (2), 53-66 (2013)

23. Sharifi, M.A., Retsios, V.: Site selection for waste disposal through spatial multiple criteria decision analysis. J. Telecommun. Inf. Technol. 3, 28–38 (2004)

24. Keeney, R.L.: Value focused thinking. Harvard University Press, Cambridge (1992)

25. Saaty, T.L.: The Analytic Hierarchy Process. McGraw Hill, New York (1980)

26. Saaty, T.L.: Theory and Applications of the Analytic Network Process. RWS Publications, Pittsburgh (2005)

27. Sharifi, M.A., Rodriguez, E.: Design and development of a planning support system for policy formulation in water resources rehabilitation: The case of Alcázar De San Juan District in the Aquifer La Mancha Spain. Int. J. Hydroinformatics 4(3), 157–175 (2002)

28. Ferretti, V., Pomarico, S.: Ecological land suitability analysis through spatial indicators: An application of the Analytic Network Process technique and Ordered Weighted Average approach. Ecol. Indic. 34, 507–519 (2013)

29. Nijkamp, P., Vindigni, G.: Impact assessment of qualitative policy scenarios. A comparative case study on land use in Sicily. Management of Environmental Quality: An International Journal 14(1), 108–133 (2003)

Air Pollution Mapping Using Nonlinear Land Use Regression Models

Alexandre Champendal[1], Mikhail Kanevski[1], and Pierre-Emmanuel Huguenot[2]

[1] Institute of Earth Surface Dynamics, University of Lausanne, Geopolis building, 1015 Lausanne, Switzerland
[2] Service de l'Air, du Bruit et des Rayonnements non Ionisant, 1211 Geneva, Switzerland
mikhail.kanevski@unil.ch

Abstract. Air pollution in cities is an important problem influencing the environment, the well-being of society as well as its economy, the management of urban zones, etc. The problem is extremely difficult due to a very complex distribution of the pollution sources, the morphology of cities and the dispersion processes leading to a multivariate nature of the pollution phenomenon and to its high spatial-temporal variability at the local scale. Therefore, the task of understanding, modelling and predicting spatial-temporal patterns of air pollution in urban zones is an interesting and challenging topic having many research axes from science-based modelling to geostatistics and data mining. Recently, the application of land use regression models (LUR) for air pollution analysis and mapping in urban zones has demonstrated their efficiency. The present research deals with a new development of nonlinear LUR models based on machine learning algorithms. A special attention is paid to the Multi-Layer Perceptron and Random Forest algorithms and their abilities to model the NO_2 pollutant in the urban zone of Geneva.

Keywords: Urban air pollution, nonlinear land use regression models, machine learning.

1   Introduction

NO_2 air pollution in cities is an important problem influencing the environment, the health of society and its economy, the management of urban zones, etc. The problem is extremely difficult due to a very complex distribution of the pollution sources, the morphology of cities and the dispersion processes leading to a multivariate nature of the pollution phenomenon and its high spatial-temporal variability at the local scale. The task of understanding, modelling and predicting the spatial-temporal patterns of air pollution in urban zones is an interesting, challenging and crucial topic subdivided into many research axes:

- Science-based modelling using diffusion models coupled with meteorological information [1];
- Geostatistical modelling – spatial or spatio-temporal [2];
- Land use regression modelling [3].

B. Murgante et al. (Eds.): ICCSA 2014, Part III, LNCS 8581, pp. 682–690, 2014.
© Springer International Publishing Switzerland 2014

Recently, land use regression (LUR) models have gained a great popularity both as a research topic and as a tool for modelling air pollution in urban zones [1-8].

In LUR models the assumption that land use has a significant impact on the concentration of air pollutants is explicitly recognized. Although a bunch of linear LUR models have already been constructed in many cities around the world (see, for example, Parenteau et al. [5], Kashima et al. [6] or Beelen [8]), this study provides an advanced approach using nonlinear modelling tools such as Multi-Layer Perceptron (MLP) and Random Forest (RF) in order to predict pollution concentrations. An important question in developing LUR models is a construction of input/feature space (measurements, land use thematic layers, etc.). This space can contain correlated or redundant features. Therefore, the task of feature extraction/feature selection during modelling should be considered as well. In the present study the input space was constructed following experts and published recommendations, and only two methods – Principal Component Analysis (feature extraction) and Random Forest (feature selection) were applied.

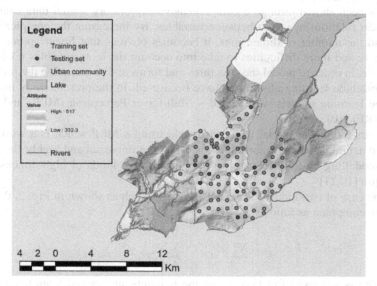

Fig. 1. The region (Geneva and Geneva Lake) of the study and the monitoring network

The real data case study deals with the spatial prediction of air pollution in canton Geneva. The yearly concentrations of Nitrogen dioxide (NO_2) were used and the studied region along with the monitoring stations are presented in Fig. 1.

2   LUR Models: Linear and Nonlinear

This study focuses on *Land Use Regression* models applied to air pollution. LUR models have been introduced in 1997 and later widely used in different countries [4].

The basic idea behind LUR is to compute a simple Multiple Linear Regression on raster or vector layers which contain land use information, such as distances to major and/or minor roads, airports, hydrographic networks, distribution of population, maximal power of heating systems, traffic, etc. Because pollution problems are not punctual, but diffuse, different buffers ranging from 100 m to 1000 m were constructed around each measurement point. Depending on the type of information, different statistics were computed.

Altogether, 110 layers were created for this study and the buffer values were chosen in order to test Henderson's model [3], considered in this research as a "benchmark model".

Through an iterative process, the significant variables are retained until a final optimal model is developed. Although, the linear approach provides good results, this work proposes a nonlinear approach of LUR concept in order to study the possible improvement of the results with a non-linear algorithms. These approaches are not very common in such studies, but have been tested with relatively good results in 2007 by Maynard [9].

It has been shown through a comprehensive exploratory spatial data analysis that non-linear relationships exist between variables. By increasing the complexity of the data and the number of dimensions, it becomes obvious that linear approaches will have more and more difficulties to take into account the inter-variables relationships. Therefore, in order to model the structures and forms and to minimize the influence of noise, machine learning algorithms have been used. In the present study mainly two machine learning models were tested – MultiLayer Perceptron (MLP) and Random Forest (RF) [10].

The first nonlinear model was constructed using a MLP which is a workhorse of artificial neural networks algorithms. The MLP is a universal and highly flexible nonlinear modelling tool which can incorporate different kinds of high-dimensional information [10,11].

In the case of two hidden layers MLP models, such as shown in Fig. 2, the predictions are computed as follows:

$$F^m(x^1, ..., x^K) = f_{out}\{\sum_{h_2=1}^{H_2} w_{h_2,m}^{(out)} f^{h_2}[\sum_{h_1}^{H_1} w_{h_2,h_1}^{(2)} f^{h_1}(\sum_{k=1}^{K} w_{k,h_1}^{(1)} x_k) + b_{h_1}] + b_{h_2}\}$$

where f is a transfer function, x_k are the K inputs, and ω and b are the connection weights and biases.

The weights between the nodes are updated using different optimization/training algorithm to minimize a mean square error (RMSE). The result of training can be interpreted as the conditional average of the target data.

In the present study, the modeling procedure was conducted as follows:

1) The 106 measurements were split into training (80 data points) and testing (26 data points) subsets;

2) Many MLP models were tested with varying numbers of neurons in the hidden layer;

3) The models were assessed using the residuals of the training and testing sets and the best ones were compared based on the error computed on the testing sets.

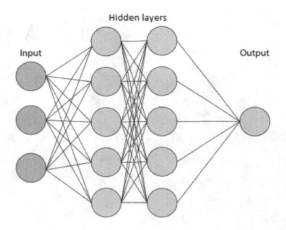

Fig. 2. The general structure of the MLP

Exploratory spatial data analysis, especially using geostatistical tools, such as variograms, confirmed that geographical X and Y coordinates are not relevant variables for modeling yearly concentrations of NO_2. Therefore, different models were trained without X and Y coordinates. Notice that the GeoMLP software described in detail in [11] was used.

Different training algorithms were applied: Levenberg-Marquardt, conjugate gradient, back propagation [11]. Due to the fact that Levenberg-Marquardt and conjugate gradient algorithms often led to over-fitting, they were not retained for the final mapping.

There are no strict rules for constructing an optimal MLP network, but regarding the RMSE error on the testing set, it turned out that the 108-2-1 network (i.e. 108 inputs, one hidden layer with 2 neurons and one output) provided the best results and was used for the final spatial predictions of the yearly concentrations of NO_2.

The resulting map produced by the best trained model is shown in Fig. 3. Its resolution is (25x25) m.

Let us note that the NO_2 predictions constructed without X and Y coordinates are greater than those taking them into account. Keeping only the model which does not contain X and Y coordinates, it is also possible to observe green spots, kinds of islands of pollution, matching the locations of different municipalities bordering the lake. These different areas are visible in Fig.4, extracted from Google Earth. Because of the high complexity of the model: 108-2-1, it is hard to interpret the weights of the link between the nodes and therefore to define the importance of each variable for this MLP model.

Therefore, another class of MLP LUR models was developed by first reducing the dimensionality of input space (feature extraction) using Principal Component Analysis (PCA). MLP modeling on PCA factors is not a very common approach. Let us note, that PCA is constructing factors/features using correlation, which is a linear measure. In a more general setting, nonlinear feature extraction/feature selection algorithms have to be considered.

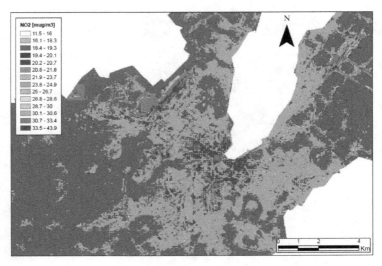

Fig. 3. The results of nonlinear LUR model using best MLP

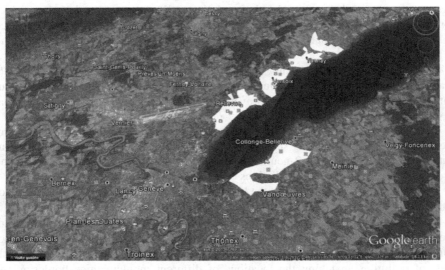

Fig. 4. Urban zones (more polluted regions) detected by the model

The number of factors introduced in different MLP models was defined on the basis of three different criteria:

- Based on the observation of the Scree Graph - the graph of the Eigen values sorted in descending order. It is used to identify and select the number of factors that explain most of the variability (Catell).
- The number of used factors is fixed once the proportion of cumulative variance has reached 90% (Joliffe).
- Using the correlation matrix to carry out the spectral decomposition is the same as choosing all eigenvalues greater than 1 (Kaiser).

Fig. 5 presents the different cuts on the Scree Graph regarding the different criteria.

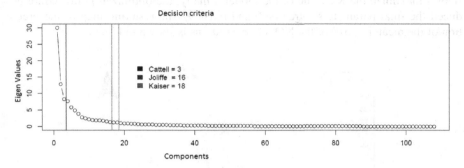

Fig. 5. PCA factors selection criteria

The factors produced by PCA are linear combinations of the original variables. Therefore, each factor contains some information about other variables. Although we have a loss of information when choosing the number of factors, a significant part of the initial information (90%) is preserved. The results produced by MLP using PCA are quite good.

The successful model gets 3.06 and 4.21 for RMSE on the training and testing sets respectively and R2 = 0.88. These values are high compared to what was obtained with the linear approache: R2 ~ 0.56.

Let us consider in short nonlinear LUR model based on the Random Forest algorithm [10]. Traditional linear LUR models select significant variables on the basis of the p-value and are therefore able to rank the variables according to their relevance. The Random Forest model gives both the results of modelling and ranks inputs according to their importance. Moreover, RF can provide also the uncertainties of the predictions.

The main idea behind RF is the construction of decision trees. N trees are created using sampling with replacements from the original dataset. On the created sample, the Gini index is used over 1/3 of the variables randomly chosen to perform the first splitting of node. At each splitting, a new set of variables is randomly chosen. This randomness takes into account the nonlinear relationship between the variables. Only two parameters are required to run a simple RF model: mtry=number of variables randomly chosen and ntree=number of trees created. Because of the bootstrapping technics, an average of 1/3 of the data points doesn't belong to the sample. This group is called Out Of Bag (OOB) and is used as a testing set in the corresponding tree which allows a quick calculation of RMSE.

The OOB group is used also to calculate the importance of the variables. After a permutation of the value of the first variable, the RMSE error is compared to the error without permutation. The greater the increase of the error is, the more important the variable is. To be able to create an optimal model using RF, we need to follow the following steps:

- Create a non-optimal model;
- Tuning: check and extract the variables of importance you want to work with;
- Optimize your parameters;
- Generate the optimal model;
- Predict and Map.

After ordering the variables by importance, it was observed that the first 11 variables maximize the R2 value of the model. Finally, the optimization procedure produced the final parameters: ntree=900 and mtry=9. The resulting map of the prediction of the mean in 2010 of the NO_2 concentrations is shown in Fig. 6.

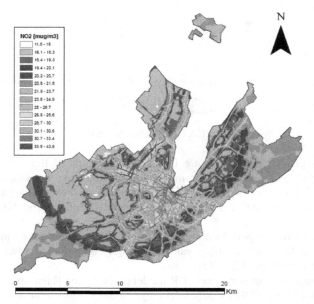

Fig. 6. Map of NO_2 pollutant produced with Random Forest algorithm

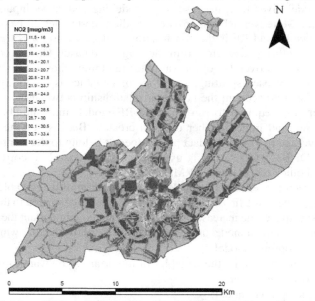

Fig. 7. Map of NO_2 pollutant produced with a Multiple Linear Regression

R2 of this model is 0.615 and RMSE on the test set is 3.85. This is an interesting result compared to the linear model. The Multiple Linear Regression model reached a R2 = 0.56 and RMSE on the test set equals 4.78. The map produced with the linear model (Fig. 7) is quite different both from RF and MLP. *It should be noted that regions outside of the measurement zone have to be treated with some cautions.* The summary of testing results is given in Table 1.

Table 1. The results of models' training

Models	RMSE Test	R2	Variables of importance
Multiple Linear Regression	4.78	56%	NO_x – 2000 m Pop – 750 m TJM – 275 m
MLP	3.88	78%	Not available
RF	3.85	61.5%	TMJ – 750 m Power_heating_system – 375 m Dist-Tarmac NO_x – 1000 m

3   Conclusions

The research presents new developments and preliminary results of air pollution in complex urban zones by nonlinear LUR models based on machine learning algorithms. Two models from machine learning algorithms - MLP and RF, were considered in detail. The algorithms belong to two different classes of machine learning, therefore it was interesting to apply and compare the results. Regarding the non-linear approaches studied in this work, it is possible to conclude that they can bring new perspectives to LUR models. Machine learning algorithms (both RF and MLP) show that they can be more efficient in comparison with linear tools and are capable to discover patterns in high dimensional data. The model with the highest R2 (78%) was produced with an MLP on 108 variables excluding X and Y coordinates.

There are several important questions to be considered in the future research. First, the construction of the input feature space (independent variables) is a challenging task, which is often based on experts' opinion and modelling experience. Features can be redundant. This poses a problem of feature selection or feature extraction. There are several fundamental approaches in ML that can be applied in nonlinear LUR modelling [10]. Second, other ML nonlinear modelling algorithms (e.g., General Regression Neural Networks, Support Vector Machines), widely used in environmental data mining, can be adapted and used to air pollution modelling in complex regions [11]. Third, monitoring network optimization (design/redesign) using geostatistics and ML algorithms can be applied in order to improve the quality of predictions and to reduce the uncertainties [12]. Finally, further studies could concern the applications of LUR models as a decision making tool in land use planning using both real and simulated data.

References

1. Briggs, D.J.: The Role of GIS: Coping with Space (and Time) in Air Pollution Exposure Assessment. J. Toxicol Env. Health 68(13-14), 1243–1261 (2005)
2. Kolovos, A., Skupin, A., Jerrett, M.: Multi-Perspective Analysis and Spatiotemporal Mapping of Air Pollution Monitoring Data. Environ. Sci. Technol. 44, 6738–6744 (2010)
3. Henderson, S.B., Beckerman, B., Jerrett, M., Brauer, M.: Application of Land Use Regression to Estimate Long-Term Concentrations of Traffic-Related Nitrogen Oxides and Fine Particule Matter. Environ. Sci. Technol. 7(41), 2422–2428 (2007)
4. Briggs, D.J., Collins, S., Elliott, P., Fisher, P., Kingham, S., Lebret, E., Pryl, K., Van Reeuwijk, H., Smallbone, K., Van der Veen, A.: Mapping Urban Air Pollution Using GIS: A Regression-Based Approach. Int. J. Geogr. Inf. Sci. 11, 699–718 (1997)
5. Parenteau, M.-P., Sawada, M.C.: The Role of Spatial Representation in the Development of a LUR Model for Ottawa, Canada. Air Qual. Atm. Health 5, 311–323 (2012)
6. Kashima, S., Yorifuji, T., Tsuda, T., Doi, H.: Application of Land Use Regression to Regulatory Air Quality Data in Japan. Science of the Total Environment 407(8), 3055–3062 (2009)
7. Ross, Z., et al.: Nitrogen Dioxide Prediction in Southern California Using Land Use Regression Modelling: Potential for Environmental Health Analyses. J. Exposure Sci. Env. Epidem. 16(2), 106–114 (2005)
8. Beelen, R., Voogt, M., Duyzer, J., Zxandveld, P., Hoek, G.: Comparison of the Performances of Land Use Regression Modelling and Dispersion Modelling in Estimating Small-Scale Variations in Long-Term Air Pollution Concentrations in a Dutch Urban Area. Atmos. Environ. 44, 4614–4621 (2010)
9. Maynard, D., Coull, B.A., Gryparis, A., Schwartz, J.: Mortality risk associated with short-term exposure to traffic particles and sulfates. Environ. Health Perspect. 115(5), 751–755 (2007)
10. Hastie, T., Tibshirani, R., Friedman, J.: The Elements of Statistical Learning, 2nd edn. Springer, New York (2009)
11. Kanevski, M., Pozdnoukhov, A., Timonin, V.: Machine Learning Algorithms for Spatial Environmental Data. Theory, Applications, and Software. EPFL Press, Lausanne (2009)
12. Tuia, D., Pozdnoukhov, A., Foresti, L., Kanevski, M.: Active Learning for Monitoring Network Optimization. In: Mateu, J., Muller, W. (eds.) Spatio-Temporal Design, pp. 285–318. John Wiley and Sons, Chichester (2013)

Towards Prospective Life Cycle Assessment: How to Identify Key Parameters Inducing Most Uncertainties in the Future? Application to Photovoltaic Systems Installed in Spain

Camille Marini and Isabelle Blanc

MINES ParisTech, O.I.E. center, Sophia Antipolis, France

Abstract. Prospective Life Cycle Assessment (LCA) is a relevant approach to assess the environmental performance of future energy pathways. Amongst different types of prospective scenarios, cornerstone scenarios meant for complex systems and long-term approaches, are of interest to assess such performance. They rely on different types of long-term projections, such as projections of technological evolutions and of energy resources. In most studies, scenarios are defined with single values for each parameter, and environmental impacts are assessed in a deterministic way. Inherent uncertainties related to these prospective assumptions are not considered and prospective LCA uncertainties are thus not addressed. In this paper we describe a methodology to account for these uncertainties and to identify the parameters inducing most of the uncertainties in the prospective LCA results. We apply this approach to prospective LCAs of photovoltaic-based electricity generation systems.

1 Introduction

World electricity consumption has multiplied by a factor of 4 in the last 40 years [18] and will keep increasing in the future due to population growth and changing lifestyle. Currently more than 68% of the total electricity production is based on fossil fuels [18]. Yet the use of fossil fuels raises growing environmental concerns, since their reserves are decreasing and their use is responsible for significant greenhouse gas (GHG) emissions [17], largely contributing to global warming (e.g., [2]). Environmental assessments of energy pathways being a critical issue, a large number of Life Cycle Assessments (LCA) have been undertaken, attempting to give a quantitative assessment of the current situation for each pathway.

An important issue we are now facing is related to the future development of energy pathways and its associated uncertainty. Such development is linked to critical environmental, technological and economical issues. Each energy pathway will have to deal with major changes in the future, such as availability and rate of depletion of resources, globalization and energy source supply, evolution of the resource potential when considering renewable energies, or technological

B. Murgante et al. (Eds.): ICCSA 2014, Part III, LNCS 8581, pp. 691–706, 2014.

developments. Uncertainties associated with these key parameters are significant and inherent to the prospective nature of the assessment.

Running scenarios is essential when handling prospective assessments. It has been applied in numerous fields and extensive research work has been undertaken to develop scenario-based LCA models [31]. Several types of scenarios can be distinguished depending on the purpose of the study [3]:

- Predictive scenarios, which answer the question "What will happen?" Predictive scenarios types are forecasts (the likely scenario occurs) and what-if (conditioned to some specific events). What-if scenarios are meant to be defined for simple objects and short-term studies.
- Explorative scenarios, which answer the question "What can happen?" They are external (related to exogenous conditions) and strategic (conditioned to some actions completed in a certain way). Cornerstone scenarios [31] are also defined as explorative scenarios and are meant for complex objects and long-term approaches.
- Normative scenarios, which answer the question "How can a specific target be reached?"

In our case, we are concerned by both predictive and explorative scenarios. Our goal is to assess the environmental performance of energy pathways, based on renewables and operating in 2050. Environmental performance is defined as the ratio of environmental impacts estimated by LCA to the electricity produced over the entire life cycle, corresponding to impacts per kWh produced. The prospective impacts depend on the future available technologies, more related to explorative scenarios. The prospective electricity production depends on future available energy resources (e.g., solar irradiation for photovoltaic systems, wind distribution for wind turbines), more related to predictive scenarios. The overall scenario needed to assess the prospective performance of energy pathways can be qualified as a cornerstone scenario [31], since it is partly explorative, meant for a long-term projection, and the considered system is complex.

In most studies, scenarios are defined with single values for each parameter over a predefined range of values and environmental impacts are assessed in a deterministic way. Inherent uncertainties related to these prospective assumptions are however not considered.

This paper describes a methodology to account for these uncertainties and to identify the key parameters inducing most of the uncertainties in these prospective performances. This methodology is based on the following steps:

1. Identification of the potential parameters to be considered (technological, geo-localization of the supply material, among others) for a given time horizon and a given area of energy production. Generation of a parametrized model of environmental performance based on these parameters.
2. Characterization of the parameter changes between the current and future situation. Definition of the uncertainties associated with future input parameter values. Different strategies to associate a distribution to these uncertainties are given.

3. Generation of the distribution of future environmental performances by applying Monte-Carlo simulations. Comparison between current and prospective environmental performance.
4. Key parameters identification and ranking with a global sensitivity analysis (GSA) based on Sobol indices.
5. Discussion and recommendations on specific key parameters inducing most uncertainties in the performance.

This methodology is applied to assess prospective environmental performance related to GHG emissions of photovoltaic (PV) systems (CdTe technology) installed in the South of Spain in 2050, given our defined cornerstone scenario. This cornerstone scenario is based on projections of developments of PV technologies, market share, and solar irradiation. Following the methodology, we are able to identify the key parameters explaining most uncertainties in the prospective environmental performance. These key parameters need to be predicted with more accuracy than the others to reduce the results' uncertainties. However, some of these key parameters may be impossible to predict, such as the manufacturing market share. Identifying them as key parameters informs us that it is essential to consider and explore their different possible developments in order not to underestimate results' uncertainties.

2   Methodology Description

2.1   Definition of a Parametrized LCA Model

We need to define as a first step a parametrized LCA model depending on a limited number of input parameters, to assess the prospective environmental performance of an electricity generation system based on renewable energy, and compare them to current performance. In this paper, the considered impacts are one dimensional, such as climate change impacts related to GHG emissions and expressed in grams of CO_2 equivalent, leading to performance in gCO_2eq/kWh. The following steps are to be taken to define this model :

1. **Definition of the objective of the study and the boundaries of the considered system.** As in any LCA, this step is essential. The functional unit must be defined and the limits of the system made explicit. The technology considered, the geographical localization of the system, the temporal horizon (for which time horizon the prospective is made), and the methodology must be specified.
2. **Identification of the input parameters of the parametrized model.** These parameters are assumed to be characteristic of the system and are likely to vary in the future. Their identification can be based on a literature review and/or discussions with experts. Their current values are known based on real observations, and their future values are predicted with more or less uncertainty. These n parameters are denoted $x_1, ..., x_n$. They must be independent (requirement for computing Sobol indices, see Sect. 2.4); if not, dependency relations must be specified to obtain a set of independent parameters.

3. **Definition of the parametrized impacts model.**
   First a parametrized Life Cycle Inventory (LCI) is realized. For parameters not made explicit, data from a LCI database, such as EcoInvent [9], are used. This parametrized LCI is converted into a parametrized LCA model by using the characterization factors corresponding to the considered impacts. This model is denoted $fi(x_1, ..., x_n)$.

4. **Definition of a parametrized model of electricity generation.**
   It estimates the electricity produced by the system over its entire production phase, and it is denoted $fe(x_1, ...x_n)$.

5. **Definition of a parametrized model of environmental performance.**
   It is obtained by combining the parametrized impacts and electricity generation models: $fp(x_1, ..., x_n) = \dfrac{fi(x_1, ..., x_n)}{fe(x_1, ..., x_n)}$.

These steps to define a parametrized LCA model have already been implemented by Padey et al (2013) [30] to assess the variability of an energy pathway.

2.2 Characterization of the Parameters Changes between the Current and Future Situation and Their Future Uncertainties

The current values $x_1^{t0}, ..., x_n^{t0}$ of the input parameters, and predicted future values $x_1^{t1}, ...x_n^{t1}$ must be identified, in order to apply the parametrized model fp and assess the current and future environmental performance.

Uncertainties consideration is a big issue in LCA. Uncertainties can affect different modeling components, the parameters, scenarios, and models; they can occur at different stages of the analysis, during the goal and scope definition, the inventory analysis, and the impact assessment; and they can have various sources (e.g., [12]). In this paper, we focus on future input parameter uncertainties, inherent to the prospective approach. We do not consider uncertainties for current input parameter values.

Uncertainties on future parameter values may be represented by considering these parameters as random variables, denoted $X_1, ..., X_n$, characterized by a distribution instead of a simple scalar value. Note that uncertainties distributions can be continuous (e.g., technology efficiency) or discrete (e.g., country of manufacture). Depending on the type of parameters and available sources of information, different strategies may be used to characterize uncertainty distributions (e.g., [12]):

Case 1 In the best case, projections of a parameter are provided with their associated uncertainty distribution. It is the case for parameters provided in the ecoSpold format [13].

Case 2 If enough projections of a parameter can be found, a distribution can be estimated by distribution fitting. A statistical test can be used to check the validity of the fit, such as a Chi-square goodness-of-fit test, or a Kolmogorov-Smirnov test.

Case 3 If the projection of a parameter is provided with qualitative information, the methodology defined by Weidema and Weismaes (1996) [40] based on the Pedigree matrix and Data Quality Indicators can be applied.

Case 4 If projections of a parameter are provided for several different scenarios, values for extreme scenarios may be used to define the uncertainty range, and depending on the likelihood, different types of distribution can be used. For instance, if there are 3 scenarios, a pessimistic, a realistic, and an optimistic, a normal or lognormal distribution may be used, with a median equal to the value provided in the realistic scenario, and a 99% confidence interval bounded by the values provided in the pessimistic and optimistic scenarios [1]. This strategy allows to transform a set of deterministic projections into an probabilistic projection associated to a distribution.

Case 5 For decadal projections of climate resources based on climate models, projections from different models can be used to represent the range of possibilities, assuming that most uncertainty is attributed to the model uncertainty [38].

Case 6 In the worst case, the projection of a parameter is provided without any other information. If it is physically plausible, the uncertainties distribution could be assumed as normal, centered on the prospective value with an 99% lower bound equal to the current parameter value.

2.3 Comparison between Current and Prospective Environmental Performances

Current and prospective performances are computed with the parametrized model defined in Sec. 2.1 applied to current and prospective input parameters values given in Sec. 2.2. The distribution of prospective performances is obtained with Monte Carlo simulations.

2.4 Identification of Parameters Inducing Most Uncertainties in the Prospective Environmental Performance

We assume that we know the uncertainties' distribution of each input parameter. The performance becomes a random variable, denoted $Y = fp(X_1, ..., X_n)$. Using Sobol sensitivity indices [37], it is possible to identify input parameters inducing most variability in the results, in our case it corresponds to parameters whose uncertainties induce most uncertainties in the performance.

More precisely, the contribution of the input parameter X_i to the total variance is quantified with the first order Sobol index, defined as:

$$S_i = \frac{Var(E(Y/X_i))}{Var(Y)} \tag{1}$$

---

[1] A lognormal distribution can be used for asymmetric distributions, and normal for symmetric distributions. For a normal distribution with median μ and standard deviation σ, 99% of values are found within $[\mu - 3\sigma; \mu + 3\sigma]$. For a lognormal distribution with median μ^* and geometric standard deviation σ^*, 99% of values are found within $[\mu^*/\sigma^{*3}; \mu^* * \sigma^{*3}]$.

where Var is the variance. The definition of Sobol indices is based on variance decomposition and requires the independence of the input parameters $X_1, ... X_n$. The first order Sobol indices can be computed using Monte Carlo simulations. The brute force method would be, for each input parameter X_i, to run a set of M Monte Carlo simulations to estimate $E(Y/X_i)$ for a fixed value of X_i and then repeat the procedure for different values of X_i, leading to a total cost of M^2 (with M being large). Saltelli (2012) [35] proposed a faster procedure, requiring $M(n + 2)$ runs (n being the number of input parameters).

Sobol indices have already been used in LCA of energy systems by Padey et al. (2013) [30], not to identify parameters inducing most performance uncertainties, but to identify parameters inducing most performance variability due to the variability of systems within one energy pathway.

3   Methodology Application

We apply the methodology described above to a residential building-integrated PV system, whose characteristics are given below. Although realistic, this application is simplified. The performance model could be refined by taking into account more input parameters, as detailed below in the boundaries of the system and in the assumptions made when identifying the input parameters of the parametrized model.

3.1   Definition of the Parametrized LCA Model

Definition of the Objective and Boundaries of the Study

Objective: Assess the current and prospective global warming performance in gCO$_2$eq/kWh of a residential PV system with the following characteristics:

- Technology: building-integrated system based on the CdTe technology and with a peak power P of $3kWp$.
- Geographical: system installed in the South of Spain at a latitude of $37°N$ and a longitude of $5°W$. The PV panel faces due South and is inclined at an angle equal to the latitude ($37°N$ in our case).
- Temporal: The prospective time horizon is 2050.
- Methodology: Except for input parameters, data come from the ecoinvent v2.2 database [23]. The characterization factor to assess the global warming impact is from IPCC 2007 with a time horizon of 100 years [19].

Boundaries: The manufacturing and production phase of the panel are considered. As stated above, it is a simplified application; so for the sake of simplicity, we do not consider the end-of-life phase and we omit the impacts related to transportation. Although transportation has been shown to have secondary effects (e.g., [1]), it should be taken into account in a more detailed study.

Identification of the Input Parameters of the Parametrized Model

A literature review mostly based on [7], [10], [23], [26], [20], and [14] led us to the identification of the following parameters:

- Life-time of the PV system (technological parameter), denoted LT and expressed in years (yr).
- Efficiency of the module (technological parameter), denoted η and expressed in percentage.
- CdTe layer thickness in the PV cell (technological parameter), denoted e and expressed in μm.
- Material utilization rate to produce the CdTe layer (technological parameter), denoted U and expressed in percentage. It corresponds to the utilisation rate of CdTe in the manufacturing process of the cell.
- Performance ratio (technological parameter), denoted PR and expressed in percentage. It corresponds to a correction factor to consider deviation from ideal conditions (dust, shadowing effects,...).
- Module manufacturing origin (parameter related to the supply chain).
- Electricity mixes in the manufacturing countries (parameter related to the supply chain).
- Irradiation, denoted Ir and expressed in kWh/(m^2.yr). It corresponds to the yearly irradiation taking into account the orientation and the tilt of the PV panel.

Figure 1 summarizes the relations between the identified characteristic parameters and the considered system. The surface S of the panel depends on the peak power P and the efficiency η of the module with $S = \dfrac{P}{\eta}$. The CdTe layer thickness e and the material utilisation rate U determine the mass of CdTe necessary for the production of $1m^2$ of module: $m_{CdTe} = \dfrac{e\mu_{CdTe}}{U}$, with $\mu_{CdTe} = 6200 kg/m^3$ being the density of semiconductor in the layer [26].

The following assumptions are made :

- Two inverters are used over the life-cycle, independently of the life-time of the system. It is the current situation, since the mean life-times of inverters and PV panel are, respectively, 15 and 30 yr. It thus assumes that these life-times will evolve in parallel, which is a realistic assumption according to experts ([20]).
- No module degradation is considered, this should be incorporated in a more realistic study.
- The electricity consumption during production processes is considered constant, since no prospective values could be found. This should be modified for a more realistic study.

Specification of the Parametrized Impacts Model

The model is built with the help of Simapro software. It is based on ecoinvent data v2.2 [23] for the processes "Photovoltaic laminate CdTe/DE", "inverter 2500W at plant/RER", "Slanted-Roof construction, integrated", and "electric installation photovoltaic plant, at plant/CH" that we parametrize as a function of the input parameters. Impacts are then obtained by combining these data with characterization factors from IPCC 2007 with a time horizon of 100 years [19]. The resulting parametrized model for the impacts is specified in Appendix 4.

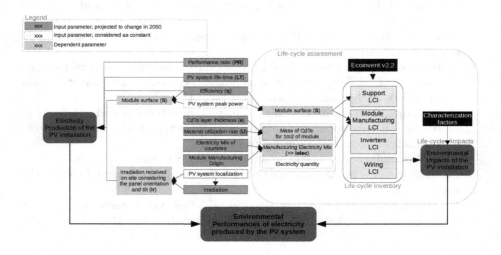

Fig. 1. Simplified parametrized model of the global warming performance for a residential PV panel based on the CdTe technology

Specification of the Parametrized Model of Electricity Generation
The electricity produced over the life cycle is based on the yearly mean production:

$$fe(\mathbf{LT}, \mathbf{PR}, \mathbf{Ir}) = \mathbf{LT}.\mathbf{PR}.\mathbf{Ir}.\eta.S = \mathbf{LT}.\mathbf{PR}.\mathbf{Ir}.P \qquad (2)$$

where $\mathbf{LT}$, $\mathbf{PR}$, $\mathbf{Ir}$, η, S and P are, respectively, the panel life-time, the performance ratio, the irradiation, the efficiency, the panel surface and the peak power, defined in Sec. 3.1 (bold letters indicate varying parameters).

Specification of the Parametrized Performance Model
The global warming performance model is obtained by combining the parametrized impacts model (specified in Appendix 4) and the parametrized model of electricity generation defined in (2):

$$fp(\eta, \mathbf{e}, \mathbf{U}, \mathbf{Ielec}, \mathbf{LT}, \mathbf{PR}, \mathbf{Ir}) = \frac{fi(\eta, \mathbf{e}, \mathbf{U}, \mathbf{Ielec})}{fe(\mathbf{LT}, \mathbf{PR}, \mathbf{Ir})} \qquad (3)$$

This model was implemented in the Python programming language.

3.2 Characterization of Parameter Changes between the Current and Future Situation and Their Future Uncertainties

The definition of the future situation, identified as a cornerstone scenario, requires the characterization of uncertainties' distributions of input parameters. This characterization is made using the different strategies summarized in Sec. 2.2.

Values for "life-time", "module efficiency", and "CdTe layer thickness" come from [20]. In this study, 3 scenarios are defined: a pessimistic one with limited improvements, a realistic scenario with reasonable improvements, and an optimistic scenario using the best available predictions for parameters identified as key for the CdTe technology. We derive a probability distribution for these parameters based on the values given for the 3 scenarios, as explained in Case 4 of Sec. 2.2.

Values and uncertainty range for the "material utilization rate" are similarly derived from [26] (Case 4 of Sec. 2.2).

Current and prospective values (for 2030 and not 2050) for the "performance ratio" were extracted from [7], but no uncertainty range could be found. The lower bound of the confidence interval is thus assumed equal to the current value, and a normal distribution with a median equal to the prospective value (Case 6 of Sec. 2.2).

Currently, module manufacturing is shared between Germany (22%), the USA (12%), and Malaysia (65%) [6]. No prospective data about module manufacturing countries could be found, as it is almost impossible to predict such parameter at such lead time. It is fundamentally associated to financial markets, regulating policies, economic measures, with time scales shorter than a decade. We thus explore 3 possibilities with equal probability: a share identical to the current situation, an occidental share (50% Germany and 50% USA), and an Asian share (50% Malaysia and 50% China).

Current estimates of the irradiation are derived from MERRA reanalysis [34].This dataset results from a combination of climate model fields and irregular observations in space and time, from 1979 to the present (satellite-area), using data assimilation. Monthly estimates of surface incident shortwave flux at a latitude of 37°N and a longitude of 5°W are converted into global tilted irradiation considering that the panel faces South and is inclined at an angle equal to the latitude (see 4 for more details). The mean for the 1985-2013 period is used as the current value for irradiation.

The percentage variation of irradiation between the present and the future situation (i.e. for a panel operating from 2050 to 2080 on average) is derived from regional climate model (RCM) simulations for the period 1950-2100 with a "moderate" emissions scenario (scenario A1B, [19]), part of the ENSEMBLE-RT2B database [5]. Contrary to MERRA data, no observations are incorporated in these simulations, they aim at modeling climate tendency, and not at providing accurate predictions for a given year, consistent with a cornerstone scenario. We use simulations from 4 different RCMs: C4I from the Swedish meteorological service [22], KNMI from the Dutch meteorological service [39], MPI from

Table 1. Current and prospective values of considered LCI parameters, as well as their associated uncertainties range (99% confidence interval). Note that no uncertainties are considered for future electricity mixes of manufacturing countries.

Parameter	Current LCI		Prospective LCI	
Life-time	30 yr	[20]	Normal distribution, median: 35 yr	[20]
		[23]	Uncertainty range:[30; 40]	[15]
Module efficiency	11.7%	[20]	Lognormal distribution, median: 19.9%	[20]
		[23]	Uncertainty range:[17.7; 22.7]	[11]
Performance ratio	78 %	[7]	Normal distribution, median: 83 % Value given for 2030 and not for 2050, without associated uncertainties	[7]
CdTe layer thickness	3μm	[20]	Lognormal distribution, median: 1μm	[20]
		[14]	Uncertainty range:[0.1; 2]	[26]
Material Utilization rate	55 %	[26]	Normal distribution, median: 85 % Uncertainty range:[70; 99]	[26]
Irradiation at 37°N, 4°W	MERRA re-analysis Mean irradiation 1985-2013 2169.7 kWh/(m^2.yr)	[34]	Based on projections from 4 Regional Climate Models (ENSEMBLES project) Irradiation from MERRA for 1985-2013 + changes between 1985-2013 and 2050-2080 estimated from the projections based 4 regional climate models. Normal distribution for changes, median: +3.3%, uncertainty range:[+1.2; +5.5]	[5]
CdTe module manufacturing origin	Germany: 22%, Malaysia:65%, USA:12%	[6]	3 explorative possibilities with equal probability: Current share, Occidental share (50% Germany, 50% USA), and Asian share (50% Malaysia, 50% China)	
Electricity mix of manufacturing countries	Data from 2009	[16]	Prospective for 2035 based on the "New Policies Scenario", No uncertainties range is considered	[16]

the Max Planck Institute for Meteorology in Hamburg [21], and SMHI from the Swedish Meteorological and Hydrological Institute [24]. Similarly to MERRA data, monthly estimates of surface incident shortwave flux at a latitude of 37°N and a longitude of 5°W are converted into global tilted irradiation, before computing 30 years averages over the periods 1985-2015 and 2050-2080. The variation percentages between averages over these two periods ranges between 1.2% and 5.5% depending on the model, reflecting small changes of future irradiation, consistent with [33]. Note that similar values are obtained when shifting the averaging periods by ±5 years. This range is used as the 99% confidence interval for the variation percentage of irradiation (Case 5 of Sec. 2.2), and a normal distribution is assumed. These variations percentages are added to the 1985-2013 mean from MERRA irradiation.

3.3   Comparison between Current and Prospective Global Warming Performances

According to the parametrized model defined in Sec. 3.1, the current global warming performance of the considered PV system reaches 23.1 gCO_2eq/kWh (represented by the dashed-dotted line in Fig.2). Although the parametrized performance model could be further refined, this figure is consistent with the literature [7] [29].

The distribution of prospective global warming performances, obtained with future parameter values specified in Table 1 (right column), is shown in Fig.2 by the histogram. The mean value of the prospective performance distribution equals 9.8 gCO_2eq/kWh, its standard deviation is 0.77 gCO_2eq/kWh, and values range between 6.9 and 13.9 gCO_2eq/kWh. The mean value is consistent with the prospective value obtained in [7] for 2030 (equals to 10 gCO_2eq/kWh). These prospective performances are significantly lower than the current value, the maximum prospective value being even 40% smaller than the current performance.

The lines on the left-hand side of Fig.2 (on the left part) indicate the prospective performances obtained with median prospective values of input parameters (right column of Table 1) and for a fixed share of manufacturing countries (dashed line for the Occidental share, dotted line for the current share, and solid line for the Asian share). Not surprisingly, a higher value is obtained for the Asian share, where the electricity mix will remain mostly based on coal; and a lower value for the Occidental share, where the electricity mix will rely on more renewable energy. Given the differences between these performances, the market share seems to considerably influence the prospective performances, as will be shown and discussed in the next section.

3.4   Identification of Parameters Inducing Most Uncertainties in the Prospective Environmental Performance

As described in Sec. 2.4, Sobol indices are computed to estimate the contributions of the different parameters to results' uncertainties, more precisely to the variance of the results that is induced by input parameters' uncertainties. The most important parameter is life time ($S_{LT} \sim 41\%$ of the total variance), followed by the share of module manufacturing origin ($S_{origin} \sim 37\%$), efficiency ($S_{\eta} \sim 18\%$), and performance ratio ($S_{PR} \sim 7\%$). Uncertainties on irradiation, CdTe layer thickness, and material utilization rate have almost no impact on the results uncertainties (their Sobol indices are smaller than 3%). If one wants to narrow uncertainties on prospective global warming performance, uncertainties on prospective life time must be reduced as a priority.

These results also show that the share of module manufacturing origin, which is almost unpredictable at such lead time, highly contributes to the variance of the results. It is, therefore, important to take into account all conceivable possibilities for this parameter in the analysis, in order not to underestimate uncertainties on the prospective performances. Indeed, if only one possibility

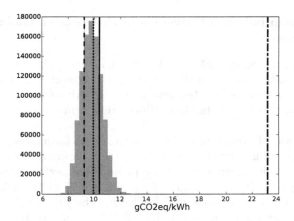

Fig. 2. Distribution of prospective global warming performances in gCO₂eq/kWh (histogram), calculated with input parameters specified in Table 1 (right column). The three lines on the left indicate prospective values obtained with median prospective values of input parameters and for a fixed share of manufacturing countries: 22% Germany, 12% USA, 65% Malaysia for the dotted line; 50% Germany and 50% USA for the dashed line; 50% Malaysia and 50% China for the solid line. The current performance is indicated by the dashed-dotted line, and obtained from input parameters specified in Table 1 (left column).

for the share of module manufacturing origin is considered (e.g, the current share), the performance standard deviation drops from 0.77 gCO₂eq/kWh to 0.61 gCO₂eq/kWh, corresponding to a decrease of 37% in the variance.

4   Summary and Discussion

In this paper, we proposed a methodology to assess the prospective environmental performance of an electricity generation system based on renewable energy, as well as the associated uncertainties, and to identify parameters contributing the most to these uncertainties based on global sensitivity analysis. The latter need to be predicted with more accuracy in order to reduce the results' uncertainties.

Prospective LCA of electricity generation systems can be found in the literature (e.g., [7], [32]) but, to our knowledge, uncertainties are not estimated, although essential to any predictive approach. Our approach is original in considering these uncertainties and in identifying parameters contributing most to performance uncertainties.

Another original feature of this work is the combination of different types of predictions (projections of technology evolutions, climate predictions to estimate future renewable resources, and projections of market share evolution), coming from different types of scenarios, both predictive and explorative.

The methodology was applied to the global warming performance of a residential PV panel based on CdTe technology, installed in the South of Spain and operating in 2050. We showed that performance decreases significantly compared with the current situation, from 23.1 gCO$_2$eq/kWh to 9.8 gCO$_2$eq/kWh on average with a standard deviation of 0.77 gCO$_2$eq/kWh due to uncertainties. The latter were mostly explained by uncertainties in the life time parameter and in the share of module manufacturing origin. Uncertainties in life time should be thus reduced to narrow performance uncertainties. It is not possible to reduce uncertainties on market share, since this parameter is hardly predictable; but it shows that possible values must be considered to avoid underestimating results uncertainties.

The methodology was illustrated for a system installed at a given location, but we could easily extend the methodology to create maps of future environmental performances at any location. Indeed, irradiation values were taken from MERRA reanalysis and regional climate models, which provide spatio-temporal irradiation that can be combined to other data to obtain maps of environmental performance. Maps allow to geographically optimize installations of PV systems with other sources of energy and to know where new installations of PV systems should be prioritized. Maps of environmental performance for PV have already been produced [27], but not yet for any prospective assessment.

It should be emphasized that parameters identified as key in this methodology correspond to parameters mostly contributing to the performance variance that is induced by input parameters uncertainties. It would also be interesting to identify another type of key parameters: those whose variations between the current and future situations induce most performance changes. However, it is not an easy task, since the performance model is non-linear and changes cannot be considered as local (so that a first order Taylor expansion would not give accurate results).

Acknowledgements. We are grateful to P. Padey, P. Blanc, and T. Ranchin for their useful comments. We acknowledge the Global Modeling and Assimilation Office and the GES DISC for the dissemination of MERRA reanalysis, as well as the European project ENSEMBLES for providing regional climate projections.

Appendix A: Specification of the Parametrized Impacts Model

From Sec. 3.1 results the following parametrized impacts model for a residential PV panel based on the CdTe technology:

$$fi(\eta, \mathbf{e}, \mathbf{U}, \mathbf{Ielec}) = Iwir + Iinv + Isup.\frac{P}{\eta} + Imod(\mathbf{e}, \mathbf{U}, \mathbf{Ielec}).\frac{P}{\eta}$$

where:

$$Imod(\mathbf{e}, \mathbf{U}, \mathbf{Ielec}) = \alpha_1 + \beta_1.\mathbf{Ielec} + \frac{\mathbf{e}.\mu_{\mathrm{CdTe}}}{\mathbf{U}}.Icdte(\mathbf{Ielec}) \qquad (4)$$

$$Icdte(\mathbf{Ielec}) = \alpha_2 + \beta_2.\mathbf{Ielec} + m_1.(\alpha_3 + \beta_3.\mathbf{Ielec} + m_3.I_{Te})$$
$$+ m_2.(\alpha_4 + \beta_4.\mathbf{Ielec} + m_4.(\alpha_5 + \beta_5.\mathbf{Ielec})) \qquad (5)$$

with:
- **Ielec** being the gCO$_2$eq/MJ of electricity, obtained by combining the module manufacturing origin and the electricity mix of the corresponding countries,
- $Iwir$, $Iinv$, $Isup$, $Imod$, and $Icdte$ being the gCO$_2$eq of, respectively, the wiring system, the 2 inverters, 1m^2 of the buiding-integrated support, 1m^2 of module, and 1kg of laminate CdTe,
- α_1 being the gCO$_2$eq/m^2 of module without considering the electricity consumption, or the production of laminate CdTe,
- α_2 being the gCO$_2$eq/kg of laminate CdTe without considering the electricity consumption, or the production of semicondutors Cd and Te,
- α_3 (α_4) being the gCO$_2$eq/kg of semiconductor Te (Cd) without considering the electricity consumption, or the production of Te (Cd),
- α_5 being the gCO$_2$eq/kg of Cd without considering the electricity consumption,
- I_{Te} being the gCO$_2$eq/kg of Te,
- β_1, β_2, β_3, β_4, and β_5 being the MJ of electricity necessary to produce 1kg of, respectively, laminate CdTe, semiconductor CdTe, semiconductor Te, semiconductor Cd, and Cd,
- m_1, m_2, m_3, and m_4 being the mass of semiconductor Te, semiconductor Cd, Te, and Cd necessary to produce 1kg of CdTe.

Appendix B: From Monthly Horizontal Irradiation to Irradiation on Inclined Surfaces

Monthly horizontal irradiation, G_m, is converted into irradiation on inclined surfaces, GTI_m, using algorithms defined in the European Solar Radiation Atlas [36]. It is based on the following steps:

1. Conversion of G_m into daily horizontal irradiation, denoted G_d, and calculation of the daily diffuse radiation D_d from G_d using the empirical model of Erbs et al (1982) [8].
2. Estimation of hourly global horizontal irradiation (G_h) and hourly diffuse radiation (D_h) from G_d and D_d using the relations defined in Collares-Pereira and Rabl (1979) [4] and Liu and Jordan (1960) [25], respectively. The hourly direct radiation, B_h, is then derived with: $B_h = G_h - D_h$.
3. Calculation of the hourly global tilted irradiation, GTI_h, as the sum of the three components: direct irradiation B_i^{tilted}, diffuse irradiation D_i^{tilted}, and reflected irradiation R_i^{tilted}. B_i^{tilted} is obtained with a geometric relation. D_i^{tilted} is derived using the Muneer algorithm (1990) [28]. R_i^{tilted} is calculated from G_h, the tilt angle, and the ground albedo (assumed equal to 0.2).
4. Conversion of GTI_h into GTI_m.

References

[1] Beloin-Saint-Pierre, D., Blanc, I.: Environmental impact of pv systems: Effects of energy sources used in production of solar panels. In: Proc. of the 24rd Eur. Photovoltac Solar Energy Conf., Hamburg, Germany, pp. 4517–4520 (September 2009)

[2] Bindoff, N., et al.: Detection and attribution of climate change: from global to regional. Tech. rep, in Climate Change 2013: The Physical Science Basis. Contribution of WG I to the 5th Assessment Report of the IPCC (2013), http://www.climatechange2013.org/images/report/WG1AR5_Chapter10_FINAL.pdf

[3] Cluzel, F., Yannou, B., Millet, D., Leroy, Y.: Exploitation scenarios in industrial system LCA. Int. J. Life Cycle Ass. 19(1), 231–245 (2014)

[4] Collares-Pereira, M., Rabl, A.: The average distribution of solar radiation-correlations between diffuse and hemispherical and between daily and hourly insolation values. Sol. Energy 22(2), 155–164 (1979)

[5] Déqué, M., Somot, S., Sanchez-Gomez, E., Goodess, C., Jacob, D., Lenderink, G., Christensen, O.: The spread amongst ensembles regional scenarios: regional climate models, driving general circulation models and interannual variability. Clim. Dynam. 38(5-6), 951–964 (2012)

[6] Dominguez-Ramos, A., Held, M., Aldaco, R., Fischer, M., Irabien, A.: Carbon footprint assessment of photovoltaic modules manufacture scenario. In: Proc. 20th Eur. Symp. Comput. Aided Process Eng. (2010)

[7] Dominguez-Ramos, A., Held, M., Aldaco, R., Fischer, M., Irabien, A.: Prospective co2 emissions from energy supplying systems: photovoltaic systems and conventional grid within spanish frame conditions. Int. J. Life Cycle Ass. 15(6), 557–566 (2010)

[8] Erbs, D., Klein, S., Duffie, J.: Estimation of the diffuse radiation fraction for hourly, daily and monthly-average global radiation. Sol. Energy 28(4), 293–302 (1982)

[9] Frischknecht, R., et al.: The ecoinvent database: Overview and methodological framework (7 pp). Int. J. Life Cycle Ass. 10(1), 3–9 (2005)

[10] Fthenakis, V., Frischknecht, R., Raugei, M., Kim, H.C., Alsema, E., Held, M., de Wild-Scholten, M.: Methodology guidelines on life cycle assessment of photovoltaic electricity. Tech. rep., IEA-PVPS Task 12 (2011)

[11] Garabedian, R.: First solar technology update. Tech. rep., First solar Inc (2013)

[12] Heijungs, R., Huijbregts, M.: A review of approaches to treat uncertainty in lca. Elsevier, Orlando (2004)

[13] Heijungs, R., Frischknecht, R.: Representing statistical distributions for uncertain parameters in lca. relationships between mathematical forms, their representation in ecospold, and their representation in cmlca (7 pp). Int. J. Life Cycle Ass. 10(4), 248–254 (2005)

[14] Houari, Y., Speirs, J., Candelise, C., Gross, R.: A system dynamics model of tellurium availability for cdte pv. Prog. in Photovoltaics 22(1), 129–146 (2014)

[15] IEA: Technology roadmap: solar photovoltaic energy. Tech. rep., International Energy Agency (2010), http://www.iea.org/publications

[16] IEA: World energy outlook. Tech. rep. (2011), http://www.iea.org/publications/freepublications/publication/name,37085,en.html

[17] IEA: co2 emissions from fuel combustion, highlights. Tech. rep. (2013), http://www.iea.org/publications/freepublications/publication/CO2EmissionsFromFuelCombustionHighlights2013.pdf

[18] IEA: Key world energy statistics 2013. Tech. rep. (2013), http://www.iea.org/publications/freepublications/publication/KeyWorld2013.pdf

[19] IPCC: Climate change 2007. ipcc fourth assessment report. the physical science basis. Tech. rep. (2007), http://www.ipcc.ch/ipccreports/ar4-wg1.htm

[20] Itten, R., Wyss, F., Frischknecht, R.: Lci of the global crystalline photovoltaics supply chain and of future photovoltaics electricity production. Tech. rep., Treeze Ltd, Uster, Switzerland (2014)

[21] Jacob, D.: A note to the simulation of the annual and inter-annual variability of the water budget over the baltic sea drainage basin. Meteorol. Atmos. Phys. 77(1-4), 61–73 (2001)

[22] Jones, C.G., Willén, U., Ullerstig, A., Hansson, U.: The rossby centre regional atmospheric climate model part i: model climatology and performance for the present climate over europe. AMBIO 33(4), 199–210 (2004)

[23] Jungbluth, N., Stucki, M., Flury, K., Frischknecht, R., Büsser, S.: Life cycle inventories of photovoltaics. ESU-services Ltd., Uster, CH (2012)

[24] Kjellström, E., Bärring, L., Gollvik, S., Hansson, U., Jones, C.: A 140-year simulation of european climate with the new version of the rossby centre regional atmospheric climate model (rca3). SMHI Reports Meteorology and Climatology 108 (2005)

[25] Liu, B.Y., Jordan, R.C.: The interrelationship and characteristic distribution of direct, diffuse and total solar radiation. Solar Energy 4(3), 1–19 (1960)

[26] Marwede, M., Reller, A.: Future recycling flows of tellurium from cadmium telluride photovoltaic waste. Resources, Conservation and Recycling 69, 35–49 (2012)

[27] Ménard, L., et al.: Benefit of geoss interoperability in assessment of environmental impacts illustrated by the case of photovoltaic systems. IEEE Journal of Selected Topics in Applied Earth Observations and Remote Sensing 5(6), 1722–1728 (2012)

[28] Muneer, T.: Solar radiation model for europe. Build. Serv. Eng. Res. T. 11(4), 153–163 (1990)

[29] Nugent, D., Sovacool, B.: Assessing the lifecycle greenhouse gas emissions from solar pv and wind energy: A critical meta-survey. Energ. Policy 65, 229–244 (2014)

[30] Padey, P., Girard, R., le Boulch, D., Blanc, I.: From lcas to simplified models: A generic methodology applied to wind power electricity. Environ. Sci. Technol. 47(3), 1231–1238 (2013)

[31] Pesonen, H.L., et al.: Framework for scenario development in lca. Int. J. Life Cycle Ass. 5(1), 21–30 (2000)

[32] Raugei, M., Frankl, P.: Life cycle impacts and costs of photovoltaic systems: current state of the art and future outlooks. Energy 34(3), 392–399 (2009)

[33] Remund, J., Müller, S.C.: Trends in global radiation between 1950 and 2100. In: 10th EMS Annual Meeting, 10th European Conference on Applications of Meteorology (ECAM) Abstracts, held September, pp. 13–17 (2010)

[34] Rienecker, M.M., et al.: Merra: Nasa's modernera retrospective analysis for research and applications. J. Climate 24(14) (2011)

[35] Saltelli, A.: Making best use of model evaluations to compute sensitivity indices. Computer Physics Communications 145(2), 280–297 (2002)

[36] Scharmer, K., Greif, J.: The European solar radiation atlas, vol. 2. Presses des MINES (2000)

[37] Sobol', I.: Sensitivity Estimates for Nonlinear Models. Mathematical Modeling and Computational Experiment 1(4), 407–414 (1993)

[38] Stainforth, D.A., et al.: Uncertainty in predictions of the climate response to rising levels of greenhouse gases. Nature 433(7024), 403–406 (2005)

[39] Van Meijgaard, E., Van Ulft, L., Van de Berg, W., Bosveld, F., Van den Hurk, B., Lenderink, G., Siebesma, A.: The knmi regional atmospheric climate model racmo version 2.1. Tech. rep (2008)

[40] Weidema, B.P., Wesnæs, M.S.: Data quality management for life cycle inventories an example of using data quality indicators. Journal of Cleaner Production 4(3), 167–174 (1996)

A New Approach to Optimization with Life Cycle Assessment: Combining Optimization with Detailed Process Simulation

Richard J. Wallace[1], Antonino Marvuglia[2], Enrico Benetto[2], and Ligia Tiruta-Barna[3]

[1] Insight Centre for Data Analytics, University College Cork, Cork, Ireland
r.wallace@4c.ucc.ie
[2] Public Research Centre Henri Tudor, Resource Centre for Environmental Technologies (CRTE), Esch-sur-Alzette, Luxembourg
{antonino.marvuglia,enrico.benetto}@tudor.lu
[3] Université de Toulouse, INSA, UPS, INP, LISBP, Toulouse, France
lbarna@insa-toulouse.fr

Abstract. In this paper we describe an approach that combines system optimization under environmental constraints with a detailed simulation of the processes of a working plant. Specifically, a flow-sheet-based model of a potable water production plant was developed and coupled with a Python script which allowed variation of operational parameters of the plant (dissolved organic carbon (DOC), UV Absorbance at 254 nm (UVA), pH of coagulation unit, etc.) and automatic determination of the values of the objectives (e.g. quantity of sludge produced as a result of the settling phase, CO_2 emissions, climate change impact), which are not known in closed form. The Python script runs an optimization algorithm, seeking the global minimum of an objective function in the parameter space. The search procedure is based on the Nelder-Mead algorithm, which does not assume smooth functions.

The script was applied to a simplified model of a potable water production plant, which includes pumping, ozonation, powdered activated carbon (PAC) addition, recirculation, coagulation, settling, filtration and disinfection. Every unit is modelled with a set of equations describing input and output mass flows of the process, as well as the chemical reactions taking place within the unit. Therefore, while simplified, the model has all the complexity of a full-fledged simulation of the plant.

Systematic exploration of the objective surfaces shows that they are irregular and multimodal in all three cases tested, although there are systematic trends in relation to parameter values.

Keywords: Life Cycle Assessment, Potable water, Optimization, Environmental constraints, Nelder-Mead.

1 Introduction

Life cycle assessment (LCA) can be effectively used to analyse any industrial process in order to improve its environmental performance. In part because of its vast promise, this area of applied research has witnessed a surge of activity in recent years.

B. Murgante et al. (Eds.): ICCSA 2014, Part III, LNCS 8581, pp. 707–720, 2014.
© Springer International Publishing Switzerland 2014

One limitation of LCA in any of its basic forms is that it does not include a systematic way of generating process alternatives for environmental improvements. This limitation can be overcome by coupling LCA with optimization tools. In this framework, LCA can be employed to assess technological solutions from an environmental perspective, whereas optimization algorithms automatically seek the best ones according to the predefined criteria.

Up to the present time, the problem of optimization under constraints, where environmental constraints are included, has been handled using techniques from mathematical programming. However, these approaches have used only generic information about production processes [1, 2]. Under these conditions, the objective functions (for which we seek minimum or maximum values) have the form of overall costs. While providing desired optimal solutions, this general approach has two important limitations:

- It is never clear to what extent the assumptions underlying the optimal solution actually hold in a specific real-world situation to which it should be applied. In particular, since, as noted by Méry et al. (2013), "life cycle inventories are traditionally based on average data ... collected on site or estimated from literature or from modelling studies performed prior to the LCA study" [3, p. 1063], there is a large degree of uncertainty in regard to the intensities of either economic and environmental processes that occur in a given real-world situation.
- It is not clear that simply considering a global cost function will be directly pertinent to the actual decisions that need to be made in a specific real-world environment. In other words, optimization is only done at a 'surface' level with regard to the actual operations.

In the present work we develop an alternative approach in which a detailed simulation model of an actual real-world operation is combined with optimization techniques working on specific parameters within each unit process. A critical feature of this simulation is that it models the operation of an actual working plant. An important part of the modelling process is the comparison of the outputs of the model with the outputs of this plant. This means that the model has been 'ground-truthed' to a sufficient degree so that optimizing its operation can lead directly to policies and prescriptions for the actual plant in question.

In the original work on the simulation model, a model-based LCA was compared with results from a conventional LCA [3, 14]. The background processes were the same; hence, any differences had to be due to foreground processes alone. These tests showed that the conventional procedure gave errors that were as high as 25%. Hence, there are very good reasons to favour and explore further a model-based approach in dealing with optimization and LCA.

The simulation model was realized in the form of Python scripts running in the LCA software Umberto® v5.5[1]. A full description of the modelling system is given in [3-5]. Since Umberto® allows us to carry out a full-fledged life cycle assessment, this

---

[1] http://www.ifu.com/en

means that we are able to incorporate a complete LCA into the plant operation that we wish to optimize.

The flow-sheet-based model described in this paper simulates a specific plant for potable water production. At the same time, the approach that we outline here can be adapted to other types of production processes. In this regard, then, the approach is quite general.

The present work will focus on the main results of this novel methodology of optimization under environmental and operational constraints, showing its manifest advantages, technical barriers and the still unexplored potential for further improvements of the model.

2   Life Cycle Assessment

LCA methodology has been standardised by the International Organisation for Standardization (ISO) [6]. In the ISO framework, a life cycle runs from raw material acquisition through product disposal. A major benefit of the ISO standard is that it sets up a well-defined division of labour in regard to the various tasks involved in assessing environmental effects. Three stages of LCA are defined: goal and scope definition, life cycle inventory (LCI) analysis, and life cycle impact assessment (LCIA). All stages, in addition, involve "interpretation"; this is a shorthand way of referring to the continuous meta-assessment that is required in an actual application of LCA, and the series of adjustments or iterations that this entails.

Two widely used methods for performing LCI are the process flow diagram (or sequential) method and the matrix method [7]. The first method represents the original approach to modelling LCA. The processes involved in the production, consumption, and disposal of a product are represented as a flow chart, and for each step in the process the associated production of some environmental emission (e.g. CO_2) is determined.

Although the process flow model is straightforward and easy to understand, it is difficult to formalise with mathematical notation and it has some fundamental shortcomings [8]. For example, feedback loops (such as the use of fuel to produce electricity and the use of electricity to produce fuel) must be solved by interrupting the feedback loop after a specified number of iterations or at the point when the change in demand is less than some specified amount, replacing the process data by data that accounts for the feedback loop, or using an infinite geometric progression.

The matrix method for the solution of the LCI problem is based on a set of linear equations (called "balance equations"). The left sides of the equations, which represent economic activities and environmental effects, are divided into two parts, the technology matrix (A), containing materials and energy inflows and outflows for all the processes involved in the life cycle of the chosen functional unit[2], and the environmental intervention matrix (B), whose elements represent the amount of pollutants

---

[2] The *functional unit* is the reference amount of the function (product or service) delivered by the system that will be addressed by the study; it is the basis for comparisons of different systems.

or natural resources respectively emitted or consumed by each of the processes solicited in the lifecycle of the chosen functional unit. Correspondingly, the right sides of the equations are separated into two vectors, the external demand vector (f), which identifies the reference flow fulfilling the chosen functional unit, and the inventory table (g). The basic matrix equations are then:

$$As = f \tag{1}$$

and

$$Bs = g \tag{2}$$

The vector s, which makes the left and right sides equal, is called the scaling vector; critically, this same vector of scaling factors is used both to calculate demand and to calculate the vector of environmental burdens, g.

The strategy for solving this system of equations is to first solve for s in equation (1), according to the following equation,

$$s = A^{-1} f, \tag{3}$$

which involves inverting the matrix A. Then s can be used in equation (2) to obtain the vector of environmental burdens, g, associated with the life cycle of the investigated product. From this vector, various environmental impacts can be derived via standard equations which make use of so-called *characterization factors*, used to translate the emissions of hazardous substances and extractions of natural resources into impact category indicators.

It will be appreciated that LCA is essentially a tool for assessing externalities quantitatively. Its strong points are that it aims to provide an exhaustive inventory of these externalities and that it attempts to do this for the entire life cycle of a product. In addition, it allows one to combine the results into a small number of impacts, thus collapsing a myriad of environmental effects into a few global measures. This further step is important for intelligent decision making.

On the other hand, by itself LCA does not indicate what one should do to alter the various environmental burdens and impacts, or how one should go about minimising them. For this reason, in recent years researchers have turned to known techniques of optimization in order to close this gap. The main approaches to the problem will be presented in section 3.

3   Combining Optimization and Environmental Concerns: Existing Approaches

As a result of the considerations just mentioned, there has been a great deal of interest in recent years in methods for combining LCA and mathematical programming techniques [9]. As in the present work, the purpose is to provide optimal solutions that incorporate both environmental and economic concerns. Here, we give some examples in order to compare these approaches with our own.

A few general themes have already emerged from this work. Most models involve multi-objective programming (MOP), so that they produce a Pareto frontier of non-dominated solutions, i.e. solutions for which the objective cannot be improved in any dimension without making it worse in other dimensions (see [1,2] for examples). Aside from this, both linear and non-linear programming models have been proposed. In many cases, the MOP involves two objectives: the first is a summary of economic costs, the second is a summary of environmental impacts. For the latter, standard translations from environmental burdens (in the inventory table) to environmental impacts on impact categories such as climate change or human health are used.

A major area of application for these techniques is process systems engineering (reviewed in [9-11]). This work has also served to expand the scope of LCA from its original product-oriented perspective to include the design of processes [12]. In most cases nonlinear programming models have been used, which can be expanded to MOP models in order to handle both environmental and economic objectives. Although some early work considered reduction of specific pollutants as part of the overall objective (see [2] for discussion), only recently has LCA been incorporated into the optimization calculations in a thoroughgoing fashion.

Pioneering work combining mathematical programming techniques and life cycle assessment for process engineering was done in the 90's by Azapagic and Clift using linear programming models [1, 13, 14]. Although this work did not employ the matrix model explicitly, it can be expressed in these terms [7]. In their later papers the multi-objective character of their approach is made explicit.

In [13] these authors describe an LP model where an objective can be defined either in terms of economic goals or environmental burdens (inventory levels). In the latter case, the goal is to minimize an aggregate function over these burdens, i.e.

$$\min \sum_{i=1}^{n} b_{ji} x_i \tag{4}$$

"where $b_{j,i}$ is the burden j from process or activity x_j" (p. 307). This is carried out under various constraints such as balance constraints between different processes, economic demands, raw materials supply, and constraints on production capacity. In later work it is made clear that each environmental burden is associated with a different point on the Pareto frontier, and it is a subsequent multi-criteria decision making problem to select one of these non-dominated points.

An example of a nonlinear programming model is the work of Guillén-Gosálbez and collaborators on selecting an "optimal flowsheet structure" for chemical synthesis [2]. (Other similar studies can be found in [15-17].) These authors use a multi-objective mixed integer nonlinear model with the following form:

$$\min_{x,y} U(x,y) = \{f_1(x,y); f_2(x)\}$$
$$\text{s.t. } h(x,y) = 0 \tag{5}$$
$$g(x,y) \le 0$$

Here, f_1 represents economic costs and f_2 environmental impacts. $h(x,y)$ represents constraints on system performance (e.g. heat and mass balances); $g(x,y)$ represents design specifications. The x variables are continuous; y are binary variables. f_2 is a weighted sum of impact factors, which are themselves weighted sums of relevant environmental burdens.

It will be noted (and, typically, it would be considered too trivial to mention) that these and other existing approaches assume that a closed-form objective is available. This, in fact, reflects a fundamental trade-off between the accuracy of the model with regard to a specific case and the power of the inferences that can be made. In other words, strong inferences, including those leading to optimization in the usual form, require a considerable degree of idealization of the situation. In most or all of the current research, this trade-off is ignored. However, especially in the field of LCA, where important prescriptive recommendations derived from the assessment will be applied to specific situations, this trade-off is of critical importance. In contrast to previous work, the present research is designed to speak to this problem directly.

4   The EVALEAU Simulator

In the framework of the project EVALEAU (Integrated environmental and economic performance assessment of drinking water production processes) funded by the French national research agency (ANR), an integrated process simulation-LCA framework was developed by INSA Toulouse and CRP Henri Tudor, to simulate and assess potable water treatment chains [3, 4]. The same framework is potentially applicable and scalable to other industrial sectors. The EVALEAU tool has unique capabilities in terms of detailed modelling of water treatment unit operations coupled to state of the art LCA, and its merits have been acknowledged by both the LCA and water communities [3].

The tool was developed in Umberto® using the Python language for code scripting. A library of unit process modules was built. Each module is a detailed and highly parameterized model of a specific water treatment process, which is further linked with the software PHREEQC® for water chemistry calculation [18]. Input data are: water composition, design, operational parameters, including those drawn from the literature and from interaction with actual plant personnel. The modules are linked to the ecoinvent® LCI database [19] for background processes. By combining the modules, water treatment chains can be modelled and evaluated in Umberto® with a high level of detail.

An important feature of the EVALEAU project is that extensive tests were performed that compared various outputs of the simulations with corresponding outputs from the actual working plant that was being modelled [4]. In almost all cases, the results produced by a simulation were in close correspondence with measured outputs of the plant. The following table shows some examples.

Table 1. Comparisons between Values Measured at a Working Plant and Values Produced by the Simulation Model of that Plant (taken from Tables 4-1 and 4-2 of [4])

	average measure (imposed limit)	simulation	difference
water quality measure			
UVA	1.26 (< 1.5)	1.31	3.9%
pH	(6.5-8.5)	7	in range
TOC [mg/L]	(< 1.5)	1.2	in range
consumption			
electricity [kWh/m^3]	0.896	0.824	-7.99%
alum. sulphate [g/m^3]	62.9	62.8	0.17%
GAC [g/m^3]	6.00	6.59	9.80%

Notes. UVA is ultra-violet absorbance, TOC is total organic carbon, GAC is granular activated carbon.

The success of these tests is strong evidence that the process modelling is accurate, and, this improves the likelihood that an LCA carried out in conjunction with the simulation is also accurate for the specific plant being modelled. This means that we can use the process simulator to determine the best values for key parameters, i.e. those that give optimal values for the objective used. There are, however, issues that must be dealt with when using this approach; these are detailed in the next section.

5   Materials and Methods

The work discussed in this paper was done in the framework of the project MILES (Multi-Objective Optimization of Industrial Processes: Lifecycle Perspectives), a research mobility project funded by the national research fund of Luxembourg (FNR). In this work we used a simplified plant model (described in [4]), which nonetheless includes all the basic water treatment processes as well as background processes, and allows a full-scale LCA to be performed based on plant requirements and outputs. The major difference from models for actual drinking water production plants is that the latter include multiple sequences of processes. Oftentimes this is due to historical developments at those plants, in which new subsystems were added with different configurations of unit properties. Hence, this limitation is not a significant restriction of the functionality of the model.

A diagram of the model's foreground processes is shown in Fig. 1. Every unit process is modelled with a set of equations describing input and output mass flows of the process, as well as chemical reactions occurring within the unit. Thus, the model has all the complexity of a full-fledged simulation.

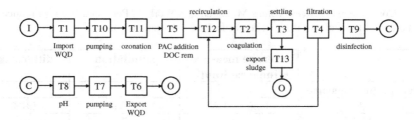

Fig. 1. Simplified model of a water treatment plant in the form of a process sheetflow. "I" indicates the starting process, "O" the final outputs, "C" the connection (path) between processes T9 and T8. Note that T1 and T6 are included in the model for bookkeeping purposes; they do not represent plant processes.

The plant model is coupled with optimization code through a top-level Python script, which still runs within the Umberto® model. This hybrid simulator-optimizer allows us to vary operational parameters of the plant and automatically determine the corresponding values of the objectives. In the present tests four parameters were included: dissolved organic carbon (DOC) used both in powdered activated carbon (PAC) addition (unit T5) and in the coagulation unit (T2 – note that here we consider these as separate parameters), UV Absorbance at 254 nm (UVA, in T11), and the pH of T2. The objectives were quantity of sludge produced as a result of the settling phase (T13), CO_2 generated, as reflected in the overall LCI, and global warming potential (GWP100) obtained from LCIA. Note that here we consider both environmental burdens and impacts simultaneously.

Because objective function values are obtained by running the simulator, the objective functions are not known in closed form. This reflects the basic trade-off described in the previous section. In fact, the mapping from parameter values to objective function values involves hundreds of chemical equations (solved by the software PHREEQC®, called by the simulator) in addition to hundreds of input-output relations in each process [3, 4]. Under these conditions, obviously, the objectives cannot be considered to be smooth (i.e. continuous) functions. Hence, the usual linear or non-linear mathematical programming approaches cannot be used directly. Instead, we rely on heuristic search techniques to explore the search space and the objective surface.

Currently, we are using a search procedure based on the Nelder-Mead method [20, 21], with different starting values derived from earlier systematic exploration (examples are given in the next section). The restarting strategy helps overcome the simple hill-climbing features of Nelder-Mead, which are not sufficient to avoid the problem of getting "stuck" in local, non-global minima.

The Nelder-Mead method uses a multi-point procedure, based on a set of points ordered by their associated objective values, to find a maximum or minimum of a complex function. The cardinality of the point set is one greater than the dimensionality of the parameter space; such as set forms a polytope called a *simplex*. As a consequence, this method is sometimes called the Simplex Method. However, because of the potential confusion with the well-known algorithm for solving linear programs that has the same name, this name is no longer commonly used.

Given an initial set of $n+1$ points, where n is the number of dimensions, the following procedure is performed repeatedly until either the range of objective values is within a certain bound or the number of repetitions of the procedure reached some pre-defined limit.

1. Calculate the centroid of the n best points, which is a point based on the mean of the values of the n points on each dimension. Then calculate a reference point, which is a weighted sum of the centroid and the worst point. (Specifically, the centroid is multiplied by $(1 + \mu_r)$, where μ_r is a small positive number, and the product of μ_r and point x_{N+1} is subtracted from the first product.)

2. Depending on the relation of the objective value for the reference point with the objective values for the existing points, choose a new point and replace the point x_{N+1} with this new point. Thus:

   a. If the objective for the reference point falls between those for points x_1 and x_N, replace point x_{N+1} with the reference point

   b. If the new objective is better than the objective for x_1, then calculate an "expansion point" based on a weighted sum of the centroid and x_{N+1} with a new weight μ_e, which is greater than μ_r (the expansion point). Then replace point x_{N+1} with the best of the reference and expansion points.

   c. If the new objective is worse than point x_N, then calculate a "contraction point" using similar methods, and replace x_{N+1} if the objective for the contraction point is better than that for x_{N+1}.

Some intuition for these various procedures can be obtained from the following example (Fig. 2), which is simplified from Figure 8.1 of [17].

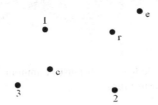

Fig. 2. Example of point locations during Nelder-Mead procedure in two dimensions

Here, there are two dimensions and, therefore three points. The μ-values are 1, 2 and $-1/2$ for μ_r, μ_e, and μ_c, respectively. The original points are labelled 1, 2, and 3. r is the reference point, e is the expansion point and c is the (inside) contraction point.

The present code differs in some respects from the standard Nelder-Mead procedure. The most important difference it that at present it does not resort to the "shrinking" step used when the (inside) contraction point gives an objective value that is worse than the current value for point x_{N+1}. This is because shrinking involves repeated calculation of the objective, which in the present case would take a large amount of time. For the present, the new point is saved as the new worst point, serving, therefore, as a kind of uphill move. In this case, no important information is lost, and accepting this new point may allow the search procedure to explore areas of the space that it would not otherwise enter.

6   Results and Discussion

Systematic exploration of the objective surface shows that it is irregular and multi-modal in all three cases, although there are systematic trends in relation to parameter values. The next figures show the surfaces obtained via a Lowess interpolation of the amount of sludge exported (Fig. 3) and overall $CO_{2\text{-eq}}$ production (Fig. 4).

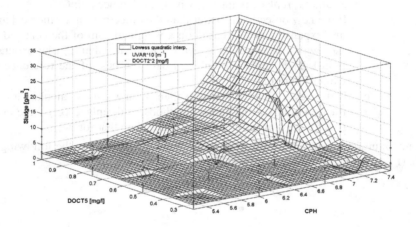

Fig. 3. Objective function for Sludge, which must be minimized, as a function of two of the four input parameters used in the test runs

These figures suggest that there are relatively few 'pockets' in the surface and that the true global optimum will be found in one of them. Of course, one must be careful not to forget that these surfaces are based on extrapolations from a sample of points. (On the other hand, this characteristic also seemed to be indicated by the raw data.) Hence, although one must be cautious at this time, there is reason to believe that this kind of survey strategy can serve to limit the search required in important ways as well as improving the chances of finding the true global optimum.

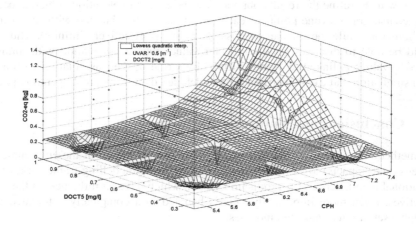

Fig. 4. Objective function for $CO_{2\text{-}eq}$, which must be minimized. Same two input parameters as in Fig. 3.

So far it has only been possible to complete a few preliminary runs with the heuristic search process. Figure 5 shows the best objective value found using all four parameters and beginning with different starting values. Each run consisted of 40 steps. These results show that there is a strong tendency for the procedure in its present form to rapidly enter a plateau; this shows that restarting from different points in the parameter space is definitely necessary with this procedure. One run did find one of the best objective values found to date, while the other two plateaued at higher values.

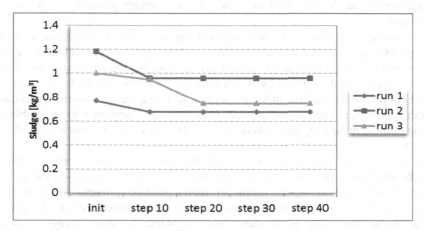

Fig. 5. Best objective value found at different stages of search on three separate runs of the Nelder-Mead procedure

Although at this point it is too early to determine how effective this procedure will be for finding optimal parameter settings, there is some indication that this approach will allow us to refine the results obtained by systematic exploration, which is necessarily coarse-grained (one point for each objective can be obtained with each run of the simulator, while the number of possible points is effectively infinite). Thus, we should be able to discover good parameter values whose association with minimal objective values is linked to "ground-truth" by virtue of their association with a detailed simulation that can be verified against detailed data from a real operating plant.

7   Conclusions and Future Work

The methodology we are using allows a flexible representation of the environmental burdens (and impacts) to be minimized and, in the same modelling framework, of the operational parameters of the production system which have an influence on the fixed objectives. Even more importantly, this is handled in a completely automated way through a set of interconnected modules.

Although the work to date is very preliminary, it serves to show that good values can be found for any given objective using a (relatively) simple heuristic search method. Since this was the basic goal of the project MILES, we can report that it has been achieved. Work is currently in progress to enhance the present methods, building on the results obtained so far.

In the future, we will consider more advanced methods of heuristic search, in particular particle swarm optimization [22]. Such methods, which can be used for many kind of optimization problem, may well be more suitable for finding or approaching true global minima than the more specialized methods from the field of global optimization used in the present work, which are often employed simply to find local minima [21]. At the same time, because these methods are computationally intensive and inherently parallel, we want to carry out tests running concurrently on multiple processors.

Still another question is whether simply doing more elaborate systematic studies will give us results as good as those obtained with heuristic search. Although we think that eventually heuristic methods should prove superior, regular systematic assessment gives us a good reference by which to judge the former methods. In addition, it should not be overlooked that in the work presented here we have used a combination of both approaches. This, in and of itself, is an interesting approach to the problem of finding optimal or near-optimal solutions in a situation where there is no closed-form objective.

We also want to explore the basis for the form of the objective surfaces obtained from the test runs of the simulator. In particular, we would like to answer the question of why certain combinations of parameter values produce 'pockets' in the objective surface, especially since this seems to be a general phenomenon. Along with this, we would like to know why at the same time the objectives seem always to show a coarse-grained regularity.

The present approach is quite general and could be adapted to other types of production processes. Input variables are all defined as parameters in Umberto®; hence, their values are attributes of Umberto®'s objects. Since the specific parameters to be used are read by the script from an input file, any parameters of any process (called a *transition* in Umberto®) can be substituted, and the code will handle them correctly. Currently, the script is being revised so that specific objectives can also be specified in an input file. Three types of objectives are of interest: outputs from specific unit processes (such as sludge from T13), outputs in the LCI, and outputs from LCIA. Hence, all that is required to generalize the code is to include the classification and specific identifying information in an input file, and the appropriate objects and values will be used by the Python script. Thus, it will be possible to run the same script with any process-model, and it will search for optimal solutions for any set of objectives based on any set of decision variables.

A final caveat is in order. In this paper we have emphasized "ground-truthing" our results and we have shown how this can be done with respect to the details of a plant operation. However, in carrying out the LCA, the simulator still uses generic data for the environmental processes themselves. So in this respect the problem of ground-truthing is still not fully resolved. Nonetheless, we are confident that the present work is a step in the right direction.

Acknowledgements. The authors thank Luxembourg's National Research Fund (FNR) for their financial support.

References

1. Azapagic, A., Clift, R.: Life Cycle Assessment and Multiobjective Optimisation. J. Clean. Prod. 7, 135–143 (1999)
2. Guillén-Gosálbez, G., Caballero, J.A., Jiménez, L.: Application of Life Cycle Assessment to the Structural Optimization of Process Flowsheets. Ind. Eng. Chem. Res. 47, 777–789 (2008)
3. Méry, Y., Tiruta-Bana, L., Benetto, E., Baudin, I.: An Integrated "Process Modelling-Life Cycle Assessment" Tool for the Assessment and Design of Water Treatment Processes. Int. J. Life Cycle Assess. 18, 1062–1070 (2013)
4. Méry, Y.: Development of an Integrated Tool for Process Modelling and Life Cycle Assessment - Ecodesign of Process Plants and Application to Drinking Water Treatment. PhD Thesis. Labora-toire d'Ingénierie des Systèmes Biologiques et deProcédés (LISBP), INSA Toulouse, France (2012)
5. Igos, E., Benetto, E., Baudin, I., Tiruta-Barna, L., Méry, Y., Arbault, D.: Cost-Performance Indicator for Comparative Environmental Assessment of Water Treatment Plants. Sci. Tot. Environ. 443, 367–374 (2013)
6. ISO 14040: Environmental Management – Life Cycle Assessment – Principles and Framework. International Organisation for Standardisation, Geneva (1998)
7. Suh, S., Huppes, G.: Methods for Life Cycle Inventory of a Product. J. Clean. Prod. 13, 687–697 (2005)
8. Heijungs, R., Suh, S.: The Computational Structure of Life Cycle Assessment. Kluwer Academic, Dordrecht (2002)
9. Grossman, I.E., Guillén-Gosálbez, G.: Scope for the Application of Mathematical Programming Techniques in the Synthesis and Planning of Sustainable Processes. Comput. Chem. Eng. 34, 1365–1376 (2010)

10. Jacquemin, L., Pontalier, P.-Y., Sablayrolles, C.: Life Cycle Assessment (LCA) Applied to the Process Industry: A Review. Int. J. Life Cycle Assess. 17, 1028–1041 (2012)
11. Pieragostini, C., Mussati, M.C., Aguirre, P.: On Process Optimization Considering LCA Methodology. J. Environ. Manag. 96, 43–54 (2012)
12. Kniel, G.E., Delmarco, K., Petrie, J.G.: Life Cycle Assessment Applied to Process Design: Environmental and Economic Analysis and Optimization of a Nitric Acid Plant. Environ. Prog. 15, 221–228 (1996)
13. Azapagic, A., Clift, R.: Linear Programming as a Tool in Life Cycle Assessment. Int. J. Life Cycle Assess. 3, 305–316 (1998)
14. Azapagic, A., Clift, R.: The Application of Life Cycle Assessment to Process Optimisation. Comput. Chem. Eng. 23, 1509–1526 (1999)
15. Eliceche, A.M., Corvalán, S.M., Martínez, P.: Environmental Life Cycle Impact as a Tool for Process Optimisation of a Utility Plant. Comput. Chem. Eng. 31, 648–656 (2007)
16. Gerber, L., Gassner, M., Maréchal, F.: Systematic Integration of LCA in Process Systems De-sign: Application to Combined Fuel and Electricity Production from Lignocelluosic Biomass. Comput. Chem. Eng. 35, 1265–1280 (2011)
17. Bernier, E., Maréchal, F., Samson, R.: Life Cycle Optimization of Energy-Intensive Processes Using Eco-Costs. Int. J. Life Cycle Asses. 18, 1747–1761 (2013)
18. Parkhurst, D.L., Appelo, C.A.J.: User's guide to PHREEQC — a computer program for speciation, reaction-path, advective-transport and inverse geochemical calculations. U.S.Geological Survey, Water Resources Investigations Report 95-4227, Lakewood, Colorado (1995)
19. Weidema, B.P., et al.: Overview and methodology. Data quality guideline for the ecoinvent database version 3. Ecoinvent Report 1 (v3), The ecoinvent centre, St. Gallen (2011)
20. Nelder, J.A., Mead, R.: A Simplex Method for Function Minimization. Comput. J. 7, 308–313 (1965)
21. Kelley, C.T.: Iterative Methods for Optimization. SIAM, Philadelphia (1999)
22. Englebrecht, A.P.: Fundamentals of Computational Swarm Intelligence. John Wiley & Sons, Chichester (2005)

The Role of Boundary Conditions in Water Quality Modeling

Vladimir J. Alarcon

Civil Engineering Department, Universidad Diego Portales, 441 Ejercito Ave., Third Floor
OOCC, Santiago, Chile
vladimir.alarcon@udp.cl

Abstract. Water quality parameters interact in nature in complex ways. An example of this complex interaction is the dependence of nutrient concentration on temperature, pH, dissolved oxygen, etc. This research presents a computational exploration of the dependence of several water quality variables on chlorophyll-a (chl-a) and phytoplankton. WASP water quality estimations for Nitrate (NO_3), Phosphate (PO_4), Ammonia (NH3), Dissolved Oxygen (DO), and biochemical oxygen demand (BOD) were produced and the results were analyzed to determine the dependence of those variables on several water quality boundary conditions. The analysis was performed for the period December 31, 1998, to April 30, 1999. The paper shows how crucial are the seaward boundary data (in particular those of chl-a and phytoplankton) on the kinetics of water quality constituents in a coastal estuary.

Keywords: Water quality modeling, WASP, nutrients, Nitrogen, Phosphorus.

1 Introduction

Several studies conclude that estuaries are strongly dependent (in terms of nutrient dynamics) from fresh water inputs, light, temperature and other factors. For example, [1] reports: "estuaries are among the most productive of marine environments, although food abundance does fluctuate greatly over space and time. The extraordinary productivity of estuaries is a product of the large amounts of nutrients that enter the estuary seasonally and of the extensive recycling of nutrients between the overlying water and the biologically active sediments". Likewise, [2] states: "fresh water drainage seasonally delivers large amounts of nutrients in dissolved and particulate form; dissolved nutrients also enter estuaries with the deep-water oceanic flow. These nutrient spikes may be the source of phytoplankton blooms in small estuaries. Infrequent storms also deliver sediment and particulate organic matter to estuaries. The microbial community in the sediments decomposes the particulate organic matter and permits the continuous recycling of nutrients between the bottom and overlying water". But, estuaries are inherently variable [3]. The National Research Council [3] reports that "it is difficult to make generalizations about nutrients in estuaries. Strong geographic variation exists in the extent of nutrient export due to variation in estuary size and morphology, variation in river input, regional variation in groundwater and precipitation inputs, and seasonal variation in nutrient cycling. The seasonal cycle of primary

B. Murgante et al. (Eds.): ICCSA 2014, Part III, LNCS 8581, pp. 721–733, 2014.
© Springer International Publishing Switzerland 2014

productivity depends upon temperature, light, and nutrient levels which are in turn affected by various interacting factors, e.g., climatic regimes, basin size and morphology, river input, etc. These factors will determine estuarine variability in both space and time."

Estuarine nutrient distribution can best be described in terms of seasonal changes in river hydrology. Nutrient behavior can be explained by two dominant factors: freshwater flushing time and biological uptake and regeneration. Superimposed on these two processes is seasonal variability in nutrient concentrations of coastal waters caused by upwelling (a process in which cold nutrient-rich water rises to the surface from the ocean depths). Freshwater flushing time determines the amount of time for the uptake of nutrients and phytoplankton, exchange with suspended particles, and interaction with the sediments. Seasonal coastal upwelling controls the time and extent of oceanic delivery of nutrients to the estuary, which, in turn, affects the magnitude and timing of the biological response within the estuary.

During winter months, fast flushing times prevent the development of significant phytoplankton populations within the estuary. Other estuarine studies have concluded that light is the dominant factor influencing phytoplankton productivity. Phytoplankton cannot significantly affect estuarine nutrient distributions during winter months even under acceptable light and high nutrient conditions. Any phytoplankton occurring within the estuary is rapidly flushed out under high flow conditions, preventing phytoplankton uptake from significantly affecting nutrient distributions in the estuary. This pattern has been observed in other estuarine systems under high flow conditions. While biological uptake would not affect nutrient concentrations during winter months, re-mineralization of organic carbon would be important for some nutrients. High NH_4 and PO_4 levels are produced during re-mineralization of organic carbon within an estuary.

During spring and summer, because river discharge rates decrease progressively from winter through summer, longer residence times occur. Under reduced flushing rates of spring and summer, the effect of biological uptake becomes evident within the estuary. Biological uptake within the estuary affects estuarine nutrient distributions.

This research presents a computational exploration on the influence of chlorophyll-a (chl-a) boundary condition in the simulation of nutrients for the estuary of St. Louis (southern USA). The exploration seeks to illustrate how critical are the establishment of reliable boundary conditions for water quality simulation of nutrient-rich estuaries.

2   Methods

2.1   Study Area

The estuary of St. Louis (Figure 1) is located in the Northern Gulf of Mexico (coastal Mississippi, USA). It is a shallow estuary (average depth is around 1.5 m) characterized by having diurnal tides (average tidal range 0.48 m), short flushing rate (residence time less than one week), and well-mixed waters [4]. It receives freshwater inputs from two main rivers: Jourdan River and Wolf River. The watersheds that these rivers drain are mostly covered by agricultural and forest lands, however there are a number of surrounding towns that impact the water quality of the estuary waters (industrial and domestic effluents).

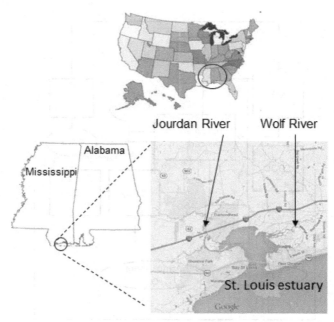

Fig. 1. Study area. Estuary of St. Luis and tributary rivers (Jourdan and Wolf rivers).

2.2   Water Quality Modeling

The water quality modeling software used in this research was the Water Quality Simulation Program-WASP [5]. This is a 3-dimensional, structured (curvilinear), dynamic water quality model based on the basic law of mass conservation. WASP solves Equation 1 for dissolved or particulate water contaminants [5].

$$\frac{\partial (V_i C_i)}{\partial t} = \sum_{j=1}^{6} \left(Q_j C_j + A_j D_j \frac{\partial C_j}{\partial x_j} \right) + \sum S_i \text{ , (i=1, ..., n).} \tag{1}$$

At a particular cell i, C_i is the concentration, V_i is the volume, A_j: area of the cell face j (j=1, ..., 6) associated with cell i. Q_j is volumetric flow rate across cell face j into i-th cell, D_j is the dispersion coefficient at cell face j, S_i are external loads and kinetic sources and sinks into i-th compartment, δx_j: distance between two neighboring cells with common interface j, δt is the time step. Figure 2 schematizes the interaction among the nutrient-related variables as modeled within WASP (EUTRO subroutine). WASP's EUTRO takes into account four major processes: phytoplankton kinetics, the phosphorus cycle, the nitrogen cycle, and the dissolved oxygen balance.

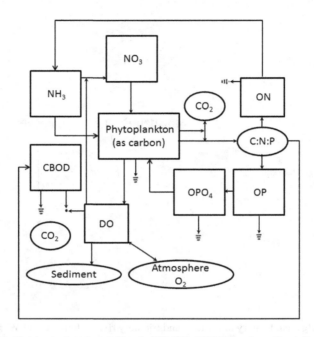

Fig. 2. Nutrient-related variables in WASP's EUTRO subroutine (after [5])

2.3   Estuary Hydrodynamics and Hydrology of Surrounding Water Bodies

This research used existing models for hydrology and hydrodynamics developed by Kieffer [6] and Hashim [7], respectively. [7] developed a hydrodynamic model for the St. Louis estuary using the Environmental and Fluid Dynamics code, EFDC [8]. EFDC is a three dimensional finite-difference code that solves the shallow-water hydrodynamics equations and it is extensively used in the USA for estuary hydrodynamics modeling. Hashim's model [7] used a structured and curvilinear mesh, with fresh water boundary conditions (from Jourdan and Wolf rivers) established with inputs from [6] watershed hydrology models for Jourdan river and Wolf river watersheds. Kieffer [6] used the Hydrological Simulation Program Fortran, HSPF [9].

2.4   Nutrient Simulation

The computational exploration was performed using the WASP code (described in 2.2). WASP has the ability of reading hydrodynamic simulation results from EFDC, using the same computational mesh with slight differences. The generation of the model application is transparent for the user provided that the EFDC output is read correctly by WASP. Figure 3 shows the resulting WASP mesh including labels of special locations within the grid (boundary conditions, sampling stations, cells where WASP output was explored, etc.).

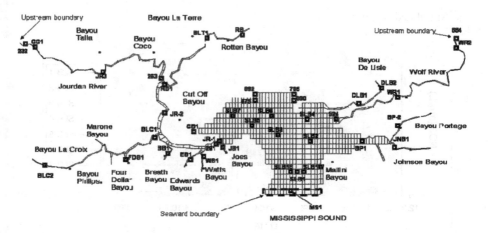

Fig. 3. WASP mesh showing boundary conditions and sampling locations

The estuary model was partially calibrated for: ammonia (NH_3), Nitrate (NO_3), Phosphate (PO_4), Dissolved Oxygen (DO), and biochemical oxygen demand (BOD). The lack of continuous measured data did not allow producing a fully calibrated and validated model. For this reason, this research explored trends of the simulated water quality variables with the purpose of identifying potential weaknesses and strengths of the estuary model.

Nutrient and other water quality variables estimations were explored at the following locations (see Figure 3): SLB1 (near the seaward boundary), SLB2 and SLB3 (middle of the estuary), SLB4 (near the confluence of Wolf River), SLB5 (near the confluence of Jourdan river), SLB6 and SLB7 (near the coast). The distribution of points also obeyed to the existence of measured water quality data (albeit scarce). WASP water quality estimations for Nitrate (NO_3), Phosphate (PO_4), Ammonia (NH_3), Dissolved Oxygen (DO), and biochemical oxygen demand (BOD) were produced and the results were analyzed. Results were produced for the period December 31, 1998, to April 30, 1999.

3   Results

3.1   Nitrate Simulated Output (NO_3)

Simulated output for Nitrate (Figure 4) is clearly dominated by the boundary conditions imposed at Wolf River. The simulated data do not show a recognizable seasonal trend. The NO$_3$ spike of 3.2 mg/L in the loading to Wolf River (on 3/14/99) produces a corresponding spike of around 2 mg/L in SLB2 and SLB4 (the nearest stations to Wolf). This NO$_3$ spike affects the simulation of most constituents included in the Saint Louis Bay model. NO$_3$ from the downstream boundary (0.05 mg/L) have a strong effect only on station SLB1.

a) Simulated results

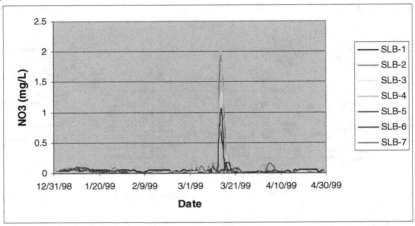

b) Boundary conditions

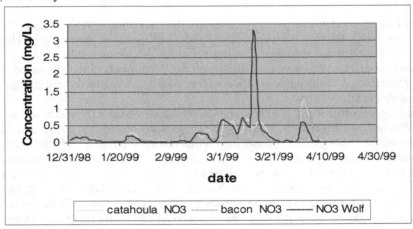

Fig. 4. Nitrate simulation

3.2   Phosphate Simulated Output (PO$_4$)

PO$_4$ values (Figure 5) at the boundaries exert some effects on the simulation results. Wolf upstream boundaries have mild effects on SLB2 and SLb4 results, and Jourdan upstream boundaries affect results on SLB5, SLB6 and SLB7. Since the WASP conceptual model of PO_4 is a function of phytoplankton phosphorus (death, growth and mineralization), phytoplankton concentrations drive the PO_4 simulation. Section 3.4 shows the trend of phytoplankton concentration through the simulation period. The similarity between the phytoplankton trend and the PO_4 trend is evident.

a) Simulated results

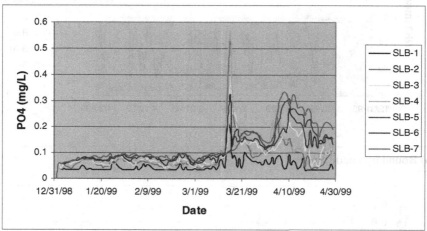

b) Boundary conditions

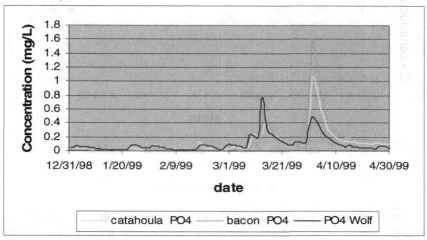

Fig. 5. Phosphate simulation

3.3   Ammonia Simulated Output (NH₃)

a) Simulated results

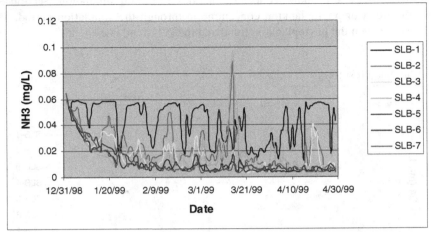

b) Boundary conditions

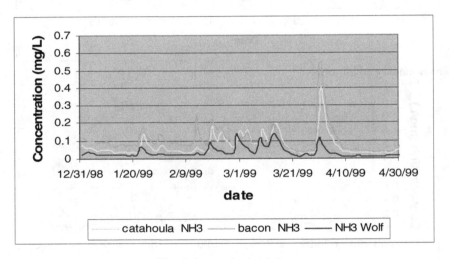

Fig. 6. Ammonia simulation

NH₃ results (Figure 6) decrease from the initial concentration values (0.065 mg/L) at the beginning of the simulation to values lower than 0.01 mg/L by the end of the simulation period. Interestingly, NH₃ boundary conditions imposed at the upstream boundaries of Wolf and Jourdan, produce a mild trend once that initial conditions are overcome, but, the most noticeable effect is the NH₃ spike that occurs at the same date of the NO₃ spike (see section 3.1). Since in the WASP formulation, the calculation of NH₃ does not directly depend on NO₃, and, on the other hand, it has a direct relationship with phytoplankton concentrations, the NH3 trend is dependent at some extent on phytoplankton. The downstream boundary value of 0.06 mg/L seems to be higher than the values trend of the stations inside the bay. This value should be reviewed.

3.4   Dissolved Oxygen Simulated Output (DO)

a) Simulated results

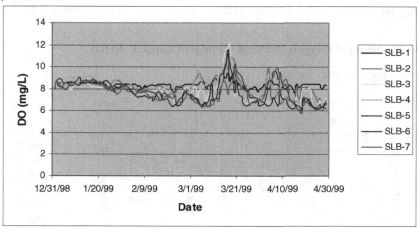

b) Boundary conditions

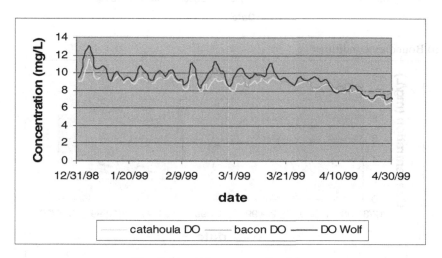

Fig. 7. Dissolved oxygen simulation

DO simulated concentrations (Figure 7) decrease from the initial concentration value of 7.92 mg/L until an average value of 6.5 mg/L by the end of the simulation. This trend is produced (again) by the trends and values specified at the upstream boundaries. The station near the downstream boundary (SLB1) is strongly affected by the DO boundary value imposed at the mouth of the estuary: 8.5 mg/L, throughout the simulation. DO simulated values, although follow the trend of the upstream boundaries, have a noticeable spike around 3/15/1999 (date where the NO_3 spike occurs) and another around 3/30/1999. These spikes don't correspond to any DO boundary (upstream or downstream). Nor it corresponds to low BOD concentrations as seen below. But they do correlate with high phytoplankton concentrations (see section 3.4).

3.5   Biochemical Oxygen Demand Simulated Output (BOD)

a) Simulated results

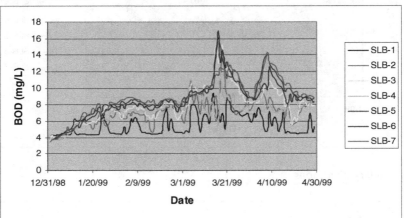

b) Boundary conditions

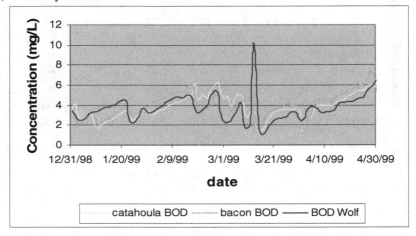

Fig. 8. BOD simulation

BOD simulated values (Figure 8) gradually increase from the initial concentration value of 2 mg/L until values higher than 8 mg/L for all stations but SLB1. The downstream boundary condition of 4 mg/L has a strong effect in SLB1 simulated concentrations and milder in those corresponding to SLB2. There are two noticeable peaks in the simulated results for SLB7, SLB6, SLB5 and SLB3. Both peaks are higher than any boundary value specified in the upstream cells of Jourdan and Wolf. Again, the general trend of the BOD curve is greatly influenced by the phytoplankton concentration during the simulation (see section 3.4).

3.6   The Role of Phytoplankton

Phytoplankton kinetics assumes a central role in eutrophication, affecting all other systems [10]. This is particularly true in the case of the estuary of St. Louis. In WASP, phytoplankton (as carbon) is calculated from chlorophyll-a values stipulated by the user. Initial and boundary chlorophyll-a values interact with environmental conditions (temperature, light, etc.) and nutrient concentration (NO_3, PO_4, NH_3), during simulation, to provide estimates of growth, death and settling of phytoplankton. The phytoplankton concentration, calculated in this manner, have effects in turn in the phosphorus and nitrogen cycles and DO-BOD calculations. Hence, it is important to include in any analysis of simulated results, an analysis of the kinetics of phytoplankton.

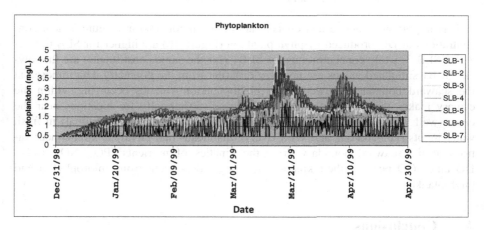

Fig. 9. Phytoplankton simulation

Figure 9 explains the two unexpected DO peaks. During the second half of March 1999 phytoplankton concentrations reach maximum levels. During the same period of time, NH_3 concentrations are lower than 0.04 mg/L and NO_3 concentrations are also low. This combination of low NH_3 and low NO_3 plus the large production of phytoplankton make the model to generate an additional source of DO. A byproduct of photosynthetic carbon fixation is the production of dissolved oxygen [10]. Figure 10 shows the amounts of DO produced by phytoplankton through the simulation period.

It is clear, in the figure below, that phytoplankton production of DO is intensive through the month of March and reaches maximum values in mid and late march. Figure above also explains the peaks of BOD. Phytoplankton death provides organic carbon which can be oxidized. In WASP, a first order kinetic reaction recycles phytoplankton carbon to BOD, constituting the BOD source that produces the trend shown in Figure 10.

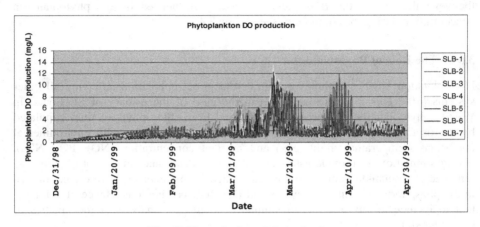

Fig. 10. Phytoplankton DO production

It is important to notice that concentrations of phytoplankton (Figure 9) and concentrations of DO produced by phytoplankton (Figure 10) are higher for SLB 6, SLB 7 and SLB 3, while concentrations for SLB1 (for the same constituents) are the lowest. SLB1 simulated values are not very different from the boundary data imposed at the seaward boundary. However, the phytoplankton kinetics occurring within the estuary makes those concentrations to increase substantially for locations near the coast (well within the estuary). Since phytoplankton is absent (or minimal) in fresh water inputs (i.e., Jourdan and Wolf rivers) this means that phytoplankton concentrations at the seaward boundary drive the kinetics of nutrients (PO_4, NO_3, NH_3), DO and BOD because their kinetics is strongly dependent from chlorophyll-a and phytoplankton.

4   Conclusions

Phytoplankton concentrations (i.e., chlorophyll-a) have an important effect in DO, BOD and PO_4 and NH_3 simulation results. The values of chl-a specified as boundary and initial conditions should be adjusted such that there are no excessive phytoplankton-produced DO and BOD during the month of March.

NO_3 is clearly dominated by boundary conditions more than any other constituent. Downstream boundary conditions for all constituents should be modified to values that range within the range of measured and simulated concentrations in the nearby.

This will avoid inconsistent jumps in the values calculated for stations near the mouth of the estuary.

Diurnal changes in DO should be interpreted in terms of all the variables involved in the calculation if a correct trend is wanted.

Since PO_4, NH_3 and BOD are strongly dependent on phytoplankton and NO_3 concentrations they will continue to respond to variations on those variables. For that reason, concentration of those constituents is not expected to have a seasonal trend.

Since phytoplankton is absent (or minimal) in fresh water boundaries, phytoplankton concentrations (i.e., chlorophyll-a concentrations since phytoplankton concentrations are calculated using chl-a values) at the seaward boundary drive the kinetics of nutrients (PO_4, NO_3, NH_3), DO and BOD because their kinetics is strongly dependent from chlorophyll-a and phytoplankton.

References

1. Yokel, B.J.: Estuarine water quality. The Conservation Fund. Study No 3. Washington DC (1975)
2. O'Connor, D.J., Thomann, R.V., DiToro, D.M.: Dynamic water quality forecasting and management. USEPA Report 660/3-73-009. Washington DC (1973)
3. National Research Council, Clean coastal waters. National Academy Press, Washington DC (2000)
4. Mojzis, A.K., Redalje, D.G.: Bacterioplankton abundances in the Saint Louis Bay. In: Northern Gulf Institute 2010 Annual Conference, May 18-20. Mobile, Alabama (2010), http://www.northerngulfinstitute.org/outreach/conferences/2010/posters/24_Mojzis_06-USM-03_Poster.pdf
5. Wool, T.A., Ambrose, R.B., Martin, J.L., Comer, E.A.: Water Quality Analysis Simulation Program, WASP (2013), http://sdi.odu.edu/mbin/wasp/win/wasp6_manual.pdf
6. Kieffer, J.: Development of a nutrient and dissolved oxygen water quality model for the Saint Louis Bay watershed. Dissertation. Mississippi State University (2002)
7. Hashim, N.B.: Watershed, hydrodynamic, and water quality models for total maximum daily load St. Louis Bay watershed Mississippi. Dissertation. Mississippi State University (2001)
8. Hamrick, J.M.: A three-dimensional environmental fluid dynamics computer code: Theoretical and computational aspects. Special Report 317 in Applied Marine Science and Ocean Engineering. The College of William and Mary, Virginia Institute of Marine Science, 63 p. (1992)
9. Environmental Protection Agency, EPA. The HSPF BMP toolkit (2013), http://www.epa.gov/athens/research/modeling/HSPFWebTools/
10. Wool, T.A., Davie, S.R., Rodriguez, H.: Development of 3-D hydrodynamic and WQ models to support total maximum daily load decision process for the Neuse River estuary. J. of Wat. Res. Plan. and Mgmt., ASCE 129(4), 295–306 (2003)

A Sequential Data Preprocessing Tool for Data Mining

Zailani Abdullah[1], Tutut Herawan[2,3], Haruna Chiroma[2], and Mustafa Mat Deris[4]

[1] School of Informatics & Applied Mathematics
Universiti Malaysia Terengganu
Gong Badak, Kuala Terengganu, Malaysia
[2] Faculty of Computer Science & Information Technology
University of Malaya
50603 Pantai Valley, Kuala Lumpur, Malaysia
[3] AMCS Research Center, Yogyakarta, Indonesia
[4] Faculty of Science Computer and Information Technology
Universiti Tun Hussein Onn Malaysia
86400 Parit Raja, Batu Pahat, Johor, Malaysia
zailania@umt.edu.my, tutut@um.edu.my, haruna.chiroma@ieee.org,
mmustafa@uthm.edu.my

Abstract. Sequential dataset is a collection of records written and read in sequential order. Information from the sequential dataset is very useful in understanding the sequential patterns and finally making an appropriate decision. However, generating of sequential dataset from log file is quite complicated and difficult. Therefore, in this study we proposed a sequential preprocessing model (SPM) and sequential preprocessing tool (SPT) as an attempt to generate the sequential dataset. The result shows that SPT can be used in generating the sequential dataset. We evaluated the performance of the developed model against the log activities captured from UMT's e-Learning System called myLearn. With the minimum modification of the dataset, it can be used by other data mining tool for further sequential patterns analysis.

Keywords: Sequential dataset; Data preprocessing; Data mining; Tool.

1   Introduction

Pattern or association rules mining is one of the important topics in data mining. Until this recently, it has been actively and widely studied [1-8] in various domain applications. Besides that, sequential pattern mining is a bit different since it involves with discrete or order of events. Sequential pattern can be defined as a consecutive or non-consecutive ordered sub-set of an events sequence [9]. A sequential pattern is an important task with broad number of applications [10]. In the last 10 years, there have been many studies on sequential pattern and its applications. Although the mining of the complete set of sequential patterns has been improved substantially, in many cases, sequential pattern still faces tough challenges in both effectiveness and efficiency. On the one hand, there could be a large number of sequential patterns in a large database. A user is often interested in only a small subset of such patterns. Presenting the complete set of sequential patterns may make the mining result hard to understand and hard to use [11]. Examples of the domain applications of sequential

B. Murgante et al. (Eds.): ICCSA 2014, Part III, LNCS 8581, pp. 734–746, 2014.
© Springer International Publishing Switzerland 2014

pattern are medicine, telecommunications, World Wide Web [12], etc. In the educational contexts, the events sequence are commonly refer to individual or group of students' actions logged by the e-Learning System. Dataset is a collection of related sets of information that is composed of separate elements but it can be manipulated as a unit [13]. In term of sequential dataset, it correspond to a collection of records that written and read in sequential order from the beginning until the end. Until this moment, educational data mining has been specifically focused on extracting the patterns in analyzing and understanding the student behavior [14]. Indeed, data mining has been considered as an integrated and functional part of e-Learning System and thus requires some adequate techniques to effectively retrieve and analyze the data. Dataset can be derived from several data sources including log file. Log file is a file that contains a list of events that goes in and out of a particular server. Its main point at keeping track of what is happening at the server. Most of the log file is saved in flat file format but for certain application systems (eg. e-Learning System) it can be also configured to channel into database system. In educational context, the information from log file is very important because it can detect regularity and deviations in groups of students. Moreover, it also can provide more information to educators about the learners' behavior, and give recommendations on how to go about with some deviation cases [15]. In order to produce a valid dataset, data processing becomes a necessary component. Data mining considered the data preprocessing as the most important technique to transform the raw data into an understandable format for further processing. There are several steps taken place in data preprocessing such as data cleaning, data integration, data transformation, data reduction and data discretization. This technique is very important because it helps in solving the problems of incomplete, noisy, inconsistent, etc. of data. In fact, data preprocessing of web log file plays an important role in web usage mining before producing the complete set of sequential patterns [16]. Educational Data Mining (EDM) refers to techniques, tools, and research designed for automatically extracting the meaning from data repositories based on the learning activities in educational settings. E-Learning is among the popular sources of data repositories in EDM. Generally, e-Learning System is referred to as Learning Management System (LMS), Course Management System (CMS), Learning Content Management System (LCMS), Managed Learning Environment (MLE), Learning Support System (LSS), Web Based Training System (WBT-System) [17]. These systems collect a lot of information that can be further processed in analyzing the students and educators' behavior. E-Learning systems offer the facilitation of communication between students and educators, sharing resources, producing content material, preparing assignments, conducting online tests [18]. From the literature, generation of sequential dataset from log file is a very interesting topic because it can be used as an input for further analysis by data mining tools. However, when dealing with the format of discrete events, the data preprocessing task becomes more complicated and difficult. Therefore in order to mitigate these problems, we proposed a Sequential Preprocessing Model (SPM) and Sequential Preprocessing Tool (SPT) with the application of e-Learning System.

The contributions of this paper are as follows. First, we do comprehensively studies about the preprocessing techniques for sequential patterns. Second, we proposed Preprocessing Model (SPM) and developed Sequential Preprocessing Tool

(SPT). Third, we evaluated the performance of the developed model against the log activities captured from UMT's e-Learning System called myLearn.

The reminder of this paper is organized as follows. Section 2 describes the related work. Section 3 describes some difinition and the proposed model. This is followed by the results and discussion in Section 4. Finally, conclusions of this work are reported in section 5.

2   Related Works

Data preprocessing is one of the important step in the data mining process. As a result, many works have been devoted to preprocess data in log file. However, only few researches have been developed in term of techniques, tools, algorithms etc. focuses on preprocessing of log file from the student log activities. Wahab *et al.* [19] elaborated the pre-processing techniques involved in extracting IIS Web Server Logs before it can be applied into data mining algorithms. Data preprocessing acts as a filter and only appropriate information from the log file will be extracted. In their experiment, the raw log files were collected from *Portal Pendidikan Utusan* Malaysia or popularly known as Tutor.com. Castellano *et al.* [20] proposed LODAP tool (log data preprocessor) to perform preprocessing of log file. LODAP takes input log file related to website and generate output based on a database that containing pages visited by user and identified user sessions. LODAP tool can reduce the size of web log file and group all the web requests into a number of user sessions. Salama *et al.* [21] introduced a new approach in preprocessing of web log files for web intrusion detection. The steps in this process are not similar and have several differences as compared to the web usage mining. The main differences between them are in the context of log files combination, user identification, session identification and after preprocessing. In this approach, two algorithms are employed to combine the log files in W3C format and NCSA format into a single file in XML format. XML file will become an input to the mining algorithms rather than the relational database. Yan Li [22] suggested path completion algorithm and an implementation of data preprocessing for web usage mining. Referer-based method is employed to append the missing pages in user access paths. The reference length of pages is modified according to the estimated average reference length of auxiliary pages. The algorithm appends the lost information and thus improves the reliability of access data for further calculations in web usage mining. Patil *et al.* [23] focused on the earlier two parts of the preprocessing which are field extraction and data cleaning. Two algorithms are specifically designed in order to clean the raw web log files and finally to insert them into a relational database. The field extraction algorithm extracts the web logs that collect the data from the web server. The data cleaning algorithm plays the function to clean the web logs and removes the redundant information. Zhang *et al.* [24] proposed a new hybrid algorithm to perform data preprocessing in web log mining based on Hadoop cluster framework. Hadoop is a distributed system infrastructure developed by Apache Foundation. It is a software platform which can run and analyze large-scale data. The experimental results show that the improved data pretreatment algorithm can improve the efficiency of web data mining. Valsamidis *et al.* [25] introduced a methodology for analyzing LMS courses and students' activity. It consists of three main steps which are logging step, preprocessing and clustering step. The first step focuses on logging of specific data from e-learning platform by considering only the fields of courseID, sessionID and Uniform Resource Locater (URL) using Apache module. The second step filters the recorded data provided by detecting the outliers and removing extreme values. The third step applies a clustering method namely Markove Clustering algorithm (MCL) to separate users into different groups according to the usage patterns. Blagojevic [26] proposed Data Mining Extensions (DMX) queries for mining student data from e-Learning System. It main aims at understanding the students' behavior more closely and plan their classes to maximize the students efficiency. There are three phases involved namely data selection, preprocessing and OnLine Analytical Processing (OLAP). The first phase is to retrieved data from the Moodle server. The second phase focuses on removing the entries that contain errors. The third phase emphasizes on generating cube and dimensions. After completing these phases, DMX query is employed to perform the analysis. Romero *et al.* [27] attempts to find that, in most e-Learning systems, all the pages accessed by students are saved in log files (either one log file for each student or just one big log file for everyone) which contain all the information about the interaction of the students with the system. Therefore, after preprocessing this information, it is possible to discover sequential patterns from these log files by using data mining algorithm. Sequential mining was first proposed by Agrawal and Srikant [17]. They have designed Apriori-based algorithm called Generalized Sequential Patterns GSP to mine all sequential patterns based on minimum support threshold. Many methods which are based on the Aprriori have been proposed to mine sequential patterns. Han [28] introduced Frequent pattern-projected Sequential pattern mining (FreeSpan) method by integrating the mining of frequent sequences and using projected sequence databases. Besides mining the complete set of patterns, FreeSpan also reduced the candidate sebsequence generation and thus outperform Apriori-based GSP algorithm by Srikant and Agrawal [29]. Pei *et al.* [30] suggested Web access pattern tree (WAP-tree) structure and WAP-Mine algorithm to efficiently mine of access patterns from Web logs. Zaki [31] proposed Sequential PAttern Discovery using Equivalence classes (SPADE) algorithm for fast discovery of Sequential Patterns. SPADE utilizes combinatorial properties to decompose the original problem into smaller sub-problems using lattice search techniques and simple join operations. Pei *et al.* [32] proposed Prefix-projected Sequential pattern mining (PrefixSpan) method to efficiently mining sequential pattern by introducing ordered growth and in the same time reducing the projected database. In most cases, PrefixSpan outperforms the apriori-based algorithm GSP, FreeSpan, and SPADE due the less memory space consumption. Shie *et al.* [33] suggested UMSP to mine high utility mobile sequential patterns. It searches for all patterns from MTS-Tree structure. Ahmed *et al.* [34] proposed UWAS-tree (utility-based web access sequence tree), and IUWAS-tree (incremental UWAS tree), for mining web access sequences in static and dynamic databases respectively. Extensive performance analyses show that our approach is very efficient for both static and incremental mining of high utility web access sequences. Yin [35] introduced USpan with lexicographic quantitative sequence to extract high utility sequence and designed concatenation mechanisms with two effective pruning strategies. USpan can defiantly identify high utility sequences in large-scale data with low minimum utility.

3   Proposed Method

There are four major components involved in generating the sequential dataset from mySQL database. All components are interrelated and the process flow is moving in one-way direction. The dataset produced in this model is in a format of flat file. A complete overview model of sequential processing model (SPM) is shown in Figure 1.

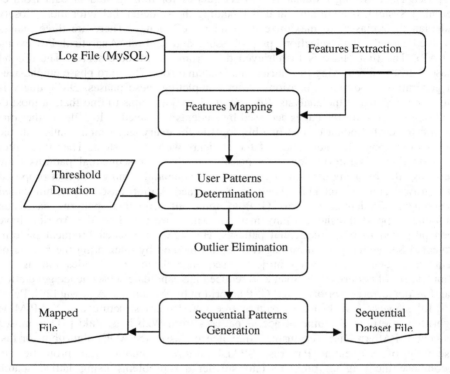

Fig. 1. The Overview of Sequential Preprocessing Model (SPM)

Log File

All the students' activities in the myLearn (e-Learning System at Universiti Malaysia Terengganu) are automatically stored in MySQL database rather than the typical text server log files. The log activities contain a great amount of information that reflects the student's learning process and academic performance. In the context of myLearn, among the crucial source of information are based on the login timestamp and requested pages. Timestamp is a sequence of characters that identifying the current time of an event occurred.

Features Extraction

SQL statement has been executed to extract only the relevant information from the MySQL database. There are about 204 tables in myLearn and the main twelve tables are mdl_config, mdl_course, mdl_course_categories, mdl_course_modules, mdl_log, mdl_course_sections, mdl_log_display, mdl_modules, mdl_user, mdl_user_admins, mdl_user_students, mdl_user_teachers. These tables are called as the meta-tables. The most important table is mdl_log which contains the attributes of id, time, userid, course, module, action, url and info. The desired attributes are obtained by merging and joining the different tables using SQL syntax. The most important attributes employed are id, userId, time and action.

Features Mapping

The unique data from attribute Action is extracted and assigned with the unique number. The attribute id, userId, time and action is transferred into a temporary table and the attribute action is mapped with the previous assigned unique number. All series of actions (replaced by the numbers) from the same userId are sorted according to the time (timestamp). The actions with the same date are grouped together and assigned. This process will continue until the last userId. Table 1 presents the mapping number of all actions.

User Patterns Formation

The sequences of actions (patterns) for each userId are generated based on the factors of date, minimum and maximum durations. For each userId, a set of sequences actions are formed and written in a single line. It means that, the single line of transaction represents the set of sequences actions that have been performed by the student in one semester for one subject. Each pattern will comply with certain threshold durations. After generating of the complete set of patterns per userId, the single transaction will be appended beta version of sequential datasets.

Outlier Elimination

Typically, the log file may contain many redundant activities (actions) within very short period of times. Unfortunately, these activities may occur many times in the same pattern. Another issue is, in one pattern it may contains only an activity. These are among the outliers that will be removed from the beta version of sequential datasets. The outlier determination and elimination process are very important because they can change the result of data analysis in the future.

Sequential Patterns Generation

The final sequential dataset (start with #data in the first line) and mapped actions file (start with #map in the first line) are produced at this phase. This dataset is generated once all the outliers are completely removed from the previous beta version of dataset. The mapped actions file contains the mapped actions in term of the associate number and the correspond action. This file can be used to interpret the actual action value during data analysis.

Sequential Dataset

The final sequential dataset and mapped actions file are ready to be used. Since not many data mining tools are available to process these inputs, the flexible tool will be developed to perform the sequential analysis in the near future based on this data format. The complete SPM algorithm is shown in Figure 2.

```
1:    Input:  Logfile, DurMin, DurMax, CourseId
2:    Output: SequentialPatterns, MappingActions
3:    for all Logfile do
4:        uAction <- ExtractUnique(Action)
5:        NumAction <- Convert2Number (uAction)
6:        Generate MappingActions
7:    end for loop
8:    foreach Action
9:        NumAllAction <- Convert2NumberAll (Action)
10:       Update Logfile
11:   end foreach loop
12:   for all Logfile do
13:       if (DurMin < Timestmp < DurMax) then
14:           DupSeqNumAction <- GenerateSeqAction(NumAllAction)
15:           SeqNumAction <- RemoveOutlier(DupSeqNumAction)
16:           AscSeqNumAction <- SortAscending(SeqNumAction)
17:           SequentialPatterns <- Append(AscSeqNumAction)
18:       end if
19:   end for loop
```

Fig. 2. SPM Algorithm

4   Results and Discussion

4.1   Experimental Setup

The experiment has been conducted on Intel® Core™ i5-3210M CPU at 2.50GHz speed with 4.00GB RAM and running on Window 7 Home Premium. SPT have been developed using Java as a programming language and NetBean IDE 7.3 with JDK 1.7 as a platform.

4.2   Dataset Description

In order to evaluate the performance of the proposed model and the tool, all log activities that related to CSC3102 course in myLearn (e-Learning System of Universiti Malaysia Terengganu) has been employed. About 17,785 records have been created of all activities from CSC3102 course between the periods of 01-Jan-2013 until 30-Apr-2013 (in one semester). The extracted data contains 61 users, 24 actions that represent the students' learning activities within 3 months and 5 minutes as sequence of duration in each pattern set. Fig. 3 shows some learning activities that have been performed by students. The main interface for the SPT is shown in Fig. 4.

☑	✎ Edit	⅔ꞓ Copy	⊝ Delete	1	1	resource view	2013-04-22 11:31:46
☑	✎ Edit	⅔ꞓ Copy	⊝ Delete	2	1	resource view	2013-04-22 11:31:44
☑	✎ Edit	⅔ꞓ Copy	⊝ Delete	3	2	course view	2013-04-22 11:30:40
☑	✎ Edit	⅔ꞓ Copy	⊝ Delete	4	1	resource view	2013-04-22 11:27:19
☑	✎ Edit	⅔ꞓ Copy	⊝ Delete	5	1	resource view	2013-04-22 11:27:13
☑	✎ Edit	⅔ꞓ Copy	⊝ Delete	6	1	resource view	2013-04-22 11:26:48
☑	✎ Edit	⅔ꞓ Copy	⊝ Delete	7	1	resource view	2013-04-22 11:26:44
☑	✎ Edit	⅔ꞓ Copy	⊝ Delete	8	1	course view	2013-04-22 11:25:04
☑	✎ Edit	⅔ꞓ Copy	⊝ Delete	9	3	course view	2013-04-22 10:45:26
☑	✎ Edit	⅔ꞓ Copy	⊝ Delete	10	3	quiz view	2013-04-22 10:45:24
☑	✎ Edit	⅔ꞓ Copy	⊝ Delete	11	3	course view	2013-04-22 10:45:17
☑	✎ Edit	⅔ꞓ Copy	⊝ Delete	12	3	quiz view	2013-04-22 10:44:55

Fig. 3. Students' Learning Activities for CSC3102 Course

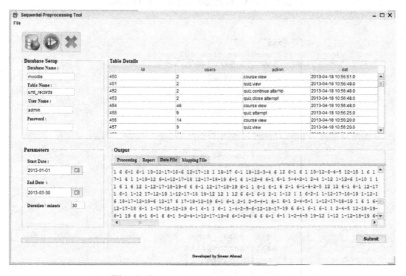

Fig. 4. Sequential Preprocessing Tool

As mentioned earlier, two flat files can be generated using the tool. Fig. 5 and Fig. 6 show sequential dataset file and mapped actions file, respectively. The first file contains 61 transactions and 24 actions to represent the activities from 61 students with the parameters of 5 minutes as duration in each pattern. From the generated mapping file, the total number of actions (students' activities) is 24.

```
#data
6-3-2-1 2-3-7 2-7-3 6-5 2-3-5 1-2 3-2 6-3 1-2 2-3 10-13-14-2 1-2-14-13 3-2 2-3-6-7 1-2 10-14-13-2 9-2 10-14-13-2 7-3-2 3-6 1-2 5-7-3
1-2 1-2 1-2 1-2 1-12 1-2 7-3-2 3-2-5-6 2-5-3 1-2 1-2 7-3-2 3-5-6 1-2 13-14-10 13-14-10 14-10-13 1-2 1-2 2-1 10-13-14-2 14-10 7-3-2
1-23 8-3-2 1-2 1-2 7-3 3-2-8 2-3-5-6 5-3-2-7 8-5-6 1-2 2-3 7-3 3-6-5 1-2 1-2-11-10 14-10-13-2 3-10-14 3-5-7 3-5-7-2 1-2-11-10 10-13-
1-2-3 1-2 2-3 6-3-5 2-3-5-6-7 1-2 1-2 7-3-1 5-1 9-19-18 2-10 10-13-14 2-11-10-14-13 2-3 3-2 1-2 2-3 1-2 1-3 3-5-7-2 2-6 1-2 14-13-10
1-2 7-3 7-2-8-5-6-3 2-3-6 1-2 2-8-3-6 1-2 2-11-10 10-14-13-2 3-2-11-14-10-13 10-14-13 13-10-14 11-10-2 2-8-6-3 2-12-15-16-17 1-2-8
1-2 1-2 2-3-1 7-3-1 3-5-6 2-1 1-2 7-3-1-2 6-5 1-2 1-2 1-2 1-2 10-13-14-2 10-13-14 14-10 14-10-2 14-10 10-14 7-2 1-2 2-3-5-6 13-14-1
1-2 1-2 7-3-1 2-3-5-6 5-6-3 1-2 3-5-7-2 10-13-14-2 14-10-2 1-2-14-10 2-10 6-7-3-2 2-3-5 2-10-14-13 1-2 1-2-13-14-10 10-14-13 7-3-2
1-2 3-1 2-3-6 3-6 1-2 3-2 2-3-5-6-7 1-2 13-14-10-2 14-10-13-2 1-2 1-2 7-3-2 2-5-6 3-2 10-2-11-14-13 1-2 2-1 1-2 1-2 13-14-10 10-14
1-2 7-3 5-3-2-6 1-2 14-10-13-2 1-2 2-5-6-7-3-1 14-13-10-2 10-14 14-10 10-14-13-2 3-6-7-2 1-2 1-2 1-2 10-14 1-2 5-7-3 2-3-6-5 1-2 1-
1-2 2-5 1-2 3-5-6-7-2 3-5-6-7-2 6-3-2 1-2 1-2 1-2 10-11-2 10-13-14 10-14-13 10-2 14-13-10-2 1-2 2-3-5-6-7 14-10-13-2 14-10-13 1-2
7-3-2 2-5-6-3-1 7-3-2 3-5-6 1-2 3-2 13-14-10-2 13-14-10 10-14 11-10-13-14 1-2 1-2 1-2 2-3-5-7 1-2 14-10-13-2 1-2 13-14-10 6-5-7 3-
1-2 1-2 7-3 2-3-5-6 3-2 1-2 1-2 1-2 10-2 10-11-13-14 10-11-14-13 14-10-2-11-14-10 2-3 2-9 1-2 3-2 1-2 7-3 5-6 2-3-1 14-10-13-2
1-2 1-2 7-3-1-2 7-3 1-2 1-2 2-1 10-2-13-14 14-10-2 14-10-13-2 1-3-2 1-2 3-1 10-2-5-6 13-10-14-2 10-14-13 10-13-14-2 2-11-10-14 18
1-2-3-6 3-2 5-3-7-2 7-3-2 1-2 1-2 10-2 1-2 1-2-10-13-14 10-13-14-2 10-14 13-10 10-13-14 10-14-13-2 2-10-14 2-6-5-7-3 10-2 10-2 1-
2-1 1-2 3-1-2 1-2 2-3-5-6-1 5-6-7-3-2 1-2 3-1-2 13-14-10-2 14-10-13-14 14-10-13 1-2 1-3-2 6-7-3-2 2-5 11-2-10-14-13 10-2 2-1 14-10-13-
1-2 1-2 7-3 2-6-3 1-2 2-1 1-2-7-10-13-14 10-14 5-3-2-10-14 10-2 14-10-13 14-10-2 2-3-5-7 10-13-14-2 1-2 1-2 14-10-13-2 3
1-2 2-1 3-6-2 3-7-2 1-2 1-2 10-2 14-10-13 1-2 7-3-2 2-3-6-5 10-2 14-10-13-2 14-10-13 1-2 10-13-14 10-2 7-3-2 3-6 1-2 1-2 10-14-2 1(
2-1 1-2 7-3 2-3-5-6 5-6-7-3-2 1-2 1-2 1-2 13-14-10 13-14-10 14-10 1-2 1-2 7-3-2-1 2-3-6-5 13-14-10-2 14-10 1-2 1-2 1
5-6-3 1-2 1-7 5-6-3 1-2 1-2 7-3-2 3-5 10-14-13-2 14-10-13 1-2 13-14-10-2 14-10 10-14-13 1-2 1-2 2-3 1-2 13-14-10-2 7-3-2 3-5 3-2 3-
1-2 1-2 1-2 1-2-7 7-3-5-6 3-5-6 1-2 1-2 10-2 10-2-13-14 10-2 2-10-14-13 10-2 1-2 1-2 10-2 14-10-13 2-14-10-13 2-3-5-6 1-2 14-10-13
1-2 7-3-2 7-3-2-5 3-6 1-19-2 1-2 3-2-5-7 2-14-10-13 1-2 1-2 1-2 10-2-8-6-5-7-3 1-14-10-13 1-2 1-2 2-12-20 2-14-10-13 7-3 3-6 1-2 1-2
1-2 1-2 1-2-7-3 7-3-2-6 2-3 1-2 3-2 2-8-5-6-3 2-10 2-13-14-10 13-14-10 1-2 13-14-10 14-10-13 1-2 1-2 2-10 10-13-14 2-3-5-6-7 1-2 1-
1-2 3-8-1 2-3-8 2-1 7-3-1 5-6-2 2-3 3-2-8 3-2 3-1 2-3 2-3 1-2 2-3-8 2-1 2-3 1-2 3-2-10 3-2 7-3-2 6-2 5-1 3-2-7 3-5-6 1(
2-1 3-2 2-3-5-6 7-3 5-6-3 1-2 1-2 10-14-13-2 10-13-14-2 10-2 14-10 2-3 1-2 1-2 10-14-13-2 3-1-2-10 5-7 1-2 14-10-13 14-10-13 7-3-1
3-5-6 2-6-3 7-3 5-3 1-23 1-3-2 1-2 7-5-6-3 1-2 10-14-2 1-3-2 2-10-13 14-10 10-14-13 1-2 6-7-3-2 2-3-5 1-2 13-14-10-2 10-14 13-14-
1-2 1-2 3-1 7-3-2 3-5-6 2-3 1-2 14-13-10 2-3-1 1-2-3 10-13-14 10-2 14-10 14-10 13-14-10 1-2 1-2 1-2-3 1-2-3 14-10-13-2 11-14-10 2-
1-2 7-3-1 7-3-5 6-3 1-3-2 10-2 10-13-14-2 10-13-14 1-14-10-13 3-2 1-2 2-10-14 10-2-13-14 3-2 3-2 3-10 3-2 1-2 10-13-14-2 1-2
1-2-23 7-3-1-2 7-3 3-5-6 1-2 1-2 1-10-2 14-10-13-2 3-2 2-3 3-2 1-3-2 1-2 10-14-13-2 10-13-14 7-3-2 10-14-13-2-5-6 1-2 10-13-14 3-2
1-2 7-3-2 7-3-5 2-5-6-3 3-5-6 14-10-13-2 7-3-2-11-10-14 1-2 1-2 2-14-10-13 7-3-2 6-3 1-2 10-14-13-2 3-5-6 2-3 7-3-2 2-3-6 1-2 1-2 1-
1-2 7-3-2 3-2-5-6 5-6 1-2-3 1-2 2-1 1-10-9-2 3-5 1-2 10-2 10-13-14 10-14 14-13-10 1-2 1-3-10-2 1-2-6-3-5-7 1-2 10-14-13-2 1-2 1-2 1(
7-3-2 3-5-6-1-2 7-3-2 2-5-6-3 1-2 1-2 3-2 3-2-5 3-5-2 7-5-6 2-3-5-6 2-3 10-2 1-2 10-2 10-13-14 1-2 3-1-2-10-13-14 2-3-5-6-7 1-2 14-10
```

Fig. 5. The Generated Sequential Dataset

```
#map
11  resource view all
2   assignment view
19  quiz attempt
7   user view
20  course unenrol
17  quiz continue attemp
12  quiz view
22  quiz view all
6   resource view
13  user update
14  course user report
1   course view
16  blog view
9   user view all
3   assignment view all
24  course editsection
18  quiz close attempt
4   assignment upload
5   upload upload
8   course recent
21  course enrol
23  forum view forum
10  forum view discussion
15  forum user report
```

Fig. 6. Mapping File of Generated Dataset

The total number of patterns and unique patterns are 2,386 and 334, respectively. The longest combination of actions in a pattern is 7 and the shortest one is 1. Meanwhile, the maximum, minimum and average numbers of patterns in a line of transaction are 140, 1 and 39, respectively. Table 1 presents top ten patterns sorted in occurrence descending order. In sequential dataset, the same pattern might appear many times within the same transaction and it is quite different with the typical dataset.

Table 1. The Occurrence and Density of Top Ten Patterns

Patterns	Occurrence	Density (%)
1-2	814	33
10-2	112	5
7-3-2	85	3
10-14	75	3
3-2	63	3
2-3	53	2
10-14-2	52	2
2-1	44	2
7-3	37	1
10-13-14	34	1

The most occurrences is pattern "1-2" which represent the combination of course view and assignment view. It indicates that many students will start with browsing the course view first and then followed by assignment view. They might be interested on finding the latest assignment should be accomplished and the due date. The second highest occurrence is pattern 10-2, which depicts the combination of forum view discussion and assignment view. During surfing the forum, they might be realized that that the new assignments are now ready and need to submit within certain dateline. The third occurrence of pattern is 7-3-2 in which it represent the combination of user view, assignment view all and assignment view. From these top three rules, most of the student still goes to page 2 (assignment view). In fact, assignment views represent 70% from the selected top ten patterns. The main reason is, the latest news of assignment will be updated at this page and if the students miss up, he or she might lose several marks and received warning from the respective lecturer.

5   Conclusion

A sequential dataset contains data items that are stored in consecutively manner [36-37,40]. This dataset can be used by data mining tools to find the complete set of subsequences patterns [38-41]. However, generating of sequential dataset especially from the log activities is quite tedious. Therefore in this study we proposed sequential preprocessing model (SPM) and sequential preprocessing tool (SPT) to generate sequential dataset from UMT e-Learning system (MyLearn) log data. The results show that SPT can be used to generate the sequential dataset from the log data in mySQL database format. Moreover, it is also expected that the dataset can be used by other data mining tools for sequential pattern analysis by a very minimum dataset modification.

Acknowledgement. This work is supported by University of Malaya High Impact Research Grant no vote UM.C/625/HIR/MOHE/SC/13/2 from Ministry of Education Malaysia.

References

1. Abdullah, Z., Herawan, T., Deris, M.M.: Detecting Definite Least Association Rules in Medical Databases. In: Herawan, T., Deris, M.M., Abawajy, J. (eds.) Proceedings of the First International Conference on Advanced Data and Information Engineering (DaEng-2013). LNEE, vol. 285, pp. 127–134. Springer, Singapore (2014)
2. Abdullah, Z., Herawan, T., Deris, M.M.: Mining Indirect Least Association Rule. In: Herawan, T., Deris, M.M., Abawajy, J. (eds.) Proceedings of the First International Conference on Advanced Data and Information Engineering (DaEng-2013). LNEE, vol. 285, pp. 159–166. Springer, Singapore (2014)
3. Abdullah, Z., Herawan, T., Deris, M.M.: Discovering Interesting Association Rules from Student Admission Dataset. In: Herawan, T., Deris, M.M., Abawajy, J. (eds.) Proceedings of the First International Conference on Advanced Data and Information Engineering (DaEng-2013). LNEE, vol. 285, pp. 135–142. Springer, Singapore (2014)
4. Herawan, T., Vitasari, P., Abdullah, Z.: Mining critical least association rules of student suffering language and social anxieties. International Journal of Continuing Engineering Education and Life-Long Learning 23(2), 128–146 (2013)
5. Abdullah, Z., Herawan, T., Deris, M.M.: Tracing significant association rules using critical least association rules model. International Journal of Innovative Computing and Applications 5(1), 3–17 (2013)
6. Herawan, T., Noraziah, A., Abdullah, Z., Deris, M.M., Abawajy, J.H.: IPMA: Indirect patterns mining algorithm. In: Nguyen, N.T., Trawiński, B., Katarzyniak, R., Jo, G.-S. (eds.) Adv. Methods for Comput. Collective Intelligence. SCI, vol. 457, pp. 187–196. Springer, Heidelberg (2013)
7. Abdullah, Z., Herawan, T., Deris, M.M.: Mining Highly-Correlated of Least Association Rules using Scalable Trie-based Algorithm. Journal of Chinese Institute of Engineers 35(5), 547–554 (2012)
8. Herawan, T., Vitasari, P., Abdullah, Z.: Mining interesting association rules on student suffering study anxieties using SLP-Growth algorithm. IGI-Global - International Journal of Knowledge and Systems Science 3(2), 24–41 (2012)
9. Martinez-Maldonado, R., Yacef, K., Kay, J., Kharrufa, A., Al-Qaraghuli, A.: Analysing frequent sequential patterns of collaborative learning activity around an interactive tabletop. In: 4th International Conference on Educational Data Mining (EDM 2011), pp. 111–120 (2011)
10. Agrawal, R., Srikant, R.: Mining Sequential Patterns. In: Proceedings of the 11th International Conference on Data Engineering (ICDE), pp. 3–14. IEEE Computer Society (1995)
11. Pei, J., Han, J., Wang, W.: Constraint-based Sequential Pattern Mining: the Pattern-Growth Methods. Journal of Intelligence and Information System 28(2), 133–160 (2007)
12. Minos, G., Hill, M., Rastogi, R., Shim, K.: Mining sequential patterns with regular expression constraints. IEEE Transactions on Knowledge and Data Engineering 14(3), 530–552
13. Kettner, J., Ebbers, M., O'Brien, W., Ogden, B.: Introduction to the New Mainframe: z/OS Basics. IBM Redbooks, NY (2011)

14. Bharadwaj, B.K., Pal, S.: Mining Educational Data to Analyze Students Performance. International Journal of Computer Science and Information Security (IJACSA) 6(2), 63–69 (2011)
15. Cocea, M., Weibelzahl, S.: Eliciting Motivation Knowledge from Log Files Towards Motivation Diagnosis for Adaptive Systems. In: Conati, C., McCoy, K., Paliouras, G. (eds.) UM 2007. LNCS (LNAI), vol. 4511, pp. 197–206. Springer, Heidelberg (2007)
16. Masseglia, F., Tanasa, D., Trousse, B.: Web Usage Mining: Sequential Pattern Extraction with a Very Low Support. In: Yu, J.X., Lin, X., Lu, H., Zhang, Y. (eds.) APWeb 2004. LNCS, vol. 3007, pp. 513–522. Springer, Heidelberg (2004)
17. Romero, C., Ventura, S., García, E.: Data mining in course management systems: Moodle case study and tutorial. Computers & Education 51(1), 368–384 (2008)
18. Lile, A.: Analyzing E-Learning Systems Using Educational Data Mining Techniques. Mediterranean Journal of Social Sciences 2(3), 403–419 (2011)
19. Wahab, M.H.A., Mohd, M.N., Hanafi, H.F., Mohsin, M.F.M.: Data Pre-processing on Web Server Logs for Generalized. Proceedings of World Academic of Science, Engineering and Technology 26, 970–977 (2008)
20. Castellano, G., Fanelli, A.M., Torsello, M.A.: Log Data Preparation for Mining Web Usage Patterns. In: IADIS International Conference Applied Computing, pp. 371–378 (2007)
21. Salama, S.E., Marie, M.I., El-Fangary, L.M., Helmy, Y.K.: Web Server Logs for Preprocessing for Web Intrusion Detection. Computer and Information Science 4(4), 123–133 (2011); Canadian Center of Science and Education
22. Li, Y., Feng, B., Mao, Q.: Research on Path Completion Technique in Web Usage Mining. In: IEEE International Symposium on Computer Science and Computational Technology, pp. 554–559 (2008)
23. Patil, P., Patil, U.: Preprocessing of web server log file for web mining. World Journal of Science and Technology 2(3), 14–18 (2012)
24. Zhang, G., Zhang, M.: The Algorithm of Data Preprocessing in Web Log Mining Based on Cloud Computing. In: Proc. of International Conference on Information Technology and Management Science, pp. 468–474 (2012)
25. Valsamidis, S., Kontogiannis, S., Kazanidis, I., Theodosiou, T., Karakos, A.: A Clustering Methodology of Web Log Data for Learning Management Systems. Educational Technology & Society 15(2), 154–167 (2012)
26. Marija Blagojevic, M., Micic, Z.: Contribution to the Creation Of DMX Queries in Mining Student Data. Int. J. Emerg. Sci. 2(3), 334–344 (2012)
27. Romero, C., Porras, A., Ventura, S., Hervas, C., Zafra, A.: Using Sequential Pattern Mining for Links Recommendation in Adaptive Hypermedia Educational Systems. In: International Conference Current Developments in Technology-Assisted Educations, pp. 1015–1020 (2006)
28. Han, J., Pei, J., Mortazavi-Asl, B., Chen, Q., Dayal, U., Hsu, M.-C.: FreeSpan: Frequent Pattern-Projected Sequential Pattern Mining. In: Proc. 2000 ACM SIGKDD Int'l Conf. Knowledge Discovery in Databases (KDD 2000), pp. 355–359 (2000)
29. Agrawal, R., Srikant, R.: Mining Sequential Patterns: Generalizations and Performance Improvements. In: Proceedings of the Fifth Int. Conference on Extending Database Technology, pp. 3–17. Avignon, France (1996)
30. Pei, J., Han, J., Mortazavi-Asl, B., Zhu, H.: Mining Access Patterns Efficiently from Web Logs. In: Proceedings of the 4th Pacific-Asia Conference on Knowledge Discovery and Data Mining (PADKK 2000), Current Issues and New Applications, pp. 396–407 (2000)

31. Zaki, M.: SPADE: An Efficient Algorithm for Mining Frequent Sequences. Machine Learning 40, 31–60 (2001)
32. Pei, J., Han, J., Mortazavi-Asl, W.J., Pinto, H., Chen, Q., Dayal, U., Hsu, M.-C.: Mining Sequential Patterns by Pattern-Growth: The PrefixSpan Approach. IEEE Transactions on Knowledge and Data Engineering 16(10), 1–17 (2004)
33. Shie, B.-E., Hsiao, H.-F., Tseng, V.S., Yu, P.S.: Mining high utility mobile sequential patterns in mobile commerce environments. In: Yu, J.X., Kim, M.H., Unland, R. (eds.) DASFAA 2011, Part I. LNCS, vol. 6587, pp. 224–238. Springer, Heidelberg (2011)
34. Ahmed, C.F., Tanbeer, S.K., Jeong, B.S.: Mining High Utility Web Access Sequences in Dynamic Web Log Data. In: Proceeding of: 11th ACIS International Conference on Software Engineering, Artificial Intelligences, Networking and Parallel/Distributed Computing, SNPD 2010, pp. 76–81 (2010)
35. Yin, J., Zheng, Z., Cao, L.: USpan: An Efficient Algorithm for Mining High Utility Sequential Patterns. In: Proceedings of the 18th ACM SIGKDD International Conference on Knowledge Discovery and Data Mining, KDD 2012, pp. 660–668 (2012)
36. Kalia, H., Dehuri, S., Ghosh, A.: A Survey on Fuzzy Association Rule Mining. International Journal of Data Warehousing and Mining 9(1), 1–27 (2013)
37. Priya, R.V., Vadivel, A.: User Behaviour Pattern Mining from Weblog. International Journal of Data Warehousing and Mining 8(2), 1–22 (2012)
38. Taniar, D., Goh, J.: On Mining Movement Pattern from Mobile Users. International Journal of Distributed Sensor Networks 3(1), 69–86 (2007)
39. Daly, O., Taniar, D.: Exception Rules Mining Based on Negative Association Rules. In: Laganá, A., Gavrilova, M.L., Kumar, V., Mun, Y., Tan, C.J.K., Gervasi, O. (eds.) ICCSA 2004. LNCS, vol. 3046, pp. 543–552. Springer, Heidelberg (2004)
40. Taniar, D., Rahayu, W., Lee, V.C.S., Daly, O.: Exception rules in association rule mining. Applied Mathematics and Computation 205(2), 735–750 (2008)
41. Ashrafi, M.Z., Taniar, D., Smith, K.A.: Redundant association rules reduction techniques. International Journal of Business Intelligence and Data Mining 2(1), 29–63 (2007)

Delivering User Stories for Implementing Logical Software Architectures by Multiple Scrum Teams[*]

Nuno Costa[1], Nuno Santos[2], Nuno Ferreira[3], and Ricardo J. Machado[1]

[1] Centro ALGORITMI, Escola de Engenharia, Universidade do Minho, Guimarães, Portugal
nunoaacosta@gmail.com, rmac@dsi.uminho.pt
[2] CCG - Centro de Computação Gráfica, Campus de Azurém, Guimarães, Portugal
nuno.santos@ccg.pt
[3] I2S Informática, Sistemas e Serviços S.A., Porto, Portugal
nuno.ferreira@i2s.pt

Abstract. In software projects, agile methodologies are based in small development cycles and in continuous communication with customers with low needs on modeling formalism for requirements elicitation and documentation. However, there are projects whose context requires formal modeling and documentation of requirements in order to raise and manage critical issues from the very beginning of the project, like architectural diagrams. This work presents an approach for deriving a list of User Stories using a logical architectural diagram as input. Derived User Stories are then delivered to multiple Scrum teams.

Keywords: Software requirements; Scrum; User Stories; Logical architectures.

1 Introduction

At the end of a requirements elicitation phase (considering traditional development processes like, for instance, the Royce's waterfall model [1], that includes Analysis, Design, Implementation, Test and Deployment phases in its lifetime), it is expectable that all information can be proper perceived by the implementation teams. Each software development methodology has its own set of artifacts for gathering requirements to be passed on to implementation teams. Amongst several artifacts, an architectural diagram is commonly used. The quality of the architecture is critically important for successful software development, since it is the key framework for all technical decisions [2]. It has a profound influence on project organizations' functioning and structure. Poor architecture can reflect poorly defined organizations with inherent project inefficiencies, poor communication, and poor decision-making. However, implementation teams may experience difficulties in understanding key information from requirements artifacts, including the architecture (namely, understanding the "who", "what" and the "why"). Even when the system requirements are (semi-)automatically derived from user requirements, like the V+V process [3], the gathered information

---

[*] This work has been supported by FCT – Fundação para a Ciência e Tecnologia in the scope of the project: PEst-OE/EEI/UI0319/2014.

B. Murgante et al. (Eds.): ICCSA 2014, Part III, LNCS 8581, pp. 747–762, 2014.
© Springer International Publishing Switzerland 2014

typically it is not yet ready to be consumed and passed on to the implementation teams. Within the V+V process, the requirements are expressed through logical architectural models and stereotyped sequence diagrams in both a process- and a product-level perspective. This approach is composed by two V-Models [4], one for each perspective, following a "Vee" shaped process. Firstly, the process-level V-Model is executed and the models (regarding a process-level perspective) are derived in succession. These models are then used as input for the first model derivation of the product-level V-Model, and then the models (regarding a product-level perspective) are derived in succession. After executing the product-level V-Model, the most important artifact is a software logical architectural diagram.

The need for Scrum implementation teams to receive the requirements information imposed by a logical architecture of large software projects resulted in an effort to develop an agile process that would easily allow mapping all these elements in User Stories. A user story is a customer-centric perspective high-level characterization of a requirement, containing only the information needed for the project developers are able to see clearly what is required to implement, enabling them to estimate its complexity your effort required to implement easily [5]. The technique described in this paper has, as a main concern, the generation of User Stories to be incorporated in the Backlog artifact during the team's own agile software development process. Additionally, the own project's context demands to share that information by multiple development teams, hence dividing the logical architecture in modules. The own project's context also arose the need for managing distributed scrum teams [6].

This paper is structured as follows: in section 2 is presented a literature review on main issues relating the problem in discussion; in section 3 is presented the V+V process that is used in requirements elicitation to derive a software logical architecture and other gathered artifacts; in section 4 is described the technique for User Story derivation; in section 5 is presented a demonstration case using the proposed technique; in section 6 is presented some suggestion of adaptation of roles, events and artifacts regarding the multiple Scrum teams; and in section 7 is presented the conclusions and future work.

2   Multiple Scrum Teams for Implementing Logical Software Architectures

The waterfall model, as proposed by Royce [1], suggests the sequential organization of activities in the process of product development software, in which one should move to a phase only when the previous phase is complete. The main activities of the waterfall model reflect the core activities of product development software through sequential execution of software product development activities: (1) analysis and requirements, (2) product (or software system) design, (3) implementation (4) testing, evaluation, and implementation, and (5) operation and maintenance of the system or software. The spiral model [7] regards an incremental development process that pays particular attention to risk management. The model explicitly suggests the development to be incremental and iterative by means of four convolutions, each designed to solve a specific sub-problem, each one resulting in documenting the results of the respective convolution. The Rational Unified Process (RUP) [8] is a well-defined object-oriented iterative development process. It provides a disciplined approach to determine tasks and responsibilities within an organization. Its main objective is to ensure, within an appropriate timetable and budget, the production of high quality software that meets users' expectations. It has a two-dimensional structure, composed by phases (inception, elaboration, construction and transition) and workflows (business modeling, requirements, analysis & design, implementation, test and deployment). Each phase may be executed iteratively, and the workflows have different importance within the phases. RUP's high formalism regarding activities, artifacts and roles has already resulted in the creation of variations to this process, one for large projects (the classic RUP), and other for small projects [9]. RUP's roles can also be tailored for small development teams, as proposed in [10].

Requirements elicitation tasks within these methodologies outputs artifacts that describe the intended system functionalities. A very common used artifact regards software architectures. A software architecture is a fundamental organization of a system embodied in its components, their relationships to each other, and to the environment, and the principles guiding its design and evolution [11]. There are approaches to support the design of software architectures, like RSEB [12], FAST [13], FORM [14], KobrA [15] and QADA [16]. The product-level perspective of the 4SRS method [17] also promotes functional decomposition of software systems. The result of the application of the 4SRS method is a logical architecture. A logical architecture can be considered a view of a system composed of a set of problem-specific abstractions supporting functional requirements [18].

These methodologies are accused of not properly involving the customer during all phases of the project, and resulted in low rate of successful rate on software projects by 1994 [19]. In opposition to the previous stated methodologies, agile development processes are based in self-organized teams for resolving their problems, dividing the implementation of complex software in small iterations periodically assessed, in order to solve eventual problems as soon as they emerge. Within these processes, eXtreme Programming (XP) [20] and Scrum [21] are the most popular methodologies [22], but there are also others, like Lean, Kanban, AUP (Agile Unified Process), Crystal Methodologies, DSDM (Dynamic Systems Development Model), FDD (Feature Driven Development) and OpenUP (Open Unified Process), amongst others.

In Scrum, temporal artifacts are produced, namely User Stories, Themes and Epics. The development is performed in a series of iterations or sprints, and progress is measured daily. Such measurement resulted in achieving more with less resources and teams are also protected against eventualities that may arise and endanger the software development (for instance, changing requirements). Even organizations with greater economic value also use agile methods.

Besides the use of Scrum in small organizations or in small projects, some techniques for adapting events, actors and artifacts arise in order to geographical distributed teams or multiple teams could work for the same product development [23]. Distributed Scrum is classified in three distributed team models [24]: (1) Isolated Scrums – teams are isolated across geographies; (2) Scrum of Scrums – multiple Scrum teams working on the same product and in the same geographical space [25]; and (3) Totally Integrated Scrum – where multiple teams are geographical distributed [26]. The Scrum methodology was also tested in projects involving different organizations trying to implement the same product [27].

3   The V+V Process and Architecture Modularization

It is absolutely necessary to assure that product-level (IT-related) requirements are perfectly aligned with process-level requirements (business processes-related), and hence, are aligned with the organization's business requirements (business mod-el-related. In the V+V process [3], technological requirements comply with the organization's business requirements. The first execution of the V-Model acts in the analysis phase and regards a process-level perspective. The first execution of the V-Model is out of the scope of this paper. The second V-Model (at product-level), is composed by *Mashed UCs* model [3], *A*- and *B-type* sequence diagrams [3], and *Use Case* models (UCs) that are used to derive (and, validate) a product-level logical ar-chitecture (*i.e.*, the software logical architecture). The execution of this V-Model is presented in Fig. 1. The logical architecture is a resulting artifact from the V-Model, and its architectural elements (AEs) and associations between them are derived from the execution of the 4SRS method [17]. An AE is a representation of the pieces from which the final logical architecture can be built. This term is used to distinguish those artifacts from the components, objects or modules used in other contexts, like in the UML structure diagrams. At the end of the product-level V-Model execution, the resulting information regards a *Context for Product Implementation* (CPI). The CPI information is constructed upon the product-level logical architecture and *B-type* se-quence diagrams, only referring to IT-related requirements. It is at the end of the V+V execution that User Stories are derived and delivered.

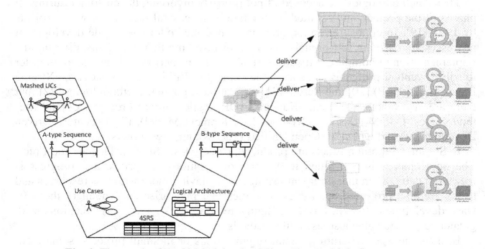

Fig. 1. The result of the V-Model to be delivered to multiple Scrum teams

Some software architectures are too complex to be directly used by Scrum teams as their requirements input artifact. Additionally, each Scrum team may be responsible for implementing only some parts of the whole architecture (*i.e.*, the CPI presented in Fig. 1). The architectural elements (representing software functionalities) that com-pose it are classified as belonging to a given software module, and thus covered by the module (Fig. 2).

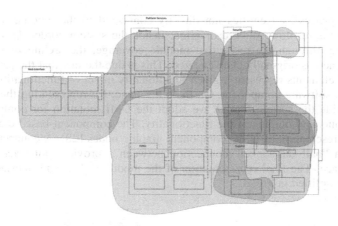

Fig. 2. Architecture modularization example

Regarding the CPI, there are two main artifacts for this task: the software logical architecture and the *B-type* sequence diagrams. The software logical architecture is the artifact typically delivered to implementation teams (Scrum or other), and it regards the architecture diagram that is modularized in this section. Besides the AEs that will compose the module, the packages are also used to identify a module. One or more packages compose a module. A given package can also be included in one or more modules. The other main artifact used regards *B-type* sequence diagrams. *B-type* sequence diagrams represent scenarios of using the software architecture, and these scenarios are the basis for identifying the required software modules. A given *B-type* sequence diagram allows identifying the need for developing a given module that responds to the execution requirements of the given scenario. However, there is not a direct relation from one *B-type* sequence diagram to a given module. A module can be composed by one or more *B-type* sequence diagrams. Additionally, a given *B-type* sequence diagram may be included in one or more modules. In its essence, a module is identified as a requirement for operationalizing a given *B-type* sequence diagram and/or the required packages.

The analyst or manager that has the overall insight on the project context and of the final product processes' must carry out the task of identifying the modules. There is no automatism for the module identification so it must be manually executed. The mindset of the analyst or manager has an impact in the resulting set of modules. The identification of a given module is based on the recognition of a specific need regarding its execution. These needs can be built on the identification of *concerns* that are dealt within the architecture execution context (in the case of aspect-oriented programming), or by the identification of software features (in the case of feature-driven development). If the mindset is oriented for agile development, the module identification will encompass a set of requirements specification (*i.e.*, user stories). Additionally, issues relating the modules' future distribution by the teams are considered, like if the teams are physically separated and/or with different working cultures and skills.

We applied filtering and collapsing techniques [17] to the border that relates to the architectural elements within the module. These techniques redefine the system boundaries, which now regards only the given module as a subsystem for design. During the filtering process, all entities not directly connected to the module must be removed from the resulting filtered diagram. Inside the system border defined for the given module, through the respective module coverage, the architectural elements were maintained as originally characterized. This is due the fact that the product-level architectural elements represent a given functionality or set of functionalities and not necessarily a software component. On the other hand, for representing the interfaces that are outside the system border, we adopt the UML notation for components, to represent inputs and outputs of the functionality. The component-based diagram uses a typical representation of UML component graphic nodes [28]. A connector may be notated by a "ball-and-socket" connection between a provided interface and a required interface. The final representation of the modules' AEs and interfaces is presented is exemplified in Fig. 3.

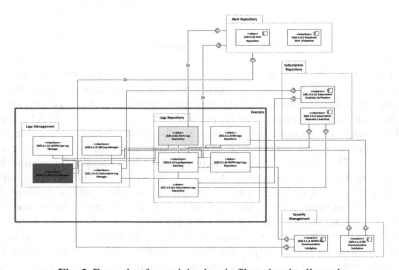

Fig. 3. Example of a module already filtered and collapsed

The architecture modularization technique (by using colored spots covering the architectural elements) also allows depicting in the diagram the existence of architectural elements that execute in more than one application (the architectural elements where there is an overlap of the module coverage). Such situation requires that their implementation must be properly managed by the development teams, *i.e.*, its development must be agreed by the teams. In order to facilitate the identification of such situations, which makes the implementation of the given architectural element more critical than the ones that are part of only one module, we identify those architectural elements by using colors. The critical architectural elements can be identified, for instance, in yellow if they are executed in two modules or red if they are executed in three modules. It is advisable that, as more critical the architectural element is considered, the "stronger" should be the color used. The information of each module is then delivered to the different Scrum teams in a project. The derived User Stories (presented in the next section) will compose Backlogs based on the modules. However, Backlogs, and any other artifacts besides User Stories, are out of the scope of this paper.

4   Scrum User Stories Derivation

The starting point for the User Stories derivation is the final product-level logical architecture diagram that results from the 4SRS method execution [17, 29]. In some cases regarding very large products, it is easy to see that these models can be extremely large and heavy to be analyzed as a whole, because these diagrams represent all the modules needed to run all desired functional requirements. Moreover, it is unlikely that, for large systems, only one Scrum team will perform all the work. In a model where there may be hundreds of modules, a Scrum team (by definition, must not have more than 10 elements) could take several years to be able to implement every detail of the model, an amount of time not feasible with the needs of a dynamic market.

Thus, deriving User Stories from the modules presented in the previous section allows that several Scrum teams can work in parallel, reducing the time required to implement and deliver the solution to the customer.

The first critical decision related to the development of our approach was to understand what should be the relation between the AEs and the User Stories. In the 4SRS method, the AEs are derived through the decomposition of Use Cases in three different types (interface, data and control). We decided to create one User Story for each AE, as depicted in Fig. 4. This decision intends to maintain the core principles for writing User Stories (*i.e.*, the INVEST characteristics – Independent, Negotiable, Valuable, Estimable, Small and Testable). Additionally, it complies with the greater flexibility for the Product Owner to follow the team's work. If the User Stories are always complex and require great effort to implement, there is the risk of diluting one of the main advantages recognized of agile methodologies: the ease of changing the direction of the team and the ability to see, in real time, which is state of commitment of the team to a Sprint. When implementing User Stories of great complexity, which may occupy the entire Sprint, only one can draw conclusions about the speed and commitment of the team at the end of the Sprint, when work (supposedly) must be completed. These arguments all meet the characteristic of having small User Stories (Small) and, thus, simplifying the implementation effort estimation (estimable).

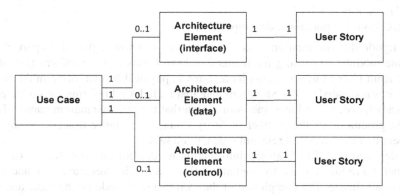

Fig. 4. Relation between Use Cases, AE's and User Stories

The information from the 4SRS method execution, mainly micro-steps 2i - "Use Case Specification", 2iii - "Architectural element naming" and 2viii - "Architectural element specification", together with the actors associated with each use case (where each AE was derived) are key elements in the generation of User Stories, since in them are encapsulated information required to write User Stories respecting the INVEST principles. The steps and micro-steps of the 4SRS method are described in [17]. The proposed technique for deriving user stories is composed by three steps, as follows:

Step 1 – Group AEs.

The first step is to group AEs and analyze their functionality and their interfaces. This step has as input the AEs from the logical architecture or, in case of an architecture modularization as the one presented in the previous section, from a given module.

Step 2 – Analyze AE specification and UC description

In this step, we gather the information from micro-steps 2i - "Use Case Specification", 2iii - "Architectural element naming" and 2viii - "Architectural element specification" and the involved actor (by reversing to the use cases that derived the AE). This step uses the traceability characteristic that the 4SRS method provides, by allowing to easily depict the original use case.

All these details of each AE should be stored with the same structure to give input to create a "card" for each User Story, containing all the information needed to carry out its estimation and subsequent implementation. Thus, for each AE is important to obtain the following information:

- · Name
- · Code
- · Type
- · Description
- · Package
- · Associations
- · Direct Associations
- · UC Associations
- · Original Use Case
- · The Actors Involved
- · UC from the functional decomposition

Alongside this information, it is also important to know if the AE is part of more than one module. Regarding the teams that have habits to keep information always visible from User Stories (placing its features in physical format, often in the form of cards), it was created a User Story template that includes all information collected and previously listed, as well as some information that the implementation team will generate in grooming, as the number of Story Points, acceptance criteria, or any other comment that the team find relevant register and save.

As depicted in Table 1, all the information needed for any stakeholder (from the customer / Product Owner to implementation teams) is presented and understood directly and simple. The completion of the card is also intended to be basic and quick as all information regarding the AE and the Use Case is available from the execution of the 4SRS method, while information on the User Story is mandatory and is defined by the implementation team during grooming.

Table 1. User story card template

User Story	US #	(name)				*Story Points*
	Acceptance criteria	• #1 • #2 • #3				
Architectural element	AE #	(num)		(nome)		
	Type	(interface/ data/ control)	*Package*			
	Specification					
	Multiple	(Yes/No)	Module			
	Association	Direct Associations				
		UC Associations				
Use Case	UC #	(num)		(name)		
	Description					
	UC Ass.	(num / name)				
	Actors					
Others		(Comments)				

Step 3 – Write the User Story

This is the final step and where the effecting output of the process is generated. The 4SRS method execution, followed by an analysis on the derived diagrams and documentation, allows triggering the procedure for mapping a logical architecture to output a set of User Stories that comply with the INVEST principles.

One of the great advantages of applying the 4SRS method to derive the AE is that it quickly allows realizing the ultimate goal of the AE: data manipulation, communication or logic operations, by reading the type of AE. This standardization of AE types simplifies the management of User Stories because, after all, they are centered on three very specific types of tasks.

The "who", the actors involved and who will perform tasks on the User Story to implement, are easily identified by analyzing their User Story Card and looking for those involved in the Use Case that derived the AE (by executing the 4SRS method). As the logic surrounding the need to represent properties of systems in Use Case and User Story is similar (to capture specific requirements in terms of interaction between users and system), it is easy to validate that those involved in the Use Cases will be benefited by the implementation actors Story of a particular User.

Regarding the "what," you can also find a direct relationship between this component and the name of AE. Firstly, it is necessary to find an action, represented by a verb, to identify what you want to implement. The division between AEs of control, and data interface simplifies this demand, since the AE interface always refer to the creation of a specific interface and is therefore an action which is fixed and constant need for the existence of a communication interface for between component and / or actors. In most cases, the name of AE only indicates what kind of interface is required. Thus, in cases of interface (*i-type*) AEs, the actors involved just need their existence in order to use them in their workflows. By using the name of the corresponding AE, and using connections want/need to have (want/need to have), the connection between the "who" (actor) and what (action) is derived. In cases of *i-type* AEs that do not have this syntax, the central part of the User Story for the "what" is simply left to the information "want/need to have an interface", and the title of the AE (the actions that will take place using that interface) used as part of the "why". For data (*d-type*) AEs the process is similar, since they usually refer to the need of the existence of repositories/storage locations or interfaces for communication with such storage spaces. Thus, the construction of the User Story follows the same rule used in *i-type* AEs.

Compared with the previous two types, control (*c-type*) AEs are quite disparate. They support the logic behind a system, representing all actions that can be performed by manipulating the data (represented by *d-type* AEs) and using interfaces for transmission (represented by *i-type* AEs). As they can represent any action on the system, typically *c-type* AEs have associated a verb that represents the action that it performs. In this case, we are deriving information of a title for a sentence. Some kind of semantic correctness of words may be required, allowing the sentence to make sense.

5   Management of Multiple Scrum Teams

Besides the User Story derivation, there are some issues that arise from the distributed Scrum teams. Each team's parallel work within the same product development must be coordinated, but breakthroughs and progress within the distributed Scrum teams are slow and hard to achieve [30]. The architecture modularization presented in section 3 promoted the identification of interface components between modules, thus identifying connection points between the distributed developments. Additionally, overlapped architectural elements identification relate to dependencies between the distributed teams.

Scrum roles and events can be easily adapted to such dependencies between teams since the agile methodologies provide this kind of flexibility. These adaptations allow distributed teams (geographical distributed or not) to work in parallel and at the same time, and are mandatory in order to prevent an increase of the time to market, that could endanger the project execution. The product backlog should be aligned by collaborating product owners [31].

Regarding the team and roles, the quantity of teams and if they are geographical distributed is irrelevant in case all teams belong to the same organization. The Scrum Master can be responsible for managing each team simultaneously. Alongside with the Product Owner, the Scrum Master can manage the team's aligned work for the implementation.

Regarding sprints and events, it is normal that all of team's sprints aren't synchronized relating its start date. The Scrum Master and the Product Owner should have total availability to work with all teams equitably. Besides, if problem reports arise from a team in a Sprint Review or a Sprint Retrospective, there is still enough time to re-schedule aspects in other teams or re-allocate resources at the end of the other team's sprints. Additionally, another issue that can be considered is that one element of each team may participate in other team's Sprint Review, and all elements should participate instead of always the same element participating in those Sprint Reviews, so all team elements have the opportunity to know other teams' work.

Finally, another issue that must be considered relates to Grooming events. A "traditional" Grooming occurs as the team elements alongside with the Scrum Master and the Product Owner elicit software requirements in form of User Stories. Since User Stories were already derived, it is still advisable that they all are revised by the entire team, its complexity (based on Story Points) estimated, and then presented to all so everyone clearly know what will be implemented. On the other hand, the technique proposed in the previous section doesn't include the elicitation of non-functional requirements, which this way can be only elicited during a Grooming.

6   Demonstration Case: The ISOFIN Project

The ISOFIN (Interoperability in Financial Software) [32] architecture aims to deliver a set of cloud-based functionalities enacting the coordination of independent services relying on private clouds. The resulting ISOFIN platform will allow the semantic and application interoperability between enrolled financial institutions (Banks, Insurance Companies and others). The global ISOFIN architecture relies on two main service types: Interconnected Business Service (IBS) and Supplier Business Service (SBS). IBSs concern a set of functionalities that are exposed from the ISOFIN core platform to ISOFIN Customers. An IBS interconnects one or more SBS's and/or IBS's exposing functionalities that relate directly to business needs. SBS's are a set of functionalities that are exposed from the ISOFIN Suppliers production infrastructure.

The ISOFIN project is executed in a consortium comprising eight entities (private companies, public research centers and universities). The initial request for the project requirements resulted in mixed and confusing sets of misaligned information. Even when a requirement found a consensus in the consortium, the intended behavior or definition was not easily understood by all the stakeholders. Our proposal of adopting a process-level perspective was agreed on and, after being executed, resulted in a set of information that the consortium is sustainably using to evolve to the traditional (product-level) development scenario. The V + V process ensured the transition between the process-level requirements of the intended system and the technological requirements that the same system must comply with. The product-level V-Model execution resulted in a CPI composed by a software logical architecture and *B-type* sequence diagrams. In the case of ISOFIN, the logical architectural diagram was very complex and difficult to analyze as a whole. It contains 105 AEs, divided into 7 different modules. As a demonstration case, we present the derivation process using the

IBS Management module, but the derivation process for the rest of the modules lfollows the same steps.

We present in Table 2 an example of the use of such a model, using the AE module for the management of IBS labeled *Test Before IBS Deployment*. Information regarding the name of the User Story presupposes the execution of the next step of this method. The acceptance criteria and the story points fields are not defined at the time of the User Story derivation. They are defined later during Sprints, so these fields were not yet defined in the card in Table 2, thus defined as "not applicable" (*N/A*) at this time.

Table 2. User Story card for "Test IBS Before Deployment"

User Story	US #1	As a Business User or a IBS Developer, I want to test IBS before deployment, in order to render IBS in pre-runtime.			
	Acceptance criteria	N/A			***Story Points***
					N/A
Architectural element	AE #	2.7.2.c	Test IBS Before Deployment		
	Type	Control	***Package***	IBS Installer	
	Specification	This AE allows testing of the IBS before deployment. This AE will be required before the execution of *{AE2.3.2.c} IBS Deployer* to verify that no problems occur during the execution of the IBS. All information need for the execution is provided by *{AE2.2.5.d} IBS Pre-Deployment Storage*			
	Multiple	No	**Module**	IBS Management	
	Association	**Direct Association**	2.7.2.i – IBS Test Generator		
		UC Association	2.3.2.c – IBS Deployer 2.2..5.d – Pre-Deployment Storage		
Use Case	UC #	2.7.2	Render IBS Pre-Runtime		
	Description	Configure and defines the pre-runtime of the IBS. This use case allows testing of the IBS before deployment.			
	UC Ass.	2.7 – Configure IBS			
	Actors	Business User IBS Developer			
Others					

The User Story must provide the "why" of a particular actor ("who") may need to perform a certain action ("what"). This information, often induced by the very title of the corresponding AE, can be complemented with a description of the Use Case from which the AE was derived. As User Stories relate to a lower-level than uses cases, the description of the use case itself can justify the need for existence of a particular User Story. In cases where the name of AEs is quite similar to use case from which the AE was derived, the significance of User Story can be found in the description of the use case. Using these rules, remaining User Stories were derived that are listed in Table 3. In Fig. 5 is represented a User Story sentence based in the derivation from Table 3.

Table 3. User Stories derived from *c-type* AEs

AE #	AE	As a(n) <actor>	I want/need <description>	In order to <outcome>
2.1.2.c1	Selected Object Configurations	ISOFIN Customer / IBS Developer	select object configurations	change (IBS Structure) configurations
2.1.4.c	Compiles IBS information	IBS Developer	compile IBS (changes and) information	create a new IBS
2.2.4.c	Define IBS Code Gaps	IBS Developer	(automatically generated code) and define IBS code gaps	create IBS code
2.2.5.c	Compile IBS code	IBS Developer	compile IBS code (and create new IBS catalog)	(keep IBS catalog and store) compile(d) IBS Code
2.2.6.c1	Selected Object Permissions	IBS Developer	select object permissions	set(/manage) permissions (and create IBS)
2.2.7.c	IBS Interface Generator	IBS Developer	(automatically) Generate IBS Interface	(store the) generate(d) IBS interface
2.3.2.c	IBS Deployer	IBS Developer	deploy IBS	execute IBS deployment
2.7.1.c	IBS Customization Filter	Business User	filter IBS (configuration and) customization	customize IBS
2.7.2.c	Test IBS Before deployment	Business User / IBS Developer	test IBS before deployment	render IBS Pre-Runtime

"As an IBS Developer, I want/need to compile IBS code, in order to create a new IBS."

Fig. 5. User Story from AE2.1.4.c

Now that the User Story derivation is complete, there are just some issues that are dealt in the multiple teams' management. In the case of the ISOFIN project, the teams were distributed but belong to the same organization and weren't geographically distributed. The quantity of teams weren't as many as the modules identified, but the total quantity of teams isn't relevant, since they belong to the same organization. Thus, the organization chose to nominate a single Scrum Master to work closely with all teams. The Product Owner was responsible for the decisions during the implementation, like detecting potential delays and decisions on critical issues across the teams.

Regarding the overlapped AEs in more than one module, it is then the Product Owner's responsibility to nominate a team to be responsible for implementation of the derived User Story and assure that the teams responsible for User Stories with dependencies with that particular one have all required documentation and provide updates on its implementation.

Overall, in the ISOFIN project, there were clear advantages in using this approach: (1) the teams experienced difficulties in interpreted the complete architecture, thus the modularization was required; (2) since the project consortium was composed by Scrum teams, they easily understood the artifacts (*i.e.*, User Stories); (3) User Stories were derived having an already designed logical architecture as input, allowing them to be properly aligned within reduced time. Connection points between modules were identified and properly covered by User Stories but, however, there was not enough time during this research work to assess that the team's efforts were in fact synched. Besides the identification of connection points, the authors believe that there is a vast area of progress in the topic of distributed Scrum teams.

7   Conclusions

In this paper, a technique for deriving software requirements compliant with Scrum teams, in the form of User Stories, from logical architectures was proposed. The input for this technique is the software logical architecture that resulted from the 4SRS method execution. The proposed technique is composed by three steps: the architectural elements from the logical architecture are grouped by their type (*c-*, *d-* or *i-type*); the information from the 4SRS method execution regarding the AEs specification and the original use case model is gathered in order to fill User Story Cards that contain the required information for the User Story derivation; and the User Stories are written based on the User Story Cards and any semantic issues are revised. Our proposal is not for building Product Backlogs based on the architecture modularization for the different Scrum teams in a project, but rather to derive User Stories regarding the module and, additionally, identify some contact points where there is a need for synchronizing efforts within distributed Scrums and effort dependencies. These points are also properly covered by User Stories. Since User Stories are not exclusive for the Scrum methodology, this approach can be used in other agile development contexts besides Scrum.

Like Scrum itself, distributed Scrum approaches were created and tested in an industrial environment, with ad hoc experiments and constant improvements. In literature, the description of how to implement Scrum on projects with distributed teams on a scientific perspective are mostly based on testimonies of organizations trying to adapt Scrum to their teams. Aspects such as synchronization sprints, platforms of understanding among various teams working on the same product, cadence meetings, among others, continue to be topics where there is a vast area of progress.

Despite there is no explicit reference regarding the need of formal documentation to initiate the product development phase, there are advantages for early defining physical and component architectures in order to define borders for the entire work. This is even more important when products are large and implemented by several teams in parallel. Such situation can be regarded as an opportunity to bring together requirements of Scrum (User Stories) and logical architectures.

In this paper, the proposed technique for deriving User Stories used as input the product-level logical architecture diagram that results from the 4SRS method execution. As future work, the proposed technique can be adapted to simpler contexts, only composed by use cases and actors, since in short and low complexity projects these are the only representation of the requirements elicited before implementation.

References

1. Royce, W.W.: Managing the development of large software systems. In: Proceedings of IEEE WESCON, Los Angeles (1970)
2. Maranzano, J.F., et al.: Architecture Reviews: Practice and Experience. IEEE Software (2005)
3. Ferreira, N., Santos, N., Soares, P., Machado, R.J., Gašević, D.: Transition from Process- to Product-level Perspective for Business Software. In: Poels, G. (ed.) CONFENIS 2012. LNBIP, vol. 139, pp. 268–275. Springer, Heidelberg (2013)
4. Haskins, C., Forsberg, K.: Systems Engineering Handbook: A Guide for System Life Cycle Processes and Activities; INCOSE-TP-2003-002-03.2. 1. INCOSE (2011)
5. Ambler, S., Lines, M.: Disciplined Agile Delivery: A Practitioner's Guide to Agile Software Delivery in the Enterprise. IBM Press (2012)
6. Woodward, E., Surdek, S., Ganis, M.: A practical guide to distributed scrum. IBM press (2010)
7. Boehm, B.W.: A spiral model of software development and enhancement. Computer 21(5), 61–72 (1988)
8. Kruchten, P.: The rational unified process: An introduction. Addison-Wesley Professional (2004)
9. Booch, G.: The Unified Modeling Language User Guide, 2/E. Pearson Education India (2005)
10. Borges, P., Monteiro, P., Machado, R.J.: Tailoring RUP to Small Software Development Teams. In: IEEE 2011 37th EUROMICRO Conference on Software Engineering and Advanced Applications, SEAA (2011)
11. IEEE Computer Society, IEEE Recommended Practice for Architectural Description of Software Intensive Systems - IEEE Std. 1471-2000 (2000)
12. Jacobson, I., Griss, M., Jonsson, P.: Software Reuse: Architecture, Process and Organization for Business Success. Addison Wesley Longman (1997)
13. Weiss, D.M., Lai, C.T.R.: Software Product-Line Engineering: A Family-Based Software Development Process. Addison-Wesley Professional (1999)
14. Kang, K.C., et al.: FORM: A feature-oriented reuse method with domain-specific reference architectures. Annals of Sw Engineering (1998)
15. Bayer, J., Muthig, D., Göpfert, B.: The library system product line. A KobrA case study. Fraunhofer IESE (2001)

16. Matinlassi, M., Niemelä, E., Dobrica, L.: Quality-driven architecture design and quality analysis method, A revolutionary initiation approach to a product line architecture, VTT Tech. Research Centre of Finland (2002)

17. Machado, R.J., Fernandes, J.M., Monteiro, P., Daskalakis, C.: Refinement of Software Architectures by Recursive Model Transformations. In: Münch, J., Vierimaa, M. (eds.) PROFES 2006. LNCS, vol. 4034, pp. 422–428. Springer, Heidelberg (2006)

18. Azevedo, S., Machado, R.J., Muthig, D., Ribeiro, H.: Refinement of Software Product Line Architectures through Recursive Modeling Techniques. In: Meersman, R., Herrero, P., Dillon, T. (eds.) OTM 2009 Workshops. LNCS, vol. 5872, pp. 411–422. Springer, Heidelberg (2009)

19. Standish, CHAOS Summary, 2009, Standish Group (1995)

20. Beck, K., Andres, C.: Extreme programming explained: Embrace change. Addison-Wesley Professional (2004)

21. Schwaber, K.: Scrum development process. In: Business Object Design and Implementation, pp. 117–134. Springer (1997)

22. VersionOne Inc., 8th Annual State of Agile Survey (2013), http://www.versionone.com/pdf/2013-state-of-agile-survey.pdf

23. Eckstein, J.: Agile software development with distributed teams: Staying agile in a global world. Addison-Wesley (2013)

24. Sutherland, J., Viktorov, A., Blount, J.: Adaptive Engineering of Large Software Projects with Distributed/Outsourced Teams. In: Proc. International Conference on Complex Systems, Boston, MA, USA (2006)

25. Cristal, M., Wildt, D., Prikladnicki, R.: Usage of Scrum practices within a global company. In: IEEE International Conference on Global Software Engineering, ICGSE 2008. IEEE (2008)

26. Paasivaara, M., Durasiewicz, S., Lassenius, C.: Using scrum in a globally distributed project: A case study. Software Process: Improvement and Practice 13(6), 527–544 (2008)

27. Dingsøyr, T., Hanssen, G.K., Dybå, T., Anker, G., Nygaard, J.O.: Developing software with scrum in a small cross-organizational project. In: Richardson, I., Runeson, P., Messnarz, R. (eds.) EuroSPI 2006. LNCS, vol. 4257, pp. 5–15. Springer, Heidelberg (2006)

28. OMG. Unified Modeling Language (UML) Superstructure Version 2.4.1 (January 2012), http://www.omg.org/spec/UML/2.4.1/

29. Machado, R.J., et al.: Transformation of UML Models for Service-Oriented Software Architectures. In: ECBS 2005, pp. 173–182. IEEE Computer Society (2005)

30. Begel, A., et al.: Coordination in large-scale software teams. In: Proceedings of the 2009 ICSE Workshop on Cooperative and Human Aspects on Software Engineering, IEEE Computer Society (2009)

31. Leffingwell, D.: Scaling software agility: Best practices for large enterprises. Pearson Education (2007)

32. ISOFIN Consortium. ISOFIN Research Project; ISOFIN Research Project (2010), http://isofincloud.i2s.pt

Monitoring the Professional Evolution
of Graduates with Multiagent Systems

Diego Fialho Rodrigues[1], Alcione de Paiva Oliveira[1], Jugurta Lisboa Filho[1],
and Alexandra Moreira[2]

[1] Universidade Federal de Viçosa, Viçosa MG, Brazil
alcione@gmail.com
http://www.dpi.ufv.br
[2] Universidade Federal de Juiz de Fora, Juiz de Fora MG, Brazil

Abstract. Several institutions need to collect and store information about people related to their activities. Among such institutions there are the educational institutions that need to monitor the professional development of its graduates. An automated system for performing this task must overcome several challenges: collect information in natural language databases, extract information through natural language processing techniques, identify the entities and the roles played by these entities in the situation described, integrate heterogeneous information and store this information on an easy handling database. Multiagent systems appear as a scalable and modular solution to accomplish this complex task that involves several steps. This paper describes a multiagent model to obtain professional information of graduates. The model was implemented and tested in a database of messages exchanged in a list of e-mails. The tests showed the feasibility of this approach in solving problems of information extraction in natural language.

Keywords: information extraction, frame semantics , ontology, tracking people.

1 Introduction

Several institutions need to collect and store information about people related to their activities. Among such institutions there are the educational institutions that need to monitor the professional development of its graduates. This information can be found on social networks, email lists, blogs and websites openly available curricula, etc.. Despite the abundance of data, there is the difficulty of aggregating all this information, since they are available in different formats, without following any particular pattern. Also, collect this information manually if an infeasible task due to the large number of sources of information. An automated system for performing this task must overcome several challenges: collect information in natural language databases, extract information through natural language processing techniques, identify the entities and the roles played by these entities in the situation described, integrate heterogeneous information and store

B. Murgante et al. (Eds.): ICCSA 2014, Part III, LNCS 8581, pp. 763–778, 2014.
© Springer International Publishing Switzerland 2014

this information on an easy handling database. Multivalent systems (MAS) [20] appear as a scalable and modular solution to accomplish this complex task that involves several steps. MAS allows systems to operate in a decentralized manner, where components can be added in a natural way, with minimal impact to the overall system due to the low coupling units.

This paper describes a system able to track information about the company people work according to the information available on public sites on the Internet. More specifically, the goal is to monitor the professional development of graduates. In Brazil, the monitoring of the professional career of the students as well as their academic development is one of the criteria for evaluation of undergraduate courses. This article is based on the model described in [14].The model was implemented as a multiagent system and tests were carried out, whose results are presented in this article (section 7). The next section addresses the problem of tracking people and what has been done to solve it.

2   Tracking People and Related Work

The search and collection of information about people has become an essential task for companies. The task of finding experts in a given field of knowledge was introduced in *The Fourteenth Text Retrieval Conference* - TREC 2005 [5]. The search for information about people was also one of the subjects of the *the Fourth International Workshop on Semantic Evaluations* [18], organized by the *Association for Computational Linguistics* (ACL). [1] presented a criterion for evaluate the performance of systems that people seek on the Internet. [6] presented a system that eliminates ambiguity in searches related to the persons name on the web by means of clustering. [13] proposed ways to mitigate the problem of establishing erroneous references in the search of people. [2,3] also sought to solve the problem of name resolution for people using clustering techniques. Most studies addressing the association of persons documents use clustering techniques to eliminate ambiguities. The basic assumption of this technique is that similar documents tend to represent the same person. Some of the major clustering methods are [2]: Single Pass, k-Means, agglomerative clustering e probabilistic latent semantic analysis (PLSA). The first three methods allow variations of the traditional method of grouping (differs in terms of efficiency and quality). These methods are also based on the fact that different documents have the same terms associated with the same individual. [4] presented a system for identifying trafficking of wild animals in social networks based on the analysis of the messages exchanged by users. Although not a system for people searchthis work also used frames and ontologies to identify the context of the utterance and the meaning of lexical items. [12] used ontologies to identify statements related to Frame TRAVEL. [16] presented a way to connect large lexical resources with the world knowledge by means of ontologies. In [17], the same author proposed a set of rules for mapping FEs with ontology elements and presented a way to connect large lexical resources with the world knowledge by means of ontologies SUMO[1].

---

[1] http://protege.stanford.edu/ontologies/sumoOntology

The monitoring of graduates can be regarded as related to the monitoring of persons task. The professional success of alumni of educational institutions can be attributed in part to the training received in these institutions.The practice commonly used by institutions is sending messages asking graduates to upgrade voluntarily their data. However, this is a practice that does not allow an efficient monitoring, since it starts from the premise that the bank address (email or physical address) are already updated and that the graduates will be willing to devote some of their time to respond and submit forms. Therefore, an automated system that seeks information on graduates from various sources would be useful. Obviously, for ethical and privacy issues, this type of system should seek information in public repositories, directly available and only with the consent of the parties involved. To implement such a system is necessary to use techniques from natural language processing and distributed processing, especially for systems-oriented agents, in order to make it a more modular and scalable system.

3 The Multiagent Approach

Multiagent System is an area within computer science that deals with aspects of Distributed Systems applied to Artificial Intelligence [20]. One of the requirements for the proposed solution to the problem of tracking graduates is that it can analyze information from various sources. The analyzers designed to extract the information from these sources should be added to the system transparently and with minimal impact on the other modules. To meet this requirement, the use of a system with agent-based architecture seems appropriate. Thus, you can change an agent specialized in extracting information from a particular base or replace a particular technique of natural language processing without interfering with another agent of the system.In addition, a multiagent platform provides the ability to add new agents, making the system scalable naturally. In the proposed architecture the data is sent to an agent that stores and manages information in a central repository. There is also the possibility of the system to benefit from the use of a hardware endowed with parallel processing, where each agent could be run on a separate processor.

The proposed system has the function of hosting several agents that will be inserted gradually. Each agent will be able to seek information in a separate source and asynchronously, and the information will be stored in a single repository. Thus the multiagent architecture is presented as a natural solution. The system is not only composed of information seekers agents. There are several types of agents, from simplest (how to communicate with a Web Service) to complex (responsible for natural language processing). The goal is to make the system to be as modular as possible, so that the insertion or removal of an agent does not compromise the functioning of the system as a whole. This should be because of the interaction between agents is made through message exchange, and thus allowing a smaller coupling between agents.

The model has six agents, namely: (i) *University Web Server*, (ii) *Email Reader*, (iii) *Questionnaire Sender*, (iv) *Questionnaire Answers Reader*, (v)

Database Manager and (vi) *NL Processor*. Each of these agents can communicate with secondary agents to assist in the execution of their tasks.

The agent *University Web Server* is responsible for fetching in the database of the educational institution the information about students who just graduated. The query is made every six months, which is the interval between each graduation. The list of graduates obtained by the agent serves as the starting point for all other searches. The list contains information such as name and e-mail that will serve to identify the graduates in the results of searches performed by other agents.

The agent *Email Reader*has the function of read messages in mailing lists groups of alumni. It was designed to search for information on the professional development of graduates. An analysis of the messages revealed that it is common for graduates to inform their current activities, such as company and town where they work and where they live and this information provided the basis for the definition of the schema that would be used for data extraction. The agent *Email Reader*has two auxiliaries. The first has the function of putting the various formats of e-mail in a single canonical form. The second agent has the function to extract the various sentences contained in a single email message.

The agent *Questionarie Sender* sends a questionnaire to the email address of each graduate.The alumni fill the information and return the form. The *Questionarie Answers Reader*, in turn, has the function to process the answers and update the database.

The agent *NL Processor* processes all the sentences provided by the agent *Mail Reader*. This agent uses techniques from natural language processing to try to identify evidence of linkage between a former student and an institution.

The agent *Database Manager* manage the data repository. None of the other agents can update the database directly. All updates are made through the agent*Database Manager*.

The Figure 1 shows the dependency between the agents. The dependency diagram for agents from Tropos methodology was used [10]. The primary and secondary agents will be detailed in the sections that follow.

4   Questionaries

Among all the ways to obtain information about people, perhaps the simplest and most direct is asking. And this is what the agent *Questionaire Sender* makes from time to time by sending an e-mail with a questionnaire to the list of ex-students of the Computer Science School. This email contains a message explaining the motivation of the questionnaire and a link to a form with all the questions in the questionnaire. The form includes questions such as: *What is your full name? What is your email address? Report a course that you have done lately. Enter the name of the institution where the course was done. Enter the name of a company you work or have worked for. When did you enter this business? When did you leave the company?,etc* .

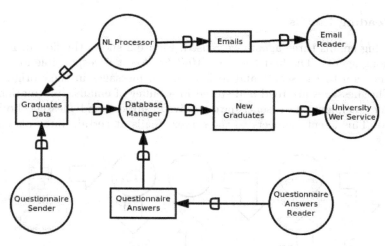

Fig. 1. Dependence between agents

Besides this information, is also stored the date on which the form was submitted. The former student answers the questionnaire and the information is automatically sent to a table in the data repository. However, the information needs to be prepared to be inserted in the repository. To accomplish this task the agent *Questionnaire Answers Reader* comes into play. This agent has the function to read the worksheet with the answers and update the data of the former student. The agent acts on time intervals, such as every month.

Questionnaires are just a way to collect more information about the graduates and it is assumed that information is reliable. However, not all people go to the trouble of answering questionnaires. Therefore, becomes necessary to collect other forms of information that will be discussed in the sections that follow.

5   Natural Language Processing

Another source of information about the graduates are the e-mails. Looking at the list of emails of former students, it is noticed that often people leave information regarding their jobs in those posts. A common example are the jobs. Generally, a person discloses a vacancy in the company where they work by posting a message to the list and email. Unlike information acquired through *Web Service* and with the questionnaires that are in a structured format, the information contained in e-mails are in natural language. Therefore, one has to perform a process using NLP techniques to be able to extract relevant information. These techniques will be described later in this article. However, before processing the messages was necessary to create mechanisms to extract e-mails from the list. Since there are various formats of e-mails and multiple communication protocols was necessary to separate this step from step that processes the sentences contained in email messages.

5.1   Reading Emails

For reading emails three agents were used. Figure 2 shows the flow of information among agents. The first agent is *Mail Reader*, it is responsible for making the connection to the server and to fetch email messages in their original formats. The messages are read and stored in a buffer of emails. Information such as sender, recipient, date sent, date received, subject and message body are captured. This agent connects to the server emails at certain intervals for new messages.

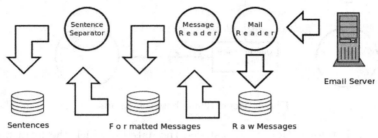

Fig. 2. Messages being processed

There are several formats of e-mail message. There are messages in plain text in html format with images, in the form of array of messages (called multipart), etc. To deal with this diversity of formats agent *Message Reader* was created. This agent reads messages from the e-mails buffer (supplied by the agent *Mail Reader*)and checks the type of the message format. This agent has at its disposal several readers to the various formats. Then it chooses the appropriate reader for the format and processes the message and translate the messages to a canonical format. These messages are placed in a second buffer to be processed by the next agent. Some formats do not matter to the system, such as images and are simply discarded.

The third agent, *Sentence Separator*, has the function of breaking messages into sentences. This is done by searching for special characters such as '.', '?', '!', etc. There are some problems in separating sentences. For example, abbreviations, such as 'Sr.',can be mistakenly identified as a sentence separator. Moreover, the emails are not always written with formality, and often people forget to put the final point. To solve the first problem, a list of abbreviations, which are consulted every time a point is found in the text was created.

The information is stored in a third buffer that will serve as input for natural language processing. Importantly, all three agents run simultaneously, in pipeline style. As soon as the agent *Mail Reader* inserts some information in the buffer, the message reader can already start processing. The same goes for *Sentence Parser*. The system allows to read e-mails from another server, just by simply instantiate another email reader, without causing major impacts on the system.

5.2   Semantic Frames

Reviewing the list of emails, it was revealed several patterns or communication schemes. Some of these patterns revealed the binding of a person with a certain organization. For example, it was noted that it is common for people to publicize job openings in their company or that they are doing a postgraduate course in any educational institution. In a certain way, the components are almost always the same: the former student, business or educational institution, city, date. The goal is to identify those patterns in the messages to capture information relevant to the problem.

To better identify these patterns, we used the Semantic Frames[8]. According to [8],Semantic frames are conceptual structures which describe a particular type of situation, object or event, along with its participants. These participants are called (*Frames Elements* ou FEs). The FEs are divided into two groups: *core* and *non-core*. The core elements always occur implicitly or explicitly. The non-core elements may or may not occur. Furthermore, each frame is evoked by a *Lexical Unit* or LU, which is a word with a meaning (form-meaning pair). The FEs act as arguments to LU, completing the the sense brought by the word.

Once the frames are defined, one can use them to recognize sentences related to frames by identifying its elements. In the case of this study, this information is related professional development of graduates. Therefore, the FEs are the information that will be analyzed and incorporated into the data repository.

The creation of the frames was done by the analysis of a *corpus* built from the messages in the mailing list of alumni. About 5000 messages sent between 1999 and 2011 were extracted. The analysis of the list of emails revealed that it is common for a graduate offer job openings in the company where he works. Along the offer, the former student can mention other important items such as the city where the company is located, the date he joined the company, the job function, etc. Usually the sentence is in the first person and the company name is mentioned. These observations were used to construct a frame, called a JOB OFFER.The frame is described in Figure 3 along with some examples. The notation used used in the description is the same adopted on the FrameNet project site[2] which proposes to build a lexical basis based on the semantics of frames for the English language..

In the proposed system, each frame would be assigned to a different agent. At the moment, only the frame JOB_OFFER was created, but in the future, other frames will be developed. The frame corresponding to each agent will search for sentences in the buffer (supplied by the agent *Sentence Separator*). The process consists of finding sentences containing a lexical unit off the frame. For a human being to read a sentence and identify the elements of frame and lexical units does not seem to be a very difficult task. However, is no simple task for a computer. The details of how a frame is identified and how the elements are found and ranked will be detailed in the subsections that follow.

---

[2] https://framenet.icsi.berkeley.edu/fndrupal/

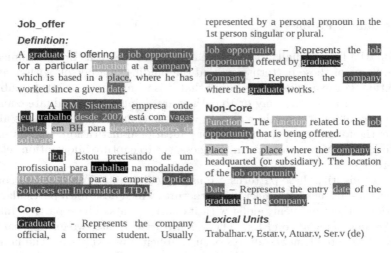

Fig. 3. The Frame JOB OFFER

5.3 The Domain Ontology

Frames can be used to make semantic analysis of the elements, but they do not necessarily define the nature or type of these elements. For example, excerpt "em São Paulo" (in São Paulo) refers to a place, however it is not possible to know whether a city or state, and which country belongs. Pto carry out the classification of the elements of the domain, we used a domain ontology [19].An ontology defines the concepts relevant to the problem, as well as all relationships between these concepts. Moreover, since ontologies allow to express a variety of relationships between concepts of the domain, you can use these relationships to make inferences about the individuals of the domain, for example, search for all companies that are located in a certain town.

The domain ontology created is meant to describe the elements that occur in the context of employment relationships between companies and graduates. This ontology has elements representing institutions where people can be employed, places (countries, states and cities) and work functions, etc. Concepts of time was also defined to represent intervals, specifically to represent the duration of a link between an institution and a person.

To help better categorize the elements, the ontology was built over two other more general ontologies (top ontologies): the ontology DUL (Dolce Ultralight)[3] and ontology $Time$[4]. As a result, there was no need to create many classes (concepts) and properties (relationships between concepts).Most classes used already defined properties in the other two ontologies, such as the relationship *is part of* between *Department* and *Organization*. The ontology developed is shown in Figure4. The parts (a), (b) and (c) contains the classes while parts (d) and (e)exhibit the properties. The elements created for this study are in bold.

---

[3] www.loa.istc.cnr.it/ontologies/DUL.owl

[4] raw.github.com/RinkeHoekstra/lkif-core/master/time.owl

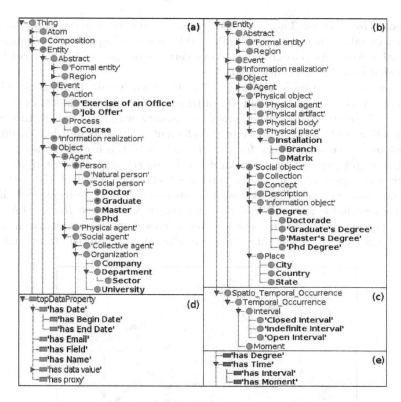

Fig. 4. The domain ontology built over the ontologies DUL and Time. A, B and C: classes. D and E: properties.

5.4 Bayesian Networks

A way to annotate the sentences with the FEs and LUs was shown. However, it must detect the background scene (frame) related to the sentence. It is the frame that attaches ultimate meaning to the sentence [8].Therefore, it is necessary to define a technique to perform the mapping between the sentence and its frame background. As already mentioned[3] used Bayes' theorem to determine the likelihood that a person is associated with a particular document. However, our work focuses on matching at the sentence level and not at the document level. Moreira [12] proposed the use of Bayesian networks to perform this mapping.

A semantic frame has several components, some mandatory and some optional. Thus, the presence of lexical units and their frame elements may indicate the occurrence of a frame. In the identification process, some elements have more weight, such as the lexical units and elements of nuclear frame. Others have a lower weight, such as non-nuclear frame elements.

In the study, we used Bayesian networks to classify sentences as proposed in [12]. According to [9],Bayesian networks are directed acyclic graphs and encoding the joint probability distribution over a set of random variables. Each variable is represented by a vertex. The correlation between variables is represented by arcs. For each variable, there is a probability table representing the local probability distribution given its parents (Departure vertices of the arcs). The model was constructed based on semantic frame.

Frames are evoked by lexical units [15]. The lexical unit, in its turn, has an argument structure, and the elements of frames behave as arguments of the LU. Based on these statements, one Bayesian Network was mounted on the frame of a job offer, and the occurrence of the elements of frames and lexical unit in the sentence serve as evidence for the occurrence of the frame. Thus, the FEs and LU are independent variables and the frame the dependent variable. The network was constructed using the software Weka [11]. A list of 150 sentences was annotated and then classified as an occurrence or not the frame. Of all the algorithms available in the Weka, the *Tree Augmented Naive Bayes* (TAN) [9] showed the best results (94.0% cases correctly classified). The network shown in Figure 5.The vertices represent each of the FEs, the LU and the frame itself.

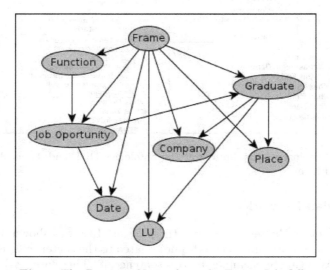

Fig. 5. The Bayesian Network on the Frame Job_Offer

The behavior of the model is always look for sentences containing LUs of the frame. Once the lexical unit is found, the task is to find frame elements. Then, the sentence is classified according to the presence or absence of each of the frame elements. The more elements are found, the higher the probability of the sentence being classified as an example of a frame.

6   Case Study

In this section we will show an example of how a sentence is processed. First, the analysis is done at the level of semantic frames. All frame elements are identified.Then these elements are connected with objects or classes of the domain ontology through the relation *hasSemanticType*. In some cases,the relationship has as its image one instance and, in other cases, a class. For example , the FEs of type *Place*possess a relationship with instances of the class *City*. It is clear that a place has a more generic meaning than a city. However, in this particular case, the important thing is to establish a connection with cities. Furthermore, it is important to know which state and country belongs this city. This is possible through discovering the relations that the instance of the domain ontology has with other instances.

In other cases, however, one need not to know the detailed characteristics of the objects, as in the case of the work performed on the job. In this case is necessary to know that only a certain segment of the sentence has the meaning of a professional occupation. Therefore, the relationship is made with the class.

The Table 1 shows the classes with which the FEs links and if this relationship is with a class or an instance.

Table 1. Frame elements and their Semantic Types

FE	Semantic Type	Image
Job_Offer.Graduate	Person	Class
Job_Offer.Job_Opportunity	Job Offer	Class
Job_Offer.Company	Company	Instance
Job_Offer.Place	City	Instance
Job_Offer.Function	Role	Class
Job_Offer.Date	Interval	Instance

In the example used is the following sentence (the company name was omitted for privacy reasons):

A X, empresa onde eu trabalho est com algumas vagas em aberto.

X, the company where I work has a few job openings.

Here, it is assumed that the entire process of searching the list of emails has already been done. Thus, is obtained a set of sentences to be subjected to the parser. Each sentence is read and the first step is to search for lexical units. There is a list of class instances *Span* that have a relation of type *evokes* with the frame Job_Offer. Once a lexical unit is found, the text is marked and follows the search for each FE.

The second step is to look for frame elements. First, we seek to the core and then the non-core elements. To do this task, one should search for classes that inherit from the class *Frame Element*. More specifically, we seek for subclasses of *Job Offer FE*. Each of these entities has a relationship *hasFE* with frame being analyzed. Each of these classes has the property *hasSemanticType*. This relationship connects a FE with some class of the domain ontology.

Once we know the type of the FE, we need to seek *Spans* whose semantic type is the same type. In this way is obtained all the candidates to fill the frame element. Now, all we have to do is check if any of these *Spans* are present in the sentence. A detail of this search is that some elements need to be preceded by a few keywords. For example, elements of type *Place* should be preceded by words such as *em* (at), *est localizada em* (is located at), etc.

When the matching of the FE is made, an instance of the frame element is created. A relation of the type *fillerOf* feita entre o *Span* and the FE. In the example, after the identification of the elements, the sentence is annotated as follows:

A *[X COMPANY]*, *empresa onde [eu GRADUATE] [trabalho LU] est [com algumas vagas em aberto JOB OPPORTUNITY]*.

[**X** COMPANY], **the company where** [**I** GRADUATE] [**work** LU] **has** [**has a few job openings** JOB OPPORTUNITY].

After the annotation of the sentence is possible to explore the instances of the domain ontology. The textual elements now have a meaning in the context of the problem. The classes to which they belong and several of its properties are identified. The element X have a relationship with an instance of the class *Company*. This object, in its turn, has several properties, such as location, owners, field of activity, etc.

After the sentence is annotated and objects of domain ontology are identified, one must decide if this sentence is really a case of a job offer. As stated in section 5.4, there is no definitive way to check if this is the case because of the polysemy inherent in natural language. Several indications may exist due to the elements that occur, however, the sentence may not be related to the frame in question. In this paper a Bayesian network was used to make such a decision, as suggested by [12]. Given a sequence of flags indicating the presence or absence of frame elements, one obtains the conditional probability indicating the chance of the sentence be related to the frame. For this example, the flags are shown in table 2. When the network was trained, were considered as true only those elements that were clearly present.

Table 2. Presence of FEs and LU

Element	Text	Value
Lexical Unit	trabalho	True
Graduate	eu	True
Job Opportunity	com algumas vagas em aberto	True
Company	X	True
Function	-	False
Place	-	False
Date	-	False

In the example sentence, the probability calculated by the Bayesian network was 97.67%. It is important to note that several frames can be added to the system, and each one must have its own Bayesian network.

7   Tests and Results

A multiagent system as described in the model was created for testing. All agents were implemented, however the tests presented here covers only the natural language processing. The objective of the tests was to compare the system performance in relation to a human being on the task to place the sentence in a frame. The analysis of both the system and those made manually generate two possible results: positive and negative. When the two ratings are confronted, there are four possible outcomes listed in the table 3. These results were plotted in a ROC graph(*Receiver Operating Characteristics*) [7].The ROC curve shows the true positive rate (y axis) and false positives (x-axis). Thus, the closer is the result of point (0, 1), better the classification. Following the same reasoning, the results near the diagonal line on the graph (*Random Guess*)represent completely random ratings.

Table 3. Possible values comparing the evaluation of the system with the human evaluation

System	Human	Evaluation
Positive	Positive	True Positive
Positive	Negative	False Positive
Negativo	Positive	False Negative
Negative	Negative	True Negative

The accuracy (ACC) in this type of test can be measured by the formula 1. TP denotes true positive whereas TN the number of true negatives. P and N are used to represent the amount of positive and negative cases respectively.

$$ACC = \frac{TP + TN}{P + N} \tag{1}$$

First, the parser searches and annotates all FE and LU. The first test deals with the evaluation of each of the frame elements. We need to know if the lexical units and FEs were correctly identified and then assess whether the frames as a whole were correctly classified.

For this test 201 samples were analyzed. The result is shown in Figure 6. Most elements obtained good results with the false positive rate below 4% and true positives above 85%. The exception was the non-core elements*City* and *Date*. Even though the false positive rate was low (less than 1%).

The second test examines the classification of frames. 546 samples were analyzed, where each sentence was received from the Bayesian network a probability that ranged from 0 to 100%of belonging or not belonging to a frame of job offer. Various limits were analyzed to determine the value below which a classification would be considered positive. The figure 7 shows the ROC graph considering limits 10%, 30%, 50%, 70% and 90%. The limits marked in 50% and 70% were closest to the point (0, 1) in the graph. Table 4 shows the true positive rate (TPR), false negatives (FPR) and accuracy (ACC).

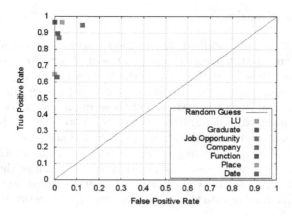

Fig. 6. Graph ROC for FEs and LU

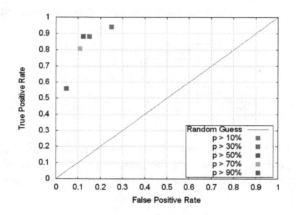

Fig. 7. ROC graph for the classification of sentences

Table 4. TPR, FPR and ACC for sentence classification

Test value	TPR	FPR	ACC
p > 10%	0,941	0,249	0,775
p > 30%	0,882	0,151	0,853
p > 50%	0,882	0,123	0,877
p > 70%	0,809	0,107	0,883
p > 90%	0,559	0,046	0,905

8   Conclusion

This paper proposed a multiagent model to monitor the professional development of graduates of higher education institutions. This is a task that involves searching for information on various sources and in different formats, requiring the use of techniques from natural language processing and distributed systems.

Thus, the use of multiagent systems with techniques PLN appears as a suitable alternative. The system was implemented following this approach and uses several techniques to extract information about the graduates. Some agents are more complex than others. However, even the simplest, as the agent that sends questionnaires, contributes with information that is stored in the data repository. One of the agents uses, besides the typical processing techniques of natural language, such as syntactic annotator, ontologies and semantic frames. The frames were used to characterize the scene which provides semantic background for the sentence. Ontologies define the nature of concepts associated with the lexical items and helps to establish the context. The results were satisfactory for a first version. The system architecture allows new frames to be used to monitor other situations related to the monitoring of graduates.

Acknowledgments. This research is supported in part by the funding agencies FAPEMIG, CNPq and by the Gapso company.

References

1. Artiles, J., Gonzalo, J., Sekine, S.: The semeval-2007 weps evaluation: Establishing a benchmark for the web people search task. In: Proceedings of SemEval 2007 Workshop. Association of Computational Linguistics (ACL), Prague (2007)
2. Balog, K., Azzopardi, L., de Rijke, M.: Resolving person names in web people search. In: King, I., Ricardo Baeza-Yates, R. (eds.) Weaving Services, Locations, and People on the WWW, pp. 301–323. Springer (2009)
3. Balog, K., de Rijke, M.: Associating people and documents. In: Macdonald, C., Ounis, I., Plachouras, V., Ruthven, I., White, R.W. (eds.) ECIR 2008. LNCS, vol. 4956, pp. 296–308. Springer, Heidelberg (2008)
4. Carrasco, R.S., Oliveira, A.P., Lisboa, J., Moreira, A., Arroyo, J.E.: Linguistic structures to support an evidence tracking system for wildlife trafficking. In: CLEI 2011 (October 2011)
5. Craswell, N., de Vries, A.: Overview of the trec-2005 enterprise track. In: The Fourteenth Text REtrieval Conference (TREC 2005) Proceedings (2006)
6. Elmacioglu, E., Tan, Y.F., Yan, S., Kan, M., Lee, D.: Psnus: Web people name disambiguation by simple clustering with rich features. Association of Computational Linguistics (ACL), Prague (2007)
7. Fawcett, T.: An introduction to roc analysis. Pattern Recognition Letters 27(8), 861–874 (2006); ROC Analysis in Pattern Recognition
8. Fillmore, C.J.: Frame semantics. In: Cognitive Linguistics: Basic Readings, pp. 373–400 (2006)
9. Friedman, N., Geiger, D., Goldszmidt, M.: Bayesian network classifiers. Machine Learning 29, 131–163 (1997)
10. Giunchiglia, F., Mylopoulos, J., Perini, A.: The tropos software development methodology: Processes, models and diagrams. In: AAMAS Conference, Bologna, Italy (2002)
11. Hall, M., Frank, E., Holmes, G., Pfahringer, B., Reutemann, P., Witten, I.H.: The weka data mining software: An update. SIGKDD Explor. Newsl. 11(1), 10–18 (2009)

12. Moreira, A.: An Ontology Grounded Framework for Frames Detection. Doctor thesis, Federal University of Juiz de Fora, Brazil (2012)
13. Popescu, O., Magnini, B.: Alleviating the problem of wrong coreferences in web person search. In: The 10th International Conference on Computational Linguistics and Intelligent Text Processing, CICLing 2009, Mexico City, Mexico, pp. 280–293 (2009)
14. Rodrigues, D.F., Oliveira, A.P., Lisboa Filho, J., Moreira, A.: Semi-automatic follow-up of graduates. In: XXXI International Conference of the Chilean Computer Science Society, SCCC 2012 (November 2012)
15. Ruppenhofer, J., Ellsworth, M., Petruck, M.R.L., Johnson, C.R., Scheffczyk, J.: FrameNet II: Extended Theory and Practice (2010)
16. Scheffczyk, J., Baker, C.F., Narayanan, S.: Ontology-based reasoning about lexical resources. In: Oltramari, A. (ed.) ONTOLEX 2006, pp. 1–8. LREC, Genoa (2006)
17. Scheffczyk, J., Pease, A., Ellsworth, M.: Linking framenet to the suggested upper merged ontology. In: Bennett, C., Fellbaum, B. (eds.) Formal Ontology in Information Systems FOIS 2006, pp. 289–300. IOS Press (2006)
18. SemEval, editor, Proceedings of SemEval, Workshop, Prague, Czech Republic. Association of Computational Linguistics, ACL (June 2007)
19. Uschold, M., Gruninger, M.: Ontologies: principles, methods and applications. The Knowledge Engineering Review 11, 93–136 (1996)
20. Wooldridge, M.: An Introduction to MultiAgent Systems. John Wiley & Sons Inc. (2002)

Topic-Based Data Merging and Routing Scheme in Many-to-Many Communication for WSNs

A.S.M. Sanwar Hosen and Gi Hwan Cho[*]

Div. of Computer Science and Engineering, Chonbuk University, Jeonju, S. Korea
{sanwar,ghcho}@jbnu.ac.kr

Abstract. This paper deals with an obvious application of the spanning tree in the many-to-many communication network design, with reflecting the optimum cost of data merging at a tuple-centroid node. Our scheme aims to achieve some degree of load balancing among the constituted tuple-nodes by defining the minimal number of tuples. Meanwhile, minimizing the number of nodes involved in data merging and forwarding towards the multiple sinks would be a fruitful mechanism to reduce the overall routing costs in the network. A mathematical evaluation shows that our method is competent to reduce the number of hops. Therefore, it is efficient in the energy minimization aspect on data routing and consequentially, fruitful to enhance the network lifetime.

Keywords: Wireless Sensor Network, Spanning Tree, Tuple, Routing Costs.

1 Introduction

Wireless Sensor Networks (WSNs) are emerging as both, an important new tier in the IT ecosystem and a rich domain of active researches involving hardware, system design, networking, security, and social factors [1][2]. Even though WSNs are contributing to be fruitful applications in different areas demanding various network strategies, it is widely agreed that network modeling is becoming important.

Sensor nodes are usually deployed randomly in a dynamic region. These deployed nodes are involved in topic-based (data-centric) sensing tasks in the environment [3]. Typically, sensor nodes participate in simultaneously sensing and routing data to the destination/sink in the network. More often, it requires other nodes as intermediate nodes in the data forwarding in source-sink bi-directionally, because of the limited transmission range of the constituted sensor nodes.

Generally, WSNs use different network topology that includes different communication scenarios: one-to-one and many-to-many communications in the network. In one-to-one communication scenario, the sensor nodes normally report its sensed data to the single sink using the linear hop-by-hop strategy [4]. It is sufficient to use the routing path by locating the shortest path towards the sink node. On the other hand, when nodes need to report data to multiple sinks in the network, (i.e.), many-to-many

---

[*] Corresponding author.

B. Murgante et al. (Eds.): ICCSA 2014, Part III, LNCS 8581, pp. 779–794, 2014.

communication, the linear one-to-one is not an optimal solution for minimizing the routing cost.

The Time Division Multiple Access (TDMA) is an efficient approach applied in the communication scenarios of WSNs in order to enhance the performance of different aspects. It is impractical to use the TDMA for avoiding collusion and saving the energy of deployed nodes in a linear one-to-one network. This is because the initiating node can randomly build multiple independent trees towards their corresponding sinks. Furthermore, the intermediate nodes are not aware of the merging mechanism in data routing. Multiple independent trees in the network may involve more nodes in data routing. This situation will cause high routing cost, and as a result, it brings out the early death of constituted nodes in the network, shortening the lifetime of the network.

Let us take as an example scenario of a WSN to monitor a forest. Sensors are deployed to measure the individual topic-based data: Temperature (TM), Humidity (HM), and Luminous (LM) around the infrastructure like Base Stations (BSs) in the forest. Thus, each sensor is a separate data source that captures information and may relay information from other nodes using multi-hops communication (merge and forward) and transmit it to the corresponding monitoring node known as BS/sink. The sink finally transmits the data to a task manager for further processing.

Problem Definition: The proposed approach is addressed for the following scenario: Consider V sensor nodes and k multiple sinks randomly deployed in an area of G. V nodes are responsible for sensing individual data. The multiple sinks may require different topics of sensed data simultaneously from the sensing node for further computation. Here, the following question should be addressed regarding what factors will be affected for the minimum routing costs: *i)* when the sink queries individual topic-based data from the constituted deployed node (in one-to-one communication), *ii)* when the multiple sinks query multi-topics of data from the constituted nodes (in many-to-many communication).

Our Solution: In order to solve the above aspects, we attempt to build n-tuples in the network based on the minimum communication cost corresponding to multiple sources and multiple sinks. Then, the nodes which have the minimum internal routing cost from a 1-hop source to a 1-hop/n-hop sink(s) node will be elected as the tuple-centroid (TC) node. These nodes act as a merging and/or an intermediate node to forward data further to any intermediate node or sink. Here, data merging is preferred at the TC node which is involved with the same topic of data generation. Otherwise, the node will simply act as a gathering and forwarding (bypassing) node.

2   Related Works

In WSN domain, the existing challenges require a well-consolidate network design based on the application implementation scenario. Meanwhile, the network needs to minimize energy consumption in every factor of its design architecture. There are many routing protocols already introduced in the last decade. For instance, some works are based on location-based protocols, as MECN [5], GEAR [6], and TBF [7], where sensor nodes are identified by means of their location. With the location information, sensor nodes can estimate the energy consumption on their routing paths.

The data centric protocol is one of the efficient techniques to reduce the extra computational costs in WSNs. For instance, in the protocols as SPIN [8], DD [9], and REEP [10], the aggregated data from multiple source nodes are routed to the sink in order to save on transmission costs. Many researchers have explored a hierarchical clustering from different perspectives as described in LEACH [11], PEGASIS [12], HEED [13], and APTEEN [14]. Clustering is known as an effective method to group the communication paths.

Considering the data transmission in-between a source and a sink, there are two routing paradigms: single-path routing and multipath routing. In single-path routing, each sensor sends its data to the sink via the shortest path. In multipath routing, each source finds the first k shortest paths to the sink and divides its load evenly among these paths as described in DP [15] and N-to-1 MD [16].

Our proposed method is very similar to the data centric, hierarchical, and multipath based protocol. To the best of our knowledge, the existing protocols are not well-suited for energy efficiency. For instance, data centric protocols emphasize the aggregation of the same topic of data, while our method permits different topics of data could be generated in the network. Therefore, it may decrease the computational costs for processing at an intermediate node to forward it to the sink.

In hierarchical protocols, a cluster header node is elected to aggregate data from its member nodes for further forwarding to the sink. Whereas, our method make use of a tuple with multiple TC nodes in order to provide the most probable scope of merging/aggregating data. It permits the constituted nodes to be involved with the different sensing tasks in a tuple. Here, the TC node(s) would be elected based on the minimum internal routing costs (communications cost among the nodes inside a tuple). Moreover, our method could select the multiple minimum routing cost paths in the network as a whole. As a result, our approach is fully capable of accomplishing the energy efficient routing in one-to-one and many-to-many communications simultaneously.

Table 1. Symbol description

Symbol	Description	Symbol	Description
G	graph/entire topology	P	data forwarding path
V	deployed node set	t	tuple/sub-tree set
u	source terminal/node	δ	number of tuples/partitions in the network (where $\delta = t$)
v	destination terminal/node	x	tuple-centroid (TC) node(s)
T	tree	α	maximum bandwidth
$SP_T(u, v)$	shortest paths between u and v	$info_i$	topic type
$d_T(u, v)$	the distance between u and v	A_i	data for a particular tuple
$E(T)$	the edge/link set of tree T	a_i	data of a particular topic
$w(e)$	assigned weight/link distance	k, n	any positive integer

3   System Modeling

Our work is inspired from the Spanning Trees (STs) [17], which encourage building a network into sub-trees with considering the shortest path edges. Firstly, minimum intercommunication routing sub-trees are constructed, and then a spanning tree is built to connect all leaf nodes to support many-to-many communication. The symbols used in our system modeling for mathematical evaluation are shown in Table 1.

3.1   Network Partition into $\delta(SP_T)$

The concept of $\delta(SP_T)$ is a partition of the network into sufficiently small n-tuples. The n-tuple is a partition of that network, which contains the number of nodes based on the balance traffic load. For instance, if $n = 3$, it is called 3-tuple with an order of nodes (u_1, u_2, u_3) as shown in Fig. 1.

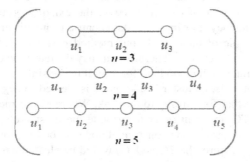

Fig. 1. An example of n-tuples

We use the terms t_i for a particular tuple, a_i for the data of individual topic from a source node u_i among the tuples, and A_i for the amount of collecting bits/data needed to forward from a tuple to the sink(s). The total data $A(t)$ among the deployed nodes in the tuples of the network, can be defined as follows:

$$A(t) = \sum_{i=1}^{n} a_i(u_i), \text{ where } A(t) \geq \alpha \text{ and } a_i \in \inf o \tag{1}$$

From the entire collected data, we can determine the average data size per node by equation (2), where $V(T)$ is the total number of deployed nodes in the network.

$$avg_a(u) = \frac{A(t)}{V(T)}, \text{ where } avg_a(u) \leq \alpha \tag{2}$$

Therefore, the number of nodes that are suitable to group into a tuple based on the maximum transmission capacity (α) of a node is shown in equation (3).

$$n - tuple = \{\frac{A(t)}{\alpha}\} / avg_a(u) \quad, \text{where} \quad A(t) \in \inf o \tag{3}$$

The amounts of data that can be generated from a particular tuple among its consti-tuted tuple-nodes are defined as follows:

$$A_i(t_i) = \sum_{i=1}^{n} a_i(u_i), \text{where} \quad u_i \in t_i \quad \text{and} \quad a_i \in \inf o \tag{4}$$

Therefore, we can partition the entire network in the following equation based on balance traffic load, where $\delta(SP_T)$ is the total number of tuples in the network.

$$\delta(SP_T) = \frac{1}{n - tuple} |V(T)| = \sum_{i=1}^{n} t_i(t) \tag{5}$$

3.2   Tuple-Centroid Node Election

In data routing, the data forwarding path from a source node to a sink should be constructed based on the n-tuple concept. This means that the number of nodes is connected at a TC node, where the connecting nodes are the member nodes (tuple-nodes) and the centroid node is the data gathering point for this tuple. For instance, Fig. 2 presents a TC node $x_1 = u_2$ of a 3-tuple (u_1, u_2, u_3) sub-tree.

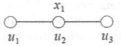

Fig. 2. An example of a TC node in a 3-tuple sub-tree

To evaluate the internal routing cost within a tuple, we use d_T to denote the shortest distance between two nodes (u,v) in a tuple, $l(t_i,e)$ to denote the load of transmitting bits/data in a message of a connecting edge e, and $w(e)$ to denote the assigned weight of the transmission cost for the bits of that distance of node u and v. Therefore, $d_T\{(u_1, u_2) + (u_2, u_3)\} = d_T\{(x_1, u_1) + (x_1, u_3)\}$, where $\forall(x_i, u_i) \in t_i$ and $d_T(u_i, u_{i+1}) > 0$. The node $x_i(u_{n-1}) = u_i$ is the TC node among the tuple-nodes, which obtain the minimum internal routing cost $Cost(t_i)$ defined as follows:

$$Cost(t_i) = \min \sum_{e \in E(t_i)} l(t_i, e) w(e) \quad,$$

$$\text{where} \quad \{l(t_i, e) \in A_i\} \le \alpha \text{ and } w(e) \in d_T(x_i, u_i) \tag{6}$$

Minimizing internal routing cost of the tuple construction leads to minimize the in-ternal communication cost of tuples in the entire network which is defined as follows:

$$C \, os \, t(t) = \min \sum_{i=1}^{n} C \, os \, t(t_i) \qquad (7)$$

3.3   Data Merging at Tuple-Centroid Node

In our method, the TC nodes have two types of forwarding strategies. The TC node x is receiving data $a_i(u_i)$ from the constituted tuple-node, and data belonging to $a_i(x_i) \in info_i$, as well as the data value is in-between the predefined thresholds would be merged together for further forwarding as in equation (4). On the other hand, the received data $a_j(u_j)$ from the constituted tuple-node which is not belonging to $a_j(x_j) \notin info_i$ (data of different topics) would be simply forwarded (bypassed) by the TC node x_i to the intermediate node x_j or directly to the sink(s). A predefined threshold of individual data of a topic value would be taken into account in the merging and the bypassing process in data routing as follows:

$$a_i(u_i) = \begin{cases} 1 & merge \; if \; threshold_1 \le a_i(u_i) \le threshold_2 \; and \; a_i(u_i) = a_i(x_i) \in \inf o_i \\ 0 & otherwise \; bypass \end{cases}$$

3.4   Routing Cost Estimation on Tuple-Connecting Edge

To ensure the minimum cost routing in the network, the minimum cost path selection in the data routing is an important factor in WSNs. Generally, when the cost of an edge represents a price for routing messages between its endpoints (e.g., transmission costs), the routing cost for a pair of vertices in a given spanning tree is defined as the sum of the costs of the edges in the unique tree path between them. The routing cost of the tree itself is the sum over all pairs of vertices in regards to the routing cost for the pair in this tree. Therefore, we can define the routing load on a connecting edge e in-between two separated tuples in the following definition.

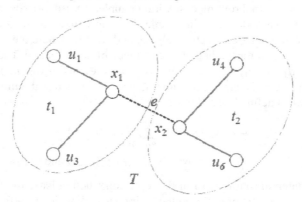

Fig. 3. An example of a tree T with two tuples

Definition 1. *Let T be a tree and $e \in E(T)$ is an edge of T. Assume t_1 and t_2 are two tuples that result by removing e from T. The routing load on the edge e is defined by $l(T,e)=2/V(t_1)//V(t_2)/$.*

Fact 1. *If the network partitioned by $\delta=1/n$-tuple, and the TDMA schedule is used, for any connecting edge $e \in E(T)$, the number of nodes in both sides (the initiating-tuples and the destination-tuple) of e are at least $\delta/V(T)/$, and the routing load on e is at least $\{2\delta^2/V(T)/^2\}/\alpha$ times that of the transmission capacity of a node. Furthermore, for any edge of the T, the routing load is upper bounded by $\{2\delta(1-\delta)/V(T)/^2\}/\alpha$ times that of the transmission capacity of a node.*

From Definition 1 and Fact 1, the routing load on the connecting edge $e \in (x_i, x_j)$ in-between two tuples t_i and t_j will be less than or equal to the transmission capacity of a node as shown in Fig. 3. In addition, the tuples are built by considering the sink node(s) as a tuple-node. The built spanning tree can establish a connecting edge from any source to any destination for bi-directional communications. Therefore, the routing cost on a connecting edge e in-between tuple is $Cost(e)=l(T,e)w(e)$.

3.5   Data Forwarding Path Selection

Intuitively, a forwarding path P is a connected general path that contains a set of TC nodes as intermediate nodes. Starting from any TC node, there are a sufficient amount of tuple-nodes which can only be reached after passing the TC node. Fig. 4 shows two nodes, u_i and u_j as the source and destination, respectively, in different tuples (t) connected by $P = \{P_1, P_2, P_3,..., P_n\}$. The path between them can be divided into three sub-paths: from edge $e(u_i, x_i) \in i$, the paths in $\{x_1, x_2,..., x_n\} \in P$, and the edge $e(x_j, u_j) \in j$.

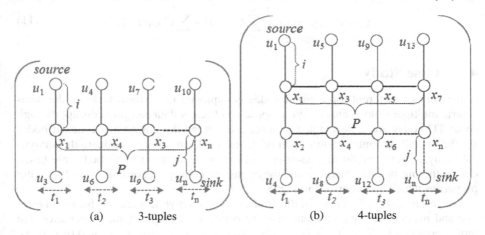

Fig. 4. An example of a path construction in 3-tuples and 4-tuples network

For each tuple contains no more than δn nodes, the distance $d_T(i, P)$ is counted at least $\{2(1-\delta)n\}/\alpha$ times that of the transmission capacity of a node as computed by the routing cost of t_i in equation (6). For each edge in P, since there are at least δn nodes in each tuple of the path, by Fact 1, the routing load on P is at least $\{2\delta(1-\delta)n^2\}/\alpha$ times that of the transmission capacity of a node.

If the equality of hop distance is $d_T(u_i, x_i) = d_T(x_j, u_j)$ in different tuples, from equation (9), the TC node can choose the minimum cost path in data routing from the initiator node towards the destination node defined as follows:

$$Cost(P_i) = \min\{2n \sum_{u \in V(t_i)} d_T(u,P) + \sum_{x_i,x_j \in V(P)} d_T^P(x_i,x_j)\} \tag{8}$$

By the definition of the routing load, we can define the minimum path cost which is obtained in path P as:

$$Cost(P_i) = \min \sum_{x_i,x_j \in V(P)} d_T^P(x_i,x_j)$$

$$= \min \sum_{e \in E(P)} l(P_i,e)w(e) \tag{9}$$

$$Cost(P) = \min \sum_{i=1}^{n} Cost(P_i) \tag{10}$$

Therefore, we can calculate the minimum routing cost from the equation (7) and (10) for the entire network as follows:

$$Cost\delta(SP_T) = \min\{\sum_{i=1}^{n} Cost(t) + \sum_{i=1}^{n} Cost(P)\} \tag{11}$$

4   Case Study

A network of different topic-based nodes is deployed to perform tasks in the data-centric multiple sinks scenario. We propose an efficient data merging concept through some TC nodes and a designated path scheme among the constituted centroid nodes for the overall minimum routing cost of the network. In order to evaluate the analytical analysis between linear one-to-one and many-to-many, we applied tuple-based data gathering with the merging and the bypassing concept according to the following properties which are described in Fig. 5.

For instance, Fig. 5(a) and (b) illustrates a 4 by 4 network scenarios for the one-to-one and many-to-many communication, respectively. In both of these scenarios, the tuple t contains $\{t_1, t_2, t_3, t_4\} = \{(u_1, x_1, x_2, u_4),(u_5, x_3, x_4, u_8),(u_9, x_5, x_6, u_{12}),(u_{13}, x_7, x_8, u_{16})\}$ and $\{(x_1, x_2),(x_3, x_4),(x_5, x_6),(x_7, x_8)\} = \{(u_2, u_3),(u_6, u_7),(u_{10}, u_{11}),(u_{14}, u_{15})\}$. We considered $\{u_1, u_4, u_{13}, u_{16}\}$ as sink nodes which can query for particular information (individual topic like TM) among the deployed nodes. It leads to a one-to-one communication as shown in Fig. 5(a). Similarly, a sink can query multi-topics of data (LM, HM, and TM) from the deployed nodes that leads to a many-to-many communication as shown in Fig. 5(b).

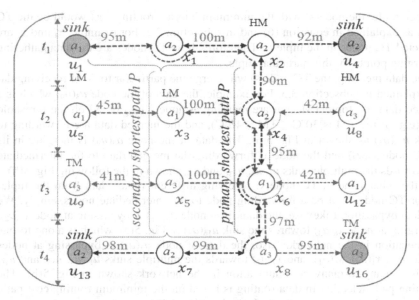

(a) one-to-one communication

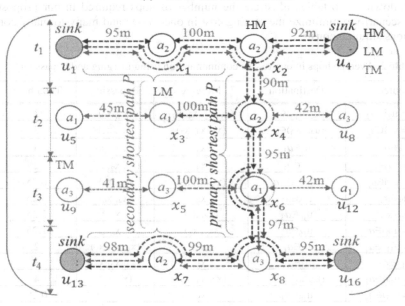

(b) many-to-many communication

Fig. 5. An example of 4-tuples complete networks

In each tuple, the node(s) with the minimum internal routing cost would be the TC node(s) as explained in equation (6) and in subsection 3.2. For example, x_1 and x_2 are two elected TC nodes in the tuple t_1. Therefore, these TC nodes are the data gathering and merging points for this particular tuple.

In the data merging, the TC node(s) will merge the particular topic of receiving data as explained in subsection 3.3. For example, the data of the node $a_3(u_9)$ which is a generated data of topic TM, that are merging at node $x_5(a_3)$ based on the threshold values (e.g., if the TM is $30^0C \leq$ TM $\leq 40^0C$), and the merged data are forwarding to the sink $u_{16}(a_3)$ as shown in Fig. 5(a). The data of the node $a_3(u_9)$ is merging in its own TC node $x_5(a_3)$ and the node x_5 forwarding the merged data to the intermediate nodes towards the multiple sinks u_1, u_4, u_{13}, and u_{16} shown in the following Fig. 5(b).

On the other hand, a TC node(s) receiving different topics data from the tuple-nodes or TC nodes will be a forwarding node to the intermediate nodes simply. We call it data bypassing. Likewise, the data from node $a_3(u_8)$ is bypassing at node $x_4(a_2)$, and merging at node $x_8(a_3)$ towards the sink $u_{16}(a_3)$ in Fig. 5(a), which is a one-to-one communication in the network. Also, the data of node $a_3(u_8)$ is bypassing at nodes $x_4(a_2)$, $x_2(a_2)$, $x_1(a_2)$, $x_6(a_1)$, and $x_7(a_2)$ towards the multiple sinks u_1, u_4, u_{13}, and u_{16}, which is a many-to-many communication in the network shown in Fig. 5(b). The forwarding path selection in data routing is based on the minimum routing cost path, which is described in equation (9) and in subsection 3.5.

The following two tables illustrate the number of hops required in our proposed network scenario to minimize the routing cost in one-to-one and many-to-many communications simultaneously.

Table 2. Average hops in one-to-one communication using merging and bypassing

Source	Destination	Merging	Bypassing	Total hops
$(x_1 = u_2) \in a_2$	$(u_4, u_{13}) \in a_2$	(x_1, x_2, x_4, x_7)	(x_6, x_8)	3
$(x_2 = u_3) \in a_2$	$(u_4, u_{13}) \in a_2$	(x_2, x_4, x_7)	(x_6, x_8)	2
$u_5 \in a_1$	$u_1 \in a_1$	x_3	(x_4, x_2, x_1)	3
$(x_3 = u_6) \in a_1$	$u_1 \in a_1$	x_3	(x_4, x_2, x_1)	2
$(x_4 = u_7) \in a_2$	$(u_4, u_{13}) \in a_2$	(x_4, x_2, x_7)	(x_8, x_6)	0.75
$u_8 \in a_3$	$u_{16} \in a_3$	x_8	(x_4, x_6)	3.25
$u_9 \in a_3$	$u_{16} \in a_3$	(x_5, x_8)	x_6	2.25
$(x_5 = u_{10}) \in a_3$	$u_{16} \in a_3$	(x_5, x_8)	x_6	1.25
$(x_6 = u_{11}) \in a_1$	$u_1 \in a_1$	x_6	(x_4, x_2, x_1)	2
$u_{12} \in a_1$	$u_1 \in a_1$	x_6	(x_4, x_2, x_1)	3
$(x_7 = u_{14}) \in a_2$	$(u_4, u_{13}) \in a_2$	(x_7, x_4, x_2)	(x_8, x_6)	4
$(x_8 = u_{15}) \in a_3$	$u_{16} \in a_3$	x_8		0.25

From Table 2, an average of 4.25 hops is required per source node corresponding to the sinks without the data merging concept in linear one-to-one communication. On the other hand, an average of 2.22 hops is required when data merging is applied. By applying data merging into the routing mechanism, the average hops are significantly reduced by 52.44% per source node in the network.

Similarly, from Table 3, an average of 9.66 hops is required per source node towards the sinks without the data merging concept in many-to-many communication. While, an average of 4.66 hops is required when it is applying the data merging. Therefore, the average hops are significantly reduced by 48.21% per source node in many-to-many communication in the network. Consequentially, we can illustrate the impact of the routing cost by considering two extreme cases in which each link has weight 1. As the proposed method is competent to reduce the number of hops, as well as average distances in its data routing, the routing cost per node would be deliberately minimized. The routing cost reduced by an average of 52.01% and 51% compared to that of linear forwarding method in one-to-one and many-to-many communication respectively.

Table 3. Average hops in many-to-many communication using merging and bypassing

Source	Destination	Merging	Bypassing	Total hops
$(x_1 = u_2) \in a_2$	$(u_1, u_4, u_{13}, u_{16})$	(x_1, x_2, x_4, x_7)	(x_6, x_8)	3.583
$(x_2 = u_3) \in a_2$	$(u_1, u_4, u_{13}, u_{16})$	(x_2, x_1, x_4, x_7)	(x_6, x_8)	2.917
$u_5 \in a_1$	$(u_1, u_4, u_{13}, u_{16})$	(x_3, x_6)	$(x_4, x_2, x_1, x_8, x_7)$	5.333
$(x_3 = u_6) \in a_1$	$(u_1, u_4, u_{13}, u_{16})$	(x_3, x_6)	$(x_4, x_2, x_1, x_8, x_7)$	4.333
$(x_4 = u_7) \in a_2$	$(u_1, u_4, u_{13}, u_{16})$	(x_4, x_2, x_1, x_7)	(x_8, x_6)	3.083
$u_8 \in a_3$	$(u_1, u_4, u_{13}, u_{16})$	x_8	$(x_4, x_2, x_1, x_6, x_7)$	7.75
$u_9 \in a_3$	$(u_1, u_4, u_{13}, u_{16})$	(x_5, x_8)	$(x_4, x_2, x_1, x_6, x_7)$	5.25
$(x_5 = u_{10}) \in a_3$	$(u_1, u_4, u_{13}, u_{16})$	(x_5, x_8)	$(x_4, x_2, x_1, x_6, x_7)$	4.25
$(x_6 = u_{11}) \in a_1$	$(u_1, u_4, u_{13}, u_{16})$	x_6	$(x_4, x_2, x_1, x_8, x_7)$	3.5
$u_{12} \in a_1$	$(u_1, u_4, u_{13}, u_{16})$	x_6	$(x_4, x_2, x_1, x_8, x_7)$	4.5
$(x_7 = u_{14}) \in a_2$	$(u_1, u_4, u_{13}, u_{16})$	(x_7, x_4, x_2, x_1)	(x_8, x_6)	4.667
$(x_8 = u_{15}) \in a_3$	$(u_1, u_4, u_{13}, u_{16})$	x_8	$(x_6, x_4, x_2, x_1, x_7)$	6.75

5   Performance Evaluation

To investigate the efficiency of the proposed method, we used a network simulator, "ns-2" [19] which is a discrete event simulator. The primary performance measurement is the increased network lifetime due to energy awareness during data routing. In the simulated scenario, a rectangle of 470m × 292m two dimensional environment was used as shown in Fig. 5(a) and (b). Approximately 52 sensor nodes of three different topics (a_1, a_2, a_3), were randomly deployed in such a way that there are always other nodes within their maximum transmission range $(R = 200m.)$.

In order to achieve the best performance in the respect to the overall data routing cost, the entire network was divided into 4 tuples of approximately 13 deployed of different topic-based nodes in each tuple based on their maximum transmission range. This method emphasizes on the minimum routing cost of the network. For this reason, we took into account only the data packets, and not the control packets for the routing cost calculations. In order to calculate the transmitter and receiver energy dissipations, equation (12) and (13) in [18] were used. We considered a maximum packet size is 552 bits. In this model, the parameter are described that the transmitter and receiver dissipate E_{elec} = 50nJ/bit to run the transmitter and receiver. To amplify the signal, amplifier dissipates E_{amp} = 100pJ/bit/m to calculate the energy in transmitting a k-bit message through a distance d between the transmitter and the receiver. This method used the transmission control protocol (TCP) as the agent of TCPSink and file transfer protocol (FTP) as the application in the simulation. We used a packet drop probability in the ratio of 0.0 to 0.2. Average 250 data packets of different topics (TM, HM and LM) of data are considered in the energy consumption in data routing per tuple-node for both of the network scenarios, one-to-one and many-to-many communications as shown in Fig. 5.

$$E_{Tx}(k,d) = E_{elec} \times k + E_{amp} \times k \times d^2 \tag{12}$$

$$E_{Rx}(k) = E_{elec} \times k \tag{13}$$

5.1   Network Model and Discussion

During the TC node selection phase of the network design procedure, all the deployed nodes broadcasted a control packet along with their status <*node_ID, data_topic*> over an empirically determined 75m transmission range. The node which received all of the control packets coming from nodes with same data topic nodes of the specific tuple were initially selected as candidate TC nodes of that specific topic of data. The average distance of receiving control packets and the cost of data packets of their own tuple were computed for each one of the initial candidate TC nodes. Finally, the candidate TC node(s) that yielded the minimum average distance as well as the minimum internal routing cost among the specific data topic nodes was selected as the TC node(s) of the particular tuple in the network.

The experimental results of the proposed network model are represented with respect to different criteria. During the simulation, the sinks (4 sinks) were located at the four corners of the designated network field as depicted in Fig. 5. As already mentioned, the sensors were deployed in a 470m × 292m rectangular field with an adjustable transmission range R not exceeding a maximum of 200m. In most of the cases, the minimum average distant TC node obtained minimum internal routing cost. It was investigated whether the internal routing cost could be different depending on the TC node(s) selection within the tuple in the network.

Fig. 6 shows that indeed different internal routing cost, in order to select the minimum cost TC node, we used 250 data packets within 5 Sec. per node, on average to route data to the TC node(s) in each tuple. The resulting internal routing costs of 13 nodes are presented for each one of the four tuples in the Figure. We observe that the routing cost relates to the average distance of the randomly selected node(s) as TC node. Specially, the internal routing cost significantly increases when the average distance of the selected TC node(s) is high. The average distance can be related to the TC node selection at the tuple in the network.

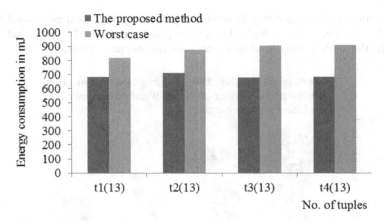

Fig. 6. Internal routing cost of tuples in the network

The proposed method selected TC nodes (two TC nodes among 13s node in a tuple) based on the average distance 58.39m from the tuple-nodes and outperforms by reducing the internal data routing cost by an average of 21.59% compared with the different cases (worst case) of randomly selected TC node(s) with the corresponding average distance (67.57m) from tuple-nodes in the tuple.

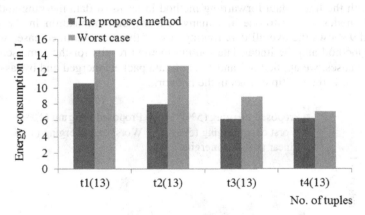

Fig. 7. Data forwarding cost

Concerning cost estimation in terms of minimum cost path selection in data routing, we performed our investigation using the same scenario for the proposed method and the worst case of TC node selection. Fig.7 shows the different path cost corresponding to the different tuples. We used the minimum cost path in the data routing towards the multiple sinks. The resulting higher path cost originated from the TC node selection in a tuple. It is proved that the minimum average distance led to the minimum path cost. The case selected by the proposed method outperforms by reduc-ing data routing cost by an average of 28.70% compared with the worst case of different corresponding distance TC nodes. Additionally, the worst case could be used to select a TC node(s) in a tuple, by selecting that candidate TC node that will be below the threshold energy level in the network reconstruction process.

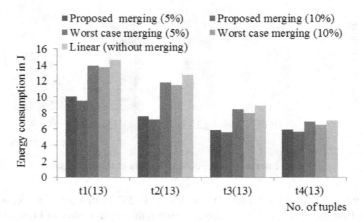

Fig. 8. Data routing cost of one-to-one communication

To measure the overall performance in the routing cost domain, we compared our method with the linear data forwarding method in terms of data merging and bypassing at TC nodes in one-to-one and many-to-many communication in the network. Fig. 8 and 9 shows the overall data routing cost of the two examined cases within the proposed method and the linear data routing method results for the same scenario. In both of the cases, we applied 5% and 10% of data packets merged and bypassed at TC node(s) towards the multiple sinks in the network.

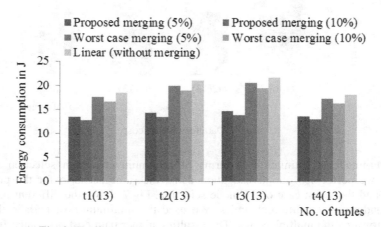

Fig. 9. The data routing cost of many-to-many communication

The results show the proposed merging and bypassing method drastically reduced by an average 32.14 %, 35.63%, 5.30%, 8.50% in one-to-one communication compared to that of corresponding linear method for two examined scenarios. Similarly, this method reduced the total routing cost by an average 29.66 %, 33.35%, 5.48%, and 10.46% in many-to-many communication compared to that of corresponding linear data routing method. Even the worst case of the proposed routing method is far better performing compared to the corresponding linear data routing method in both of the network scenarios. We observed that, when the data merging ratio are increased, the packet received delay and computation cost of data merging also significantly increased at TC nodes.

6   Conclusion and Future Work

In this paper, we introduced a novel network model inspired from the spanning trees and an optimization problem, where it partitions the network into the optimum number of tuples. The network partitioning is based on the load balance in data gathering and in merging at the TC nodes as well as the overall communication costs, which rely on the minimum routing costs for tuple formation. The TC node(s) election in each tuple and minimum cost path selection in data routing play an important role to be reflected on designing an efficient routing mechanism in regards to network lifetime.

As a future work, we plan to solve the TDMA schedule to slot the tuple-nodes to connect to the TC node in order to avoid the unnecessary delay in regards to the centroid node to merge the gathered data from the same type of data involved nodes. Also, an efficient TDMA schedule is required in data routing to merge data or bypass an intermediate node (TC node) for further forwarding towards the sinks.

References

1. Akyildiz, I., Su, W., Sankarasubramaniam, Y., Cayirci, E.: Wireless Sensor Networks: A Survey. Computer Networks 38, 393–422 (2002)
2. Dai, S., Jing, X., Li, L.: Research and Analysis on Routing Protocols for Wireless Sensor Networks. In: International Conference on Communication, Circuits and Systems, vol. 1, pp. 407–411 (2005)
3. Yang, Z., Liu, Y.: Quality of Trilateration: Confidence-Based Iterative Localization. IEEE Transaction on Parallel and Distributed Systems 21, 631–640 (2010)
4. Mottola, L., Picco, G.: MUSTER: Adaptive Energy-Aware Multisink Routing in Wireless Sensor Networks. IEEE Transaction on Mobile Computing 10, 1694–1709 (2011)
5. Rodoplu, V., Meng, T.: Minimum Energy Mobile Wireless Networks. IEEE Journal on Selected Areas in Communications 17, 1333–1344 (1999)
6. Yu, Y., Govindan, R., Estrin, D.: Geographical and Energy Aware Routing: A Recursive Data Dissemination Protocol for Wireless Sensor Networks. Technical Report UCLA/CSDA-TR-01-0023, UCLA Computer Science Department (2001)
7. Nath, B., Niculescu, D.: Routing on a Curve. ACM SIGCOMM Computer Communication Review 33, 155–160 (2003)
8. Heinzelman, W., Kulik, J., Balakrishnan, H.: Adaptive Protocols for Information Dissimination in Wireless Sensor Networks. In: The 5th ACM/IEEE International Conference on Mobile Computing and Networking, pp. 174–185 (1999)

9. Intanagonwiwat, C., Govindan, R., Estrin, D.: Directed Diffusion: A Scalable and Robust Communication Paradigm for Sensor Networks. In: ACM MobiCom 2000, pp. 56–67 (2000)
10. Zabin, F., Misra, S., Woungang, I., Rashvand, H., Ma, N., Ali, M.: REEP: Data Centric, Energy-Efficient and Reliable Routing Protocol for Wireless Sensor Networks. IET Communication 2(8), 995–1008 (2008)
11. Heinzelman, W., Chandrakasan, A., Balakrishnan, H.: Energy-Efficient Communication Protocol for Wireless Microsensor Networks. In: The 33rd International Conference on System Sciences, p. 8020 (2000)
12. Lindsey, S., Raghavendra, C.: PEGASIS: Power-Efficient Gathering in Sensor Information Systems. In: International Conference on Aerospace, pp. 1125–1130 (2002)
13. Younis, O.: HEED: A Hybrid, Energy-Efficient, Distributed Clustering Approach for Ad Hoc Sensor Networks. IEEE Transaction on Mobile Computing 3(4), 366–379 (2004)
14. Manjeshwar, A., Agarwal, D.: APTEEN: A Hybrid Protocol for Efficient Routing and Comprehensive Information Retrieval in Wireless Sensor Networks. In: Parallel and Distributed Processing Symposium, pp. 195–202 (2002)
15. Lindsey, S., Raghavendra, C., Sivalingam, K.: Data Gatering in Sensor Networks using the Energy Delay Metric. In: The 5th International Conference on Parallel and Distributed Processing Symposium, pp. 2001–2008 (2001)
16. Chu, M., Haussecker, H., Zhao, F.: Scalable Information-Driven Sensor Querying and Routing for Ad Hoc Heterogeneous Sensor Networks. Journal of High Performance Computing Application 16(3), 293–313 (2002)
17. Wu, B., Chao, K.: Spanning Trees and Optimization Problems. Chapman & Hall (2004)
18. Awwad, S., Ng Noordin, N., Rasid, M.: Cluster based routing protocol for mobile nodes in wireless sensor network. In: International Symposium on Collaborative Technologies and Systems, pp. 233–241. Baltimore, MD (2009)
19. The network simulator (ns-2), http://www.isi.edu/nsnam/ns/

A Physical-Geometric Approach to Model Thin Dynamical Structures in CAD Systems

Vitalis Wiens[1], J.P.T. Mueller[1], Andreas G. Weber[1], and Dominik L. Michels[2]

[1] Multimedia, Simulation, Virtual Reality Group, Institute of Computer Science II, University of Bonn, Friedrich-Ebert-Allee 144, 53113 Bonn, Germany
{muellerp,wiens,weber}@cs.uni-bonn.de
[2] Department of Computing and Mathematical Sciences, California Institute of Technology, 1200 E. California Blvd., MC 305-16, Pasadena, CA 91125-2100, USA
dominik@caltech.edu

Abstract. The efficient accurate modeling of thin, approximately one-dimensional structures, like cables, fibers, threads, tubes, wires, etc. in CAD systems is a complicated task since the dynamical behavior has to be computed at interactive frame rates to enable a productive workflow. Traditional physical methods often have the deficiency that the solution process is expensive and heavily dependent on minor details of the underlying geometry and the configuration of the applied numerical solver. In contrast, pure geometrical methods are not able to handle all occurring effects in an accurate way.

To overcome this shortcomings, we present a novel and general hybrid physical-geometric approach: the structure's dynamics is handled in a physically accurate way based on the special Cosserat theory of rods capable of capturing effects like bending, twisting, shearing, and extension deformations, while collisions are resolved using a fast geometric sweep strategy which is robust under different numerical and geometric resolutions.

As a result, fast editable high quality tubes can easily be designed including their dynamical behavior.

Keywords: CAD, Cosserat Theory of Rods, Fiber Modeling, Fiber Simulation, Geometric Collision Handling.

1 Introduction

Thin, approximately one-dimensional structures, like cables, fibers, threads, tubes, and wires are ubiquitous objects and occur in a variety of virtual scenes in computer-aided design. Such scenarios are for example ranging from advertising design to the construction of complex column 1 wiring harnesses.

If multiple slender structures occur in the same scene, collisions have to be resolved accurately to avoid interpenetrating configurations in order to achieve realistic results. To reach a productive workflow, the CAD system should handle this collisions automatically instead of leaving this legacy almost impossible to solve for the user in complex scenes.

B. Murgante et al. (Eds.): ICCSA 2014, Part III, LNCS 8581, pp. 795–808, 2014.

In dynamical scenarios the requirements are significantly higher, since in addition to the collisions, different deformation modes like bending, twisting, shearing, and extension have to be handled. For example cables typically exhibit a very high resistance against tensile deformations, whereas their bending and twisting stiffnesses are comparatively low, which leads to a dynamical process which takes place on different time scales. From a mathematical point of view so-called "stiff" differential equations have to be handled, see [1]. According to the classical definition of Curtiss and Hirschfelder the term "stiff" indicates situations in which "...certain implicit methods perform better than explicit ones" (cf. [2]). Stiffness as such is not a characteristic of differential equations nor is it a property of the numerical methods applied to solve these equations, rather it is an issue how efficient the solution process can be realized. Of course, computer-aided design is an interactive process for which reason efficiency plays an important role to enable a productive workflow.

To reach user interaction at interactive frame rates, up-to-date CAD systems typically increase the resolution of the underlying geometry as well as the numerical accuracy immediately prior to the computation of the final results whereas low resolutions are used during the interactive design process. Physics-based collision handling is one aspect, which can profoundly be dependent on different numerical and geometric resolutions.

We introduce an efficient hybrid physical-geometric algorithm to model dynamical approximately one-dimensional objects using CAD system, in which the structure's dynamics is handled in a physically accurate way, while collisions are resolved by employing a geometric sweep along the objects. The final framework is robust under different numerical and geometric resolutions and can be integrated in existing CAD systems.

In this regard our specific contributions are as follows.

- A general formulation of the special Cosserat theory of rods is described. It is able to physically accurate describe the motion of one-dimensional structures covering bending, twisting, shearing, and extension deformations.
- This formulation is extended with an advanced geometric sweep strategy to efficiently resolve collisions and interpenetrations.
- High numerical stability of the solution process of the resulting inherently "stiff" equations of motion is achieved efficiently due to the application of the generalized α-method.
- The usability and advantageous behavior of the resulting algorithm in the context of computer-aided design is successfully demonstrated by presenting examples from the design and engineering section.

2   Related Work

We provide a brief overview of the literature concerned with the simulation of slender approximately one-dimensional structures.

In 1991, Rosenblum et al. [3] presented one of the first methods to simulate the dynamics of a single hair fiber by employing a particle system. Furthermore, several approximate solutions to the Cosserat equations have been published. For example, in [4] each fiber is represented by a Super-Helix, i.e. a piecewise helical rod which is animated by employing the principles of Lagrangian mechanics. A Cosserat model for rods that is very suitable for being used in CAD systems is presented in [5]. The method allows the integration of contact forces but has no provision for efficient handling of interactions in complex collision scenarios. A thorough mathematical derivation of the Euler-Lagrange equations for nonlinearly elastic rods with self-contact is presented in [6]. The authors of [7] developed a fiber simulation framework in which the dynamical behavior is independent of the discretization. This is realized by computing energies per element using a finite element methods. In contrast, a symbolic-numeric method for solving Kirchhoff rods is presented in [8].

The authors of [9] adapted concepts from differential geometry to realize a discrete geometric model of thin flexible rods. In [10], the cantilever beam equation from mechanical engineering is employed to model single hair fibers on a human head. There is no treatment of the interactions between the individual fibers, as it was considered computationally too challenging for such an amount of objects.

Frameworks that successfully handle interactions in large complex fiber-based systems like in a shake of the head were amongst others presented in [11,12]. The authors of [11] employ a contact handling method based on a complementarity problem. Such kinds of formulations have also been addressed in [13,14,15]. In [12] fiber-fiber interactions including sticking and clumping behavior as well as collisions with objects are resolved by representing the object collision geometry as a level set signed distance function.

3 A Hybrid Physical-Geometric Approach

We devise a hybrid physical-geometric approach, in which the dynamics of single objects is handled physically accurately by employing a continuum mechanics formulation based on the special Cosserat theory of rods. In contrast, collision handling is realized in a geometric way by performing a sweep strategy along the objects.

3.1 Dynamics of One-dimensional Structures

Cables, fibers, threads, tubes, and wires can approximately be considered as one-dimensional continua that undergo bending, twisting, shearing, and longitudinal dilation deformation. Following the special Cosserat theory of rods (cf. [16]), the motion of a special rod is given by

$$(s,t) \mapsto (r(s,t), d_1(s,t), d_2(s,t)), \tag{1}$$

where $r(s,t)$ is its center line. It is furnished with a set of right-handed orthonormal basis directors $\{d_1(s,t), d_2(s,t), d_3(s,t)\}$ with $d_3 := d_1 \times d_2$ as illustrated in Fig. (1). The directors d_1 and d_2 span the cross-section plane. The rod's deformation is obtained by its motion defined by (1) is related to some reference configuration $\{r^\circ(s,t), d_1^\circ(s,t), d_2^\circ(s,t)\}$.

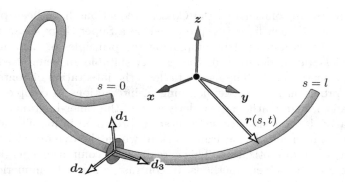

Fig. 1. The vector set $\{d_k\}$ forms a right-handed orthonormal basis at each point of the center line. The directors d_1 and d_2 span the local material cross-section, whereas d_3 is perpendicular to the material cross-section. Note that in the presence of shear deformations d_3 is not tangent to the center line of the fiber.

Further, there exist vector values functions $\boldsymbol{\kappa}$ and $\boldsymbol{\omega}$ such that the directors evolve according to the kinematic relations

$$\partial_s d_k = \boldsymbol{\kappa} \times d_k,$$
$$\partial_t d_k = \boldsymbol{\omega} \times d_k,$$

where $\boldsymbol{\kappa}$ is the Darboux and $\boldsymbol{\omega}$ the twist vector. Their components are given with respect to the orthonormal basis, i.e. $\boldsymbol{\kappa} = \sum_{k=1}^{3} \kappa_k d_k$ and $\boldsymbol{\omega} = \sum_{k=1}^{3} \omega_k d_k$.

The linear strains of the rod are given by $\boldsymbol{\nu} = \sum_{k=1}^{3} \nu_k d_i = \partial_s r$ and the velocity of a cross-section material plane by $\boldsymbol{v} = \partial_t r$. The triples $(\kappa_1, \kappa_2, \kappa_3)$, $(\omega_1, \omega_2, \omega_3)$, (ν_1, ν_2, ν_3), and (v_1, v_2, v_3) are denoted by $\mathbf{k}$, $\mathbf{w}$, $\mathbf{v}$, and $\mathbf{p}$ respectively. In particular, $\mathbf{k} := (\kappa_1, \kappa_2, \kappa_3)$ and $\mathbf{v} := (\nu_1, \nu_2, \nu_3)$ are the strain variables that uniquely determine the motion of the rod Eqn. (1) at every instant in time t (except for a rigid body motion). Their components have a physical meaning: they describe the bending of the rod with respect to the two major axes of the cross section (κ_1, κ_2), the torsion (κ_3), shear (ν_1, ν_2) and extension (ν_3). Moreover, since $\partial_t \partial_s d_k = \partial_s \partial_t d_k$ we obtain the compatibility equation

$$\partial_t \boldsymbol{\kappa} = \partial_s \boldsymbol{\omega} - \boldsymbol{\kappa} \times \boldsymbol{\omega}.$$

In the same sense we have

$$\partial_t \boldsymbol{\nu} = \partial_s \boldsymbol{v}.$$

From this the equations of motion can be derived, which read

$$\partial_s \boldsymbol{n} + \boldsymbol{f} = \rho A \partial_t \boldsymbol{v} + \rho \left(I_1 \partial_{tt} d_1 + I_2 \partial_{tt} d_2 \right),$$
$$\partial_s \boldsymbol{m} + \boldsymbol{\nu} \times \boldsymbol{n} + \boldsymbol{l} = \rho \left(I_1 d_1 + I_2 d_2 \right) \times \partial_t \boldsymbol{v} + \partial_t \left(\rho \boldsymbol{J} \boldsymbol{\omega} \right),$$

where $\boldsymbol{n} = n_k d_k$ and $\boldsymbol{m} = m_k d_k$ are the internal stresses and $\boldsymbol{f}$ and $\boldsymbol{l}$ are the external forces and torques acting on the rod. I_1 and I_2 are the first mass moments of inertia of cross section per unit length and $\boldsymbol{J}$ is the inertia tensor of cross section per unit length. Further, we define $\mathbf{n} := (n_1, n_2, n_3)$ and $\mathbf{m} := (m_1, m_2, m_3)$. n_1 and n_2 are the shear forces, and $\boldsymbol{n} \cdot \boldsymbol{\nu} / \|\boldsymbol{\nu}\|$ the tension. m_1 and m_2 are the bending moments and m_3 the twisting moment.

In order to relate the internal stresses $\boldsymbol{n}$ and $\boldsymbol{m}$ to the kinematic quantities $\boldsymbol{\nu}$ and $\boldsymbol{\kappa}$ we introduce constitutive equations of the form

$$\mathbf{n}(s,t) = \hat{\mathbf{n}}\left(\mathbf{k}(s,t), \mathbf{v}(s,t), s\right),$$
$$\mathbf{m}(s,t) = \hat{\mathbf{m}}\left(\mathbf{k}(s,t), \mathbf{v}(s,t), s\right).$$

For fixed s, the common domain $\mathcal{V}(s)$ of these constitutive functions is a subset of $(\mathbf{k}, \mathbf{v})$ describing orientation preserving deformations. $\mathcal{V}(s)$ consists at least of all $(\mathbf{k}, \mathbf{v})$ that satisfy $\nu_3 = \boldsymbol{\nu} \cdot \boldsymbol{d}_3 > 0$. The rod is called hyper-elastic, if there exists a strain-energy-function

$$W : \{(\mathbf{k}, \mathbf{v} \in \mathcal{V})\} \to \mathbb{R},$$

such that

$$\hat{\mathbf{n}}\left(\mathbf{k}, \mathbf{v}, s\right) = \partial W\left(\mathbf{k}, \mathbf{v}, s\right) / \partial \mathbf{v},$$
$$\hat{\mathbf{m}}\left(\mathbf{k}, \mathbf{v}, s\right) = \partial W\left(\mathbf{k}, \mathbf{v}, s\right) / \partial \mathbf{u}.$$

The full system of partial differential equations governing the deformation of an elastic rod is thus given by the following first order system,

$$\partial_t \boldsymbol{d}_k = \boldsymbol{\omega} \times \boldsymbol{d}_k, \tag{2}$$
$$\partial_t \boldsymbol{\kappa} = \partial_s \boldsymbol{\omega} - \boldsymbol{\kappa} \times \boldsymbol{\omega},$$
$$\partial_t \boldsymbol{\nu} = \partial_s \boldsymbol{v},$$
$$\partial_t \left(\rho \boldsymbol{J} \boldsymbol{\omega}\right) = \partial_s \left(\hat{m}_k(\mathbf{k}, \mathbf{v}) \boldsymbol{d}_k\right) + \boldsymbol{\nu} \times \hat{n}_k(\mathbf{k}, \mathbf{v}) \boldsymbol{d}_k,$$
$$\rho A \partial_t \boldsymbol{v} = \partial_s \left(\hat{n}_k(\mathbf{k}, \mathbf{v}) \boldsymbol{d}_k\right).$$

This system can be decoupled from the kinematic equation (2) by decomposing it with respect to the basis $\{\boldsymbol{d}_k\}$ which yields

$$\partial_t \mathbf{k} = \partial_s \mathbf{w} - \mathbf{w} \times \mathbf{k}, \tag{3}$$
$$\partial_t \mathbf{v} = \partial_s \mathbf{p} + \mathbf{u} \times \mathbf{p} - \mathbf{w} \times \mathbf{v}, \tag{4}$$
$$\partial_t \left(\rho \mathbf{J} \mathbf{w}\right) = \partial_s \hat{\mathbf{m}} + \mathbf{k} \times \hat{\mathbf{m}} + \mathbf{v} \times \hat{\mathbf{n}} - \mathbf{w} \times \left(\rho \mathbf{J} \mathbf{w}\right), \tag{5}$$
$$\rho A \partial_t \mathbf{p} = \partial_s \hat{\mathbf{n}} + \mathbf{k} \times \hat{\mathbf{n}} - \mathbf{w} \times \left(\rho A \mathbf{p}\right). \tag{6}$$

The dynamical behavior of the object is influenced by the density ρ, the cross-section area A, the area moments of inertia I_1 and I_2 of the cross section, the polar moment of inertia I_3, and the inertia tensor $\mathbf{J}$. The constitutive laws for elastic material behavior become

$$\hat{\mathbf{n}}(s,t) = \left(GA\left(\nu_1 - \nu_1^\circ\right), GA\left(\nu_2 - \nu_2^\circ\right), EA\left(\nu_3 - \nu_3^\circ\right)\right),$$

with the initial strain vector field $\boldsymbol{\nu}^{\circ}(s)$, shear modulus G, Young's modulus E, and

$$\hat{\mathbf{m}}(s,t) = \left(E_b I_1 \left(\kappa_1 - \kappa_1^{\circ}\right), E_b I_2 \left(\kappa_2 - \kappa_2^{\circ}\right), GI_3 \left(\kappa_3 - \kappa_3^{\circ}\right)\right),$$

with the initial bending and torsion vector field $\boldsymbol{\kappa}^{\circ}(s)$ and Young's modulus E_b of bending.

For example in the case of a fiber with a circular cross section, the bending stiffnesses $E_b I_{\{1,2\}}$ and torsional stiffnesses $G_b I_3$ change with the fourth power of the fiber diameter. They are usually orders of magnitude smaller than the tensile stiffnesses EA and shearing stiffnesses GA scaling with the second power. This renders the problem inherently "stiff" and hence numerically challenging. Therefore we solve the system (3-6) using the generalized α-method (cf. [17]). This numerical solver was originally introduced for "stiff" mechanical problems. For this purpose it is desirable to achieve a maximum damping of the high-frequency parts to reach high numerical stability while the low-frequency parts are damped as less as possible. Its use is motivated due to the fact that the ratio of this damping intensities is optimal (cf. [18]).

Since external forces like gravity are to be considered as well, they have to be added to the right-hand side of (6) after transforming them into the local basis. For this purpose the kinematic equation (2) has be solved additionally. This realized using its analytical solution, see [19].

3.2   Geometric Collision Handling

In this section, we briefly describe our geometric collision handling strategy which can be used to resolve collisions and interpenetrations in the fiber set $\boldsymbol{F}$ whose configuration is designed from an user input or is computed by a physical simulation.

The geometry of a single fiber

$$(s,t_0) \mapsto (\boldsymbol{r}(s,t_0), \boldsymbol{d}_1(s,t_0), \boldsymbol{d}_2(s,t_0)),$$

parametrized with $s \in [0, N\Delta s]$ at a fixed point in time t_0 can be described in a discrete version using

$$\boldsymbol{f} = (\boldsymbol{s}_1 := [\boldsymbol{v}_1, \boldsymbol{v}_2], \boldsymbol{s}_2 := [\boldsymbol{v}_2, \boldsymbol{v}_3], \cdots, \boldsymbol{s}_N := [\boldsymbol{v}_N, \boldsymbol{v}_{N+1}]),$$

with $\boldsymbol{v}_i := \boldsymbol{r}\left((i-1)\Delta s, t_0\right)$ and

$$\left(\boldsymbol{\kappa}(\boldsymbol{v}_i, t_0), \boldsymbol{\omega}(\boldsymbol{v}_i, t_0)\right)_{i \in \{1,2,\cdots,N\}},$$

in the case equidistant lengths of the segments $\boldsymbol{s}_1, \boldsymbol{s}_2, \cdots, \boldsymbol{s}_N$.

Algorithm 1. A geometric sweep method to resolve collisions and interpenetrations in sets of slender structures.

Input: Set of objects F at time t, number of tries m, and functions $\alpha(\cdot,\cdot)$, $g(\cdot)$.
Output: Interpenetration free set of fibers F.

1: $\mathcal{K} \leftarrow \mathsf{initKdTree}(F)$
2: $r \leftarrow \mathsf{longestSegmentLength}(F)/2 + 2\,\mathsf{largestSemimajor}(F)$
3: **for each** $f \leftarrow ([v_1, v_2], [v_2, v_3], \cdots, [v_N, v_{N+1}]) \in F$ **do**
4:     **for each** $s \leftarrow [v_p, v_n] \in f$ **with indices** p, n **do**
5:        $S \leftarrow \mathsf{sphericalLookupAroundVertex}(\mathcal{K}, v_p, r)$
6:        $S \leftarrow S \setminus \{[v_{p-1}, v_p], [v_p, v_n]\}$
7:        $d_{\mathsf{stick}} \leftarrow \mathsf{normalize}(v_n - v_p)$
8:        $d_{\mathsf{push}} \leftarrow \mathsf{normalize}(\sum_{[s_1, s_2] \in S} \mathsf{normalize}(v_n - s_1) + \mathsf{normalize}(v_n - s_2))$
9:        $\alpha^* \leftarrow \alpha(\kappa(v_p, t), \omega(v_p, t))$
10:       $p \leftarrow \mathsf{normalize}(\mathsf{lerp}(d_{\mathsf{push}}, d_{\mathsf{stick}} \times d_{\mathsf{push}}, \alpha^*))$
11:       $v'_n \leftarrow v_n + p$
12:       $v'_n \leftarrow v_p + \|v_n - v_p\|_2\, \mathsf{normalize}(v'_n - v_p)$
13:       **if** (cylindrical segment of $[v_p, v'_n]) \cap S \neq \varnothing$ **then**
14:          $V \leftarrow \varnothing$
15:          $u \leftarrow \mathsf{normalize}(d_{\mathsf{stick}} \times p)$
16:          $v \leftarrow \mathsf{normalize}(d_{\mathsf{stick}} \times u)$
17:          **for** $i \leftarrow 1$ **to** m **do**
18:             $R \leftarrow [u, v, d_{\mathsf{stick}}]$
19:             $\theta \leftarrow \arccos(1 - \sqrt{\mathsf{random}(0,1)})$
20:             $\varphi \leftarrow 2\pi\,\mathsf{random}(0,1)$
21:             $v'_n \leftarrow v_p + \|v_n - v_p\|_2\, R\,[\sin\theta \cos\varphi, \sin\theta \sin\varphi, \cos\theta]^\mathsf{T}$
22:             **if** (cylindrical segment of $[v_p, v'_n]) \cap S = \varnothing$ **then**
23:                $V \leftarrow V \cup \{v'_n\}$
24:             **end if**
25:          **end for**
26:          **if** $|V| > 0$ **then**
27:             $v'_n \leftarrow \underset{v'_n \in V}{\mathrm{argmax}}\left(\mathsf{lerp}\big(\mathsf{normalize}(v'_n - v_p) \cdot p,\; \mathsf{normalize}(v'_n - v_p) \cdot d_{\mathsf{stick}}, \alpha^*\big)\right)$
28:          **else**
29:             ▷ **Backtracking**
30:          **end if**
31:       **end if**
32:       **for** $i \leftarrow 1$ **to** N **do**
33:          $\delta \leftarrow v'_n - v_n$
34:          $v_i \leftarrow v_i + g\big(\mathrm{sgn}(i - n) \sum_{k=\min\{i,n\}+1}^{\max\{i,n\}} \|v_k - v_{k-1}\|_2\big)\,\delta$
35:       **end for**
36:       $\mathcal{K} \leftarrow \mathsf{updateKdTree}(\mathcal{K}, F)$
37:     **end for**
38: **end for**

In F, several collision scenarios can occur like full collisions with intersecting border cross sections between the cylindrical segments and partial collisions, as illustrated in Fig. (2). Therefore, a fast detection of the cylindrical segment-segment collision is needed. Since the cross sections are of elliptic shape, computing collisions between primitives is sufficient. For a detailed explanation, we refer to [20].

For collision handling, we modify each fiber in place to ensure an intersection free result, as listed in Alg. (1). In order to perform an efficient collision detection, we build up a k-d tree $\mathcal{K}$ from all segments in the fibers $f \in F$. The lookups into $\mathcal{K}$ are performed with spherical regions around vertices with fixed radii determined from the longest segment length and the largest semimajor, which is sufficient to find all intersecting elliptical segments.

In the following, we consider each segment $s = [v_p, v_n]$ of all $f \in F$ and manipulate the location of v_n and its successors if needed to resolve intersections.

In order to shift v_n we have to compute the moving direction p. Let S be the result of the spherical lookup with radius r around v_p in $\mathcal{K}$. This also includes the segments $s_{p-1} = [v_{p-1}, v_p]$ and $s_p = [v_p, v_n]$ adjacent to v_p. They are always included in S for radius r, and have to be excluded from further collision testing for this segment.

In order to determine an appropriate vector p we have to model two effects that are present when handling intersections: Firstly, we want to let p point into the mean direction d_{push} that points away from all segments in S. Secondly, we want to introduce a twist direction that leads to a twisting of the current segment around the other segments in the local neighborhood. This twist vector is orthogonal to the current segment direction d_{stick}, as well as to d_{push}.

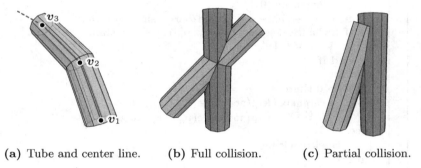

(a) Tube and center line.   (b) Full collision.   (c) Partial collision.

Fig. 2. Illustration of a tube with its center line (2a) and two different collision types occurring in sets of slender objects like cables, fibers, threads, tubes, and wires (2b,2c)

The twisting of the current segment is dependent on the Darboux vector κ and twist vector ω given by the result of the physical simulation. It is furthermore customizable by a twist adjustment function α, that is used for linear interpolating between a no twisting at all (d_{push}) and a full twisting scenario ($d_{\text{stick}} \times d_{\text{push}}$) as illustrated in Fig. (3). The result of this adjusted twist direction is p in which v_n is moved.

Before using more elaborate measures, we shift v_n by p, renormalize to the original segment length and check if an intersection-free configuration is obtained. This is an optimization since in many cases no other choice of v_n is needed.

However, if we did not obtain an intersection-free configuration, resorting to different means of calculating the new location of v_n is necessary. For that we calculate the transformation matrix $R = [u, v, d_{\text{stick}}]$ of an orthonormal basis system located at v_n, whose local vertical axis points into the direction of d_{stick}, and is arbitrarily rotated in the remaining dimension with axes u and v. After that we employ a cosine-weighted sampling of the hemispherical directions θ, φ in the orthogonal frame R, and check if moving v_n into this direction while maintaining the current segment length yields an intersection-free configuration. If this is the case, we save this location into an set of possible locations V.

If multiple locations in the last step are found, we take the one whose direction is both closest to the adjusted twist direction p and closest to the original segment direction d_{stick}. This trade-off is also adjustable using α. Only if no location could be found in the previous step we cannot resolve the intersections of this time step. In this case a backtracking approach is employed.

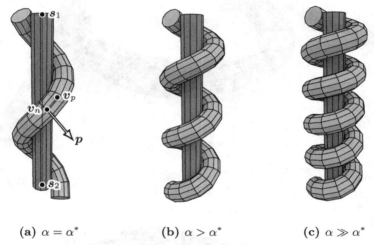

(a) $\alpha = \alpha^*$ **(b)** $\alpha > \alpha^*$ **(c)** $\alpha \gg \alpha^*$

Fig. 3. Illustration of different resolved collision scenarios. In each scene a tube is striped around another one using different twist adjustments α. Moreover, the collision resolving process is visualized (3a).

Both in the case that we have directly found a displaced location for v_n or that we have found such a location by hemispherical sampling, we end up with a new location v'_n. This new location does preserve the length of the current segment. The location of the vertex v_n should be set to v'_n, and the vertices of f in the local neighborhood of v_n should be moved proportionally. This proportional editing step employs a falloff function $g : \mathbb{R} \to \mathbb{R}_+$ that is dependent on the signed distance between an vertex v_i of f and the currently moved vertex v_n. Thus $g(\cdot)$ is subject to the following conditions:

Firstly, the currently manipulated vertex v_n should be moved directly onto v'_n, i.e. $g(0) = 1$ and secondly, all previously manipulated vertices should not be moved to avoid reintroducing collisions between already resolved segments, i.e. $g(x) = 0$ for all $x < 0$.

This introduces a smooth editing falloff and yields better results for the following segments. Finally we update the k-d tree $\mathcal{K}$ with these new changed segments.

4   Evaluation and Results

We design several simulation scenarios of coiled cables, tangled earphones, and parts of a wiring harness with multiple cables. In this context, we assume a circular

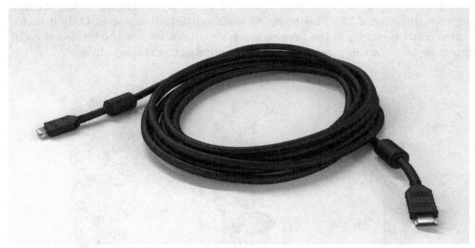

Fig. 4. Photo-realistic rendering of a coiled HDMI cable with photo comparison below. A radius of 3.75 mm is used, the cable's length is set to 5.00 m, and the twist adjustment α is uniformly sampled between 0 and 2.5.

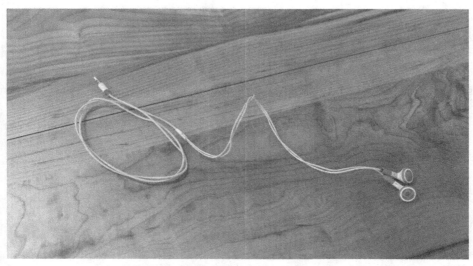

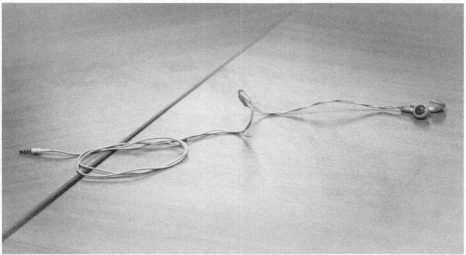

Fig. 5. Photo-realistic rendering of the final result of the tangled earphones falling on a table simulation scenario with photo comparison below. A radius of 0.75 mm for the lower part of length 73.50 cm (respectively a radius 1.00 mm for the upper part of length 33.00 cm) is used. The twist adjustment is set to $\alpha = 1$.

cross section with diverse radii, different cable lengths, and material parameters $E = E_b = 124.00\,\mathrm{GPa}$ and $G = 47.00\,\mathrm{GPa}$. A mass per volume ratio of $8.92\mathrm{g/cm}^3$ is used according to the density of copper.

We choose a falloff function given by the truncated and scaled Gaussian $g(\cdot) = \sqrt{2\pi}\sigma\,\mathcal{G}_{0,\sigma}(\cdot)$ for $x \geq 0$ and $g(\cdot) = 0$ else, with mean 0 and standard deviation $\sigma = 1$ normalized for object size. Different twist adjustments α are being used.

Fig. 6. Photo-realistic rendering of a part of a wiring harness with photo comparison below. This part contains nine cables. For each a radius of 1.25 mm is used and the lengths of the visible parts are 11.40 cm. The twist adjustment is set to $\alpha = 1$.

The algorithm has an expected runtime of $\mathcal{O}(n^{5/3})$, with n being the number of all segments of all fibers. The initialization of the k-d tree $\mathcal{K}$ takes $\mathcal{O}(n \log n)$ time. Each spherical lookup for the n segments has a complexity of $\mathcal{O}(n^{2/3} + q)$ for q returned segments in the lookup radius; these lookups are the dominant time sink of this algorithm. The update of the k-d tree can run in $\mathcal{O}(\log n)$ for each segment: The number of changed vertices due to the proportional editing is bounded by $\lceil s/l \rceil$, with l being the minimal segment length of $\boldsymbol{F}$ and s being the length of the support of g. As our algorithm leaves the segment lengths invariant, the maximum number of changed vertices is constant and the update step is not the limiting factor of this algorithm. Overall, we get a sub-quadratic runtime of $\mathcal{O}(n^{5/3})$.

All modeling processes were realized at interactive frame rates. The final results are illustrated in Fig. (4-6) including comparisons with real photos. Since all these design and engineering example scenes have been created in minutes and optically correspond to reality, advanced usability as well as a high degree of realism is successfully demonstrated.

5 Conclusion and Future Work

In this contribution an extended formulation of the special Cosserat theory of rods with a geometric sweep strategy to resolve collisions and interpenetrations is introduced. This hybrid physical-geometric framework is able to physically accurate capture effects like bending, twisting, shearing, and extension, and at the same time it handles collision and interpenetrations successfully and efficiently. Overall numerical stability is achieved by applying the generalized α-method although the differential equations behaves "stiff".

The usability and advantageous behavior in the context of CAD systems is demonstrated with design and engineering example scenes.

In our future work, we aim to apply the presented general framework for the physically based modeling of slender structures to further application scenarios like the molecular modeling of DNA strands.

Acknowledgements. We are grateful to Christoph Peters and Ralf Sarlette for helpful suggestions and discussions. Moreover our thanks go to the anonymous reviewers for many critical comments and feedback.

Finally, the partial financial support of the German National Merit Foundation is also gratefully acknowledged.

References

1. Hairer, E., Wanner, G.: Solving Ordinary Differential Equations II: Stiff and Differential-Algebraic Problems. Springer (1996)
2. Curtiss, C.F., Hirschfelder, J.O.: Integration of stiff equations. Proceedings of the National Academy of Sciences of the United States of America 38(3), 235–243 (1952)
3. Rosenblum, R.E., Carlson, W.E., Tripp, E.: Simulating the structure and dynamics of human hair: Modeling, rendering and animation. The Journal of Visualization and Computer Animation 2(4), 141–148 (1991)
4. Bertails, F., Audoly, B., Cani, M.-P., Querleux, B., Leroy, F., Lévêque, J.-L.: Super-Helices for Predicting the Dynamics of Natural Hair. ACM Transactions on Graphics (TOG) 25(3), 1180–1187 (2006)
5. Grégoire, M., Schömer, E.: Interactive simulation of one-dimensional flexible parts. Computer-Aided Design 39(8), 694–707 (2007)

6. Schuricht, F., Von der Mosel, H.: Euler-Lagrange equations for nonlinearly elastic rods with self-contact. Archive for Rational Mechanics and Analysis 168(1), 35–82 (2003)
7. Spillmann, J., Teschner, M.: CoRdE: Cosserat rod elements for the dynamic simulation of one-dimensional elastic objects. In: Proceedings of the 2007 ACM SIGGRAPH/Eurographics Symposium on Computer Animation, pp. 63–72. Eurographics Association (2007)
8. Sobottka, G., Weber, A.: A symbolic-numeric approach to tube modeling in CAD systems. In: Ganzha, V.G., Mayr, E.W., Vorozhtsov, E.V. (eds.) CASC 2006. LNCS, vol. 4194, pp. 279–283. Springer, Heidelberg (2006)
9. Bergou, M., Wardetzky, M., Robinson, S., Audoly, B., Grinspun, E.: Discrete elastic rods. ACM Transactions on Graphics (TOG) 27(3), 63:1–63:12 (2008)
10. Anjyo, K.-I., Usami, Y., Kurihara, T.: A simple method for extracting the natural beauty of hair. ACM SIGGRAPH Computer Graphics 26, 111–120 (1992)
11. Daviet, G., Bertails-Descoubes, F., Boissieux, L.: A hybrid iterative solver for robustly capturing coulomb friction in hair dynamics. ACM Transactions on Graphics (TOG) 30, 139:1–139:12 (2011)
12. Selle, A., Lentine, M., Fedkiw, R.: A mass spring model for hair simulation. ACM Transactions on Graphics (TOG) 27, 64:1–64:11 (2008)
13. Fukushima, M., Luo, Z.-Q., Tseng, P.: Smoothing functions for second-order-cone complementarity problems. SIAM Journal on Optimization 12(2), 436–460 (2002)
14. Jiang, H.: Global convergence analysis of the generalized Newton and Gauss-Newton methods of the Fischer-Burmeister equation for the complementarity problem. Mathematics of Operations Research 24(3), 529–543 (1999)
15. Silcowitz, M., Niebe, S., Erleben, K.: Nonsmooth Newton method for Fischer function reformulation of contact force problems for interactive rigid body simulation. In: Proceedings of the Sixth Workshop on Virtual Reality Interactions and Physical Simulations, pp. 105–114. Eurographics Association (2009)
16. Antman, S.S.: Nonlinear Problems of Elasticity. Applied Mathematical Sciences, vol. 107. Springer (2005)
17. Sobottka, G., Lay, T., Weber, A.: Stable Integration of the Dynamic Cosserat Equations with Application to Hair Modeling. Journal of WSCG 16, 73–80 (2008)
18. Chung, J., Hulbert, G.M.: A Time Integration Algorithm for Structural Dynamics With Improved Numerical Dissipation: The Generalized-α Method. Journal of Applied Mechanics 60(2), 371–375 (1993)
19. Murray, R.M., Li, Z., Sastry, S.S.: A Mathematical Introduction to Robotic Manipulation. CRC press (1994)
20. Eberly, D.H.: 3D Game Engine Design: A Practical Approach to Real-Time Computer Graphics, 2nd edn. Morgan Kaufmann (2006)

Author Index

Printed in the United States
By Bookmasters